VIRUSES AND HUMAN
DISEASE

VIRUSES AND HUMAN DISEASE

SECOND EDITION

JAMES H. STRAUSS
ELLEN G. STRAUSS
Division of Biology
California Institute of Technology
Pasadena, California

AMSTERDAM • BOSTON • HEIDELBERG • LONDON
NEW YORK • OXFORD • PARIS • SAN DIEGO
SAN FRANCISCO • SINGAPORE • SYDNEY • TOKYO
Academic Press is an imprint of Elsevier

ELSEVIER

30 Corporate Drive, Suite 400, Burlington, MA 01803, USA
525 B Street, Suite 1900, San Diego, California 92101-4495, USA
84 Theobald's Road, London WC1X 8RR, UK

This book is printed on acid-free paper. ∞

Library of Congress Cataloging-in-Publication Data
Application submitted

British Library Cataloguing in Publication Data
A catalogue record for this book is available from the British Library

ISBN: 978-0-12-373741-0

For all information on all Elsevier Academic Press publications
visit our Web site at www.books.elsevier.com

Printed in Canada
08 09 10   9 8 7 6 5 4 3 2 1

Table of Contents

8. Emerging and Reemerging Viral Diseases

9. Subviral Agents

10. Host Defenses against Viral Infection and Viral Counterdefenses

11. Gene Therapy

Appendix

Preface to the Second Edition

Our knowledge of viruses continues to expand rapidly and this book has been thoroughly revised to incorporate new information that has appeared since the first edition. In this process, virtually all of the figures and tables have been redrawn to include the latest information and the text has been extensively rewritten. New or refined approaches to the study of viruses, especially detailed structural studies but also more extensive sequencing studies, has resulted in a deeper understanding of virus evolution leading to many changes in virus taxonomy. New advances in cellular biology that highlight the importance of Toll-like receptors and RNAi in the response of vertebrates to viral infection have changed our understanding of how viruses interact with their hosts. The appearance of new global threats from previously unknown viruses such as SARS, the spread of viruses into new areas such as the spread of West Nile virus across the Americas, the continuing spread of Nipah virus in Southeast Asia, the outbreaks of filoviruses that are threatening endangered primates in Africa, as well as the justifiable worry about bird flu, reminds us that viruses have not been conquered but continue to threaten humans worldwide and has induced us to write a separate chapter on emerging and reemerging viral diseases.

We would once again like to thank the many people who read various chapters in the course of preparing the two editions of this book. We gratefully acknowledge the contributions of Elaine Bearer, Tom Benjamin, Pamela Bjorkman, Tara Chapman, Bruce Chesbro, Marie Csete, Diane Griffin, Jack Johnson, Bill Joklik, Minnie McMillan, Dennis O'Callaghan, James Ou, Ellen Rothenberg, Gail Wertz, Eckard Wimmer, and William Wunner. We are also grateful to the students in our course during the past few years for feedback on the text and figures as they evolved.

1

Overview of Viruses and Virus Infection

INTRODUCTION

The Science of Virology

The science of virology is relatively young. We can recognize specific viruses as the causative agents of epidemics that occurred hundreds or thousands of years ago from written descriptions of disease or from study of mummies with characteristic abnormalities. Furthermore, immunization against smallpox has been practiced for more than a millennium. However, it was only approximately 100 years ago that viruses were shown to be filterable and therefore distinct from bacteria that cause infectious disease. It was only about 60 years ago that the composition of viruses was described, and even more recently before they could be visualized as particles in the electron microscope. Within the last 20 years, however, the revolution of modern biotechnology has led to an explosive increase in our knowledge of viruses and their interactions with their hosts. Virology, the study of viruses, includes many aspects: the molecular biology of virus replication; the structure of viruses; the interactions of viruses and hosts and the diseases they cause in those hosts; the evolution and history of viruses and viral diseases; virus epidemiology, the ecological niche occupied by viruses and how they spread from victim to victim; and the prevention of viral disease by vaccination, drugs, or other methods. The field is vast and any treatment of viruses must perforce be selective.

Viruses are known to infect most organisms, including bacteria, blue-green algae, fungi, plants, insects, and vertebrates, but we attempt here to provide an overview of virology that emphasizes their potential as human disease agents. Because of the scope of virology, and because human viruses that cause disease, especially epidemic disease, are not uniformly distributed across virus families, the treatment is not intended to be comprehensive. Nevertheless, we feel that it is important that the human viruses be presented in the perspective of viruses as a whole so that some overall understanding of this fascinating group of agents can emerge. Thus, we consider many nonhuman viruses that are important for our understanding of the evolution and biology of viruses.

Viruses Cause Disease but Are Also Useful as Tools

Viruses are of intense interest because many cause serious illness in humans or domestic animals, and others damage crop plants. During the last century, progress in the control of infectious diseases through improved sanitation, safer water supplies, the development of antibiotics and vaccines, and better medical care have dramatically reduced the threat to human health from these agents, especially in developed countries. This is illustrated in Fig. 1.1, in which the death rate from infectious disease in the United States during the last century is shown. At the beginning of the twentieth century, 0.8% of the population died each year from infectious diseases. Today the rate is less than one-tenth as great. The use of vaccines has led to effective control of the most dangerous of the viruses. Smallpox virus has been eradicated worldwide by means of an ambitious and concerted effort, sponsored by the World Health Organization, to vaccinate all people at risk for the disease. Poliovirus and measles virus have been eliminated from the Americas by intensive vaccination programs. There is hope that these two viruses will also be eradicated worldwide in the near future. Vaccines exist for the control of many other viral diseases including, among others, mumps, rabies, rubella, yellow fever, Japanese encephalitis, rotaviral gastroenteritis, and, very recently, papillomaviral disease that is the primary cause of cervical cancer.

The dramatic decline in the death rate from infectious disease has led to a certain amount of complacency. There is a

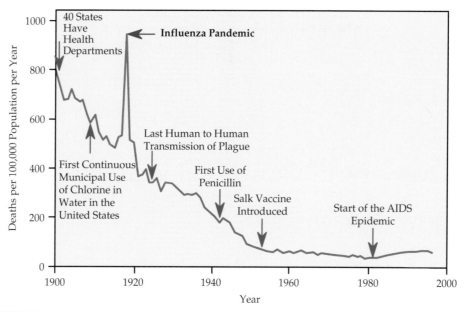

FIGURE 1.1   Death rate from infectious diseases in the United States, 1900–1996. The death rate dropped over the twentieth century from around 800 deaths per 100,000 population per year to about 50. Significant milestones in public health are shown. After dropping steadily for 80 years, interrupted only by the influenza pandemic of 1918–1919, the death rate began to rise in 1980 with the advent of the AIDS (*a*cquired *i*mmuno*d*eficiency *s*yndrome) epidemic. From *Morbidity and Mortality Weekly Report (MMWR)* (1999), Vol. 48, #29, p. 621.

small but vocal movement in the United States and Europe to eliminate immunization against viruses, for example. However, viral diseases continue to plague humans, as do infectious diseases caused by bacteria, protozoa, fungi, and multicellular parasites. Deaths worldwide due to infectious disease are shown in Fig. 1.2, divided into six categories. In 2002 more than 3 million deaths occurred as a result of acute respiratory disease, much of which is caused by viruses. More than 2 million deaths were attributed to diarrheal diseases, about half of which are due to viruses. AIDS killed 3 million people worldwide in 2002, and measles is still a significant killer in developing countries. Recognition is growing that infectious diseases, of which viruses form a major component, have not been conquered by the introduction of vaccines and drugs. Viral diseases and disease caused by other pathogens continue to resist elimination. Furthermore, the overuse of antibiotics has resulted in an upsurge in antibiotic-resistant bacteria, which has exacerbated the problems caused by them.

The incidence of disease in various parts of the world caused by a number of widespread viruses is illustrated in Fig. 1.3. In the Americas and, for most viruses, Europe as well, widespread use of vaccines has almost eliminated disease caused by viruses for which vaccines exist. In developing countries, measles, poliovirus, yellow fever virus, and rabies virus, as well as others not shown in the figure, still cause serious problems although good vaccines exist. However, developed countries as well as developing countries suffer from viruses for which no vaccines exist to the

current time. Human immunodeficiency virus (HIV), illustrated in the figure, is a case in point.

The persistence of viruses is in part due to their ability to change rapidly and adapt to new situations. HIV is the most striking example of the appearance of a virus that has recently entered the human population and caused a plague of worldwide importance. The arrival of this virus in the United States caused a noticeable rise in the total number of deaths from infectious disease, as seen in Fig. 1.1. Other, previously undescribed viruses also continue to emerge as serious pathogens. Sin Nombre virus, a previously unknown virus associated with rodents, caused a 1994 outbreak in the United States of hantavirus pulmonary syndrome with a 50% case fatality rate, and it is now recognized as being widespread in North America. Junin virus, the cause of Argentine hemorrhagic fever, as well as related viruses have become a more serious problem in South America with the spread of farming. Ebola virus, responsible for several small African epidemics with a case fatality rate of 70%, was first described in the 1970s. Nipah virus, a previously unknown virus of bats, appeared in 1998 and caused 258 cases of encephalitis, with a 40% fatality rate, in Malaysia and Singapore. The SARS virus, also a previously unknown virus of bats, caused an epidemic that killed more than 700 humans worldwide in 2002–2003. The H5N1 strain of influenza, known as "bird flu," has killed more than 150 humans in the last few years and there is fear that it might eventually cause a worldwide pandemic with hundreds of millions of deaths. It is obvious that the potential for rapid spread of all viruses is increasing as faster and

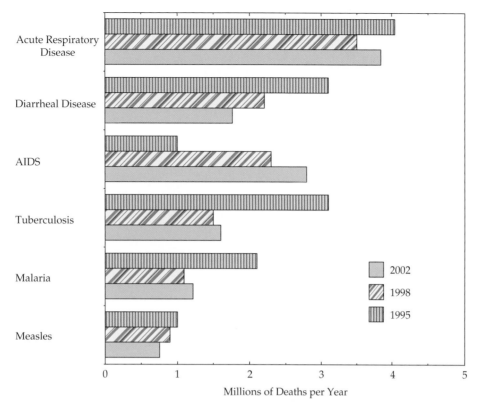

FIGURE 1.2   Six leading infectious diseases as causes of death. Data are the totals for all ages worldwide in 1995, 1998, and 2002. The data came from the World Health Organization Web site: http://www.who.int/infectious-disease-report/pages/graph 5.html, and the World Health Report 2003 at: (http://www.who.int/whr/2003/en/).

more extensive travel becomes ever more routine. The possibility exists that any of these viruses could become more widespread, as has HIV since its appearance in Africa perhaps half a century ago, and as has West Nile virus, which spread to the Americas in 1999. A discussion of emerging and reemerging viral diseases is found in Chapter 8.

Newly emerging viruses are not the only ones to plague humans, however. Many viruses that have been known for a long time continue to cause widespread problems. Respiratory syncytial virus, as an example, is a major cause of pneumonia in infants. Despite much effort, it has not yet been possible to develop an effective vaccine. Even when vaccines exist, problems may continue. For example, influenza virus changes rapidly and the vaccine for it must be reformulated yearly. Because the major reservoir for influenza is birds, it is not possible to eradicate the virus. Thus, to control influenza would require that the entire population be immunized yearly. This is a formidable problem and the virus continues to cause annual epidemics with a significant death rate (Chapter 4). Although primarily a killer of the elderly, the potential of influenza to kill the young and healthy was shown by the worldwide epidemic of influenza in 1918 in which 20–100 million people died worldwide. In

the United States, 1% of the population died during the epidemic and perhaps half of all deaths were due to influenza (Fig. 1.1). Continuing study of virus replication and virus interactions with their hosts, surveillance of viruses in the field, and efforts to develop new vaccines as well as other methods of control are still important.

The other side of the coin is that viruses have been useful to us as tools for the study of molecular and cellular biology. Further, the development of viruses as vectors for the expression of foreign genes has given them a new and expanded role in science and medicine, including their potential use in gene therapy (Chapter 11). As testimony to the importance of viruses in the study of biology, numerous Nobel Prizes have been awarded in recognition of important advances in biological science that resulted from studies that involved viruses (Table 1.1). To cite a few examples, Max Delbrück received the prize for pioneering studies in what is now called molecular biology, using bacteriophage T4. Cellular oncogenes were first discovered from their presence in retroviruses that could transform cells in culture, a discovery that resulted in a prize for Francis Peyton Rous for his discovery of transforming retroviruses, and for Michael Bishop and Harold Varmus, for showing that a

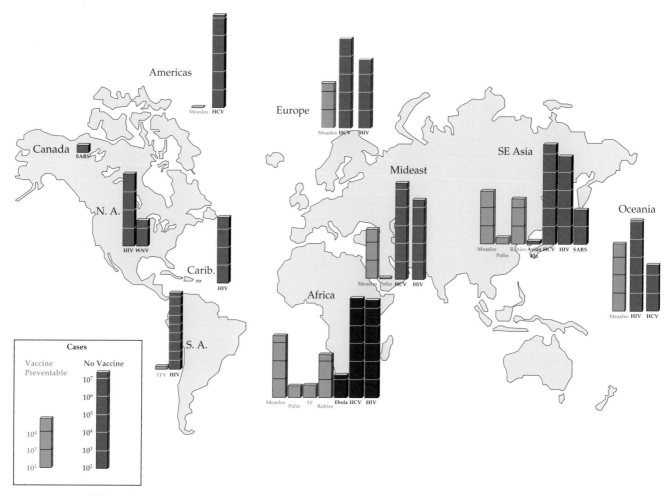

FIGURE 1.3   Incidence of selected infectious diseases worldwide and the effect of vaccination. The number of cases is shown on a log scale such that each division represents 10-fold more cases than the division below it. The diseases for which vaccines exist are shown in red. Adapted from Lattanzi *et al.* (2006), Figure 1. SARS, severe acute respiratory syndrome; HIV, human immunodeficiency virus; WNV, West Nile virus; YFV, yellow fever virus; HCV, hepatitis C virus.

transforming retroviral gene had a cellular counterpart. As a third example, the development of the modern methods of gene cloning have relied heavily on the use of restriction enzymes and recombinant DNA technology, first developed by Daniel Nathans and Paul Berg working with SV40 virus, and on the use of reverse transcriptase, discovered by David Baltimore and Howard Temin in retroviruses. As another example, the study of the interactions of viruses with the immune system has told us much about how this essential means of defense against disease functions, and this resulted in a prize for Rolf Zinkernagel and Peter Doherty. The study of viruses and their use as tools has told us as much about human biology as it has told us about the viruses themselves.

In addition to the interest in viruses that arises from their medical and scientific importance, viruses form a fascinating evolutionary system. There is debate as to how ancient are viruses. Some argue that RNA viruses contain remnants of the RNA world that existed before the invention of DNA. All would accept the idea that viruses have been present for hundreds of millions of years and have helped to shape the evolution of their hosts. Viruses are capable of very rapid change, both from drift due to nucleotide substitutions that may occur at a rate 10^6-fold greater than that of the plants and animals that they infect, and from recombination that leads to the development of entirely new families of viruses. This makes it difficult to trace the evolution of viruses back more than a few millennia or perhaps a few million years. The development of increasingly refined methods of sequence analysis, and the determination of more structures of virally encoded proteins, which change far more slowly than do the amino acid sequences that form the structure, have helped identify relationships among viruses that were not at first obvious. The coevolution of viruses and their hosts remains a study that is intrinsically interesting and has much to tell us about human biology.

TABLE 1.1   Nobel Prizes Involving Virology[a]

Year	Names	Nobel citation; *virus group or family*
1946 [Chemistry]	Wendell Stanley	Isolation, purification and crystallization of tobacco mosaic virus; *Tobamovirus*
1951	Max Theiler	Development of yellow fever vaccine; *Flaviviridae*
1954	John F. Enders, Thomas Weller, Frederick C. Robbins	Growth and cultivation of poliovirus; *Picornaviridae*
1958	Joshua Lederberg	Transforming bacteriophages
1965	Francois Jacob, André Lwoff, Jacques Monod	Operons; bacteriophages
1966	Francis Peyton Rous	Discovery of tumor-producing viruses; *Retroviridae*
1969	Max Delbrück, Alfred D. Hershey, Salvador E. Luria	Mechanism of virus infection in living cells; bacteriophages
1975	David Baltimore, Howard M. Temin, Renato Dulbecco	Discoveries concerning the interaction between tumor viruses and the genetic material of the cell; *Retroviridae*
1976	D. Carleton Gajdusek, Baruch S. Blumberg	New mechanisms for the origin and dissemination of infectious diseases; B with *Hepadnaviridae*, G with prions
1978[b]	Daniel Nathans	Application of restriction endonucleases to the study of the genetics of SV40; *Polyomaviridae*
1980 [Chemistry]	Paul Berg	Studies of the biochemistry of nucleic acids, with particular regard to recombinant DNA (SV40); *Polyomaviridae*
1982 [Chemistry]	Aaron Klug	Development of crystallographic electron microscopy and structural elucidation of biologically important nucleic acid–protein complexes; *Tobamovirus and Tymovirus*
1988[b]	George Hitchings, Gertrude Elion	Important principles of drug treatment using nucleotide analogues (acyclovir)
1989	J. Michael Bishop, Harold E. Varmus	Discovery of the cellular origin of retroviral oncogenes; *Retroviridae*
1993	Phillip A. Sharp, Richard J. Roberts	Discoveries of split (spliced) genes; *Adenoviridae*
1996	Rolf Zinkernagel, Peter Doherty	Presentation of viral epitopes by MHC
1997	Stanley Prusiner	Prions
2006	Andrew Fire, Craig Mello	Discovery of RNAi

[a] All prizes listed are in Physiology or Medicine except those three marked [Chemistry].
[b] In these two instances, the prize was shared with unlisted recipients whose work did not involve viruses.

The Nature of Viruses

Viruses are subcellular, infectious agents that are obligate intracellular parasites. They infect and take over a host cell in order to replicate. The mature, extracellular virus particle is called a virion. The virion contains a genome that may be DNA or RNA wrapped in a protein coat called a capsid or nucleocapsid. Some viruses have a lipid envelope surrounding the nucleocapsid (they are "enveloped"). In such viruses, glycoproteins encoded by the virus are embedded in the lipid envelope. The function of the capsid or envelope is to protect the viral genome while it is extracellular and to promote the entry of the genome into a new, susceptible cell. The structure of viruses is covered in detail in Chapter 2.

The nucleic acid genome of a virus contains the information needed by the virus to replicate and produce new virions after its introduction into a susceptible cell. Virions bind to receptors on the surface of the cell, and by processes described later the genome is released into the cytoplasm of the cell, sometimes still in association with protein ("uncoating"). The genome then redirects the cell to the replication of itself and to the production of progeny virions. The cellular machinery that is in place for the production of energy (synthesis of ATP) and for macromolecular synthesis, such as translation of mRNA to produce proteins, is essential.

It is useful to think of the proteins encoded in viral genomes as belonging to three major classes. First, most viruses encode enzymes required for replication of the genome and the production of mRNA from it. RNA viruses must encode an RNA polymerase or replicase, since cells do not normally replicate RNA. Most DNA viruses have access to the cellular DNA replication machinery in the nucleus, but even so, many encode new DNA polymerases for the replication of their genomes. Even if they use cellular DNA polymerases, many DNA viruses encode at least an initiation protein for genome replication. An overview of the replication strategies used by different viruses is presented later, and details of the replication machinery used by each virus are given in the chapters that describe individual viruses. Second, viruses must encode proteins that are used in the assembly of progeny viruses. For simpler viruses, these may consist of only one or a few structural proteins that assemble with the genome to form the progeny virion. More complicated viruses may encode scaffolding proteins that are required for assembly but are not present in the virion. In some cases, viral proteins required for assembly may have proteolytic activity. Assembly of viruses is described in Chapter 2. Third, many (most?) viruses encode proteins that interfere with defense mechanisms of the host. These defenses include, for example, the immune response and the interferon response of vertebrates, which are highly evolved and effective methods of controlling and eliminating virus infection; and the DNA restriction system in bacteria, so useful in molecular biology and genetic engineering, that prevents the introduction of foreign DNA. Vertebrate defenses against viruses, and the ways in which viruses counter these defenses, are described in Chapter 10.

It is obvious that viruses that have larger genomes and encode larger numbers of proteins, such as the herpesviruses (family *Herpesviridae*), have more complex life cycles and assemble more complex virions than viruses with small genomes, such as poliovirus (family *Picornaviridae*). The smallest known nondefective viruses have genomes of about 3 kb (1 kb = 1000 nucleotides in the case of single-stranded genomes or 1000 base pairs in the case of double-stranded genomes). These small viruses may encode as few as three proteins (e.g., the bacteriophage MS2). At the other extreme, the largest known RNA viruses, the coronaviruses (family *Coronaviridae*), have genomes somewhat larger than 30 kb, whereas the largest DNA viruses, poxviruses belonging to the genera *Entomopoxvirus A* and *C* (family *Poxviridae*), have genomes of up to 380 kb. These large DNA viruses encode hundreds of proteins and can finely regulate their life cycle. Further, as stated before, many or even most viruses interfere with host defenses. In the smaller viruses this may involve only one or two proteins that interfere with limited aspects of host defense, whereas the large viruses have the luxury of encoding more than a dozen proteins that can finely regulate the host defense mechanisms. It is worthwhile remembering that even the largest viral genomes are small compared to the size of the bacterial genome (2000 kb) and miniscule compared to the size of the human genome (2×10^6 kb).

There are other subcellular infectious agents that are even "smaller" than viruses. These include satellite viruses, which are dependent for their replication on other viruses; viroids, small (~300 nucleotide) RNAs that are not translated and have no capsid; and prions, infectious agents whose identity remains controversial, but which may consist only of protein. These agents are covered in Chapter 9.

CLASSIFICATION OF VIRUSES

The Many Kinds of Viruses

Three broad classes of viruses can be recognized, which may have independent evolutionary origins. One class, which includes the poxviruses and herpesviruses among many others, contains DNA as the genome, whether single stranded or double stranded, and the DNA genome is replicated by direct DNA → DNA copying. During infection, the viral DNA is transcribed by cellular and/or viral RNA polymerases, depending on the virus, to produce mRNAs for translation into viral proteins. The DNA genome is replicated by DNA polymerases that can be of viral or cellular origin. Replication of the genomes of most eukaryotic DNA viruses and assembly of progeny viruses occur in the nucleus, but the poxviruses replicate in the cytoplasm.

A second class of viruses contains RNA as their genome and the RNA is replicated by direct RNA → RNA copying. Some RNA viruses, such as yellow fever virus (family *Flaviviridae*) and poliovirus (family *Picornaviridae*), have a genome that is a messenger RNA, defined as plus-strand RNA. Other RNA viruses, such as measles virus (family *Paramyxoviridae*) and rabies virus (family *Rhabdoviridae*), have a genome that is anti-messenger sense, defined as minus strand. The arenaviruses (family *Arenaviridae*) and some of the genera belonging to the family *Bunyaviridae* have a genome that has regions of both messenger and anti-messenger sense and are called ambisense. The replication of these viruses follows a minus-sense strategy, however, and they are classified with the minus-sense viruses. Finally, some RNA viruses, for example, rotaviruses (family *Reoviridae*), have double-strand RNA genomes. In the case of all RNA viruses, virus-encoded proteins are required to form a replicase to replicate the viral RNA, since cells do not possess (efficient) RNA → RNA copying enzymes. In the case of the minus-strand RNA viruses and double-strand RNA viruses, these RNA synthesizing enzymes also synthesize mRNA and are packaged in the virion, because their genomes cannot function as messengers. Replication of the genome proceeds through RNA intermediates that are complementary to the genome in a process that follows the same rules as DNA replication.

The third class of viruses encodes the enzyme reverse transcriptase (RT), and these viruses have an RNA → DNA step in their life cycle. The genetic information encoded by these viruses thus alternates between being present in RNA and being present in DNA. Retroviruses (e.g., HIV, family *Retroviridae*) contain the RNA phase in the virion; they have a single-stranded RNA genome that is present in the virus particle in two copies. Thus, the replication of their genome occurs through a DNA intermediate (RNA → DNA → RNA). The hepadnaviruses (e.g., hepatitis B virus, family *Hepadnaviridae*) contain the DNA phase as their genome, which is circular and largely double stranded. Thus their genome replicates through an RNA intermediate (DNA → RNA → DNA). Just as the minus-strand RNA viruses and double-strand RNA viruses package their replicase proteins, the retroviruses package active RT, which is required to begin the replication of the genome in the virions. Although in many treatments the retroviruses are considered with the RNA viruses and the hepadnaviruses with the DNA viruses, we consider these viruses to form a distinct class, the RT-encoding class, and in this book references to RNA viruses or to DNA viruses are not meant to apply to the retroviruses or the hepadnaviruses.

All viruses, with one exception, are haploid; that is, they contain only one copy of the genomic nucleic acid. The exception is the retroviruses, which are diploid and contain two identical copies of the single-stranded genomic RNA. The nucleic acid genome may consist of a single piece of DNA or RNA or may consist of two or more nonidentical fragments. The latter can be considered analogous to chromosomes and can reassort during replication. In the case of animal viruses, when a virus has more than one genome segment, all of the different segments are present within a single virus particle. In the case of plant viruses with multiple genome segments, it is quite common for the different genome segments to be separately encapsidated into different particles. In this case, the infectious unit is multipartite: Infection to produce a complete replication cycle requires simultaneous infection by particles containing all of the different genome segments. Although this does not seem to pose a problem for the transmission of plant viruses, it must pose a problem for the transmission of animal viruses since such animal viruses have not been found. This difference probably arises because of different modes of transmission, the fact that many plant viruses grow to exceptionally high titers, and the fact that many plants grow to very high density.

The ICTV Classification of Viruses

The International Committee on Taxonomy of Viruses (ICTV), a committee organized by the Virology Division of the International Union of Microbiological Societies, is attempting to devise a uniform system for the classification and nomenclature of all viruses. Viruses are classified into species on the basis of a close relationship. The decision as to what constitutes a species is arbitrary because a species usually contains many different strains that may differ significantly (10% or more) in nucleotide sequence. Whether two isolates should be considered as being the same species rather than representing two different species can be controversial. Virus species that exhibit close relationships are then grouped into a genus. Species within a genus usually share significant nucleotide sequence identity demonstrated by antigenic cross-reaction or by direct sequencing

of the genome. Genera are grouped into families, which can be considered the fundamental unit of virus taxonomy. Classification into families is based on the type and size of the nucleic acid genome, the structure of the virion, and the strategy of replication used by the virus, which is determined in part by the organization of the genome. Groupings into families are not always straightforward because little or no sequence identity is present between members of different genera. However, uniting viruses into families attempts to recognize evolutionary relationships and is valuable for organizing information about viruses.

Higher taxonomic classifications have not been recognized for the most part. To date only three orders (*Caudovirales*, *Nidovirales*, *Mononegavirales*) have been established that group together a few families. Taxonomic classification at higher levels is difficult because viruses evolve rapidly and it can be difficult to prove that any two given families are descended from a common ancestor, although it is almost certain that higher groupings based on common evolution do exist and will be elucidated with time. Viral evolution involves not only sequence divergence, however, but also the widespread occurrence of recombination during the rise of the modern families, a feature that blurs the genetic relationships between viruses. Two families may share, for example, a related polymerase gene but have structural protein genes that appear unrelated; how should such viruses be classified?

The ICTV has recognized 5450 viruses as species (more than 30,000 strains of viruses exist in collections around the world), and classified these 5450 species into 287 genera belonging to 73 families plus a number of "floating" genera that have not yet been assigned to a family. An overview of these families, in which viruses that cause human disease are emphasized, is shown in Table 1.2. Included in the table is the type of nucleic acid that serves as the genome, the genome size, the names of many families, and the major groups of hosts infected by viruses within each grouping. For many families the names and detailed characteristics are not shown here, but a complete listing of families can be found in the reports of the ICTV on virus taxonomy or in *The Encyclopedia of Virology* (2nd ed.). Tables that describe the members of families that infect humans are presented in the chapters that follow in which the various virus families are considered in some detail.

AN OVERVIEW OF THE REPLICATION CYCLE OF VIRUSES

Receptors for Virus Entry

The infection cycle of an animal virus begins with its attachment to a receptor expressed on the surface of a susceptible cell, followed by penetration of the genome, either

TABLE 1.2   Major Virus Families

Nucleic acid	Genome size	Segments	Family	Genera	Major hosts (number of members infecting that host)[a]	
DS DNA	130–375 kbp	1	Poxviridae	8 + 3	Vertebrates (35 + 9T)[b], insects (27), plus 15 U[c]	
	170–190 kbp	1	Asfariviridae	1	Vertebrates (1)	
	170–400 kbp	1	Iridoviridae	3 + 2	Vertebrates (2 + 5T), insects (6 + 11T)	
	120–220 kbp	1	Herpesviridae	9	Vertebrates (61 + 7T + 56U)	
	80–180 kbp	1	Baculoviridae	2	Insects (36 + 8T)	
	28–48 kbp	1	Adenoviridae	4	Vertebrates (32 + 9T)	
	5 kbp	1	Polyomaviridae	1	Vertebrates (12)	
	6.8–8.4 kbp	1	Papillomaviridae	16	Vertebrates (7 + 88T + 13U)	
	Various	1	Several families	—	Bacteria (42 + 368T)	
SS DNA	4–6 kbp	1	Parvoviridae	5 + 4	Vertebrates (33 + 3T), insects (6 + 18T)	
	Various	1	Several families	—	Bacteria (43 + 38T), plants (98 + 11T)	
DS RNA	20–30 kbp	10–12	Reoviridae	6 + 2 + 4	Vertebrates (52 + 24T), insects (1 + 7T), plants (13 + 1T)	
	5.9 kbp	2	Birnaviridae	2 + 1	Vertebrates (3), insects (1)	
	4.6–7.0 kbp	1 or 2	Three families	—	Fungi (7 + 7T), plants (30 + 15T), protozoans (14)	
SS (+)RNA	28–33 kb	1	Coronaviridae	2	Vertebrates (17 + 1T)	
	13–16 kb	1	Arteriviridae	1	Vertebrates (4)	
	10–13 kb	1	Togaviridae	2	Vertebrates (insect vectors)(28)	
	10–12 kb	1	Flaviviridae	3	Vertebrates (some insect vectors) (59 + 4T)	
	7–8.5 kb	1	Picornaviridae	9	Vertebrates (30 + 1T + 23U)	
	7–8 kb	1	Astroviridae	2	Vertebrates (9)	
	8 kb	1	Caliciviridae	4	Vertebrates (6 + 1T)	
	7.2 kb	1	Hepeviridae	1	Vertebrates (1)	
	Various	1 to 3	Many families	—	Plants (496 + 84T + 5U)	
SS (−)RNA	15–16 kb	1	Paramyxoviridae	7	Vertebrates (34 + 2U)	
	19 kb	1	Filoviridae	2	Vertebrates (5)	
	11–16	1	Rhabdoviridae	4 + 2	Vertebrates (23 + 25T + 40U), invertebrates (20U), plants (15)	
	6 kb	1	Bornaviridae	1	Vertebrates (1)	
	10–15 kb	8	Orthomyxoviridae	5	Vertebrates (7)	
	12–23 kb	3	Bunyaviridae	4 + 1	Vertebrates and insect vectors (86 + 20T), plants (9 + 7T)	
	11 kb	2	Arenaviridae	1	Vertebrates (19 + 1T)	
SS RNA RT	7–10 kb	dimer	Retroviridae	7	Vertebrates (53 + 2T)	DNA Intermediate
DS DNA RT	3 kb	1	Hepadnaviridae	2	Vertebrates (5 + 1T)	RNA intermediate
	8 kb	1	Caulimoviridae	6	Plants (26 + 10T)	RNA intermediate

[a] Vertebrates in red indicate humans are among the vertebrates infected. Vertebrates in blue indicate non-human hosts only; plant hosts are in green; insect hosts in yellow; bacterial hosts are black.
[b] T = tentatively assigned to a particular genus.
[c] U = assigned to the family, but not to any particular genus within the family.
Source: Data for this table is from Fauquet et al. (2005).

naked or complexed with protein, into the cytoplasm. Binding often occurs in several steps. For many viruses, the virion first binds to an accessory receptor that is present in high concentration on the surface of the cell. These accessory receptors are usually bound with low affinity and binding often has a large electrostatic component. Use of accessory receptors seems to be fairly common among viruses adapted to grow in cell culture but less common in primary isolates of viruses from animals. This first stage binding to an acces-

sory receptor is not required for virus entry even where used, but such binding does accelerate the rate of binding and uptake of the virus.

Binding to a high-affinity, virus-specific receptor is required for virus entry, and virus may be transferred to its high-affinity receptor after primary binding to an accessory receptor, or may bind directly to its high-affinity receptor. Cells that fail to express the appropriate receptor cannot be infected by the virus. These receptors are specifically bound by one or more of the external proteins of a virus. Each virus uses a specific receptor (or perhaps a specific set of receptors) expressed on the cell surface, and both protein receptors and carbohydrate receptors are known. In some cases, unrelated viruses make use of identical receptors. A protein called CAR (*Coxsackie-adenovirus receptor*), a member of the immunoglobulin (Ig) superfamily, is used by the RNA virus Coxsackie B virus (*Picornaviridae*) and by many adenoviruses (*Adenoviridae*), which are DNA viruses. Sialic acid, a carbohydrate attached to most glycoproteins, is used by influenza virus (family *Orthomyxoviridae*), human coronavirus OC3 (family *Coronaviridae*), reovirus (*Reoviridae*), bovine parvovirus (*Parvoviridae*), and many other viruses. Conversely, members of the same viral family may use widely

disparate receptors. Fig. 1.4 illustrates a number of receptors used by different retroviruses (family *Retroviridae*). These receptors differ widely in their structures and in their cellular functions. Where known, the region of the cellular receptor that is bound by the virus is indicated. Table 1.3 lists receptors used by different herpesviruses (*Herpesviridae*) and different coronaviruses.

In addition to the requirement for a high-affinity or primary receptor, many viruses also require a coreceptor in order to penetrate into the cell. In the current model for virus entry, a virus first binds to the primary receptor and then binds to the coreceptor. Only on binding to the coreceptor can the virus enter the cell. The best studied example is HIV, which uses the cell surface molecule called CD4 as

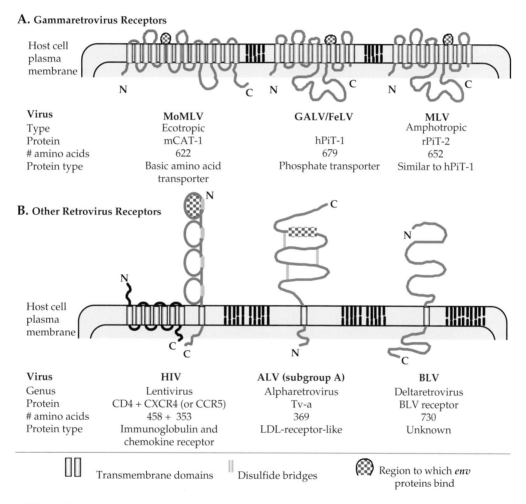

FIGURE 1.4 Cellular receptors for retroviruses. The structures of various retrovirus receptors are shown schematically to illustrate their orientation in the cell plasma membrane. The receptors for the gammaretroviruses contain multiple transmembrane domains and have known cellular functions. The HIV receptor consists of a molecule of CD4 plus a chemokine receptor such as CXCR4. The receptor for alpharetroviruses is a Type II membrane protein similar to the LDL receptor, with the N terminus in the cytoplasm. Little is known about the cellular function of the BLV receptor, other than its orientation as a Type I membrane protein. Abbreviations: MLV, murine leukemia virus; GALV, gibbon/ape leukemia virus; FeLV, feline leukemia virus; HIV, human immunodeficiency virus; ALV, avian leukosis virus; BLV, bovine leukemia virus; LDL, low density lipoprotein. Adapted from Fields *et al.* (1996) p. 1788 and Coffin *et al.* (1997) pp. 76–82.

TABLE 1.3   Viruses Within a Family that Use Unrelated Receptors

Family	Virus	High-affinity receptor	Accessory receptor
Herpesviridae			
Alpha	Herpes simplex	HIgR (CD155 family)	Heparan sulfate
		HVEM (TNF receptor family)	
	Pseudorabies	140 kD heparan sulfate proteoglycan	
		85 kD integral membrane protein	
		CD155 and related proteins	
Beta	Cytomegalovirus	protein?? unidentified	Heparan sulfate
Gamma	Bovine herpesvirus	56 kD protein	Heparan sulfate
	Epstein Barr	CD21 (CR2 receptor)	
Coronaviridae			
Group 1			
Porcine	TGEV[a]	Porcine APN: aminopeptidase N	
Feline	FIPV[a]	Feline APN: aminopeptidase N	
Human	229e	Human APN: human aminopeptidase N	
Human	NL63	ACE2: Human angiotensin-converting enzyme 2	
Group 2			
Human	SARS	ACE2: Human angiotensin-converting enzyme 2	
Murine	Mouse hepatitis	CEACAM1: Carcinoembryonic antigen-related cell adhesion molecule 1[b]	
Bovine	Bovine coronavirus	Sialic acid residues on glycoproteins and glycolipids	

[a] Virus abbreviation: TGEV, transmissible gastroenteritis virus (of swine); FIPV, feline infectious peritonitis virus.

[b] Note that entry of mouse hepatitis variants of extended host range is independent of CEACAM1, and instead uses heparan sulfate as an entry receptor.

a primary receptor and various chemokines as coreceptors (see later).

The nature of the receptors utilized by a virus determines in part its host range, tissue tropism, and the pathology of the disease caused by it. Thus, the identification of virus receptors is important, but identification of receptors is not always straightforward.

Primary (High-Affinity) Receptors

Many members of the Ig superfamily are used by viruses as high-affinity receptors, as illustrated in Fig. 1.5. The Ig superfamily contains thousands of members, which play important roles in vertebrate biology. The best known members are found in the immune system (Chapter 10), from which the family gets its name. Members of this superfamily contain one or more Ig domains of about 100 amino acids that arose by duplication of a prototypical gene. During evolution of the superfamily, thousands of different proteins arose by a combination of continuing gene duplication, sequence divergence, and recombination. Many proteins belonging to this superfamily are expressed on the surface of cells, where they serve many functions, and many have been usurped by animal viruses for use as receptors.

Other surface proteins used as receptors include the vibronectin receptor $\alpha_v\beta_3$, used by several members of the *Picornaviridae*; aminopeptidase N, used by some coronaviruses; CD55, used by Coxsackie A21 virus; the different proteins illustrated in Fig. 1.4; and other proteins too numerous to describe here. The receptors used by four viruses are described in more detail as examples of the approaches used to identify receptors and their importance for virus pathology.

One well-characterized receptor is that for poliovirus, which attaches to a cell surface molecule that is a member of the Ig superfamily (Fig. 1.5). The normal cellular function of this protein is unknown. It was first called simply the poliovirus receptor or PVR, but has now been renamed CD155, following a scheme for the designation of cell surface proteins. Poliovirus will bind only to the version of this molecule that is expressed in primates, and not to the version expressed in rodents, for example. Thus, in nature, poliovirus infection is restricted to primates. Although chicken cells or most mammalian cells that lack CD155 are resistant to poliovirus infection, they can be transfected with the viral RNA by a process that bypasses the receptors. When infected in this way, they produce a full yield of virus, showing that the block to replication is at the level of entry.

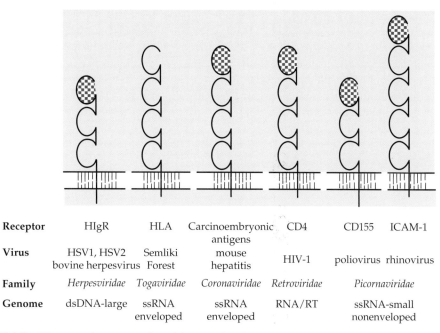

Receptor	HIgR	HLA	Carcinoembryonic antigens	CD4	CD155	ICAM-1
Virus	HSV1, HSV2 bovine herpesvirus	Semliki Forest	mouse hepatitis	HIV-1	poliovirus	rhinovirus
Family	*Herpesviridae*	*Togaviridae*	*Coronaviridae*	*Retroviridae*	*Picornaviridae*	
Genome	dsDNA-large	ssRNA enveloped	ssRNA enveloped	RNA/RT	ssRNA-small nonenveloped	

FIGURE 1.5   Diagrammatic representation of immunoglobulin superfamily membrane proteins that are used as receptors by viruses. The domains indicated by cross-hatching have been shown to be required for receptor activity. ssRNA, single-strand RNA; dsDNA, double-strand DNA; RNA/RT, RNA reverse transcribed into DNA.

Cells lacking CD155 have been transfected with expression clones so that they express CD155, and these modified cells are sensitive to infection with poliovirus. This was, in fact, the way the receptor was identified. Such a system also allows the testing of chimeric receptors, in which various domains of CD155 come from the human protein and other parts come from the homologous mouse protein, or even from entirely different proteins like CD4. In this way it was shown that only the distal Ig domain from human CD155 (cross-hatched in Fig. 1.5) is required for a chimeric protein to function as a receptor for poliovirus.

In humans, CD155 is expressed on many cells, including cells of the gut, nasopharynx, and the central nervous system (CNS). Infection begins in the tonsils, lymph nodes of the neck, Peyer's patches, and the small intestine. In more than 98% of cases, the infection progresses no further and no illness, or only minor illness, results. In some cases, however, virus spreads to the CNS, probably both by passing through the blood–brain barrier and through retrograde axonal transport. Once in the CNS, the virus expresses an astounding preference for motor neurons, whose destruction leads to paralysis or even death via a disease called poliomyelitis. This preference for motor neurons, and the failure of the virus to grow in other tissues, is not understood. Athough CD155 is required for virus entry, other factors within the cell are also important for efficient virus replication.

Making use of the CD155 gene, transgenic mice have been generated in which the syndrome of poliomyelitis can be faithfully reproduced. Although these transgenic mice can be infected only by injection of virus and not by ingestion, the normal route of poliovirus infection in humans, a small animal model for poliomyelitis is valuable for the study of virus pathology or for vaccine development. To date, our information on the pathology of poliovirus in the CNS was obtained only from experimental infection of nonhuman primates, which are very expensive to maintain, or from humans naturally infected with the virus.

As a second example of virus–receptor interactions, HIV utilizes as its receptor a cell surface molecule known as CD4, which is also a member of the Ig superfamily (Figs 1.4 and 1.5). As described later, a coreceptor is also required. CD4 is primarily expressed on the surface of certain lymphocytes (described in more detail in Chapter 10). Furthermore, the virus has a narrow host range and will bind with high efficiency only to the human version of CD4 (Fig. 1.4). Thus, humans are the primary host of HIV. Immune function is impaired over time as helper CD4+ T cells, which are required for an immune response directed against infectious agents, are killed by virus infection, leading to the observed syndrome of AIDS (*a*cquired *i*mmuno*d*eficiency *s*yndrome). The virus can also infect cells of the monocyte-macrophage lineage, and possibly other cells in the CNS, leading to neurological manifestations.

As a third example of virus–receptor interaction, among the receptors used by Sindbis virus (family *Togaviridae*) is the high-affinity laminin receptor. Sindbis virus is an arbovirus, that is, it is arthropod-borne. In nature it alternates between replication in mosquitoes, which acquire the virus when they take a blood meal from an infected vertebrate, and higher vertebrates, which acquire the virus when bitten by an infected mosquito. The high-affinity laminin receptor is a cell adhesion molecule that binds to laminin present in basement membranes. It has been very highly conserved during evolution, and Sindbis virus will bind to both the mosquito version and the mammalian version of this protein. Viruses with broad host ranges, such as arboviruses, must use receptors that are highly conserved, or must have evolved the ability to use different receptors in different hosts.

Finally, as a fourth example, the receptor for influenza virus (family *Orthomyxoviridae*) is sialic acid covalently linked to glycoproteins or glycolipids present at the cell surface. Because sialic acid is expressed on many different cells and in many different organisms, the virus has the potential to have a very wide host range. The virus infects many birds and mammals, and its maintenance in nature depends on its ability to infect such a broad spectrum of animals. The epidemiology of influenza virus will be considered in Chapter 4.

Accessory Receptors and Coreceptors

The process by which a virus binds to a cell and penetrates into the cytoplasm may be complicated by the participation of more than one cellular protein in the process. Some viruses may be able to use more than one primary receptor, which thus serve as alternative receptors. Second, many viruses appear to first bind to a low-affinity receptor or accessory receptor before transfer to a high-affinity receptor by which the virus enters the cell. Third, many viruses absolutely require a coreceptor, in addition to the primary receptor, for entry.

Many viruses, belonging to different families, have been shown to bind to glycosoaminoglycans such as heparan sulfate (Table 1.4), which are expressed on the surface of many cells. In at least some cases, however, such as for human herpes simplex virus (HSV) (family *Herpesviridae*), heparan sulfate is not absolutely required for the entry of the virus. Cells that do not express heparan sulfate or from which it has been removed can still be infected by HSV. Heparan sulfate does dramatically increase the efficiency of infection, however. The current model is that HSV first binds to heparan sulfate with low affinity and is then transferred to the primary receptor for entry. In this model, heparan sulfate serves an accessory function, which can be dispensed with.

The primary receptor for HSV has now been identified as a protein belonging to the Ig superfamily (Fig. 1.5). This protein is closely related to CD155, and, in fact, CD155 will serve as a receptor for some herpesviruses, but not for HSV. The story is further complicated by the fact that more than one protein can serve as a receptor for HSV. Two of these

proteins, one called HIgR (for *h*erpesvirus *Ig*-like *r*eceptor) and the other called either PRR-1 (for *p*oliovirus *r*eceptor *r*elated) or HveA (for *h*erpes*v*irus *e*ntry mediator *A*), appear to be splice variants that have the same ectodomain.

Heparan sulfate may serve as an accessory receptor for the other viruses shown in Table 1.4, or it may serve as a primary receptor for some or all. It was thought that it may be a primary receptor for dengue virus (family *Flaviviridae*), but recent work has identified other candidates as the primary receptor. In the case of Sindbis virus, the situation is complex and interesting. Primary isolates of the virus do not bind to heparan sulfate. Passage of the virus in cultured cells selects for viruses that do bind to heparan sulfate, and which infect cultured cells more efficiently. It is thought that selection for heparin sulfate binding upon passage of the virus in the laboratory speeds up the process of infection in cultured cells because virus bound to the cell surface by binding to heparin sulfate can diffuse in two dimensions rather than three to encounter its high-affinity receptor. In infected animals, however, heparin sulfate binding attenuates the virus, perhaps allowing the animal to clear the virus more quickly.

Many viruses absolutely require a coreceptor for entry, in addition to the primary receptor to which the virus first binds. The best studied example is HIV, which requires one of a number of chemokine receptors as a coreceptor. Thus a

TABLE 1.4   Viruses That Bind to Heparin-Like Glycosaminoglycans

Virus	Family	High affinity receptor
RNA viruses		
Sindbis	*Togaviridae*	High affinity laminin receptor
Dengue	*Flaviviridae*	???
Hepatitis C	*Flaviviridae*	CD81
Foot and mouth disease	*Picornaviridae*	$\alpha_v\beta_3$ integrin
Respiratory syncytial	*Paramyxoviridae*	???
Retroviruses		
HIV-1	*Retroviridae*	CD4 (Ig superfamily)
DNA viruses		
Vaccinia	*Poxviridae*	EGF receptor ???
Human papillomavirus	*Papillomaviridae*	Syndecan-1[a]
Herpes simplex	*Herpesviridae*	HIgR (CD155 family)
Adeno-associated type 2	*Parvoviridae*	FGFR1

[a] In this case the heparan sulfate proteoglycan appears to be the primary receptor protein.

Abbreviations used: EGF receptor, epidermal growth factor receptor; HIgR, herpes immunoglobulin-like receptor; CD155, the poliovirus receptor; FGFR1, human fibroblast growth factor receptor 1.

mouse cell that is genetically engineered to express human CD4 will bind HIV, but binding does not lead to entry of the virus into the cell. Only if the cell is engineered to express both human CD4 and a human chemokine receptor can the virus both bind to and enter into the cell. It is thought that binding to the first or primary receptor induces conformational changes in the virion that allow it to bind to the second or coreceptor.

The requirement for a coreceptor has important implications for the pathology of HIV. Chemokines are small proteins, secreted by certain cells of the immune system, that serve as chemoattractants for lymphocytes. They are important regulators of the immune system and are described in Chapter 10. Different classes of lymphocytes express receptors for different chemokines at their surface. To simplify the story, macrophage-tropic (M-tropic) strains of HIV, which is the virus most commonly transmitted sexually to previously uninfected individuals, require a coreceptor called CCR5 (a receptor for β chemokines). Human genetics has shown that two mutations can block the expression of CCR5. One is a 32-nucleotide deletion in the gene, the second is a mutation that results in a stop codon in the CCR5 open reading frame (ORF). The deletion mutation is fairly common, present in about 20% of Caucasians of European descent, whereas the stop codon mutation has been reported in only one individual. Individuals who lack functional CCR5 because they are homozygous for the deleted form, or in the case of one individual, heterozygous for the deletion but whose second copy of CCR5 has the stop codon, are resistant to infection by HIV. Heterozygous individuals who have only one functional copy of the CCR5 gene appear to be partially resistant. Although they can be infected with HIV, the probability of transmission has been reported to be lower, and once infected, progression to AIDS is slower. During the course of infection by HIV, T-cell-tropic strains (T-tropic) of HIV arise that require a different coreceptor, called CXCR4 (a receptor for α chemokines). After the appearance of T-tropic virus, both M-tropic and T-tropic strains cocirculate. The requirement for a new coreceptor is associated with mutations in the surface glycoprotein of HIV. The presence of T-tropic viruses is associated with more rapid progression to severe clinical disease.

Entry of Plant Viruses

Many plant viruses are important pathogens of food crops and have been intensively studied. No specific receptors have been identified to date, and it has been suggested that virus penetration of plant cells requires mechanical damage to the cell in order to allow the virus entry. Such mechanical damage can be caused by farm implements or by damage to the plant caused by insects such as aphids or leafhoppers that feed on the plants. Many plant viruses are transmitted

by insect or fungal pests, in fact, with which the virus has a specific association. There remains the possibility that specific receptors will be identified in the future, however, for at least some plant viruses.

Penetration

After the virus binds to a receptor, the next step toward successful infection is the introduction of the viral genome into the cytoplasm of the cell. In some cases, a subviral particle containing the viral nucleic acid is introduced into the cell. This particle may be the nucleocapsid of the virus or it may be an activated core particle. For other viruses, only the nucleic acid is introduced. The protein(s) that promotes entry may be the same as the protein(s) that binds to the receptor, or it may be a different protein in the virion.

For enveloped viruses, penetration into the cytoplasm involves the fusion of the envelope of the virus with a cellular membrane, which may be either the plasma membrane or an intracellular membrane. Fusion is promoted by a fusion domain that resides in one of the viral surface proteins. Activation of the fusion process is thought to require a change in the structure of the viral fusion protein that exposes the fusion domain. For viruses that fuse at the plasma membrane, interaction with the receptor appears to be sufficient to activate the fusion protein. In the case of viruses that fuse with intracellular membranes, the virus is internalized via various cellular vesicular pathways, which may differ depending upon the virus. The best studied internalization process is endocytosis into clathrin-coated vesicles and progression through the endosomal pathway. During transit, the clathrin coat is lost and the endosomes become progressively acidified. On exposure to a defined acidic pH (often ~5–6), activation of the fusion protein occurs and results in fusion of the viral envelope with that of the endosome. In either case, the nucleocapsid of the virus is present in the cytoplasm after fusion.

A dramatic conformational rearrangement of the hemagglutinin glycoprotein (HA) of influenza virus, a virus that fuses with internal membranes, has been observed by X-ray crystallography of HA following its exposure to low pH. HA, which is cleaved into two disulfide-bonded fragments HA_1 and HA_2, forms trimers that are present in a spike on the surface of the virion. The atomic structure of an HA monomer is illustrated in Fig. 1.6. HA_1 (shown in blue) is external and derived from the N-terminal part of the precursor. It contains the domain (indicated with a star in the figure) that binds to sialic acid receptors. HA_2 (shown in red) is derived from the C-terminal part of the precursor and has a C-terminal anchor that spans the viral membrane. The fusion domain (yellow) is present at the N terminus of HA_2, hidden in a hydrophobic pocket within the spike near the lipid bilayer of the virus envelope. Exposure to low pH results in a dramatic rearrangement

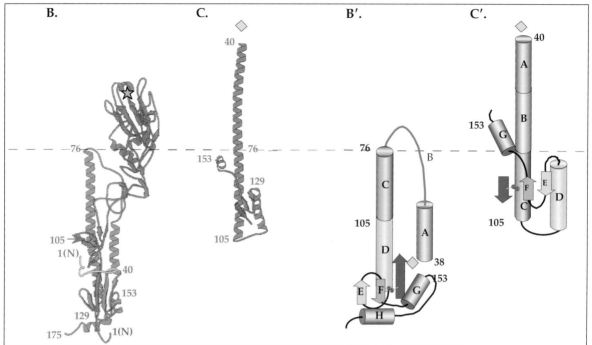

FIGURE 1.6   The folded structure of the influenza hemagglutinin and its rearrangement when exposed to low pH. (A) A schematic of the cleaved HA molecule. S is the signal peptide, TM is the membrane-spanning domain. HA_1 is in blue, HA_2 is in red, and the fusion peptide is shown in yellow. The same color scheme is used in (B) and (C). (B) X-ray crystallographic structure of the HA monomer. TM was removed by proteolytic digestion prior to crystallization. The receptor-binding pocket in HA_1 is shown with a green star. In the virion HA occurs as a trimeric spike. (C) The HA_2 monomer in the fusion active form. The fragment shown is produced by digesting with thermolysin, which removes most of HA_1 and the fusion peptide of HA_2. Certain residues are numbered to facilitate comparison of the two forms. The approximate location of the fusion peptide before thermolysin digestion is indicated with a yellow diamond. (B′) Diagrammatic representation of the HA_2 shown in (B), with α helices shown as cylinders and β sheets as arrows. The disulfide link between HA_1 and HA_2 is shown in ochre. The domains of HA_2 are color coded from N terminus to C terminus with a rainbow. (C′) Diagrammatic representation of the fusion-active form shown in (C). Redrawn from Fields *et al.* (1996) p. 1361, with permission.

of HA that exposes the hydrophobic peptide and transports it more than 100 Å upward, where it is thought to insert into the cellular membrane and promote fusion. It is assumed that similar events occur for all enveloped viruses, whether fusion is at the cell surface or with an internal membrane.

Studies with HIV have further refined our understanding of the fusion process. The external glycoproteins of HIV are also synthesized as a precursor that is cleaved into an N-terminal protein (called gp120) and a C-terminal, membrane-spanning protein (called gp41). Like the case for influenza (and many other enveloped viruses), the glycoproteins form trimers. A model for the process of fusion is shown in Fig. 1.7. The external gp120 binds to the receptor CD4 and then to the coreceptor chemokine. The fusion domain at the N terminus of gp41 rearranges and penetrates the host cell membrane. Two trimeric helical bundles in gp41 then

rearrange to form a hexameric helical bundle, which forces the cellular membrane and the viral membrane together, resulting in fusion. Fusion can be blocked by peptides that bind to one or the other of the trimeric bundles, preventing the formation of the hexameric bundle.

For nonenveloped viruses, the mechanism by which the virus breaches the cell membrane is less clear. After binding to a receptor, somehow the virus or some subviral component ends up on the cytoplasmic side of a cellular membrane, the plasma membrane for some viruses, or the membrane of an endosomal vesicle for others. It is believed that the interaction of the virus with a receptor, perhaps potentiated by the low pH in endosomes for those viruses that enter via the endosomal pathway, causes conformational rearrangements in the proteins of the virus capsid that result in the formation of a pore in the membrane. In the case of poliovirus, it

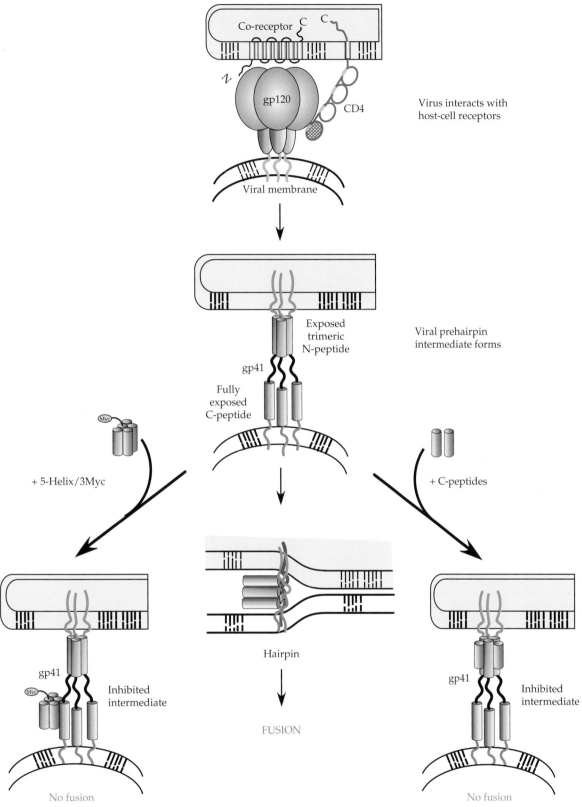

FIGURE 1.7   A model for HIV-1 membrane fusion and two forms of inhibition. In the native state gp120 partially shields gp41. When gp120 interacts with receptors and coreceptors on the host cell surface, gp41 undergoes a configurational rearrangement to the transient prehairpin intermediate, in which both the N and C peptides of gp41 are exposed. Fusion can be inhibited either by binding of C peptides to the trimeric N-peptide bundle or by binding of 5-helix/3Myc to a gp41 C peptide. Figure is adapted from Figure 6 in Koshiba and Chan (2003).

is known that interactions with receptors *in vitro* will lead to conformational rearrangements of the virion that result in the release of one of the virion proteins, called VP4. The N terminus of VP4 is myristylated and thus hydrophobic [myristic acid = $CH_3(CH_2)_{12}COOH$]. It is proposed that the conformational changes induced by receptor binding result in the insertion of the myristic acid on VP4 into the cell membrane and the formation of a channel through which the RNA can enter the cell. It is presumed that other viruses also have hydrophobic domains that allow them to enter. A number of other viruses also have a structural protein with a myristilated N terminus that might promote entry. In some viruses, there is thought to be a hydrophobic fusion domain in a structural protein that provides this function.

The entry process may be very efficient. In the case of enveloped viruses, there is evidence that at least for some viruses the specific infectivity in cultured cells can be one (all virions can initiate infection), and successful penetration is thought to be efficient for all enveloped viruses. For non-enveloped viruses, the situation varies. The specific infectivity of reoviruses assayed in cultured cells can be almost one but for other viruses, entry may be less efficient. For example, the specific infectivity of poliovirus in cultured cells is usually less than 1%. In general it is not known how such specific infectivites assayed in cultured cells relate to the infectivity of the virus when infecting host animals.

During entry of at least some viruses it is known that cellular functions must be activated and it is thought that binding of the virus to its receptor signals the cell to do something that is required for virus penetration. For example, binding of adenoviruses activates a pathway that results in polymerization of actin and endocytosis of the virus. As a second example, internalization of the polyomavirus SV40 is regulated by at least five different kinases. These activations of cellular pathways are only beginning to be unraveled.

Following initial penetration into the cytoplasm, further uncoating steps must often occur. It has been suggested that, at least in some cases, translation of the genomic RNA of plus-strand RNA viruses may promote its release from the nucleocapsid. In other words, the ribosomes may pull the RNA into the cytoplasm. In other cases, specific factors in the host cell, or the translation products of early viral transcripts, have been proposed to play a role in further uncoating.

It is interesting to note that bacteriophage face the problem of penetrating a rigid bacterial cell wall, rather than one of simply penetrating a plasma membrane or intracellular membrane. Many bacteriophage have evolved a tail by which they attach to the cell surface, drill a hole into the cell, and deliver the DNA into the bacterium. In some phage, the tail is contractile, leading to the analogy that the DNA is injected into the bacterium. Tailless phage are also known that introduce their DNA into the bacterium by other mechanisms.

Replication and Expression of the Virus Genome

The replication strategy of a virus, that is, how the genome is organized and how it is expressed so as to lead to the formation of progeny virions, is an essential component in the classification of a virus. Moreover, it is necessary to understand the replication strategy in order to decipher the pathogenic mechanisms of a virus and, therefore, to design strategies to interfere with viral disease.

DNA Viruses

A simple schematic representation of the replication of a DNA virus is shown in Fig. 1.8. After binding to a receptor and penetration of the genome into the cell, the first event in

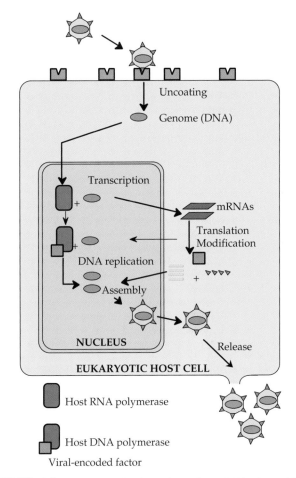

FIGURE 1.8 General replication scheme for a DNA virus. After a DNA virus attaches to a cellular membrane receptor, the virus DNA enters the cell and is transported to the cell nucleus. There it is transcribed into mRNA by host RNA polymerase. Viral mRNAs are translated by host ribosomes in the cytoplasm, and newly synthesized viral proteins, both structural and nonstructural, are transported back to the nucleus. After the DNA genome is replicated in the nucleus, either by the host DNA polymerase or by a new viral-encoded polymerase, progeny virus particles are assembled and ultimately released from the cell. Adapted from Mims *et al.* (1993) p. 2.3.

the replication of a DNA virus is the production of mRNAs from the viral DNA. For all animal DNA viruses except poxviruses, the infecting genome is transported to the nucleus where it is transcribed by cellular RNA polymerase. The poxviruses replicate in the cytoplasm and do not have access to host cell polymerases. Therefore, in poxviruses, early mRNA is transcribed from the incoming genome by a virus-encoded RNA polymerase that is present in the virus core. For all animal DNA viruses, translation of early mRNA is required for viral DNA replication to proceed. Early gene products may include DNA polymerases, proteins that bind to the origin of replication and lead to initiation of DNA replication, proteins that stimulate the cell to enter S phase and thus increase the supply of materials required for DNA synthesis, or products required for further disassembly of subviral particles.

The initiation of the replication of a viral genome is a specific event that requires an origin of replication, a specific sequence element that is bound by cellular and (usually) viral factors. Once initiated, DNA replication proceeds, catalyzed by either a cellular or a viral DNA polymerase. The mechanisms by which replication is initiated and continued are different for different viruses.

DNA polymerases, in general, are unable to initiate a polynucleotide chain. They can only extend an existing chain, following instructions from a DNA template. Replication of cellular DNA, including that of bacteria, requires the initiation of polynucleotide chains by a specific RNA polymerase called DNA polymerase α-primase, or primase for short. The resulting RNA primers are then extended by DNA polymerase. The ribonucleotides in the primer are removed after extension of the polynucleotide chain as DNA. Removal requires the excision of the ribonucleotides by a $5' \rightarrow 3'$ exonuclease, fill-in by DNA polymerase, and sealing of the nick by ligase. Because DNA polymerases can synthesize polynucleotide chains only in a $5' \rightarrow 3'$ direction, and cannot initiate a DNA chain, removal of the RNA primer creates a problem at the end of a linear chromosome. How is the 5′end of a DNA chain to be generated? The chromosomes of eukaryotic cells have special sequences at the ends, called telomeres, that function in replication to regenerate ends. The telomeres become shortened with continued replication, and eukaryotic cells that lack telomerase to repair the telomeres can undergo only a limited number of replication events before they lose the ability to divide.

Viruses and bacteria have developed other mechanisms to solve this problem. The chromosomes of bacteria are circular, so there is no 5′ end to deal with. Many DNA viruses have adopted a similar solution. Many have circular genomes (e.g., poxviruses, polyomaviruses, papillomaviruses). Others have linear genomes that cyclize before or during replication (e.g., herpesviruses). Some DNA viruses manage to replicate linear genomes, however. Adenoviruses use a virus-encoded protein as a primer, which remains covalently linked to the 5′ end of the linear genome. The single-stranded parvovirus DNA genome replicates via a foldback mechanism in which the ends of the DNA fold back and are then extended

by DNA polymerase. Unit sized genomes are cut from the multilength genomes that result from this replication scheme and are packaged into virions.

Once initiated, the progression of the replication fork is different in different viruses, as illustrated in Fig. 1.9. In SV40 (family *Polyomaviridae*), for example, the genome is circular. An RNA primer is synthesized by primase to initiate replication, and the replication fork then proceeds in both directions. The product is two double-strand circles. In the herpesviruses, the genome is circular while it is replicating but the replication fork proceeds in only one direction. A linear double-strand DNA is produced by what has been called a rolling circle. For this, one strand is nicked by an endonuclease and used as a primer. The strand displaced by the synthesis of the new strand is made double stranded by the same mechanism used by the host cell for lagging strand synthesis. In adenoviruses, in contrast, the genome is linear and the replication fork proceeds in only one direction. A single-strand DNA is displaced during the progression of the fork and coated with viral proteins. It can be made double stranded by an independent synthesis event. These different mechanisms will be described in more detail in the discussions of the different DNA viruses in Chapter 7.

As infection proceeds, most DNA viruses undergo a regular developmental cycle, in which transcription of early genes is followed by the transcription of late genes. Activation of the late genes may result from production of a new RNA polymerase or the production of factors that change the activity of existing polymerases so that a new class of promoters is recognized. The developmental cycle is, in general, more elaborate in the larger viruses than in the smaller viruses.

Plus-Strand RNA Viruses

A simple schematic of the replication of a plus-strand RNA virus is shown in Fig. 1.10. The virus example shown is enveloped and gives rise to subgenomic RNAs (see later). Although the details of RNA replication and virus release are different for other viruses, this scheme is representative of the steps required for gene expression and RNA replication.

Following entry of the genome into the cell, the first event in replication is the translation of the incoming genomic RNA, which is a messenger, to produce proteins required for synthesis of antigenomic copies, also called minus strands, of the genomic RNA. Because the replication cycle begins by translating the RNA genome to produce the enzymes for RNA synthesis, the naked RNA is infectious, that is, introduction of the genomic RNA into a susceptible cell will result in a complete infection cycle. The antigenomic copy of the genome serves as a template for the production of more plus-strand genomes. For some plus-strand viruses, the genomic RNA is the only mRNA produced, as illustrated schematically in Fig. 1.11A. It is translated into a polyprotein, a long, multifunctional protein that is cleaved by viral proteases, and sometimes also by

A. Bidirectional DNA Replication in SV40

B. Rolling Circle DNA Replication in Herpesvirus

C. Adenovirus DNA Replication by Displacement Synthesis

D. Parvovirus DNA Synthesis by Rolling Hairpin Mechanism

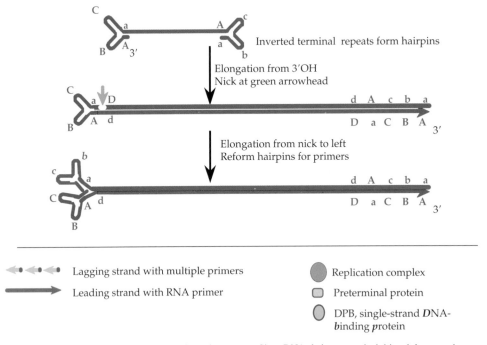

FIGURE 1.9   Models for DNA replication in various virus groups. Since DNA chains cannot be initiated *de novo*, viruses have used a variety of ways to prime new synthesis, such as (A) using RNA primers generated by a primase, (B) elongation from a 3'OH formed at a nick in a circular molecule, (C) priming by an attached protein, and (D) priming by hairpins formed of inverted terminal repeats. Adapted from Flint *et al.* (2000) Figures 9.8, 9.16, 9.10, and 9.9, respectively.

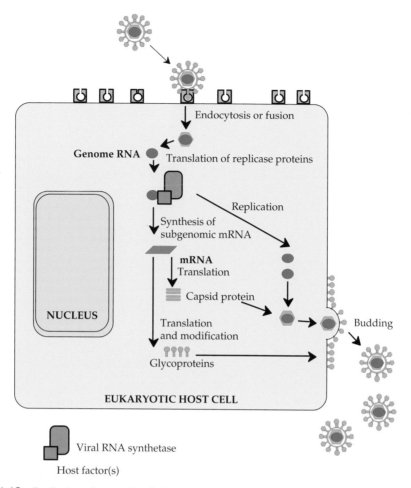

FIGURE 1.10   Replication of an enveloped, plus-strand RNA virus. After the virus attaches to a cellular receptor, fusion of the virus envelope with the cell plasma membrane or with an endocytic vesicle releases the nucleocapsid into the cytoplasm. The genome RNA is an mRNA, and is translated on cytoplasmic ribosomes into the proteins required for RNA synthesis. The synthetase complex can both replicate the RNA to produce new genomes and synthesize viral subgenomic mRNAs from a minus-strand copy of the genome. The viral structural proteins are then translated from these subgenomic mRNAs. In the example shown, the capsid protein assembles with the genome RNA to form a capsid, while the membrane glycoproteins are transported to the cell plasma membrane. In the final maturation step the nucleocapsid buds out through areas of modified membrane to release the enveloped particle. Adapted from Mims *et al.* (1993) p. 2.3 and Strauss and Strauss (1997) Figure 2.2.

cellular proteases, to produce the final viral proteins. For other plus-strand RNA viruses, one or more subgenomic mRNAs are also produced from the antigenomic template (Fig. 1.11B). For these viruses, the genomic RNA is translated into a polyprotein required for RNA replication (i.e., the synthesis of the antigenomic template and synthesis of more genomic RNA) and for the synthesis of the subgenomic mRNAs. The subgenomic mRNAs are translated into the structural proteins required for assembly of progeny virions. Some viruses, such as the coronaviruses (family *Coronaviridae*), which produce multiple subgenomic RNAs, also use subgenomic RNAs to produce nonstructural proteins that are required for the virus replication cycle but not for RNA synthesis.

The replication of the genome and synthesis of subgenomic RNAs require recognition of promoters in the viral RNAs by the viral RNA synthetase. This synthetase contains several pro-

teins encoded by the virus, one of which is an RNA polymerase. Cellular proteins are also components of the synthetase.

All eukaryotic plus-strand RNA viruses replicate in the cytoplasm. There is no known nuclear involvement in their replication. In fact, where examined, plus-strand viruses will even replicate in enucleated cells. However, it is known that for many viruses, virus-encoded proteins are transported to the nucleus, where they may inhibit nuclear functions. For example, a poliovirus protein cleaves transcription factors in the nucleus.

Minus-Sense and Ambisense RNA Viruses

The ambisense RNA viruses and the minus-sense viruses are closely related. One family, the *Bunyaviridae*, even contains both types of viruses as members. The ambisense strategy is, in fact, a simple modification of the minus-sense strategy,

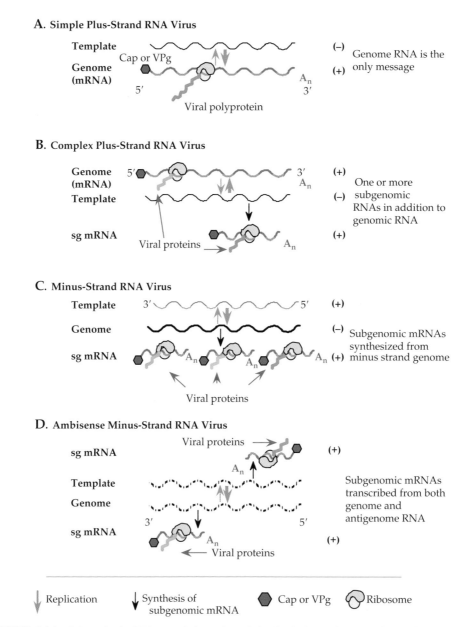

FIGURE 1.11   Schematic of mRNA transcription and translation for the four major types of RNA viruses.

and these viruses are generally lumped together as "negative-strand" or "minus-strand" RNA viruses (Table 1.2).

A simple schematic of the replication of a minus-sense or ambisense RNA virus is shown in Fig. 1.12. All of these viruses are enveloped. After fusion of the virus envelope with a host cell membrane (some enter at the plasma membrane, some via the endosomal pathway), the virus nucleocapsid enters the cytoplasm. The nucleocapsid is helical (Chapter 2). It remains intact and the viral RNA is never released from it. Because the viral genome cannot be translated, the first event after entry of the nucleocapsid must be the synthesis of mRNAs. Thus, the minus-sense or ambisense strategy requires that the viral RNA synthetase be an integral compo-

nent of an infectious virion and the naked RNA is not infectious if delivered into a cell.

Multiple mRNAs are synthesized by the enzymes present in the nucleocapsid. Each mRNA is usually monocistronic in the sense that it is translated into a single protein, not into a polyprotein (illustrated schematically in Fig. 1.11C). mRNAs are released from the nucleocapsid into the cytoplasm, where they are translated. The newly synthesized proteins are required for the replication of the genome.

Replication of the RNA requires the production of a complementary copy of the genome, as is the case for all RNA viruses, but the antigenomic or vcRNA (for virion complementary) is distinct from mRNA (Fig. 1.11C). Although

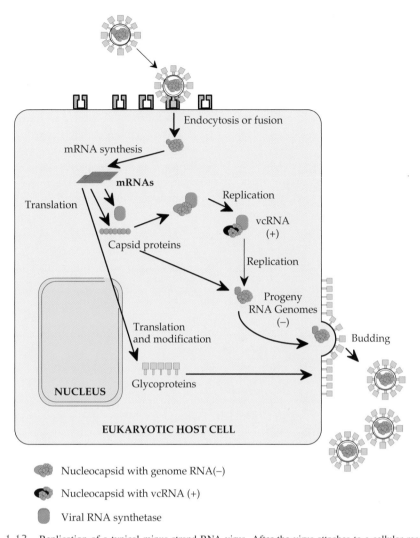

FIGURE 1.12   Replication of a typical minus-strand RNA virus. After the virus attaches to a cellular receptor, the nucleocapsid, containing the viral RNA synthetase, is released into the cytoplasm. The viral synthetase first synthesizes mRNAs, which are translated into the viral proteins required for synthesis of full-length complementary RNAs (vcRNAs). These vcRNAs are the templates for minus-strand genome RNA synthesis. Throughout replication, minus-strand genomes and plus-strand vcRNAs are present in nucleocapsids. Viral mRNAs are also translated into membrane glycoproteins that are transported to the cell plasma membrane (or in some cases specialized internal membranes). In the final maturation step, the nucleocapsid buds out through areas of modified membrane to release the enveloped particle. Adapted from Strauss and Strauss (1997) Figure 2.3 on p. 77.

technically plus sense, it is not translated and is always present in nucleocapsids with the associated RNA synthetic machinery. Replication requires ongoing protein synthesis to supply protein for encapsidation of the nascent antigenomic RNA during its synthesis. In the absence of such protein, the system defaults to the synthesis of mRNAs. The antigenomic RNA in nucleocapsids can be used as a template to synthesize genomic RNA if proteins for the encapsidation of the nascent genomic RNA are available.

In the ambisense viruses, the antigenomic RNA can also be used as a template for mRNA (Fig. 1.11D). Thus, ambisense viruses modify the minus-sense strategy by synthesizing mRNA from both the genome and the antigenome. Neither the genome nor the antigenome serves as mRNA.

The effect is to delay the synthesis of mRNAs that are made from the antigenomic RNA and thus to introduce a timing mechanism into the virus life cycle.

The mRNAs synthesized by minus-sense or ambisense viruses differ in several key features from their templates. First, the mRNAs lack the promoters required for encapsidation or replication of the genome or antigenome. Thus, they are not encapsidated and do not serve as templates for the synthesis of minus strand. Second, as befits their function as messengers, the mRNAs of most of these viruses are capped and polyadenylated, whereas genomic and antigenomic RNAs are not. Third, the mRNAs of the viruses in the families *Orthomyxoviridae*, *Arenaviridae*, and *Bunyaviridae* have 5′ extensions that are not present in the genome or antigenome, which, where well studied, are obtained from cellular mRNAs. Fourth, although most minus-strand and ambisense RNA viruses replicate in the cytoplasm, influenza virus and bornavirus RNA replication occurs in the nucleus. Thus, these RNAs have access to the splicing enzymes of the host. Two of the mRNAs of influenza viruses are exported in both an unspliced and a singly spliced version, and bornaviruses produce a number of spliced as well as unspliced mRNAs.

Double-Stranded RNA Viruses

The *Reoviridae*, the best studied of the double-strand RNA viruses, comprise a very large family of viruses that infect vertebrates, insects, and plants (Table 1.2). The genome consists of 10–12 pieces of double-strand RNA. The incoming virus particle is only partially uncoated. This partial uncoating activates an enzymatic activity within the resulting subviral particle or core that synthesizes an mRNA from each genome fragment. These mRNAs are extruded from the subviral particle and translated by the usual cellular machinery. Thus, the reoviruses share with the minus-strand RNA viruses the attribute that the incoming virus genome remains associated with virus proteins in a core that has the virus enzymatic machinery required to synthesize RNA, and the first step in replication, following entry into a cell, is the synthesis of mRNAs.

The mRNAs also serve as intermediates in the replication of the viral genome and the formation of progeny virions. After translation, the mRNAs become associated with virus proteins. At some point, complexes are formed that contain double-stranded forms of the mRNAs; in these complexes, the 10–12 segments are found in equimolar amounts. These complexes can mature into progeny virions. In other words, mRNAs eventually form the plus strands of the double-strand genome segments.

Retroviruses

An overview of the replication cycle of a retrovirus is shown in Fig. 1.13. The retroviruses are enveloped and enter

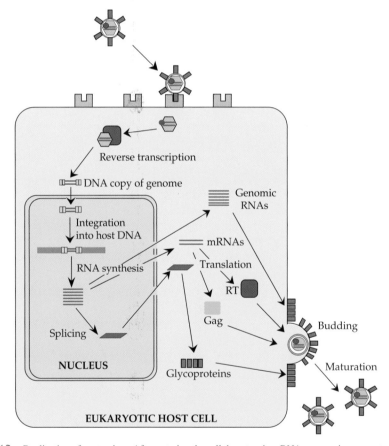

FIGURE 1.13   Replication of a retrovirus. After entering the cell the retrovirus RNA genome is reverse transcribed into double-stranded DNA by RT present in the virion. The DNA copy migrates to the cell nucleus and integrates into the host genome as the "provirus." Viral mRNAs are transcribed from proviral DNA by host cell enzymes in the nucleus. Both spliced and unspliced mRNAs are translated into viral proteins in the cytoplasm. The capsid precursor protein, "Gag," and RT are translated from full-length RNA. The glycoproteins are translated from spliced mRNA and transported to the cell plasma membrane. Immature virions containing Gag, RT, and the genome RNA assemble near the modified cell membrane. The final maturation step involves proteolytic cleavage of Gag by the viral protease and budding to produce enveloped particles. Adapted from Fields *et al.* (1996) p. 1786, and Coffin *et al.* (1997) p. 8.

the cell by fusion, some at the plasma membrane, some at an internal membrane. After entry, the first event is the production of a double-strand DNA copy of the RNA genome. This requires the activities of the enzymes RT and RNase H, which are present in the virion. RT synthesizes DNA from either a DNA or an RNA template. RNase H degrades the RNA strand of a DNA–RNA hybrid and is essential for reverse transcription of the genome. The mechanism by which the genome is reverse transcribed is complicated and is described in detail in Chapter 6.

The double-strand DNA copy of the genome is transported to the nucleus, where it integrates into host DNA. Integration is essentially random within the host genome and requires a recombinational event that is catalyzed by another protein present in the virus, called integrase. The integrated DNA copy, called a provirus, is transcribed by cellular RNA polymerases to produce an RNA that is identical to the viral genome. This RNA is exported to the cytoplasm either unspliced or as one or more spliced mRNAs.

The genomic RNA is a messenger for the translation of a series of polyproteins. These polyproteins contain the translation products of genes called *gag*, *pro*, and *pol*. Gag (group-specific *an*tigen) proteins form the capsid of the virus. Pro is a protease that processes the polyprotein precursors. Pol contains RT, RNase H, and integrase. The three genes are immediately adjacent in the genomic RNA, separated by translation stop codons whose arrangement and number depend on the virus. The arrangement of stop codons is described in detail in Chapter 6, and the mechanisms by which readthrough of stop codons occurs to produce longer polyproteins are described later in this chapter. A simple diagram of one retrovirus arrangement is shown in Fig. 1.14 as an example. In this example, the polyproteins translated from the genomic RNA are Gag and Gag–Pro–Pol. These two polyproteins assemble with two copies of the virus genome to form the capsid of the virus, usually at regions of the plasma membrane where virus glycoproteins are present. During and immediately after virus assembly by budding,

the viral protease cleaves Gag into several components and also separates the enzymes. Cleavage is essential for the assembled virion to be infectious. Thus, current inhibitors of HIV target the protease of the virus as well as the RT, both of which are required for replication, but neither of which is present in the uninfected cell.

The simple retroviruses also produce one spliced mRNA, which is translated into a precursor for the envelope glycoproteins. In some retroviruses, notably the lentiviruses, of which HIV is a member, differential splicing can also lead to the production of mRNAs for a number of regulatory proteins.

Hepadnaviruses

A schematic of the replication of an hepadnavirus is shown in Fig. 1.15. Hepadnaviruses, which are enveloped, have a life cycle that also involves alternation of the information in the genome between DNA and RNA. The incoming genome is circular, partially double-stranded DNA. The genome is transported to the nucleus where it is converted to a covalently closed, circular, double-strand DNA (cccDNA). Unlike the retroviruses, the DNA does not integrate into the host genome but persists in the nucleus as a nonreplicating episome. It is transcribed by cellular RNA polymerases, using several different promoters in the cccDNA, to produce a series of RNAs. These RNAs are exported to the cytoplasm where they serve as mRNAs. One of these RNAs, called the pregenomic (pg) RNA, is slightly longer than unit length and serves as a template for reverse transcription into DNA. Reverse transcription is performed by RT and RNase H that are translated from the viral mRNAs. It occurs in a core particle assembled from viral capsid proteins, the viral enzymes, and pgRNA. Reverse transcription, described in detail in Chapter 6, resembles that which occurs in the retroviruses, but differs in details. A complete minus-sense DNA is first synthesized by reverse transcription. Second strand plus-sense DNA is then initiated but only partially completed, so that the core contains partially double-stranded,

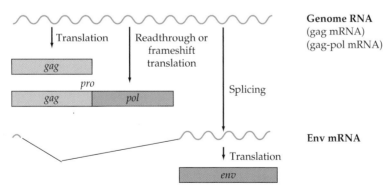

FIGURE 1.14 Transcription and translation of the retroviral genome. Three major polyproteins (shown as colored blocks) are produced. Gag is processed to form the nucleocapsid proteins, Pol contains RT, RNase H, and integrase, and Env is the precursor to the membrane glycoproteins. The protease, "pro," lies between *gag* and *pol* and may be in either the *gag* reading frame or the *pol* reading frame.

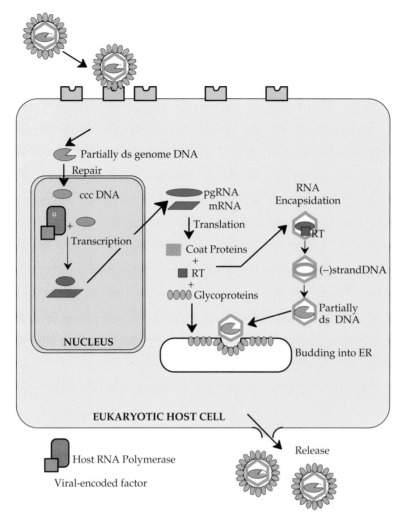

FIGURE 1.15   Simplified scheme of hepadnavirus replication. In the virion, the partially ds DNA genome consists of one full-length minus-strand DNA, and a plus-strand DNA of variable length. After the virus enters the cell, the genome is repaired to a closed circular coiled form (cccDNA) in the nucleus. RNAs, of several sizes, including pregenomic (pg) RNA of greater than unit length, are transcribed by host RNA polymerases. These mRNAs are exported and translated into viral proteins in the cytoplasm. The pgRNA is encapsidated, then reverse transcribed into minus-strand DNA. The last step before budding into the endoplasmic reticulum is the partial synthesis of (+) strand DNA. Scheme derived from Fields *et al.* (1996) p. 2709.

circular DNA (i.e., the genome). The core with its DNA can proceed through one of two pathways. Early in infection, newly assembled cores may serve to amplify the cccDNA present in the nucleus. These cores contain genomic DNA and are essentially indistinguishable from cores that enter the cytoplasm upon infection by a virion, and their genomic DNA can be transported to the nucleus and converted into cccDNA. Amplification of cccDNA occurs only through an RNA intermediate, using the pathway just described; there is no direct replication of the DNA in the nucleus. Later in infection, the cores mature into virions by budding through the endoplasmic reticulum. The switch to budding appears to be driven by the presence of viral envelope proteins.

Cellular Functions Required for Replication and Expression of the Viral Genome

The relationship between a virus and its host is an intimate one, shaped by a long history of coevolution. Viruses have small genomes and cannot encode all the functions required for successful replication and have borrowed many cellular proteins as components of their replication machinery. The nature of the interactions between virus proteins and cellular proteins is an important determinant of the host range and pathology of a virus.

All animal DNA viruses, with the exception of the poxviruses, replicate in the nucleus. They make use of the cellular machinery that exists there for the replication of their DNA and the transcription of their mRNAs. Some viruses use this machinery almost exclusively, whereas others, particularly the larger ones, encode their own DNA or RNA polymerases. However, almost all DNA viruses encode at least a protein required for the recognition of the origin of replication in their DNA. The interplay between the viral proteins and the cellular proteins can affect the host range of the virus. The monkey virus SV40 (family *Polyomaviridae*) will replicate in monkey cells but not in mouse cells, whereas the closely related mouse polyomavirus (also family *Polyomaviridae*) will replicate in mouse cells but not in monkey cells. The basis for the host restriction is an incompatibility between the DNA polymerase α-primase of the nonpermissive host and the T antigen of the restricted virus. T antigens are large multifunctional proteins, one of whose functions is to bind to the origin of replication. The T antigens of the viruses form a preinitiation complex on the viral origin of replication, which then recruits the primase into the complex. Because the preinitiation complex containing SV40 T antigen cannot recruit the mouse primase to form an initiation complex, SV40 DNA replication does not occur in mouse cells. However, replication will occur in mouse cells if they are transfected with the gene for monkey primase. Similarly, monkey primase is not recruited into the complex containing mouse polyoma virus T antigen and mouse polyoma virus does not replicate in monkey cells.

In the case of RNA viruses, there is no preexisting cellular machinery to replicate their RNA, and all RNA viruses must encode at least an RNA-dependent RNA polymerase. This RNA polymerase associates with other viral and host proteins to form an RNA replicase complex, which has the ability to recognize promoters in the viral RNA as starting points for RNA synthesis. Early studies on the RNA replicase of RNA phage Qβ showed that three cellular proteins were associated with the viral RNA polymerase and were required for the replication of Qβ RNA. These three proteins, ribosomal protein S1 and two translation elongation factors EF-Ts and EF-Tu, all function in protein synthesis in the cell. The virus appropriates these three proteins in order to assemble an active replicase complex. Recent studies on animal and plant RNA viruses have shown that a variety of cellular proteins also appear to be required for their transcription and replication. One interesting finding is that the animal equivalents of EF-Ts and EF-Tu are required for the activity of the replicase of vesicular stomatitis virus. This suggests that the association of these two translation factors with viral RNA replicases is ancient. Several other cellular proteins have also been found to be associated with viral RNA polymerases or with viral RNA during replication, but evidence for their functional role is incomplete.

Although our knowledge of the nature of host factors involved in the replication of viral genomes and the interplay between virus-encoded and host cell proteins is incomplete, it is clear that such factors can potentially limit the host range of a virus. The restriction of SV40 in cells that do not make a compatible primase was cited earlier in this section. As a second example, the replication of poliovirus in a restricted set of cells in the gastrointestinal tract and its profound tropism for neurons if it reaches the CNS was also described earlier. These tropisms exhibited by poliovirus do not correlate with the distribution of receptors for the virus, and are a result of restrictions on growth after entry of the virus. Thus in addition to a requirement for a specific receptor for the virus to enter a cell, there may be a need for specific host factors to permit replication once a virus enters a cell. The permissivity of a cell for virus replication after its entry, as well as the distribution of receptors for a virus, are major determinants of viral pathogenesis.

Translation and Processing of Viral Proteins

Viral mRNAs are translated by the cellular translation machinery. Most mRNAs of animal viruses are capped and polyadenylated. Thus, the translation pathways are the same as those that operate with cellular mRNAs, although many viruses interfere with the translation of host mRNAs to give the viral mRNAs free access to the translation machinery. However, there are mechanisms of translation and processing used by some viruses that have no known cellular counterpart. These appear to have evolved because of the special problems faced by viruses and are described below.

Cap-Independent Translation of Viral mRNAs

A 5'-terminal cap on an mRNA is normally required for its translation. A cellular cap-binding protein binds the cap as part of the translation initiation pathway. The mRNA is then scanned by the initiation complex, starting at the cap, and translation begins at a downstream AUG start codon that is present in a favorable context. However, some viruses, such as poliovirus and other members of the *Picornaviridae* and hepatitis C virus (genus *Hepacivirus*, family *Flaviviridae*), have uncapped mRNAs and use another mechanism for the initiation of translation. The 5' nontranslated regions (NTRs) of the RNAs of two picornaviruses, poliovirus and encephalomyelitis virus (EMCV), are illustrated schematically in Fig. 1.16. These 5' NTRs are long—more than 700 nucleotides. Within this 5' region is a sequence of about 400 nucleotides called an IRES (*i*nternal *r*ibosome *e*ntry *s*ite). Ribosomes bind to the IRES and initiate translation in a cap-independent fashion. It is known that the secondary structure of the IRES is critical for its function and that the position of the initiating AUG with respect to the IRES is important. Interestingly, IRES elements may have entirely different sequences in different viruses and, hence, different apparent higher order structures. Comparison of the IRES elements of poliovirus and EMCV in Fig. 1.16 illustrates the differences

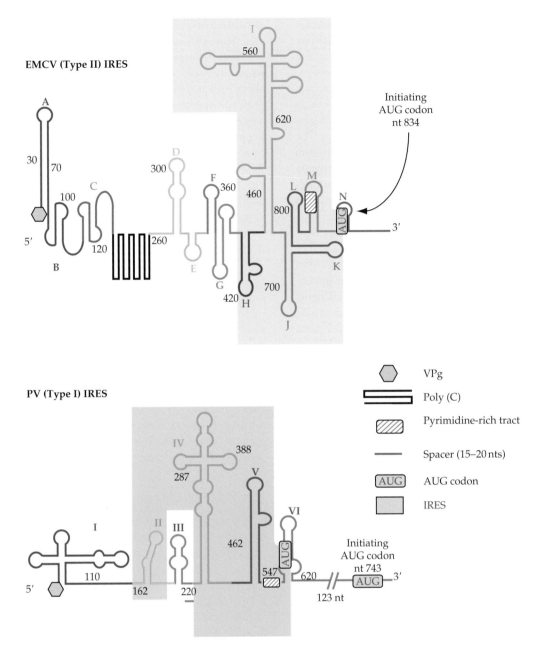

FIGURE 1.16 Schematic diagram of the secondary structures in the 5'-nontranslated regions of encephalomyelocarditis virus (EMCV) and poliovirus (PV). The *internal ribosome entry* sites, or IRES elements, are shaded. There is an element at the 3' border of the IRES consisting of a pyrimidine-rich tract (box with diagonal hatching), a spacer (magenta line) (usually 15–20 nucleotides) and an AUG codon. In the Type II IRES this AUG is the initiating codon, but in a Type I IRES a second codon, in this case 123 nucleotides further down, is used for initiation. Adapted from Wimmer *et al.* (1993) p. 374 with permission.

in apparent structure. Yet they promote internal initiation in a similar fashion, and the IRES elements of poliovirus and EMCV can be exchanged to yield a viable virus.

The IRES elements of viruses are always found within the 5' NTR. However, if an IRES is placed in the middle of an mRNA, it will function to initiate translation. True polycistronic mRNAs have been constructed by placing multiple IRES elements into mRNAs, or by combining cap-dependent translation of a 5' gene with an IRES-dependent translation of a 3' gene. Since it is possible to construct polycistronic mRNAs using IRES elements, it is somewhat puzzling that animal viruses have never used them to do so.

Using an IRES allows a virus to preferentially translate viral mRNAs. Thus, for example, poliovirus blocks cap-dependent translation after infection by cleaving the cellular cap-binding protein. By blocking cap-dependent translation, most host mRNAs cannot be translated (although IRESs are known to exist in some eukaryotic mRNAs), whereas viral mRNAs are not affected. This reserves the translation machinery for translation of viral mRNAs and also blocks many host defense mechanisms (Chapter 10).

Ribosomal Frameshifting

Retroviruses, many plus-strand RNA viruses, and certain other viruses expand the range of polyproteins produced by having long open reading frames (ORFs) interrupted by stop codons. Termination of the polyprotein at the stop codon leads to production of a truncated polyprotein having certain functions, whereas readthrough leads to the production of a longer polyprotein with additional functions. This was illustrated in Fig. 1.14 for a retrovirus. Two mechanisms are used by different viruses to ignore the stop codon. The first is simply to read the stop codon as sense. In this case the downstream sequences are in the same reading frame as the upstream sequences. The mechanism of readthrough is thought to involve wobble in the third codon position that allows a tRNA to bind to and insert an amino acid at the stop codon position. Both amber (UAG) and opal (UGA) codons have been found in different readthrough positions. Readthrough efficiency is variable, although usually between 5 and 20%.

The second mechansim for ignoring the stop codon is ribosomal frameshifting in which the reading frame is shifted into the plus 1 or minus 1 frame upstream of the stop codon. In this case, the downstream sequences are in a different reading frame from the upstream sequences. Fig. 1.17

shows the −1 frameshifting that occurs between the *gag* and *pol* genes of avian leukosis virus (ALV). Frameshifting occurs at a precise sequence known as a "slippery" sequence, shown in green. Slippery sequences have short strings of the same nucleotide and are often rich in A and U. For ALV, termination at the UAG indicated with the asterisk produces a polyprotein that contains the sequences for Gag and Pro. Frameshifting results in movement into the −1 frame at this point so that the codon following UUA (leucine) becomes AUA (isoleucine) rather than UAG (stop), and translation of the downstream gene (*pol*) follows to produce the polyprotein Gag–Pro–Pol.

Frameshifting usually requires a structural feature, such as a hairpin, downstream of the slippery sequence in order to slow the ribosome at this point and allow more time for the frameshifting to take place. The structure that is required for frameshifting in a yeast RNA virus called L-A is shown in Fig. 1.18. The sizable hairpin just downstream of the slippery sequence is further stabilized by a pseudoknot. In this case the slippery sequence, in which frameshifting takes place, occurs about 110 nucleotides upstream of the stop codon for the upstream reading frame (Gag protein reading frame). Thus, these 110 nucleotides are translated in two different reading frames: the Gag reading frame if frameshifting does not occur, or the Pol reading frame if frameshifting does occur. The sequence illustrated in Fig. 1.18 is the minimum sequence required for frameshifting. It can be placed in the middle of an unrelated mRNA and −1 frameshifting will occur.

Ribosomal Frameshift in Avian Leukosis Virus

Mechanism of the (−1) Frameshift

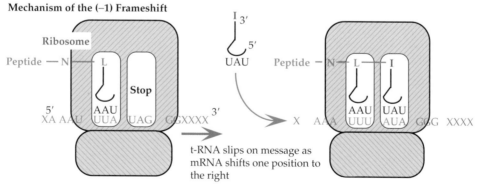

FIGURE 1.17   Proposed mechanism of the (−1) ribosomal frameshift that occurs in ALV (avian leukosis virus). The slippery sequence is shown in green. The asterisk identifies the UAG codon that terminates the upstream (Gag–Pro) ORF. Frameshifting is thought to require a pseudoknot downstream of the "slippery sequence" (illustrated in Fig. 1.18). Adapted from Fields *et al.* (1996) p. 577 and Goff (1997) Figure 3.15 on p. 156.

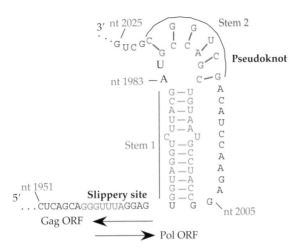

FIGURE 1.18 The site in L-A RNA that promotes –1 ribosomal frameshifting to form the Gag–Pol fusion protein. The pseudoknot makes the ribosome pause over the slippery site (nt 1958–1964 shown in green). The *gag* and *pol* ORFs overlap for 130 nucleotides and in the absence of frameshifting the Gag protein terminates at nt 2072. Adapted from Fields *et al.* (1996) p. 566.

The efficiency of frameshifting is variable and depends on the frameshifting sequences. Thus, the efficiency of frameshifting can be modulated by mutations in the virus. For example, the frameshifting sequence in Fig. 1.18 results in frameshifting about 2% of the time. Changing the slippery sequence from GGGUUU to UUUUUU results in frameshifting 12% of the time. Frameshifting efficiencies of from 2% to more than 20% have been observed at different frameshift sites in different viruses.

Processing of Viral Polyproteins

Many viruses produce polyproteins, as described earlier in this chapter. These polyproteins are cleaved into individual proteins by viral enzymes, with the exception of precursors to viral envelope proteins, which have access to cellular enzymes present in subcellular compartments. The use of viral enzymes to process polyproteins may be due in part to a lack of appropriate proteases in the cytoplasm of eukaryotes. However, the use of a viral protease also allows the virus to fine-tune the processing events, and this control is often used to regulate the viral replication cycle.

The viral proteases are components of the translated polyproteins. Some cleavages catalyzed by these proteases can occur *in cis*, in a monomolecular reaction in which the polyprotein cleaves itself. Other cleavages occur *in trans*, whereby a polyprotein, or a cleaved product containing the protease, cleaves another polyprotein in a bimolecular reaction. In many cases the series of cleavages effected by a protease proceeds in a defined manner and serves to regulate the virus life cycle.

Three types of virus proteases have been found in different viruses. Proteases related to the serine proteases of animals are common in animal RNA viruses. The animal serine pro-

teases, of which chymotrypsin is a well-studied example, have a catalytic triad composed of histidine, aspartic acid, and serine, which form the active site of the enzyme. Where structures have been determined, the viral proteases possess a fold related to that of chymotrypsin and possess an active site with geometry identical to that of chymotrypsin. The structure of protease 3C^pro of a rhinovirus (family *Picornaviridae*) is shown in Fig. 1.19A as an example (another example is shown in Chapter 2, Fig. 2.14B). Although most viral serine-type proteases have a catalytic triad composed of histidine, aspartic acid, and serine, the rhinovirus protease has an active site composed of histidine, glutamic acid, which replaces aspartic acid, and cysteine, which replaces serine. The replacement of the active site serine by cysteine causes problems with nomenclature. Here we use serine protease (or serine-type protease) to refer to a protease whose structure and active site geometry are related to those of the animal serine proteases, rather than as a description of the catalytic amino acid. These similarities in structure make it likely that the viruses acquired the protease from a host and modeled it to fit their own needs.

A second group of proteases is related to the papain-like enzymes of plants and animals. These proteases have a catalytic dyad consisting of histidine and cysteine. A third residue is sometimes considered a part of the active site (as described later) and the active site is then referred to as a catalytic triad. Model folding studies have suggested that these viral papain-like proteases are related to the cellular counterparts. Such proteases are found in many plus-strand RNA viruses and in adenoviruses. The structure of the adenovirus papain-like protease is shown in Fig. 1.19B. Notice that the structure of this enzyme is very different from the serine-type protease of rhinoviruses in Fig. 1.19A, even though the active site is composed of the same three residues, Cys, His, and Glu. In addition, the sequence of the active site residues in the linear amino acid sequence of serine-type proteases is His, Glu, Cys, whereas the order is Cys, His, Glu in the adenovirus protease and in papain (where the active site contains Asn rather than Glu).

The retroviruses encode a protease to process the Gag and Gag–Pol polyproteins during virus maturation. The protease is related to the aspartate proteases of animals, which include pepsin, renin, and cathepsin D. The structure of the HIV protease is shown in Fig. 1.19C. The active site of aspartate proteases consists of two aspartate residues. In the animal enzymes, the two aspartate residues are present in a single polypeptide chain that folds to bring the aspartate residues together to form the active site. In HIV, the active site is formed by dimerization of two Pro protein monomers of about 100 residues, each of which contributes one of the aspartate residues to the active site.

Assembly of Progeny Virions

The last stage in the virus life cycle is the assembly of progeny virions and their release from the infected cell. The

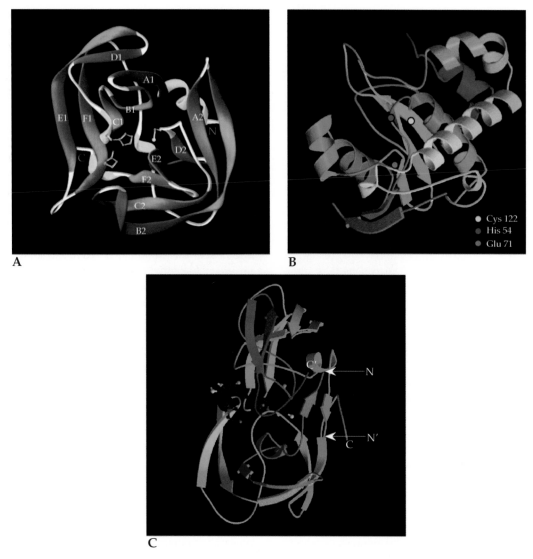

FIGURE 1.19 Divergent structures of viral proteases. (A) Ribbon diagram of human rhinovirus 3C^{pro}. The β strands are shown in blue and the helical secondary structure is shown in yellow. The side chains that make up the catalytic triad are shown: Cys-146 (with the sulfur atom shown in yellow), His-40, and Glu-71 (with the atoms of the charged carboxyl group shown in red). The N termini and C termini are indicated "N" and "C," respectively, in red. From Matthews *et al.* (1994) with permission. (B) Ribbon diagram of the papain-like protease of adenovirus complexed with its 11 amino acid cofactor (red arrow at bottom). The protein trace is colored from N terminus to C terminus as the visible spectrum from red to violet. The locations of the amino acids making up the catalytic triad (Cys-122, His-54, and Glu-71) are indicated with dots. From Ding *et al.* (1996), with permission. (C) The HIV protease, a dimer of identical subunits, is shown complexed with a non–amino acid inhibitor in the active site. One monomer is colored from red to green (N terminus to C terminus) while the second one is colored from green to dark blue. Adapted from Rutenber *et al.* (1993) with permission.

assembly of viruses will be discussed in Chapter 2, after the structure of viruses is described.

EFFECTS OF VIRUS INFECTION ON THE HOST CELL

Cells can be described as permissive, semipermissive, or nonpermissive for virus replication. Semipermissive or nonpermissive cells lack factors required for a complete replication cycle. The term *nonpermissive* usually refers to a cell in which no progeny virus are produced. A cell may be nonpermissive because it lacks receptors for the virus or because it lacks factors required by the virus after entry. In the latter case, an abortive infection may occur in which virus replication begins but does not result in the production of progeny virus. The term *semipermissive* usually refers to a cell in which a small yield of progeny virus may be produced.

Several types of viral infection cycle can be distinguished. The infection may be lytic, latent, persistent, or chronic. In some cases, virus infection results in the transformation of a cell.

Lytic Infection or Latent Infection

In a lytic infection, the virus replicates to high titer, host cell macromolecular synthesis is shut down, and the host cell dies. Bacterial cells are usually actively lysed by the elaboration of a specific lysis product during bacteriophage infection. Animal viruses, in contrast, usually cause cell death by inducing apoptosis or programmed cell death. Apoptosis is a suicide pathway in which the mitochrondria cease to function, the cell destroys its DNA, and the cell fragments into small vesicles (Chapter 10). Cell death may also be due to necrosis, a generalized loss of cell integrity caused by virus interference with activities necessary for the upkeep of the cell. Membrane integrity is lost during necrotic cell death and cytoplasmic contents leak out of the cell. Apoptosis is a normal event in animals and does not result in an inflammatory response. In contrast, necrosis does result in an inflammatory response.

During lytic infection, profound changes in the condition of the cell occur well before it dies and fragments. These changes may result in alterations that are observable in the light microscope, such as changes in the morphology of the cell, the formation of vacuoles within the cell, or the fusion of cells to form syncytia. Such changes are given the name cytopathic effect, or CPE. CPE is often an early sign that the cell is infected.

In latent infections, no virus replication occurs. The best understood case of latent infection is that of temperate bacteriophage, which express genes that repress the replication of the virus. Once the lysogenic state is established, in which viral replication is repressed, it can persist indefinitely. Among vertebrate viruses, many of the herpesviruses are capable of latently infecting specialized cells that are nonpermissive or semipermissive for virus replication. As one example, herpes simplex virus type 1 establishes a lifelong, latent infection of neurons of the trigeminal ganglia. In this case it is thought that latent state arises because the neuron lacks cellular factors required for the transcription and replication of the viral DNA, rather than because of the production of a herpes protein that suppresses replication. Reactivation of the virus at times leads to active replication of the virus in epithelial cells innervated by the infected neuron, resulting in fever blisters, usually at the lip margin.

Persistent Versus Chronic Infection

Persistent infection and chronic infection are often used interchangeably, but these two terms will be distinguished here in order to describe two types of infections that persist by different mechanisms. In what is here referred to as a persistent infection, an infected cell lives and produces progeny virus indefinitely. The retroviruses represent the best studied case of persistent infection. During infection by retroviruses, the DNA copy of the genome is integrated into the host cell genome. Continual transcription of the genome and assembly of progeny virus, which bud from the cell surface, occur without apparent ill effects on the host cell. The infected cell retains its normal functions and can divide. However, although infection by most retroviruses does not lead to cell death, active replication of HIV can result in cell death.

Chronic infection is a property of a group of cells or of an organism in which lytic infection is established in many cells, but many potentially susceptible cells escape the infection at any particular time, for whatever reason. The infection is not cleared and the continual appearance of susceptible cells in the population leads to the continued presence of replicating virus. One well-known example of a chronic infection in humans is HIV, in which the infection cannot be cleared by the immune system and the virus continues to replicate. AIDS results when the immune system is finally overwhelmed by the virus. Hepatitis B and hepatitis C viruses are also well known for their ability to establish chronic liver infections that can persist for life.

Transformation of Cells

The normal outcome of the infection of a cell by a virus is the death of the cell and the release of progeny virus. The major exceptions are the persistent infection of cells by retroviruses and the latent infection of cells by viruses such as herpesviruses, in which the cell survives with its properties little altered except for the new ability to produce virus. However, another possible outcome is the transformation of the cell, which involves not only the survival of the cell but an alteration in its growth properties caused by deregulation of the cell cycle. Transformed cells may be able to induce the formation of a tumor if they are produced within an animal or are injected into an animal after formation *ex vivo*. Transformation of a cell needs to be distinguished from tumorigenicity, the ability of the transformed cell to cause a tumor. Transformed cells may fail to cause a tumor because they are rejected by the host's immune system or because the transformed cells lack some properties required for the growth of a tumor in an animal, in which case additional mutations may eventually allow tumors to form.

The avian and mammalian sarcoma viruses, specialized retroviruses that arise when a cellular oncogene is incorporated into the retroviral genome, can transform cells in culture and cause tumors in animals. It was this feature that led to their discovery in the first place and resulted in intensive study of the retroviruses. Cellular oncogenes encode proteins that regulate the cell cycle. They induce the cell to enter S phase, in which DNA replication occurs, on receipt of appropriate signals. Following infection by a sarcoma virus and intergration of the provirus into the host genome, the overexpression of the incorporated oncogene, or expression of a mutated oncogene that continuously induces the cell to multiply, results in transformation of the infected cell. The incorporation of cellular oncogenes into a retrovirus is an accident that results from recombination of the viral genome with a cellular mRNA encoding an oncogene. The oncogene replaces viral genes in the genome and almost all sarcoma viruses are defective, unable to undergo a complete replication cycle without a helper. In nature sarcoma viruses are able to cause a tumor in the animal in which they arise but are not passed on, and thus die out. The subject of sarcoma viruses will be covered in more detail in Chapter 6, after a discussion of the genome organization of retroviruses and the details of their replication.

Many DNA viruses also encode proteins that are capable of transforming cells. These viral oncogenes induce cycling in infected cells, providing an environment suitable for the replication of the viral DNA. Whereas the cellular oncogenes present in sarcoma viruses serve no function in viral replication, the viral oncogenes in DNA viruses are essential for viral replication. If the cell is not induced to enter S phase, the virus replicates poorly or not at all. Infection of a cell by a DNA virus normally leads to the death of the cell caused by replication of the virus. Thus, although transformed, the infected cell does not survive. However, if the virus is unable to undergo a complete lytic cycle, either because it is defective or because the cell infected is nonpermissive or semipermissive, the cell may survive as a transformed cell if the early (transforming) genes continue to be expressed. We return to the subject of viral oncogenes in Chapter 7, because they are an ingredient of the replication cycle of DNA viruses.

Transformation of cells is accompanied by a number of phenotypic changes. Cells maintained in culture are normally derived from solid tissues and require anchorage to a solid substrate as well as a supply of nutrients, including growth factors, for growth. Transformed cells have a decreased requirement for a solid substrate and may have lost such a requirement entirely, and thus may grow in soft agar or in liquid culture. They have decreased requirements for growth factors and will continue to divide in medium with lowered amounts of such factors. They grow to higher densities in culture than do normal cells, and tend to pile up in multilayers on the surface of the culture dish (loss of contact inhibition or density-dependent growth). Transformed cells may be immortal and able to divide indefinitely in culture, whereas normal cells stop dividing after a limited number of divisions. Many of these changes in growth properties are reflected in changes in their cytoskeleton, their metabolism, and in their interactions with the extracellular matrix.

Lymphocytes, which do not require anchorage for growth, can also be transformed by the appropriate viruses.

They may become immortal and able to divide indefinitely in culture. In the animal this may lead to leukemias or lymphomas.

EPIDEMIOLOGY: THE SPREAD OF VIRUSES FROM PERSON TO PERSON

Viruses must be able to pass from one infected organism to another if they are to persist. The spread of specific viruses will be considered together with their other attributes in the chapters that follow, but it is useful to consider virus epidemiology in overview at this point. The tissues infected by a virus and the seriousness of the disease caused by it are attributes that determine in part the mechanism of spread of a virus. Thus, knowledge of the epidemiology of a virus is important for understanding the biology of its replication and pathology.

We can discriminate several general ways in which animal viruses are spread: oral–fecal, airborne, blood-borne (including viruses that are spread by bloodsucking arthropods), sexual, and congenital. We can also distinguish human viruses that have humans as their major or only host (referred to here as human viruses), and viruses that are also associated with other animals (referred to as zoonoses).

Viruses spread by an oral–fecal route are disseminated by ingestion of contaminated food or water. Infection begins in the gut, and it may or may not spread to other organs. Many of these viruses cause gastroenteritis. Virus is excreted in feces or urine to continue the cycle. Such viruses are usually fairly stable outside the organism because they may have to persist in an infectious form for long periods of time before being ingested by the next victim.

Airborne or respiratory viruses are spread when virus present in the respiratory tract is expelled as aerosols or in mucus. Infection begins when contaminated air is inhaled or when virus present in mucosal secretions, for example, on doorknobs or on a companion's hands, is contacted and the virus is transferred to mucosal surfaces in the nose, mouth, or eyes. These viruses are often unstable outside the body and spread usually requires close person-to-person contact. Infection begins on mucosal surfaces in the nose, the upper respiratory tract, or the eye. Many of these viruses are restricted to growth in the upper respiratory tract and cause respiratory disease, but some are able to spread to other organs and cause disseminated disease.

Blood-borne viruses establish a viremia in which infectious virus circulates in the blood. Some are transmitted by bloodsucking arthropods, which act as vectors, whereas others are transmitted by exposure to contaminated blood or other bodily fluids. Arboviruses (e.g., yellow fever virus, genus *Flavivirus*, family *Flaviviridae*) can replicate in both arthropods, such as ticks or mosquitoes, and in vertebrates. The arthropod may become infected when it takes a blood meal from a viremic vertebrate. After replication of the virus in the arthropod, it can be transmitted to a vertebrate when the arthropod takes another blood meal. Although arboviruses tend to have broad host ranges, a virus is usually maintained in only one or a few vertebrate hosts and vectored by a limited set of arthropods.

Therapeutic blood transfusion, use of hypodermic injections, and intravenous drug use are methods of spread of many blood-borne viruses. HIV, hepatitis B virus (family *Hepadnaviridae*), and hepatitis C virus (genus *Hepacivirus*, family *Flaviviridae*), for example, are commonly spread among drug users through sharing of contaminated needles. Transfusion with contaminated blood is still possible despite diagnostic tests to identify infected blood products. In developed countries, the blood supply is screened for HIV and hepatitis B and C viruses, as well as other viral agents for which tests exist, but in developing countries contaminated blood is often still a major problem. Blood-borne viruses that are not arboviruses are often spread sexually as well as by the methods stated earlier, but in some cases it is not clear how the viruses were spread before the introduction of blood transfusion and hypodermic needles.

Because of the need to establish a significant viremia, which requires extensive viral production in organs that can shed virus into the bloodstream, blood-borne viruses often cause serious disease. Furthermore, because spread is direct, these viruses need not be stable outside the body and usually have a short half-life outside an organism.

Many viruses are transmitted by sexual contact. Virus may be present in warts in the genital area (e.g., herpes simplex virus type 2 and human papillomaviruses) or in semen or vaginal secretions (e.g., HIV, hepatitis B virus). Infection begins in the genital mucosa but may spread to other organs. Because the opportunity for spread by sexual contact is much more restricted than for spread by other routes, viruses spread by sexual transmission almost invariably set up long-term chronic infections that cause only mild disease, at least early in infection. This allows the virus to be disseminated over long periods of time.

Many viruses can be spread vertically. Congenital infection of the fetus *in utero* or during passage of the infant through the birth canal occurs with viruses such as HIV, cytomegalovirus (family *Herpesviridae*), and rubella virus (genus *Rubivirus*, family *Togaviridae*). Vertical transmission can also occur shortly after birth, by breast-feeding, for example. HTLV I (family *Retroviridae*), which causes leukemia in humans, is such a virus.

Many of the viruses that cause human disease infect only humans in nature, or are maintained only in humans (e.g., all of the human herpesviruses, HIV, hepatitis B, poliovirus). Thus, spread is from one person to another.

Many others are associated with wildlife and spread is often from animal to man (e.g., most of the arboviruses, rabies virus, the hantaviruses, and the arenaviruses). The hantaviruses (family *Bunyaviridae*) and arenaviruses (family *Arenaviridae*) are associated with small rodents, in which they cause little disease. Humans, in which these viruses (e.g., Lassa fever virus; Junin virus, the causative agent of Argentine hemorrhagic fever; and Sin Nombre virus, the causative agent of hantavirus pulmonary syndrome) cause serious illness, can become infected by inhaling aerosols containing excreta from infected rodents. Rabies virus (family *Rhabdoviridae*) is associated with unvaccinated domestic dogs and with many species of wildlife, including foxes, coyotes, skunks, raccoons, and bats. It is spread by the bites of infected animals. The virus is present in salivary fluid of the infected animal, and the disease induces an infected animal to become aggressive and bite potential hosts. Interestingly, although many humans die worldwide of rabies each year contracted from the bites of rabid animals, human-to-human transmission does not occur. Some other human infections have been contracted from wildlife only indirectly. Nipah virus (family *Paramyxoviridae*) is associated with flying foxes, large fruit-eating bats. The virus recently caused an epidemic of disease in pigs in Malaysia and Singapore, and pig farmers contracted the disease from the pigs. Before the epidemic died out, 258 humans developed encephalitis, of which 40% died.

FURTHER READING

Effects of Infectious Disease on Human History

Crosby, A. W. (1989). *America's Forgotten Pandemic: The Influenza of 1918*. Cambridge, England, Cambridge University Press.

Gould, T. (1995). *A Summer Plague: Polio and Its Survivors*. New Haven, CT, Yale University Press.

Hopkins, D. R. (1983). *Princes and Peasants: Smallpox in History*. Chicago, University of Chicago Press.

Kolata, G. (1999). *Flu: The Story of the Great Influenza Pandemic of 1918 and the Search for the Virus that Caused It*. New York, Farrar, Straus, and Giroux.

Koprowski, H., and Oldstone, M. B. A. (Eds.) (1996). *Microbe Hunters Then and Now*. Bloomington, IL, Medi-Ed Press.

Oldstone, M. B. A. (1998). *Viruses, Plagues, and History*. New York, Oxford University Press.

Reid, A. M., Fanning, T. G., Janczewski, T. A., *et al*. (2000). Characterization of the 1918 "Spanish" influenza virus neuraminidase gene. *Proc. Natl. Acad. Sci. U.S.A.* **97**: 6785–6790.

Sharp, P. M. (2002). Origins of human virus diversity. *Cell* **108**: 305–312.

Virus Taxonomy

Fauquet, C. M., Mayo, M. A., Maniloff, J., Desselberger, U., and Ball, L. A. (Eds.) (2005). *Virus Taxonomy: Eighth Report of the International Committee on Taxonomy of Viruses*. Burlington, MA, Elsevier Academic Press.

Virus Receptors and Coreceptors, Entry, and Fusion

Barnard, R. J. O., Elleder, D., and Young, J. A. T. (2006). Avian sarcoma and leukosis virus-receptor interactions: From classical genetics to novel insights into virus-cell membrane fusion. *Virology* **344:** 25–29.

Barth, H., Schnober, E. K., Zhang, F., Linhardt, R. J., Depla, E., *et al.* (2006). Viral and cellular determinants of the hepatitis C virus envelope-heparan sulfate interaction. *J. Virol.* **80:** 10579–10590.

Bella, J., and Rossmann, M. G. (1999). Review: Rhinoviruses and their ICAM receptors. *J. Struct. Biol.* **128:** 69–74.

Belting, M. (2003). Heparan sulfate proteoglycan as a plasma membrane carrier. *Trends Biochem. Sci.* **28:** 145–151.

Bullough, P. A., Hughson, F. M., Skehel, J. J., *et al.* (1994). Structure of influenza hemagglutinin at the pH of membrane-fusion. *Nature* **371:** 37–43.

Clapham, P. R., and McKnight, A. (2002). Cell surface receptors, virus entry and tropism of primate lentiviruses. *J. Gen. Virol.* **83:** 1809–1829.

Cocchi, F., Menotti, L., Mirandola, P., *et al.* (1998). The ectodomain of a novel member of the immunoglobulin subfamily related to the poliovirus receptor has the attributes of a bona fide receptor for herpes simplex virus types 1 and 2 in human cells. *J. Virol.* **72:** 9992–10002.

Fuller, A. O., and Perez-Romero, P. (2002). Mechanisms of DNA virus infection: Entry and early events. *Front. Biosci.* **7:** 390–406.

de Haan, C. A. M., Li, Z., te Lintelo, E., *et al.* (2005). Murine coronavirus with an extended host range uses heparan sulfate as an entry receptor. *J. Virol.* **79:** 14451–14456.

Harrison, S. C. (2005). Mechanism of membrane fusion by viral envelope proteins. *Adv. Virus Res.* **64:** 231–261.

Kielian, M., and Rey, F. A. (2006). Virus membrane-fusion proteins: more than one way to make a hairpin. *Nature Rev. Microbiol.* **4:** 67–76.

Koshiba, T., and Chan, D. C. (2003). The prefusogenic intermediate of HIV-1 gp41 contains exposed C-peptide regions. *J. Biol. Chem.* **278:** 7573–7579.

Li, W., Moore, M. J., Vasilieva, N., *et al.* (2003). Angiotensin-converting enzyme 2 is a functional receptor for the SARS coronavirus. *Nature* **426:** 450–454.

Lopez, S., and Arias, C. F. (2006). Early steps in rotavirus cell entry. *Curr. Top. Microbiol. Immunol.* **309:** 39–66.

Marsh, M., and Helenius, A. (2006). Virus entry: open sesame. *Cell* **124:** 729–740.

Rey, F. A. (2006). Molecular gymnastics at the herpesvirus surface. *EMBO Rep.* **7:** 1000–1005.

Ryman, K. D., Klimstra, W. B., and Johnston, R. E. (2004). Attenuation of Sindbis virus variants incorporating uncleaved PE2 glycoprotein in correlated with attachment to cell-surface heparan sulfate. *Virology* **322:** 1–12.

Saragovi, H. U., Rebai, N., Roux, E., *et al.* (1998). Signal transduction and antiproliferative function of the mammalian receptor for type 3 reovirus. *Curr. Top. Microbiol. Immunol.* **233:** 155–166.

Sommerfelt, M. (1999). Retrovirus receptors. *J. Gen. Virol.* **80:** 3049–3064.

Spillmann, D. (2001). Heparan sulfate: anchor for viral intruders? *Biochimie* **83:** 811–817.

Stiasny, K., and Heinz, F. X. (2006). Flavivirus membrane fusion. *J. Gen. Virol.* **87:** 2755–2766.

Vivès, R. R., Lortat-Jacob, H., and Fender, P. (2006). Heparan sulphate proteoglycans and viral vectors: Ally or foe? *Curr. Gene Ther.* **6:** 35–44.

Wang, K.-S., Kuhn, R. J., Strauss, E. G., *et al.* (1992). High affinity laminin receptor is a receptor for Sindbis virus in mammalian cells. *J. Virol.* **66:** 4992–5001.

Weissenhorn, W., Dessen, A., Calder, L. J., *et al.* (1999). Structural basis for membrane fusion by enveloped viruses. *Mol. Membr. Biol.* **16:** 3–9.

Yanagi, Y., Takeda, M., and Ohno, S. (2006). *Measles virus*: cellular receptors, tropism, and pathogenesis. *J. Gen. Virol.* **87:** 2767–2779.

Viral Proteases

Byrd, C. M., and Hruby, D. E. (2006). Viral proteinases: targets of opportunity. *Drug Dev. Res.* **67:** 501–510.

de Francesco, R., and Steinkuhler, C. (2000). Structure and function of the hepatitis C virus NS3-NS4A proteinase. *Curr. Top. Microbiol.* **242:** 149–169.

Ding, J., McGrath, W. J., Sweet, R. M., *et al.* (1996). Crystal structure of the human adenovirus proteinase with its 11 amino acid cofactor. *EMBO J.* **15:** 1778–1783.

Matthews, D. A., Smith, W. W., Ferre, R. A., *et al.* (1994). Structure of the human rhinovirus 3C protease reveals a trypsin-like polypeptide fold, RNA-binding site, and means for cleaving precursor polyprotein. *Cell* **77:** 761–771.

Rutenber, E., Fauman, E. B., Keenan, R. J., *et al.* (1993). Structure of a nonpeptide inhibitor complexed with HIV-1 protease. Developing a cycle of structure-based drug design. *J. Biol. Chem.* **268:** 15343–15346.

Strauss, J. H., Ed. (1990). Viral Proteinases. *Semin.Virol.* **1:** 307–384.

Frameshifting, Pseudoknots, and IRES

Baird, S. D., Turcotte, M., Korneluk, R. G., and Holcik, M. (2006). Searching for IRES. *RNA—a publication of the RNA Society* **12:** 1755–1785.

Deiman, B. A. L. M., and Pleij, C. W. A. (1997). Pseudoknots: a vital feature in viral RNA. *Semin. Virol.* **8:** 166–175.

Dinman, J. D., Icho T., and Wickner, R. B. (1991). A-1 ribosomal frameshift in a double-stranded RNA virus of yeast forms a gag-pol fusion protein. *Proc. Natl. Acad. Sci. U.S.A.* **88:** 174–178.

Jang, S. K. (2006). Internal initiation: IRES elements of picornaviruses and hepatitis C virus. *Virus Res.* **119:** 2–15.

Kim, Y.-G., Su, L., Maas, S., *et al.* (1999). Specific mutations in a viral RNA pseudoknot drastically change ribosomal frameshifting efficiency. *Proc. Natl. Acad. Sci. U.S.A.* **96:** 14234–14239.

Stewart, S. R., and Semler, B. L. (1997). RNA determinants of picornavirus cap-independent translation initiation. *Semin. Virol.* **8:** 242–255.

Vagner, S., Galy, B., and Pyronnet, S. (2001). Irresistible IRES: Attracting the translation machinery to internal ribosome entry sites. *EMBO Rep.* **2:** 893–898.

2

The Structure of Viruses

INTRODUCTION

Virus particles, called virions, contain the viral genome encapsidated in a protein coat. The function of the coat is to protect the genome of the virus in the extracellular environment as well as to bind to a new host cell and introduce the genome into it. Viral genomes are small and limited in their coding capacity, which requires that three-dimensional virions be formed using a limited number of different proteins. For the smallest viruses, only one protein may be used to construct the virion, whereas the largest viruses may use 30 or more proteins. To form a three-dimensional structure using only a few proteins requires that the structure must be regular, with each protein subunit occupying a position at least approximately equivalent to that occupied by all other proteins of its class in the final structure (the principle of quasi-equivalence), although some viruses are now known to violate the principle of quasi-equivalence. A regular three-dimensional structure can be formed from repeating subunits using either helical symmetry or icosahedral symmetry principles. In the case of the smallest viruses, the final structure is simple and quite regular. Larger viruses with more proteins at their disposal can build more elaborate structures. Enveloped viruses may be quite regular in construction or may have irregular features, because the use of lipid envelopes allows irregularities in construction.

Selected families of vertebrate viruses are listed in Table 2.1 grouped by the morphologies of the virions. Also shown for each family is the presence or absence of an envelope in the virion, the triangulation number (defined later) if the virus is icosahedral, the morphology of the nucleocapsid or core, and figure numbers where the structures of members of a family are illustrated. Electron micrographs of five DNA viruses belonging to different families and of five RNA viruses belonging to different families are shown in Fig. 2.1. The viruses chosen represent viruses that are among the largest known and the smallest known, and are all shown to the same scale for comparison. For each virus, the top micrograph is of a virus that has been negatively stained, the middle micrograph is of a section of infected cells, and the bottom panel shows a schematic representation of the virus. The structures of these and other viruses are described next.

HELICAL SYMMETRY

Helical viruses appear rod shaped in the electron microscope. The rod can be flexible or stiff. The best studied example of a simple helical virus is tobacco mosaic virus (TMV). The TMV virion is a rigid rod 18 nm in diameter and 300 nm long (Fig. 2.2B). It contains 2130 copies of a single capsid protein of 17.5 kDa. In the right-hand helix, each protein subunit has six nearest neighbors and each subunit occupies a position equivalent to every other capsid protein subunit in the resulting network (Fig. 2.2A), except for those subunits at the very ends of the helix. Each capsid molecule binds three nucleotides of RNA within a groove in the protein. The helix has a pitch of 23 Å and there are $16\frac{1}{3}$ subunits per turn of the helix. The length of the TMV virion (300 nm) is determined by the size of the RNA (6.4 kb).

Many viruses are constructed with helical symmetry and often contain only one protein or a very few proteins. The popularity of the helix may be due in part to the fact that the length of the particle is not fixed and RNAs or DNAs of different sizes can be readily accommodated. Thus the genome size is not fixed, unlike that of icosahedral viruses.

TABLE 2.1   Listing of Selected Vertebrate Virus Families by Morphology

Morphology of particle/virus family	Enveloped	Triangulation number	Morphology of nucleocapsid	Figure numbers
Icosahedral				
Adenoviridae	No	$T = 25$	Not applicable	Figs. 2.1, 2.12
Reoviridae	No	$T = 13l^a$	Icosahedral	Figs. 2.1, 2.5, 2.11
Papillomaviridae	No	$P = 7d^{a,b}$	Not applicable	Figs. 2.1,[c] 2.5
Polyomaviridae	No	$P = 7d^{a,b}$	Not applicable	Figs. 2.1,[c] 2.5, 2.10
Parvoviridor	No	$T = 1$	Not applicable	Figs. 2.1, 2.5
Picornaviridae	No	$P = 3^b$	Not applicable	Figs. 2.1,[d] 2.5, 2.7, 2.8
Astroviridae	No		Not applicable	
Caliciviridae	No	$T = 3$	Not applicable	
Herpesviridae	Yes	$T = 16$	Icosahedral	Figs. 2.1, 2.5, 2.20
Togaviridae	Yes	$T = 4$	Icosahedral	Figs. 2.5, 2.14, 2.18
Flaviviridae	Yes	$T = 3$	Icosahedral	Figs. 2.5, 2.18
Irregular				
Poxviridae (ovoid)	Yes		Dumbbell	Figs. 2.1, 2.24
Rhabdoviridae (bacilliform)	Yes		Coiled helix	Figs. 2.1, 2.23
Spherical				
Retroviridae	Yes	?	Icosahedral?	Figs. 2.1, 2.21
Round[e]				
Coronaviridae	Yes		Helical or tubular	
Paramyxoviridae	Yes		Helical	Fig. 2.22
Orthomyxoviridae	Yes		Helical	Figs. 2.1, 2.22
Bunyaviridae	Yes		Helical	
Arenaviridae	Yes		Helical	
Hepadnaviridae	Yes	$T = 3$ & 4	Icosahedral	
Filamentous[c]				
Filoviridae	Yes		Helical	Fig. 2.19

[a]Two mirror image structures can be formed using $T = 13$ or $T = 7$, a symmetry referred to as "d" or "l."

[b]Because the subunits are not exactly equivalent, *Papillomaviridae*, *Polyomaviridae*, as well as poliovirus, have "pseudo-triangulation numbers" so are referred to as $P=7$, $P=7$, and $P=3$ symmetries respectively.

[c]*Papovaviridae*, referred to in Fig. 2.1 have now been separated into *Polyomaviridae* and *Papillomaviridae*.

[d]Enterovirus, referred to in Fig. 2.1 is a genus of the *Picornaviridae*.

[e]Virions are often pleiomorphic.

ICOSAHEDRAL SYMMETRY

Virions can be approximately spherical in shape, based on icosahedral symmetry. Since the time of Euclid, there have been known to exist only five regular solids in which each face of the solid is a regular polygon: the tetrahedron, the cube, the octahedron, the dodecahedron, and the icosahedron. The icosahedron has 20 faces, each of which is a regular triangle, and thus each face has threefold rotational symmetry (Fig. 2.3A). There are 12 vertices where 5 faces meet, and thus each vertex has fivefold rotational symmetry. There are 30 edges in which 2 faces meet, and each

edge possesses twofold rotational symmetry. Thus the icosahedron is characterized by twofold, threefold, and fivefold symmetry axes. The dodecahedron, the next simpler regular solid, has the same symmetry axes as the icosahedron and is therefore isomorphous with it in symmetry: the dodecahedron has 12 faces which are regular pentagons, 20 vertices where three faces meet, and 30 edges with twofold symmetry. The three remaining regular solids have different symmetry axes. The vast majority of regular viruses that appear spherical have icosahedral symmetry.

In an icosahedron, the smallest number of subunits that can form the three-dimensional structure is 60 (5 subunits at

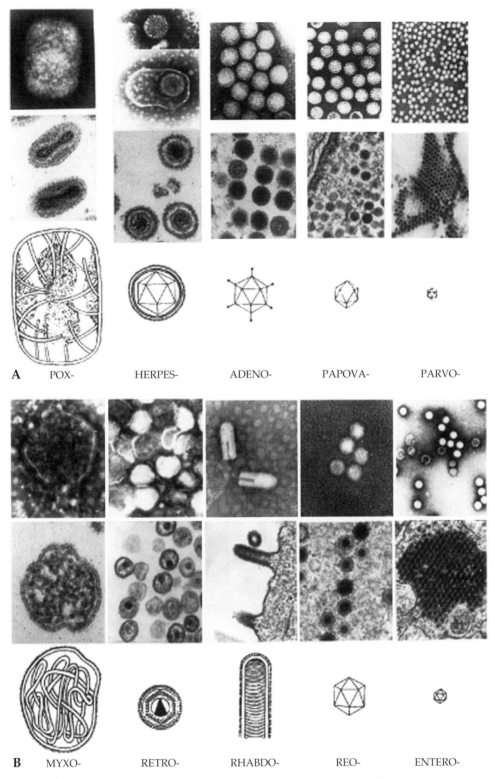

A   POX-   HERPES-   ADENO-   PAPOVA-   PARVO-

B   MYXO-   RETRO-   RHABDO-   REO-   ENTERO-

FIGURE 2.1   Relative sizes and shapes of representative (A) DNA- and (B) RNA- containing viruses. In each panel the top row shows negatively stained virus preparations, the second row shows thin sections of virus-infected cells, and the bottom row illustrates schematic diagrams of the viruses. Magnification of the electron micrographs is 50,000. From Granoff and Webster (1999), p. 401.

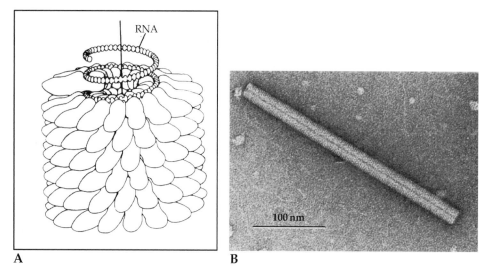

FIGURE 2.2   Structure of TMV, a helical plant virus. (A) Schematic diagram of a TMV particle showing about 5% of the total length. From Murphy *et al.* (1995), p. 434. (B) Electron micrograph of a negatively stained TMV rod. From www.ncbi.nlm.nih.gov.

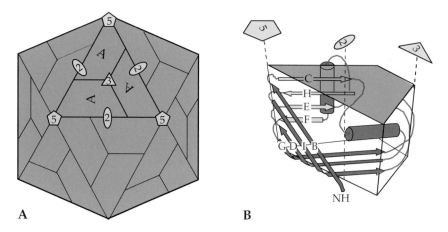

FIGURE 2.3   A simple icosahedral virus. (A) Diagram of an icosahedral capsid made up of 60 identical copies of a protein subunit, shown as blue trapezoids labeled "A." The twofold, threefold, and fivefold axes of symmetry are shown in yellow. This is the largest assembly in which every subunit is in an identical environment. (B) Schematic representation of the subunit building block found in many RNA viruses, known as the eightfold β barrel or β sandwich. The β sheets, labeled B through I from the N terminus of the protein, are shown as yellow and red arrows; two possible α helices joining these sheets are shown in green. Some proteins have insertions in the C–D, E–F, and G–H loops, but insertions are uncommon at the narrow end of the wedge (at the fivefold axis). From Granoff and Webster (1999) Vol. 3, color plate 31. [Originally from J. Johnson (1996)].

each of the 12 vertices, or viewed slightly differently, 3 units on each of the 20 triangular faces). Some viruses do in fact use 60 subunits, but most use more subunits in order to provide a larger shell capable of holding more nucleic acid. The number of subunits in an icosahedral structure is $60T$, where the permissible values of T are given by $T = H^2 + HK + K^2$, where H and K are integers and T is called the triangulation number. Permissible triangulation numbers are 1, 3, 4, 7, 9, 12, 13, 16, and so forth. A subunit defined in this way is not necessarily formed by one protein molecule, although in most cases this is how a structural subunit is in fact formed. Some viruses that form regular structures that are constructed using icosahedral symmetry principles do not possess true icosahedral symmetry. In such cases they are said to have pseudo-triangulation numbers. Examples are described later.

Structural studies of viruses have shown that the capsid proteins that form the virions of many plant and animal icosahedral viruses have a common fold. This fold, an eight-stranded antiparallel β sandwich, is illustrated in Fig. 2.3B. The presence of a common fold suggests that these capsid proteins have a common origin even if no sequence identity is detectable. The divergence in sequence while maintaining this basic fold is illustrated in Fig. 2.4, where capsid proteins of three viruses are shown. SV40 (family *Polyomaviridae*), poliovirus (family *Picornaviridae*), and bluetongue virus (family *Reoviridae*) are a DNA virus, a single-strand RNA virus, and a double-strand RNA virus, respectively. Their capsid proteins have insertions into the basic eight-stranded antiparallel β-sandwich structure that serve important functions in virus assembly. However, they all possess a region exhibiting the common β-sandwich fold and may have originated from a common ancestral protein. Thus, once a suitable capsid protein arose that could be used to construct simple icosahedral particles, it may ultimately have been acquired by many viruses. The viruses that possess capsid proteins with this fold may be related by descent from common ancestral viruses, or recombination may have resulted in the incorporation of this successful ancestral capsid protein into many lines of viruses.

Because the size of the icosahedral shell is fixed by geometric constraints, it is difficult for a change in the size of a viral genome to occur. A change in size will require a change in the triangulation number or changes in the capsid proteins sufficient to produce a larger or smaller internal volume. In either case, the changes in the capsid proteins required are relatively slow to occur on an evolutionary timescale and the size of an icosahedral virus is "frozen" for long periods of evolutionary time. For this reason, as well as for other reasons, most viruses have optimized the information content in their genomes, as will be clear when individual viruses are discussed in the following chapters.

Comparison of Icosahedral Viruses

Cryoelectron microscopy has been used to determine the structure of numerous icosahedral viruses to a resolution of 7 to 25 Å. For this, a virus-containing solution on an electron microscope grid is frozen very rapidly so that the sample is embedded in amorphous frozen water. The sample must be maintained at liquid nitrogen temperatures so that ice crystals do not form and interfere with imaging. Unstained, slightly out-of-focus images of the virus are captured on

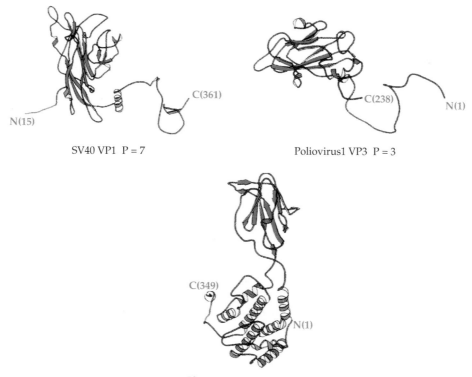

SV40 VP1 P = 7

Poliovirus1 VP3 P = 3

Bluetongue virus VP7 T = 13

FIGURE 2.4   Structure of three vertebrate virus protein subunits that assemble into icosahedral shells. The N termini and C termini are labeled with the residue number in parenthesis. The β barrels are shown as red arrows, α helices are gray coils, and the subunit regions involved in quasi-symmetric interactions that are critical for assembly are colored green. SV40 and PV have triangulation numbers of "pseudo-T=7" or P = 7 and "pseudo-T=3" or P = 3, respectively. Adapted from Granoff and Webster (1999), Vol. 3, plate 32.

film, or more recently captured electronically, using a low dose of electrons. These images are digitized and the density measured. Mathematical algorithms that take advantage of the symmetry of the particle are used to reconstruct the structure of the particle.

A gallery of structures of viruses determined by cryoelectron microscopy is shown in Fig. 2.5. All of the images are to scale so that the relative sizes of the virions are apparent. The largest particle is the nucleocapsid of herpes simplex virus, which is 1250 Å in diameter and has $T=16$ symmetry (the virion is enveloped but only the nucleocapsid is regular). The rotavirus and reovirus virions are smaller and have $T=13$. Human papillomavirus and mouse polyomavirus are pseudo-$T=7$. Ross River virus (RRV) (family *Togaviridae*) is enveloped but has regular symmetry, with $T=4$. Several examples of viruses with $T=3$ or pseudo-$T=3$ are shown (dengue 2, flock house, rhino-, polio-, and cowpea mosaic viruses, of which dengue 2 is enveloped but regular and the

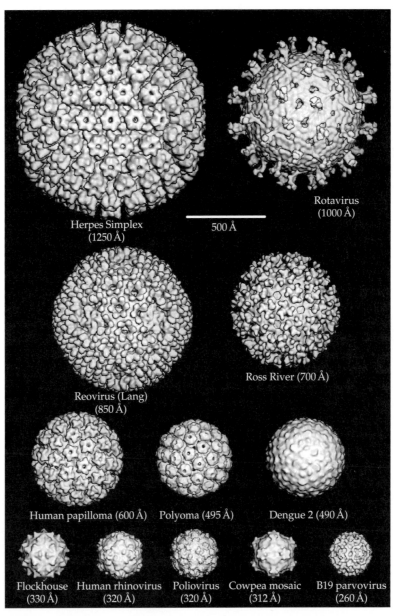

FIGURE 2.5 Gallery of three-dimensional reconstructions of icosahedral viruses from cryoelectron micrographs. All virus structures are surface shaded and are viewed along a twofold axis of symmetry except Ross River, which is viewed along a three-fold axis. All of the images are of intact virus particles except for the herpes simplex structure, which is of the nucleocapsid of the virus. Most of the images are taken from Baker *et al.* (1999), except the images of Ross River virus and of dengue virus, which were kindly provided by Drs. R. J. Kuhn and T. S. Baker.

rest are not enveloped). B19 parvovirus has $T=1$. The general correlation is that larger particles are constructed using higher triangulation numbers, which allows the use of larger numbers of protein subunits. Larger particles accommodate larger genomes.

Atomic Structure of $T=3$ Viruses

Because the simplest viruses are regular structures, they will often crystallize, and such crystals may be suitable for X-ray diffraction. Many viruses formed using icosahedral symmetry principles have been solved to atomic resolution, and a discussion of representative viruses that illustrate the principles used in construction of various viruses is presented here.

Among $T=3$ viruses, the structures of several plant viruses, including tomato bushy stunt virus (TBSV) (genus *Tombusvirus*, family *Tombusviridae*), turnip crinkle virus (TCV) (genus *Carmovirus*, family *Tombusviridae*), and Southern bean mosaic virus (SBMV) (genus *Sobemovirus*, not yet assigned to family), have been solved. All three of these viruses have capsid proteins possessing the eight-stranded antiparallel β sandwich. $T=3$ means that 180 identical molecules of capsid protein are utilized to construct the shell. The structures of two insect viruses that are also simple $T=3$ structures have also been solved. As an example of these simple structures, the $T=3$ capsid of the insect virus, flock house virus (family *Nodaviridae*), is illustrated in Fig. 2.6.

The 180 subunits in these $T=3$ structures interact with one another in one of two different ways, such that the protein shell can be thought of as being composed of an assembly of 60 AB dimers and 30 CC dimers (Fig. 2.6A). The bond angle between the two subunits of the dimer is more acute in the AB dimers than in the CC dimers (Figs. 2.6B and C). For the plant viruses, there are N-terminal and C-terminal extensions from the capsid proteins that are involved in interactions between the subunits and with the RNA. The N-terminal extensions have a positively charged, disordered domain for interacting with and neutralizing the charge on the RNA and a connecting arm that interacts with other subunits. In the case of the CC dimers, the connecting arms interdigitate with two others around the icosahedral three-fold axis to form an interconnected internal framework. In the case of the AB conformational dimer, the arms are disordered, allowing sharper curvature. For flock house virus, the RNA plays a role in controlling the curvature of the CC dimers, as illustrated in Fig. 2.6.

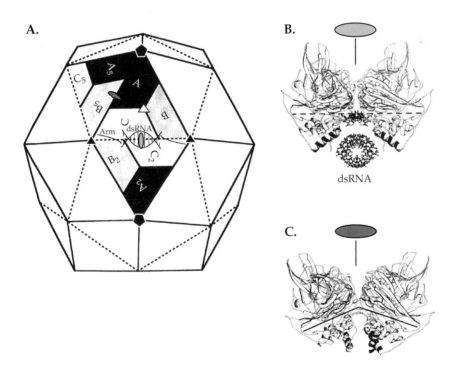

FIGURE 2.6   (A) Diagrammatic representation of a $T=3$ virus, flock house virus. The positions of the three identical proteins that make up a triangular face are only quasi-equivalent. The angle between the A and B_5 units (shown with a red oval and in diagram (C) is more acute than that along the C–C_2 edge, shown with a blue oval, and diagram (B). This difference in the angles is due to the presence of an RNA molecule located under the C–C_2 edge. From Johnson (1996), with permission.

Atomic Structure of Viruses Having Pseudo-*T*=3 Symmetry

The structures of several picornaviruses and of a plant comovirus (cowpea mosaic virus) have also been solved to atomic resolution. The structures of these viruses are similar to those of the plant *T*=3 viruses, but the 180 subunits that form the virion are not all identical. A comparison of the structure of a *T*=3 virus with those of poliovirus and of cowpea mosaic virus is shown in Fig. 2.7. Poliovirus has 60 copies of each of three different proteins, whereas the comovirus has 60 copies of an L protein (each of which fills the niche of two units) and 60 copies of an S protein. All three poliovirus capsid proteins have the eight-stranded antiparallel β-sandwich fold. In the comoviruses, the L protein has two β-sandwich structures fused to form one large protein, and the S protein is formed from one sandwich. The structures of the picornavirus and comovirus virions are called pseudo-*T*=3 or *P*=3, since they are not true *T*=3 structures.

The picornavirus virion is 300 Å in diameter. The 60 molecules of each of the three different proteins have different roles in the final structure, as illustrated in Fig. 2.8, in which the structure of a rhinovirus is shown. Notice that five copies of VP1 are found at each fivefold axis (compare Fig. 2.7 with Fig. 2.8). VP1, VP2, and VP3 are structurally related to one another, as stated, all possessing the common β-sandwich fold. There exists a depression around each fivefold axis of rhinoviruses that has been termed a "canyon." This depression is believed to be the site at which the virus interacts with the cellular receptor during entry, as illustrated in Fig. 2.9. This interaction is thought to lead to conformational changes that open a channel at the fivefold axis, through which VP4 is extruded, followed by the viral RNA.

FIGURE 2.8 Three-dimensional space-filling model of the human rhinovirus 14 virion, based upon X-ray crystallographic data. VP1 is shown in blue, VP2 in green, and VP3 in red. VP4 is interior and not visible in this view. This figure was kindly provided by Dr. Michael Rossmann.

Atomic Structure of Polyomaviruses

The structures of both mouse polyomavirus and of SV40 virus, two members of the family *Polyomaviridae*, have been solved to atomic resolution. Both viruses possess pseudo-*T*=7 icosahedral symmetry. Although *T*=7

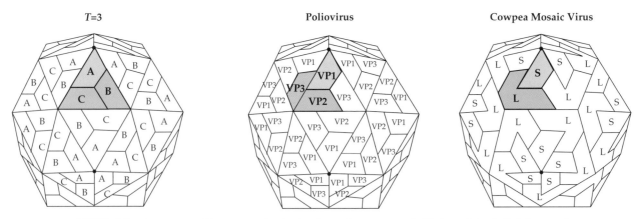

FIGURE 2.7 Arrangement of the coat protein subunits of comoviruses compared with those of simple *T*=3 viruses and picornaviruses. In simple viruses, the asymmetric unit contains three copies of a single protein β sandwich, labeled A, B, and C in order to distinguish them. In picornaviruses such as poliovirus the asymmetric unit is made up of three similar but not identical proteins, all of which have the β-sandwich structure. In comoviruses such as cowpea mosaic virus, two of the β-sandwich subunits are fused to give the L protein. Adapted from Granoff and Webster (1999), p. 287.

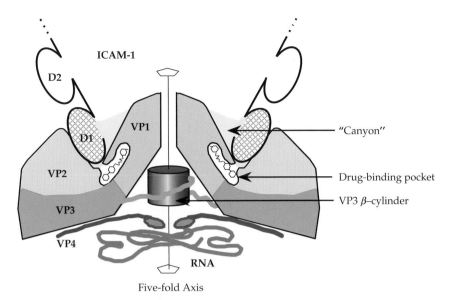

FIGURE 2.9   Binding of the rhinovirus receptor, ICAM-1, to the "canyon" at a fivefold vertex of a virion of a major group human rhinovirus. The colors of the three virion proteins are the same as those shown in the surface view in Fig. 2.8. The distal two domains of ICAM-1 are represented schematically (cross-hatched) as they were in Fig. 1.5. The amino-terminal domains of the five VP3 molecules around the fivefold axis form a five-stranded β cylinder on the virion's interior and are thought to stabilize the pentamer. Below the canyon is the hydrophobic pocket where certain antiviral drugs (indicated schematically in black) are known to bind. Adapted from Kolatkar *et al.* (1999).

symmetry would require 420 subunits, these viruses contain only 360 copies of a major structural protein known as VP1. These 360 copies are assembled as 72 pentamers. Twelve of the 72 pentamers lie on the fivefold axes and the remaining 60 fill the intervening surface in a closely packed array (Fig. 2.10). These latter pentamers are thus sixfold coordinated and the proteins in the shell are not all in quasi-equivalent positions, a surprising finding for our understanding of the principles by which viruses can be constructed. The pentamers are stabilized by interactions of the β sheets between adjacent monomers in a pentamer (Fig. 2.10C). The pentamers are then tied together by C-terminal arms of VP1 that invade monomers in an adjacent pentamer (Figs. 2.10B and C). Because each pentamer that is sixfold coordinated has five C-terminal arms to interact with six neighboring pentamers, the interactions between monomers in different pentamers are not all identical (Fig. 2.10B). Flexibility in the C-terminal arm allows it to form contacts in different ways.

Atomic Structure of Bluetongue Virus

Members of the reovirus family are regular *T*=13 icosahedral particles. They are composed of two or three concentric protein shells. Cryoelectron microscopy has been used to

solve the structure of one or more members of three genera within the *Reoviridae*, namely *Reovirus, Rotavirus,* and *Orbivirus*, to about 25-Å resolution. Structures of a reovirus and of a rotavirus are shown in Fig. 2.5. The complete structure of virions has not been determined because of their large size, but in a remarkable feat the atomic structure of the core of bluetongue virus (genus *Orbivirus*) has now been solved. This is the largest structure determined to atomic resolution to date. Solution of the structure was possible because the virus particle had been solved to 25 Å by cryoelectron microscopy, and the structures of a number of virion proteins had been solved to atomic resolution by X-ray diffraction. Fitting the atomic structure of the proteins into the 25-Å structure gave a preliminary reconstruction at high resolution, which allowed the interpretation of the X-ray data to atomic resolution.

Core particles are formed following infection, when the outer layer is proteolytically cleaved (described in more detail in Chapter 5). The structure of the inner surface of the bluetongue virus core is shown in Fig. 2.11A and of the outer surface in Fig. 2.11B. The outer surface is formed by 780 copies of a single protein, called VP7, in a regular *T*=13 icosahedral lattice. The inner surface is surprising, however. It is formed by 120 copies of a single protein, called VP3. These 120 copies have been described as forming a *T*=2

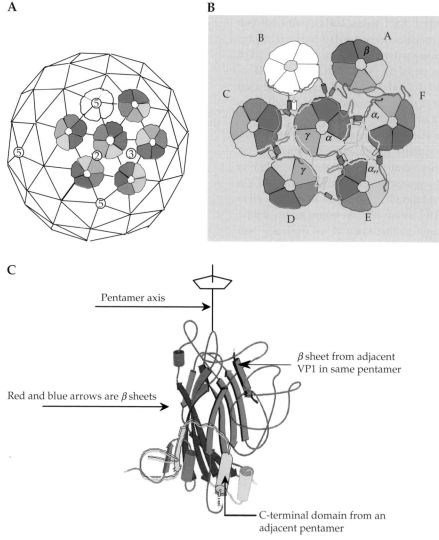

FIGURE 2.10 Organization of the capsid of the polyomavirus SV40. (A) Arrangement of the strict pentamers (white) and quasi pentamers (colored) on the T=7d icosahedral lattice. (B) Schematic showing the pattern of interchange of arms in the virion. The central pentamer shares "arms" with six neighboring pentamers. (C) A single VP1 subunit, viewed normal to the pentamer axis. The N-terminal domain is green, the C-terminal domain is yellow, and the β sandwich is shown as arrows of blue and red. The central yellow C-terminal domain (outlined in black) comes from an adjacent pentamer and the β sheet outlined in black comes from the neighboring VP1 within the same pentamer. From Figure 1 Stehle *et al.* (1996) and Fields *et al.* (1996), Color Plate 4.

lattice. Because T=2 is not a permitted triangulation number, these 120 copies, strictly speaking, form a T=1 lattice in which each unit of the lattice is composed of two copies of VP3. However, the interactions are not symmetrical, leading to the suggested terminology of T=2.

It has been suggested that the inner core furnishes a template for the assembly of the T=13 outer surface. The reasoning is that a T=13 structure may have difficulty in forming, whereas the T=2 (or T=1) structure could form readily. In this model, the threefold symmetry axis of the inner surface could serve to nucleate VP7 trimers and organize the T=13 structure.

Structure of Adenoviruses

Cryoelectron microscopy has also been applied to adenoviruses, which have a triangulation number of 25 or pseudo-25. Various interpretations of the structure of adenoviruses, both schematic and as determined by microscopy or crystallography, are shown in Fig. 2.12. Three copies of a protein called the hexon protein associate to form a structure called a hexon (Fig. 2.12C). The hexon is the basic building block of adenoviruses. Five hexons, called peripentonal hexons, surround each of the 12 vertices of the icosahedron (which, as has been stated, have fivefold rotational symmetry). Between the

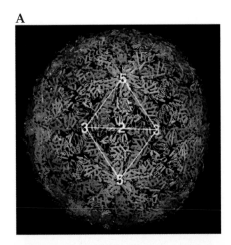

FIGURE 2.11   The essential features of the orbivirus native core particle. The asymmetric unit is indicated by the white lines forming a triangle and the fivefold, threefold, and twofold axes are marked. (A) The inner capsid layer of the bluetongue virus (BTV) core is composed of 120 molecules of VP3, arranged in what has been called $T=2$ symmetry. Note the green subunit A and the red subunit B which fill the asymmetric unit. (B) The core surface layer is composed of 780 copies of VP7 arranged as 260 trimers, with $T=13$ symmetry. The asymmetric unit contains 13 copies of VP7, arranged as five trimers, labeled P, Q, R, S, and T, with each trimer a different color. Trimer "T" in blue sits on the icosahedral threefold axis and thus contributes only a monomer to the asymmetric unit. From Granoff and Webster (1999), Color Plate 17.

groups of peripentonal hexons are found groups of 9 hexons, which are sixfold coordinated. Each group of 9 hexons forms the surface of one of the triangular faces. Thus, there are 60 peripentonal hexons and 180 hexons in groups of nine.

The structure of the hexon trimer has been solved to atomic resolution by X-ray crystallography (Fig. 2.12C).

The hexon protein has two eight-strand β sandwiches to give the trimer an approximately sixfold symmetry (and each hexon protein fills the role of two symmetry units). There are long loops that intertwine to form a triangular top. These structures can be fitted uniquely into the envelope of density determined by cryoelectron microscopy, which produces a structure refined to atomic resolution for most of the capsid. The minor proteins can be fitted into this structure.

From the 12 vertices of the icosahedron project long fibers that are anchored in the surface by a unit called the penton base. Each fiber terminates in a spherical extension that forms an organ of attachment to a host cell (Fig. 2.12A). The length of the fiber differs in the different adenoviruses.

NONENVELOPED VIRUSES WITH MORE COMPLICATED STRUCTURAL FEATURES

In addition to the nonenveloped viruses that possess relatively straightforward icosahedral symmetry or helical symmetry, many viruses possess more complicated symmetries made possible by the utilization of a large number of structural proteins to form the virion. The tailed bacteriophages are prominent examples of this (Fig. 2.13). Some of the tailed bacteriophages possess a head that is a regular icosahedron (or, in at least one case, an octahedron) connected to a tail that possesses helical symmetry. Other appendages, such as baseplates, collars, and tail fibers, may be connected to the tail. Other tailed bacteriophages have heads that are assembled using more complicated patterns. For example, the T-even bacteriophages have a large head, which can be thought of as being formed of two hemi-icosahedrons possessing regular icosahedral symmetry, which are elongated in the form of a prolate ellipsoid by subunits arranged in a regular net connecting the two icosahedral ends of the head of the virus.

ENVELOPED VIRUSES

Many animal viruses and some plant viruses are enveloped; that is, they have a lipid-containing envelope surrounding a nucleocapsid. The lipids are derived from the host cell. Although there is some selectivity and reorganization of lipids during virus formation, the lipid composition in general mirrors the composition of the cellular membrane from which the envelope was derived. However, the proteins in the nucleocapsid, which may possess either helical or icosahedral symmetry, and the proteins in the envelope are encoded in the virus. The protein–protein interactions that are responsible for assembly of the mature enveloped virions differ among the different families and the structures of the resulting virions differ. The virions of alphaviruses, and of flaviviruses, are uniform

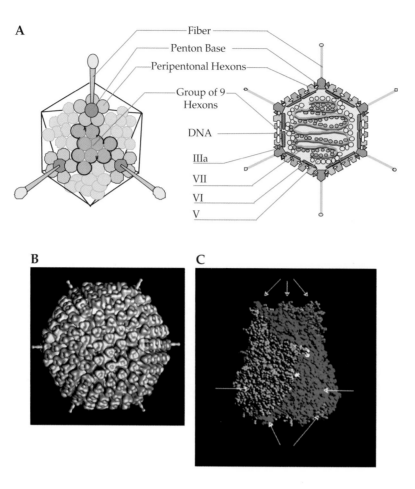

FIGURE 2.12   Structure of adenovirus particles. (A) Schematic drawing of the outer shell of an adenovirus (left), and a schematic cross section through an adenovirus particle, showing the locations of minor polypeptide components (right). The virus is composed of 60 peripentonal hexons at the bases of the fibers at the fivefold vertices, and groups of 9 hexons, one on each triangular face of the icosahedron. (B) Cryoeletron microscopic reconstruction of an adenovirus virion, viewed down the threefold axis. (C) Space-filling model of the hexon trimer, with each subunit in a different color. The atomic structure of the hexon has been solved and fitted into the cryoelectron microscopic reconstruction. The locations of the minor constitutents, indicated schematically in A, were deduced by subtraction. (A) is from Fields *et al.* (1996) p. 80, (B) is from Stewart *et al.* (1991), and (C) is from Athappilly *et al.* (1994).

structures that possess icosahedral symmetry. Poxviruses, rhabdoviruses, and retroviruses also appear to have a regular structure, but there is flexibility in the composition of the particle and the mature virions do not possess icosahedral symmetry. The herpesvirus nucleocapsid is a regular icosahedral structure (Fig. 2.5), but the enveloped herpesvirions are not regular. Other enveloped viruses are irregular, often pleiomorphic, and are heterogeneous in composition to a greater or lesser extent. The structures of different enveloped viruses that illustrate these various points are described next.

The Nucleocapsid

The nucleocapsids of enveloped RNA viruses are fairly simple structures that contain only one major structural protein, often referred to as the nucleocapsid protein or core protein. This protein is usually quite basic or has a basic domain. It binds to the viral RNA and encapsidates it to form the nucleocapsid. For most RNA viruses, nucleocapsids can be recognized as distinct structures within the infected cell and can be isolated from virions by treatment with detergents that dissolve the envelope. The nucleocapsids of alphaviruses, and probably flaviviruses and arteriviruses as well, are regular icosahedral structures, and there are no other proteins within the nucleocapsid other than the nucleocapsid protein. In contrast, the nucleocapsids of all minus-strand viruses are helical and contain, in addition to the major nucleocapsid protein, two or more minor proteins that possess enzymatic activity. As described, the nucleocapsids of minus-strand RNA viruses remain intact within the cell during the entire infection cycle and serve as machines that make viral RNA. The coronaviruses also have helical nucleocapsids, but being plus-strand RNA viruses they do not need to carry enzymes in the virion to initiate infection. The helical nucleocapsids

A. Enterobacteria phage T2

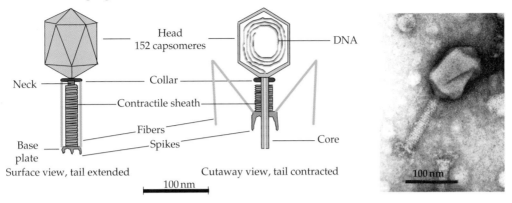

Head
152 capsomeres

DNA

Neck

Collar

Contractile sheath

Fibers

Spikes

Base
plate

Core

Surface view, tail extended

Cutaway view, tail contracted

100 nm

Electron micrograph

B. Enterobacteria phage T7

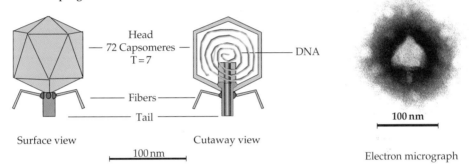

Head
72 Capsomeres
T = 7

DNA

Fibers

Tail

Surface view

Cutaway view

100 nm

100 nm

Electron micrograph

C. Lambda-like Phage

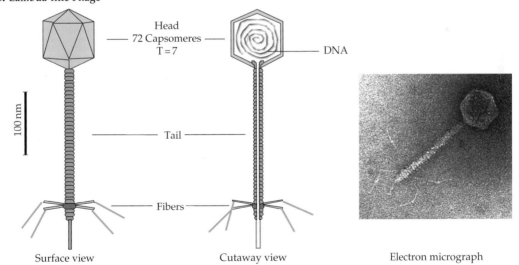

Head
72 Capsomeres
T = 7

DNA

100 nm

Tail

Fibers

Surface view

Cutaway view

Electron micrograph

FIGURE 2.13 Morphology of some bacteriophages (members of the *Caudovirales*). (A) Enterobacteria phage T2, in the family *Myoviridae*. The head is an elongated pentagonal structure. (B) Enterobacteria phage T7, a member of the *Podoviridae*. (C) Enterobacteria phage λ, a member of the *Siphoviridae*. All electron micrographs are stained with uranyl acetate, and all bars shown are 100 nm. Phage diagrams are adapted from Murphy *et al*. (1995) pp. 51, 60, 55. Electron micrographs of T2 and T7 were kindly provided by Dr. H.-W. Ackermann, Laval University, Quebec. The electron micrograph of Ur-Lambda (note the long kinked tail fibers) was kindly provided by Dr. Roger Hendrix.

of (−) RNA viruses appear disordered within the envelope of all viruses except the rhabdoviruses, in which they are coiled in a regular fashion (see later).

The nucleocapsids of retroviruses also appear to be fairly simple structures. They are formed from one major precursor protein, the Gag polyprotein, that is cleaved during maturation into four or five components. The precursor nucleocapsid is spherically symmetric but lacks icosahedral symmetry. The mature nucleocapsid produced by cleavage of Gag may or may not be spherical symmetric. The nucleocapsid also contains minor proteins, produced by cleavage of Gag–Pro–Pol, as described in Chapter 1. These minor proteins include the protease, RT, RNase H, and integrase that are required to cleave the polyprotein precursors, to make a cDNA copy of the viral RNA, and to integrate this cDNA copy into the host chromosome.

The two families of enveloped DNA viruses that we consider here, the poxviruses and the herpesviruses, contain large genomes and complicated virus structures. The nucleocapsids of herpesviruses are regular icosahedrons but those of poxviruses are complicated structures containing a core and associated lateral bodies.

Envelope Glycoproteins

The external proteins of enveloped virions are virus-encoded proteins that are anchored in the lipid bilayer of the virus or whose precursors are anchored in the lipid bilayer. In the vast majority of cases these proteins are glycoproteins, although examples are known that do not contain bound carbohydrate. These proteins are translated from viral mRNAs and transported by the usual cellular processes to reach the membrane at which budding will occur. When budding is at the cell plasma membrane, the glycoproteins are transported via the Golgi apparatus to the cell surface. Some enveloped viruses mature at intracellular membranes, and in these cases the glycoproteins are directed to the appropriate place in the cell. Both Type I integral membrane proteins, in which the N terminus of the protein is outside the lipid bilayer and the C terminus is inside the bilayer, and Type II integral membrane proteins, which have the inverse orientation with the C terminus outside, are known for different viruses. Many viral glycoproteins are produced as precursor molecules that are cleaved by cellular proteases during the maturation process.

Following synthesis of viral glycoproteins, during which they are transported into the lumen of the endoplasmic reticulum (ER) in an unfolded state, they must fold to assume their proper conformation, and assume their proper oxidation state by formation of the correct disulfide bonds. This process often occurs very quickly, but for some viral glycoproteins it can take hours. Folding is often assisted by chaperonins present in the endoplasmic reticulum. It is believed that at least one function of the carbohydrate chains attached to the protein is to increase the solubility of the unfolded glycoproteins in the lumen of the ER so that they do not aggregate prior to folding. During folding, the solubility of the proteins is increased by hiding hydrophobic domains within the interior of the protein and leaving hydrophilic domains at the surface.

The glycoproteins possess a number of important functions in addition to their structural functions. They carry the attachment domains by which the virus binds to a susceptible cell. This activity is thought to be related to the ability of many viruses, nonenveloped as well as enveloped, to bind to and agglutinate red blood cells, a process called hemagglutination. The protein possessing hemagglutinating activity is often called the hemagglutinin or HA. The viral glycoproteins also possess a fusion activity that promotes the fusion of the membrane of the virus with a membrane of the cell. The protein possessing this activity is sometimes called the fusion protein, or F. The glycoproteins, being external on the virus, are also primary targets of the humoral immune system, in which circulating antibodies are directed against viruses; many of these are neutralizing antibodies that inactivate the virus.

The glycoproteins of some enveloped viruses also contain enzymatic activities. Many orthomyxoviruses and paramyxoviruses possess a neuraminidase that will remove sialic acid from glycoproteins. The primary receptor for these viruses is sialic acid. The neuraminidase may allow the virus to penetrate through mucus to reach a susceptible cell. It also removes sialic acid from the viral glycoproteins so that these glycoproteins or the mature virions do not aggregate, and from the surface of an infected cell, thereby preventing released virions from binding to it. The viral protein possessing neuraminidase activity may be called NA, or in the case of a protein that is both a neuraminidase and hemagglutinin, HN.

The structure of most enveloped viruses is not as rigorously constrained as that of icosahedral virus particles. The glycoproteins are not required to form an impenetrable shell, which is instead a function of the lipid bilayer. They appear to tolerate mutations more readily than do proteins that must form a tight icosahedral shell and appear to evolve rapidly in response to immune pressure. However, the integrity of the lipid bilayer is essential for virus infectivity, and enveloped viruses are very sensitive to detergents.

Other Structural Proteins in Enveloped Viruses

In some enveloped viruses, there is a structural protein that underlies the lipid envelope but which does not form part of the nucleocapsid. Several families of minus-strand RNA viruses possess such a protein, called the matrix protein. This protein may serve as an adapter between the nucleocapsid and the envelope. It may also have regulatory functions in viral RNA replication. The herpesviruses also have proteins underlying the envelope that form a thick layer called the tegument. The thickness of the tegument is not uniform within a virion, giving rise to some irregularity in its structure. The tegument proteins perform important functions early after infection of a cell by a herpesvirus (see Chapter 7).

Structure of Alphaviruses

The alphaviruses, a genus in the family *Togaviridae*, are exceptional among enveloped RNA viruses in the regularity of their virions, which are uniform icosahedral particles. Virions of two alphaviruses have been crystallized and the crystals are regular enough to diffract to 30–40-Å resolution. Higher resolution has been obtained from cryoelectron microscopy, which has been used to determine the structures of several alphaviruses to 7–25 Å (Fig. 2.5).

More detailed reconstructions of Sindbis virus and Ross River virus (RRV) have been derived from a combination of cryoelectron microscopy of the intact virion and X-ray crystallography of alphavirus structural proteins. A cutaway view of RRV at about 25-Å resolution is shown in Fig. 2.14A. The nucleocapsid, shown in red and yellow, has a diameter of 400 Å, and is a regular icosahedron with $T=4$ symmetry. It is formed from 240 copies of a single species of capsid protein of size 30 kDa. Note the fivefold and sixfold

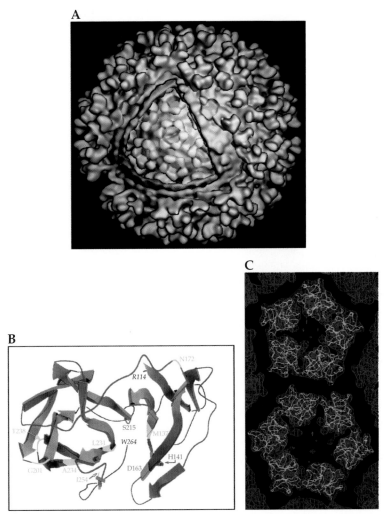

FIGURE 2.14   Structure of Ross River virus reconstructed from cryoelectron microscopy. (A) Cutaway view of the cryoelectron reconstruction illustrating the multilayered structure of the virion. Envelope glycoproteins are shown in blue, the lipid bilayer in green, the ordered part of the nucleocapsid in yellow, and the remainder of the nucleocapsid in orange. (B) Ribbon diagram of the X-ray crystallographic structure of the Sindbis virus capsid protein, with β sheets represented by large arrows. Only the carboxy-terminal domain, which starts at Arg-114, is ordered in crystals. The active site residues of the autoprotease, Ser-215, His-141, and Asp-163, are shown in red. The carboxy-terminal Trp-264, which is the P1 residue of the cleavage site, lies within the active site of the enzyme. The seven residues shown in yellow-green may interact with the cytoplasmic domain of glycoprotein E2 during budding of the nucleocapsid. (C) Fit of the Sindbis capsid protein Cα trace (yellow) into the electron density of Ross River virus (blue) determined by cryoelectron microscopy. (A) and (B) were adapted from Strauss *et al.* (1995), Figures 4 and 3, respectively, and (C) was kindly provided by Richard J. Kuhn.

coordinated pedestals in yellow that rise above the red background of RNA and unstructured parts of the protein. Each of these pedestals is formed by the ordered domains of one capsid protein molecule. The lipid bilayer is shown in green and is positioned between the capsid and the external shell of glycoproteins, shown in blue. The glycoproteins are also icosahedrally arranged with $T=4$ symmetry. The complete structure is therefore quite regular and the virion has been described as composed of two interacting protein shells with a lipid bilayer sandwiched between.

The structure of the ordered part of the capsid protein of Sindbis virus has been solved to atomic resolution by conventional X-ray crystallography and this structure is shown in Fig. 2.14B. The first 113 residues are disordered and the structure is formed by residues 114–264. This ordered domain has a structure that is very different from the eight-fold β sandwich described earlier (compare Fig. 2.14B with Figs. 2.3B and 2.4). Instead, its fold resembles that of chymotrypsin, and it has an active site that consists of a catalytic triad whose geometry is identical to that of chymotrypsin. The capsid protein is an active protease that cleaves itself from a polyprotein precursor. After cleavage, the C-terminal tryptophan-264 remains in the active site and the enzymatic activity of the protein is lost.

The interactions between the capsid protein subunits that lead to formation of the $T=4$ icosahedral lattice have been deduced by fitting the electron density of the capsid protein at 2.5-Å resolution into the electron density of the nucleocapsid found by cryoelectron microscopy. Such a reconstruction, based on a cryoEM structure of Sindbis virus at a resolution of better than 10 Å, is shown in Fig. 2.14C. The fit of the capsid protein is unique and the combined approaches of X-ray crystallography and cryoelectron microscopy thus define the structure of the shell of the nucleocapsid to atomic resolution.

The envelopes of alphaviruses contain 240 copies of each of two virus-encoded glycoproteins, called E1 and E2. E2 is first produced as a precursor called PE2. E1 and PE2 form a heterodimer shortly after synthesis, and both span the lipid bilayer as Type I integral membrane proteins (having a membrane-spanning anchor at or near the C terminus). The C-terminal cytoplasmic extension of PE2 interacts in a specific fashion with a nucleocapsid protein so that there is a one-to-one correspondance between a capsid protein and a glycoprotein heterodimer. The 240 glycoprotein heterodimers form a $T=4$ icosahedral lattice on the surface of the particle by interacting with one another and with the capsid proteins. Because of the glycoprotein–capsid protein interactions, the icosahedral lattices of the nucleocapsid and the glycoproteins are coordinated.

At some time during transport of the glycoprotein heterodimers to the cell surface, PE2 is cleaved by a cellular protease called furin to form E2. E1 and E2 remain associated as a heterodimer. If cleavage is prevented, noninfectious particles are produced that contain PE2 and E1.

In the virion, three glycoprotein heterodimers associate to form a trimeric structure called a spike, easily seen in Figs. 2.5 and 2.14A. It is not known if the spike assembles during virus assembly or if heterodimers trimerize during their transport to the cell surface. A reconstruction of a spike of Sindbis virus at a resolution better than 10 Å is shown in Fig. 2.15A. In this reconstruction, the electron density of E1 has been replaced by the E1 structure of the related Semliki Forest virus determined to atomic resolution by X-ray crystallography. The three copies of E1 project upwards at an angle of about 45° and are shown in three colors because they have slightly dif-

A. Sindbis Virus   **B.** Dengue Virus

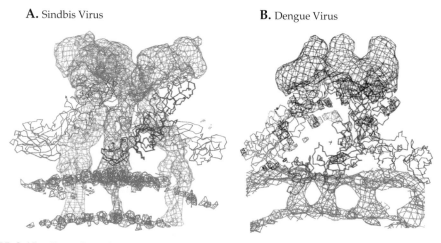

FIGURE 2.15   Comparison of (A) the spike structure of mature Sindbis virus (an alphavirus) with (B) the spike of immature dengue virus (a flavivirus). The Cα backbones of the three E1 (Sindbis) and E (dengue) glycoprotein ectodomains are shown in red, green, and blue, as they were fitted into the cryoelectron density envelope. The E1 and E densities have been zeroed out, leaving the gray envelope that corresponds to E2 for Sindbis and prM for dengue. The density corresponding to the lipid bilayer is shown in bright green. Adapted from Figure 5 in Y. Zhang *et al.* (2003) with permission.

ferent environments. The electron density in gray that remains after subtracting the density due to E1 is thus the electron density of E2. E2 projects further upward than does E1 and covers the apex of E1, which has the fusion peptide. Thus, E2 covers the fusion peptide with a hydrophobic pocket so that it does not interact with the hydrophilic environment. The apex of the E2 spike contains the domains that attach to receptors on a susceptible cell. Both E1 and E2 have C-terminal membrane-spanning anchors that traverse the lipid bilayer shown in green. The C-terminal domain of E1 is not present in the protein whose structure has been determined because hydrophobic domains do not easily crystallize. Thus, the electron density shown traversing the lipid bilayer arises from both E1 and E2 and shows that the two membrane spanning anchors go through as paired α helical structures (Fig. 2.16).

Upon entry of an alphavirus into a cell, the acidic pH of endosomal vesicles causes disassembly of E2/E1 heterodim-

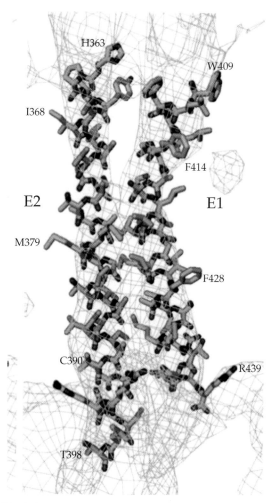

FIGURE 2.16 The E1 and E2 transmembrane helices of Sindbis (an alphavirus) determined from a 9Å resolution cryoelectron microscopy reconstruction. Shown are E1 residues from 409 to 439 and E2 residues 363 to 398 fitted into the transmembrane density. This is Figure 6 from Mukhopadhyay *et al.* (2006), reprinted with permission.

ers and trimerization of E1 to form homotrimers. The fusion peptide is exposed and penetrates the target bilayer of the host endosomal membrane. Fusion follows by methods discussed in Chapter 1.

Structure of Flaviviruses

Flaviviruses also possess a regular icosahedral structure (Fig. 2.5) that has been solved by methods similar to those used to determine the structure of alphaviruses. The structures of alphaviruses and flaviviruses are related and have descended from a common ancestral structure. Like alphaviruses, flaviviruses produce two structural glycoproteins, called E and prM (for precursor to M). E is homologous to E1 of alphaviruses. Although no sequence identity is detectable, the structures of the two proteins are virtually identical and are formed with a similar fold (Fig. 2.17). prM and E form a heterodimer and immature particles can be formed if cleavage of prM is prevented. The glycoprotein heterodimers in these immature particles trimerize to form spikes whose structure is very similar to the spikes of alphaviruses (Fig. 2.15B). The arrangement of the glycoproteins on the immature virus particle is illustrated in Fig. 2.18A and a cryoEM reconstruction that illustrates the surface of the immature dengue virus particle is shown in Fig. 2.18D. The major differences between the immature flavivirus particle and the alphavirus particle are that there are 180 copies of the heterodimer in the flavivirus particle arranged in a *T*=3 icosahedral structure rather than 240 heterodimers arranged in a *T*=4 structure in alphaviruses; that prM is a smaller molecule than PE2 so that in the flavivirus spike there is but a thin trace of density that projects downward, parallel to E, from the cap that shields the fusion peptide (Fig. 2.15B) rather than a substantial trace of density in the alphavirus spike (Fig. 2.15A); and that the C-terminal regions of prM and E that enter the membrane do so independently and do not emerge from the internal side of the membrane (illustrated in Fig. 2.19B), unlike the alphavirus membrane spanning regions (Fig. 2.16). There is no evidence that the membrane glycoproteins interact with the nucleocapsid in flaviviruses, and the flavivirus nucleocapsid, assuming it is a regular icosahedral structure, is not coordinated with the icosahedral structure formed by the spikes, which is again different from alphaviruses where the C-terminal domain of PE2 interacts with the nucleocapsid.

Following cleavage of prM by furin to form M, there is a dramatic rearrangement of the flavivirus glycoproteins such that the final virion structure is very different from the alphavirus structure. The heterodimers dissociate and E-E homodimers are formed that collapse over the lipid bilayer (Fig. 2.19). The E-E homodimers occur in two totally different environments, either perpendicular to the twofold axis where they interact with homodimers in side-to-side interactions, forming a herringbone pattern, or parallel to a twofold axis where they interact at fivefold and threefold axes (Fig. 2.18B). A comparison of the immature flavivirus particle,

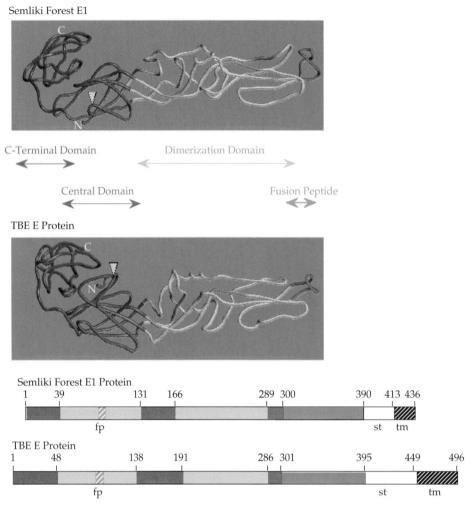

FIGURE 2.17 Comparison of the folds of the Semliki Forest (alphavirus) E1 protein and the tick-borne encephalitis (flavivirus) E protein. At the top are shown the X-ray crystal structures of the two proteins. At the bottom is a schematic of the linear amino acid sequences, color coded to indicate which amino acids contribute to each of the domains. Adapted from Figure 2 in Lescar *et al.* (2001) with permission.

the mature flavivirus particle, and the alphavirus particle is shown in Fig. 2.18. The mature flavivirus is smooth, with no surface projections, and is 50 nm in diameter (Fig. 2.18E). The immature particle is ragged in appearance and is 60 nm in diameter (Fig. 2.18D). The alphavirus virion shows conspicuous spikes and is 70 nm in diameter (Fig. 2.18F). The arrangement of E or E1 in the three particles is also shown (Figs. 2.18A, B, C), illustrating the differences in their arrangement.

Entry of flaviviruses follows pathways similar to those used by alphaviruses. The acidic pH of the endosome causes the E-E homodimers to reorganize to form E homotrimers. These trimers must reorient so that the exposed fusion peptide is projected upwards where it penetrates the host membrane.

Structure of Other Enveloped Viruses with Round Nucleocapsids

The arteriviruses possess icosahedral nucleocapsids, but the mature virion does not appear to be regular in structure. Detailed structures of these particles are not available.

The herpesviruses are large DNA viruses that have a *T*=16 icosahedral nucleocapsid (Fig. 2.5). A schematic diagram of an intact herpesvirion is shown in Fig. 2.20A. Underneath the envelope is a protein layer called the tegument. The tegument does not have a uniform thickness, and thus the virion is not uniform. Two electron micrographs of herpesvirions are shown in Figs. 2.20B and 2.20C that illustrate the irregularity of the particle and the differing thickness of the tegument in different particles.

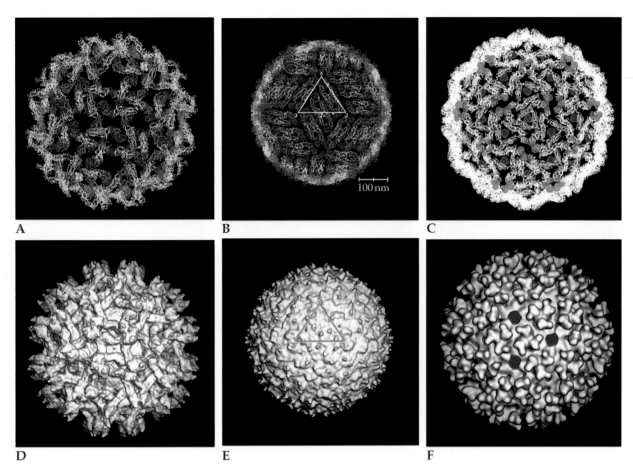

FIGURE 2.18   Fitting the X-ray crystal structures of flavivirus E protein and alphavirus E1 protein into the respective cryoelectron density envelopes of the virions. (A) Dengue E protein fitted into the cryoelectron density of immature prM-containing particle; (B) dengue E protein fitted into the envelope of the mature virion; (C) the fit of Sindbis alphavirus E1 into the Sindbis virus envelope; (D) cryoelectron reconstruction of the immature dengue prM-containing particle at 16-Å resolution; (E) cryoelectron reconstruction of the mature dengue virion at 12-Å resolution; (F) cryoelectron reconstruction of Ross River alphavirus at 25-Å resolution. Panels A and C were provided by Richard J. Kuhn; panel B is adapted from Figure 3c in Kuhn *et al.* (2002) with permission; panel D is adapted from Figure 3b in Y. Zhang *et al.* (2003) with permission; panel E is reprinted from Figure 1a in W. Zhang *et al.* (2003) with permission; panel F is adapted from Figure 4 in Strauss *et al.* (1995) with permission.

The retroviruses have a nucleocapsid that forms initially using spherical symmetry principles. Cleavage of Gag during virus maturation results in a nucleocapsid that is not icosahedral and that is often eccentrically located in the virion. Fig. 2.21A presents a schematic of a retrovirus particle that illustrates the current model for the location of the various proteins after cleavage of Gag and Gag–Pol. Figs. 2.21B, C, and D show electron micrographs of budding virus particles and of mature extracellular virions for three genera of retroviruses. Betaretrovirus particles usually mature by the formation of a nucleocapsid within the cytoplasm that then buds through the plasma membrane. This process is shown in Fig. 2.21B for mouse mammary tumor virus. In the top micrograph in Fig. 2.21B, preassembled capsids are seen in the cytoplasm. In the middle micrograph, budding of the capsid through the plasma membrane is illustrated. In the bottom micrograph, a mature virion with an eccentrically located capsid is shown.

In gammaretroviruses, the capsid forms during budding, and the nucleocapsid is round and centrally located in the mature virion. This process is illustrated in Fig. 2.21C for murine leukemia virus. The top micrograph shows a budding particle with a partially assembled capsid. The bottom micrograph shows a mature virion.

In the lentiviruses, of which HIV is a member, the capsid also forms as a distinct structure only during budding. In the top panel of Fig. 2.21D is shown a budding particle of bovine immunodeficiency virus. After cleavage of Gag to form the mature virion, the capsid usually appears cone shaped or bar shaped (bottom panel of Fig. 2.21D).

Mason–Pfizer monkey virus is a betaretrovirus whose capsid is cone shaped and centrally located in the mature virion. A single amino acid change in the matrix protein MA determines whether the capsid preassembles and then buds, or whether the capsids assemble during budding.

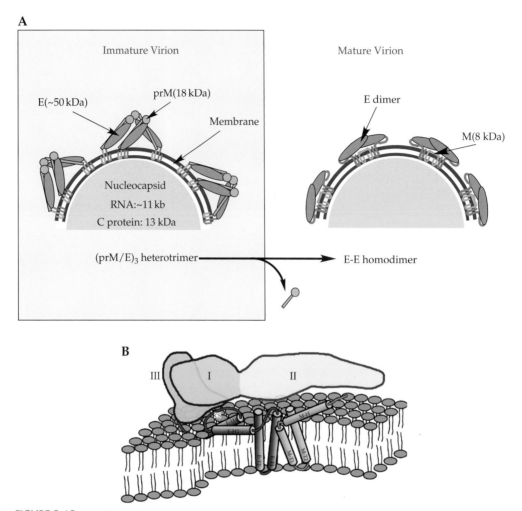

FIGURE 2.19 (A) Schematic representation of maturation of dengue virus. In the immature particle, as shown in panels (A) and (D) of Fig. 2.18, three heterodimers of E and prM come together to form a heterotrimer. Upon cleavage of prM to M and pr, E collapses onto the surface of the mature particle (panels B and E in Fig. 2.18) as homodimers. (B) Diagram of the dengue virus E protein in the mature particle as derived from cryoEM. This shows the three domains colored in the same way as in Figs. 2.17 and 2.18 as well as the locations of the transmembrane and intramembrane helices in blue and orange for E and M, respectively. Reprinted from Figure 4a in W. Zhang *et al.* (2003) with permission.

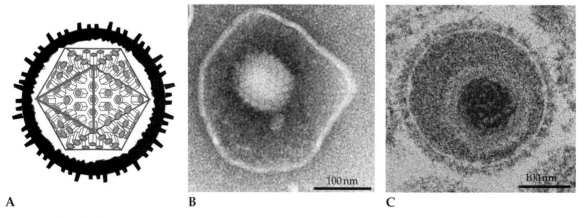

FIGURE 2.20 Two views of herpes simplex virus. (A) Cutaway schematic representation showing the outer envelope with projecting spikes, the irregular inner margin of the envelope due to the tegument, and the icosahedral core containing 162 capsomeres in a *T*=16 arrangement. One of the triangular faces of the icosahedron is outlined. Adapted from Murphy *et al.* (1995) p. 114. (B) Negatively stained electron micrograph of an intracellular particle of bovine herpesvirus. (C) Section through a bovine herpesvirion. Images in (B) and (C) were kindly provided by Dr. Peter Wild.

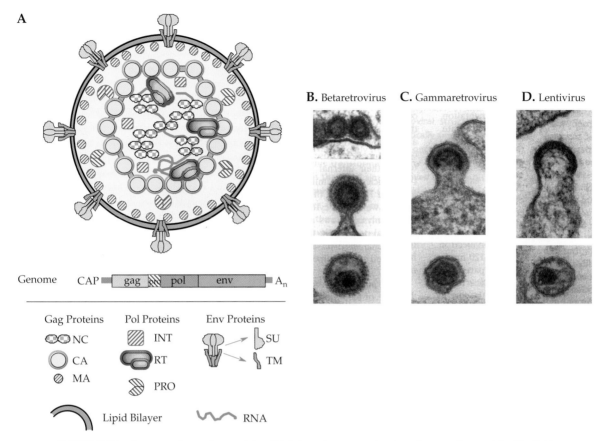

Genome   CAP   gag   pol   env   A_n

Gag Proteins   Pol Proteins   Env Proteins
NC   INT   SU
CA   RT   TM
MA   PRO

Lipid Bilayer   RNA

FIGURE 2.21   Structure of retrovirus particles. (A) Schematic cross section through a retrovirus particle. The lipid bilayer surrounds the particle and has imbedded in it trimeric spikes composed of surface (SU) and transmembrane (TM) envelope proteins. The internal nonglycosylated proteins are encoded by the *gag* gene and include NC, the nucleocapsid protein complexed with the genomic RNA, CA, the major capsid protein, and MA, the matrix protein that lines the inner surface of the membrane. Other components include RT, the reverse transcriptase, IN, the integrase, and PR, the protease. Adapted from Coffin *et al.* (1997), *Retroviruses*, p. 2. (B) Electron micrographs of mouse mammary tumor virus particles. Top: intracytoplasmic particles; middle: budding particles; bottom: mature extracytoplasmic particles. (C) Electron micrographs of murine leukemia virus particles. Top: budding particles; bottom: mature extracytoplasmic particles. (D) Electron micrographs of bovine immunodeficiency virus. Top: budding particles; bottom: mature extracytoplasmic particles. Adapted from Coffin *et al.* (1997), *Retroviruses*, p. 30.

Thus, the point at which capsids assemble does not reflect a fundamental difference in retroviruses. Preassembly of capsids or assembly during budding appears to depend on the stability of the capsid in the cell. Stable capsids can pre-assemble. Unstable capsids require interactions with other viral components to form as a recognizable structure.

Enveloped Viruses with Helical Nucleocapsids

The coronaviruses and the minus-strand RNA viruses have nucleocapsids with helical symmetry. The structures of the mature virions are irregular, with the exception of the rhabdoviruses, and the glycoprotein composition is not invariant. Because of the lack of regularity in these viruses, as well as the lack of symmetry, detailed structural studies of virions have not been possible. The lack of regularity arises in part because in these viruses there is no direct interaction between the nucleocapsid and the glycoproteins. The lack of such interactions permits these viruses to form pseudotypes, in which glycoproteins from other viruses substitute for those of the virus in question. Pseudotypes are also formed by retroviruses.

The structures of paramyxoviruses and orthomyxoviruses are illustrated schematically in Fig. 2.22. The helical nucleocapsids contain a major nucleocapsid protein called N or NP, and the minor proteins P (NS1) and L (PB1, PB2, PA), as shown. There is a matrix protein M (M1) lining the inside of the lipid bilayer and also two glycoproteins anchored in the bilayer that form external spikes. The two glycoproteins, called F and HN in paramyxoviruses and HA and NA in orthomyxoviruses, do not form heterodimers but rather form homooligomers so that there are two different kinds of spikes on the surface of the virions. HA in the orthomyxoviruses forms homotrimers whereas NA forms homotetramers, and the two types of spikes can be distinguished in the electron microscope if the

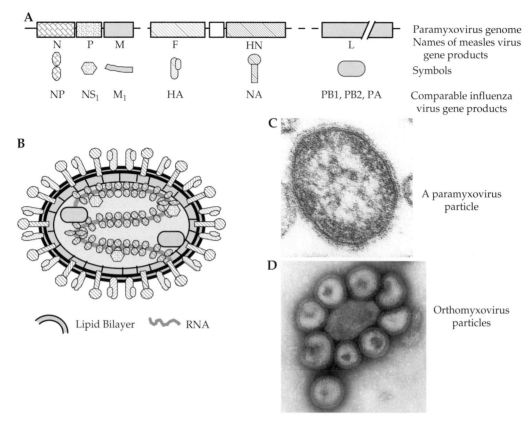

FIGURE 2.22 Morphology of orthomyxoviruses and paramyxoviruses. (A) Schematic of the genome organization of a paramyxovirus, Sendai virus. The names of the gene products and symbols to be used in the diagram are indicated. Also shown are the comparable gene products of influenza virus, an orthomyxovirus. (B) Schematic cutaway view of an orthomyxovirus or paramyxovirus particle. The nucleocapsid consists of a helical structure made up of the RNA complexed with many copies of the nucleocapsid protein. This internal structure also contains a few molecules of the RNA polymerase L (or PB1, PB2, PA in influenza virus), and P (or NS1). The nucleocapsid is enveloped in a lipid bilayer derived from the host cell in which are embedded two different glycoproteins, F and HN (or HA and NA in influenza virus) and which is lined on the inner surface with the matrix protein M. (C) Electron micrograph of a thin section of a measles particle. This photo was taken by Cynthia S. Goldsmith and obtained from the Public Health Image Library (PHIL). (D) Electron micrograph of a negatively stained influenza virus particles. This photo was taken by Fred Murphy and obtained from PHIL.

resolution is high enough. As occurs in many enveloped RNA viruses, F and HA are produced as precursors that are cleaved by furin during transport of the proteins. Cleavage is required to activate the fusion peptide of the virus, which is found at the N terminus of the C-terminal product (see Fig. 1.6). Electron micrographs of virions are shown in Figs. 2.22C and D. The particles in the preparations shown are round and reasonably uniform, but in other preparations the virions are pleomorphic baglike structures that are not uniform in appearance. In fact, clinical specimens of some orthomyxoviruses and paramyxoviruses are often filamentous rather than round, illustrating the flexible nature of the structure of the virion. The micrograph of the paramyxovirus measles virus shown in Fig. 2.22C is a thin section and illustrates the lack of higher order structure in the internal helical nucleocapsid. The micrograph of the orthomyxovirus influenza A virus shown in Fig. 2.22D is of a negative-stained preparation and illustrates the spikes that decorate the virus particle.

The structures of rhabdoviruses and filoviruses are illustrated in Fig. 2.23. The rhabdoviruses assemble into bullet-shaped or bacilliform particles in which the helical nucleocapsid is wound in a regular elongated spiral conformation (Figs. 2.23B and C). The virus encodes only five proteins (Fig. 2.23A), all of which occur in the virion (Fig. 2.23B). The nucleocapsid contains the major nucleocapsid protein N and the two minor proteins L and NS. The matrix protein M lines the inner surface of the envelope, and G is an external glycoprotein that is anchored in the lipid bilayer of the envelope. Budding is from the plasma membrane (Fig. 2.23D).

The filoviruses are so named because the virion is filamentous. A schematic diagram of a filovirus is shown in Fig. 2.23E, and electron micrographs of two filoviruses, Marburg virus and Ebola virus, are shown in Figs. 2.23F and G. Notice that in the electron microscope, filovirus virions often take the shape of a shephard's crook or the number 6.

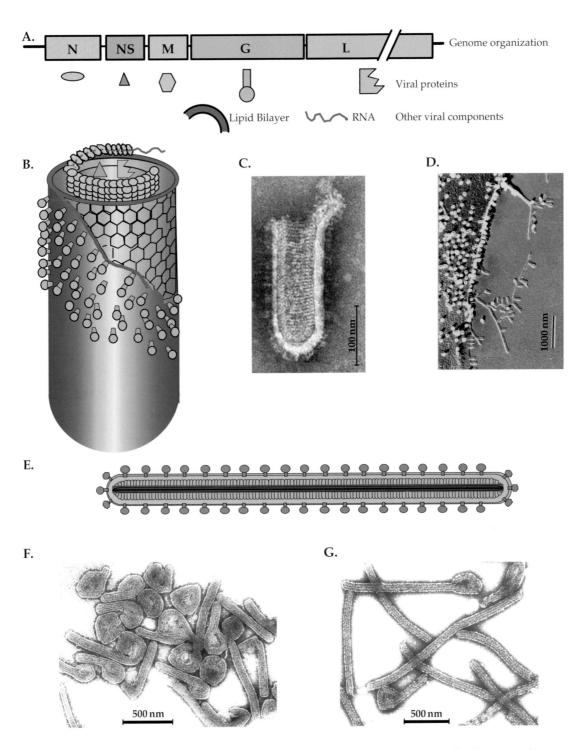

FIGURE 2.23 Morphology of the *Rhabdoviridae* and *Filoviridae*. (A) Genome organization of vesicular stomatitis virus (VSV), a *Vesiculovirus*, with the symbols for the various viral components shown below. (B) Cutaway diagram of a VSV particle. (C) A negatively stained electron micrograph of VSV virions. (D) Surface replica of a chicken embryo fibroblast infected with VSV. Note that the magnification is approximately 1/10 of that shown in (C). (E) Diagram of a filovirus, using the same color code for the components. (F) Negatively stained preparation of Marburg virus. (G) Filamentous forms of Ebola (Reston) virus. (B) is from Simpson and Hauser (1966); (F) and (G) are from Murphy *et al.* (1995) p. 289; and (D) is from Birdwell and Strauss (1975).

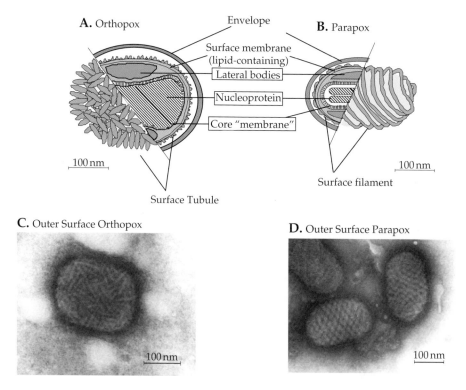

FIGURE 2.24   Morphology of orthopox and parapox virions. (A) and (B) Diagrams of orthopox and parapox virions. At the far left and right are shown the surfaces of the particles as they are isolated from infected cells, with the outer tubules or the outer filament shown in turquoise. The inner parts of each diagram show the enveloped particle in cross section illustrating the core membrane, the lateral bodies, and the nucleoprotein. Adapted from Fenner and Nakano (1988). (C) Purified vaccinia virus negatively stained with phosphotungstate; (D) Image of the outer surface of Orf parapox virus. Images in (C) and (D) were kindly provided by Prof. Stewart McNulty, Veterinary Sciences Division, Queens University of Belfast.

Vaccinia Virus

The poxviruses, large DNA-containing viruses, also have lipid envelopes. In fact, they may have two lipid-containing envelopes. The structures of poxviruses belonging to two different genera, *Orthopox* and *Parapox,* are illustrated in Fig. 2.24. Electron micrographs of the orthopox virus vaccinia virus and of a parapox virus are also shown.

Vaccinia virus has been described as brick shaped. The interior of the virion consists of a nucleoprotein core and two (in vertebrate viruses) proteinaceous lateral bodies. Surrounding these is a lipid-containing surface membrane, outside of which are several virus-encoded proteins present in structures referred to as tubules. This particle is called an intracellular infectious virion. As its name implies, it is present inside an infected cell, and if freed from the cell it is infectious. A second form of the virion is found outside the cell and is called an extracellular enveloped virion. This second form has a second, external lipid-containing envelope with which is associated five additional vaccinia proteins. This form of the virion is also infectious.

Parapox virions are similar to orthopox virions. However, their morphology is detectably different, as illustrated in Fig. 2.24.

ASSEMBLY OF VIRIONS

Self Assembly

Virions self-assemble within the infected cell. In most cases, assembly appears to begin with the interaction of one or more of the structural proteins with an encapsidation signal in the viral genome, which ensures that viral genomes are preferentially packaged. After initiation, encapsidation continues by recruitment of additional structural protein molecules until the complete helical or icosahedral structure has been assembled. Thus, packaging of the viral genome is coincident with assembly of the virion, or of the nucleocapsid in the case of enveloped viruses. The requirement for a packaging signal may not be absolute. In many viruses that contain an encapsidation signal, RNAs or DNAs lacking such a signal may be encapsidated, but with much lower efficiency. For some viruses, there is no evidence for an encapsidation signal.

Assembly of the TMV rod (Fig. 2.2) has been well studied. Several coat protein molecules, perhaps in the form of a disk, bind to a specific nucleation site within TMV RNA to initiate encapsidation. Once the nucleation event occurs, additional protein subunits are recruited into the structure and assembly proceeds in both directions until the RNA is completely encapsidated. The length of the virion is thus determined by the size of the RNA.

The assembly of the icosahedral turnip crinkle virion has also been well studied. Assembly of this $T=3$ structure is initiated by formation of a stable complex that consists of six capsid protein molecules bound to a specific encapsidation signal in the viral RNA. Additional capsid protein dimers are then recruited into the complex until the structure is complete.

It is probable that most other viruses assemble in a manner similar to these two well-studied examples. At least some viruses deviate from this model, however, and assemble an empty particle into which the viral genome is later recruited. It is also known that many viruses will assemble empty particles if the structural proteins are expressed in large amounts in the absence of viral genomes, even if assembly is normally coincident with encapsidation of the viral genome in infected cells.

Enveloped Viruses

The nucleocapsids of most enveloped viruses form within the cell by pathways assumed to be similar to those described above. They can often be isolated from infected cells, and for many viruses the assembly of nucleocapsids does not require viral budding or even the expression of viral surface glycoproteins. After assembly, the nucleocapsids bud through a cellular membrane, which contains viral glycoproteins, to acquire their envelope. Budding retroviruses were illustrated in Fig. 2.21 and budding rhabdoviruses in Fig. 2.23. A gallery of budding viruses belonging to other families is shown in Fig. 2.25. The membrane chosen for budding depends on the virus and depends, in part if not entirely, on the membrane to which the viral glycoproteins are directed by signals within those glycoproteins. Many viruses bud through the cell plasma membrane (Figs. 2.25B–F); in polarized cells, only one side of the cell may be used. Other viruses, such as the coronaviruses and the bunyaviruses, use the endoplasmic reticulum or other internal membranes. The herpesviruses replicate in the nucleus and the nucleocapsid assembles in the nucleus; in this case, the first budding event is through the nuclear membrane (Fig. 2.25A).

Although the nucleocapsid of most enveloped viruses assembles independently within the cell and then buds to acquire an envelope, exceptions are known. The example of retroviruses, some of which assemble a nucleocapsid during virus budding, was discussed earlier. In these viruses, morphogenesis is a coordinated event.

The forces that result in virus budding are not well understood for most enveloped viruses. In the case of the alphaviruses, there is evidence for specific interactions between the cytoplasmic domains of the glycoproteins and binding sites on the nucleocapsid proteins. The model for budding of these viruses is that the nucleocapsid first binds to one or a few glycoprotein heterodimers at the plasma membrane. By a process of lateral diffusion, additional glycoprotein heterodimers move in and are bound until a full complement is achieved and the virus is now outside the cell. Additional free energy for budding is furnished by lateral interactions between the glycoproteins, which form a contiguous layer on the surface (Fig. 2.18C). This model accounts for the regularity of the virion, the one-to-one ratio of the structural proteins in the virion, and the requirement of the virus for its own glycoproteins in order to bud.

In other enveloped viruses, however, there is little evidence for nucleocapsid–glycoprotein interactions. The protein composition of the virion is usually not fixed, but can vary within limits. In fact, glycoproteins from unrelated viruses can often be substituted. In the extreme case of retroviruses, noninfectious virus particles will form that are completely devoid of glycoprotein. The matrix proteins appear to play a key role in the budding process, as do other protein–protein interactions that are yet to be determined.

Maturation Cleavages in Viral Structural Proteins

For most animal viruses, there are one or more cleavages in structural protein precursors during assembly of virions that are required to activate the infectivity of the virion. Interestingly, these cleavages may either stabilize or destabilize the virion in the extracellular environment, depending on the virus. Many of these cleavages are effected by viral proteases, whereas others are performed by cellular proteases present in subcellular organelles. Virions are formed by the spontaneous assembly of components in the infected cell, sometimes with the aid of assembly factors ("scaffolds") that do not form components of the mature virion. For most nonenveloped viruses, complete assembly occurs within the cell cytoplasm or nucleoplasm. For enveloped viruses, final assembly of the virus occurs by budding through a cellular membrane. In either case, the virion must subsequently disassemble spontaneously on infection of a new cell. The cleavages that occur during assembly of the virus potentiate penetration of a susceptible cell after binding of the virus to it, and the subsequent disassembly of the virion on entry into the cell. A few examples will be described that illustrate the range of cleavage events that occur in different virus families.

A. Herpes

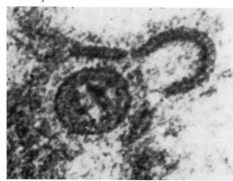

B. Machupo

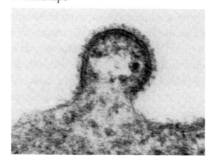

C. Sindbis

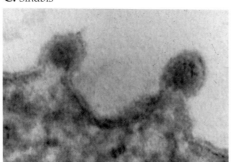

D. Rubella

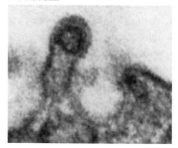

F. SV5 Round Particle Bud

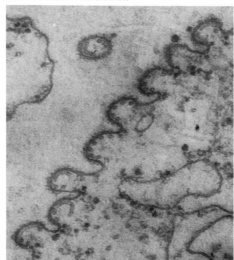

E. SV5 Filamentous Bud

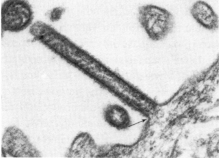

FIGURE 2.25   Gallery of budding figures of viruses representing several different families. (A) Thin section of a herpes simplex virion (*Herpesviridae*) in an infected Hep-2 cell. The particle is apparently coated with an inner envelope, and is in the process of acquiring its outer envelope from the nuclear membrane. From Roizman (1969). (B) Machupo virus (*Arenaviridae*) budding from a Raji cell. From Murphy *et al.* (1969). (Magnification: 120,000×). (C) Sindbis virus (*Togaviridae*) budding from the plasma membrane of an infected chicken cell. From Strauss *et al.* (1995). (160,000×). (D) Rubella virions (*Togaviridae*) budding from the surface of a BHK cell. From Higashi *et al.* (1976). (190,000×). (E) A portion of the cell surface with SV5 filaments (*Paramyxoviridae*) in the process of budding. From Compans and Choppin (1973). (45,000×). (F) A row of SV5 virions budding from the surface of a monkey kidney cell. Cross sections of the nucleocapsid can be seen within several of the particles. From Compans *et al.* (1966). (87,000×).

In the picornaviruses, a provirion is first formed that is composed of the viral RNA complexed with three viral proteins, called VP0, VP1, and VP3. During maturation to form the virion, VP0 is cleaved to VP2 and VP4. No protease has been found that performs this cleavage, and it has been postulated that the virion RNA may catalyze it. Cleavage to produce VP4, which is found within the interior of the capsid shell, as illustrated schematically in Fig. 2.9, is required for the virus to be infectious. As described in Chapter 1, VP4 appears to be required for entry of the virus into the cell. This maturation cleavage has another important consequence. Whereas the provirion is quite unstable, the mature virion is very stable. The poliovirus virion will survive treatment with proteolytic enzymes and detergents, and survives exposure to the acidic pH of less than 2 that is present in the stomach. Only on binding to its receptor (Figs. 1.5 and 2.9) is poliovirus destabilized such that VP4 can be released for entry of the viral RNA into the host cell.

Similarly, the insect nodaviruses first assemble as a procapsid containing the RNA and 180 copies of a single protein species called α (44 kDa). Over a period of many hours, spontaneous cleavage of α occurs to form β (40 kDa) and γ (4 kDa). This cleavage is required for the particle to be infectious. These events in nodaviruses have been well studied because it has been possible to assemble particles *in vitro,* and the structures of both cleaved and uncleaved particles have been solved to atomic resolution.

Rotaviruses, which form a genus in the family *Reoviridae,* must be activated by cleavage with trypsin after release from an infected cell in order to be infectious. Trypsin is present in the gut, where the viruses replicate, and activation occurs normally during the infection cycle of the virus in animals. When the viruses are grown in cultured cells, however, trypsin must be supplied exogenously.

A different type of cleavage event occurs during assembly of retroviruses and adenoviruses, as well as of a number of other viruses. During assembly of retroviruses, the Gag and Gag–Pol precursor polyproteins are incorporated, together with the viral RNA, into a precursor nucleocapsid. These polyproteins must be cleaved into several pieces by a protease present in the polyprotein if the virus is to be infectious. These cleavages often visibly alter the structure of the particle as seen in the electron microscope (Fig. 2.21). An analogous situation occurs in adenoviruses, where a viral protease processes a protein precursor in the core of the immature virion.

In most enveloped viruses, one of the envelope proteins is produced as a precursor whose cleavage is required to activate the infectivity of the virus. This cleavage may occur prior to budding, catalyzed by a host enzyme called furin, or may occur after release of the virus, catalyzed by other host enzymes. The example of the hemagglutinin of influenza virus was described in Chapter 1. This protein is produced as a precursor called HA0, which is cleaved to HA1 and HA2

(Fig. 1.6). Cleavage is required to potentiate the fusion activity present at the N terminus of HA2. As a second example, alphaviruses produce two envelope glycoproteins that form a heterodimer and one of the glycoproteins is produced as a precursor, as described before. The heterodimer containing the uncleaved precursor is quite stable, so that a particle containing uncleaved heterodimer is not infectious. The cleaved heterodimer, which is required for virus entry, is much less stable and dissociates readily during infection. Thus, in contrast to the poliovirus maturation cleavage, the alphavirus cleavage makes the virion less stable rather than more stable. Maturation cleavages also occur in the envelope glycoproteins of retroviruses, paramyxoviruses, flaviviruses, and coronaviruses.

Neutralization of Charge on the Virion Genome

DNA or RNA has a high net negative charge, and there is a need for counterions to neutralize this charge in order to form a virion. In many viruses, positively charged polymers are incorporated that neutralize half or so of the nucleic acid charge. The DNA in the virions of the polyomaviruses is complexed with cellular histones. The viral genomes in these viruses have been referred to as minichromosomes. In contrast, the adenoviruses encode their own basic proteins that complex with the genome in the core of the virion. Another strategy is used by the herpesviruses, which incorporate polyamines into the virion. Herpes simplex virus has been estimated to incorporate 70,000 molecules of spermidine and 40,000 molecules of spermine, which would be sufficient to neutralize about 40% of the DNA charge. Among RNA viruses, the nucleocapsid proteins are often quite basic and neutralize part of the charge on the RNA. As one example, the N-terminal 110 amino acids of the capsid protein of Sindbis virus have a net positive charge of 29. The positive charges within this domain of the 240 capsid proteins in a nucleocapsid would be sufficient to neutralize about 60% of the charge on the RNA genome. This charged domain is thought to penetrate into the interior of the nucleocapsid and complex with the viral RNA.

STABILITY OF VIRIONS

Virions differ greatly in stability, and these differences are often correlated with the means by which viruses infect new hosts. Viruses that must persist in the extracellular environment for considerable periods, for example, must be more stable than viruses that pass quickly from one host to the next. As an example of such requirements, consider the closely related polioviruses and rhinoviruses, members of two different genera of the family *Picornaviridae.* These viruses shared a common ancestor in the not too distant past and have structures that are very similar. The polioviruses are spread by an oral–fecal route and have the ability to persist in a hostile extracellular environment for some time where they may contaminate drinking water or food. Furthermore, they must pass through the stomach, where the pH is less than 2, to reach the intestinal tract where they begin the infection cycle. It is not surprising, therefore, that the poliovirion is stable to storage and to treatments such as exposure to mild detergents or to pH < 2. In contrast, rhinoviruses are spread by aerosols or contaminated mucus, and spread normally requires close contact. The rhinovirion is less stable than the poliovirion. It survives for only a limited period of time in the external environment and is sensitive to treatment with detergents or exposure to pH 3.

FURTHER READING

General Structure of Viruses

Johnson, J. E., and Speir, J. A. (1999). Principles of virus structure. In *Encyclopedia of Virology* (A. Granoff and R. G. Webster, Eds.), San Diego, Academic Press, pp, 1946–1956.

Johnson, J. E. (1996). Functional implications of protein–protein interactions in icosahedral viruses. *Proc. Natl. Acad. Sci. U.S.A.* **93**: 27–33.

Harrison, S. C. (2006). Principles of Virus Structure, Chapter 3 in: *Fields Virology, Fifth Edition* (D. M. Knipe and P. M. Howley, Eds. in chief), Philadelphia, Lippincott Williams & Wilkins, pp. 59–98.

Chiu, W., Burnett, R. M., and Garcea, R. L. (1997). *Structural Biology of Viruses*. Oxford, Oxford University Press.

Dalton, A. J., and F. Haguenau, Eds. (1973). *Ultrastructure of Animal Viruses and Bacteriophages: An Atlas*. Ultrastructure in Biological Systems. New York, Academic Press.

Cryoelectron Microscopy

Baker, T. S., Olson, N. H., and Fuller, S. D. (1999). Adding the third dimension to virus life cycles: Three-dimensional reconstruction of icosahedral viruses from cryo-electron micrographs. *Microbiol. Mol. Biol. Rev.* **63**: 862–922.

Forsell, K., Xing, L., Kozlovska, T., Cheng, R. H., and Garoff, H. (2000). Membrane proteins organize a symmetrical virus. *EMBO J.* **19**: 5081–5091.

Mukhopadhyay, S., Zhang, W., Gabler, S., *et al.* (2006). Mapping the structure and function of the E1 and E2 glycoproteins of Alphaviruses. *Structure* **14**: 63–73

Zhang, Y., Corver, J., Chipman, P. R., *et al.* (2003). Structures of immature flavivirus particles. *EMBO J.* **22**: 2604–2613.

Zhang, Y., Zhang, W., Ogata, S., *et al.* (2004). Conformational changes of the flavivirus E glycoprotein. *Structure* **12**: 1607–1618.

Zhang, W., Chipman, P. R., Corver, J., *et al.* (2003). Visualization of membrane protein domains by cryo-electron microscopy of dengue virus. *Nature Struct. Biol.* **10**: 907–912.

X-ray Crystallography

Athappilly, F. K., Murali, R., Rux, J. J., *et al.* (1994). The refined crystal structure of hexon, the major coat protein of adenovirus type 2, at 2.9Å resolution. *J. Mol. Biol.* **242**: 430–455.

Cheng, R. H., Kuhn, R. J., Olson, N. H., *et al.* (1995). Nucleocapsid and glycoprotein organization in an enveloped virus. *Cell* **80**: 621–630.

Hogle, J. M., Chow, M., and Filman, D. J. (1985). Three-dimensional structure of poliovirus at 2.9Å resolution. *Science* **229**: 1358–1365.

Lescar, J., Roussel, A., Wien, M. W., *et al.* (2001). The fusion glycoprotein shell of Semliki Forest virus: an icosahedral assembly primed for fusogenic activation at endosomal pH. *Cell* **105**: 137–148.

Modis, Y., Ogata, S. A., Clements, D. E., and Harrison, S. C. (2003). A ligand-binding pocket in the dengue virus envelope glycoprotein. *Proc. Natl. Acad. Sci. U.S.A.* **100**: 6986–6991.

Prasad, B. V., Hardy, M. E., Dokland, T., *et al.* (1999). X-ray crystallographic structure of the Norwalk virus capsid. *Science* **286**: 287–290.

Rossmann, M. G., Arnold, E., Erickson, J. W., *et al.* (1985). Structure of a human common cold virus and functional relationship to other picornaviruses. *Nature* (*London*) **317**: 145–153.

Stehle, T., Gamblin, S. J., Yan, Y., *et al.* (1996). The structure of simian virus 40 refined at 3.1Å resolution. *Structure* **4**: 165–182.

Canyon Hypothesis

Rossmann, M. G. (1989). The canyon hypothesis. *Viral Immunol.* **2**: 143–161.

Virus Assembly

Condit, R. C., Moussatche, N., and Traktman, P. (2006). In a Nutshell: structure and assembly of the vaccinia virion. *Adv. Virus. Res.* **66**: 31– 35.

Kuhn, R. J., Zhang, W., Rossmann, M. G., *et al.* (2002). Structure of dengue virus: implications for flavivirus organization, maturation, and fusion. *Cell* **108**: 717–725

Mukhopadhyay, S., Kuhn, R. J., and Rossmann, M. G. (2005). A structural perspective of the *Flavivirus* life cycle. *Nature Rev. Microbiol.* **3**: 13–22.

Pesavento, J. B., Crawford, S. E., Estes, M. K., and Prasad, B. V. V. (2006). Rotavirus proteins: structure and assembly. *Curr. Top. Microbiol. Immunol.* **309**: 189–219.

Roy, P., and Noad, R. (2006). Bluetongue virus assembly and morphogenesis. *Curr. Top. Microbiol. Immunol.* **309**: 87–116.

Strauss, J. H., Strauss, E. G., and Kuhn, R. J. (1995). Budding of alphaviruses. *Trends Microbiol.* **3**: 346–350.

3

Plus-Strand RNA Viruses

INTRODUCTION

The plus-strand RNA [(+)RNA] viruses comprise a very large group of viruses belonging to many families. Among these are viruses that cause epidemic disease in humans, including encephalitis, hepatitis, polyarthritis, yellow fever, dengue fever, poliomyelitis, and the common cold. The number of cases of human disease caused by these viruses each year is enormous. As examples, dengue viruses infect an estimated 50 to 100 million people each year; most humans suffer at least one rhinovirus-induced cold each year, with the cases therefore numbering in the billions; and most humans during their lifetime will suffer several episodes of gastroenteritis caused by astroviruses or caliciviruses. In terms of frequency and severity of illness, the (+)RNA viruses contain many serious human pathogens, and we will begin our description of viruses with this group.

The human (+)RNA viruses belong to seven families (Table 3.1). These seven families also contain numerous viruses that infect other vertebrates, of which many are important pathogens of domestic animals. Large numbers of (+)RNA viruses that infect plants are also known; in fact, most plant viruses contain (+)RNA genomes. The plant viruses, however, belong to different families and are currently classified by the International Committee on Taxonomy of Viruses (ICTV) into nine families plus many unassigned genera. Because of their importance as disease agents of domestic crops, much is known about these viruses. Other families of (+)RNA viruses include two families of bacterial viruses, one of fungal viruses, and four families of insect viruses (the nodaviruses, in particular, have been intensively studied). Thus, the (+)RNA viruses have evolved into many distinctly different families and must have arisen long ago. In this chapter, the seven families of viruses that include human viruses as members are considered, followed by a brief discussion of

the relationships of these viruses to the plant viruses and what this means in terms of virus evolution.

FAMILY *PICORNAVIRIDAE*

The picornaviruses are so named because they are small (*pico* = small), *RNA*-containing viruses. Nine genera of picornaviruses, five of which contain human pathogens, are currently recognized (Table 3.2), and more will probably be recognized as further studies of the known viruses occur and as new viruses are described. A dendrogram that illustrates the relationship of the nine genera to one another, as well as the relationships of a number of viruses within the various genera, is shown in Fig. 3.1. This dendrogram makes clear that all picornaviruses are closely related. They share significant nucleotide and amino acid sequence identity and form a well-defined taxon. The dendrogram also illustrates the rationale for grouping these viruses into nine genera.

As described in Chapter 2, the structures of several picornaviruses have been solved to atomic resolution by X-ray crystallography. The picornavirus virion is composed of 60 copies of each of four different proteins (called VP1–4) that form an icosahedral shell having $T=3$ symmetry (or pseudo-$T=3$) and a diameter of approximately 30 nm (see Figs. 2.1, 2.5, 2.7, and 2.8).

Organization and Expression of the Genome

The structure of the genome of poliovirus and comparison of it with the genomes of the other genera are shown in Fig. 3.2. The picornaviral genome is a single RNA molecule of about 7.5 kb. It contains one open reading frame (ORF) and is translated into one long polyprotein. This polyprotein is

TABLE 3.1   Families of Plus-Strand RNA Viruses That Contain Human Pathogens

Family	Size of genome (nucleotides)	Other vertebrate hosts	Representative human pathogens
Picornaviridae	~7500	Cattle, primates, mice	Poliovirus Human rhinovirus Hepatitis A
Caliciviridae	~7500	Rabbits, swine, cats	Norwalk
Hepeviridae	7200	Primates, swine	Hepatitis E
Astroviridae	6800–7900	Cattle, ducks, sheep, swine	Human astrovirus
Togaviridae	~11,600	Mammals, birds, horses	Semliki Forest, Ross River, WEE, VEE, EEE, Mayaro, rubella
Flaviviridae	9500–12,500	Swine, cattle, primates, birds	Dengue, yellow fever, JE, MVE, TBE, WNV, hepatitis C
Coronaviridae	20,000–30,000	Mice, birds, swine, cattle, bats	SARS coronavirus

Virus name abbreviations: WEE, VEE, EEE, Western, Venezuelan, Eastern equine encephalitis viruses; JE, Japanese encephalitis virus; MVE, Murray Valley encephalitis virus; TBE, tick-borne encephalitis virus; WNV, West Nile virus; SARS, severe acute respiratory syndrome.

TABLE 3.2   *Picornaviridae*

Genus/members	Virus name abbreviation[a]	Usual host(s)	Transmission	Disease	World distribution
Enterovirus					
Human enterovirus A	HEV-A	Humans	Oral–fecal, contact	See Table 3.5	Worldwide
Human enterovirus B	HEV-B	Humans	Oral–fecal, contact	See Table 3.5	Worldwide
Human enterovirus C	HEV-C	Humans	Oral–fecal, contact	See Table 3.5	Worldwide
Human enterovirus D	HEV-D	Humans	Oral–fecal, contact	See Table 3.5	Worldwide
Poliovirus (Types 1, 2, and 3)	PV	Humans	Oral–fecal, contact	See Table 3.5	Originally worldwide, extirpated in Americas
Bovine enterovirus, porcine enteroviruses A and B; Unassigned enteroviruses of humans and monkeys					
Rhinovirus					
Human rhinoviruses (>100 serotypes)	HRV-A, HRV-B	Humans	Aerosols, contact	Common cold	Worldwide
Cardiovirus					
Encephalomyocarditis Theilovirus	EMCV	Mice	Oral–fecal, contact	Encephalitis, myocarditis	Worldwide
Aphthovirus					
Foot and mouth disease	FMDV	Cattle, swine	Oral–fecal, contact	Lesions on mouth and feet	Worldwide (except United States, Australia)
Equine rhinitis A	ERAV	Horses			Worldwide
Hepatovirus					
Hepatitis A	HAV	Humans	Oral–fecal	Hepatitis	Endemic worldwide
Parechovirus					
Human parechovirus	HPeV	Humans	Oral–fecal	Gastroenteritis	Worldwide
Erbovirus					
Equine rhinitis B	ERBV	Horses	?	?	?
Kobuvirus					
Aichi	AiV	Humans	Oral–fecal	Gastroenteritis	Isolated in Japan (oysters)
Teschovirus					
Porcine teschoviruses (10 species recognized)	PTV-1	Swine	Oral–fecal	Paralysis, porcine encephalomyelitis	Britain, central and Eastern Europe

[a] Standard abbreviations are given for either the virus listed (such as poliovirus) or for the type member of the genus.

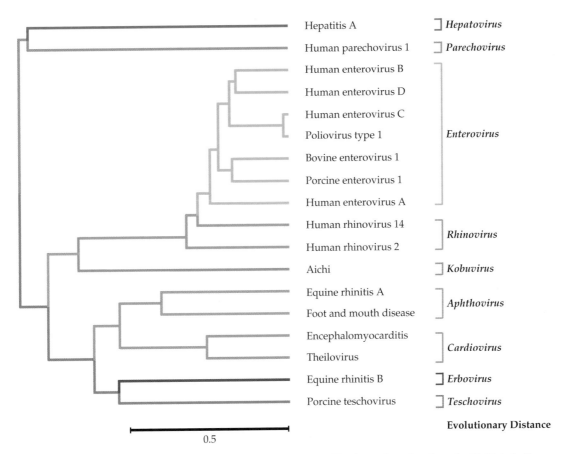

Hepatitis A —] *Hepatovirus*

Human parechovirus 1 —] *Parechovirus*

Human enterovirus B

Human enterovirus D

Human enterovirus C

Poliovirus type 1

Bovine enterovirus 1

Porcine enterovirus 1

Human enterovirus A

— *Enterovirus*

Human rhinovirus 14

Human rhinovirus 2

—] *Rhinovirus*

Aichi —] *Kobuvirus*

Equine rhinitis A

Foot and mouth disease

—] *Aphthovirus*

Encephalomyocarditis

Theilovirus

— *Cardiovirus*

Equine rhinitis B —] *Erbovirus*

Porcine teschovirus —] *Teschovirus*

Evolutionary Distance

0.5

FIGURE 3.1   Relationships between 18 representative picornaviruses. The viruses shown have been classified into the 9 recognized genera. This tree was generated from the amino acid sequences of the 3D^{pol} proteins. Adapted from Yamashita et al. (2003), and updated with the taxonomy found in Fauquet *et al.* (2005). Evolutionary distance, calculated by the UPGMA (unweighted pair group method with averages) method, is the number of residue substitutions that have occurred between two sequences since their divergence from a common ancestor and is defined as D = number of base mismatches/total alignment length in nucleotides.

cleaved by one or more virus-encoded proteinases to form more than 25 different polypeptides, including processing intermediates (not all of which are shown in the figure) as well as final cleavage products. The ORF in the genome contains three regions, called P1 (the 5′ region), P2 (the middle region), and P3 (the 3′ region). Region 1 encodes the structural proteins and regions 2 and 3 encode proteins required for RNA replication. The genome organization of all picornaviruses is similar, but each genus differs in important details. For example, the aphthoviruses and the cardioviruses have a poly(C) tract near the 5′ end of the RNA that is important for virus replication. These two genera also have a leader polypeptide that precedes the structural protein region. The aphthovirus leader peptide is a papain-like protease that cleaves itself from the polyprotein and has a role in the shutoff of cellular protein synthesis. The function of the cardiovirus leader is not known. Hepatitis A virus, Aichi virus, and echovirus 22, representatives of three other genera, also have leaders.

The picornaviral genome has a small protein, VPg, covalently bound to the 5′ end, which is the primer for initiation of RNA synthesis. VPg is normally removed from RNA that serves as mRNA by a cellular enzyme, but its removal is not required for its translation. The 3′ end of the RNA is poly-adenylated. As described in Chapter 1, the 5′ nontranslated region of a picornaviral RNA possesses an IRES (internal ribosome entry site) and the RNA is translated by a cap-independent mechanism. The translation of picornaviral RNA is greatly favored in the infected cell because picornaviruses interfere with host cell macromolecular synthesis and, in particular, interfere with host protein synthesis. Infection with entero-, rhino-, and aphthoviruses leads to proteolytic cleavage of a cellular protein called eIF4G that is a component of the cap-binding complex. Cleavage of this protein by 2A^{pro} of entero- and rhinoviruses or by the leader protease of aphthoviruses results in inhibition of the translation of RNAs that require the cap-binding protein complex, that is, capped host cell mRNAs. The cardioviruses, which are also cap independent, interfere with translation of host mRNAs in a different way, by interfering with phosphorylation of cap-binding protein. Poliovirus also interferes with host protein synthesis by cleavage of poly(A)-binding protein by the viral 3C^{pro}, but the mechanism by which this interference operates on cellular protein synthesis and not viral protein synthesis, since poliovirus RNA is also polyadenylated, is not yet clear. In addition, 3C^{pro} of apthoviruses cleaves initiation factors eIF4A and eIF4GI, and it is thought that these cleavages lead to a decrease in the level of viral protein synthesis later in infection, which facilitates packaging of the viral RNA.

The viral 3C^{pro} and its precursor 3CD^{pro} make multiple cleavages in the polyprotein translated from the genome, as illustrated for poliovirus in Fig. 3.2. Some cleavages occur *in cis* and some *in trans*. The crystal structures of 3C^{pro} of poliovirus and of a rhinovirus have been solved to atomic resolution, and their core structure resembles that of chymotrypsin (Fig. 1.19A). The catalytic center has the same geometry as that of chymotrypsin, but in 3C^{pro} the catalytic serine has been replaced by cysteine. Moreover, in many, but not all, picornaviruses the aspartic acid in the catalytic triad has been replaced by glutamic acid. Thus, 3C^{pro} is related to cellular serine proteases and may have originated by the capture of a cellular serine protease during the evolution of the viruses.

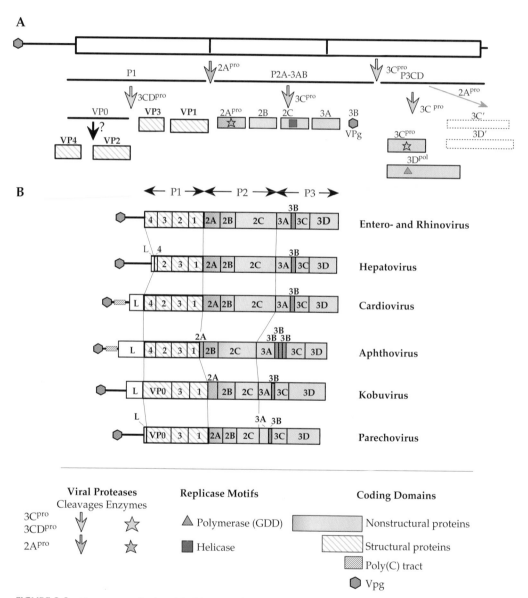

FIGURE 3.2   Genome organization of the *Picornaviridae*. (A) Genome organization of poliovirus showing the proteolytic processing steps. Both the 3C^{pro} and 2A^{pro} proteases are "serine-type proteases" with cysteine in the catalytic site. (B) Comparative genome organizations of representatives of seven of the nine genera of *Picornaviridae*. The key to the different shadings of coding domains and the symbols for various enzymatic motifs used in both (A) and (B) is given below. Adapted from Murphy *et al.* (1995) p. 300 and Yamashita *et al.* (1998).

Protease 3C^pro is present in all picornaviruses whereas protease 2A^pro is present in only a subset of picornaviruses. In poliovirus, 2A^pro, like 3C^pro, is a serine protease in which the active site serine has been replaced by cysteine. 2A^pro catalyzes one essential cleavage in the polyprotein of poliovirus, that between P1 and P2. This cleavage occurs *in cis*. The proteolytic activity of 2A^pro is also required for other functions during poliovirus replication, the nature of which have not been established. An interesting experiment is illustrated in Fig. 3.3 because it illustrates the power of molecular genetics and the tricks that modern virologists can play with viruses. This experiment will serve as a prelude to the discussion of the uses of viruses as vectors in Chapter 11. A poliovirus was constructed in which a stop codon was placed after the structural protein domain (region 1), so that 2A^pro was not needed to remove P1 from the polyprotein precursor. The stop codon was followed by an IRES and a new AUG start codon, so that P2A and the rest of the genome could be translated from the polycistronic RNA. This virus was viable. However, when the 2A proteinase was inactivated by changing the active site cysteine, the resulting virus was dead, showing that the proteolytic activity of P2A is required not only to separate regions 1 and 2 of the polyprotein but also for other function(s).

Protein 2A of rhinoviruses is also a protease. The crystal structure of protein 2A of human rhinovirus type 2 reveals that this protease is unrelated to 2A^pro of polioviruses, however. Thus, these two proteases in closely related viruses have different origins, and the viruses have solved the problem of how to separate regions 1 and 2 in the polyprotein in different ways. Furthermore, this finding illustrates that recombination to introduce new functions into viral genomes has been important in the evolution of these viruses, a theme to which we will refer many times in this book. Still another solution to the problem of separating regions 1 and 2 has been adopted by the cardio- and aphthoviruses. Protein 2A is not a protease in these viruses. Indeed, 2A is only 18 residues long, and cleavage between P1 and P2 is catalyzed by 3C^pro. Another interesting feature of these viruses is that the cleavage at the 2A–2B junction occurs spontaneously during translation, catalyzed by the specific amino acid sequence at the scissile bond. This cotranslational cleavage occurs only during translation on eukaryotic ribosomes, and it has been proposed that no cleavage actually occurs, but that the 2A sequence prevents synthesis of the specific peptide bond at the 2A–2B junction.

In addition to these cleavages catalyzed by 2A^pro, 3C^pro, and the leader protease of aphthoviruses, VP0 is cleaved during virion maturation to VP2 and VP4 in most, but not all, picornaviruses. Available evidence suggests that this cleavage is not catalyzed by a protease.

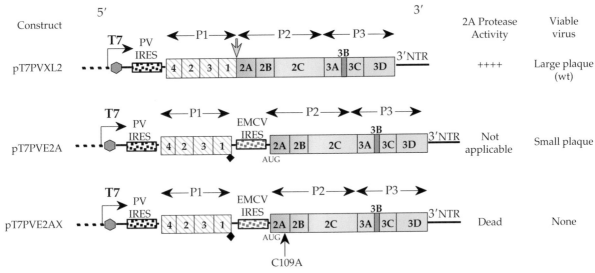

FIGURE 3.3 Diagramatic illustration of constructs used to unravel the functions of protein 2A in poliovirus replication. cDNA copies of the virus RNA can be manipulated by genetic engineering to insert IRES elements or make specific mutations. RNA can be transcribed from the clones *in vitro* and used to infect cells, which is possible for plus-strand RNA viruses because the first event after infection is translation of the genomic RNA. The wild-type construct pT7PVXL2 is shown in the top line. The 2A proteolytic activity normally cleaves the bond between domains P1 and P2 of the translated polyprotein. If this function is rendered nonessential, as in construct pT7PVE2A, by the insertion of a stop codon at the C terminus of P1 (solid diamond), followed by a second IRES and an initiation AUG at the beginning of 2A, virus is still produced, but forms small plaques. Thus separation of the structural region and the nonstructural region in this way results in viable virus. However, if the proteolytic activity of 2A is inactivated by mutation of the catalytic cysteine to alanine as in pT7PVE2AX, no virus is produced, demonstrating that the proteolytic activity of 2A^pro is necessary for other functions in addition to the P1/P2 cleavage. The pink hexagon is the VPg encoded in 3B and linked to the 5′ end of the RNA. Adapted from Lu *et al.* (1995) and Molla *et al.* (1993).

Functions of the Picornavirus Proteins

The cleavage product P1 consists of a polyprotein precursor for the four structural proteins of the virus, VP1–4. P1 is first cleaved *in trans* to VP0, VP1, and VP3 by 3CD^{pro} (Fig. 3.2A). VP0 is later cleaved to VP2 and VP4 during assembly of most picornaviruses.

The cleavage products of P2 and P3 are required for RNA replication. 2A^{pro} has been described. Protein 2B from a Coxsackie virus has been shown to induce the influx of extracellular Ca^{2+} and the release of Ca^{2+} from stores in the endoplasmic reticulum, and it is proposed that this protein induces lesions in cellular membranes that allow release of progeny virions. 2C^{ATPase} has been shown to be an ATPase, not a GTPase, and contains sequence motifs characteristics of helicases. Many, but not all, RNA viruses encode helicases to unwind duplex RNA during replication, and it is assumed that 2C^{ATPase} performs such a function. The precursor to 2B and to 2C^{ATPase}, a protein called 2BC^{ATPase}, has a different role in RNA replication. It is required for proliferation of membranous structures in poliovirus-infected cells that serve as sites for RNA replication.

Region 3 encodes VPg, 3CD^{pro}/3C^{pro}, and the viral RNA polymerase 3D^{pol}. Cleavages effected by 3C^{pro} are illustrated in Fig. 3.2. 3C^{pro} may also have a regulatory role in the virus life cycle, because the cleavage intermediate 3CD^{pro}, which is fairly long lived, has properties that differ from 3C^{pro}. One function of 3CD^{pro} is to bind the viral RNA in conjunction with 3AB, the precursor for VPg, or with a cellular protein, poly(C)-binding protein. Formation of a complex with the viral RNA is essential for its replication, and differential cleavage of the 3C–3D bond during the infection cycle may regulate replication. A strategy in which precursor proteins perform different functions than those performed by the final cleavage products, such as those illustrated by 2BC^{ATPase} and 3CD^{pro}, allows the virus to optimize the coding capacity of its small genome, because a given sequence is used for more than one function.

Replication of Picornaviruses

The replication of poliovirus has been particularly well studied and the virus has served as a model for the replication of eukaryotic RNA viruses. All nonstructural poliovirus proteins, including cleavage intermediates, have been purified and studied for their possible function as enzymes or RNA-binding proteins. These studies have been complemented by studies of replication complexes isolated from infected cells, studies using replicons in which the luciferase gene replaces the P1 coding region (see, for example, Fig. 3.3 and Chapter 11), and studies of processes that occur in infected cells.

Replication of poliovirus RNA is associated with cellular membranes, as appears to be true of all eukaryotic plus-strand RNA viruses. The 3A protein has hydrophobic sequences that may be involved in this association. During replication, a full-length complementary copy of the genomic RNA is produced that serves as a template for the synthesis of genomic RNA (illustrated schematically in Fig. 1.11A). This complementary RNA template has been variously called minus-strand RNA [abbreviated (−)RNA], antigenomic RNA, or virion-complementary (vc) RNA. Much more (+)RNA than (−)RNA is produced, since (+)RNA is needed for translation and encapsidation into progeny as well as for replication, whereas (−)RNA is needed only as a template for making (+)RNA. It is probable that disproportionate amounts of (+) and (−) strands are synthesized because the promoters in the viral RNA recognized by the viral replication machinery for (+) and (−)RNA synthesis (which might also be called origins of replication) differ in their strength, but other mechanisms are known to be used in at least some RNA viruses.

The RNA-dependent RNA polymerase 3D^{pol} is strictly primer dependent. In the presence of template, 3D^{pol} can uridylate VPg on a specific tyrosine residue. This nucleotidyl peptide, VPgpU or VPgpUpU, then functions as a primer for the initiation of RNA synthesis. It is of interest that several viruses belonging to other families, such as hepatitis B virus (a virus that uses reverse transcription during the replication of its genome) and adenovirus (a DNA virus), have also adopted the strategy of using a protein primer for initiation of nucleic acid synthesis.

The nature and function of the promoters in the poliovirus genome that are involved in the initiation of RNA replication are incompletely understood. One essential element has been called a *c*is-acting *r*eplication *e*lement, abbreviated *cre*, or 3B-uridylation site, abbreviated *bus*. *cre* is a stem-loop structure that contains a motif in the loop, AAACA, that is conserved in all picornaviruses. This motif serves as a template for the uridylation of VPg described before, which is required for the initiation of RNA synthesis. The description of this element as *cis* acting is a misnomer because the element can act *in trans*, and there is a pool of VPgpUpU within poliovirus-infected cells that can be used to initiate RNA synthesis. The *cre* element is found in different locations in different picornaviruses. In poliovirus it is found in the coding sequence for 2C, in rhinovirus in the coding region for VP1, and in FMDV it is in the 5′ NTR. Furthermore, the element can be moved to other regions within a viral genome and still function normally.

A second sequence element required for RNA replication, present in polioviruses and rhinoviruses if not in all picornaviruses, is located within the 5′-terminal NTR. This element forms a cloverleaf that binds protein complexes containing 3CD^{pro}.

In addition to the various viral proteins just described, a number of cellular proteins are also required for viral RNA replication. In fact, cellular proteins appear to be required for replication of all (+)RNA virus RNAs, but the identity of these proteins and their function in viral RNA replication is only poorly understood. One such protein in the case of poliovirus is a cellular protein called heterogeneous nuclear ribonucleoprotein C1 (hnRNP C1), which interacts with RNA synthesis initiation complexes and appears to be required for the initiation of positive-strand RNA. A second protein is the poly(A)-binding protein. Efficient replication of polio RNA requires a poly(A) tract at the 3′ end of the RNA that is at least 20 residues in length, and it is believed that the poly(A)-binding protein binds this poly(A) tract and participates in the initiation of minus-strand RNA synthesis.

It has been possible to achieve a complete replication cycle of poliovirus in an extract of uninfected HeLa cells. RNA from poliovirus virions added to such an extract will direct the synthesis of all the poliovirus proteins, and these in turn will replicate the input RNA and encapsidate the progeny genomes. This cell-free, *de novo* synthesizing system for poliovirus, is as yet unique in virology.

In cell culture, most picornaviruses complete their replication cycle in about 6 hours. The infection is cytolytic, and large quantities of virus are produced. An exception is hepatitis A virus, which establishes chronic infections in cell culture and grows to very low titers.

Genus *Enterovirus*

Enteroviruses replicate primarily in the enteric tract where they usually cause only mild disease. More serious enteroviral disease may develop after spread to other organs, such as the central nervous system or the heart. Enteroviruses are normally contracted though ingestion of the virus, either in contaminated food or water or by exposure to the virus through contacts with individuals that are excreting the virus. The epidemiology of poliovirus has been the most intensively studied among the enteroviruses. Poliovirus is present in oropharyngeal secretions early after infection and is excreted in feces over a period of weeks following infection. The virus spreads readily and rapidly through households, which demonstrates the importance of close contacts in virus spread. The virus also has the ability to persist in the external environment for weeks under favorable conditions, and this may represent another source of infection during epidemics. Sewage surveys, for example, have been used to follow poliovirus epidemics, and poliovirus has been found in lakes and swimming pools.

In general, enteroviruses have a fairly narrow host range. Most of the well-studied viruses are human viruses, because humans take a particular interest in the viruses that cause them the most trouble, but enteroviruses of nonhuman primates, pigs, cattle, and insects are known. The more than 65 known human enteroviruses, many of which are important pathogens, normally infect only humans, but poliovirus will infect Old World monkeys. It has been suggested that the virus may be a natural pathogen of these monkeys but it is unlikely that nonhuman primates constitute a reservoir for it, which is important in relation to efforts spearheaded by the World Health Organization to eradicate poliovirus globally.

The classification of human enteroviruses has recently undergone extensive revision, based upon the wealth of sequence information that is increasingly available. Previously, classification was based upon the symptomology of disease caused or upon the characteristics of the growth of a virus in experimental animals or in cultured cells. Poliovirus has been known for more than a century as the causative agent of epidemic poliomyelitis. It was first shown to be a filterable virus in 1908. However, early experiments could only be conducted in monkeys, because the virus will only infect primates. Thus, the amount of information that could be obtained was limited, but such studies eventually showed that more than one poliovirus serotype existed. The development of methods for the cultivation of viruses in cell culture in the 1940s made it possible to screen human stool samples in an effort to type poliovirus isolates, which was necessary if a vaccine was to be produced. Such screening resulted not only in the identification of three serotypes of poliovirus, but also in the discovery of many other enteroviruses as well. The study of virology in the United States owes much to the campaign to develop a vaccine against poliomyelitis. This campaign generated a great deal of public support, which led to funding through private as well as governmental agencies, and the successful development of a vaccine reinforced this support.

The first of these other enteroviruses to be found were two Coxsackie viruses, found by screening patients in Coxsackie, New York, who were suffering from paralysis during a polio epidemic. Coxsackie viruses will infect mice and are classified into two subgroups, called A and B, which differ in their biological properties in mice. They were simply given serial numbers in the order of their isolation—23 Coxsackie A viruses and 6 Coxsackie B viruses are now recognized. Another series of enteroviruses that were first identified in these early studies were called echoviruses (enteric cytopathic human orphan virus), because these viruses infected the enteric tract of humans, caused cytopathology in cultured cells, and were orphans, not known to cause disease. Echoviruses were distinguished from Coxsackie viruses by their inability to infect suckling mice. Currently 29 echoviruses are recognized in the genus *Enterovirus*. The latest human viruses to be isolated are now simply called enteroviruses and given serial numbers. The first four such viruses to be recognized were thus called human enterovirus 68, 69, 70, and 71. Numbering started with 68 because at the time there were thought to be 67 polio, Coxsackie, and echoviruses. However, 5 of these (one Coxsackie A virus and 4 echoviruses) were subsequently found to be misidentified, and one (echovirus 22) is sufficiently distinct that it has been renamed human parechovirus and classified into the genus *Parechovirus* (Table 3.2).

Thus, from such studies, a total of 65 human enteroviruses was isolated and, as indicated, classified according to their biological properties. As the genomes of these various viruses were sequenced, it became apparent that these viruses fell into five lineages or clades whose members are closely related to one another. As described in Chapter 1, the definition of a virus species is somewhat arbitrary, but the purpose of classification is to recognize evolutionary relationships, and 63 of these 65 human enteroviruses have now been reclassified into five species, called *poliovirus* and *human enterovirus A, B, C, D*. These assignments are shown in Table 3.3, and the various members of a species are now considered serotypes. This table contains information on serotypes accepted as of the 2005 ICTV report. More than 80 serotypes are now known and as new serotypes continue to be identified and characterized, it is to be expected that this number will continue to grow. The extensive sequence data have also uncovered examples of recombination that have occurred during the evolution of these viruses, both within species and between species.

Sequence information has also been used to identify one species of bovine enterovirus and two species of pig enteroviruses (*Porcine Enterovirus A and B*) in the genus *Enterovirus* (Table 3.3). There are 2 serotypes assigned to *Simian enterovirus A* and 17 other known monkey enteroviruses have as yet to be classified into species. The monkey viruses form a distinct clade related to porcine enterovirus 8 (*Porcine Enterovirus A*) and will probably be classified into one or two species.

Polioviruses

The best known of the enteroviruses are the three serotypes of poliovirus. These viruses are the causative agents of poliomyelitis, a disease characterized by the death of motor neurons in the spinal cord. Most poliovirus infections of susceptible humans are inapparent or result in a mild febrile illness in which cells of the pharynx and the gut are infected and recovery is uncomplicated. However, a transient viremia is established following infection (viremia = virus present in the blood), and in a small percentage (<2%) of infections the virus invades the central nervous system (CNS), where it infects motor neurons in the spinal cord and, in severe cases, other regions of the CNS. The mechanism by which the virus enters the CNS is still controversial. Current information supports the hypothesis that viremia allows the virus to enter by penetrating through the blood–brain barrier, but entry via retrograde transport in axons that serve the periphery may also be involved. In any event, infection of the CNS can result in paralysis, which can be severe enough to be fatal because of paralysis of respiratory muscles. The name *poliomyelitis* comes from the Greek words *polio* = gray and *myelo* = spinal cord, from the pathology caused by damage to the motor neurons in the spinal cord, which are located in the gray matter.

Polioviruses readily undergo recombination with other polioviruses and with at least some other enteroviruses. The distinguishing feature of a poliovirus, what makes it a poliovirus, is the structural protein module and not the nonstructural protein module associated with the structural proteins. This has importance implications for vaccines that protect against poliomyelitis, as described later.

TABLE 3.3   Current Taxonomy of the Genus *Enterovirus*

Species	Strains, subtypes, and serotypes
Human enterovirus A	Human Coxsackie viruses A2–8, 10, 12, 14, 16, human enterovirus 71, 76
Human enterovirus B	Human Coxsackie virus A9
	Human Coxsackie viruses B1–6 (including swine vesicular disease virus)
	Human echoviruses 1–7
	21 other human echoviruses
	Human enterovirus 69, 73–78
Human enterovirus C	Human Coxsackie viruses A1, 11, 13, 15, 17, 19–22, 24
Human enterovirus D	Human enteroviruses 68, 70
Poliovirus	Human poliovirus types 1, 2, and 3
Bovine enterovirus	Bovine enteroviruses 1, 2
Porcine enterovirus A	Porcine enterovirus 8
Porcine enterovirus B	Porcine enterovirus 9, 10
Simian enterovirus A	Simian enterovirus A1, A2-plaque
Unassigned viruses	17 simian enteroviruses

Epidemic Poliomyelitis

Polioviruses appear to have been important pathogens of humans for a very long time. The depiction of a lame priest on an Egyptian stele that dates from 3500 years ago suggests that poliovirus was present in ancient Egypt, and references to clubfoot in ancient Greek and Roman writings probably signifies that polio was present at these early times. However, although it is very likely that poliovirus has been widespread in humans for thousands of years, there is no firm evidence for poliomyelitis in human populations until about 200 years ago, when the virus appears to have been (or to have become) widespread. Serosurveys in the United States in the 1930s and 1940s, before the introduction of the Salk and Sabin vaccines, indicated that 80–100% of adults had been infected by poliovirus at some time in their lives. Studies in other areas of the world, including studies of lameness in populations, also suggest that, at least in the 1900s, the majority of the world's population had been infected with poliovirus.

Paradoxically, even though poliovirus was surely widespread earlier, poliomyelitis epidemics of large proportions evolved only during the twentieth century and they were concentrated at first in countries practicing the highest standards of hygiene. This startling phenomenon has been explained as resulting from changes in human behavior. Originally, the highly infectious virus was contracted by infants shortly after birth when they were still protected by maternal antibodies (see Chapter 10 for a discussion of maternal antibodies). This natural infection served to immunize the infant, protecting it from poliomyelitis for life. However, when the chain of immunization was interrupted upon removal of the virus from the environment by the development of hygienic conditions, unprotected children grew up, giving rise to susceptible populations. If the virus invades such populations, epidemics rapidly evolve.

Notice that this scenario requires that infants be infected very early, while still protected by maternal antibodies. After these antibodies wane, the infant is susceptible to poliomyelitis, although it has been thought that infection of susceptible but very young children is less likely to cause poliomyelitis. Statistics of the fraction of young children who contract poliomyelitis in societies in which the virus is endemic, rather than epidemic, are not well defined, in part because of the high death rate of children in such societies due to many infectious diseases. However, surveys conducted in the twentieth century of lameness in populations, most of which is probably due to paralytic polio, found similar extents of lameness whether the virus was endemic or epidemic.

In any event, it is clear that changes in human behavior can bring about serious complications relating to infectious disease, and such scenarios have recurred many times during the last century. However, it is important to note that although higher standards of hygiene led eventually to epidemics of poliomyelitis, these standards also led to a reduction in diseases caused by numerous other infectious agents, both viral and bacterial (see Fig. 1.1).

Control of Epidemic Poliomyelitis

Before it was controlled with vaccines, epidemic poliomyelitis was greatly feared, and it is hard now for people to realize the extent of fear that the disease induced. It was not simply that the disease could be fatal, but the specter of the iron lung and the wheelchair hanging over teenagers or young adults who were the most likely to contract the disease. Furthermore, the epidemics struck during the summer, during the summer breaks of schools or universities. Many human pathogenic viruses are known to prefer a season for attack on humans: influenza during the winter, measles in early spring, enteroviruses during the summer. It is thought that this phenomenon relates to air temperature and humidity. For example, poliovirus infections are correlated with humidity in the Americas and in Europe.

In the United States, there were huge poliovirus epidemics every summer in the 1950s in which more than 50,000 people, mostly children or adolescents, became ill. Of these cases, about 20,000 were paralytic and 2000–3000 people died (Fig. 3.4). Death was often the result of the paralysis of the muscles required for breathing, and iron lungs were introduced for mechanical ventilation of poliomyelitis patients until their muscles recovered sufficiently that they could breathe on their own. Wards containing dozens of patients in iron lungs became a common sight in the large epidemics of the 1950s (Fig. 3.5), and there were fears that larger wards containing still more iron lungs would be required as the epidemics became more virulent. Of the survivors of poliomyelitis, many were permanently paralyzed and confined to wheelchairs or required the use of crutches for walking. One of the best known poliomyelitis cases is that of Franklin D. Roosevelt, who contracted poliovirus in 1921 at the age of 39 and was in a wheelchair for the rest of his life, although he continued to lead an active political life.

Introduction of the Salk and Sabin vaccines in the 1950s and 1960s led to the elimination of poliovirus in the United States over a period of about 2 decades (Fig. 3.4) and more recently has led to the elimination of poliovirus throughout the Americas. The Salk vaccine, which was the first to be developed, is an inactivated virus vaccine that is given as a series of injections. Introduction of this vaccine resulted in a rapid decrease in the number of poliovirus cases. However, because the vaccine induces circulating antibodies but little in the way of mucosal immunity (see Chapter 10), it prevents poliomyelitis, the disease, by preventing spread of the virus from the gastrointestinal (GI) tract to the CNS, but not infection of the GI tract by the virus. The virus thus remained in circulation. The Sabin vaccine, introduced shortly thereafter,

Cases of Poliomyelitis in the United States from 1951–2004

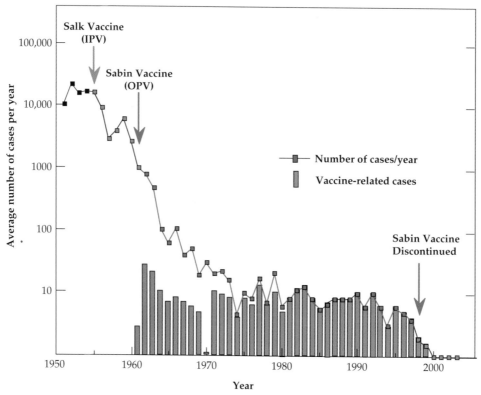

FIGURE 3.4   Total number of cases of poliomyelitis in the United States from 1951 to 2004 and the number of vaccine-related cases after the introduction of the live virus Sabin vaccine. IPV is *i*nactivated *p*olio *v*accine; OPV is *o*ral *p*olio *v*accine. Data from N. Nathanson *et al.* (1996) p. 556 and from *Morbidity and Mortality Weekly Report (MMWR)*. Note that the scale is logarithmic, with each division portraying 10 times as many cases as the one below.

is a live attenuated vaccine that is given orally. Attenuation was achieved by blind passage of the virus followed by testing of the resulting virus in monkeys. The changes resulting from passage are now known and are shown in Table 3.4; two changes are sufficient to make the virus avirulent in the case of types 2 and 3. The introduction of the Sabin vaccine led to a further rapid decline in paralytic poliomyelitis. This vaccine has the drawback that it induces a very small number of cases of paralytic disease, termed vaccine associated paralytic poliomyelitis (VAPP), that result from reversion of the attenuated virus to virulence. The incidence rate is about 1 per million persons inoculated, and there were about 10 such cases per year in the United States until use of this vaccine was discontinued in the year 2000 (Fig. 3.4). The efficacy of the Sabin vaccine is very high, however, because it induces mucosal immunity as well as other forms of immunity. It prevents subsequent infection by the wild-type virus, thus allowing eradication of the wild-type virus if coverage is sufficiently broad. In addition, it is much cheaper and simpler to manufacture and administer than the Salk vaccine

(oral administration of relatively small doses of live virus versus injection of large amounts of inactivated virulent virus), making it suitable for widespread use in developing countries. Worldwide use of this vaccine has resulted in the eradication of wild-type poliovirus in the United States and throughout the Americas (the last case of indigenous poliovirus infection in the Americas occurred in Peru in 1991). Poliovirus is in the process of being eradicated in other parts of the world, although it is still endemic in areas of Africa and Asia. With the extirpation of polio in the United States, the use of Sabin vaccine was discontinued in this country and it has been replaced with the Salk vaccine, in order to eliminate VAPP.

As described before, polioviruses undergo recombination with other enteroviruses. It is important, therefore, to note that the mutations in the Sabin vaccines that render the virus attenuated are all found in the structural region of the genome (Fig. 3.6). Thus, recombination with other enteroviruses cannot restore the virulence of the virus. Reversion to virulence requires the back mutation of the attenuating

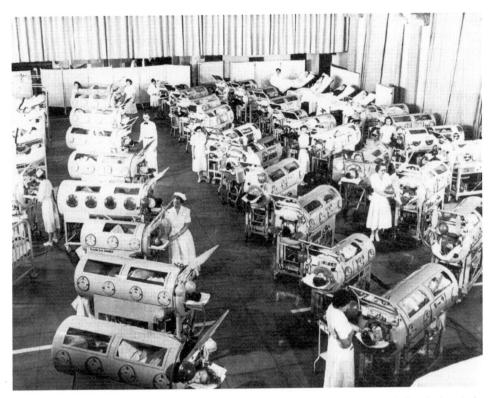

FIGURE 3.5 Ward of iron lungs and rocking beds at the poliomyelitis rehabilitation center in Rancho Los Amigos, California. From Halstead (1998) with the permission of the author and the publisher.

mutations in the vaccine viruses or, conceivably, recombination between two attenuated poliovirus strains that eliminates the attenuating mutations. In the latter case, however, it is unlikely that most recombinants would be virulent because of the incompatibility of the various nonstructural proteins with one another.

Development of the Polio Vaccine

The polio vaccine has been enormously successful in controlling this virus scourge, but the history of its development and the current difficulties in complete eradication of polio have important lessons for us. The original Salk vaccine was incompletely inactivated because the science of virology was insufficiently developed to assay for minute amounts of residual live virus in solutions containing very high concentrations of virus. The result was that this vaccine caused a small number of cases of poliomyelitis, but in retrospect the risk–reward ratio was favorable because of the significant decline in natural infections (Fig. 3.4). As soon as the problem was recognized, more stringent methods of inactivation

TABLE 3.4 Characteristics of Poliovirus Vaccines

Salk vaccine	Inactivated wild-type poliovirus (three types)					
Sabin vaccine	Live poliovirus, attenuated by mutations in:					
	Type 1		Type 2		Type 3	
	nt	aa	nt	aa	nt	aa
5′NTR	A480G	—	G481A	—	C472U	—
VP1	G2795A	A106T	C2909U	T143I	U2 493C	I6T
	C2879U	L134F				
VP3	U2438A	L225M			C2034U	S91F
VP4	G935U	A65S				

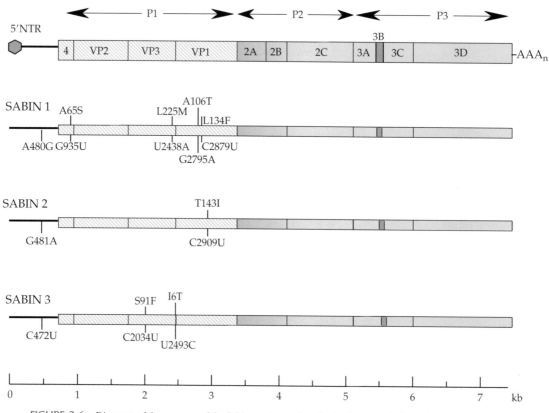

FIGURE 3.6   Diagrams of the genomes of the Sabin vaccine strains of poliovirus types 1, 2, and 3 showing the locations of the attenuating mutations.

were quickly developed that resulted in complete inactivation of the infectivity of the virus, solving this problem and serving as an example for development of other vaccines. The introduction of the Salk vaccine, although enormously successful in controlling polio, also suffered from early problems when it was found that early lots were contaminated with the monkey virus SV40 (described in Chapter 7). Infection of humans by this virus appears to be benign, although there is some evidence that, very rarely, brain tumors may be associated with infection. In any event, this episode brought to light the issue of adventitious contamination of cell cultures with viruses that infect the host supplying tissues for culture. A third problem that arose during development of polio vaccines was the infection of a number of laboratory workers in Germany with Marburg virus. These workers were employed in the isolation of cells from the kidneys of wild-caught monkeys that were to be used in propagating polioviruses, and some of the monkeys were infected with Marburg virus, at that time an unknown virus. Several people died in the ensuing epidemic (described in Chapter 4).

Eradication of Polioviruses

Introduction of the Sabin vaccine led to the eradication of poliovirus from the Americas, and in 1988 the World Health Organization (WHO) initiated a campaign, the Global Polio Eradication Initiative, to eradicate poliovirus worldwide by the year 2000. Although falling short of this goal, significant progress has been made. The number of polio cases worldwide fell from an estimated 300,000+ cases in the mid 1980s to fewer than 3000 by 2000 and subsequently to still lower levels (Fig. 3.7). In 2000 the Centers for Disease Control and Prevention (CDC) said that 2971 cases were reported of which 719 were confirmed by laboratory analysis. In 2001, 537 cases were reported of which 473 were confirmed by laboratory analysis, and these cases occurred in just 10 countries. It appeared that eradication would be achieved soon. The eradication campaign hit a snag recently, however, when Muslim clerics in Nigeria claimed that the vaccine could cause AIDS or infertility. In 2003, officials in some parts of Nigeria suspended local vaccination programs, and an epidemic of poliomyelitis in Nigeria resulted that then spread to neighboring countries that had been free of polio (Fig. 3.8). By 2005 polio had spread to a total of 16 countries that had previously been polio free. Further setbacks in the polio vaccination initiative have resulted from civil unrest in Sudan and other countries that resulted in interference with vaccine campaigns and the reestablishment of poliovirus transmission. Health ministers from Africa are stepping up vaccination

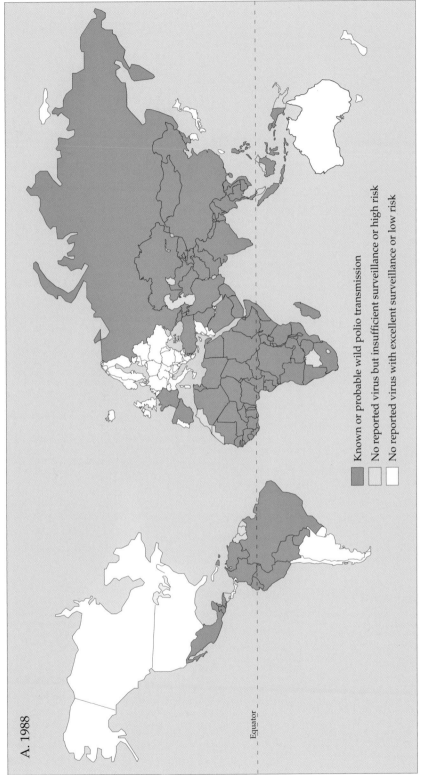

A. 1988

Equator

Known or probable wild polio transmission

No reported virus but insufficient surveillance or high risk

No reported virus with excellent surveillance or low risk

FIGURE 3.7   Maps showing the worldwide distribution of wild poliovirus and the effects of the global eradication efforts. (A) Wild poliovirus transmission in 1988. This is from the Web site www.who.int/gpv-surv/graphics/NY_graphics/ global_polio_98.htm.

Continued

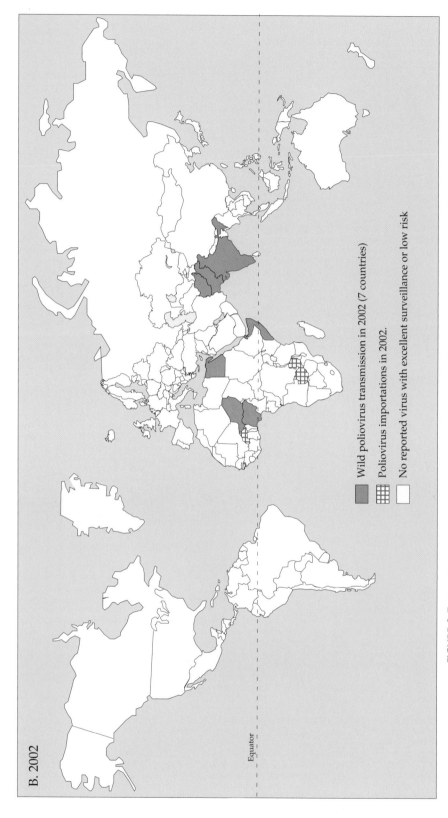

B. 2002

Equator

Wild poliovirus transmission in 2002 (7 countries)

Poliovirus importations in 2002.

No reported virus with excellent surveillance or low risk

FIGURE 3.7 (Cont'd) (B) Wild poliovirus transmission as of 13 March 2002 (from *MMWR* Vol. 49, #16, p. 352).

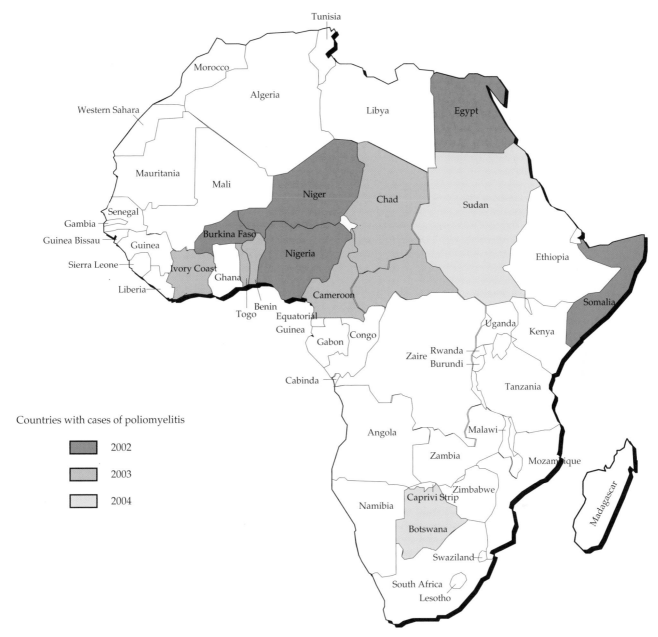

FIGURE 3.8　Reemergence of poliomyelitis in Africa 2002–2004. Data from *MMWR*.

programs to reestablish control of poliovirus transmission, but these situations illustrate problems that result from the continuing conflicts among societies.

Should Routine Poliovirus Immunization Be Eliminated?

If poliovirus is finally eradicated worldwide, should vaccination against the virus be scaled back? Limiting vaccination would be important because of the residual virulence of the vaccine virus. Virulent revertants of vaccine virus not only cause a small number of individual cases (Fig. 3.4), but have led to small epidemics of poliomyelitis when the virulent virus, derived by reversion, circulates in nonimmunized or incompletely immunized contacts (e.g., 21 cases in Hispaniola in 2000–2001), illustrating the potential for continued outbreaks arising from vaccination. The Salk vaccine could still be used in developed countries, but it seems unlikely that routine administration of Salk vaccine would be used in developing countries. It should be possible to design new attenuated viruses for vaccine purposes that would be safer than the Sabin vaccine, but in the absence of poliovirus epidemics no drug company would want to undertake the very expensive development of a new vaccine, especially in view of legal problems that would be sure to arise.

Associated problems are the difficulties in being sure that poliovirus is truly eradicated and the possibility that virulent poliovirus could reemerge. It is known that immunodeficient children who have received the vaccine virus can continue to secrete virus for long periods of time, during which the virus may revert to virulence. Even wild-type poliovirus can circulate silently because most infections are inapparent. In regions where poliomyelitis has reemerged following interruption of vaccination campaigns, it has been found in some cases that the virus had been circulating for at least 2 years before reemergence. In addition, poliomyelitis can be caused by enteroviruses other than poliovirus (see later), a possible source of confusion in diagnosis.

If vaccination were terminated and poliovirus were to reemerge in a naïve population, it could lead to a widespread epidemic. Possible sources of reemerging virus include new strains that might arise from other enteroviruses, circulating strains of wild poliovirus or vaccine-derived virus that have escaped detection, or the inadvertent or deliberate release of wild poliovirus by escape from a laboratory or introduction by terrorists. It would be necessary to maintain stockpiles of polio vaccine to counter such threats, and in the case of the Salk vaccine this stockpiling itself could serve as a possible source of accidental escape of the wild virus. Since the events of 9/11, it is also obvious that terrorists would have no compunctions about releasing a virulent virus into the U.S. population if such a virus could be obtained. Thus, even if (when?) poliovirus is eradicated, the policies with respect to vaccination will require careful consideration.

Post-Polio Syndrome

Although poliovirus has been eradicated from developed countries, there is a large cohort of people infected in the 1950s who are or were paralyzed. Many paralyzed poliomyelitis patients were ultimately able to resume almost normal activities. Through a process of axonal sprouting and reenervation of muscles by the motor neurons that survived the infection, many learned to walk and use their previously paralyzed limbs. In many, recovery was effectively complete. However, a syndrome called post-polio syndrome has emerged to plague a significant fraction, perhaps 40%, of the survivors of paralytic poliomyelitis. This syndrome appears 30–40 years after polio infection and is characterized by fatigue, pain, and weakness. The weakness may be severe enough to require the use of a wheelchair. The syndrome results from the degeneration of motor neurons, but the reasons for the degeneration are not clear. The favored hypothesis is that it is the result of overuse of the surviving motor neurons, which are forced to do the work of many. A second possibility is that the surviving neurons were damaged by the original poliovirus infection and fail prematurely. A third, albeit unlikely, possibility is that poliovirus persists in neurons and is somehow reactivated, even in the presence of anti-polio antibody. In model studies using Sindbis virus infection of mice, it has been found that the virus can persist in neurons in a latent state for at least 1–2 years. There is no evidence that poliovirus might similarly persist in humans for 40 years, however, and such persistence seems unlikely. Other possible explanations for the failure of motor neurons in post-polio syndrome have also been suggested. Fortunately, paralytic poliomyelitis and its sequelae may soon be a thing of the past.

Other Enteroviral Diseases of Humans

As described, 62 human enteroviruses other than polioviruses are currently recognized in the ICTV catalog and classified as serotypes of 4 different species. Although most of these have been known for 50 years, it is only recently that the association of many of these viruses with significant human illness has been shown. In fact, it has now been established that most enteroviruses do cause disease, and many of them cause significant episodes of serious disease (Table 3.5). Study of disease caused by these viruses has been complicated by the fact that there are so many enteroviruses, of which at least some have multiple strains that may differ in disease-causing potential, and by the fact that serious disease is an uncommon complication of infection by most enteroviruses (even for poliovirus most infections do not result in significant disease). This has made it difficult to ascribe any particular disease to infection by any particular virus. However, even though serious disease is an uncommon complication, enteroviral infections are very common, and the total number of cases of disease caused by these viruses is large. These illnesses include very infrequent paralytic disease essentially indistinguishable clinically from that caused by poliovirus; myocarditis and pericarditis (caused especially by the Coxsackie B viruses) that is usually subclinical but can be acute and result in significant cardiac compromise; aseptic meningitis; encephalitis; hepatitis; the common cold (perhaps a quarter of summer colds are due to enteroviruses); diarrheal disease; febrile illnesses; rash; hand-foot-and-mouth disease (a common childhood illness caused by several serotypes in human enterovirus A); and epidemic acute hemorrhagic conjunctivitis (an epidemic disease caused by enterovirus 70 that appeared recently and spread around the world). The Coxsackie B viruses are also associated epidemiologically with juvenile onset diabetes in humans but how (or even whether) they cause diabetes is still unresolved. There are no vaccines for any of these viruses.

Genus *Rhinovirus*

The human rhinoviruses are the causative agents of about half of human colds, the most characteristic symptom of which is rhinitis (inflammation of the nasal mucous membrane and characterized by a runny nose). Other viruses that

TABLE 3.5 Clinical Syndromes Associated with Human Enteroviruses

Clinical syndrome	Poliovirus	Enterovirus A	Enterovirus B	Enterovirus C	Enterovirus D
Paralysis	Types 1, 2, 3	Coxsackie A7, A9 Enterovirus 71	Coxsackie B2–B5 Echoviruses 4, 6, 9, 11, 30	—	Enterovirus 70
Aseptic meningitis	—	Coxsackie A2, A4, A7, A9, A10	Coxsackie B1–B6 All echoviruses except 12, 24, 26, 29, 32, 33	—	—
Pericarditis, myocarditis	—	—	Coxsackie B1–B5 Echoviruses 1, 6, 9, 19	—	—
Encephalitis	—	Enterovirus 71	Coxsackie B1–B5 Echoviruses 2, 6, 9, 19	—	Enterovirus 70
Hepatitis	—	Coxsackie A4	Coxsackie A9, B5 Echovirus 4, 9	—	—
Upper respiratory disease, pneumonia	—	—	Coxsackie B4, B5	Coxsackie A21, A24	Enterovirus 68
Hand, foot, and mouth disease	—	Enterovirus 71 Coxsackie A5, A10, A16	—	—	—
Acute hemorrhagic conjunctivitis	—	—	—	Coxsackie A24	Enterovirus 70
Undifferentiated febrile illness	Types 1, 2, 3	—	Coxsackie B1–B6	—	—

serve as major causes of the common cold include some of the enteroviruses, just described, and the coronaviruses, described later. One hundred serotypes of human rhinoviruses are currently recognized. Eighteen of these have been assigned to the species *human rhinovirus A* and three of them to *human rhinovirus B*. The remaining 79 serotypes have not yet been assigned to a species. There are also three serotypes of bovine rhinovirus known to exist, and there are rhinoviruses for other animals that have as yet to be well characterized. In general, rhinoviruses are specific for a particular species or for a limited range of species, and this restriction appears to work at the level of receptors required for virus entry (see Chapter 1).

The 100 serotypes of human rhinoviruses are not cross protective and the result is that we are subject to many rhinovirus colds during our lifetimes. Young children, not having been exposed to rhinoviruses and other viruses that cause colds, contract many colds a year. Adults, having become immune to many of these viruses through hard experience, have fewer colds per year, usually only about one. However, the extent and duration of immunity to a particular rhinovirus induced by infection are not well established. There are so many rhinoviruses (and although rhinoviral disease may be miserable it is not life threatening) that detailed studies on cohorts of people over many

years have not been done to establish whether immunity to a particular rhinovirus following infection is long lived. For the same reasons, there are no vaccines for any of these viruses.

Rhinoviruses replicate in the upper respiratory tract and are transmitted by direct person-to-person contact. Coughing and sneezing, common syndromes of rhinovirus infection, help spread the virus to nearby contacts. It is not clear how much of the spread is due to aerosolization of the virus on coughing or sneezing followed by inhalation of the aerosolized virus by a susceptible contact, and how much is due to contact with mucus that contains virus, such as by handshake or contact with contaminated doorknobs, followed by transmission of the virus to mucosal membranes in the nose or the mouth.

It is an interesting and informative historical fact that early attempts to isolate rhinoviruses using standard cell culture techniques were unsuccessful. Most cells in the body are maintained at 37°C at a pH of 7.4, and cells in culture are normally maintained under these conditions. However, cells in the upper respiratory tract are maintained at a lower temperature, about 33°C, because the inhalation of outside air through the upper respiratory tract keeps this area cool, and at a pH significantly less than 7.4 because of the high concentration of CO_2 in expired air. Rhinoviruses replicate well in cultured cells under these altered conditions and appear to require the lower temperature and lower pH for efficient growth. In part because of this, rhinovirus infection is limited to the upper respiratory tract, and rhinoviruses almost never cause lower respiratory tract infections.

It is also of interest that rhinoviruses are sensitive to very low pH, and infectivity is destroyed by exposure to pH 3. The related polioviruses, however, survive exposure to pH 2, which is necessary because, being enteroviruses, they must survive passage through the stomach in order to infect an animal.

Genus *Cardiovirus*

The *Cardiovirus* genus consists of several viruses of mice of which encephalomyocarditis virus (EMC) has been extensively studied as a model picornavirus. It is closely related to other picornaviruses (Fig. 3.1) although differing in certain important characteristics. The EMC IRES has proved more useful than the poliovirus IRES in experiments that require polycistronic mRNAs or that express proteins in a cap-independent fashion in vertebrate expression systems. Theiler's virus, another member of this genus, causes demyelinating disease in mice and has been extensively studied as a model for multiple sclerosis in humans.

Genus *Hepatovirus*

Hepatitis in Humans

Many different viruses, belonging to several virus families, are known to cause hepatitis (inflammation of the liver) in humans. These different viruses have different modes of transmission and cause illness of different degrees of severity (although all hepatitis is serious) that results from destruction of liver cells caused by growth of these viruses in the liver as a target organ. Hepatitis is characterized by fatigue and other symptoms that result from inadequate liver function, and it may be fatal if sufficient destruction of the liver takes place. A characteristic feature of acute hepatitis is the presence of elevated levels of liver enzymes circulating in the blood that results from the destruction of liver cells. Many cases of hepatitis are accompanied by jaundice (turning yellow) because of the destruction of the liver, which is responsible for clearing bilirubin from the blood.

The viruses whose primary disease syndrome in humans is hepatitis, and which therefore target the liver as the principal or only organ infected, or viruses that are closely related to viruses that cause such hepatitis, have historically been named hepatitis virus followed by a letter, in the order of isolation. Thus we have hepatitis A virus, the first to be isolated, hepatitis B virus, the second, and so forth. Because these viruses belong to a number of different families, confusion can arise because of the similar names even though the viruses are unrelated. For reference, Table 3.6 presents a description of the currently known viruses whose name includes hepatitis. Figure 3.9 shows the incidence of hepatitis in the United States in 1997 caused by hepatitis viruses A, B, and C, which are the most important causes of viral hepatitis in the United States. Identification of which hepatitis virus is responsible for any specific case of hepatitis requires immunologic tests or virus isolation, because symptoms are similar.

Hepatitis A virus is a picornavirus and will be considered here. The other viruses will be considered when their respective families are introduced.

Hepatitis A Virus

Hepatitis A virus (HAV) is a causative agent of infectious hepatitis in humans. The virus is worldwide in distribution. Only one serotype is known, but isolates from different areas or different times can be grouped into different genotypes or strains. The most distantly related HAV isolates share about 75% nucleotide sequence identity, but most isolates are much more closely related. HAV is a typical picornavirus but is an outlier in the family (Fig. 3.1). It shares only 28% amino acid identity in its structural proteins with any other picornavirus, whereas most picornaviruses are more closely related to one another.

The number of cases of hepatitis A in the world has been estimated to be more than 1.4 million each year. In 1998 in the United States, for example, ~37,000 cases of hepatitis were reported, of which two-thirds were diagnosed as caused by HAV. HAV is spread through contaminated food and water. Filter-feeding shellfish like oysters are known to concentrate the virus, and consumption of raw shellfish has been the cause of many epidemics of hepatitis A. A 2003 epidemic of more that 500 cases in Pennsylvania was caused by consumption of green onions that are believed to have been contaminated during harvest. Infection by HAV usually results in a self-limited illness in which the patient recovers with relatively few sequelae. The illness can be quite serious, even fatal, however, because 90% of the liver tissue can be destroyed by virus infection, and liver function is severely impaired until the liver recovers. The seriousness of disease is age dependent. Very young children suffer little disease but with advancing age infection by the virus becomes more serious. The mortality rate in children younger than 14 is only 0.1%, but HAV infection in people older than 40 results in a fatality rate of 2.1%.

Before the introduction of a vaccine, the only prophylaxis for HAV was injection of immune gamma globulin, which provided protection from the virus for a few weeks. Two inactivated virus vaccines against HAV were licensed in the mid 1990s that have been found to give long-lived protection. Results from a clinical trial in Thailand that showed the

TABLE 3.6   Causative Agents of Viral Hepatitis in Humans

Virus	*Family/ genus*	Genome type/ *size in kb*	Transmission	Chronicity?	Long-term effects	Annual[a] U.S. acute cases/*deaths* in 2004	Chronic hepatitis (millions of cases) U.S./*World*
Hepatitis A	*Picornaviridae/ Hepatovirus*	ss (+) RNA/ *7.5 kb*	Fecal–oral	Very little	Few if any	5683/*76*	0/0
Hepatitis B	*Hepadnaviridae/ Orthohepadno-virus*	ds DNA (RT)/[b] *3.2 kb*	Parenteral, sexual, vertical	10% of adults, 90% of neonates	HCC[c], cirrhosis	6212/*659*	1.2/*300–400*
Hepatitis C	*Flaviviridae/ Hepacivirus*	ss (+) RNA/ *9.4 kb*	Parenteral, sexual, vertical	>50%	HCC[c], cirrhosis	720/*4321*	3/*50–100*
Hepatitis D	*Deltavirus*	ss, circular RNA/*1.7 kb*	Parenteral, (sexual, vertical?)	Yes	Exacerbates symptoms of Hep B	7500/ *1000*	0.07/*?*
Hepatitis E	*Hepeviridae/ Hepevirus*	ss (+) RNA/ *7.5 kb*	Fecal–oral	No	Few if any	Very rare	0/0
Hepatitis F[d]	?						
Hepatitis G	*Flaviviridae/ Hepacivirus*	ss (+) RNA/ *9.4 kb*	Parenteral, other?	Yes	??	??/*none*	??

[a] Data from MMWR Summary of Notifiable Diseases—2004. It is noteworthy that acute cases of both hepatitis A and hepatitis B have declined significantly since 1990 with the introduction of vaccines that are now in widespread use; see Figure 3.9 below.

[b] RT is reverse transcriptase. Nucleic acid in virion is partially ds DNA, consisting of a full-length minus-strand DNA of 3.2 kb, and an incomplete plus-strand DNA that is variable in length.

[c] HCC, Hepatocellular carcinoma.

[d] Isolate from a fulminant case of hepatitis, not further characterized.

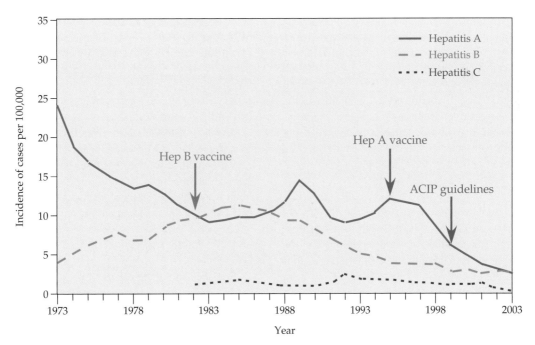

FIGURE 3.9   Incidence (cases per 100,000 population) of viral hepatitis in the United States between 1973 and 2003. Hepatitis A incidence was the lowest ever in 2004, but there has been a trend for cyclic increases every decade, and future increases could occur. However, with the expansion of recommended vaccination to include children in all communities where the incidence was consistently above the national average (1999 ACIP guidelines), the incidence of HAV has continued to plummet. No vaccine for hepatitis C exists, but an antibody test for hepatitis C was first introduced in May 1990. This graph is adapted from *MMWR* Vol. 52, #54, *Summary of Notifiable Diseases United States 2003.*

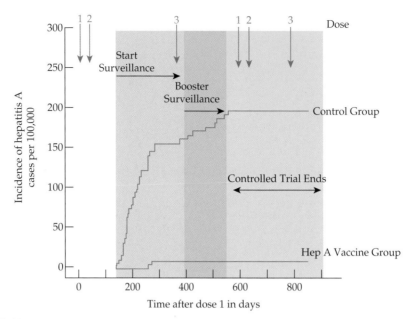

FIGURE 3.10   Controlled trial of a hepatitis A vaccine in Thailand. Children were divided into two groups, and at the times indicated by the blue arrows the vaccine group was given hepatitis A vaccine and the control group was given hepatitis B vaccine. There were two periods of surveillance for cases of hepatitis A, indicated by the pink overlays. The controlled trial ended at 540 days when the control group was given hepatitis A vaccine at the times indicated by the magenta arrows and the vaccine group was given hepatitis B vaccine. Adapted from Figure 3 in Innis *et al.* (1994).

efficacy of this vaccine are shown in Fig. 3.10, as an example of the type of data that can be obtained in clinical trials. The introduction of these vaccines and the recommendation in 1996 of routine childhood vaccination has resulted in a steady decline in the rate of hepatitis A in the United States. In 2003 there were 7653 cases reported (2.7 per 100,000 population), and childhood vaccination appears to have been important in the control of this disease. Of interest is the finding that hepatitis A rates were much higher in the western United States before the introduction of the vaccines, but the rates are now similar across the United States, and an increasing proportion of cases occur in adults.

Genus *Kobuvirus*

In Aichi, Japan in 1989, a stool specimen from a patient suffering from gastroenteritis associated with consumption of raw oysters was found to contain a new picornavirus, subsequently called Aichi virus. Since then the virus has been found not only in Japan but also in Pakistani children with gastroenteritis and in Japanese travelers returning from Southeast Asia with gastroenteritis, and the virus thus appears to be widespread and to be one of the many viruses that are causative agents of human gastroenteritis. The virus is sufficiently distinct in its sequence from other picornaviruses to be classified in a separate genus, *Kobuvirus* (Fig. 3.1). The name is derived from the Japanese word for bump, because the virion appears bumpy in the electron microscope. There are

interesting differences in the replication of this virus from those of other picornaviruses, such as the presence of VP0 in virus particles rather than the cleaved products VP2 and VP4 (see Fig. 3.2). The worldwide burden of human gastroenteritis caused by Aichi virus or closely related viruses is unknown at present. As described later, many viruses belonging to a number of virus families cause epidemic gastroenteritis in humans, and sorting out the causative agents is difficult and requires time.

A virus related to Aichi virus appears to be widespread in cattle in the Aichi area of Japan. This virus, classified for now as a member of the *Kobuvirus* genus, apparently causes inapparent infection in cattle and has been called bovine kobuvirus. Thus, it is possible that kobuviruses are widely distributed in the world and infect a number of species.

Genus *Aphthovirus*

Foot-and-mouth disease viruses (FMDV) belong to seven currently recognized serotypes. They cause a debilitating disease, foot-and-mouth disease, in cattle and other animals, and are economically important pathogens. FMDV was eliminated from the United States many years ago by the simple expedient of killing all infected animals until such time as the virus was extirpated. The last epidemic in the United States occurred in 1929. The virus still circulates in Europe, South America, and other parts of the world, and the U.S. Department of Agriculture maintains strict quarantines in order to prevent the virus from reappearing in this country. In the United States, work with the virus is allowed only on Plum Island in Long Island Sound, in order to prevent its accidental release. Because of its agricultural importance, the molecular biology of the virus has been intensively studied.

In 2001, a large epidemic of FMDV occurred in Western Europe. The epidemic began in February in British sheep and spread to cattle and pigs in Britain and on the continent. By June more than 2000 infected animals had been detected. The epidemic was controlled by restricting the movement of sheep, cattle, and swine, and culling of herds in which FMDV was found. Almost 4 million animals were destroyed in this process. The damage to the British cattle industry was particularly distressful because this epidemic occurred only a few years after widespread culling of cattle to control an epidemic of a prion disease called "mad cow disease" (Chapter 9). Beginning in March, many rural areas often visited by tourists were closed to prevent the spread of the virus, and the U.S. Department of Agriculture was especially vigilant in examining travelers returning from affected countries. British authorities considered vaccinating cattle with commercial vaccines to control the epidemic, but vaccines have not been used by the British to date. Vaccination for FMDV is used in some parts of the world, but is controversial because it is then difficult to distinguish between vaccinated animals and infected animals. Thus, for example, the United States does not allow the importation of beef from areas where vaccination is practiced because it is not possible to rule out the presence of FMDV infection. Although FMDV was thought to have been eradicated by June 2001, and restrictions were about to be lifted, 16 new cases of FMDV were then found in Britain that delayed lifting of the restrictions in that country. The last new case was found on September 30, 2001, and Britain was declared free of FMDV on January 15, 2002.

FAMILY *CALICIVIRIDAE*

The caliciviruses are nonenveloped viruses possessing icosahedral symmetry and having a diameter of about 30 nm. The name comes from the Latin word for cup or goblet (source of the English word chalice) because there are cuplike depressions in the surface of the virion when viewed in the electron microscope. The characteristics of a number of caliciviruses are shown in Table 3.7. Four genera of caliciviruses are currently recognized, two of which, genus *Norovirus* (whose members were formerly called the Norwalk-like viruses) and genus *Sapovirus* (formerly called the Sapporo-like viruses), contain agents that cause human gastroenteritis. The genera *Vesivirus* and *Lagovirus* contain viruses of cats, rabbits, pigs, and sea lions. Additional caliciviruses of cattle, dogs, mink, walrus, and chickens remain to be classified as to genus. Most or all caliciviruses are host specific, infecting only a single animal species.

TABLE 3.7 *Caliciviridae*

Genus/members	Virus name abbreviation	Usual host(s)	Transmission	Disease	World distribution
Vesivirus					
Vesicular exanthema of swine (includes San Miguel sea lion virus)	VESV	Swine	Oral, contact	Fever, lesions on snout and feet	California
Feline calicivirus	FCV	Cats	Contact	Rhinitis, pneumonia, fever	Worldwide
Rabbit vesivirus		Rabbits			
Lagovirus					
Rabbit hemorrhagic disease	RHDV	Rabbits	Water-borne, oral–fecal	Hemorrhages	China, Europe, Australia[a]
European brown hare syndrome	EBHSV	Hares			Europe
Norovirus					
Norwalk	NV	Humans	Water-borne, oral–fecal	Epidemic gastroenteritis	Worldwide
Sapovirus					
Sapporo	SV	Humans	Water-borne, oral–fecal	Epidemic gastroenteritis	Worldwide

Unassigned members of the family include caliciviruses of cattle, dogs, fowl, mink, and walrus

[a] RHDV was inadvertently introduced into Australia.

The human caliciviruses have been difficult to study because it has not been possible to grow them in cell culture or to infect experimental animals. As a consequence, we know less about them than we do about such well-studied viruses as the picornaviruses. The first calicivirus described was a virus of sea lions (San Miguel sea lion virus, genus *Vesivirus*), and most of what we know of the molecular biology of caliciviruses comes from studies of this virus and of other nonhuman viruses such as feline calicivirus, which will replicate in cultured cells. The complete sequences of a number of the human viruses have now been obtained, which has greatly expanded our knowledge of these viruses, but details of the molecular biology of their replication are still lacking.

The caliciviruses are distant relatives of the picornaviruses. Where known, the (+)RNA genome of about 8 kb has a 5′-terminal VPg and a 3′-terminal poly(A), as do the picornaviruses, and calicivirus proteins share sequence identity

with the picornavirus 2C^{ATPase}, 3D^{pol}, and 3C^{pro}. Unlike the picornaviruses, however, the nonstructural proteins are 5′ terminal and the structural protein(s) 3′ terminal (Fig. 3.11). Translation of the genomic RNA produces a polyprotein that contains the sequences of the nonstructural proteins. Cleavage of the polyproteins is presumably effected by 3C^{pro} (also called 3C^{pro}-like or 3CL^{pro}). In those caliciviruses that have been studied in cell culture, a subgenomic RNA is produced that is assumed to be an mRNA for the major capsid protein and, at least in one case, a minor capsid protein as well, and it is assumed that all caliciviruses produce a subgenomic RNA in order to produce the structural protein(s). There is evidence that this subgenomic RNA, once produced, can replicate independently of the viral genomic RNA. The organization of the calicivirus genome differs slightly among the genera as shown in Fig. 3.11. In the lagoviruses and sapoviruses, the major capsid protein is in frame with the nonstructural proteins and contiguous with them. Thus,

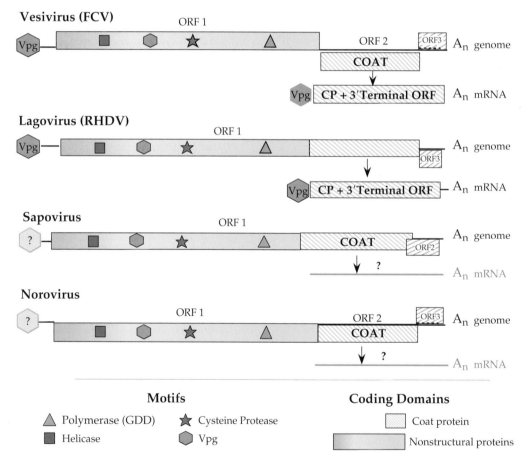

FIGURE 3.11   Diagrammatic representation of the genome organization of the four genera of *Caliciviridae*. Notice that the coat protein is encoded in ORF2 of vesiviruses and noroviruses but as the C-terminal portion of ORF1 in lagoviruses and sapoviruses. Subgenomic mRNAs have been identified for vesiviruses and lagoviruses. FCV, feline calicivirus; RHDV, rabbit hemorrhagic disease virus. Adapted from Fauquet *et al.* (2005), p. 847.

it is possible that the nonstructural polyprotein contains the sequences of the structural proteins as well, but is assumed that the major source of the structural proteins is translation of a subgenomic mRNA, although details of the viral expression strategy are still lacking. In the noroviruses and vesiviruses, the major capsid protein is in a different reading frame.

Another, fairly short, ORF is found in the 3′-terminal region of each calicivirus genome. The translation product of this ORF is a minor structural protein.

Caliciviruses have been shown to interfere with host protein synthesis. The mechanism used is the same as one of the mechanisms used by poliovirus, cleavage of poly(A)-binding protein by 3C^{pro}. The norovirus protease cleaves this host protein at the same site as does the poliovirus 3C^{pro}, whereas the feline calicivirus protease cleaves it at a site 24 residues downstream of the poliovirus site.

Noroviruses

Norwalk virus is a human virus that causes gastroenteritis. As described, there is no cultured cell line in which to propagate the viruses, nor even an animal model in which to grow it, and studies have relied on human volunteers for virus propagation. This has severely limited the amount of information (and material) that can be obtained. However, because of the power of modern gene cloning technology, the entire genome of Norwalk has been cloned and sequenced, starting from stools of experimentally infected human volunteers. These genome sequences can be aligned with the genomes of other caliciviruses (Fig. 3.11), and the mechanisms of the expression of the genome and the functions of the encoded proteins can be predicted from studies of the animal caliciviruses. Thus, for example, it is assumed that the Norwalk viruses possess a 5′-terminal VPg as does San Miguel sea lion virus, but to date it has not been possible to prove this. However, the genomic sequence of Norwalk virus contains a sequence related to the VPg of other caliciviruses, providing evidence that Norwalk virus RNA does in fact carry a 5′-terminal VPg. Clones of various regions of the viral RNA can be expressed in cultured cells to study the expression and function of various domains. For example, the capsid protein of Norwalk virus has been expressed in insect cells and these proteins spontaneously assemble into T=3 virus-like particles of 38-nm diameter whose structure has been determined to atomic resolution by X-ray diffraction techniques. Of interest is the recent discovery of a mouse norovirus. The virus causes only silent infections in immunocompetent mice but, importantly, it will replicate in cultured cells, and thus has the potential to serve as a surrogate virus to work out details of the molecular biology of replication of noroviruses. Another recent development is the finding that human noroviruses will replicate in a cell

system in which the viral RNA is introduced by transcription from cDNA clones.

Several isolates of the Norwalk virus have been studied, all of which share more than 50% sequence identity and which are named after the location where they were first isolated. These include Norwalk virus, Hawaii virus, Snow Mountain virus, Southampton virus, Lordsdale virus, and Desert Shield virus. These viruses are extraordinarily infectious. In one epidemic investigated by the Centers for Disease Control and Prevention and local health authorities, a baker preparing food for a wedding was ill with gastroenteritis. After using the toilet, he washed his hands thoroughly before handling food, but his hands were still contaminated with virus, perhaps under the fingernails. He used his hands to stir a very large pot of icing used to glaze cakes and doughnuts that were distributed at the wedding reception, and managed to contaminate the icing with virus. Every guest at the reception who ate as much as a single doughnut contracted gastroenteritis.

The Norwalk viruses regularly cause epidemics of gastroenteritis. The incubation period is short (24 hours on average) and the course of disease is also short (1–2 days). Following recovery, immunity is established to the virus, but the duration of immunity appears to be fairly short (perhaps one or a few years). Studies of immunity are made difficult by the finding that some fraction of the human population seems to be resistant to any particular Norwalk virus studied, perhaps because of a lack of receptors for the virus, and by the fact that there are so many viruses that cause gastroenteritis.

The Norwalk group of viruses is worldwide in its distribution but epidemiological studies of these viruses have been difficult in the absence of a cell culture system or an experimental animal that can be infected by the virus. Food-borne infections are estimated to cause 76 million cases of human illness in the United States each year, of which 300,000 lead to hospitalization and 5000 are fatal. A significant fraction of these illnesses, perhaps one-third, are caused by bacteria, for which ready tests are available for use by health authorities and these epidemics have received wide attention. Until recently, however, little testing for Norwalk virus was performed. With the determination of the sequence of Norwalk RNA followed by the development of RT–PCR techniques to rapidly screen for Norwalk virus, the CDC has now attempted to assess the importance of Norwalk virus in food-borne outbreaks of gastroenteritis. Analysis of available data suggests that perhaps 50% of food-borne outbreaks in the United States are attributable to Norwalk virus, and these viruses are therefore responsible for the majority of outbreaks of nonbacterial gastroenteritis in the United States. Further, the Norwalk outbreaks are larger than those associated with bacteria, a median of 25 cases per outbreak versus 15 cases in bacterial outbreaks and 7 cases in outbreaks of unknown etiology. More than half of the Norwalk outbreaks were associated with eating salads, sandwiches, or fresh produce, and in 40% of Norwalk outbreaks restaurants or caterers were associated. Some outbreaks have been associated with the consumption of raw oysters for reasons described earlier. Given the total number of cases of gastroenteritis in humans, it is clear that Norwalk virus is a very important disease pathogen.

Norwalk virus has also established itself as a bane of the cruise ship industry. In 2004, for example, there were 38 outbreaks of gastrointestinal illness on cruise ships, defined as at least 3% of the passengers and crew aboard the ship developing diarrhea or vomiting not due to seasickness and accompanied by other symptoms such as fever or aching muscles. Most of such outbreaks are due to noroviruses. In one series of episodes studied by the CDC, a cruise ship suffered outbreaks during six consecutive cruises, despite extensive efforts to cleanse the ship. After the second outbreak, the ship was taken out of service for a week and scrubbed exhaustively, but the outbreaks continued. The CDC investigation showed that one particular genotype of virus was present in all of the outbreaks, but that new genotypes were also introduced into the ship from time to time.

Sapoviruses

Many isolates of Sapporo virus have been made in association with outbreaks of gastroenteritis. Sapporo virus also appears to be a widespread virus that is an important cause of human gastroenteritis, but much less is known about the epidemiology of Sapporo virus than about Norwalk virus.

Rabbit Hemorrhagic Disease Virus

Rabbit hemorrhagic disease virus (RHDV), a member of the genus *Lagovirus*, causes an often fatal illness in European rabbits. It has been used in Australia and New Zealand in an attempt to control large and destructive populations of introduced rabbits, a topic that will be covered in more detail in Chapter 7 when we discuss rabbit myxoma virus. RHDV was being studied on an island off the south coast of Australia as a possible rabbit control agent when it was "inadvertently" introduced onto the mainland. Once introduced it spread rapidly, probably aided by local farmers. The virus was later introduced, at first illegally and then legally, into New Zealand. The virus has been only moderately successful in controlling the rabbit populations. When first introduced, it was highly pathogenic, causing high fatality rates. After some time, however, an increasing proportion of rabbits survived infection. In New Zealand, the virus appears to have developed the ability to establish persistent or latent infections, perhaps allowing the rabbit host to survive challenge with virulent virus. There may also have been a background level of immunity to the virus caused by the circulation of a similar virus in the rabbit population of New Zealand for many years before introduction of the virulent virus.

FAMILY *HEPEVIRIDAE*

Hepatitis E virus is similar in size and genome organization (Fig. 3.12) to the caliciviruses and until recently was considered to be a member of the *Caliciviridae*. It differs in important details, however, and classification as a distinct family was deemed appropriate. The viral RNA is capped rather than possessing a VPg, and the virus-encoded protease, whose sequence has been deduced from the sequence of the viral genome, appears to be a papain-like protease rather than a $3C^{pro}$-like enzyme, another difference from the caliciviruses. Like the caliciviruses, the capsid protein is 3′ of the nonstructural proteins and is translated from a subgenomic mRNA.

HEV is one of several viruses that cause human hepatitis (Table 3.6), in this case epidemic water-borne hepatitis. Infection is by the oral–fecal route and contaminated water is often the source of the infection. Unlike picornaviruses

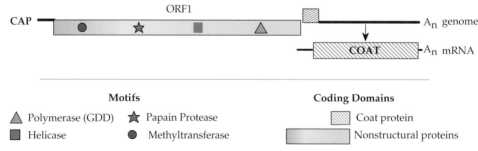

FIGURE 3.12   Genome organization of hepatitis E virus, adapted from Fauquet *et al.* (2005) p. 854.

and caliciviruses, the virus has a wide host range, infecting a variety of mammals. The virus is widespread in pigs, both domestic and wild, and pigs appear to be the major reservoir host for the virus. Thus, transmission of the virus is zoonotic, from animals to humans. In one particularly interesting study in Japan, four humans were infected by the virus upon eating raw deer meat from a wild deer. The nucleotide sequence of the virus isolated from the humans differed from the sequence of virus isolated from the deer by fewer than 8 nucleotides, clearly showing that the deer served as the source of the infection. From a wild boar captured in the same area a year later, virus was isolated that differed by up to 23 nucleotides from the deer–human viruses. It has been found that in Japan only about 1% of deer are positive for HEV antibodies, whereas 3 of 7 wild boars were found to be positive and almost 100% of Japanese pigs are positive by 6 months of age. Thus, pigs are the major reservoir and spread the virus to other mammals as well as to humans.

HEV is found in Asia, Africa, Southern Europe, and Mexico, where it causes thousands of cases of hepatitis each year (Fig. 3.13). A dendrogram of various geographical isolates of HEV illustrates that the New World strain has diverged significantly from the Asian isolates (Fig. 3.14). Thus, there is little or no circulation of virus between different geographic regions. The disease is severe but the fatality rate is low (<1%), with the prominent exception that the fatality rate in pregnant women can be 20%. There is no vaccine or treatment for the virus at present.

An avian HEV is also known. This virus shares about 50% nucleotide sequence identity with the mammalian virus. It is an important pathogen of commercial broiler chickens, but does not appear to infect mammals.

FAMILY *ASTROVIRIDAE*

Astroviruses constitute a recently described family of animal viruses. Some of these are human viruses that cause gastroenteritis, but astroviruses for cattle, pigs, sheep, and ducks are also known (Table 3.8). The name comes from the Greek word for star, from starlike structures on the surface of the virion. Unlike the human caliciviruses, the human astroviruses will grow in cultured cells, and progress on understanding their molecular biology has been more rapid. They are small viruses, 30 nm, with icosahedral symmetry, and a genome of 7 kb.

The replicase proteins of astroviruses are translated from the genomic RNA as two polyproteins (Fig. 3.15). The smaller translation product (1a) terminates at a stop codon; ribosomal frameshifting (Chapter 1) at a retrovirus-like slippery

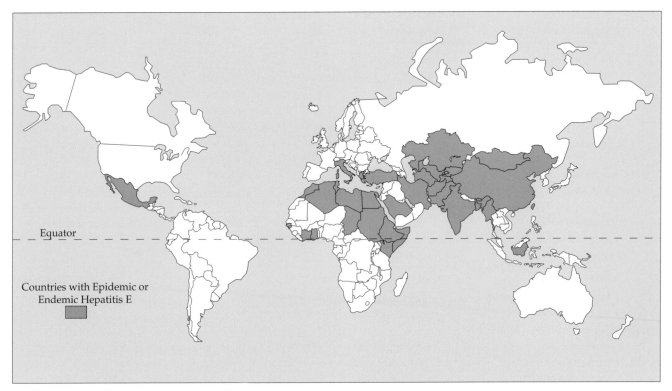

FIGURE 3.13   Worldwide incidence of epidemic and endemic hepatitis E. In addition, serosurveys indicate that 1–2% of blood donors in United States and Western Europe have detectable IgG antibodies to hepatitis E. Adapted from Fields *et al.* (1996) p. 2838.

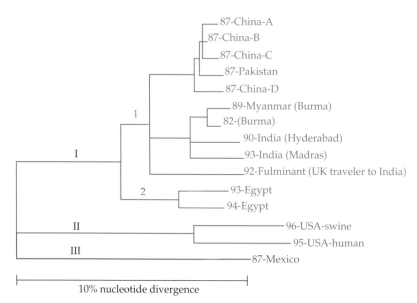

FIGURE 3.14   Phylogenetic tree of HEV isolates based on the complete sequences of ORF2 (1983 nt), encoding the coat protein. Isolates are listed by year and location of isolation. Branch lengths are proportional to the evolutionary distance between sequences. Roman numerals are used to denote genotypes (<85% nucleotide sequence identity), and arabic numbers denote subgenotypes (<92.5% nucleotide sequence identity). The genetic groupings of the HEV strains reflect the geographic relationships of the places from which they were isolated, showing that the viruses are geographically isolated and do not circulate over wide areas. Adapted from Figure 1 in Tsarev *et al.* (1999).

TABLE 3.8   *Astroviridae*

Genus/members[a]	Virus name abbreviation	Usual host(s)	Transmission	Disease
Mamastrovirus				
Human astrovirus (5)	HAstV	Humans	Water-borne, oral–fecal	Gastroenteritis
Bovine astrovirus (2)	BAstV	Cattle	Water-borne, oral–fecal	Gastroenteritis
Feline astrovirus	FAstV	Cats	Water-borne, oral–fecal	Gastroenteritis
Ovine astrovirus 1	OAstV	Sheep	Water-borne, oral–fecal	Gastroenteritis
Porcine astrovirus 1	PAstV	Swine	Water-borne, oral–fecal	Gastroenteritis
Avastrovirus				
Duck astrovirus	DAstV	Ducks	Water-borne, oral–fecal	Fatal hepatitis in ducklings
Turkey astrovirus	TAstV	Turkeys	Water-borne, oral–fecal	Gastroenteritis

[a] All astroviruses identified are worldwide in distribution.

sequence allows readthrough into a second reading frame (1b) to produce a longer polyprotein (1ab). A subgenomic mRNA is translated into the capsid protein of the virus.

Astroviruses encode a 3C-like serine protease (with serine in the active site) and an RNA polymerase related to other viral RNA polymerases, but there is no evidence for a helicase or a capping enzyme. It has been suggested that astroviruses contain a VPg, consistent with the lack of a capping enzyme, but no domain encoding a VPg has been identified in the genome.

Epidemiological studies suggest that astroviruses are an important cause of acute human gastroenteritis worldwide. A summary of virus families known to contain viruses that cause gastroenteritis and acute diarrhea in humans as well as other vertebrates is shown in Table 3.9. In addition to the picornaviruses, caliciviruses, and astroviruses described earlier, some adenoviruses and reoviruses also cause human gastroenteritis, and members of additional families may also cause human gastroenteritis, as shown in the table. In addition to the viruses listed, there may yet

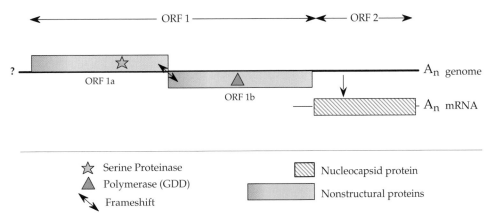

FIGURE 3.15   Genome organization of a human astrovirus. ORF 1a is in a different reading frame than ORF 1b. Ribosomal frameshifting (arrows) during translation is required to produce the ORF 1b protein. Adapted from Fauquet *et al.* (2005) p. 860.

TABLE 3.9   Viruses Causing Acute Diarrhea

Nucleic acid	Family	Virus[a]	Host range
Single-stranded, plus-sense RNA	*Caliciviridae*	Sapporo, Norwalk feline calicivirus	Humans, cattle, swine, chickens, dogs, cats
	Astroviridae	Numerous astroviruses	Humans, cattle, swine, cats, dogs, avian species
	Coronaviridae	PEDV, TGEV and others	Swine, cattle, foals, mice, rabbits, dogs, cats turkeys, (humans?)[b]
		Numerous toroviruses	Cattle, horses (goats, sheep, swine, rabbits, mice, humans?)
	Flaviviridae	Pestivirus BVDV	Cattle
	Picornaviridae	Aichi, human parechovirus 1	Humans
Single-stranded, minus-sense RNA	*Paramyxoviridae*	Canine distemper	Dogs
		Newcastle disease	Chickens, fowl
Double-stranded RNA	*Reoviridae*	Rotavirus A	Mammalian and avian species, humans
		Rotavirus B	Swine, cattle, sheep, rodents, humans
		Rotavirus C	Swine, ferrets, humans
		Rotaviruses D, F, G	Avian species
		Rotavirus E	Swine
Single-stranded DNA	*Parvoviridae*	Numerous parvoviruses	Cattle, cats, dogs, mink, (humans?)
Double-stranded DNA	*Adenoviridae*	Human Ad40,41	Humans

[a] Abbreviations: PEDV, porcine epidemic diarrhea virus; TGEV, transmissible gastroentieritis virus; BVDV, bovine viral diarrhea virus.
[b] The role of the listed viruses in causing diarrhea has not been proven for species listed in parentheses.
Source: Adapted from Granoff and Webster (1999), p. 442.

be other small viruses that cause gastroenteritis in humans. Virus particles have been seen in stools of humans suffering from gastroenteritis that have not as yet been characterized and are referred to simply as SRVs (small, round viruses). Some of these may be members of families not yet characterized.

FAMILY *TOGAVIRIDAE*

The family *Togaviridae* contains two genera, genus *Alphavirus* and genus *Rubivirus*. The family name comes from the Latin word for cloak, and the name was given to them because they are enveloped. The 30 alphaviruses have

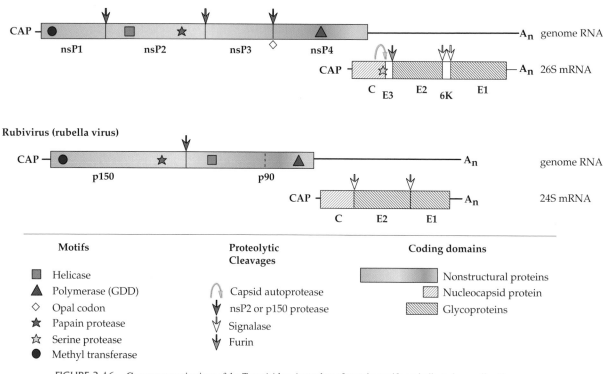

FIGURE 3.16   Genome organizations of the *Togaviridae*. A number of protein motifs are indicated, as well as the enzymes responsible for the proteolytic cleavages. The opal codon shown between nsP3 and nsP4 is leaky and is not present in all alphavirus isolates; readthrough produces small amounts of nsP4. Adapted from Strauss and Strauss (1994) Figure 34.

a (+)RNA genome of about 12 kb, whereas rubella virus, the only member of the *Rubivirus* genus, has a genome of 10 kb. The genomes of alphaviruses and of rubella virus are organized in a similar fashion, as illustrated in Fig. 3.16. The virions of the two groups are also roughly similar in size (70 nm for alphaviruses, 50 nm for rubiviruses) and structure (icosahedral nucleocapsids surrounded by a lipoprotein envelope). Structures of alphaviruses are illustrated in Figs. 2.5, 2.14, and 2.18. However, although the two genera exhibit similarities, they are only distantly related. As an historical footnote, the flaviviruses, described after the togaviruses, were once classified as a genus within the *Togaviridae*. When sequences of alphaviruses and flaviviruses were determined, however, they were found to be unrelated and the flaviviruses were placed into a new family.

Genus *Alphavirus*

The alphaviruses have a worldwide distribution. They get their name from the Greek letter alpha because they were once known as the Group A arboviruses. Many cause important illnesses in humans, and information for a representative selection of these viruses is presented in Table 3.10. Because of their importance as disease agents and aided by the fact that alphaviruses grow well in cultured cells and in animal models, this group of viruses has been well studied. The genomes of many of them have been sequenced in their entirety. All members of the genus are closely related and share extensive amino acid sequence identity, with the most distantly related alphaviruses sharing about 40% amino acid sequence identity on average and viruses belonging to the same lineage sharing higher sequence identity. A dendrogram that illustrates their relationships is shown in Fig. 3.17. A number of lineages or clades are present, including a clade of aquatic viruses, a clade of encephalitic viruses (EEE, VEE), the Sindbis clade (which includes Aura virus and many strains of Sindbis virus), the SFV clade (which includes many viruses which cause arthritis including polyarthritis), and a clade of recombinant viruses (the WEE lineage). The dendrogram illustrates the interesting fact that during evolution of alphaviruses, there was a singular recombination event between Eastern equine encephalitis virus and a Sindbis-like virus to produce Western equine encephalitis virus, which subsequently evolved into a number of different viruses (the WEE lineage). Recombination events

TABLE 3.10   *Togaviridae*

Genus/members	Virus name abbreviation	Usual host(s)	Transmission	Disease	World distribution
Alphavirus					
Sindbis	SINV	Mammals[a], Birds	Mosquito-borne	Arthralgia, rash, fever	Old World
Semliki Forest	SFV	Mammals[a]	Mosquito-borne	Arthralgia, fever	Africa
Ross River, Barmah Forest	RRV, BFV	Mammals[a]	Mosquito-borne	Polyarthritis, fever, rash	Australasia
Ft. Morgan	FMV	Birds	Vectored by swallow bug	?	North America
Chikungunya, O'Nyong-nyong	CHIKV, ONNV	Humans	Mosquito-borne	Arthralgia, fever	Africa
Mayaro	MAYV	Mammals[a]	Mosquito-borne	Arthralgia, fever	South America
Eastern, Western, Venezuelan equine encephalitis	EEEV, WEEV, VEEV	Mammals[a], Birds	Mosquito-borne	Encephalitis	Americas
Salmon pancreas disease virus	SPDV	Fish	No arthropod vector		
Rubivirus					
Rubella	RUBV	Humans	No arthropod vector	Rash, congenital abnormalities	Americas, Europe

[a] Humans can be infected by these viruses, but humans are not the primary mammalian reservoir.

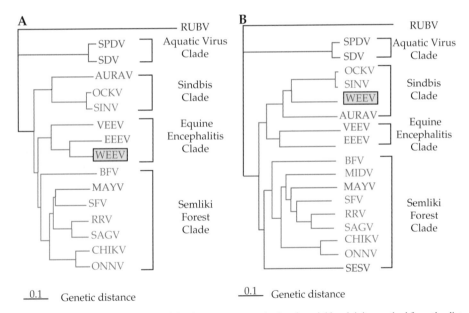

FIGURE 3.17   Phylogenetic trees of the alphaviruses, constructed using the neighbor-joining method from the distances computed using Clustal X v 1.81 software, with the sequences from rubella virus as outgroup. The Old World viruses are shown in blue, the New World viruses in red. The vertical distances are arbitrary, but the lengths of the horizontal branches indicate the number of amino acid substitutions along the branch. (A) A tree derived from the amino acid sequences of nonstructural proteins nsP1 through nsP4, comprising roughly 2475 amino acids (B) A tree derived from the sequences of the virion structural proteins or roughly 1245 amino acids. The trees are very similar in architecture with the exception of the position of WEE (boxed), whose nonstructural proteins are most closely related to EEE, but whose structural proteins resemble SIN, indicating that a recombinational event has taken place to produce WEE. Note that Fort Morgan, Buggy Creek, and Highlands J (not shown) are three recombinant New World viruses closely related to WEE. Most virus abbreviations are found in Table 3.10; AURAV, Aura; MIDV, Middelburg; SDV, sleeping disease; OCKV, Ockelbo; SAGV, Sagiyama; SESV, Southern elephant seal virus. These trees were adapted from Luers *et al.* (2005) with permission.

in which the recombinant virus persists and prospers as a distinct virus appear to be uncommon to rare, but there is much evidence that recombination has played an important role in the evolution of viruses. The dendrogram also illustrates the fact that there have only been a limited number of transfers of alphaviruses between the Americas and the Old World. In fact, three transfers between the Americas and the Old World are sufficient to explain the dendrogram, and the majority of members of the three major lineages are restricted to either the Americas or to the Old World. This contrasts with many virus families, where individual viruses are often worldwide in distribution and evidence is abundant for the mixing of lineages between the two hemispheres.

Expression of the Genome

Alphaviruses enter a cell via endosomal uptake and fuse with the endosomal membrane upon exposure to a suitably low pH that differs somewhat from virus to virus. The capsid protein has an affinity for ribosomes and there is evidence that ribosomes help disassemble the nucleocapsid upon its entry into the cytoplasm by binding the capsid protein. Release of the genomic RNA, which is capped and polyadenylated, is followed by its translation into a nonstructural polyprotein that is cleaved into four polypeptides by a viral protease (Fig. 3.16). Activities present in these proteins include a capping activity in the N-terminal protein (nsP1); helicase, papain-like protease, and RNA triphosphatase activities in nsP2; and RNA polymerase in nsP4. A viral encoded capping activity is required to cap the viral mRNAs (the genomic RNA and a subgenomic RNA) because replication occurs in the cytoplasm and the virus does not have access to cellular capping enzymes. The RNA triphosphatase in nsP2 is required to remove the terminal phosphate in the 5′ triphosphate on the RNA in order to prepare the RNA for capping by the viral enzyme, the RNA helicase is needed to unwind the RNAs during replication, and the protease is required to process the precursor polyprotein. The RNA polymerase is needed to synthesize viral RNA. All four nonstructural proteins are required to synthesize the viral RNA. Replication of the RNA and synthesis of a subgenomic RNA follow the pattern illustrated schematically in Figure 1.9B.

The function of the phosphoprotein nsP3 in RNA replication is unknown. It is phosphorylated on several threonines and serines in a nonconserved domain in the C-terminal region of the protein. The N-terminal domain of nsP3 is a conserved domain (often referred to as the X domain) that is also present in a number of other viruses (rubella virus, hepatitis E virus, coronaviruses) as well as being widely distributed in bacteria, archae, and eukaryotes. The virus domain shares up to 35–40% amino acid sequence identity with the cellular homologues, whose function is unknown.

Replication of the viral RNA takes place in association with cellular membranes. Small spherical invaginations form in the membranes, induced by viral proteins, in which replication occurs. Protein nsP1 has been shown to interact with membranes by means of a specific domain within the protein, and it is assumed that this association is important for the membrane association of the replication machinery.

Studies of the viral nonstructural protease have shown that the cleavages that process the polyprotein control viral RNA replication. During or shortly after translation, the full-length polyprotein precursor (called P1234) cleaves itself in cis to produce P123 and nsP4. These form an RNA synthetase, probably together with cellular proteins, that can make complementary (−)RNA from the genomic RNA template, but which cannot make (+)RNA efficiently. Subsequent cleavage of P123 in trans, between nsP1 and nsP2, gives rise to a synthetase that can make both (+)RNA and (−)RNA. A second cleavage between nsP2 and nsP3 gives rise to a synthetase that can make only (+)RNA. Thus, (−)RNA templates are made early, but as infection proceeds and the concentration of protease builds up, trans cleavage occurs and (−)RNA synthesis is shut down (Fig. 3.18). After this time, plus-strand RNA synthesis continues from the preformed minus-strand templates but no further increase in the rate of RNA synthesis takes place. This control mechanism may have evolved not only to make the infection process more efficient, since genomic RNA for progeny virions and subgenomic RNAs for translation of structural proteins required

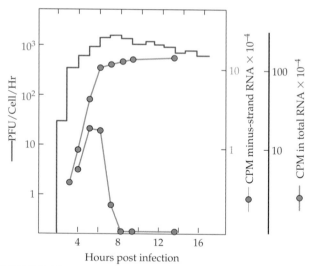

FIGURE 3.18 Growth curve of Sindbis virus infection in chicken cells at 30°C. At the times shown, the medium was harvested and replaced with fresh medium, and the titer was determined. The red line shows release of infectious virus into the culture fluid. For determining the rate of RNA synthesis, cells infected as for virus assay were pulsed with radioactive uridine for 1 hour at the times shown. Monolayers were harvested and incorporation into total RNA (green line) and minus-strand RNA (blue line) were determined. Adapted from Strauss and Strauss (1994) Figure 5.

for assembly of progeny virions are needed in much larger quantities than minus-strand templates, but also to control the virulence of the virus. In the case of alphaviruses, it is particularly important to control the virulence of the virus in the mosquito vector. As discussed in more detail later, it is important that the infection process in the mosquito be tightly controlled, and shutting down minus-strand RNA synthesis not only prevents further exponential increase in virus replication but also allows downregulation of virus replication as minus-strand templates decay. It also has the effect that the infected cell becomes resistant to superinfection by the same or a related virus because no (−)RNA templates can be made, which could be especially important in the mosquito vector. The resistance to superinfection by the same or related viruses is called superinfection exclusion or homologous interference.

The rate of cleavage of the early synthetase that makes (−)RNA to convert it into one that can make (+)RNA is thought to be controlled in part by a leaky stop codon between nsP3 and nsP4 that is present in most, but not all, alphaviruses (Fig. 3.16). Termination at this codon produces P123, which can act *in trans* as a protease but cannot act as a synthetase because it lacks the nsP4 RNA polymerase. In addition to a more rapid buildup of protease that accelerates the rate of conversion to (+)RNA synthesis, this additional

P123 and its cleavage products may serve to accelerate the rate of RNA synthesis. There is evidence from genetic studies that domains in nsP1 and nsP2, among others, are required for the recognition of viral promoters and the initiation of RNA synthesis, and additional helicase activity could also speed up the rate of RNA synthesis.

During infection by alphaviruses, a subgenomic mRNA is produced that serves as the message for the production of the structural proteins of the virus, which consist of a capsid protein and two glycoproteins. The 4.1-kb subgenomic RNA is synthesized by the viral replicase from the (−)RNA template using an internal promoter of 24 nucleotides. The activity of this core 24-nucleotide promoter is increased by enhancer sequences present in the 100 nucleotides upstream of the core promoter. The structural proteins are translated as a polyprotein and cleaved by a combination of viral and cellular enzymes. The capsid protein is itself a serine autoprotease that cleaves itself from the N terminus of the nascent polyprotein. It has a fold that is similar to that of chymotrypsin (Fig. 2.14B), suggesting that it was derived from a cellular serine protease during evolution of the virus. After release of the capsid protein, N-terminal signal sequences and internal signal sequences in the glycoprotein polyprotein precursor lead to its insertion into the endoplasmic reticulum (Figure 3.19). This precursor is cleaved

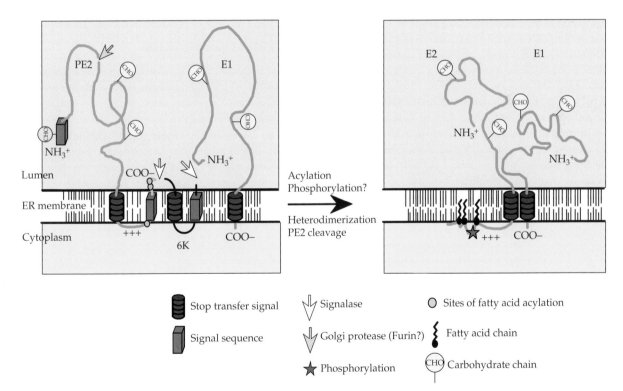

FIGURE 3.19 Schematic diagram of the configurations of glycoproteins PE2 and E1 in membranes. The left panel shows the configuration of PE2 immediately after signalase cleavage from 6K, with the C terminus in the lumen of the ER. The right panel shows the configuration of the E2-E1 heterodimer after phosphorylation–dephosphorylation with the C terminus in the cytoplasm. Adapted from Strauss and Strauss (1994), Figures 7 and 9.

by signalase, a cellular enzyme that resides in the lumen of the endoplasmic reticulum, to produce glycoprotein PE2 (a precursor to glycoprotein E2), 6K (a small hydrophobic peptide located between E2 and E1), and glycoprotein E1. PE2 and E1 quickly form a heterodimer. At some time, the C terminus of PE2/E2 is withdrawn from the lipid bilayer into the cytoplasm, where it can interact with the capsid during virus assembly. During transport of the heterodimer to the cell plasma membrane, PE2 is cleaved by another cellular enzyme, furin or a furin-like enzyme, to form E2 and E3. E3 is a small glycoprotein that in most alphaviruses is lost into the culture fluid.

Viral Promoters

The replication of alphaviral RNA requires the recognition of specific promoters in the viral RNA by the viral RNA synthetase. These promoters act *in cis*, that is, they must be present in the RNA to be used as a template, and both viral and cellular proteins may be involved in the recognition of these promoters. Four alphavirus promoters, or components of promoters, are illustrated in Figure 3.20. The best understood of these promoters is that for the production of the subgenomic mRNA for the structural proteins. The basal promoter consists of 24 nucleotides, of which 19 are upstream of the transcription start site and 5 are copied into the subgenomic RNA. This subgenomic promoter can be placed in front of any RNA sequence and the viral synthetase will use it to synthesize a subgenomic mRNA. This property of the promoter has made alphaviruses useful as expression vectors (Chapter 11).

The promoters for synthesis of full-length genomic RNA and the antigenomic (−)RNA template are less well understood. A linear sequence element at the 3′ end of the (+)RNA and two elements at or near the 5′ end of the genomic RNA that can form stem-loop structures (Fig. 3.20) are required for the efficient synthesis of both the antigenomic RNA from the genomic RNA and for genomic RNA synthesis from the antigenomic template. The genomic RNA (and presumably the antigenomic RNA as well) is known to cyclize in the absence of protein (Fig. 3.21), thus bringing the sequence elements at the two ends of the molecule into close proximity, allowing the viral replicase to interact with both ends of the RNA at once when initiating synthesis of new RNA. It seems likely that this mechanism evolved so that only full-length RNA molecules can be replicated. This eliminates replication not only of the subgenomic RNA but also of any broken RNA molecules.

It is noteworthy that the RNAs of many, perhaps most, RNA viruses, both plus stranded and minus stranded, cyclize at some stage of RNA replication or translation. In some cases, as in the alphaviruses, cyclization occurs by means of complementary sequences near the ends of the RNA and requires no protein to stabilize the interaction. In other cases, the interactions of the complementary sequences near the ends of the RNA are stabilized by the binding of viral or cellular proteins. And in still other cases, cyclization is effected by cellular proteins. In one scenario, a cellular protein such as the poly(A)-binding protein binds to the 3′ poly(A) tract and another cellular protein, such as one that binds to translation initiation signals, binds near the 5′ end of the molecule. These two cellular proteins then interact with each other to hold the two ends of the RNA near one another.

Assembly of Progeny Virions

The structure of alphaviruses has been described in Chapter 2. PE2-E1 heterodimers form in the endoplasmic reticulum and are transported to the cell surface. During transport, they are cleaved to form E2-E1 heterodimers. The PE2-E1 heterodimer is more stable to acidic pH than is the E2-E1 heterodimer and survives the mildly acidic pH of transport vesicles. Cleavage is necessary to prime the E2-E1 heterodimer for disassembly upon entry into a susceptible

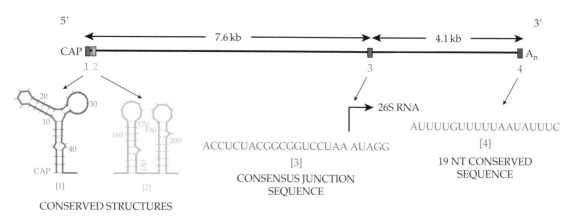

FIGURE 3.20   Promoters in the alphavirus genome. Adapted from Strauss and Strauss (1994), Figure 18.

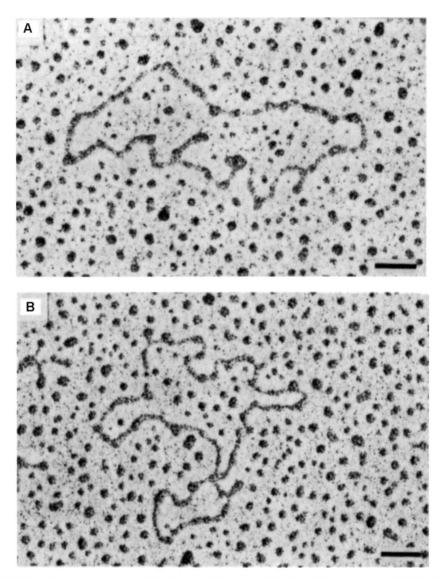

FIGURE 3.21 Electron micrographs of Sindbis virus 49S genomic RNA. The RNA was treated with 0.5M glyoxal in 0.1M phosphate buffer for either 30min (A) or 40min (B) at 35°C before spreading. Scale bar is 100nm. Adapted from Frey *et al.* (1979) Figure 3.

cell, which is required for formation of E1 homotrimers that are responsible for fusion. During transport or virus assembly, three E2-E1 heterodimers trimerize to form the spikes found in the virion. Virions normally mature when a preassembled nucleocapsid consisting of the genomic RNA and 240 copies of capsid protein buds through the cell plasma membrane to acquire a lipoprotein envelope containing 240 copies of the E1-E2 heterodimer (Fig. 2.25C). The free energy for budding is provided by lateral interactions between the viral glycoproteins as they assemble around the nucleocapsid and by the interaction of the C-terminal domain of glycoprotein E2 with a docking site on each nucleocapsid protein molecule. The assembled virion has icosahedral symmetry in which the symmetry axes of the nucleocapsid are coordinated with those of the glycoprotein shell by the E2–capsid interactions. The diameter of the assembled virion is 70nm (Figs. 2.5 and 2.14A).

The 6K protein is required for efficient assembly of virions. Virions that appear to be normal in every way will assemble in the complete absence of 6K but only inefficiently and virus that lacks the 6K gene produces only low yields. It has been shown that 6K expressed alone will form ion channels, but whether this is important for virus assembly is not known.

Alphaviruses Are Arboviruses

All alphaviruses are arboviruses (**ar**thropod-**bo**rne animal viruses), with the probable exception of the newly described fish viruses, and were once referred to as

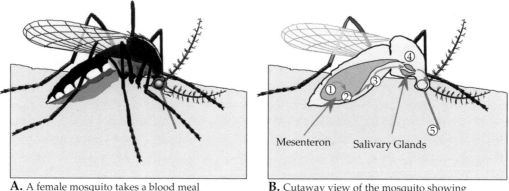

A. A female mosquito takes a blood meal

B. Cutaway view of the mosquito showing steps in the replication and transmission of an arbovirus

FIGURE 3.22   Sequential steps necessary for a mosquito to transmit an arbovirus. (1) A female mosquito ingests an infectious blood meal and virus enters the mesenteron. (2) Virus infects and multiplies in mesenteronal epithelial cells. (3) Virus is released across the basal membrane of the epithelial cells and replicates in other tissues. (4) Virus infects salivary glands. (5) Virus is released from the epithelial cells of the salivary glands and is transmitted in the saliva during feeding. Adapted from Monath (1988).

the Group A arboviruses. In nature, they alternate between replication in arthropod vectors, usually mosquitoes, and higher vertebrates. A mosquito may become infected on taking a blood meal from a viremic vertebrate, which can have 10^8 or more infectious particles per milliliter of blood. The infection in the mosquito, which is almost asymptomatic and lifelong, begins in the midgut and spreads to the salivary glands, as illustrated in Fig. 3.22. After infection of the salivary glands, the mosquito can transmit the virus to a new vertebrate host when it next takes a blood meal, injecting saliva in the process. Infection in the vertebrate begins in the tissues surrounding the bite or in regional lymph nodes, but then spreads to other organs. The infection is usually self-limited and the vertebrate is capable of infecting mosquitoes for only a brief time after viremia is established but before an immune response limits circulating virus. The necessity to alternate between two such different hosts has constrained the evolution of arboviruses—changes that adapt the virus to one host or that are neutral in one host are often deleterious in the alternate host. Thus, the evolutionary pressures on arboviruses are different from those on viruses such as poliovirus, which infects only primates.

Different alphaviruses infect different spectra of mosquitoes and vertebrates in nature. It is useful to distinguish between reservoir hosts in which the virus is maintained in nature and dead-end hosts in which infection normally does not lead to continuity of the infection cycle. We can also distinguish between enzootic cycles, in which the virus is continuously maintained in nature and which may or may not result in disease in the enzootic host, and epizootic cycles, in which the virus breaks out and causes epidemics of disease that may die out with time. Two types of natural transmission cycles are illustrated in Fig. 3.23. In Fig. 3.23A, a simple transmission cycle is illustrated, such as that of urban yellow fever infection of humans (see the section on flaviviruses later). In this cycle, humans are the only vertebrate hosts and the virus alternates between infection of a human and infection of the mosquito vector *Aedes aegypti*. Figure 3.23B illustrates a complex transmission pattern, using as an example the transmission of Eastern equine encephalitis virus in North America. This virus has a vertebrate reservoir consisting primarily of migratory songbirds and is transmitted by the mosquito, *Culiseta melanura*, a common inhabitant of freshwater swamps in eastern North America. However, the virus is capable of infecting other mosquitoes and has even been isolated from naturally infected chicken mites. It also infects mammals, including humans. On occasion, the virus breaks out of its enzootic cycle to cause epidemics of disease in pheasants, transmitted and maintained by an epizootic vector mosquito. Either enzootic or epizootic vectors are capable of infecting humans or domestic animals if they invade the areas in which these mosquitoes are present, but these hosts are usually dead-end hosts and do not further spread the virus.

Most alphaviruses are capable of infecting both mammals and birds, and the nature of the vertebrate reservoir depends on the virus or even the strain of virus, which may differ by geographic location. Thus, for example, Sindbis virus in nature is normally maintained in birds, which are its usual vertebrate reservoir. However, the virus is capable of infecting mammals, including humans, and has also been isolated from amphibians and reptiles. Numerous species of mosquitoes form its insect reservoir, but it has also been isolated from other hematophagous arthropods, including mites. In contrast, Ross River virus is maintained in small marsupial mammals in Australia and does not appear to use birds as hosts.

A. Simple Transmission Cycle

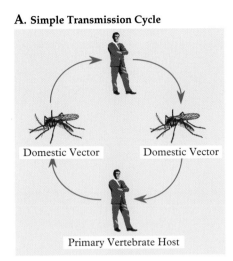

B. Complex Transmission Cycle

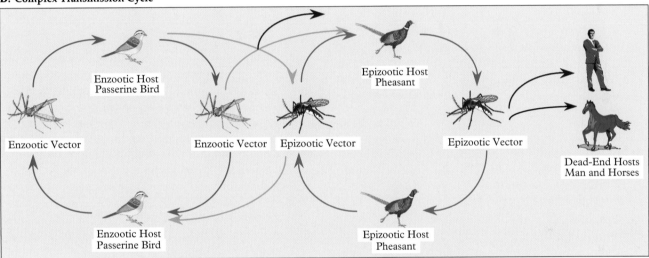

FIGURE 3.23 Two generalized transmission cycles of arboviruses. (A) Simple cycle, such as urban yellow fever, involving a single vector (*Aedes aegypti* mosquitoes) and a single vertebrate host (man). (B) Complex cycle, such as that for Eastern equine encephalitis, where the virus is maintained in an enzootic host (passerine birds) with an enzootic vector (*Culiseta melanura*), but can enter an epizootic vector (another insect) and be transmitted to epizootic hosts, and tangentially to dead-end hosts like man. An intermediate type of cycle is illustrated in Figure 5.12 for Colorado tick fever. Adapted from Monath (1988) p. 129.

Arboviruses that are transmitted by mosquitoes, ticks, sand flies, or other bloodsucking arthropods are known from several families of viruses. A selection of arboviruses from three virus families, most of which cause human disease, is listed in Table 3.11, together with the diseases they cause.

Pathology in the Mosquito Vector

The mosquito vector suffers little pathology upon infection by an arbovirus. This must be so if the virus is to persist, because transmission of the virus to a new vertebrate requires

that the female mosquito survive long enough and be healthy enough to take another blood meal, which is required for egg development. The time between infection of the female and the time at which the mosquito is capable of transmitting the virus is called the extrinsic incubation period. This period varies widely with temperature, humidity, the virus, and the mosquito. It can be as short as 2 days or as long as 2 weeks, although in alphaviruses it is seldom longer than 1 week.

Although infection of the mosquito is relatively benign, recent studies have shown that there is some pathology associated with arboviral infection. There is limited pathology

TABLE 3.11   Representative arboviruses that cause human disease

Family/virus	Nonfatal systemic febrile illness	Encephalitis Frequency[b]	% Mortality[c]	Hemorrhagic Fever (HF) Frequency[b]	% Mortality[d]
Togaviridae					
Chikungunya	Most cases, Ep			Rare, Ep	rare
Mayaro	Most cases, Ep				
O'nyong-nyong	Most cases, Ep				
Ross River	Most cases, Ep				
Sindbis	Most cases, Ep				
EEE	Most cases	Rare	50–70		
VEE	Most cases, Ep	Rare	0.1–20[e]		
WEE	Most cases	Rare	5–10		
Flaviviridae					
Dengue (1–4)	Most cases, Ep			Rare, Ep	3–12
West Nile	Ep	En	~9%[f]		
Japanese encephalitis		<1%	30–40[g]		
Kyasanur Forest		En	5	En	5
Murray Valley		Ep	20–70		
Rocio		Ep	13		
St. Louis encephalitis		Ep	4–20		
Tick-borne encephalitis					
Eastern		Rare	30		
Central European		Rare	1–10		
Omsk hemorrhagic fever				En	1–2
Yellow fever				Most cases	5–20
Bunyaviridae					
Bunyamwera	En				
Germiston	En				
Sand fly fever	Ep				
Rift Valley fever	Ep			En	1–5
California encephalitis		En	1		
Crimean hemorrhagic fever				En	15–20

[a] In addition to the disease manifestations listed, most viruses in this table can cause mild febrile illness; some viruses are endemic (En) but others cause occasional outbreaks or epidemics (Ep).

[b] Frequency relates to the relative number of cases exhibiting encephalitis or HF relative to the total number of infections. This number can be difficult to estimate, since only the most seriously ill (e.g., hospital patients in an epidemic) may be reported as infections. Mortality ≥ 10% is highlighted in red.

[c] Percent mortality is the percent of those *with encephalitis* who succumb.

[d] Percent mortality is the percent of those *with HF* who succumb.

[e] Mortality in children is at the high end of the range given.

[f] Mortality from recent epidemics in the United States.

[g] Mortality generally lower in children.

in the gut, which might be important for virus spread from the gut into the hemolymph, from which it spreads to the salivary glands, in some cases after an increase in titer following replication in fat body cells. More importantly, however, are studies showing that arbovirus infection results in decreased survival and reproductive capacity. Thus, the persistence of an arbovirus requires a delicate balance. The virus must replicate to sufficiently high titer in the vertebrate to cause sufficient viremia to infect a mosquito, which usually results in symptomatic disease.

But in the mosquito the infection must be controlled so as to produce sufficiently high titers of virus in salivary fluid without damaging too extensively the ability of the mosquito to survive and reproduce.

Overwintering by Arboviruses

In humid tropical or subtropical areas in which mosquitoes are active throughout the year, an arbovirus can be maintained by continuous transmission between invertebrate and vertebrate hosts. Virus activity may vary during the year, for example it may be much greater during a rainy season when the number of mosquito vectors is higher, but the virus is active throughout the year and human infection can occur at any time. However, in temperate zones in which adult mosquitoes die off in the winter, or in very dry areas in which mosquitoes are only active after sporadic rains, the virus must have a mechanism by which it overwinters. Mosquitoes survive winters (or extended droughts) by suspending development of the young at some stage. In some mosquitoes, eggs are laid but do not develop until conditions are favorable. In others, developing young diapause, suspending development at some stage of embryonic development or larval development, until conditions are favorable, such as the return of spring. Thus, one mechanism for overwintering by some arboviruses is transovarial transmission, in which the virus infects oocytes in the infected female at an early stage of development; the replication cycle of the virus is suspended during diapause and the animal develops normally. When the newly hatched mosquito begins to fly, it is already infected. One way to search for transovarial transmission in the field is to look for virus in male mosquitoes. Since these do not feed on blood, the only way for them to become infected is by transovarial transmission. Some alphaviruses are known to use this mechanism whereas other alphaviruses do not. A second mechanism used by some arboviruses is persistent infection of a vertebrate host so that infected vertebrates are present when the mosquitoes emerge once again. Some alphaviruses are known to persist in humans or other vertebrates for extended periods, but such persistence is rare and it is not known whether this could be a means of overwintering. In some arboviruses, however, such as the coltiviruses (see Chapter 5), persistence is known to be essential to the maintenance of the virus in nature and the virus has evolved specific mechanisms to persist. A third mechanism is the reintroduction of the virus into an area from regions where it persists year-round, for example, by the return of migratory birds or by infected mosquitoes being blown over large distances by storms. The whole subject of overwintering is an interesting evolutionary and ecological study, and in the case of many arboviruses, including most alphaviruses, overwintering is only poorly understood.

Seasonality of Disease

It is obvious from the preceding discussion that mosquito-borne diseases are seasonal. In temperate areas, disease is absent in the winter. In spring when mosquitoes first arise, disease is also absent or rare. It is only with time that the intensity of virus transmission builds up to the point that humans become infected with some frequency. There have also been suggestions that the arboviruses that first appear may not be as virulent as viruses that appear later in the season, perhaps because they are adapted to the mosquito vector and must readapt to the vertebrate host. In any event, arboviral epidemics characteristically occur in mid to late summer and early fall in temperate regions. In other regions, epidemics may be associated with heavy rainfall that results in an increase in the mosquito population.

Encephalitic Alphaviruses

Most, perhaps all, alphaviruses are neurotropic. They readily infect neurons in culture or in experimental animals. Infection of neurons is dependent upon the age of the animal as well as dependent upon the strain of virus. Sindbis virus infection of the mouse has been used as an experimental model to study virus induced encephalitis (inflammation of the brain, from *enceph* = brain, *itis* = inflammation) and encephalomyelitis (inflammation of the brain and spinal cord, from the preceding plus *myel* = medulla or marrow). Some strains of the virus will invade the central nervous system after peripheral inoculation and cause encephalitis, whereas other strains of the virus do not invade the CNS following peripheral inoculation but will cause encephalitis upon direct inoculation of the virus into the brain, and still other strains do not cause overt encephalitis even though neurons may be infected. In all cases, the infection of neurons to cause encephalitis is age related. Very young mice are susceptible to most strains of the virus, whereas infection of older mice does not result in encephalitis for many strains of the virus. Manipulation of the viral genome in the laboratory has resulted in the identification of specific genes that are important for neurovirulence.

It is important to distinguish neurotropism, the ability to infect neurons or other cells of the CNS, from neuroinvasion, the ability of the virus to cross the blood–brain barrier and invade the CNS, from neurovirulence, the ability to cause brain disease once the CNS has been invaded. Although most alphaviruses studied are neurotropic, only three viruses regularly cause encephalitis in humans or domestic animals. These three viruses, Eastern (EEE), Western (WEE), and Venezuelan equine encephalitis (VEE) viruses, are New World alphaviruses that cause fatal encephalitis in horses. WEE and EEE regularly cause encephalitis in humans as

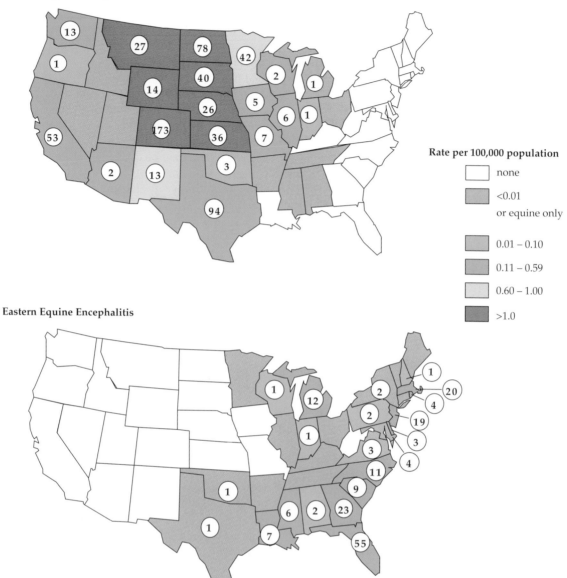

FIGURE 3.24 Geographic distribution of reported human cases of alphavirus encephalitis in the United States from 1964 to 2003. Colors indicate the rate in cases per 100,000 population by state and the actual numbers of cases are shown. Note that there were two human cases of WEE in 1994 and one in 1999, but none have been reported since then. Reported cases of EEE from 1994 through 1998, for which no state locations were given, were 1 in 1994, 4 in 1995, 5 in 1996, 14 in 1997, and 4 in 1998. There were 5 cases of EEE in 1999, 3 in 2000, 9 in 2001, 10 in 2002, and 14 in 2003. Since 1998 EEE cases are reported by state and the cases for 1999–2003 have been added to the cumulative state totals. Adapted from Fields *et al.* (1996) p. 875 and additional data from *MMWR, Summary of Notifiable Diseases, 1998, 1999, 2000, 2001, 2002, 2003.*

well, although the number of cases is small (Table 3.11 and Fig. 3.24). The mechanisms by which the CNS is invaded by alphaviruses or by viruses belonging to other families that cause encephalitis (poliovirus, certain flaviviruses, and others) are imperfectly known. Model studies in mice have indicated that Sindbis virus invades the CNS by retrograde axonal transport from the peripheral site of infection where primary replication occurs, one of the mechanisms that also

may be important for the invasion of the CNS by poliovirus (see earlier). Infection of the nasal neuroepithelium, whose neurons project directly to the CNS, may also lead in infection of the CNS. Sindbis virus appears to use this mechanism as well as transport from the peripheral sites, and transport from the neural epithelium may be particularly important in VEE infection. A third possible mechanism is the establishment of viremia followed by infection of choroid plexus epithelial cells such that the virus replicates across the blood–brain barrier. As described, this mechanism appears to be particularly important in poliovirus invasion of the CNS, but does not appear to be used during invasion by Sindbis virus. Thus, the primary mechanism used appears to differ among different viruses, and more than one mechanism may be important for any individual virus.

Alphaviruses have been used as a model system to study recovery of mice from virally induced encephalitis. The CNS is immunologically privileged (see Chapter 10) because neurons are not replaceable once destroyed, and the mechanisms used to cure the CNS of viruses differ from those used in other organs such as the liver or intestinal tract. It has been shown that humeral antibodies are required for clearance from the CNS but the mechanism is unknown. Cytotoxic T cells are also required for recovery from virally caused CNS disease, but by means of secretion of interferon-γ rather than by killing of infected cells.

Eastern Equine Encephalitis Virus

The majority of human infections by EEE are inapparent or result in febrile illness that is usually mild. Encephalitis results in only about 4% of infected adults but in more than 10% of children less than 10 years old, demonstrating the age-related aspects of the disease. EEE-caused encephalitis is fatal about half the time, with the highest fatality rates in the very young and the very old. Survivors usually have neurological deficits. There have been an average of seven cases of EEE encephalitis per year in the United States over the last 50 years. Horses are more sensitive to the virus and more likely to become infected, and the virus is of major concern to horse breeders in the eastern United States. The case fatality rate is 80–90%, and in years past there have been epidemics involving thousands of horses; the largest such epidemic on record killed more than 11,000 horses in Louisiana and Texas in 1947.

As described before, in North America EEE is maintained in an enzootic cycle in birds that inhabit freshwater swamps, vectored by *Culiseta melanura*. Perhaps because many of these birds are migratory and therefore capable of spreading the virus over large distances, the virus is fairly uniform throughout its North American range. Viruses isolated over a period of more than 60 years from many different areas of North America show less than 2% nucleotide sequence divergence. As illustrated in Fig. 3.23, epizootic cycles in other birds can occur. Mosquitoes vectoring either the enzootic cycle or an epizootic cycle are capable of transmitting the disease to horses or humans.

EEE in South America is a distinct virus. It is vectored primarily by *Culex* species and is not associated with severe disease in humans. In addition to birds, small mammals, especially rodents, which are sedentary, are an important reservoir for the virus. Perhaps because of this, the virus has evolved into a number of strains that differ by up to 25% in nucleotide sequence.

Western Equine Encephalitis Virus

WEE is less virulent for humans than is EEE. Encephalitis occurs in only 1 of 1000 adults infected by WEE, but in about 2% of children younger than 5, and the encephalitis produced by WEE is less severe, with an overall case fatality rate of about 3%, but about 8% in persons older than 50. The virus is also less virulent in horses, in which the case fatality rate is about 40%. The virus is a recombinant between EEE and a Sindbis-like virus, probably Aura virus present in South America (Fig. 3.17). Because the glycoproteins were derived from the Aura parent, the virus is serologically related to the Sindbis lineage. The RNA replication proteins are derived from EEE, as is the encephalitic potential of the virus. The fact that the virus is less virulent than EEE is consistent with laboratory studies that chimerization of a virus, a technique being used to produce live virus vaccines, usually results in lowered virulence (see, e.g., Chapter 11).

WEE is endemic from western Canada discontinuously to southern South America. In the western United States, the primary vector is *Culex tarsalis* and the vertebrate reservoir is again birds, although jackrabbits and possibly other mammals may be important in some areas. In the past, infection by the virus was common. An epizootic of WEE in 1912 killed an estimated 25,000 horses in the western United States. In 1960, 34% of humans tested in rural areas of California were found to be positive for antibodies to WEE, indicating past infection by the virus. Infection is now less common in the United States, and only about 2% of humans in similar areas of California were found to be seropositive for WEE in the mid 1990s. Confirming this trend, an average of 34 cases of human WEE encephalitis occurred per year in the United States from 1955 to 1984, but this number has declined thereafter and there has been only one case in the United States since 1994 (Fig. 3.24). This dramatic decline may have resulted from mosquito control measures, the widespread use of insect repellents, the adoption of air-conditioning and window screens that has resulted in fewer bites by mosquitoes, especially night flying mosquitoes, and because fewer horses, which are amplifying hosts for WEE, are used in farming. No vaccines for this virus are available for widespread use, although experimental vaccines exist that are given to laboratory personnel who work with the virus.

In Central and South America, WEE remains a widespread problem because the horse is still in widespread use as a farm animal. However, the virus is not associated with significant human disease in South America, for reasons unknown. In South America, small mammals appear to be an important vertebrate reservoir.

As described, WEE arose by recombination, an event that occurred at some unknown time but probably more that 1000 years ago. Since its origin, it has diverged into a number of different viruses, including Fort Morgan virus found in Colorado and Oklahoma and Highlands J virus present along the east coast of the United States. Fort Morgan virus is transmitted by swallow bugs present in the nests of swallows to the young birds of the year, an interesting adaptation. Highlands J virus is maintained in a cycle similar to that for EEE. Neither of these viruses has been associated with human disease.

Venezuelan Equine Encephalitis Virus Complex

The VEE complex consists of a number of closely related viruses that are endemic to tropical and subtropical areas of the Americas. These viruses were first classified by serology into six subtypes of VEE, numbered with Roman numerals from I to VI. Some of these subtypes were further subdivided; in particular, subtype I was subdivided into IAB, IC, ID, IE, and IF. Partial or complete sequences of most of these viruses are now available and these subtypes are now considered to be full species (Fig. 3.25). As a species, VEE virus consists of IAB, IC, and ID; IE, also considered a strain of VEE, will be considered in more detail later. Subtype II is now named Everglades virus and is endemic to Florida. Subtype IIIA is called Mucambo virus, IIIB is Tonate virus (of which Bijou Bridge virus is another isolate), IV is Pixuna virus, V is Cabassou virus, and VI is Rio Negro virus; most of these are South American viruses. Subtype IF (not shown in Fig. 3.25) is also a South American virus but is only distantly related to other subtype I viruses, and is now considered a separate species, named to the present by its isolate designation 78V353I virus. These eight virus species are maintained in an endemic cycle in which the principal mosquito vectors are species belonging to the genus *Culex*, subgenus *Melanoconin*, and the major vertebrate reservoirs are small mammals, primarily cotton rats and opossums. These endemic viruses are able to infect horses but replicate to only low titers in them, causing no disease and not establishing an epidemic cycle. Similarly, these endemic viruses are not associated with significant human disease.

The endemic VEE ID gives rise to epizootic viruses IAB and IC by a small number of mutations in envelope glycoprotein E2 that allow the virus to grow to high titers in horses. The horse serves as an important amplifying host that allows the virus to become epidemic and spread over wide areas. *Aedes taeniorhynchus* and *Psorophora confinnis* are important mosquito vectors in these outbreaks. In these epidemics, which occur at intervals of 10–20 years in Venezuela, Columbia, Peru, and Ecuador, large numbers of horses die of encephalitis and significant episodes of human illness occur. For example, an epidemic in Venezuela and Columbia in 1995 resulted in disease in an estimated 75,000 to 100,000 people, including 3000 cases of neurological disease and 300 deaths. As with the other encephalitic alphaviruses, children and older people are more at risk for serious illness.

Because of the importance of VEE epidemics, an inactivated virus vaccine was produced many years ago from a

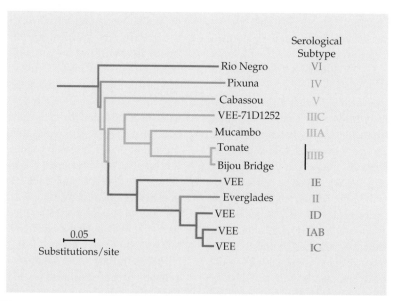

FIGURE 3.25 Phylogenetic tree of the various strains of Venezuelan equine encephalitis virus (VEE) and species of viruses formerly considered to be serological subtypes of VEE. This tree was generated from partial E1 envelope gene sequences using a neighbor-joining algorithm. Adapted from Fauquet *et al.* (2005), Figure 4 on p. 1007.

strain of IAB virus called Trinidad Donkey for use in horses and for laboratory personnel who work with the virus. This vaccine suffered from poor immunogenicity and insufficient inactivation. Use of the vaccine, for example, led to a wide-ranging epidemic of VEE in 1969–1973 that spread from South America up through Central America to southern Texas, causing many thousands of cases of disease in horses and humans. Because of these problems, use of this vaccine was abandoned and a new attenuated vaccine strain called TC-83, derived from the Trinidad Donkey virus by passage in culture, was developed. This vaccine is effective but is reactogenic and causes mild illness in many recipients. An inactivated version of this attenuated virus has also been produced as a vaccine but is not as effective as the live virus vaccine.

VEE virus can spread by aerosols as well as by mosquito transmission. Many cases of laboratory acquired infection have resulted from aerosols. Because of this property and the incapacitating disease caused by VEE, this virus was weaponized by the Russians in the past as a potential biowarfare agent.

Virtually all epidemics of epizootic VEE have been due to IAB and IC viruses. However, a recent epizootic of VEE in Mexico resulted from an IE strain that had a mutation in envelope glycoprotein E2. Unlike the IAB and IC strains, the epizootic IE virus grows to only low titers in horses. It is believed that the mutation in IE virus allowed a more efficient interaction with mosquito vectors that are widespread and numerous, allowing more efficient transmission of the virus. As shown in Fig. 3.25, IE virus is more closely related to Everglades virus than it is to ID virus, and the virus may be reclassified in the future.

Alphaviruses That Cause Arthritides

A number of alphaviruses cause disease in humans characterized by fever, rash, and joint involvement. The pain from arthritis (joint inflammation, from *arth* = joint and *itis* = inflammation) or arthralgia (joint pain, from *arth* and *algia* = pain) following infection by these viruses can be so severe as to be disabling and can last for a year or more with relapses of severe arthritis being common during this period. The names of some of these viruses come from the crippling pain caused by the disease resulting from viral infection.

Ross River Virus

Ross River virus is widespread in Australia, New Guinea, and the Solomon Islands. The disease induced by virus infection is known as epidemic polyarthritis and is characterized by pain, often accompanied by frank swelling, in the small joints of the hands and feet and in the knees. The principal vectors are *Aedes camptorhynchus* and *Aedes vigilax* in coastal regions, which breed in salt marshes, and *Culex*

annulirostris and *Aedes notoscriptus* in other areas, which breed in freshwater. The vertebrate reservoir consists of various macropods. Because mammals constitute the vertebrate reservoir, different strains of the virus have evolved in different geographic regions for reasons discussed earlier. In some areas of Australia in which the virus is endemic, a majority of the population may be seropositive for RRV, indicating a high attack rate. It has been estimated that 2 to 30% of infected humans develop clinical symptoms following infection. As noted, recovery may be prolonged, with arthritic symptoms recurring over a period of a year or more.

Of interest was a wide-ranging epidemic of Ross River polyarthritis that swept through the South Pacific in 1979–1980. The epidemic began when a single viremic traveler from Australia landed in Fiji. The epidemic began near the airport and eventually spread throughout the island. From there it jumped to other islands having air contact with Fiji. During this epidemic, it is believed that humans were the primary or only vertebrate host, with the disease being transmitted from mosquito to human to mosquito without the intervention of another animal reservoir (the cycle illustrated in Fig. 3.23A). Sequencing studies have suggested that a mutation in glycoprotein E2 may have allowed this cycle to become established. The mosquito vector during this epidemic was *Aedes polynesiensis*. This epidemic was explosive with most of the people on the affected islands becoming infected by the virus and about 10% of them becoming ill. The epidemic eventually burned itself out because all the humans had become immune and the virus failed to establish an endemic cycle in other animals in the region. After 20 years without RRV in Fiji, three recent cases of RRV disease have been reported in travelers to Fiji. It is thought that RRV has been reintroduced into the island on these occasions and susceptible tourists infected, but that the native population is now largely immune to the virus and no new epidemic has occurred.

Sindbis Virus

The prototype alphavirus is Sindbis virus, named after the town of Sindbis, Egypt, where it was first isolated in 1953 from mosquitoes. It has the widest distribution of any alphavirus, endemic from northern Europe through the Middle East and India to South Africa, Southeast Asia, Indonesia, the Philippines, New Guinea, and Australia. As might be expected from this broad range, strains of virus isolated from different regions may differ by 20% or so in nucleotide sequence and differ in their epidemiology and disease potential. As far as known, birds are the vertebrate reservoir for the virus over its entire range with different mosquitoes serving as vectors in different areas. Over most of its range no human illness is associated with infection or illness is very rare, even though in regions such as the Nile Delta seroprevalence rates may be fairly high. However, strains of the virus in northern Europe, South Africa, and Australia cause significant episodes of arthritic illness.

In northern Europe the strains of Sindbis virus that cause arthritic disease are called Ockelbo virus in Sweden (present between the 60th and 64th parallels), Pogosta virus, widespread in Finland, and Karelian Fever virus, present in far Western Russia. The virus is maintained in migratory birds or in game birds and the mosquito vectors are various species of *Culex* and *Culiseta*. *Aedes* species may spread the virus to humans. The disease in humans is typical of alphavirus arthritic disease with fever, rash, and joint inflammation and the number of cases can be large during epidemic years. Children are less likely to develop disease upon infection. In South Africa strains of Sindbis virus also cause human disease, but the number of cases appears to be small and the virus has been less well studied. Strains of Sindbis virus that are present in Australia also cause human disease, primarily in northeastern Australia. The virus appears to be fairly uniform throughout Australia but to be replaced every so often by new strains that presumably invade from the north.

Mayaro Virus

Mayaro virus is present in the northern half of South America (Trinidad, Surinam, Brazil, Columbia, Bolivia). It is maintained by *Haemagogus* mosquitoes and humans usually contract the virus while in humid tropical forests. Rubber workers are at risk of infection and the polyarthritis caused by the disease can be debilitating, preventing the workers from gainful employment. Mayaro belongs to the Semliki Forest virus clade (Fig. 3.17) and causes a disease that is similar to that caused by the related Ross River virus. It is the only known representative of this clade in the Americas and represents one of the very few transfers of alphaviruses between the Old and the New World.

Chikungunya Virus and Related Viruses

Chikungunya (CHIK) virus, a member of the Semliki Forest virus clade, is endemic or epidemic from sub-Saharan Africa through India and Southeast Asia to the Philippines. The name comes from Swahili meaning "that which bends up," from the intense arthralgia that causes patients to lie with joints flexed. In Africa the virus is maintained in an endemic cycle that is similar to that for yellow fever virus (see later). The mosquito vectors are *Aedes africanus* and *Aedes furcifer* and subhuman primates are the vertebrate reservoir. During explosive epidemics in urban areas of Africa and Asia, *Aedes aegypti* is the vector and human–mosquito–human cycles maintain the virus. During an epidemic, a large fraction of the susceptible human population may contract the disease. The epidemic then dies out, to return when reintroduced after a period of time that allows a susceptible population to build up. Epidemics often occur during the rainy season when the population of *Aedes aegypti* is highest.

In 2006, a hugh epidemic of CHIK, accompanied in many locations by infections of dengue virus as well (see later) whose disease symptoms can be similar to those caused by CHIK, occurred throughout the Indian Ocean region. The virus appears to have been imported from East Africa, and many thousands of cases occurred in the region, affecting the islands of Reunion, Mauritius, Madagascar, Mayotte, the Seychelles, and the Maldives as well as the Indian subcontinent. In Reunion there occurred more than 200,000 cases in a population of under 800,000. Many of these islands are popular tourist destinations, and many cases were imported into Europe. The primary mosquito vector was *Aedes albopictus* or *Aedes aegypti*, depending upon the location, both of which are also efficient vectors for dengue virus.

O'nyong-nyong (ONN) virus is a close relative of CHIK that has caused very large epidemics of disease in East Africa similar to that caused by CHIK. The name also comes from the very painful arthralgia accompanying the disease. Epidemics affecting up to two million people have occurred followed by the disappearance of the virus for long periods. In these epidemics the virus is transmitted by *Anopheles funestrus* and *Anopheles gambiae*, mosquitoes that are major vectors of malaria, and these epidemics represent the only known cases of epidemic transmission of an alphavirus by anopheline mosquitoes. An endemic cycle presumably maintains the virus during interepidemic periods but, if so, this cycle is unknown. A strain of ONN called Igbo-Ora virus, from the name of the Nigerian village in which the virus was first isolated, is present in West Africa.

Barmah Forest Virus

Barmah Forest virus is an Australian virus that is an outlier in the Semliki Forest clade (Fig. 3.17). It also causes polyarthritis in humans. It is probably maintained in a cycle similar to that for Ross River virus.

Other Alphaviruses

Other alphaviruses are known that infect higher vertebrates including humans, but most are not associated with disease in humans. Semliki Forest virus, named after the Semliki Forest in Uganda, has been extensively characterized as a model system to study the molecular biology of alphaviruses. Most strains cause no human illness, but strains from central Africa cause a disease characterized by exceptionally severe headache, fever, and rash. One case of fatal human encephalitis caused by this virus occurred in a laboratory worker, who is believed to have contracted the virus via aerosols. Getah virus, widespread in Asia, causes a mild febrile illness in humans.

Recent isolates of new alphaviruses include Southern elephant seal virus, currently unclassified. It is spread by a louse, *Lepidophthirus macrorhini*, that infests the seals.

Two fish alphaviruses have been recently isolated, salmon pancreas disease virus and sleeping disease virus, which are now considered to be strains of the same species, to be called salmonid alphavirus. A third strain of the virus, Norwegian salmonid alphavirus, has been found in western Norway. Salmonid alphavirus is an important pathogen of Atlantic salmon and rainbow trout in Norway, Britain, Ireland, and France. It causes significant economic losses in the farmed fish industry and is the leading cause of economic loss in Ireland. The viruses can be spread by direct fish-to-fish infection and are not known to have an arthropod vector, but since all other alphaviruses are vectored by arthropods there has been speculation that a sea louse *Lepeophtheirus salmonis* might be involved in the transmission of the virus. It is not known if the primary location of infection is saltwater or freshwater.

Rubella Virus

Rubella virus is less well understood than alphaviruses because it grows very poorly in cultured cells and its genome possesses an extraordinarily high GC content (70%), which retarded efforts to sequence and express the viral genome. The complete sequences of several strains are now known and detailed molecular studies are under way. The genome is approximately 10 kb in size and is expressed similarly to the alphavirus genome: The genomic RNA is translated into a polyprotein cleaved by a papain-like protease into two pieces, and a subgenomic mRNA is translated into structural proteins consisting of a capsid protein and two envelope glycoproteins (Fig. 3.16). Interestingly, minus-strand RNA

synthesis is controlled in the same way as in alphaviruses. The uncleaved nonstructural polyprotein synthesizes minus-strand RNA, whereas the cleaved products synthesize only plus-strand RNA.

Rubella virus infects only humans and there is no other reservoir. Infection is by person-to-person contact, primarily through aerosols. It causes a relatively benign illness, sometimes called German measles, with a characteristic rash and is (or was) one of the typical childhood diseases. However, infection of a pregnant woman in the first trimester of pregnancy can have devastating effects on the developing fetus. The virus sets up a long-lived infection in the fetus that often causes developmental abnormalities resulting in children with severe handicaps (congenital rubella syndrome). An attenuated virus vaccine against rubella has been developed that is now routinely given to children as part of mumps–measles–rubella vaccination (MMR). Since the vaccine was introduced, there has been a drastic reduction in the number of cases of rubella in the United States (Fig. 3.26) and other developed countries.

Because the postnatal disease caused by rubella virus is trivial, the purpose of the rubella vaccine is to protect against future birth defects rather than to protect the individual vaccinated. This raises interesting ethical questions about its use. In some societies, only females were vaccinated, since they would want to protect their future children from the effects of rubella. However, because males remained susceptible to the virus, it continued to circulate in the population and rubella-caused birth defects continued to occur. To protect against this, the only solution is to vaccinate the entire population so as to eliminate the virus from the society.

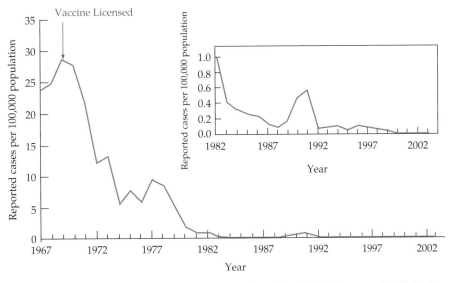

FIGURE 3.26   Incidence of rubella virus, by year, in the United States. From *MMWR Summary of Notifiable Diseases in the United States for 1997, 1998, 1999, 2000, 2001, 2002, 2003.*

The rubella vaccine has now been in use for many years and is generally safe and effective when given to children. The present vaccine has a high incidence of side effects in adults, however, especially arthralgia and arthritis. Nonetheless, vaccination is recommended for women of childbearing age, as well as for certain health care personnel, who have never been vaccinated and who are seronegative. The need exists to improve the vaccine, and current efforts to understand the molecular biology of the virus in more detail will hopefully lead to the development of a better vaccine.

FAMILY *FLAVIVIRIDAE*

The *Flaviviridae* are named after the prototype virus, yellow fever virus, *flavus* being the Latin word for yellow. The *Flaviviridae* are divided into three genera, the genus *Flavivirus*, the genus *Pestivirus*, and the genus *Hepacivirus*. A partial listing of viruses in the three genera is given in Table 3.12. In the following discussion, the term flavivirus refers only to members of the genus *Flavivirus* unless otherwise specified.

The genome organizations of members of the three genera are shown in Fig. 3.27. The genomes of the three genera are similar in size (11 kb for flaviviruses, 12.5 kb for pestiviruses, 9.4 kb for hepaciviruses) and organization. These viruses, like the picornaviruses, have a genome that contains only a single ORF. This ORF is translated into a long polyprotein that is processed by cleavage into 10 or more polypeptides. Processing of the precursor polyprotein is complicated. Cleavage is effected by a combination of one or two or three (depending on the virus) virus encoded proteases and two or more cellular proteases. The structural proteins are encoded in the 5′-terminal region of the genome (like picornaviruses). However, all members of the *Flaviviridae* are enveloped, unlike the picornaviruses, and the structural proteins consist of a nucleocapsid protein and two or three envelope glycoproteins. Cellular proteases make the cleavages that separate the glycoproteins, but the cleavages in the nonstructural region of the polyprotein, which is required for RNA replication, are made by one or two virus-encoded proteases. Even so, cellular signalase makes at least one of the cleavages in the nonstructural domain of flaviviruses. The cleavage pathways in this genus are described in detail next.

All members of the *Flaviviridae* encode a serine protease with a catalytic triad consisting of serine, histidine, and aspartic acid. The protease resides in the nonstructural

TABLE 3.12   *Flaviviridae*

Genus/species	Virus name abbreviation	Usual host(s)	Transmission	Disease	World distribution
Flavivirus					
Dengue (Types 1–4)	DENV	Humans	Mosquito-borne	Dengue fever, shock, hemorrhage,	Worldwide
Yellow fever	YFV	Primates[a]	Mosquito-borne	Hemorrhage, liver destruction	Africa, Americas
Japanese encephalitis	JEV	Mammals,[a] especially swine	Mosquito-borne	Encephalitis	Widespread in Asia
St. Louis encephalitis	SLEV	Mammals,[a] birds	Mosquito-borne	Encephalitis	North America
Murray Valley encephalitis	MVEV	Mammals,[a] birds	Mosquito-borne	Encephalitis	Australia
Tick-borne encephalitis	TBEV	Mammals[a]	Tick-borne	Encephalitis	Europe, Asia
West Nile	WNV	Mammals,[a] birds	Mosquito-borne	Encephalitis	Europe, Africa, North America
Hepacivirus					
Hepatitis C	HCV	Humans	Parenteral, transfusion	Hepatitis, liver cancer	Worldwide
Pestivirus					
Classical swine fever	CSFV	Swine	Contact	Fever, acute gastroenteritis	Europe, Americas
Bovine viral diarrhea	BVDV	Cattle	Contact	Usually none[b]	Worldwide

[a] Including humans.
[b] Calves infected *in utero* develop persistent infections that can lead to mucosal disease.

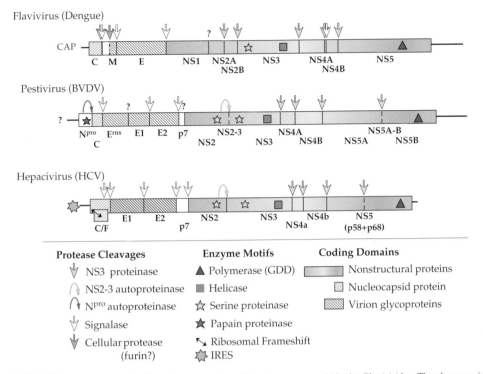

FIGURE 3.27 Genome organization of representatives of the three genera within the *Flaviviridae*. The cleavage sites indicated with dashed lines have not been precisely localized. Adapted from Figures 2, 4, 5 of the *Flaviviridae* in Fauquet *et al.* (2005) pp. 981–998. The data on the F protein of hepatitis C virus, which is produced from the core protein sequence by ribosomal frameshifting, came from Xu *et al.* (2004).

region called NS3, just upstream of a helicase. The crystal structure of the dengue virus protease and of the hepatitis C virus (HCV) protease has been solved to atomic resolution and they possess a fold similar to chymotrypsin, as is the case for other viral serine proteases whose structures have been solved. The enzyme is interesting in that a second polypeptide is required for activity, NS2B in flaviviruses and NS4A in HCV. From the atomic structure of the HCV protease complexed with the region of NS4A required for activity, it is clear that NS4A forms an integral part of the folded protease. Thus, it is puzzling that the protease consists of two cleaved products rather than one continuous polypeptide chain.

Flaviviruses encode only the NS3 protease. Hepaciviruses and pestiviruses encode a second protease in the NS2 region that cleaves between NS2 and NS3. Pestiviruses also encode a third protease at the N terminus of the polyprotein whose only known function is to cleave itself from the polyprotein precursor.

Flaviviruses have capped genomes whose translation is cap dependent. In contrast, the hepacivirus and pestivirus genomes are not capped and have an IRES in the 5′ non-translated region. Members of family *Flaviviridae* do not have a poly(A) tail at the 3′ end of the RNA. Instead, a stable stem-loop structure is present at the 3′ end of the genome that is required for replication of the genomic RNA and for its

translation, as described in more detail later. No nucleotide or amino acid sequence identity can be detected between members of different genera except for isolated motifs that are signatures of various enzymatic functions.

Viruses in the family are enveloped. As described later and in Chapter 2, members of the genus *Flavivirus* have a structure that is related to that of alphaviruses. Details of the structures of pestiviruses and hepaciviruses are lacking. Flaviviruses mature at intracytoplasmic membranes rather than at the plasma membrane.

Genus *Flavivirus*

There are about 53 species of flaviviruses currently recognized, and many species have important subtypes that are also named. A representative sample is given in Tables 3.11 and 3.12. The relationships of the viruses to one another is illustrated by the dendrogram in Fig. 3.28. All members of the genus are closely related and share significant amino acid sequence identity in their proteins, which results in serological cross-reactivity. Historically, members of this genus were assigned to it on the basis of these cross-reactions. Most are arthropod-borne, and they were once referred to as Group B arboviruses. They can be divided into three major groups

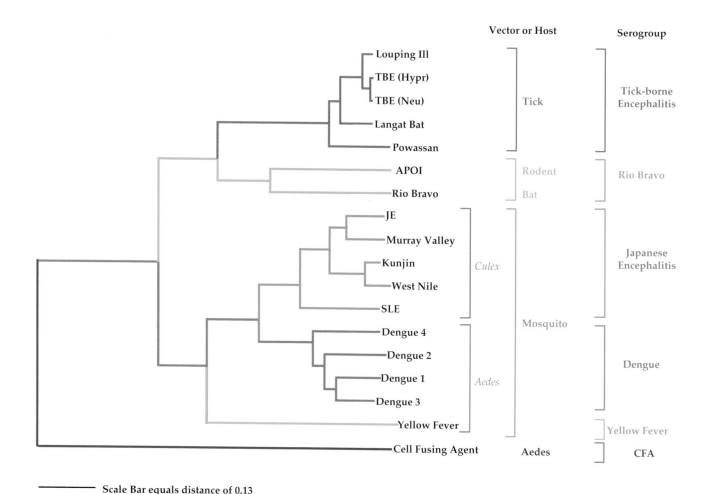

Vector or Host

Serogroup

Louping Ill
TBE (Hypr)
TBE (Neu)
Langat Bat
Powassan

Tick

Tick-borne
Encephalitis

APOI
Rio Bravo

Rodent
Bat

Rio Bravo

JE
Murray Valley
Kunjin
West Nile
SLE

Culex

Japanese
Encephalitis

Dengue 4
Dengue 2
Dengue 1
Dengue 3
Yellow Fever

Aedes

Mosquito

Dengue

Yellow Fever

Cell Fusing Agent

Aedes

CFA

Scale Bar equals distance of 0.13

FIGURE 3.28 Phylogenetic tree of the flaviviruses based on NS3 polyprotein region using the neighbor-joining method. Data are from Billoir *et al.* (2000).

based on the vector utilized: the mosquito-borne group (which includes yellow fever, the dengue complex, and the Japanese encephalitis complex), the tick-borne encephalitis group (the TBE complex), and a group that lacks an arthropod vector. The last are of limited medical importance. Notice that in the phylogenetic tree in Fig. 3.28, the tick-borne viruses and the mosquito-borne viruses belong to different lineages. These viruses are adapted to a tick vector or to a mosquito vector, and interchange of vectors does not occur. Further, the tree indicates that the mosquito-borne viruses separate out into a lineage vectored primarily by mosquitoes belonging to the genus *Culex* and lineages vectored by mosquitoes belonging to the genus *Aedes*. However, in these lineages the restriction on the mosquito vector is not firm and mosquitoes belonging to other genera may vector many of these viruses. As two examples, the ancestral yellow fever virus in Africa is vectored by *Aedes* mosquitoes in both sylvan and urban cycles, but in the Americas it is vectored by *Hemagogous* mosquitoes in a sylvan cycle and *Aedes* mosquitoes in an urban cycle, as described later. West Nile virus, recently

introduced into the United States, is vectored by a very wide variety of mosquitoes, and this has been in part responsible for the rapid spread of the virus across the United States.

Expression of the Viral Genome

The genome organization of a typical flavivirus is illustrated in Fig. 3.27. As for all plus-strand RNA viruses, the genomic RNA is a messenger and in the case of flaviviruses serves as the messenger for all of the virus encoded proteins. The RNA is capped but lacks 3′ poly(A). There is a stem-loop structure at the 3′ end which serves the same function as poly(A) in other messengers. This structure increases the efficiency of translation of the RNA by about 10-fold and will substitute for poly(A) in model systems. Viral proteins are not required for this effect and therefore cellular proteins must interact with this structure in order to increase the efficiency of translation. It is known that during translation of mRNAs that are capped and polyadenylated, there is an initiation complex formed that contains both cap-binding protein and poly(A)-binding protein. Thus, the complex interacts with both ends of the mRNA to initiate translation. It is assumed that a cellular protein binds the 3′ stem-loop of flaviviruses and interacts with the initiation complex so as to perform the same function as the poly(A)-binding protein. Formation of this complex in the case of flavivirus RNAs could be enhanced by cyclization of the viral RNA described later, although the primary function of cyclization appears to be in the replication of the viral RNA.

The processing of the long polyprotein produced from the genome is complicated and is illustrated in Fig. 3.29 as an example of complex processing events that can occur in viral polyproteins associated with lipid bilayers. The nucleocapsid protein is 5′ terminal in the genome and is removed from the precursor polyprotein by the viral NS2B–NS3 protease. Two envelope proteins, prM (precursor to M) and E (envelope), follow. Both are anchored in the endoplasmic reticulum by C-terminal membrane-spanning domains and are usually, but not always, glycoproteins. A series of inter-

nal signal sequences is responsible for the multiple insertion events required to insert prM, E, and the following protein, NS1, into the endoplasmic reticulum. After separation of these three proteins by signalase, prM and E form a heterodimer. prM is cleaved to M by furin during transport of the heterodimer or during virus assembly. Assembly of virions is described in more detail later.

Following the E protein is NS1 (NS for nonstructural). NS1 is a glycoprotein and has multiple functions that are only poorly understood. It is found as dimers and higher multimers in three locations in mammalian cells: intracellular; anchored in the plasma membrane by a GPI (glycosyl-phosphotidylinositol) anchor; and as a soluble protein secreted from the infected cell. It is required for RNA replication, presumably a function of the intracellular form of the protein. For this function, it interacts with NS4A. The cell surface-anchored form is capable of antibody-induced signal transduction that may play a role in cell activation. The function of the secreted form is unknown but it has

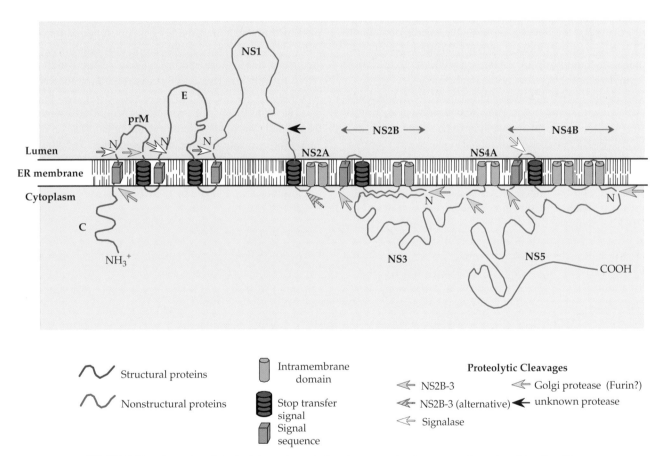

FIGURE 3.29 Processing of the flavivirus polyprotein into the structural and nonstructural proteins of the virus. The structural proteins (blue) at the N terminus of the polyprotein are processed primarily by signalase, with one late cleavage in prM due to furin. The nonstructural proteins (green) are mostly processed by the viral NS2B–NS3 protease. As indicated in the figure, the central 40 amino acids of NS2B interact with NS3, tying NS3 to the membrane, and this interaction is essential for proteolytic function. The striped arrow shows the alternative site of cleavage within NS2A that may lead to an anchored form of NS1. Adapted from Figures 3, 4, and 6 in Strauss and Strauss (1996).

been speculated that it has a role in counteracting immune responses to the virus.

Next in the polyprotein precursor are two hydrophobic polypeptides called NS2A and NS2B. These proteins are cleaved by the viral NS2B–NS3 protease. They are associated with membranes and may serve to anchor parts of the replication machinery to internal membranes in the cell. NS2A has multiple functions. It inhibits the production of interferon-α/β by infected cells. As described in more detail in Chapter 10, the interferons are potent inhibitors of virus replication and most, perhaps all, viruses encode products to block interferon action. NS2A also has a role in the production of infectious particles from the infected cell, since certain mutations in this protein block virus assembly but do not affect other aspects of the virus life cycle. These mutants can be suppressed by changes in the NS3 helicase domain, suggesting an interaction between NS2A and NS3. NS2B also interacts with NS3, but with the protease domain. It is a cofactor required for the NS3 protease activity and the central domain of NS2B forms a complex with NS3, which follows NS2B in the polyprotein precursor. The NS2B–NS3 serine protease cleaves many bonds in the polyprotein. NS3 also has at least two other activities—the middle domain of NS3 is a helicase, required for RNA replication, and the C-terminal domain has RNA triphosphatase activity, an activity required for the capping of the viral genome.

NS4A and NS4B are hydrophobic polypeptides that are associated with membranes. They may function in assembly of the viral replicase on intracellular membranes. Both the viral NS2B–NS3 protease and cellular signalase are required to produce the final cleaved products.

NS5 is the viral RNA polymerase. It appears to be a soluble cytoplasmic protein that associates with membranes through association with other viral peptides. It also has methyltransferase activity and thus is the capping enzyme that caps the viral genome. Thus, capping requires two flaviviral proteins, NS3 (RNA triphosphatase) and NS5 (capping enzyme). Note the similiarity to alphaviruses where the RNA triphosphatase activity is also on the helicase-protease protein (nsP2, the analogue of flaviviral NS3), and the methyltransferase or capping activity is a different protein (nsP1). However, in alphaviruses the capping enzyme and the RNA polymerase (nsP4) are distinct proteins, whereas in flaviviruses they are present in the same polypeptide.

Replication of the Viral RNA

RNA replication is associated with the nuclear membrane. The composition of the replicase complex is not understood but is assumed to consist of many (most? all?) of the viral nonstructural proteins with associated cellular proteins. Cyclization of the RNA is required for replication. Sequences from the 5′ and 3′ regions of dengue virus RNA that form a number of stem-loop structures and that also cyclize the RNA are illustrated in Fig. 3.30, where two possible structures are shown. Experimental data have shown that the RNA sequence in the capsid protein downstream of the AUG start codon is involved in cyclization (region marked CS1). Sequences upstream of the start codon are also known to be required for cyclization, and the two structures show different ways that these might be used for cyclization. This figure also illustrates the long stem-loop structure at the 3′ end of the RNA, discussed earlier. A stem-loop structure in the 5′ region just upstream of the CS1 region has also been shown to be important in translation of the RNA, in this case for recognition of the AUG start codon, which is found in a poor context for a start codon.

The sequences surrounding CS1 are illustrated for a number of mosquito-borne flaviviruses in Fig. 3.31. This eight nucleotide sequence is invariant among the mosquito-borne flaviviruses, and experimental studies have shown that this sequence is important for cyclization and replication of the RNA. The 3′ sequences complementary to this region are found in the 3′ nontranslated region (see also Fig. 3.30). Changes in these sequences that eliminate cyclization prevent the RNA from replicating, even in model systems in which translation of the RNA is not required for expression of the replicase. Compensating mutations in the partner sequence that restore cyclization restore the ability of the RNA to replicate. Thus, cyclization is required for RNA replication.

The identities of the promoters recognized by the RNA replication machinery are as yet unknown, but the requirement for cyclization suggests that sequences at both ends of the RNA are required. The conservation of the 8-nucleotide core sequence suggests that these sequences might be part of the promoter recognized by the RNA replicase.

Formation of the Virion

Most flaviviruses mature at intracellular membranes. Budding figures have been described only rarely and assembly may be associated with the complex processing of the polyprotein. West Nile virus is an exception to this general rule. It grows to higher titers in cultured cells than other flaviviruses and budding of preassembled nucleocapsids at the plama membrane is readily seen. Even in this case, however, intracellular assembly of virions is also seen.

The processing of the structural proteins from the precursor polyprotein was described earlier. prM and E form a heterodimer shortly after synthesis. The assembly of flaviviruses has clear parallels with that of alphaviruses, as described in Chapter 2. E of flaviviruses and E1 of alphaviruses are homologous proteins, having the same structure and function (see Fig. 2.17). A heterodimer is first formed, between prM and E in flaviviruses and PE2 and E1 in alphaviruses. Immature virus particles can be isolated that have uncleaved PE2 or prM whose infectivity is very low. The trimeric spikes

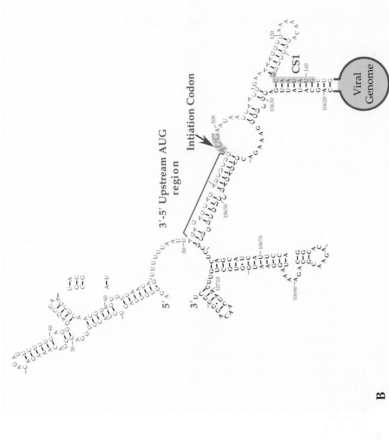

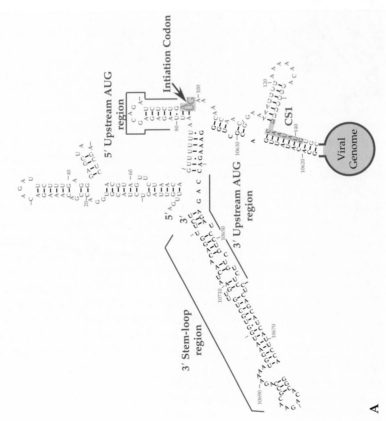

FIGURE 3.30   Computer-generated secondary structures of the last 106 nucleotides (nts 10618–10723 in black) and the first 144 nucleotides (in red) of the dengue 2 genome of strain 16681. The rest of the genome is indicated schematically by the green circle. The two predicted secondary structures shown in (A) and (B) are those with the lowest ΔG. The 3′ stem-loop structure, the 3′ and 5′ upstream AUG regions, the initiating AUG codon, and the conserved sequence (CS1, illustrated in Fig. 3.30) are labeled. Note that the stem-loop structure just upstream of CS1 (nt 114ff) is important for initiation of translation at the AUG start codon. This figure is adapted from Figure 5 in Alvarez *et al.* (2005).

111

                                                    CS1
             5'   ┌─────────────┐
    YF   (147) CCCUGGGC G │ UCAAUAUG │ GUACGACGAG (173)   3'

    MVE   (126) CC CCGGGUCG │ UCAAUAUG │ CUAAAACGCG (153)
    JE   (126) AA CCGGGCUA │ UCAAUAUG │ CUGAAACGCG (153)
    WN   (127) AA CCGGGCUG │ UCAAUAUG │ CUAAAACGCG (154)
    SLE   (129) AA CCGGGUUG │ UCAAUAUG │ CUAAAACGCG (156)

    DEN4   (127) GACCAC CUU │ UCAAUAUG │ CUGAAACGCG (153)
    DEN2   (125) ACACGC CUU │ UCAAUAUG │ CUGAAACGCG (151)
                            └─────────────┘

FIGURE 3.31   Conserved nucleotide sequence elements in the 5' region encoding the capsid protein in six different mosquito-borne flaviviruses. The number of the first and last nucleotides shown is given in parentheses. The boxed nucleotides in red are those postulated to be important for cyclization of the RNA. Residues shaded in green are complementary to sequences at the 3' end and those given in blue are conserved but probably not involved in cyclization. Adapted from Figure 7 of Hahn *et al.* (1987).

in these immature particles are quite similar in structure (see Fig. 2.15). Flaviviruses have a triangulation number of 3 and therefore 60 trimeric spikes each consisting of 3 heterodimers of prM and E. Alphaviruses have a triangulation number of 4 and therefore there are 80 trimeric spikes each consisting of 3 heterodimers of PE2 and E1. After cleavage of prM or of PE2, however, the structures are quite different. Alphaviruses retain 80 trimeric spikes with E2-E1 heterodimers. In flaviviruses, however, the M-E heterodimer dissociates and there is a dramatic rearrangement whereby 90 E-E homodimers are formed and the particle shrinks from 60 nm in diameter to 50 nm. Upon infection of a cell and exposure of the mature flavivirion to acidic pH there is another dramatic rearrangement and 60 E-E-E homotrimers are formed that tilt up so that the fusion peptide at the extremity of domain 2 is inserted into the cellular target membrane and fusion results. It seems clear that the structures of alphaviruses and flaviviruses had a common origin, and that recombination during their evolution resulted in this common structure becoming associated with different suites of RNA replication enzymes. It is possible that this structure, which allows enveloped viruses to have a regular icosahedral structure, evolved only once. It is also of note that both groups of these viruses are primarily arboviruses and perhaps this common structure is important for this.

The flavivirus nucleocapsid is thought to be icosahedral in symmetry, perhaps having a triangulation number of 3. There appears to be no interaction between the envelope proteins and the capsid proteins in flaviviruses, however, unlike the situation for alphaviruses, so that the icosahedral structure of the nucleocapsid, if it exists, is not coordinated with the icosahedral arrangement of the glycoproteins forming the outer surface of the virion.

Yellow Fever Virus

Many flaviviruses are important pathogens of humans. Different viruses may cause encephalitis, hemorrhagic fever with shock, fulminant liver failure, or disease characterized

by fever and rash. Several important viruses and their diseases are listed in Tables 3.11 and 3.12. We begin the discussion of these viruses with yellow fever virus, the prototype flavivirus and a virus whose history was important in the development of the science of virology and of vaccinology.

Yellow fever virus (YFV) was once greatly feared and is still capable of causing large epidemics. The virus is viscerotropic in primates, the only natural hosts for it. The growth of the virus in the liver, a major target organ, causes the major symptoms of disease and the symptoms from which the name of the virus derives, jaundice following destruction of liver cells. The virus also replicates in other organs, such as kidney and heart, and causes hemorrhaging. Illness is accompanied by high fever. Death occurs in 20–50% of serious infections, usually on days 7–10 of illness and usually as a result of extensive liver necrosis.

YFV is present today in Africa and Latin America. It originated in Africa and spread to the Americas with European colonization and the introduction of slaves. The virus is maintained in two different cycles. In an endemic or sylvan cycle, it is maintained in *Aedes africanus* and other *Aedes* mosquitoes in Africa and in *Haemogogus* mosquitoes in the Americas. Monkeys form the vertebrate reservoir. In this cycle, forest workers and other humans who enter deep forests are at risk. Infection of humans can lead to the establishment of an epidemic or urban cycle in which the virus is transmitted by the mosquito *Aedes aegypti* and humans are the vertebrate reservoir. In this cycle, all urban dwellers are at risk. *Aedes aegypti* is a commensal of man, breeding around human habitation. It is widespread in the warmer regions of the world, including the southern United States, Central America and the Caribbean, large regions of South America, sub-Saharan Africa, the Indian subcontinent, Southeast Asia, Indonesia, and northern Australia.

History of Yellow Fever

In the 1800s, YFV was continuously epidemic in the Caribbean region, where it had a pronounced influence on the development and settlement of the Americas by the Europeans. Caucasians and Native Americans are very sensitive to yellow fever, usually suffering a serious illness with a high death rate. Black Africans, who were brought as slaves to the New World to replace Native American slaves who had died in large numbers from European diseases, in general suffer less severe disease following yellow fever infection, presumably having been selected for partial resistance by millennia of coexistence with the virus. Their relative resistance to yellow fever resulted in the importation of even more black slaves into yellow fever zones. The high death rate among French soldiers sent to the Caribbean region to control black slaves was probably responsible for the decision by Napoleon to abandon the Louisiana territory by selling it to the United States, by which the United States underwent a huge territorial expansion. The high death rate among French engineers and workers in the 1880s under de Lesseps, who had previously supervised the construction of the Suez Canal, led to the abandonment of the attempt by the French to build a canal through Panama. The Panama Canal through Panama was built by the United States only after yellow fever was controlled.

From its focus in the Caribbean, yellow fever regularly spread to port cities in the southern and southeastern United States and as far north as Philadelphia, New York, and Boston. Epidemic yellow fever even reached London. The virus also spread up the Mississippi River from New Orleans. The virus was transported from its focus in the Caribbean by ships, which carried freshwater in which mosquitoes could breed. If there was yellow fever on the ship, the disease was maintained and could be transmitted by the mosquitoes or by infected individuals to ports at which the ships called. Yellow fever epidemics could afflict most of the population of a city and result in death rates of 20% or more of the city's original population.

One telling account of an epidemic in Norfolk, Virginia, in 1855 is described in the report of a committee of physicians established to examine the causes of this epidemic. Quarantine procedures to prevent the introduction of yellow fever were often thwarted by captains who concealed the presence of the disease to avoid a lengthy quarantine, even going to the extreme of secretly burying crew members who died while in quarantine. On June 6, 1855, the steamer *Ben Franklin* arrived from St. Thomas and anchored at the quarantine ground. There had been three cases of yellow fever on the ship during the voyage, of whom two died and were buried, one on land and one at sea. There was yet another case on board during quarantine who died and was buried ashore. Yet when the health officer, Dr. Gordon, visited the ship, he was told by the captain that there was no disease on the ship. The captain did admit that there had been two deaths during the voyage but ascribed them to other causes. After 13 days in quarantine and continued inspection by Dr. Gordon, who finding nothing amiss believed the captain's report that

there was no disease aboard, the ship was allowed to dock. Yellow fever soon appeared in Norfolk. A number of early cases among the citizens of the town were ascribed to the ship passing within a half mile of their homes, and it is possible that infected mosquitoes were blown ashore, although it is also possible that workmen visiting the ship while laid up for repairs may have brought the disease into the town. The disease then spread in all directions at a uniform rate of about 40 yards per day until it encompassed the whole city. The epidemic peaked at the end of August and died out after a frost in October. During the epidemic, an estimated 10,000 cases of yellow fever occurred in a population of 16,000, and 2000 died of the disease. The report established two other facts about the disease: Persons who had had yellow fever previously were immune, and the epidemic was not spread by person-to-person contact.

The Walter Reed Investigation

At the turn of the twentieth century, there was much debate as to the mechanism by which yellow fever spread. The Department of the Army sent an expedition, under the command of Walter Reed, to Cuba, recently acquired by the United States from Spain, to study the disease. The commission undertook to test the thesis that the virus was transmitted by mosquitoes, using themselves as human volunteers. Mosquitoes were allowed to feed on yellow fever patients and then on members of the commission. At first there was a lack of understanding about the fact that mosquitoes are infected only by feeding on patients early in their disease, before an effective immune response arises, and about the necessity for an extrinsic incubation period in the mosquito, during which the virus establishes an infection in the salivary glands, before it can transmit the virus. Ultimately, however, the investigation team did succeed in infecting themselves by mosquito transmission and one member of the commission, Dr. Jesse Lazear, died of it. Fortunately, his was the only death recorded in these experiments. It is of note that in the days before the introduction of a vaccine, most researchers who studied yellow fever in the field or in the laboratory ultimately contracted the disease and many of them died.

With the discovery that the virus was mosquito borne, the U.S. Army began a campaign in Havana to eliminate mosquito breeding places by eliminating sources of water around human habitation. It was (and still is) common for drinking water to be stored around houses throughout Latin America in large pots that served as excellent breeding places for *Ae. aegypti*. The campaign, which included smashing such water containers, succeeded in breaking the mosquito transmission cycle and yellow fever as an epidemic agent disappeared from Havana within months. This approach was later exported to other areas with great success, including Panama. These successes led to the belief that yellow fever could be eradicated, but the discovery of the endemic cycle of yellow fever dispelled this idea. Forest workers who cut down trees and brought the mosquitoes down from the upper canopy, where they transmit the disease to monkeys, were particularly at risk. Once infected, a person is able to bring the disease back to town where it can get into the *Ae. aegypti* population and start an urban epidemic.

The Yellow Fever Vaccine

In the late 1920s, yellow fever virus was successfully propagated in rhesus monkeys, in which it causes a lethal disease and in which it can be experimentally passed from monkey to monkey. One such strain was derived from an infected human named Asibi. Theiler and Smith passed the Asibi strain of yellow fever in chicken cells, and after approximately 100 passages, it was found that the resulting virus was no longer virulent for rhesus monkeys. After additional passages, this virus, called 17D, was ultimately used as a live virus vaccine in humans and has proved to be one of the best and most efficacious vaccines ever developed. The vaccine virus has been given to about 350 million people. It causes very few side reactions, although three recent vacinees developed full-blown yellow fever and died. The vaccine is essentially 100% effective in providing long-lasting protection against yellow fever. This vaccine is routinely given to travelers to regions where yellow fever is endemic and is used to control the spread of epidemic yellow fever in Latin America and, with less success, in Africa. The success of this vaccine has served as a model for the development of other live virus vaccines, namely, passing the virus in cultured cells from a nonnative host. Recent sequencing studies have found that the 17D vaccine differs from the parental Asibi strain at 48 nucleotides that result in 22 amino acid substitutions. The substitutions responsible for the attenuation of the virus are not known, but it is suggestive that 8 of the amino acid substitutions are found in the E protein, where they might alter host range.

Yellow Fever Today

Although not as wide ranging as previously, yellow fever continues to cause epidemics in Africa and South America as illustrated in Fig. 3.32. On an annual basis, 50–300 cases are officially reported in South America and up to 5000 cases in Africa, but these figures are significantly underreported and the World Health Organization estimates that there are 200,000 cases of yellow fever each year with 30,000 deaths. Between 1986 and 1991, annual outbreaks of yellow fever occurred in Nigeria that probably resulted in hundreds of thousands of cases. An intense campaign beginning in 1992 to vaccinate the population of Nigeria has resulted in the virtual disappearance of yellow fever in Nigeria, but epidemics

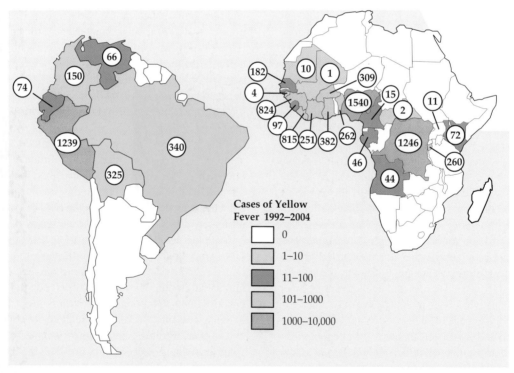

FIGURE 3.32 Cumulative number of cases of yellow fever reported to the World Health Organization for the years 1992 through 2004, by country. It is suspected that cases in Africa may be underreported by a factor of 10 or more. Immunization coverage in Africa has remained low and the disease has continued to spread. Major epidemics (>250 cases) have occurred in Liberia, Burundi, and Peru in 1995, Guinea in 2000, Burkina Faso in 2002, and the Democratic Republic of Congo in 2004. Note that while Nigeria had 19,891 cases between 1980 and 1991, since 1994 there have been only 12 cases. Data from: http://www.who.int/immunization_monitoring/data/data_subject/en/index.html.

continue to occur in other African countries. Epidemics of yellow fever also continue to occur in Peru, Bolivia, Brazil, Ecuador, Columbia, and Venezuela, perhaps in part due to the reemergence of *Ae. aegypti* in South America as described in more detail later. There was one imported case of yellow fever in the United States in 1996, in which an American who visited the jungles of Brazil along the Amazon River without being immunized returned to the United States with yellow fever and died of the disease. Because of the endemic cycle in which monkeys are the reservoir, it is probably impossible to eradicate the virus as has been done with smallpox and as is planned for poliovirus and measles virus.

Dengue Viruses

The four dengue viruses, now considered by the ICTV to be serotypes of a single viral species, have recently undergone a dramatic expansion in range. The incidence of dengue fever is estimated to have increased 30-fold over the last 40 years and dengue viruses now infect an estimated 50–100 million humans each year. Infection may be subclinical or may result in dengue fever, which is usually uncomplicated but which can progress to dengue hemorrhagic fever (DHF) or dengue shock syndrome (DSS). Uncomplicated dengue fever is characterized by headache, fever, rash, myalgia (muscle pain, from *myo* = muscle and *algia* = pain), bone pain, and prostration. The disease may be mild or it may be extremely painful (an old name for the disease is break-bone fever which dramatically describes the joint pain that can occur), but it is almost never fatal. However, progression to DHF or DSS is associated with a significant mortality rate. In the absence of medical care, mortality can be as high as 20%, but with good medical care the mortality rate is a few percent. Up to 250,000 cases of DHF and DSS are recorded each year, most of them in children, and in Southeast Asia DHF and DSS are a leading cause of mortality in children. DHF and DSS have also become important in Latin America. It is thought that DHF and DSS are caused by immune enhancement in which infection by one serotype of dengue virus expands the population of cells that can be infected by a second serotype. Many antibodies induced by the four dengue viruses are cross-reactive, reacting not only with the infecting virus but with the other dengue viruses as well. Immediately following infection, in fact, a person is immune to all four serotypes. With time this cross protection fades, and in less than a year the person remains immune only to the infecting virus, probably for life. After this time, infection with another serotype can occur and cause disease. Still present, however, are cross-reactive antibodies that can react with the newly infecting virus but which cannot neutralize the virus to provide protection. The nonneutralizing antibodies are thought to enable the virus to infect a larger number of lymphocytes by means of Fc receptors (see Chapter 10) than would otherwise result from infection only of cells expressing the dengue receptor. This expanded pool of infected cells results in increased cytokine production and can result in capillary leakage and shock. Although second infections are common, infection by a third serotype is rare. Evidently the boost to the immune system from the second infection results in an increase in the amount and avidity of cross-reactive antibodies.

Although infection by a second serotype is important for the development of DHF, it is also known that the probability of contracting DHF is in part a function of the virulence of the virus that is responsible for the second infection. Some dengue strains grow better than others and are more likely to cause DHF upon a second infection than other strains of the same virus. As one example, there was very little DHF in Sri Lanka before 1989 despite the continuing circulation of all four dengue serotypes. In 1989, however, a new strain of DEN-3 appeared in Sri Lanka that caused a large number of cases of DHF. A second example, DEN-2 in the Americas, is described in Chapter 8.

Because second infections by a different serotype are much more likely to lead to DHF than primary infections, the development of vaccines against dengue has progressed slowly. The possibility is real that immunizing against one serotype might put a person at risk for a more serious illness. Current efforts in Thailand are directed toward developing a quadrivalent attenuated virus vaccine that would immunize against all four serotypes simultaneously. U.S. scientists are independently attempting to develop vaccines for the viruses, based either on attenuated dengue viruses or on the development of chimeric flaviviruses that express dengue envelope antigens in a yellow fever vaccine background (see Chapter 11). These various vaccine candidates are in clinical trials as of this writing. A major problem has been the tendency of vaccinated humans to respond strongly to one of the four serotypes in live virus vaccines, often to DEN-3, while responding only weakly or not at all to other serotypes. Changing the ratios of the four viruses in the mix and use of multiple inoculations are being tested as possible ways to overcome this problem.

Dengue viruses are maintained in *Ae. aegypti* in urban settings in most of the world, but also in *Ae. albopictus* in Asia, and humans are the vertebrate reservoir. Part of the difficulty in developing vaccines is that there is no animal model for the disease. The virus will infect monkeys but does not cause disease in them. Small animal models of infection exist but the infection process is artificial and the resulting disease is not dengue fever (DF) or DHF. DF and DHF are exclusively human diseases and the dengue viruses that infect humans are exclusively human viruses.

The Spread of Dengue Viruses

The four serotypes of the dengue viruses arose in Old World monkeys and jumped to humans an estimated 200–1000 years ago. As noted before, the human viruses are now strictly human viruses. Although they will infect monkeys under laboratory conditions (without causing disease), the reservoir in nature is exclusively humans. However, the monkey viruses still exist as monkey viruses in a sylvatic cycle in Asia and Africa. The human viruses have been continuously active over large areas of Asia and the Pacific region for many years. In areas of Thailand where the viruses are endemic, for example, most people are infected by multiple serotypes in childhood and DHF is a leading cause of mortality in children.

The viruses have recently dramatically expanded their range in the Americas. Before 1970 there was very little dengue activity in the Americas, probably because of mosquito control efforts that were abandoned about that time. Following this, dengue activity increased dramatically, especially upon the introduction of new strains of the virus from Asia. By the 1990s there occurred widespread epidemics that affected many millions of people every year. Epidemics have resulted in an estimated 100 million cases of dengue infection in Brazil alone. The introductions of Asian viruses included more virulent strains of dengue that together with the circulation of multiple serotypes led to epidemics of DHF. Dengue virus is now a major health problem in Latin America. Dengue has also become more active in the Pacific region. Recent epidemics in Hawaii, the first in 50 years, have resulted in more than a hundred documented cases of dengue fever. This topic of the origin and spread of dengue to the Americas is discussed in more detail in Chapter 8.

Japanese Encephalitis Virus

The Japanese encephalitis (JE) complex of flaviviruses includes a large number of related viruses, many of which cause encephalitis in humans. For these viruses, the majority of human infections are inapparent and, for most, fewer than 1% of infections result in neurological disease. However, when encephalitis develops it is often serious with case fatality rates as high as 50% and neurological sequelae are frequent among survivors. In addition to JE, these include St. Louis encephalitis (SLE), Murray Valley encephalitis (MVE), and West Nile viruses. The close relationships of these viruses are illustrated in Fig. 3.28. Some of these viruses are widespread whereas others are much more local in their distribution. West Nile virus, for example, is now virtually worldwide (the Australian strain is often called Kunjin virus), whereas MVE virus is only found in Australia. Thus, circulation of at least some of these viruses has occurred over widespread areas, as is the case for the dengue viruses described in the preceding section. For most of the viruses in this lineage, birds form the major vertebrate reservoir and culicine mosquitoes are the major vectors.

JE virus is distributed throughout Asia, including Japan, India, Southeast Asia, Indonesia, the Philippines, and Borneo (Fig. 3.33). Reported cases of JE encephalitis average 30,000–50,000 per year with 10,000 deaths, but the disease is greatly underreported. Only one JE virus infection in 200 or 300 results in encephalitis, with children and the elderly being at higher risk. The fatality rate following JE encephalitis is 2–40% in different outbreaks, but 45–70% of survivors have neurological sequelae. In endemic areas, virtually all people have been infected by the time they reach adulthood. Bird–mosquito–bird transmission is the normal transmission cycle, but domestic pigs are particularly important amplifying hosts for transmission to humans because they are found in proximity to their human owners. Various species of *Culex* mosquitoes transmit the virus. During peak transmission seasons, up to 1% of *Culex* mosquitoes around human habitations may be virus infected. Travelers to endemic regions have a probability of about 10^{-4}/week of contracting JE, and 24 cases of JE encephalitis in travelers were reported between 1978 and 1992. Inactivated virus vaccines are in use in different regions of Asia. The Japanese have long used such a vaccine to eliminate JE encephalitis from their population, and the Chinese have recently developed a vaccine that is being used in China and Thailand. The Japanese vaccine is available in the United States for travelers to endemic regions. Of considerable interest is the finding that JE virus infection may reactivate in mice after the immune system first damps it out. Reactivation in other animals may also occur and could be important for persistence of the virus in nature.

For the dengue viruses, immune enhancement is important for the disease caused in humans. There is no evidence that immune enhancement plays a role in the disease caused by JE virus or other flaviviruses such as MVE virus. It is interesting, however, that in a mouse model system, prior treatment with subneutralizing concentrations of anti-JE serum resulted in an increase in virus growth and in mortality in the mice following infection by MVE virus. This suggests that the potential for immune enhancement exists for other flaviviruses but does not occur in humans other than the dengue viruses because the immune reaction to flaviviruses is normally strong and not cross-reactive and subneutralizing concentrations of antiviral antibodies do not exist. It does raise a warning flag for vaccines, however, and vaccines that increase the seriousness of disease caused by subsequent infection have occurred in the case of measles virus and respiratory syncytial virus (see Chapters 4 and 10).

As indicated before, most human infections by JE virus do not result in invasion of the nervous system and encephalitis does not occur. Two laboratory experiments are of interest in this regard. In one experiment, JE virus variants were selected that failed to bind to mouse brain membrane receptor preparations. These mutants were attenuated for neuroinvasiveness and neurovirulence because their receptor-binding preferences were altered. In a second experiment, passage of JE virus and of MVE virus in cultured cells selected for variants that bound to glycosaminoglycans (GAG). These

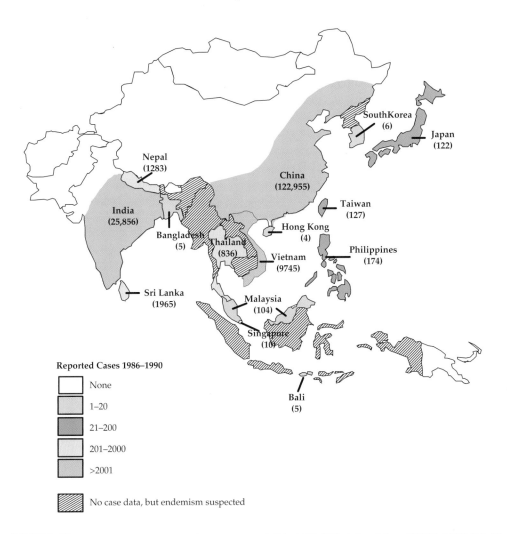

FIGURE 3.33 Range and reported cases of Japanese encephalitis, 1986–1990. Adapted from *MMWR* (1993) Vol. 42, RR-1, p. 2. Since this detailed report, the first human cases were reported in Papua New Guinea in 1997, there were two fatal cases on islands in the Torres Strait in 1995, and the virus was detected in mainland Australia (the Cape York Peninsula) in 1998.

variants were rapidly removed from the bloodstream when inoculated into mice and were attenuated. It was suggested that GAGs present on cells and extracellular matrices result in the removal of these variants from blood and tissues before replication and neural invasion can take place. Evidence was presented that the attenuation of the live JE virus used as a vaccine in China, called SA14-14-2, may have resulted, at least in part, from such an effect.

West Nile Virus

West Nile (WN) virus was first isolated in 1937 from the blood of an infected woman in the West Nile province of Uganda. Until 1999 it was not considered an important pathogen, causing only sporadic cases of encephalitis in parts of Africa, Asia, and Europe and having little effect on wildlife.

But with the recent occurrence of outbreaks of encephalitis in Europe, the Middle East, and North America the situation changed. Not only has infection by the virus resulted in large numbers of cases of neurological disease in humans, but domestic animals, especially horses, and wildlife, particularly birds, have been severely impacted. The effect of the virus in North America has been especially dramatic, as described in more detail in Chapter 8. Since its arrival in 1999, more than 20,000 Americans have become ill from West Nile infection and more than 800 have died. The virus has also had severe effects upon horses and many species of birds.

Two lineages of WN virus are recognized. The lineage present in North America, Europe, the Middle East, India, Australia (where the subtype present has been called Kunjin virus), and parts of Africa, called lineage 1, contains both virulent and attenuated strains and is responsible for WN disease. Lineage 2 is present only in sub-Saharan Africa and Madagascar and is mostly maintained in an enzootic cycle. The association of the virus with significant outbreaks of disease in humans, domestic animals, and birds, and the widespread dispersion of the virus in Eurasia and the Americas, appears to be the result of the emergence of a more virulent strain of the virus.

Various species of culicine mosquitoes are the principal vectors of WN virus, although the virus has also been isolated from species of *Aedes*, *Coquillettidia*, *Culiseta*, and *Ochlerotatus*, among others. In Europe and Africa the principal vectors are *Culex pipiens*, *Cx. univittatus*, and *Cx antennatus*, in India *Cx. vishnui*, in Australia *Cx. annulirostris*, and in North America *Cx. pipiens*, *Cx. quinquefasciatus*, and *Cx. tarsalis*, among others. During epidemics, from 0.1% to as many as 15% of *Culex* mosquitoes were found to be infected.

WN virus infects a wide spectrum of vertebrates. The reservoir of the virus is various species of birds. More than 150 species of birds were shown to be infected by WN virus and only in birds, with few exceptions, does the virus produce high enough viremia titers to infect mosquitoes. Laboratory studies showed that viremia titers of 10^5 to 10^7 are required to infect mosquitoes; at the lower titer fewer than 15% of feeding mosquitoes become infected, whereas at the higher titer more than 70% become infected. Mammals in general do not generate such high viremia titers after infection by WN virus. The maximum viremic titer in humans and horses, for example, appears to be about 10^3. Among birds, grackles, corvids (crows, ravens, jays, magpies), house finches, house sparrows, shorebirds, hawks, and owls are most susceptible to the virus and show high mortality rates upon infection (25–100% in various studies). They develop sufficiently high viremia to efficiently infect mosquitoes that feed upon them. WN virus has also been isolated from amphibians and reptiles, and the lake frog of Russia develops sufficiently high viremia that it might serve as a reservoir.

There are reports that WN virus can persist in infected animals for a considerable time. Virus could be isolated from experimentally infected ducks and pigeons for more than 3 months, for example. Such persistence could be important in the persistence of the virus in nature. Also important is transovarial transmission, which occurs in many flaviviruses, and West Nile is no exception. During transovarial transmission, the eggs laid by the mosquito are infected and the emergent mosquito is thus infected. This mechanism is especially important in temperate climates where adult mosquitoes die off during the winter and the species persists as diapausing embryos or larvae.

WN virus can also be spread by means other than by mosquitoes, although the importance of such spread in the maintenance and spread of the virus is unknown. Birds excrete virus in their feces, which can serve as a source of infection of contacts. Alligators were accidentally infected by feeding them infected birds. In humans, transmission of the virus by blood transfusion or via breast milk has occurred.

Pathology of West Nile Disease

About 80% of human infections appear to be asymptomatic. In the 20% of infections that result in clinical disease, most result in a self-limited illness characterized by fever, headache, fatigue, malaise, muscle pain, and weakness. In fewer than 1% of infected humans does the virus cross the blood–brain barrier and cause neurological disease such as meningitis, encephalitis, or paralysis. In about 13% of patients experiencing neurological disease, infection of anterior horn cells of spinal motor neurons causes an acute flaccid paralysis very similar to poliomyelitis resulting from poliovirus infection. The fatality rate for neuroinvasive disease is about 10%, and many survivors, especially those with poliomyelitis type disease, never fully recover.

Infection with WN virus may be serious because the virus interferes with the innate immune system (see Chapter 10). Several of the nonstructural proteins of the virus interfere with phosphorylation of the Janus kinases JAK1 and Tyk2. This prevents the activation of the transcription factors STAT1 and STAT2 and their transport to the nucleus. Activation of these factors is required for the cell to respond to the signaling of the interferons, normally a first line of defense against viral infection.

Vaccines to protect horses have been developed. One is an inactivated virus vaccine. The other is a recombinant vaccine that uses canarypox virus to express WN virus antigens. Two viruses for human use are also being developed. The first is an inactivated virus vaccine. The second is a chimeric YF–WN live virus vaccine in which the prM and E proteins of the 17D YF virus vaccine strain have been replaced with those of WN virus (see Chapter 11).

Other Flaviviruses of the JE Complex

MVE virus is an Australian virus that is closely related to JE virus. It causes encephalitis in humans, but the number of cases is small. Birds are the primary vertebrate reservoir, and epidemics of MVE have been associated with wet years when the mosquito population expands and nomadic waterfowl invade regions that are normally too dry to support them. *Culex annulirostris* is the primary vector for MVE.

SLE virus is a North American virus belonging to the JE complex that causes regular epidemics of encephalitis in the United States. The virus is widely distributed and cases of SLE encephalitis have been recorded in every state, with the majority of cases occurring in the Mississippi River valley, Texas, California, and Florida. Data for the years 1964–2003 are shown in Fig. 3.34. In the epidemic year 1975 there were 1815 cases of SLE encephalitis officially reported in the United States, but in nonepidemic years there may be fewer

FIGURE 3.34 Distribution of cases of St. Louis encephalitis occurring between 1964 and 2003 in the United States, shown by state. The large number of cases in Florida includes the most recent U.S. epidemic, which occurred in 1990, during which Florida reported 223 cases and 11 deaths. Data came from Fields *et al.* (1996) p. 981, and *MMWR, Summary of Notifiable Diseases, 1996, 1997, 1998, 1999, 2000, 2001, 2002, 2003*. St. Louis encephalitis was not a notifiable disease nationwide until 1998. Recent reported cases were 24 in 1998, 4 in Florida in 1999, 2 in Texas in 2000, 79 in 2001 of which 71 were in Louisiana, 28 in 2002, and 41 in 2003. These have been incorporated into the state totals shown.

that 50 cases. The most recent epidemic occurred in 1990 in Florida with 223 cases and 11 deaths. The case fatality rate is about 7% overall, but is higher in the elderly. Most infections by SLE are inapparent, as is the case for many encephalitis viruses. The ratio of inapparent to clinical infection is age dependent and varies from 800 to 1 in children to 85 to 1 in the elderly. The virus is transmitted by *Culex* mosquitoes, and the primary vertebrate reservoirs are wild birds.

Tick-Borne Encephalitis Viruses

The tick-borne encephalitis (TBE) viruses are important pathogens in Europe and Asia, and there is also a representative in North America. The viruses include Central European encephalitis (CEE), louping ill, Russian spring-summer encephalitis (RSSE), Kyasanur Forest disease, Omsk hemorrhagic fever, and Powassan viruses. Members of the TBE complex form a distinct group within the flaviviruses (Fig. 3.28), but share 40% amino acid sequence identity with the mosquito-borne flaviviruses, showing their close relationship to other flaviviruses. Most TBE viruses are transmitted by *Ixodes* ticks and can cause a fatal encephalitis in humans. An inactivated virus vaccine is widely used in Central Europe to protect people exposed to ticks. Even so, several thousand cases of TBE encephalitis occur each year. The case fatality rate is 1–2%, with 10–20% of survivors having sequelae in the RSSE form. RSSE, and perhaps other TBE viruses, can also be contracted by drinking raw goat's milk and possibly other forms of raw milk. The virus has a tendency to set up persistent infection in experimental animals and possibly in humans as well. Although *Ixodes* ticks are the primary vector, *Dermacentor* ticks and ticks of other genera are also capable of transmitting the virus. The distributions of two species of *Ixodes* ticks that are important vectors of TBE are shown in Fig. 3.35 together with the geographic range of endemic TBE disease.

Powassan virus is a member of the complex found in North America and in Russia. In North America, 20 cases of Powassan encephalitis have been reported since 1958.

All known TBE viruses cause encephalitis in humans with the exception of Omsk hemorrhagic fever virus, which causes hemorrhagic fever in humans, as its name implies, in the absence of encephalitis. Two other members of this complex, Kyasanur Forest disease virus and Alkhurma virus, which are closely related and may represent isolates of the same virus, also cause hemorrhagic fever in humans but it is associated with encephalitis. Omsk hemorrhagic fever virus also differs from other TBE viruses in that its principal tick vector is *Dermacentor reticulates* rather than an *Ixodes* tick.

Cell Fusing Agent

A flavivirus called cell fusing agent was discovered in laboratory cultures of *Ae. aegypti* cells in 1975. The relationship of this virus to other flaviviruses is shown in Fig. 3.28.

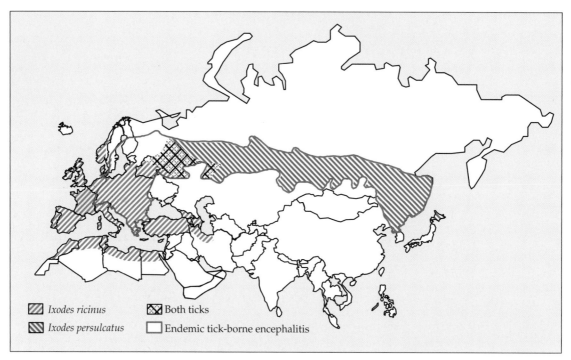

FIGURE 3.35   Geographic distribution of two major tick vectors of tick-borne encephalitis. Also shown is the major region in which TBE is endemic. Adapted from Porterfield (1995) p. 207.

This virus is an insect only virus and is not known to infect vertebrates. Very recently, new isolates of a strain of cell fusing agent have been made from wild-caught mosquitoes in Puerto Rico belonging to at least two genera, *Aedes* and *Culex*. Remarkably, DNA sequences related to cell fusing agent have been identified in the genomes of wild-caught mosquitoes, presumably having integrated into the mosquito genome at some time in the distant past. This and other recent isolates of new flaviviruses has led to the suggestion that there are as yet many flaviviruses in nature that remain to be identified.

Genus *Hepacivirus*

Hepatitis C virus (HCV) forms a second genus in the *Flaviviridae*. The virus was discovered in 1989 as a causative agent of nonA-nonB hepatitis in humans. Despite the inability to grow the virus in culture or in a small animal model, the complete genome sequence of the virus was established using the methods of modern biotechnology and verified by injection of viral RNA produced from cDNA clones into the liver of a chimpanzee, the only animal other than humans that is infectible by the virus. The HCV genome, which is slightly smaller than those of the flaviviruses and pestiviruses, has an organization similar to those of the other members of the family (Fig. 3.27). It has a number of important differences from the genome of members of the genus *Flavivirus*, some of which are illustrated in the figure. One

is the presence of two proteases rather than one. The NS3 protease is shared with the flaviviruses but requires NS4A as a cofactor rather than NS2B. The second protease, which has a catalytic cysteine, is present in NS2 and its only known cleavage function in viral replication is to cleave the NS2–NS3 bond.

A second difference is the lack of a 5′ cap and the possession instead of an IRES, so that initiation of translation of the plus-strand genome is not cap-dependent but uses an IRES as does poliovirus. A third difference is the production of a small (17 kDa), short-lived protein called F (for frame shift) or ARFP (for *a*lternative *r*eading *f*rame *p*rotein) that is encoded within the C protein gene in a different reading frame. Translation of this protein requires initiation at the 5′ end of the polyprotein followed by a frameshift near residue 11 of the capsid protein. There is evidence that it is produced in infected humans but it is not known if this protein plays a role in virus replication. Of note is the fact that no other plus-strand RNA virus is known to produce two different proteins from two different reading frames in the same nucleotide sequence, although this phenomenon occurs in several other classes of viruses.

HCV also differs in the way that RNA replication is anchored to a membrane. RNA replication in plus-strand RNA viruses occurs in association with membranes. In flaviviruses, the RNA polymerase is thought to associate with membranes by means of its association with membrane bound proteins such as NS4A or NS4B. In HCV, the RNA polymerase NS5B is itself anchored in the membrane by a C-terminal transmembrane anchor. Interestingly, this anchor is required for RNA replication and the HCV sequence cannot be substituted with that from the pestivirus bovine viral diarrhea virus. Thus, this anchor plays a role in RNA replication other than simply anchoring the polymerase in the membrane.

Another interesting difference is the cleavage of the N-terminal capsid protein from the polyprotein precursor. The capsid protein is anchored in the membrane by a C-terminal transmembrane anchor, as described earlier for flaviviruses. In flaviviruses the capsid protein is cleaved from this signal sequence anchor by the NS3 protease, but in HCV it is cleaved by a cellular protein, signal peptide peptidase.

Because of its importance as a human disease agent, HCV has been the subject of intensive study. Progress has been relatively slow because the only animal model for the disease is the chimpanzee, which are rare and expensive, limiting the number of experiments that can be performed, and because until very recently there was no cell culture system in which the virus would undergo a complete replication cycle and release infectious virus. Studies in cultured cells have relied upon the expression of parts of the genome in expression vectors, and more recently upon the replication of truncated versions of the genome called replicons. Replicons encode all of the genes required for RNA replication but lack the genes encoding structural proteins. Thus, only part of the virus life cycle can be studied using these reagents. The very recent development of systems using cultured cells that allow a complete virus replication cycle with the release of infectious virus will allow more rapid progress in the future. A second advance has been the use of immunodeficient mice (*s*evere *c*ombined *i*mmuno*d*eficiency or SCID mice) into which have been grafted human liver cells. These can be infected by HCV and although the numbers of such animals is limiting, they possess obvious advantages over the use of chimps.

Natural History of HCV

HCV is a causative agent of blood-borne hepatitis in man. In the United States, HCV was once spread primarily through transfusion of contaminated blood, but the development of a diagnostic screen for the virus has virtually eliminated this source of infection in the developed world. However, the virus continues to be transmitted through the sharing of needles by drug users. The virus can also be transmitted sexually or from mother to child but these mechanisms are inefficient. There are additional mechanisms of transmission that are not well understood. In some developing countries, circumcision or scarification practices may be important in the spread of the virus.

HCV is worldwide in distribution, as illustrated in Fig. 3.36. It has been estimated that ~3% of the world's population, 170 million people, are infected by the virus. The highest infection rate found was among Egyptian blood donors, where up to 19% were seropositive for HCV, which may have resulted in part from past treatment for bilharziasis

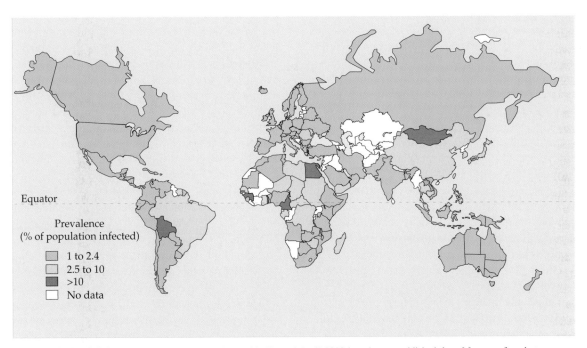

FIGURE 3.36 Worldwide prevalence of hepatitis C as of April 2003 based upon published data. Map was found at: http://www.reliefweb.int/rw/RWB.NSF/db900LargeMaps/SKAR-64GDV4?OpenDocument.

using inadequately sterilized needles. There are six different clades or genotypes of the virus, which differ by more than 30% in nucleotide sequence and are numbered from 1 to 6. In turn, each clade has many isolates that may differ by up to 25% in nucleotide sequence, so that the clades can be subdivided into subclades called a, b, etc. These different viruses all cause the same disease but differ in the severity of the disease caused and in their ease of cure. Genotype 1 is the clade commonly found in the United States and probably became widespread only with the introduction of blood transfusion in the 1940s. In Africa and Asia, the virus has been endemic for a long time, and the different clades have different geographic distributions. Thus, for example, clade 5 is commonly found only in South Africa, clade 4 is widely distributed in the Middle East, clade 6 in eastern Asia, and clade 2b in the Mediterranean and the Far East.

HCV Disease and Its Treatment

Infection with HCV can be extremely serious. The initial infection may cause no disease or may result in hepatitis accompanied by jaundice, but fulminant liver failure is rare. However, in 70–80% of infections, the infection becomes chronic. During chronic infection, up to 10^{12} viruses are produced each day and turn over with a half-life of about 3 hours, and the more or less constant viral load in the blood is 10^3–10^7 per ml. This chronic infection is well tolerated by some and in a minority of cases spontaneous remission may occur in the absence of medical intervention. However, in many persons chronic hepatitis results. Most seriously, in about 20% of chronic infections liver cirrhosis develops after a long lag, usually more than 20 years, and hepatocellular carcinoma develops in up to 2.5%. Liver failure due to HCV infection is the leading cause of liver transplantation in the United States.

The current treatment for chronic HCV infection is injection of interferon-α conjugated to polyethylene glycol, which increases its stability, together with the inhibitor ribavirin. This treatment results in curing the infection in about half the cases but the cure rate depends upon the genotype of the virus. In one trial, 42% of patients chronically infected with genotype 1 HCV were cured whereas patients chronically infected with genotype 2 or 3 virus exhibited a cure rate of 80%. This treatment is not only expensive but relatively toxic and many patients tolerate it poorly. This consideration, as well as the fact that half the patients show no effect or only transient relief from this treatment, has led to intense efforts to develop new treatments. These include efforts to develop vaccines as well as efforts to develop antiviral agents that will interrupt virus replication or prevent the virus from interfering with the host defenses against the virus. Antivirals currently in clinical trials include nucleoside analogues that when incorporated into viral RNA result in chain termination, two compounds that bind to the viral

RNA polymerase NS5B and inhibit its activity, two inhibitors of the viral NS3–4A protease, and three compounds that modulate the immune system. Other drugs are also being studied as possible antivirals.

HCV Suppression of the Immune Response

In order to establish a chronic infection, HCV interferes with many aspects of both the innate and adaptive immune responses of the host. The importance of such interference for chronicity and the persistence of the virus in nature is illustrated by the fact that the virus interferes in so many different ways. The immune system is described in some detail in Chapter 10. Here we note that the first line of defense against viral infection is the production of type 1 interferons (IFN) α and β, components of the innate immune system. The NS3–4A protease of HCV interferes with the induction of IFN by cleaving two intermediates, called MAV5 and TRIF, in two different but overlapping activation pathways. MAV5 is required in the pathway that starts from an intracellular sensor of double-strand RNA called RIG-1, whereas TRIF is required in the pathway that starts from a membrane bound sensor of double-strand RNA called Toll-like receptor 3 (see Chapter 10). The result is that both pathways are disabled.

The HCV core protein interferes with the activity of any IFN that might be produced. It induces the expression of cellular proteins called SOCS1 and SOCS3. These block the JAK–STAT pathway by which IFN induces the production of hundreds of proteins required for defense against viral infection (Chapter 10). Protein NS5A independently interferes with the IFN system in at least two ways. It induces the production of IL-8, which attenuates the expression of genes induced by the activity of IFN. It also binds to a protein called PKR that is induced by IFN, thereby inhibiting its activity. Protein E2 also inhibits PKR. Other HCV proteins are also known to interfere with the activity of IFN.

HCV also interferes with the adaptive immune system. Interestingly, instead of a general interference with the adaptive system, as happens with HIV, for example, that cripples immune responses against all pathogens, the modulation by HCV is limited to HCV-specific responses, leaving the immune system free to control other viral infections. The mechanisms by which this occurs are incompletely understood. What is known is that successful clearance of HCV infection in humans is associated with a strong T-cell response, both CD8+ and CD4+, and that immunologic memory results such that although reinfection by HCV can occur, it does not lead to chronic infection. In humans in which the infection becomes chronic, CD8+ cytotoxic T cells are relatively few and these T cells recognize fewer epitopes. In one study, CD4+ helper T cells from persistent infections recognized very few epitopes whereas those from humans who cleared the infection recognized up to 14 different epitopes.

HCV-Related Viruses

Viruses related to HCV, called GB viruses (from the initials of a surgeon with hepatitis from which they were first isolated), are known. GBV-A and GBV-B viruses have a genome organization very similar to that of HCV, but share little amino acid sequence identity with HCV or with each other. They may eventually be classified as new genera within the *Flaviviridae*, more closely related to genus *Hepacivirus* than to genus *Flavivirus* or genus *Pestivirus*. A third virus, GBV-C, also called hepatitis G virus or HGV, is related to GBV-A. These three viruses appear to be widely distributed and establish chronic infections in humans, but there is no evidence that they cause disease.

Genus *Pestivirus*

Three closely related viruses belonging to the genus *Pestivirus* are important pathogens of domestic animals and have been well characterized. These are bovine viral diarrhea virus (BVDV), classical swine fever virus (CSFV) (also called hog cholera virus), and border disease virus of sheep (BDV). These three viruses share more than 70% amino acid sequence identity and exhibit extensive serological cross-reactivity. Their genome organization is similar to those of other viruses in the family (Fig. 3.37). Pestiviruses have also been isolated from a number of other mammals including giraffe, deer, bison, bongo, alpaca, and reindeer. The taxonomic status of these isolates is still unclear. Some are classified as strains of one of the three viruses just listed but some, at least, may represent other species of pestivirus

BVDV exhibits an important and interesting disease syndrome in cattle. Animals infected as adults by the virus may exhibit no disease or may have symptoms that include diarrhea, but they recover uneventfully. However, when a pregnant cow is infected by the virus, infection of the fetus may cause the fetus to become immunologically tolerant to the virus, resulting in a chronic infection that lasts for the life of the animal. Such *in utero* infection may lead to developmental abnormalities or runting in the calf, and may render the calf sensitive to infection by other microorganisms, all of which have serious economic effects. A more interesting effect of the chronic infection, however, is the development in some animals of fatal mucosal disease at the age of 1–2

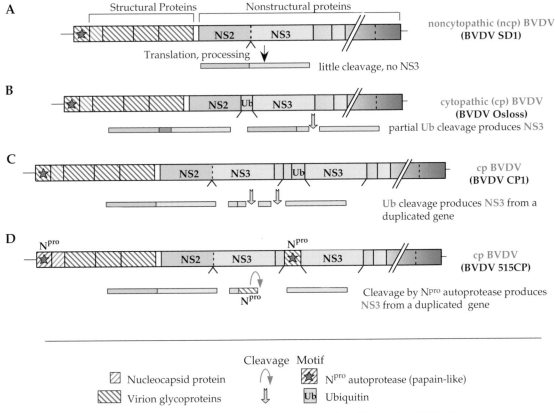

FIGURE 3.37   Genome organization of cytopathic and noncytopathic strains of the pestivirus BVDV. (A) In noncytopathic (wild-type) strains little cleavage occurs between NS2 and NS3. In cytopathic strains, NS3 is produced either by an upstream insertion of ubiquitin (see B), insertion of multiple ubiquitin sequences plus duplication of NS3 sequences (see C), or duplication of the N^pro proteinase and its insertion immediately upstream of a duplicated NS3 (see D). Data for this figure came from Meyers and Thiel (1996).

years; once symptoms appear, the animal dies within weeks. Animals that die of this disease are found to be infected by two types of BVDV. One is the normal wild-type virus, which is noncytopathic in cultured cells. The second type of BVDV is a new strain that is cytopathic in cultured cells. The cytopathic BVDV strain is derived from the wild-type strain by recombination, which occurs during the persistent infection. Several different cytopathic BVDV strains have been sequenced, and they all have in common that NS2–3 (formerly called p125) is cleaved to produce NS3 (also called p80). It is the production of NS3 that renders the virus cytolytic in culture and causes lethal mucosal disease in cattle. As illustrated in Fig. 3.37, the cleavage to produce NS3 can be induced in several different ways. In at least three cytopathic BVDV strains, cellular ubiquitin sequences were inserted (in different ways) within the sequence encoding this protein, such that a cellular enzyme that cleaves specifically after ubiquitin cleaves the BVDV polyprotein to produce NS3. Another mechanism to produce NS3 was the insertion of the BVDV N^{pro} autoprotease immediately upstream of the NS3 sequence. A third mechanism, not illustrated in the figure, is the insertion of cellular sequences derived from a protein called Jiv.

Why the production of NS3 renders the virus cytopathic and capable of causing lethal disease in cattle is a fascinating story of self-imposed limitation on virus growth not unlike the story of alphavirus downregulation described earlier in this chapter. For BVDV to persist in nature it must be able to establish persistent infection because persistently infected animals that continue to shed the virus are an important reservoir for the virus. Cytopathic viruses are not able to establish persistent infection and come to a dead end when they arise. It turns out that cleavage to form NS3 is essential for virus replication, and cleavage occurs early after infection by all BVDV strains, noncytopathic as well as cytopathic. Cleavage is effected by a protease in NS2, and this protease, like the NS3 protease of all members of the family, requires a cofactor for function. This cofactor is a cellular protein, however, not a virally encoded protein. This cellular cofactor is the protein Jiv, which forms a stable (but noncovalent) complex with NS2–3. The amounts of Jiv in the infected cell are limited; however, and it is soon titrated out. Once no free Jiv remains, NS2–3 cleavage cannot occur and no further increase in viral replication is possible, allowing the establishment of persistent infection with only limited amounts of virus being produced. If NS2–3 cleavage continues to occur because new protease sites have been introduced or because the virus encodes its own Jiv, virus replication continues to accelerate until it overwhelms the cell and the cell dies. As an aside, it is possible that the virus host range is controlled by the presence or absence in cells of sufficient Jiv able to act as a cofactor.

Thus, pestiviruses encode three proteases. The NS3 protease common to all *Flaviviridae* makes many cleavages in the polyprotein and functions both *in cis* and *in trans*. This protease requires a virally encoded cofactor, NS2B in flaviviruses and NS4A in pestiviruses and hepaciviruses. N^{pro} is an autoprotease whose only known cleavage in normal infection is to release itself from the polyprotein. The NS2 protease described here also functions as an autoprotease that makes only one cleavage, that between NS2 and NS3.

CSFV is epidemic in pig populations and causes serious illness, with different isolates differing in their virulence. Infection of pregnant sows can lead to abortion or to birth of persistently infected piglets, which soon die. BDV also can cause congenital infection, which can lead to abortion or to birth of animals that display a number of defects.

FAMILY *CORONAVIRIDAE*

The name *Coronaviridae* comes from the Latin word meaning crown, from the appearance of the array of spikes around the enveloped virion. The family is composed of a number of RNA-containing animal viruses currently classified into two genera, the genus *Coronavirus* (whose members will here be called coronaviruses) and the genus *Torovirus* (whose members will be referred to as toroviruses). A representative listing of viruses in the two genera is found in Table 3.13. The family is classified together with the *Arteriviridae* and the *Roniviridae* (described later) in the Order *Nidovirales*, after the Latin word *nido* meaning nest, because they produce a nested set of mRNAs. Coronaviruses are somewhat larger in size (120–160 nm) than the toroviruses (120–140 nm) and have a larger genome (about 30 kb compared to 20 kb). In contrast to other (+)RNA viruses, the nucleocapsids of *Coronaviridae* are constructed using helical symmetry. The coronaviruses have a helical nucleocapsid 10–20 nm in diameter, whereas the toroviruses have a tubular nucleocapsid that appears toroidal in shape in the virion. The coronavirus virion is roughly spherical, whereas the torovirus virion is disk shaped or rod shaped. The viruses mature by budding through intracytoplasmic membranes. The coronaviruses have been well studied, whereas the toroviruses, which are composed of one pathogen of horses, one pathogen of cattle, a presumptive human torovirus, and a possible torovirus of swine, have attracted less attention.

Genus *Coronavirus*

The coronaviruses have the largest RNA genome known, 27–32 kb in size. The genome size of RNA viruses is thought to be limited by the mutation rate during RNA synthesis. Because there is no proofreading during RNA synthesis, an inherent mistake frequency results that is in the order of 10^{-4}. Thus, error-free replication of an RNA genome becomes impossible once the genome becomes too large. The 30-kb genome of coronaviruses may represent this upper limit. It is

TABLE 3.13   *Coronaviridae*

Genus/members	Virus name abbreviation	Usual host(s)	Transmission	Disease	World distribution
Coronavirus					
Group 1					
Transmissable gasteroenteritis	TGEV	Swine	Contact	Gastroenteritis	United States, Europe
Human coronaviruses 229, NL63	HCoV	Humans	Aerosols	Common cold	Americas, Europe
Group 2A					
Human coronaviruses OC43, HKU-1	HCoV	Humans	Aerosols	Common cold	Americas, Europe
Murine hepatitis	MHV	Mice	Aerosols, contact	Gastroenteritis, hepatitis	Laboratory mouse colonies worldwide
Group 2B					
Severe acute repiratory syndrome	SARS	Bats[a], Humans	Aerosols, contact	Fever, pneumonia, severe respiratory disease	Asia, Americas
Group 3					
Infectious bronchitis	IBV	Birds	Mechanical, oral–fecal	Bronchitis	Worldwide
Torovirus					
Berne (equine torovirus)	EqTV	Horses	Oral–fecal	Diarrhea	Europe, Americas
Breda (bovine torovirus)	BoTV	Cattle	Oral–fecal	Diarrhea	?
Human torovirus	HuTV	Humans	?	Diarrhea	?

[a] Bats have been identified as the vertebrate reservoir, but disease is primarily in humans.

also possible that because the coronaviruses undergo high-frequency recombination, as described later, they may be able to accommodate these large genomes because recombination offers a possible mechanism for correcting defective genomes. Intriguingly, coronaviruses and other members of the *Nidovirales* encode a number of RNA-processing enzymes including a 3′-to-5′ exonuclease that could conceivably make proofreading possible during RNA replication. However, there is as yet no evidence that the mutation frequency during coronaviral RNA replication is less than that occurring during replication of other RNA viruses.

The coronaviruses are grouped into three clades called groups 1, 2, 3, and examples are given in Table 3.13. Assignments were first based on serological cross-reactivity but more recently on sequence relatedness. Group 1 viruses include porcine epidemic diarrhea virus, porcine transmissible gastroenteritis virus, canine coronavirus, feline infectious peritonitis virus, and two human viruses, human coronaviruses 229E and NL63. Group 2 viruses are subdivided into two clades. Group 2A contains murine hepatitis virus (MHV), bovine coronavirus, rat sialodacryoadenitis virus, porcine hemagglutinating encephalomyelitis virus, canine respiratory coronavirus, equine coronavirus, and one human virus, human coronavirus OC43. Group 2B contains severe acute respiratory syndrome coronavirus (SARS HCoV). Group 3 contains a number of avian viruses, avian

infectious bronchitis virus, turkey coronavirus, and recently described viruses of geese, pigeons, and ducks. Where known, the viruses in these different groups use different receptors to enter cells (see Table 1.3). A number of group 1 viruses use aminopeptidase N, also called CD13. Several group 2A viruses are known to use carcinoembryonic antigen-related adhesion molecules, which are members of the Ig superfamily. SARS virus uses angiotensin-converting enzyme 2.

Translation of the Viral Genome: The Nonstructural Proteins

The coronavirus genome is, as in the case of all plus-strand RNA viruses, a messenger, and the naked RNA is infectious. The organization of the 27.6-kb genome of avian infectious bronchitis virus (IBV) is shown in Fig. 3.38 as an example for the genus. The RNA, which is capped and polyadenylated, is translated into two polyproteins required for the replication of the viral RNA and the production of subgenomic mRNAs. The first polyprotein terminates at a stop codon 12.4 kb from the 5′ end of the RNA. Ribosomal frameshifting occurs frequently, however, and in the shifted frame, translation continues to the end of the RNA replicase-encoding region at 20.4 kb. The resulting polyproteins are cleaved by virus-encoded proteases, as illustrated in

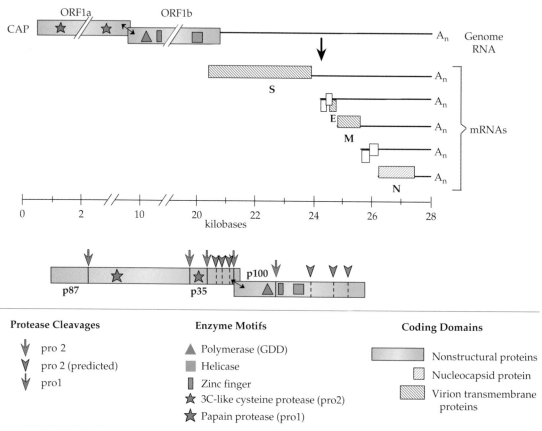

FIGURE 3.38   Upper panel: genome organization of the coronavirus, avian infectious bronchitis virus (IBV). ORF1a and ORF1b encode components of the viral replicase, and are translated as two polyproteins, with ribosomal frameshifting at the double-headed arrow. The remaining viral components are encoded in a nested set of mRNAs. The hatched proteins are polypeptides found in virions. White boxes are open reading frames of unknown function. E is a minor virion component, but essential for virus assembly. Lower panel: proteolytic processing of the IBV ORF1ab polyprotein. Motifs of papain-like proteases (pro1), 3C-like cysteine protease (pro2), RNA polymerase (GDD), zinc finger, and helicase are indicated with various symbols. Arrows at cleavage sites are color coded according to the protease responsible. Green arrowheads are predicted cleavage sites for pro2. Adapted from de Vries *et al.* (1997) with permission.

Fig. 3.38B. All coronaviruses possess at least two proteases, one papain-like and the other serine-like (but with cysteine at the active site as in poliovirus), and some encode a second papain-like protease so that they encode three proteases. Processing is complicated, as indicated in the figure.

Nidoviruses differ fundamentally from other RNA viruses in the number of nonstructural enzymes that they encode for the synthesis of the viral RNAs or for the purpose of enabling vigorous viral replication. The size of the RNA devoted to encoding these proteins in the coronaviruses is 20–30 kb, larger than the entire genome of other RNA viruses, and the number of cleaved products produced from the polyprotein precursors is large, on the order of 16. Perhaps the large size of the genome requires this. It is known that at least some of these proteins are devoted to countering host defenses against viral infection, which is surely important for the persistence of the viruses in nature.

The nonstructural proteins encoded in this domain of the genome include the two or three proteases described before,

an RNA polymerase, an RNA helicase, and enzymes involved in capping that must perform functions similar to the corresponding enzymes in other (+)RNA viruses. The papain-like protease, however, has another function in addition to processing some of the sites in the nonstructural polyprotein. It is a deubiquitinating enzyme (DUB) whose precise role in virus infection is unknown. Ubiquitin and ubiquitin-like proteins (UBLs) are small proteins that are covalently attached to other proteins by ubiquitinating enzymes, either as single molecules or as branched chains. The role of ubiquitination is only incompletely understood but plays an important role in many cellular processes. One role of ubiquitination is to target proteins for degradation by the proteosome, a well-studied phenomenon. Ubiquination is also involved in membrane protein trafficking, in the activation of the transcription factor NFκB, in DNA repair, and in autophagy, a response to starvation in which double membrane structures are assembled that might serve as viral replication sites. Thus a viral DUB might stabilize proteins that enhance viral replication, or might be important for the induction of NFκB, an important transcription factor during viral infection, or it might be important for constructing viral replication sites. There are also at least 10 UBLs derived from the same common ancestor as ubiquitin that are also conjugated to proteins to control cellular activities in ways that are but incompletely understood. One of these is the product of interferon-stimulated gene 15 (ISG15). This protein is induced by interferon and plays an unknown role in regulating the immune response to viral infection. It is not known if the viral DUB might also remove conjugated ISG15, but it is known that some other viruses target ISG conjugation. Influenza B virus produces a protein that binds to ISG, preventing the ISGlation of proteins. In addition, African swine fever virus, a large DNA virus, has a DUB that is thought to block the production of interferon by unknown mechanisms.

In addition to these gene products, nidoviruses encode distant homologues of at least five cellular enzymes associated with RNA processing. These are an endoribonuclease that cleaves after uridine residues, the 3'-to-5' exonuclease mentioned earlier, a methyltransferase that might be part of the capping complex, an adenosine diphosphate-ribose 1'-phosphatase, and cyclic phosphodiesterase. Most coronaviruses encode all five of these enzymes whereas roniviruses encode only three and arteriviruses only one. The functions of these enzymes in the virus life cycle are unknown.

Production of Subgenomic RNAs

The members of the *Nidovirales* produce a nested set of subgenomic mRNAs (Fig. 3.38), which are capped and polyadenylated. The number produced depends on the virus but is 5 to 8 for most. Each subgenomic RNA is a messenger that is translated into one to three proteins from the 5' ORF(s) in the mRNA. The five subgenomic mRNAs of IBV and the proteins translated from them are illustrated in Fig. 3.38A. Four of the subgenomic mRNAs are translated into the structural proteins in the virion, S, E, M, and N, found in that order in the genomes of all coronaviruses. Four small accessory proteins of unknown function are also produced, two from the E mRNA and two from RNA 5. Coronaviruses encode variable numbers of such accessory proteins which are not conserved as to sequence or to number among the various members of the family and whose function in unknown. It is also not known how multiple proteins are translated from a single mRNAs in the case of the coronaviruses.

Two mechanisms have been proposed for the production of these subgenomic RNAs. The first mechanism proposed was primer-directed synthesis from the (−)RNA template (i.e., from the antigenome produced from the genomic RNA). In this model, a primer of about 60 nucleotides is transcribed from the 3' end of the template, which is therefore identical to the 5' end of the genomic RNA. The primer is proposed to dissociate from the template and to be used by the viral RNA synthetase to reinitiate synthesis at any of the several subgenomic promoters in the (−)RNA template. Evidence for this model includes the fact that each subgenomic RNA has at its 5' end the same 60 nucleotides that are present at the 5' end of the genomic RNA, and that there is a short sequence element present at the beginning of each gene that could act as an acceptor for the primer (this sequence, e.g., is ACGAAC in the SARS CoV). A recent model proposes that the bulk of the subgenomic mRNAs are produced by independent replication of the subgenomic RNAs as replicons. Such replication is thought to be possible because the mRNAs contain both the 5' and 3' sequences present in the genomic RNA, and therefore possess the promoters required for replication. Evidence for this model includes the fact that both plus-sense and minus-sense subgenomic RNAs are present in infected cells. The model favored is that the subgenomic RNAs are first produced during synthesis of minus-strand RNA from the genomic RNA. In this model, synthesis initiates at the 3' end of the genome and then jumps to the 5' leader at one of the junctions between the genes. Once produced, the subgenomic RNAs begin independent replication.

Coronaviruses undergo high-frequency recombination in which up to 10% of the progeny may be recombinant. It is proposed that the mechanism for generation of the subgenomic RNAs, which requires the polymerase to stop at defined sites and then reinitiate synthesis at defined promoters, may allow the formation of perfect recombinants at high frequency.

Envelope Glycoproteins

Coronaviruses possess three envelope proteins—a spike protein (S), a membrane protein (M), and an envelope protein (E). The spike protein is a large protein (e.g., 1255 residues in the SARS CoV) that is heavily glycosylated (more than 10 carbohydrate chains attached) and anchored in the membrane of the virion by a transmembrane domain near the C terminus, with a C-terminal cytoplasmic tail of about 40 residues. It forms trimers that project from the surface of the membrane and give coronaviruses their characteristic corona. These spikes possess the receptor-binding activity, the major neutralizing epitopes, and the fusion activity of the virion. S contains two domains of about equal size called S1 (N terminal) and S2 (C terminal), and in some, but not all, coronaviruses these two domains are separated into different proteins by proteolytic cleavage of S. S1 contains the receptor-binding region and S2 contains the fusion domain. S is not well conserved, with only about 30% sequence identity among S proteins of coronaviruses belonging to different groups.

The M protein is smaller, 221 residues in SARS CoV, and spans the lipid bilayer three times such that it has only a small fraction of its mass exposed outside the bilayer. The E protein is quite small, only 76 residues in SARS CoV, and has one membrane-spanning region. These two proteins are important for virion morphogenesis.

Some coronaviruses belonging to group 2 also possess a fourth envelope protein, a hemagglutinin-esterase (HE). Remarkably, this protein appears to be homologous to the H-E of influenza C virus (described in the next chapter). It appears that recombination between a coronavirus and an influenza C virus occurred that led to exchange of this protein. Because only some coronaviruses possess HE, whereas all influenza C viruses possess it, the simplest hypothesis is that HE was an influenza C protein that was acquired by a coronavirus. Presumably, this acquisition was maintained because it extended the host range of the coronavirus by allowing it to infect cells by binding to 5-N-acetyl-9-O-acetyl-N-neuraminic acid, a type of sialic acid, or to related sialic acids, depending on the specificity of the HE. Maintaining the HE protein has a cost for the virus. Mouse hepatitis virus loses HE when passed in culture, demonstrating that it is not needed for replication in cultured cells and that virus without the gene outcompete virus with the gene. In mice, MHV with HE is more virulent than virus without this gene and can spread more easily to the nervous system. Importantly, the HE gene is conserved in MHV strains isolated in the field, showing that this gene confers a selective advantage upon the virus.

Diseases Caused by Coronaviruses

Until recently, coronaviruses were considered to cause only mild disease in humans. Two human coronaviruses were known, HCoV OC43 (group 2A) and HCoV 229E (group 1). These viruses are responsible for about 25% of human colds and are spread by a respiratory route. Unlike rhinoviruses, they cause not only upper respiratory tract infections but sometimes lower respiratory tract infections as well, which are more serious. There is weak evidence that coronaviruses might also cause gastroenteritis in humans, because there have been reports of coronaviruses in the stools of people suffering from gastroenteritis. The status of coronaviruses as human disease agents changed with the recent isolation of two new human coronaviruses, NL63 (group 1) and HKU1 (group 2A), and with the 2003 epidemic of SARS (group 2B). NL63 is an important cause of severe lower respiratory tract infections in both adults and children. HKU1 has been isolated from adults with pneumonia. SARS causes an atypical pneumonia that carries a 10% fatality rate. It is a bat virus that jumped to humans in China, causing an epidemic of SARS that began in 2002. In 2003 it was spread around the world by air travelers, eventually causing more than 8000 cases of human disease and almost 800 deaths. It was eventually controlled by culling of animals that served as intermediates in passing the virus from bats to humans, and by quarantine procedures. There is concern that epidemics will recur since the virus is widely distributed in China. This topic is covered in more detail in Chapter 8.

Coronaviruses for many other animals are known, including mice, chickens, pigs, and cats. Diseases associated with various coronaviruses in these animals include respiratory disease, gastroenteritis, hepatitis, and a syndrome similar to multiple sclerosis of humans, as well as other illnesses. Mouse hepatitis virus has been particularly well studied as a model for the genus. Feline infectious peritonitis (FIP) coronavirus has also been intensively studied. This virus causes a severe infection of cats that is often fatal. It is immunosuppressive and the high fatality rate results from an inability to control the infection such that viral replication eventually reaches very high levels. Vaccination of cats with either structural proteins or nonstructural proteins did not protect the animals. In fact, vaccination with structural proteins made subsequent infection with live virus more severe. Persistent infection was observed in most animals, and there is evidence that virus remains even in animals that eventually control the infection since virus replication can resume if the animals are immunosuppressed. There are some parallels with SARS infection of humans, in that T-cell lymphopenia and viral persistence have been reported.

FAMILY *ARTERIVIRIDAE*

The family *Arteriviridae* contains four viruses, which are listed in Table 3.14. There are no known human viruses in the family, but it is of interest because it represents an intermediate between the coronaviruses and other enveloped (+)RNA viruses. The genome of equine arterivirus is illustrated in Fig. 3.39. The arteriviruses have a 13-kb genome that is very similar in organization and expression strategy to that of coronaviruses. The virion (60 nm) is enveloped, as are the coronaviruses, but the nucleocapsid, which is poorly defined, is probably icosahedral rather than helical. The arteriviruses could have arisen by the acquisition of new structural proteins by a coronavirus (or vice versa). The existence of this family, which appears to be a coronavirus with structural proteins that lead to icosahedral symmetry rather than helical symmetry, illustrates a problem for taxonomy. The ICTV has classified these viruses as a distinct family, but created the order *Nidovirales* to indicate their relation to the coronaviruses.

The four arteriviruses are lactate dehydrogenase-elevating virus of mice (LDV), equine arteritis virus (EAV), simian hemorrhagic fever virus (SHFV), and porcine reproductive and respiratory syndrome virus (PRRSV). The primary target cells in their respective hosts are macrophages, and all are associated with persistent, long-term infections. LDV causes a lifelong infection of mice that requires special care to detect. EAV causes epizootics of subclinical or mild respiratory diseases in adult horses. Infection can lead to abortions in pregnant mares, and infection of young horses causes a more serious illness. The virus persists for long

TABLE 3.14   *Arteriviridae* and *Roniviridae*

Genus/members	Virus name abbreviation	Usual host(s)	Transmission	Disease	World distribution
Arteriviridae					
Arterivirus					
Equine arteritis	EAV	Horses	Aerosols, contact	Fever, necrosis of arteries, abortion	Worldwide
Porcine reproductive and respiratory syndrome	PRRSV	Pigs	Oral–fecal?	Infertility, respiratory distress	?
Lactic dehydrogenase-elevating	LDV	Mice	Biting	?	?
Simian hemorrhagic fever	SHFV	Monkeys	Biting	Hemorrhage	?
Roniviridae					
Okavirus					
Gill-associated virus	GAV	Invertebrates (prawns)	Vertical, horizontal	Chronic subclinical, also acute necrosis of lymphoid organ	Asia and Australia

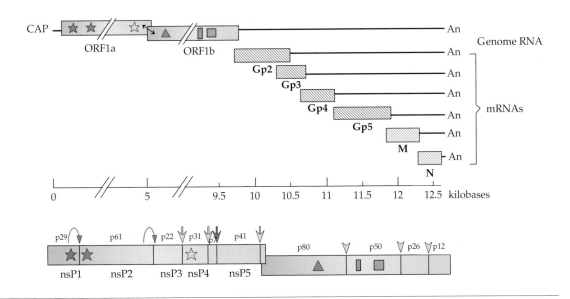

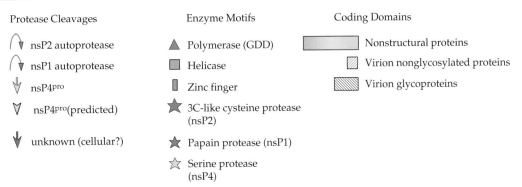

FIGURE 3.39   Upper panel: genome organization of an arterivirus, equine arteritis virus. ORF1a and ORF1b encode components of the viral replicase and are translated as a polyprotein with ribosomal frameshifting at the arrow. The remaining viral components are encoded in a nested set of mRNAs. The hatched proteins are polypeptides found in virions. Lower panel: proteolytic processing of the equine arteritis virus ORF1ab polyprotein. Positions of motifs of proteases, polymerase, zinc finger, and helicase are indicated with various symbols. Arrows are color coded to indicate cleavage by the corresponding protease. Arrowheads are predicted cleavages. Blue arrowhead is a cleavage site possibly cleaved by a cellular protease. Adapted from de Vries *et al.* (1997) and den Boon *et al.* (1991).

periods, and in stallions the virus may be secreted in semen for the life of the animal.

PRRSV causes respiratory distress in pigs of all ages and abortions and stillbirths in pregnant sows. SHFV is an African virus that causes persistent, inapparent infections in African monkeys. When introduced into colonies of Asian monkeys, however, it causes fatal hemorrhagic fever.

FAMILY *RONIVIRIDAE*

The *Roniviridae*, from **ro**d-shaped **ni**dovirus, are represented by a single known virus, gill-associated virus, which infects shrimp (Table 3.14). Its genome organization presents yet another permutation of how ancestral genes become associated with one another. The nonstructural genes, which occupy 20 kb, are translated from the genomic RNA by mechanisms that are very similar as those used by other members of the *Nidovirales* (Fig. 3.40). However, the structural proteins are translated from only two subgenomic mRNAs, one that is translated into the nucleocapsid protein, and one that is translated into a polyprotein precursor for the envelope proteins, which are separated from one another by signalase. The assembled virion is bacilliform in shape, 150–200 nm long and 40–60 nm in thickness. The virion thus resembles that of the rhabdoviruses (Chapter 4) rather than those of other nidoviruses.

THE PLUS-STRAND RNA VIRUSES OF PLANTS

Most plant viruses possess (+)RNA as their genome. Some have as their genome a single RNA molecule and produce subgenomic mRNAs, whereas in others the viral genome is divided into two or three or more segments. In plant viruses in which the genome is present in more than one segment, each segment is packaged separately into different particles and infection requires the introduction into the same cell of at least one of each genome segment. It is of interest that such an arrangement is common in plant viruses but nonexistent in animal viruses, presumably because of differences in the mechanisms by which plant and animal viruses spread and infect new cells or new hosts. Many (+)RNA plant viruses are rod shaped, formed using helical symmetry (e.g., tobacco mosaic virus, Fig. 2.2), while others are icosahedral (e.g., the comovirus cowpea mosaic virus, Figs. 2.5 and 2.7). No (+)RNA plant viruses are enveloped. Many of these viruses are major agricultural pathogens responsible for a great deal of crop damage worldwide. Although important as plant pathogens, plant viruses will not be covered here except for a description of the genomes of certain families that are of particular interest because of what they tell us about the evolution of viruses.

Several families of (+)RNA plant viruses share sequence identity with one another and with the alphaviruses. This collection of viruses, sometimes referred to as the Sindbis superfamily or the alphavirus superfamily, includes the alphaviruses, the tobamoviruses, the bromoviruses, and other families of plant viruses. The genomes of the tobamovirus tobacco mosaic virus (TMV), the bromovirus brome mosaic virus (BMV), and the alphavirus Sindbis virus are compared in Fig. 3.41. The genome of TMV is one molecule of (+)RNA and two subgenomic RNAs are produced. The genome of BMV consists of three molecules of (+)RNA and one subgenomic RNA is made. The alphaviruses have been described. Notice that a characteristic of this superfamily is that all viruses in it produce at least one subgenomic mRNA. The members of this superfamily all share three proteins (or protein domains) with demonstrable sequence homology, as indicated in the figure. These three are a viral

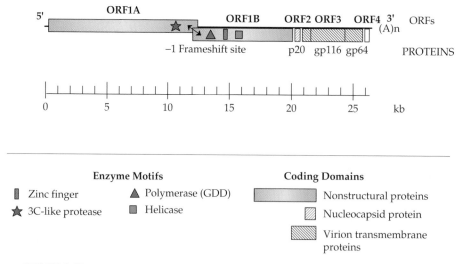

FIGURE 3.40 Genome organization of the *Roniviridae*. Redrawn from Cowley and Walker (2002).

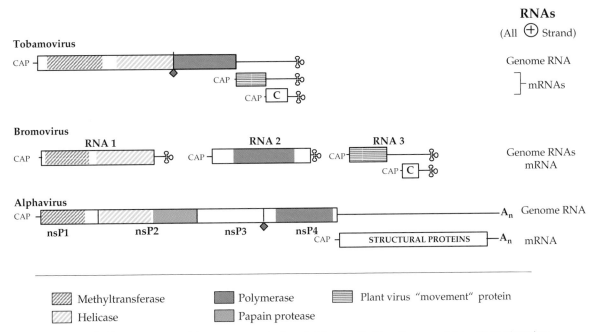

FIGURE 3.41   Comparison of the genome organization of alphavirus Sindbis with representatives of two plant virus families. Three shaded domains illustrate regions of low but significant sequence homology, which extend over hundreds of amino acids, within the methyltransferase, helicase, and polymerase proteins. The blue diamond is a leaky termination codon that is read through to produce the downstream blue-shaded domains in the tobamoviruses and the alphaviruses. C is the coat protein. The plant viruses have no module corresponding to the protease in nsP2 nor to protein nsP3. The alphaviruses have no domain corresponding to the "movement" protein of plant viruses. Adapted from Strauss and Strauss (1994), Figure 35.

RNA polymerase, a helicase, and a capping enzyme (characterized by methyltransferase activity). In the case of the alphaviruses and the tobamoviruses, all three domains are found on one genome segment and readthrough is required to translate the polymerase. In the bromoviruses, the capping enzyme and the helicase are encoded on one segment, but the polymerase is encoded on a different segment. The alphaviruses encode a protease to separate the three domains from one another, but the plant viruses do not. While these three shared proteins have clearly diverged from a common ancestral source, other domains within the nonstructural proteins are different from family to family. The alphavirus protease and nsP3 are not shared with the plant viruses, while the plant viruses possess movement proteins that are not shared with the alphaviruses. The structural proteins of the different families are also distinct. These observations clearly point to the occurrence of extensive recombinational events during the evolution of this group of viruses from a common ancestral source. Recombination has brought together new combinations of genes appropriate to the different lifestyles of the various members of the superfamily. In addition, the structural proteins differ among these three families so that the structures of the virions are very different from one another. The alphaviruses are enveloped, icosahedral particles (Figs. 2.5, 2.14A, and 2.25C). TMV is rod shaped (Fig. 2.2). BMV is icosahedral

but is not enveloped. Thus, recombination has brought together different RNA replication modules with different structural protein modules to give rise to the current families of viruses.

Similar considerations pertain to two families of plant viruses (the *Comoviridae* and the *Potyviridae*) and the animal picornaviruses, which are all related to one another and are sometimes referred to as the picornavirus superfamily. The *Comoviridae* have a bipartite genome, whereas the *Potyviridae* and the *Picornaviridae* have a single molecule of RNA as their genome. Characteristics of this superfamily include the absence of subgenomic RNAs, the presence of 5' VPg and 3' poly(A) on the viral RNAs, and the production of at least one protease. The genome organizations of two members of the *Comoviridae* that belong to different genera, tomato black ring virus (genus *Nepovirus*) and cowpea mosaic virus (genus *Comovirus*), are compared with that of picornavirus poliovirus in Fig. 3.42. The members of this superfamily have demonstrable homologies in their RNA polymerases, 2C helicases, and 3Cpro proteases. Further, the RNA genomes have a 5' VPg and are polyadenylated, as noted. Proteases, VPg's, and poly(A) are very unusual in plant viruses, found only in members of this superfamily. It is clear that these viruses are all related to one another, and that multiple recombination events have taken place to give rise to the current families.

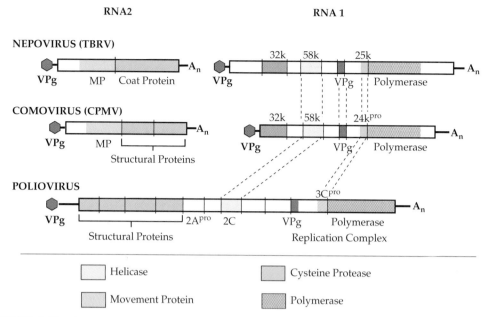

FIGURE 3.42   Comparison of the genomes of bipartite como- and nepoviruses and monopartite poliovirus. Domains in the helicase, polymerase, and protease that share sequence homology over long stretches of amino acids are identified with differently colored patterns. The related 32k proteins of como- and nepoviruses and the movement proteins encoded in RNA 2 (MP) have no counterpart in poliovirus. TBRV, tomato black ring virus; CPMV, cowpea mosaic virus. The structural proteins of the three viruses show no sequence similarity. Adapted from Strauss and Strauss (1997) Figure 2.12.

ORIGIN AND EVOLUTION OF PLUS-STRAND RNA VIRUSES

A reasonable hypothesis for the origin of the RNA viruses is that they began as an mRNA that encoded an RNA polymerase. The acquisition of an origin of replication that allowed the mRNA itself to be replicated by its encoded product would give rise to a self-replicating RNA and could have represented the first step in the development of a virus. Subsequent recombination with an mRNA encoding an RNA-binding protein that could be modeled into a capsid would give rise to a very simple virus. This protovirus could then evolve through continued mutation and recombination into something more complex. In support of this idea is the fact that the capsid protein of a large number of viruses, including bacterial viruses, plant viruses, and animal viruses that are otherwise unrelated to one another, share a common fold, suggesting that once a proper capsid protein arose it was retained during the evolution of many viruses while being modeled into new shapes.

Examples of the importance of recombination in the evolution of RNA viruses have been discussed. Computer-aided studies that have attempted to align the amino acid sequences of the proteins of different (+)RNA viruses have suggested that all these viruses share core functions that have common ancestral origins. These results are summarized in Fig. 3.43. All RNA viruses possess an RNA polymerase and these all appear to have derived from a common ancestral source. However, three lineages of RNA polymerases can

be distinguished that probably diverged from one another early in the evolution of RNA viruses. Most RNA viruses also possess an RNA helicase that is required to unwind the RNA during replication. These helicases also appear to have diverged from a single source, but three lineages can be distinguished here as well. A third shared function in those RNA viruses with capped mRNAs is a methyltransferase gene (an activity required for capping), and two methyltransferase lineages can be distinguished. Finally there are the viral proteases that process polyproteins. The two distinct types of proteases with independent origins are the proteases derived from serine proteases (which may possess serine or cysteine at the active site) and the papain-like proteases. The different lineages of these four core activities have been reassorted in various ways during the evolution of the RNA viruses, as shown in the figure.

The second mechanism for divergence among viruses is mutation. Lack of proofreading in RNA replication means that the mistake frequency during replication is very high, on the order of 10^{-4}. Most mistakes are deleterious and do not persist in the population. However, because the mistake frequency is so high, many different sequences can be tried rapidly because of the rapid replication rate of viruses. The net result is that the rate of sequence divergence in RNA viruses is very high, up to 10^{6}-fold faster than their eukaryotic hosts. Three studies of the rate of sequence divergence in RNA viruses are illustrated in Fig. 3.44. In these studies, regions of the genomes of viruses isolated over a period of many years were compared. The rates of sequence divergence in a

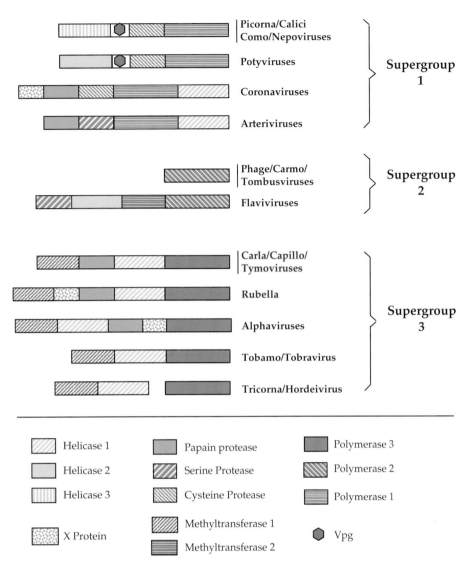

FIGURE 3.43 Genome organizations of plus-strand RNA viruses, grouped into three supergroups on the basis of sequence relationships among the nonstructral proteins. The RNA polymerases (blue), the proteinases (green), and the helicases (yellow) are each divided into three groups; the methyltransferases (red) are divided into two groups. From a relatively small number of building blocks, it is possible to arrive at the genomes of all of these viruses by divergence of individual domains and by recombination to reassemble them into different plans. Adapted from Strauss and Strauss (1994) Figure 36.

picornavirus and in influenza virus (Chapter 4) were found to be 0.5–1% per year. Changes in third codon positions, which are usually silent, occur more rapidly than changes in first or second codon positions, which usually result in an amino acid substitution. In alphaviruses, which alternate between insect and vertebrate hosts, the rate of divergence was significantly less, 0.03% per year, because changes that might be neutral or positively selected in one host are often deleterious in the other host. One of the apparent paradoxes of such studies is the observation that despite rapid sequence divergence, the properties of most viruses appear to remain stable for centuries or millennia. This is due in part to the fact that although the sequence may drift, the virus continues to

fill the same niche and selection ensures that the properties of the virus change only slowly. A second factor is that different domains of the genome, or even different nucleotides or amino acids, diverge at very different rates because of differences in selection pressure. Studies of the rates of divergence of viruses perforce will measure the rates of domains that diverge most rapidly. There is no fossil record to tell us when currently extant viruses might have diverged from one another, and viruses in collections have all been isolated within the last 70 years. Thus, all studies of divergence in nature examine only the divergence that has occurred within the last 70 years. Such considerations have two practical implications. Vaccines developed against most viruses

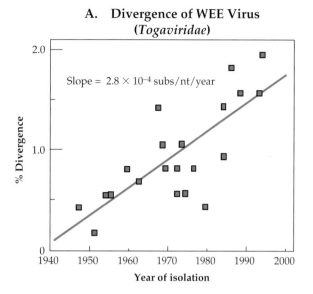

A. Divergence of WEE Virus
(*Togaviridae*)

Slope = 2.8×10^{-4} subs/nt/year

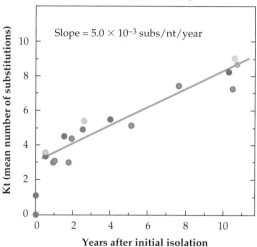

B. Divergence of Enterovirus 70
(*Picornaviridae*)

Slope = 5.0×10^{-3} subs/nt/year

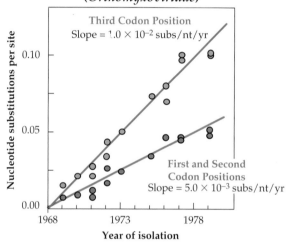

C. Divergence of Influenza HA (H3 subtype)
(*Orthomyxoviridae*)

Third Codon Position
Slope = 1.0×10^{-2} subs/nt/yr

First and Second
Codon Positions
Slope = 5.0×10^{-3} subs/nt/yr

FIGURE 3.44   Divergence plots for two different plus-strand RNA viruses, which differ by more than an order of magnitude in their divergence rates, and a minus-strand virus. Data for (A) are from Weaver *et al.* (1997) for 477 nucleotides of the C terminus of the E1 protein-coding region, for (B) from Takeda *et al.* (1994) for the 918 nucleotides which encode the VP1 protein, and for (C) from Saitou and Nei (1986) for the ~1700 nt encoding the hemagglutinin. In (B) the dots are color coded by location of isolation: red, North Africa; orange, Europe; yellow, Pakistan; green, South East Asia; blue, Japan; purple, Honduras.

continue to be effective for long periods of time, but ultimately may have to be reformulated as the virus drifts. Second, viruses can in principle change very rapidly. Human immunodeficiency virus (Chapter 6) and influenza virus (Chapter 4) do change rapidly in response to immune pressure, and viruses that jump to a new host have been known to change rapidly (e.g., the DNA virus canine parvovirus, Chapter 7, and SARS coronavirus, this chapter and Chapter 8).

FURTHER READING

General

Mackenzie, J. (2005). Wrapping things up about virus RNA replication. *Traffic* **6**:967–977.

Nathanson, N., Ahmed, R., Gonzalez-Scarano, F., *et al.* (Eds.) (1996). *Viral Pathogenesis*, Philadelphia, Lippincott-Raven.

Porterfield, J. S. (Ed.) (1995). *Exotic Viral Infections*, Kass Handbook of Infectious Diseases. London, Chapman & Hall Medical.

Picornaviridae

Bedard, K. M., and Semler, B. L. (2004). Regulation of picornavirus gene expression. *Microbes Infect.* **6**: 702–713.

Belsham, G. J. (2005) Translation and replication of FMDV RNA. *Curr. Top. Microbiol. Immunol.* **288**: 43–70.

Carrillo, C., Tulman, E. R., Delhon, G., *et al.* (2005). Comparative genomics of foot-and-mouth disease virus. *J. Virol.* **79**: 6487–6504.

Cherkasova, E. A., Yakovenko, M. L., Rezapkin, G. V., *et al.* (2005). Spread of vaccine-derived poliovirus from a paralytic case in an immunodeficient child: an insight into the natural evolution of oral polio vaccine. *J. Virol.* **79**: 1062–1070.

Halstead, L. S. (1998). Post-polio syndrome. *Scientific American* **April**: 42–47.

Heymann, D. L., Sutter, R. W., and Aylward, R. B. (2005). A global call for new polio vaccines. *Nature* **434**: 699–700.

Hollinger, B., and Emerson, S. U. (2006). Hepatitis A virus. Chapter 27 in *Fields Virology, Fifth Edition* (D. M. Knipe and P. M. Howley, Eds. in chief), Philadelphia, Lippincott Williams & Wilkins, pp. 911–948.

Minor, P. D. (2004). Polio eradication, cessation of vaccination and re-emergence of disease. *Nature Rev. Microbiol.* **2**: 473–482.

Oshinsky, D. M. (2005). *Polio: An American Story*. New York, Oxford University Press.

Pallansch, M., and Roos, R. (2006). Enteroviruses: Polioviruses, coxsackievirus, echoviruses, and newer enteroviruses. Chapter 25 in: *Fields Virology, Fifth Edition* (D. M. Knipe and P. M. Howley, Eds. in chief), Philadelphia, Lippincott Williams & Wilkins, pp. 839–894.

Racaniello, V. R. (2006). *Picornaviridae*: The viruses and their replication. Chapter 24 in: *Fields Virology, Fifth Edition* (D. M. Knipe and P. M. Howley, Eds. in chief), Philadelphia, Williams & Wilkins, pp. 795–838.

Turner, R. B., and Couch, R. B. (2006). Rhinoviruses. Chapter 26 in *Fields Virology, Fifth Edition* (D. M. Knipe and P. M. Howley, Eds. in chief), Philadelphia, Lippincott Williams & Wilkins, pp. 895–910.

Whitton, J. L., Cornell, C. T., and Feuer, R. (2005). Host and virus determinants of picornavirus pathogenesis and tropism. *Nature Rev. Microbiol.* **3**: 765–776.

Yamashita, T., Ito, M., Kabashima, Y., Tsuzuki, H., Fujiura, A., and Sakae, K. (2003). Isolation and characterization of a new species of kobuvirus associated with cattle. *J. Gen. Virol.* **84**: 3069–3077.

Zell, R., Krumbholz, A., Dauber, M., Hoey, E., and Wutzler, P. (2006). Molecular-based reclassification of the bovine enteroviruses. *J. Gen. Virol.* **87**: 375–385.

Caliciviridae

Bertolotti-Ciarlet, A., White, L. J., Chen, R., Prasad, B. V. V., and Estes, M. K. (2002). Structural requirements for the assembly of Norwalk virus-like particles. *J. Virol.* **76**: 4044–4055.

Green, K. Y. (2006). Human caliciviruses. Chapter 28 in: *Fields Virology, Fifth Edition* (D. M. Knipe and P. M. Howley, Eds. in chief), Philadelphia, Lippincott Williams & Wilkins, pp. 949–980.

Katayama, K., Shirato-Horikoshi, H., Kojima, S., *et al.* (2002). Phylogenetic analysis of the complete genome of 18 Norwalk-like viruses. *Virology* **299**: 225–239.

Kuyumcu-Martinez, M., Belliot, G., Sosnovtsev, S. V., et al. (2004). Calicivirus 3C-like proteinase inhibits cellular translation by cleavage of poly(A)-binding protein. *J. Virol.* **78**: 8172–8182.

Martín-Alonso, J. M., Skilling, D. E., González-Molleda, L., *et al.* (2005). Isolation and characterization of a new *Vesivirus* from rabbits. *Virology* **337**: 373–382.

Matson, D. O., and Szucs, G. (2003). Calicivirus infections in children. *Curr. Opin. Infect. Dis.* **16**: 241–246.

Thiel, H. J., and Konig, M. (1999). Caliciviruses: an overview. *Vet. Microbiol.* **69**: 55–62.

Widdowson, M.-A., Sulka, A., Bulens, S. N., *et al.* (2005). Norovirus and foodborne disease, United States, 1991–2000. *Emerg. Infect. Dis.* **11**: 95–102.

Hepatitis E

Emerson, S. U., and Purcell, R. H. (2006). Hepatitis E virus. Chapter 78 in: *Fields Virology, Fifth Edition* (D. M. Knipe and P. M. Howley, Eds. in chief), Philadelphia, Lippincott Williams & Wilkins, pp. 3047–3056.

Takahashi, K., Kitajima, N., Abe, N., and Mishiro, S. (2004). Complete or near-complete nucleotide sequences of hepatitis E virus genome recovered from a wild boar, a deer, and four patients who ate the deer. *Virology* **330**: 510–505.

Tsarev, S. A., Binn, L. N., Gomatos, P. J., *et al.* (1999). Phylogenetic analysis of hepatitis E virus isolates from Egypt. *J. Med. Virol.* **57**: 68–74.

Astroviridae

Glass, R. I., Noel, J., Mitchell, D., *et al.* (1996). The changing epidemiology of astrovirus-associated gastroenteritis: A review. *Arch. Virol.* **141**: 287–300.

Hart, C. A., and Cunliffe, N. A. (1999). Viral gastroenteritis. *Curr. Opin. Infect. Dis.* **12**: 447–457.

Kiang, D., and Matsui, S. M. (2002). Proteolytic processing of a human astrovirus nonstructural protein. *J. Gen. Virol.* **83**: 25–34.

Méndez, E., and Arias, C. F. (2006). Astroviruses. Chapter 29 in: *Fields Virology, Fifth Edition* (D. M. Knipe and P. M. Howley, Eds. in chief), Philadelphia, Lippincott Williams & Wilkins, pp. 981–1000.

Togaviridae

Calisher, C. H. (1994). Medically important arboviruses in the United States and Canada. *Microbiol. Rev.* **7**: 89–116.

Chikungunya and dengue in the south west Indian Ocean. (2006). A news bulletin from the WHO at http://www.who.int/csr/don/2006-03-17/en/index.html.

Ehrengruber, M. U. (2002). Alphaviral gene transfer in neurobiology. *Brain Res. Bull.* **59**: 13–22.

Garoff, H., Sjöberg, M., and Cheng, R. H. (2004). Budding of alphaviruses. *Vir. Res.* **106**: 103–116.

Gorchakov, R., Hardy, R., Rice, C. M., and Frolov, I. (2004). Selection of functional 5′ cis-acting elements promoting efficient Sindbis virus genome replication. *J. Virol.* **78**: 61–75.

Griffin, D. E. (2006). Alphaviruses. Chapter 31 in: *Fields Virology, Fifth Edition* (D. M. Knipe and P. M. Howley, Eds. in chief), Philadelphia, Lippincott Williams & Wilkins, pp. 1023–1068.

Hobman, T. C., and Chantler, J. K. (2006). Rubella virus. Chapter 32 in: *Fields Virology, Fifth Edition* (D. M. Knipe and P. M. Howley, Eds. in chief), Philadelphia, Lippincott Williams & Wilkins, pp. 1069–1100.

Kääriäinen, L., and Ahola, T. (2002). Functions of alphavirus nonstructural proteins in RNA replication. *Prog. Nucl. Acid. Res.* **71**: 187–222.

Klapsing, P., MacLean, J. D., Glaze, S., *et al.* (2005). Ross River virus disease reemergence, Fiji, 2003–2004. *Emerg. Infect. Dis.* **11**: 613–615.

Kuhn, R. J. (2006). *Togaviridae*: the viruses and their replication. Chapter 30 in: *Fields Virology, Fifth Edition* (D. M. Knipe and P. M. Howley, Eds. in chief), Philadelphia, Lippincott Williams & Wilkins, pp. 1001–1022.

Laine, M., Luukkainen, R., and Toivanen, A. (2004). Sindbis virus and other alphaviruses as cause of human arthritic disease. *J. Intern. Med.* **256**: 457–471.

Luers, A. J., Adams, S. D., Smalley, J. V., and Campanella, J. J. (2005). A phylogenomic study of the genus *Alphavirus* employing whole genome comparison. *Comp. Funct. Genomics* **6**: 217–227.

Mackenzie, J. S., Poindinger, M., Lindsay, M. D., Hall, R. A., and Sammels, L. M. (1996). Molecular epidemiology and evolution of mosquito-borne flaviviruses and alphaviruses enzootic in Australia. *Virus Genes* **11**: 225–237.

Monath, T. P. (Ed.) (1988). *The Arboviruses: Epidemiology and Ecology.* Boca Raton, FL, CRC Press..

Sawicki, D. L., Perri, S., Polo, J. M., and Sawicki, S. G. (2006). Role for nsP2 proteins in the cessation of alphavirus minus-strand synthesis by host cells. *J. Virol.* **80**: 360–371.

Strauss, J. H., and Strauss, E. G. (1994). The alphaviruses: Gene expression, replication, and evolution. *Microbiol. Rev.* **58**: 491–562.

Strauss, J. H., and Strauss, E. G. (1997). Recombination in alphaviruses. *Semin. Virol.* **8**: 85–94.

Weston, J., Villoing, S., Brémont, M., *et al.* (2002). Comparison of two aquatic alphaviruses, salmon pancreas disease virus and sleeping disease virus, by using genome sequence analysis, monoclonal reactivity, and cross-infection. *J. Virol.* **76**: 6155–6163.

Flaviviridae

Alvarez, D. E., Lodiero, M. F., Ludueña, S. J., Pietrasanta, L. I.., and Gamarnik, A. V. (2005). Long range RNA–RNA interactions circularize the dengue virus genome *J. Virol.* **79**: 6631–6643.

Bartholomeusz, A., and Thompson, P. (1999). *Flaviviridae* polymerase and RNA replication. *J. Viral Hepatitis* **6**: 261–270.

Bryant, J. E., Vasconcelos, P. F. C., Rijnbrand, R. C. A., *et al.* (2005). Size heterogeneity in the 3′ noncoding region of South American isolates of yellow fever virus. *J. Virol.* **79**: 3807–3821.

Chambers, T. J., Hahn, C. S., Galler, R., *et al.* (1990). Flavivirus genome organization, expression, and replication. *Annu. Rev. Microbiol.* **44**: 649–688.

Cook, S., Bennett, S. N., Holmes, E. C., *et al.* (2006). Isolation of a new strain of the flavivirus cell fusing agent in a natural mosquito population from Puerto Rico. *J. Gen. Virol.* **87**: 735–748.

Elshuber, S., and Mandl, C. W. (2005). Resuscitating mutations in a furin cleavage-deficient mutant of the flavivirus tick-borne encephalitis virus. *J. Virol.* **79**: 11813–11823.

Goncalvez, A. P., Escalante, A. A., Pujol, F. H., *et al.* (2002). Diversity and evolution of the envelope gene of dengue virus type 1. *Virology* **303**: 110–119.

Gritsun, T. S., Lashkevich, V. A., and Gould, E. A. (2003). Tick-borne encephalitis. *Antiviral Res.* **57**: 129–146.

Gubler, D., Kuno, G., and Markoff, L. (2006). Flaviviruses. Chapter 34 in: *Fields Virology, Fifth Edition* (D. M. Knipe and P. M. Howley, Eds. in chief), Philadelphia, Lippincott Williams & Wilkins, pp. 1153–1252.

Hall, R. A., Nisbt, D. J., Pham, K. B., *et al.* (2003). DNA vaccine coding for the full-length infectious Kunjin virus RNA protects mice against the New York strain of West Nile virus. *Proc. Natl. Acad. Sci. U.S.A.* **100**: 10460–10464.

Hayes, E. B., Komar, N., Nasci, R. S., *et al.* (2005). Epidemiology and transmission dynamics of West Nile virus disease. *Emerg. Infect. Dis.* **11**: 1167–1173.

Jones, C. T. Patkar, C. G., and Kuhn, R. J. (2005). Construction and applications of yellow fever virus replicons. *Virology* **331**: 247–259.

Lindenbach, B. D., Thiel, H.-J., and Rice, C. M. (2006). *Flaviviridae*: The viruses and their replication. Chapter 33 in: *Fields Virology, Fifth Edition* (D. M. Knipe and P. M. Howley, Eds. in chief), Philadelphia, Lippincott Williams & Wilkins, pp. 1101–1152.

Lanciotti, R. S., Ebel, G. D., Deubel, V., *et al.* (2002). Complete genome sequence and phylogenetic analysis of West Nile virus strains isolated from the United States, Europe, and the Middle East. *Virology* **298**: 96–105.

Lorenz, I. C., Kartenbeck, J., Mezzacasa, A., *et al.* (2003). Intracellular assembly and secretion of recombinant subviral particles from tick-borne encephalitis virus. *J. Virol.* **77**: 4370–4382.

Meyers, G., and H.-J. Thiel (1996). Molecular characterization of pestiviruses. *Adv. Virus Res.* **47**: 53–118.

Mukhopadhyay, S., Kuhn, R. J., and Rossmann, M. G. (2005). A structural perspective of the flavivirus life cycle. *Nature Revs. Microbiol.* **3**: 13–22.

Rey, F. A., Heinz, F. X., Mandl, C. W., *et al.* (1995). The envelope glycoprotein from tick borne encephalitis virus at 2Å resolution. *Nature* **375**: 291–298.

Rigau-Pérez, J. G., Clark, G. G., Gubler, D. J., *et al.* (1998). Dengue and dengue haemorrhagic fever. *Lancet* **352**: 971–977.

Stiasny, K., and Heinz, F. X. (2006). Flavivirus membrane fusion. *J. Gen. Virol.* **87**: 2755–2766.

Van der Meulen, K. M., Pensaert, M. B., and Nauwynck, H. J. (2005). West Nile virus in the vertebrate world. *Arch. Virol.* **150**: 637–657.

Zhang, Y., Zhang, W., Ogata, S., *et al.* (2004). Conformational changes of the flavivirus E glycoprotein. *Structure* **12**: 1607–1618.

Hepatitis C

Bartenschlager, R., and Lohmann, V. (2000). Replication of hepatitis C virus. *J. Gen. Virol.* **81**: 1631–1648.

Bukh, J., and Purcell, R. H. (2006). A milestone for hepatitis C virus research: a virus generated in cell culture is fully viable *in vivo. Proc. Natl. Acad. Sci. U.S.A.* **103**: 3500–3501.

"Hepatitis C," (a Nature Insight encompassing several review articles). (2005). *Nature* **436**: 929–978.

Lemon, S. M., Walker, C. M., Alter, M. J., and Yi, M. K. (2006). Hepatitis C viruses. Chapter 35 in: *Fields Virology, Fifth Edition* (D. M. Knipe and P. M. Howley, Eds. in chief), Philadelphia, Lippincott Williams & Wilkins, pp. 1253–1304

Macdonald, A., and Harris, M. (2004). Hepatitis C virus NS5A: tales of a promiscuous protein. *J. Gen. Virol.* **85**: 2485–2502.

Simmonds, P. (2004). Genetic diversity and evolution of hepatitis C virus— 15 years on. *J. Gen. Virol.* **85**: 3173–3188.

Wieland, S. F., and Chisari, F. V. (2005). Stealth and cunning: hepatitis B and hepatitis C viruses. *J. Virol.* **79**: 9369–9380.

Coronaviridae and Arteriviridae

de Vries, A. A. F., Horzinek, M. C., Rottier, P. J. M., *et al.* (1997). The genome organization of the Nidovirales: Similarities and differences between arteri-, toro-, and coronaviruses. *Semin. Virol.* **8**: 33–47.

Gorbalenya, A. E., Enjuanes, L., Ziebuhr, J., and Snijder, E. J. (2006). Nidovirales: Evolving the largest RNA virus genome. *Vir. Res.* **117**: 17–37.

Kahn, J. S. (2006). The widening scope of coronaviruses. *Curr. Opin. Pediatr.* **18**: 42–47.

Lai, M. M. C., Perlman, S., and Anderson, L. J. (2006). *Coronaviridae.* Chapter 36 in: *Fields Virology, Fifth Edition* (D. M. Knipe and P. M. Howley, Eds. in chief), Philadelphia, Lippincott Williams & Wilkins, pp. 1305–1336.

Pasternak, A. O., Spaan, W. J. M., and Snijder, E. J. (2006). Nidovirus transcription: how to make sense …? *J. Gen. Virol.* **87**: 1403–1421.

Sawicki, S. G., Sawicki, D. L., Younker, D., *et al.* (2005). Functional and genetic analysis of coronavirus replicase-transcriptase proteins. *PLOS* **1**: 310–322.

Snijder, E. J., and Spaan, W. J. M. (2006). Arteriviruses. Chapter 37 in: *Fields Virology, Fifth Edition* (D. M. Knipe and P. M. Howley, Eds. in chief), Philadelphia, Lippincott Williams & Wilkins, pp. 1337–1356.

Evolution

Gallei, A., Rümenapf, T., Thiel, H.-J., and Becher, P. (2005). Characterization of helper virus-independent cytopathogenic classical swine fever virus generated by an in vivo RNA recombination system. *J. Virol.* **79**: 2440–2448.

Koonin, E. V., and Dolja, V. V. (1993). Evolution and taxomony of positive-strand RNA viruses—implications of comparative analysis of amino-acid sequences. *Crit. Rev. Biochem. Mol. Biol.* **28**: 375–430.

Worobey, M., and Holmes, E. C. (1999). Evolutionary aspects of recombination in RNA viruses. *J. Gen. Virol.* **80**: 2535–2543.

4

Minus-Strand RNA Viruses

INTRODUCTION

Seven families of viruses contain minus-strand RNA [(−)RNA], also called negative-strand RNA, as their genome. These are listed in Table 4.1. Included in the table are the names of the genera belonging to these families and the hosts infected by these viruses. Six of the families are known to contain members that cause epidemics of serious human illness. Diseases caused by these viruses include influenza (*Orthomyxoviridae*), mumps and measles (*Paramyxoviridae*), rabies (*Rhabdoviridae*), encephalitis (several members of the *Bunyaviridae*), upper and lower respiratory tract disease (numerous viruses in the *Paramyxoviridae*), and hemorrhagic fever (many viruses belonging to the *Bunyaviridae*, the *Arenaviridae*, and the *Filoviridae*), as well as other diseases. Bornavirus, the sole representative of the *Bornaviridae*, also infects humans and may cause neurological illness, but proof of causality is lacking. Many of the (−)RNA viruses presently infect virtually the entire human population at some point in time (e.g., respiratory syncytial virus, influenza virus), whereas others did so before the introduction of vaccines against them (e.g., measles virus and mumps virus). These viruses are thus responsible for a very large number of cases of human illness. The diseases caused by such widespread viruses are usually serious but have a low (although not insignificant) fatality rate. In contrast, some (−)RNA viruses, such as rabies and Ebola viruses, cause illnesses with high fatality rates but (fortunately) infect only a small fraction of the human population. The (−)RNA viruses are major causes of human suffering, and all seven families and the viruses that belong to these families will be described here.

OVERVIEW OF THE MINUS-STRAND RNA VIRUSES

Viruses belonging to four families of (−)RNA viruses, the *Paramyxoviridae*, the *Rhabdoviridae*, the *Filoviridae*, and the *Bornaviridae*, contain a nonsegmented RNA genome having similar organization. They are grouped into the order *Mononegavirales* (*mono* because the genome is in one piece, *nega* for negative-strand RNA). This was the first order to be recognized by the International Committee on Taxonomy of Viruses and still is one of only three orders currently recognized. Viruses belonging to the other three families, the *Arenaviridae*, *Bunyaviridae*, and *Orthomyxoviridae*, possess segmented genomes with two, three, and six to eight segments, respectively. Regardless of whether the genome is one RNA molecule or is segmented, the genomes of all (−)RNA viruses possess a similar suite of genes, as illustrated in Fig. 4.1. In the *Mononegavirales*, the order of genes along the genome is conserved among the viruses (although the number of genes may differ). In the viruses with segmented genomes, the genes can be ordered in the same way if the segments are aligned as shown. In addition, many features of virion structure and of replication pathways are shared among the (−)RNA viruses.

Structure of the Virions

All (−)RNA viruses are enveloped and have helical nucleocapsids. The different families encode either one or two glycoproteins (called G in most of the families but called HA, NA, F, or HN in some, after hemagglutinating, neuraminidase, or fusion properties). These glycoproteins are present in the viral envelope. In most cases, cleavages

TABLE 4.1   Negative-strand RNA Viruses

Family/genus	Genome size (in kb)	Type virus[a]	Host(s)[b]	Transmission
Mononegavirales (nonsegmented)				
Rhabdoviridae	13–16			
Vesiculovirus		VSIV	Vertebrates	Some arthropod-borne
Lyssavirus		Rabies	Vertebrates	Contact with saliva
Ephemerovirus		BEFV	Cattle	Arthropod-borne
Novirhabdovirus		IHNV	Fish	
Two genera of plant viruses				Arthropod-borne
Filoviridae	13			
Marburgvirus		Marburg	Vertebrates	?
Ebolavirus		Zaire Ebola	Vertebrates	?
Paramyxoviridae	16–20			
Respirovirus		Sendai	Vertebrates	Airborne
Morbillivirus		Measles	Vertebrates	Airborne
Rubulavirus		Mumps	Vertebrates	Airborne
Henipavirus		Hendra	Vertebrates	Airborne
Avulavirus		Newcastle disease	Birds	Airborne
Pneumovirus		HRSV	Vertebrates	Airborne
Metapneumovirus		TRTV	Turkeys	Airborne
Bornaviridae				
Bornavirus	~9	BDV	Vertebrates	Contaminated forage
Segmented Negative Strand RNA Viruses				
Orthomyxoviridae	10–14.6			
Influenzavirus A	8 segments	Influenza A	Vertebrates	Airborne
Influenzavirus B		Influenza B	Vertebrates	Airborne
Influenzavirus C	7 segments	Influenza C	Vertebrates	Airborne
Thogotovirus	6 segments	Thogoto	Vertebrates	Arthropod-borne
Isavirus	8 segments	ISAV	Fish	Waterborne
Bunyaviridae	11–20 in 3 segments			
Orthobunyavirus		Bunyamwera	Vertebrates	Mosquito-borne
Hantavirus		Hantaan	Vertebrates	Feces–urine–saliva
Nairovirus		Dugbe	Vertebrates	Tickborne
Phlebovirus		Rift Valley fever	Vertebrates	Arthopod-borne
Tospovirus		TSWV	Plants	Thrips
Arenaviridae	10–14 in 2 segments			
Arenavirus		LCMV	Vertebrates	Urine–saliva

[a] Abbreviations of virus names: VSIV, vesicular stomatitis Indiana virus; BEFV, bovine ephemeral fever virus; IHNV, infectious hematopoietic necrosis virus; HRSV, human respiratory syncytial virus; TRTV, turkey rhinotracheitis virus; BDV, Borna disease virus; ISAV, infectious salmon anemia virus; TSWV, tomato spotted wilt virus; LCMV, lymphocytic choriomeningitis virus.
[b] In all cases, "Vertebrates" includes humans.

are required to produce the mature glycoproteins, such as cleavage to release a signal peptide, cleavage to separate two glycoproteins produced as a common precursor, or cleavage to activate viral infectivity. The glycoproteins project from the lipid bilayer as spikes that are visible in the electron microscope (see, e.g., Fig. 2.18D).

All (−)RNA viruses have a single major nucleocapsid protein (called N) that encapsidates the virion RNA to form the helical nucleocapsid. Also present in the nucleocapsid is a phosphorylated protein that is required for RNA synthesis, variously called P (for phosphoprotein) or NS (for nonstructural protein because it was not originally known to

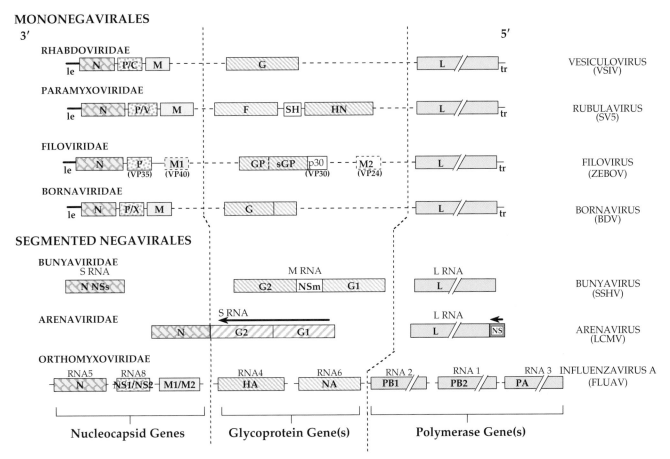

FIGURE 4.1 Genome organizations of the *Negavirales*. The genomes of representatives of the four families of *Mononegavirales* have been aligned to illustrate functional similarity between gene products. The individual gene segments of the representatives of the three families with segmented genomes, *Bunyaviridae*, *Arenaviridae*, and *Orthomyxoviridae*, have been aligned according to similarity of function with those of the *Mononegavirales* above. Gene expression strategies for the other genera of *Bunyaviridae* vary (see Fig. 4.21). Abbreviations of virus names are as follows: VSIV, vesicular stomatitis Indiana virus; SV5, simian virus 5; ZEBOV, Zaire ebolavirus; BDV, Borna disease virus; SSHV, snowshoe hare virus; LCMV, lymphocytic choriomeningitis virus; FLUAV, influenza A virus. The gene products are abbreviated as follows: le is a leader sequence; N is the nucleoprotein; P is the phosphoprotein; M (M1, M2) are matrix proteins; G (G1, G2) are membrane glycoproteins; F is the fusion glycoprotein; HN is the hemagglutinin-neuraminidase glycoprotein; L is the RNA polymerase; NA is the neuraminidase glycoprotein; HA is the hemagglutinin glycoprotein; NS (NV, SH, NSs, NSm) are nonstructural proteins; PB1, PB2, and PA are components of the influenza RNA polymerase; tr is the trailer sequence. Within a given genome, the genes are drawn approximately to scale. mRNAs for most genes would be synthesized left to right; however, an arrow over a gene means that it is in the opposite orientation (ambisense genes). Redrawn from Strauss *et al.* (1996), Figure 5.

be a component of the virion), as well as a few molecules of an RNA-dependent RNA polymerase. The polymerase is a large, multifunctional protein called L in most families but is present as three proteins in the *Orthomyxoviridae*. L and P form a core polymerase that replicates the viral genome and synthesizes mRNAs.

A matrix protein (M) is present in all of the viruses except the bunyaviruses and the arenaviruses. M underlies the lipid bilayer where it interacts with the nucleocapsid. M also inhibits host transcription and shuts down viral RNA synthesis prior to packaging.

The (−)RNA virions are heterogeneous to a greater or lesser extent. Members of five families often appear roughly

spherical in the electron microscope. The example of influenza virus is shown in Figs. 2.1 and 2.22D, and the paramyxovirus measles virus is shown in Fig. 2.22C. The compositions of these virions are not rigorously fixed and some variability in the ratios of the different components, particularly in the glycoprotein content, is present. The rhabdoviruses are bullet shaped or bacilliform and appear more regular (Fig. 2.23), but even here variations in the composition of the glycoproteins in the envelope can occur. The filoviruses are filamentous (Fig. 2.23). Orthomyxoviruses and paramyxoviruses also produce filamentous forms as well as round virions (see Fig. 2.25E). In fact, clinical isolates of influenza viruses and human respiratory syncytial virus are predominantly filamentous.

Synthesis of mRNAs

For all (−)RNA viruses, the first event in infection is the synthesis of mRNAs from the minus-strand genome by the RNA polymerase present in the nucleocapsid. Because this polymerase is necessary for the production of the mRNAs, and because the proteins translated from the mRNAs are required for replication of the genome, the naked genomes of (−)RNA viruses are not infectious, nor are complementary RNA copies of the genomes. It has been possible, nonetheless, to rescue virus from cDNA clones of viral genomes by using special tricks, as described in Chapter 11.

Multiple mRNAs are produced from minus-strand genomes. By definition, each region of the genome from which an independent mRNA is synthesized is called a gene. In (−)RNA viruses with segmented genomes, it is obvious that multiple mRNAs are produced (the number of mRNAs produced actually exceeds the number of segments, as described later). In the *Mononegavirales*, multiple mRNAs arise from the use of a single polymerase entry site at the 3′ end of the genome. The polymerase then recognizes conserved start and stop signals at the beginning and end of each gene to generate discrete mRNAs. The amount of mRNA produced for any given gene is controlled by the location of the gene relative to the single polymerase entry site, because mRNA synthesis is obligatorily sequential and attenuation occurs at each gene junction. Thus, more mRNA for the proteins encoded 3′ in the genome is made and more protein is thus translated from these genes. The N protein, required for encapsidation of both genome and antigenome, is thereby produced in the largest quantities and the RNA polymerase, needed in the smallest quantities, is made in the smallest quantities. The synthesis of mRNAs is described in more detail in the sections on *Rhabdoviridae*.

Most of the mRNAs are translated into a single protein, but a few of the genes produce mRNAs that are translated into more than one product. Multiple products can be produced from the same gene by the use of alternative translation initiation codons during translation of an mRNA; by the introduction of nontemplated nucleotides during mRNA synthesis, which results in a shift in the reading frame; or by splicing of an mRNA. The P genes, in particular, of most of the (−)RNA viruses are translated into multiple products, and two of the segments of influenza virus, which replicates in the nucleus, can be spliced to produce a second mRNA encoding a different product. In no case are the mRNAs exact complements of virion RNAs. This is obvious in the case of the *Mononegavirales*, where as many as 7–10 mRNAs are produced from a single long genomic RNA, but is also true of the segmented (−)RNA viruses, where the mRNAs lack *cis*-active sequences required for encapsidation and replication that are present near the ends of the antigenome segments. Thus, the mRNAs of (−)RNA viruses do not replicate nor are they packaged into virions.

In contrast to the translation strategy used by the (+)RNA viruses, the (−)RNA viruses do not produce polyproteins that require processing by virally encoded enzymes, and virus-encoded proteases are unknown among them. However, most of the glycoproteins of the (−)RNA viruses are produced as precursors that are processed by cellular enzymes, and some of these precursors can be considered to be polyproteins.

Replication of the Genome

Replication of the (−)RNA genome requires the production of a complementary copy of the genome, called an antigenome or virus-complementary RNA (vcRNA), which is distinct from the mRNAs (schematically illustrated in Figs. 1.11C and D). Neither the genomic (−)RNA nor the antigenomic template produced during replication is ever free in the cytoplasm. Instead, replication of the genome, as well as the synthesis of mRNAs, takes place in nucleocapsids (sometimes referred to as ribonucleoprotein or RNP), which always contain the phosphoprotein and the polymerase as well as N and the viral RNA. Replication can only occur in the presence of ongoing protein synthesis to produce the new proteins required to encapsidate the genome or antigenome. The mRNAs can be synthesized in the absence of viral protein synthesis and lack encapsidation signals, so that they are released into the cytoplasm where they can associate with ribosomes and be translated. Thus, early after infection, mRNAs are synthesized. After translation of the mRNAs, which leads to production of sufficient amounts of viral proteins, a switch to the production of antigenomes for use as templates occurs, followed by production of genomic RNA from the antigenomic templates.

The genomes (or genome segments) of all (−)RNA viruses have sequences at the ends that are complementary (so-called inverted terminal repeats). In the bunyaviruses, the RNAs form panhandles, circular structures that are visible in the electron microscope. Panhandles have also been reported for influenza A virus. In other viruses, circles have not been seen but may form transiently during replication. It is possible that these complementary sequences exist to promote cyclization of the RNA, which may be required for replication of the genome or synthesis of mRNAs. It has been shown for influenza A virus that the viral RNA replicase interacts with both ends of the RNA during synthesis of RNA, similar to the story for alphaviruses and flaviviruses described in Chapter 3. Another possible explanation for the complementary sequences is that the promoter at the 3′ end of the genomic RNA that is recognized by the viral RNA synthetase for the production of antigenomes is the same, at least in part, as the promoter at the 3′ end of the antigenomic RNA that is used to initiate the production of genomic RNA.

In this event, the sequences at the two ends of the genome or antigenome that encompass these promoters would be complementary.

Host Range of the (–)RNA Viruses

All seven families contain members that infect higher vertebrates, including humans. For five of the families, only vertebrate hosts are known. The rhabdoviruses and bunyaviruses, however, have a broader host range. Some are arboviruses that replicate in an arthropod vector as well as in a vertebrate host, and others infect only insects. In addition, some genera of rhabdoviruses and bunyaviruses consist of plant viruses. Some of these are transmitted to the plants by insect vectors in which the viruses also replicate.

FAMILY *RHABDOVIRIDAE*

The genome organization of the rhabdoviruses is the simplest of the (–)RNA viruses and it is useful to begin our coverage with this group. The genome is a single piece of minus-strand RNA 11–15 kb in size. The genomes of all rhabdoviruses contain five core genes, called N, P, M, G, and L in that order in the genome reading 3' to 5', which result in the production of five to seven proteins, five of which are present in the virion. Some rhabdoviruses contain only these five genes, but others contain one to five extra genes inserted in various regions of the genome. The animal rhabdoviruses are bullet shaped, approximately 200 nm long and 75 nm in diameter (Fig. 2.23), whereas some of the plant viruses are bacilliform, being rounded at both ends. The rhabdoviruses infect mammals, birds, fish, insects, and plants, and are presently divided into six genera. A listing of these genera and a representative sample of the viruses in each genus, together with several characteristics of each virus, are shown in Table 4.2. Members of three genera infect mammals, namely, the vesiculoviruses (type virus: vesicular stomatitis Indiana virus), lyssaviruses (type virus: rabies virus), and ephemeroviruses (type virus: bovine ephemeral fever virus). The novirhabdoviruses infect fish, and the cytorhabdoviruses and nucleorhabdoviruses infect plants. Some or all of the members of four genera are transmitted by arthropods (Table 4.2). In addition, a large number of the more than 175 currently known rhabdoviruses have not been assigned to a genus. The animal rhabdoviruses replicate in the cytoplasm, but certain of the plant rhabdoviruses may replicate in the nucleus.

Genus *Vesiculovirus*

Vesicular stomatitis virus (VSV) has been extensively studied and serves as a model for the replication of (–)RNA viruses in general and rhabdoviruses in particular. Three serotypes have been recognized, Indiana (VSIV), New Jersey (VSNJV), and Alagoas. VSIV is the prototype virus of the genus and has a genome size of 11,161 nt. The genome is neither capped nor polyadenylated, consistent with the fact that it is minus-strand RNA.

Synthesis of mRNAs

The VSV nucleocapsid has about 1250 copies of N protein as its major structural component, leading to the conclusion that each N protein interacts with 9 nucleotides of RNA. The nucleocapsid also contains about 470 molecules of P and 50 copies of L. It can synthesize RNA, and P, L, and N are all required for this activity. The organization of the genome and the production of five mRNAs from it are illustrated in Fig. 4.2. There is a single polymerase entry site at the 3' end of the genome, and production of mRNAs is obligatorily sequential. Synthesis begins at the exact 3' end of the genome and a leader RNA of 48 nucleotides is first synthesized. The leader is released and synthesis of the first mRNA, that for N, is initiated. The RNA polymerase complex has capping activity, and the mRNA is capped during or shortly after initiation. At the end of the gene for N, the transcriptase reaches a conserved sequence AUACUUUUUUU, where it begins to stutter and produces a poly(A) tract at the 3' end of the mRNA. The polymerase complex will not terminate or stutter unless the conserved AUAC is present immediately upstream of the U_7 tract, and the sequence $AUACU_7$ is therefore a consensus termination-polyadenylation signal. The capped and polyadenylated mRNA for N is terminated and released, the transcriptase skips the next two nucleotides, which are referred to as the intergenic sequence, and initiates synthesis of the second mRNA, that for P, at the conserved gene start signal UUGUC. Following synthesis of this mRNA, the polymerase again stutters at the oligo(U) tract in the $AUACU_7$ signal to produce a poly(A) tract, releases the capped and polyadenylated mRNA, skips the next two nucleotides, and begins synthesis of the third gene, that for M. The process continues in this way through the fourth gene (the G protein) and the fifth gene (the L protein, L for large because it comprises about 60% of the genome). In this way, five capped and polyadenylated mRNAs are produced. In VSV, the intergenic sequence is always two nucleotides. After releasing the L mRNA, the polymerase complex terminates synthesis some 50 nucleotides before the 5' end of the genome is reached.

As described earlier, synthesis of the mRNAs proceeds in strict sequential order and the attenuation that occurs at each initiation step results in a gradient in the amounts of mRNAs produced. This attenuation appears to be important for regulation of the virus life cycle, so that the mRNAs for proteins needed in most abundance are produced in most abundance. Reorganization of the genome to change the order of genes gives rise to viable virus, but the yield of such virus during an infection cycle in cultured cells, and thus the fitness of the virus, is reduced.

TABLE 4.2   *Rhabdoviridae*

Genus/members[a]	Virus name abbreviation	Usual host(s)	Transmission/ vector?	Disease	World distribution
Vesiculovirus					
Vesicular stomatitis Indiana	VSIV	Humans, horses, ruminants, swine	Airborne, Insects?	Vesicles on tongue and lips	Americas
Chandipura Virus	CHPV	Mammals, including humans	Sandflies	Febrile illness	India, Asia?
Piry	PIRYV	Mice, humans	Sandflies	Febrile illness	Brazil
Lyssavirus					
Rabies	RABV	Humans, dogs, skunks, foxes, raccoons	Infectious saliva	Malaise, then delirium, then coma and death	Worldwide except some islands, and Australia
Bat lyssaviruses	ABLV,[b] EBLV, LBV	Bats, humans	Infectious saliva	Like rabies	Europe, Africa, Australia
Mokola	?	Humans, dogs, cats, shrews	?	Like rabies	Africa
Ephemerovirus					
Bovine ephemeral fever	BEFV	Cattle, water buffalo	Hematophagous arthropods	Fever, anorexia	Africa, Asia, Australia
Adelaide River	ARV	Cattle			
Berrimah	BRMV	Cattle			
Novirhabdovirus					
Infectious hematopoietic necrosis (and other fish viruses)	IHNV	Salmonid fish	Waterborne, contaminated eggs	Hemorrhage	Pacific Northwest of North America
Cytorhabdovirus					
Lettuce necrotic yellows		Plants	Aphids	?	? ?
Northern cereal mosaic		Plants	Leafhoppers		
Strawberry crinkle		Plants	Aphids		
Nucleorhabdovirus					
Potato yellow dwarf		Plants	Leafhoppers		
Maize mosaic		Plants	Leafhoppers		
Sonchus yellow net		Plants	Aphids		

[a] Representative members of each genus are shown, and the first virus listed is the type species.

[b] Virus name abbreviations: ABLV, Australian bat lyssavirus; EBLV, European bat lyssaviruses -1 and -2; LBV, Lagos bat virus.

The mRNAs for N, M, G, and L are each translated into a single protein. That for P is translated into three proteins. The major translation product of this mRNA is P, which is produced using an initiation codon near the 5′ end of the mRNA. Initiation of translation also occurs at two downstream AUGs. These two downstream AUGs are in frame with one another but in a different reading frame from P. Use of these alternative AUGs leads to the synthesis of short proteins of 55 and 65 amino acids (of which the shorter protein is a truncated version of the longer one). The functions of these small proteins have not been established for VSV, but extra products translated from the P gene of the paramyxoviruses are known to interfere with host defense mechanisms, as described later.

Replication of the Genome

Synthesis of viral proteins, in particular of the N protein, allows the enzymatic activity present in the genomic nucleocapsid to switch from synthesis of messengers to replication of the genome. Replication requires producing a full-length antigenomic template, and the immediate encapsidation of the newly synthesized (+)RNA into plus-strand RNP containing N, P, and L during synthesis is required. In the absence of N for encapsidation, the system defaults to synthesis of mRNAs. The M protein also appears to regulate RNA synthesis. In the replication mode, the polymerase complex ignores all of the initiation, termination, and polyadenylation signals utilized to produce mRNAs, and instead produces a perfect

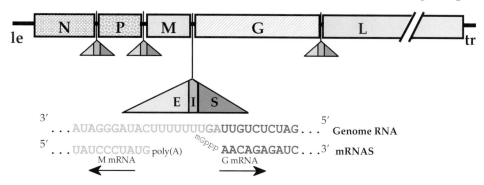

A Location of intergenic sequences of VSV (a rhabdovirus), and detailed view of the M/G intergenic region

3′ ...AUAGGGAUACUUUUUUUGAUUGUCUCUAG... 5′ Genome RNA

5′ ...UAUCCCUAUG poly(A) mGppp AACAGAGAUC...3′ mRNAS

M mRNA   G mRNA

B Genomic sequences at other intergenic regions in the VSV genome

3′   5′

N/P ...CGAUGUAUACUUUUUUUGAUUGUCUAUAG...

P/M ...CAUCUGAUACUUUUUUUCAUUGUCUAUAG...

G/L ...UUAAAAAUACUUUUUUUGAUUGUCGUUAG...

FIGURE 4.2 (A) Schematic diagram of the VSV genome. le is the leader sequence; tr is the trailer sequence. The 5 genes N, P, M, G, and L were defined in the legend to Fig. 4.1 and are described in more detail in the text. The positions of the conserved regulatory sequences at the gene boundaries are shown by the triangles. Each of these intergenic sequences is composed of E (end), I (intergenic), and S (start) domains. (B) Sequences in VSV at the other three gene boundaries. Data for this figure came from Rose and Schubert (1987).

complementary copy of the genome. The antigenomic RNA can be copied by the polymerase activity in the (+)RNP to produce more genomic RNA. This also requires that the RNA be immediately encapsidated. The new genomic RNP can be used to amplify the replication of viral RNA or, later in infection, can bud to produce progeny virions.

Maturation of Virus

The G protein has a 16-residue N-terminal signal sequence that leads to its insertion into the endoplasmic reticulum during translation. The signal is removed by cellular signalase. The resulting 495-residue protein is anchored near the C terminus by a 20-residue transmembrane anchor, with the 29 C-terminal residues forming a cytoplasmic domain (i.e., it is a type 1 integral membrane protein). G is glycosylated on two asparagine residues and transported to the plasma membrane, where progeny viruses are formed by budding (Fig. 2.23D). The M protein appears to form an adaptor between the glycoprotein present in the plasma membrane and the nucleocapsids assembled inside the cell. M also acts to repress RNA synthesis by the viral nucleocapsid. The G protein contains the fusion activity and receptor recognition activities of the virus, and it is the only protein present on the surface of the virion. The assembled virion contains about 1200 molecules of G, present as trimers that form spikes visible in the electron microscope, and about 1800 molecules of M.

Vesiculovirus Diseases

VSV causes nonfatal but economically important and debilitating disease in cattle, pigs, and horses. The name of the virus comes from the vesicles that it induces on the tongue and lips. These symptoms resemble those caused by foot-and-mouth disease virus and epidemics of VSV disease in domestic animals result in disruptive quarantines as well as complications in efforts for control of FMDV. Human infection is common in rural areas where VSV is endemic in domestic animals; 25–90% of farmers in such areas may have anti-VSV antibodies, showing past infection by the virus. Human infection is largely asymptomatic or associated with a mild febrile illness, sometimes accompanied by herpeslike lesions in the mouth or on the lips or nose. Serological surverys also show that the virus infects bats, deer, and monkeys in endemic areas. The virus also replicates in numerous arthropods and has been isolated from mosquitoes, sand flies, black flies, culicoides, houseflies, and eye gnats. The natural cycle of VSV in nature is not understood and the epidemiological importance of mosquitoes or other hematophagous arthropods in transmission of the virus is not clear.

VSV is endemic in Latin America from Mexico to northern South America where outbreaks of disease occur every year. VSNJV accounts for the majority of the clinical cases in this region. Sporadic outbreaks occur both north and south of this endemic area. In the United States, sporadic outbreaks occur in the Southwest at intervals of about 10 years, caused by both VSIV and VSNJV. In the Southeast, VSNJV was endemic until the 1970s. After this, VSNJV remained endemic only on Ossabaw Island off Georgia, where it is transmitted to feral pigs by sand flies. In the rest of the Southeast, no clinical disease caused by VSV has been reported since 1976 and there have been only occasional findings of seropositive wild animals.

Chandipura virus, another member of the genus *Vesiculovirus* (Table 4.2), is widespread in India, where it infects humans and domestic animals. It is also present in Senegal. It has been isolated from sand flies, which are believed to serve as vectors of the virus. Until recently it was thought to cause no disease or only mild febrile illness in humans. However, recent epidemics of encephalitis in children in India in 2003 and 2004 have been traced to the virus, showing that it has the potential to be a significant human pathogen. In the 2003 epidemic in Andhra Pradesh, for example, 183 of 329 affected children died.

More than 20 other vesiculoviruses are known. As one example, Isafahan virus has been isolated from sand flies in Iran. There is serological evidence of human infection in several central Asian countries but no definite evidence for human illness caused by it.

Genus *Lyssavirus*

The rhabdovirus of greatest medical interest is rabies virus, which belongs to the genus *Lyssavirus*. Seven genotypes or species of lyssavirus are currently recognized and two additional genotypes have been proposed. Genotype 1 is classical rabies virus and is virtually worldwide in distribution. It is the only lyssavirus found in the United States where it infects a wide range of hosts, notably raccoons, wolves, skunks, and bats. Mokola virus is an African virus that is known to infect dogs and cats as well as shrews and humans. The remaining five genotypes are bat-associated lyssaviruses. Lagos bat virus and Duvenhage virus are African, there are two European bat lyssaviruses called type 1 and type 2, and Australian bat lyssavirus is Australian as its name implies. Two additional genotypes of bat viruses present in central Asia have been proposed.

The genome of rabies closely resembles that of the VSV, although very little sequence identity exists between the genomes of the viruses belonging to the two genera. One difference is the lack of a second protein encoded in the P gene of lyssaviruses.

Rabies Virus

Most lyssaviruses cause the disease called rabies in humans and other mammals. It is a uniformly fatal disease of man and of other mammals, and has been known since the twenty-third century B.C. Rabies virus is present in the saliva of a rabid animal and is transmitted by its bite. Infection begins in tissues surrounding the site of the bite. Without treatment the virus may be transmitted to the brain, where replication of the virus leads to the disease called rabies. It is believed that the virus enters neurons by using acetylcholine receptors as a receptor, followed by transport up the axon until it reaches the cell body. The probability that rabies will develop following the bite of a rabid animal depends on the location of the bite, the species doing the biting, and the virus strain. In the absence of treatment, bites on the face and head result in rabies in 40–80% of cases, whereas bites on the legs result in rabies in 0–10% of cases. The incubation period to development of symptomatic rabies can vary from less than a week to several years. Once the virus reaches the brain, it spreads from there to a variety of organs. To be transmitted, it must spread to the salivary glands. Infection of neurons in the brain may result in behavioral changes that cause the animal to become belligerent and bite other animals, so that the virus present in salivary fluid is transmitted. In humans, the disease may be paralytic or may result in nonspecific neurological symptoms including anxiety, agitation, and delirium. Biting behavior is not a consequence of rabies-induced neurological disease in humans, and human-to-human transmission does not occur. Two to 7 days after symptoms of rabies appear, coma and death ensue. Only three cases of humans recovering from symptomatic rabies have been recorded.

For centuries, the saliva of a rabid dog was thought to be the source of rabies infection, but it was only in 1804 that Zinke succeeded in transmitting rabies from it. In the late 1800s, Pasteur adapted rabies virus to laboratory animals and developed the concept of protective vaccination against rabies. The dessicated spinal cords from rabies-infected rabbits became the first rabies vaccine. On July 6, 1885, this vaccine was used to immunize Joseph Miester, who had been bitten 14 times by a rabid dog. Because of the multiplicity of bites, he would almost surely have died, but the Pasteur vaccine saved him. A vaccine grown in nervous system tissue and inactivated by phenol rather than drying was the accepted rabies vaccine for decades. In the 1960s, a safer inactivated virus vaccine derived from virus grown in cultured human cells was introduced. The rabies vaccine is unique in that it is normally given after exposure to the virus, in conjunction with anti-rabies antiserum. This is possible because there is a window of time following the bite of a rabid animal before rabies develops, during which a protective immune response can be induced. Veterinarians and wildlife workers who are potentially exposed to rabid animals, as well as biologists who work with rabies virus in the laboratory, are immunized prophylatically, but the protective immune response can be of short duration and immunity must be tested at regular intervals.

In the United States, Canada, and Western Europe, where vaccination of domestic dogs is widely practiced, wild animals such as raccoons and skunks maintain the virus and transmit it to humans or their domestic animals. Fig. 4.3 shows the decline in number of cases of rabies in dogs and humans in the United States since the 1940s, the result of compulsory vaccination of pet dogs. Fig. 4.3 also shows the increase in rabies in wild animals since 1940. Fig. 4.4 illustrates the explosive spread of rabies in raccoons on the eastern seaboard in the last 20 years. The focus of this spread in Virginia arose from the import of 3500 raccoons from Florida into the Washington, D.C., area by members of the cabinet of President Carter, for the purpose of raccoon hunting. These imported animals ignited an epidemic of rabies in raccoons that has slowly spread up and down the Atlantic seaboard, as shown in Fig. 4.4. In other parts of the United States, foxes and skunks are important hosts for rabies and disease also occurs at times in wolves and coyotes. Bats are also an important carrier of rabies. In other parts of the world, where licensing and immunization of pets is not required, domestic dogs continue to be the principal vectors that transmit rabies to humans. Rabies remains a significant global health problem. More than one million people annually undergo antirabies treatment following exposure to the virus, and 50,000 people die of rabies each year.

Efforts to control rabies in wildlife in the United States and Western Europe have met with some success. These efforts involve vaccinating wildlife with attenuated rabies virus or with recombinant vaccinia viruses that express the rabies G protein, using bait containing one of these viruses that is dispersed by hand or by airplane. In the eastern United States, spread of baits has been used to slow or prevent the further spread of rabies up the eastern seaboard. In Europe, baits have been used to set up barriers to halt the spread of rabies in foxes. The European efforts have been more successful than those in the United States.

The perpetuation of rabies in nature is somewhat of a mystery because of the fact that it can be maintained only in rabid animals who die quickly of the infection. How is it that the virus manages to persist? One possibility arises from recent findings that rabies virus can establish a latent infection in humans. Five cases have been documented in which people did not develop symptoms for 7 or more years after infection with the virus. In at least some of these cases, progression to rabies appeared to be triggered by hormonal changes during puberty. If the virus can establish a latent infection in other animals that is later followed by reemergence of the virus and its transmission to new susceptibles, this could serve as a reservoir of the virus.

Bats may also be an important reservoir of (classical) rabies virus (bat lyssaviruses are discussed separately later). Rabies virus infection of bats seems to take longer to kill the animal, during which time the virus may be transmissible through the bite of an infected bat or through aerosols from infected bat feces or saliva spray. However, rabies virus in bats is distinguishable from rabies virus strains in other wildlife by nucleotide sequence analysis. Thus, mixing of bat rabies and rabies in other wildlife is infrequent. Bats can transmit rabies to humans, and cases of human rabies transmitted by bats in the United States have been documented. In fact, in the United States in recent years, cases of human

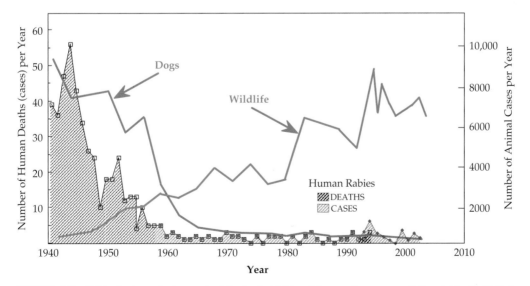

FIGURE 4.3 Rabies in domestic dogs and wild animals (right scale) versus human cases (left scale) in the United States 1940–2003. Note that untreated rabies in humans is uniformly fatal. Data from Smith *et al.* (1995), and *MMWR Summaries of Notifiable Diseases, 1996* and *2003*.

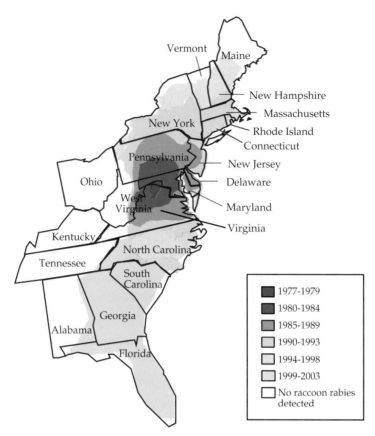

FIGURE 4.4   Spread of raccoon rabies over 3- to 5-year increments in the states of the Atlantic seaboard. Over 25 years the virus has spread from a small focal area in northern Virginia to encompass much of the entire region from southern Maine to Florida. From *Morbidity and Mortality Weekly Report* (*MMWR*) (1997) Vol. 45, p. 1119, updated with data from CDC found on www.rabavert.com/caserc.html.

rabies resulting from infection with bat-associated rabies virus have been more numerous than cases resulting from infection by bites of other rabid wildlife. In many cases of bat-associated rabies, the mechanism by which the virus was transmitted to the human is not known, because no exposure to bats, rabid or otherwise, could be shown.

Bat Lyssaviruses

Australia was long believed to be completely free of rabies. However, it has recently been found that many Australian bats carry a virus known as Australian bat lyssavirus (ABLV). Two cases of fatal human rabies that were caused by infection with this bat virus have occurred in the last few years. In one case, a woman caring for injured bats was bitten by a bat in her care. In the second incident, a woman was bitten while trying to remove a bat that had landed on a child. No rabies has been found in other animals, presumably because there is no efficient mechanism for transmission of the virus among other mammals present in the continent. However, the disease could potentially spread to dogs and cats that have been introduced into Australia over the years.

There are two major types of bats. Bats belonging to the suborder *Megachiroptera*, of which 40 genera are recognized, are large and feed on fruit and nectar in flowers. Members belonging to the genus *Pteropus* are often called flying foxes and are found from Australia across India to Madagascar. There are four species of flying foxes in Australia. Bats belonging to the suborder *Microchiroptera* are smaller, feed on insects, and have developed echolocation to find their prey in the dark. Both types of bats carry ABLV in Australia, and of the two cases of human infection that have resulted, one was from a flying fox and the second was from an insectivorous bat. The strains of virus in the two types of bats are distinguishable, differing by about 20% in nucleotide sequence.

Two different European bat lyssaviruses exist. Four human deaths resulting from infection by these viruses have occurred, and there is concern that more cases might occur. Mokola virus has also infected humans in Africa.

Other Genera of Rhabdoviruses

Bovine ephemeral fever virus, genus *Ephemerovirus*, is an arbovirus that causes economically important disease in domestic cattle and water buffalo in tropical areas of the Old World. This virus, as well as other members of the genus, are not known to be human disease agents. The genome encodes five additional genes located between G and L. One of these encodes an additional glycoprotein whose significance is unknown. The other four genes encode five small proteins of unknown function.

The novirhabdoviruses infect salmon and other fish and are responsible for economic losses in fish farming operations. One additional gene is present in infectious hematopoietic necrosis virus, located between G and L. The nucleorhabdoviruses and cytorhabdoviruses are plant viruses that are transmitted by arthropods. They replicate in both the arthropod vectors as well as in plants and are the plant equivalent of arboviruses. Two (in maize fine streak virus, a nucleorhabdovirus) or four (in northern cereal mosaic virus, a cytorhabdovirus) additional genes are positioned between P and M. There are in addition many plant rhabdoviruses that have not been assigned to genus.

FAMILY *PARAMYXOVIRIDAE*

The family *Paramyxoviridae* has seven genera, listed in Table 4.3 together with representative viruses in each genus. The relationships among the genera are illustrated in the tree shown in Fig. 4.5. Each genus represents a distinct lineage. Furthermore, *Respirovirus*, *Morbillivirus*, *Henipavirus*, and *Rubulavirus* are more closely related to one another than to *Pneumovirus* and *Metapneumovirus*, and the family is divided into two subfamilies, *Paramyxovirinae* and *Pneumovirinae*. Many of the viruses belonging to this family are very important human pathogens. Some, such as measles virus and mumps virus, have been known for a long time—the infectious diseases caused by these viruses were known to the ancients. At the other extreme, Hendra virus, first described as an "equine morbillivirus of Australia," and the related Nipah virus of Southeast Asia have been known for less than a decade. These two new viruses have been classified into a new genus, *Henipavirus*.

Replication of the *Paramyxoviridae*

The genome organizations of five viruses representing five genera of the paramyxoviruses are shown in Fig. 4.6. The paramyxovirus genome is larger than that of the rhabdoviruses, 15–20 kb, and encodes more proteins, 8–11 or more. A core of six genes is present in paramyxoviruses, N, P, M, F, H, L (different names are used for some of the genes as shown in the figure). The P gene uses more than one reading frame to encode multiple proteins in most of these viruses. Rubulaviruses possess a seventh gene, encoding a protein called SH, and pneumoviruses possess an even larger constellation of genes, 10 in number. It is possible that still other genes are hidden within some of these large genomes. For example, the SH gene, encoding a very small protein, was discovered only recently.

The N or NP (= N in rhabdoviruses), P, M, and L proteins serve the same functions as their counterparts in rhabdoviruses. G of rhabdoviruses is replaced by two glycoproteins in paramyxoviruses, one called F and the other H or HN or G, depending on the virus. The order of genes in the paramyxoviruses is the same as in the rhabdoviruses, and the genome of the ancestral paramyxoviruses could have arisen from that of a rhabdovirus by insertion of extra genes (or vice versa by deletion of genes). Of interest is the fact that the rhabdovirus N protein binds nine nucleotides whereas the paramyxovirus N protein binds six nucleotides, and paramyxoviral RNAs, where studied, contain a number of nucleotides divisible by six.

Virus replication occurs in the cytoplasm. Like the rhabdoviruses, paramyxovirus mRNAs are transcribed sequentially beginning at the 3 end of the genome and the mechanisms to produce these mRNAs are similar to those employed by rhabdoviruses. A leader is first transcribed, poly(A) tracts are added by stuttering at oligo(U) stretches at the end of each gene, intergenic nucleotides are skipped by the polymerase during synthesis of mRNAs, and attenuation of mRNA synthesis occurs at each junction. The intergenic nucleotides are variable among paramyxoviruses, however. They are GAA or GGG for some viruses, but are variable in sequence and in length, from 1 to 60 nucleotides, for others. The mechanisms by which the virus switches from synthesis of mRNAs to replication of the genome are the same as those used by the rhabdoviruses.

The Viral Glycoproteins

Paramyxovirus virions are 150–350 nm in diameter and contain a helical nucleocapsid that is 8–12 nm in diameter. Virions are usually round but pleomorphic (Fig. 2.22C), and filamentous forms are also produced by some viruses that are more common in clinical isolates (Fig. 2.25E). Virions are produced by budding from the plasma membrane (Figs. 2.25E and F). The virion size differs even within a single species and the composition of the virion is not as well defined as for some enveloped viruses. There are two glycoproteins on the surface. One is a fusion protein that is required for the fusion of the viral membrane with the cell plasma membrane. Paramyxoviruses fuse with the plasma membrane, not with endosomal membranes, and fusion does not require exposure to low pH. The fusion protein is produced as a precursor called F_0. When first synthesized, F_0 has an N-terminal signal sequence that results in its insertion into the endoplasmic reticulum during translation. The signal sequence is removed by signal peptidase and the resulting type 1 integral membrane protein is anchored by a membrane-spanning region near the C terminus. F_0 is cleaved either by cellular furin (many paramyxoviruses)

TABLE 4.3   *Paramyxoviridae*

Genus/members[a]	Virus name abbreviation	Usual host(s)	Transmission	Disease	World distribution
Paramyxovirinae					
Respirovirus					
Human parainfluenza 1,3	HPIV-1,3	Humans	Airborne	Respiratory disease	Worldwide
Bovine parainfluenza 3	BPIV-3	Cattle, sheep	Airborne	Respiratory disease	Worldwide
Sendai	SeV	Mice	Airborne	Respiratory disease	Worldwide
Rubulavirus					
Mumps	MuV	Humans	Airborne	Parotitis, orchitis, meningitis	Worldwide
Human parainfluenza 2, 4a,4b	HPIV-2,4	Humans	Airborne	Respiratory disease	Worldwide
Simian virus 5	SV-5	Monkeys, canines	Airborne	Respiratory disease	Worldwide
Menangle	?	Bats/swine	???	Reproductive abnormalities	Australia
Morbillivirus					
Measles	MeV	Humans, monkeys	Airborne	Fever, rash, SSPE[b], immune suppression	Worldwide
Rinderpest	RPV	Cattle, swine	Airborne	Gastroenteritis	Worldwide
Distemper	CDV, PDV	Dogs, marine mammals	Airborne	Immune suppression, gastroenteritis, CNS disease	Worldwide
Henipavirus					
Hendra (equine morbillivirus)	HeV	Humans, equines, *Pteropus* fruit bats	Body fluids?	Respiratory disease, encephalitis	Australia
Nipah	NiV	Humans, swine, cats, dogs	Body fluids?	Respiratory disease, encephalitis	Malaysia, Singapore
Avulavirus					
Newcastle disease, avian paramyxoviruses 2–9	NDV	Gallinaceous birds	Airborne	Respiratory distress, diarrhea	Worldwide
Pneumovirinae					
Pneumovirus					
Human respiratory syncytial	HRSV	Humans	Airborne	Respiratory disease	Worldwide
Bovine respiratory syncytial	BRSV	Cattle	Airborne	Respiratory disease	Worldwide
Pneumonia virus of mice	PVM	Mice	Airborne	Respiratory disease	Worldwide
Metapneumovirus					
Turkey rhinotracheitis	TRTV	Turkeys	Airborne	Respiratory disease	Worldwide

[a] Representative members of each genus are shown, and the first virus listed is the type species.
[b] Abbreviations: SSPE, subacute sclerosing panencephalitis; CNS, central nervous system; CDV, canine distemper virus; PDV, phocine distemper virus.

or cathepsin L (henipaviruses) within the cell or by other cellular enzymes after release of the virion from the cell, depending on its sequence. Cleavage is required for the virus to be infectious and the cleavage products, F_1 (the N-terminal part of the precursor) and F_2 (the C-terminal part which is anchored in the membrane), remain covalently linked through a disulfide bond. The fusion domain consists of the N-terminal 20 amino acids of F_2, but this domain is not fusogenic until cleavage of F_0 has occurred. Those strains whose F_0 can be cleaved intracellularly by furin (which recognizes the sequence RXRR or RXKR) or cathepsin L are able to spread systemically and in general cause serious disease. In contrast, viruses that require cleavage of F_0 by proteases after the release of (noninfectious) virions from the cell, usually at a single basic residue by trypsin-like enzymes such as Clara or miniplasmin that are limited in their distribution in the animal, cannot spread systemically and are usually restricted to the respiratory tract. F oligomerizes to form trimers that are visible as spikes on the surface of the virion.

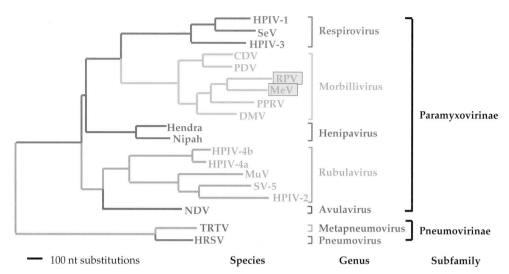

FIGURE 4.5   Phylogenetic tree of the *Paramyxoviridae* derived from the nucleotide sequences of the N ORF. Most of the virus abbreviations are found in Table 4.3. CDV, canine distemper; PDV, phocine distemper; PRRV, peste-des-petits-ruminants; DMV, dolphin morbillivirus. Notice that the closest relative of measles (MeV, boxed), a human virus, is rinderpest (RPV, boxed), a virus of cattle and pigs. Adapted from Chua *et al.* (2000).

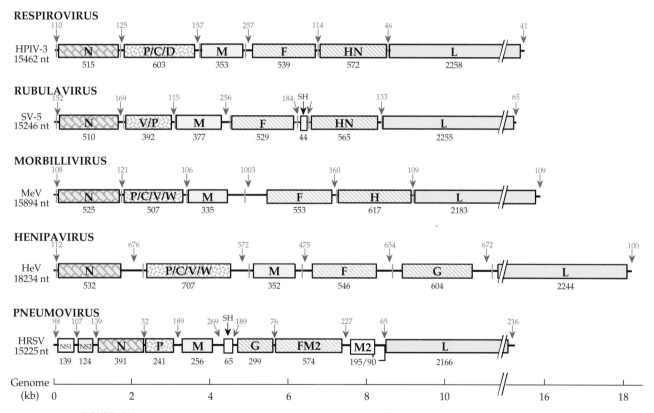

FIGURE 4.6   Genome organizations of the five genera of the *Paramyxoviridae* that infect mammals. The genome is shown 3′ to 5′ for the minus-strand RNA. For the top four genera, each gene begins with the vertical pink bar marking the intergenic sequence. The untranscribed intergenic sequences of respiroviruses, morbilliviruses, and henipaviruses are 3 nt in length. Those of rubulaviruses and pneumoviruses vary in length from 1 to more than 60 nucleotides. The boxes are the ORFs encoding the nucleocapsid (N), the P (V, C, W, D) complex, the matrix protein (M), the fusion protein (F), the glycoprotein (G, HN, or H), and the polymerase (L). Numbers above the arrows are the total number of nucleotides between the ORFs; numbers below the boxes are the number of amino acids in the protein. HPIV-3, human parainfluenza virus 3; SV5, simian virus 5; MeV, measles virus; HeV, Hendra virus; and HRSV, human respiratory syncytial virus.

The second glycoprotein is called the hemagglutinin-neu-raminidase (HN), the hemagglutinin (H), or simply G (for glycoprotein), depending on the virus. This protein is a type 2 integral membrane protein. The signal sequence at the N terminus is not removed but instead serves as the transmembrane anchor for the protein, so that it has its N terminus inside and its C terminus outside. This protein contains the receptor-binding activity of the virus.

Many paramyxoviruses belonging to the genera *Respirovirus* and *Rubulavirus* use sialic acid (*N*-acetylneuraminic acid) bound to protein or lipids as a receptor. Because this receptor is also present on red blood cells, these viruses can cause red blood cells to clump or agglutinate, a process called hemagglutination (*heme* = the red compound in red blood cells that binds oxygen). In paramyxoviruses that use sialic acid as a receptor, this second glycoprotein is also a neuraminidase, in which case it is called HN. Neuraminidase removes sialic acid from potential receptors and from virus glycoproteins. By removing sialic acid from the virus glycoproteins and from the cell surface, released virus is prevented from aggregating with itself or sticking to infected cells. It also increases the probability that the virus will successfully initiate infection of a suitable animal. Mucus, which lines the respiratory tract where the viruses begins infection, contains sialic acid and might otherwise bind virus, preventing its entry into cells, if the virus could not release from mucous by destroying these receptors.

Other receptors used by paramyxoviruses include, among others, CD46 (measles virus), Ephrin B2 (henipaviruses), and glycosaminoglycans (respiratory syncytial virus). If the receptor used by the virus is found on red blood cells, the viruses will hemagglutinate, but if the receptor is not sialic acid, the virus will not contain a neuraminidase. In this case, the second glycoprotein is called H. If the viruses are not known to hemagglutinate, the second glycoprotein is simply called G, for glycoprotein. In any event, this second glycoprotein of paramyxoviruses, best studied in the case of the HN of some paramyxoviruses, oligomerizes to form tetrameric spikes on the surface of the virion.

Some paramyxoviruses belonging to the genera *Rubulavirus* and *Pneumovirus* encode a third integral membrane protein. This small (44–64 residues) protein is called SH or 1A and is glycosylated in the pneumovirus respiratory syncytial virus but not in the rubulaviruses SV5 and mumps. In the rubulaviruses the protein is a type 1 integral membrane protein whose gene is positioned between F and H (or HN). In SV5 and, probably, in mumps virus also, the protein interferes with the TNF-α mediated apoptosis pathway. Mutants lacking this protein will replicate in cell culture but cause extensive apoptosis, and are attenuated in animals.

The P Gene

Expression of the P gene of paramyxoviruses belonging to the subfamily *Paramyxovirinae*, sometimes called the P/V or P/C/V gene, is remarkable, as illustrated in Fig. 4.7. The P gene, or its equivalent, of most (−)RNA viruses is used to make more than one protein, as was described earlier for the vesiculoviruses and as will be described later for other viruses, but the translation strategies used by some paramyxoviruses result in maximal use of the potential information contained within this gene. In some paramyxoviruses, alternative AUG start codons are used to produce two different proteins translated from different reading frames, similar to what occurs in the vesiculoviruses. A second strategy used by paramyxoviruses is to add nontemplated nucleotides to the mRNA during synthesis in order to shift the reading frame downstream of the added nucleotides. The ultimate use of the paramyxovirus P gene occurs in some viruses in which all three reading frames are translated by using one or both of these strategies to produce four or more proteins.

In respiroviruses, morbilliviruses, and henipaviruses, translation of P mRNA can start at one of two different AUGs that are in different reading frames. One of the two proteins produced is called C and the other P (Fig. 4.7). In addition, during transcription of P mRNA in most members of the *Paramyxovirinae*, nontemplated G residues are added at a specific site in the gene. In measles or Sendai viruses, addition of one G shifts the reading frame after this point to produce a new protein called V, which is rich in cysteine residues. Thus P and V share their N-terminal sequence but diverge after the site where the extra G is added. In the case of parainfluenza virus 3, addition of two G's leads to mRNA translated into a protein called D. Another respirovirus, HPIV-1, lacks editing in the P gene (Fig. 4.7). In the rubulaviruses, the V protein is translated from the unmodified transcript, and production of mRNA for P requires addition of two nontemplated G residues. In mumps virus, addition of 4 G residues also occurs to produce a third protein.

During translation of these P mRNAs, the situation becomes even more complicated. In some viruses, multiple in-frame start codons are used to initiate translation of C. Thus, different forms of C are produced that are variously truncated at their N terminus. For example, Sendai virus expresses four C proteins, called C′, C, Y1, and Y2, of which C′ starts at an ACG codon. The many different protein products produced from the P gene have not been fully characterized, and perhaps not yet fully enumerated, and the various functions of this wealth of proteins are only partially unraveled.

Addition of the nontemplated G residues is thought to involve a mechanism similar to the stuttering that produces a poly(A) tract opposite a string of U's in the template. The nontemplated G's are always added at a specific, unique place in the genome characterized by a string of C's. There must be some signal within the genome that is recognized by the viral polymerase for the addition of the extra G's, similar to the case for the addition of the poly(A) tract at the end of mRNAs.

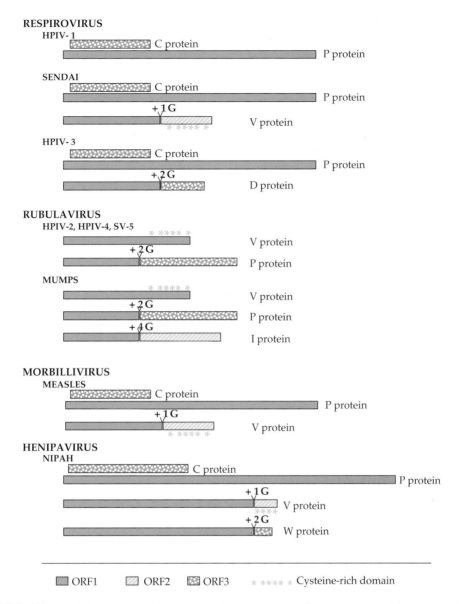

FIGURE 4.7 Translation strategy of the P gene of paramyxoviruses. In most paramyxoviruses, nontemplated nucleotides are inserted during transcription of P to shift the translation frame. Alternative translation start codons are also used. The result is the production of up to four proteins from this one gene. Adapted from Strauss and Strauss (1991) and Chua *et al.* (2000).

An important function of the V and C proteins, perhaps the primary function, is to block the action of the hosts interferon (IFN) system. Such an activity is called a luxury function because it is not needed for virus replication in cultured cells but is needed for a successful infection of an animal. Production of IFN is the first line of defense of birds and mammals against virus infection and the system is described in detail in Chapter 10. The importance of IFN in controlling viral infection is shown by the fact that most if not all viruses interfere with its action in some way. In paramyxoviruses, the V and C proteins act in different ways to block IFN production or its activity once induced. A potent inducer of IFN

production is double-strand RNA (dsRNA), and there are at least two cellular sensors that detect dsRNA and induce the production of IFN (described in more detail in Chapter 10). Intracellular dsRNA can be sensed by a helicase called RIG-1. Through a complicated pathway involving caspase-recruitment domains (CARD), two transcription factors, IRF-3 and NFκB, are activated and transported to the nucleus. These form a complex that leads to the transcription of the mRNA for IFN-β. An overlapping pathway can start from extracellular dsRNA, which is bound by a cellular receptor called Toll-like receptor 3. The resulting signal cascade results in the production of the same two activated transcription factors. The V proteins of several paramyxoviruses interfere with either activation pathway, preventing the induction of IFN. In addition, the W protein of Nipah virus blocks the activity of IRF-3 in the nucleus. The cysteine-rich C-terminal domains of the V proteins are highly conserved and presumably work in the same way in this pathway.

The V proteins, and in some cases the P proteins, also block the activity of IFN-β once it is produced. IFN-β is exported from the cell where it can be bound by Type I IFN receptors at the surface of the same cell or other cells. Once IFN-β is bound, the receptor, which consists of two different subunits, heterodimerizes and associated adaptor proteins phosphorylate one another and phosphorylate transcription factors called STAT1 and STAT2. Once phosphorylated, the STATs heterodimerize and are transported to the nucleus where they form part of a transcription complex that transcribes mRNA from hundreds of IFN-responsive genes and the products of these genes establish an antiviral state. The V protein and P protein of Nipah virus cause the STATs to aggregate into large, inactive complexes. The V proteins of several other paramyxoviruses cause STAT1 or STAT2 to be degraded by proteasomes. The W proteins of Nipah virus also blocks the activity of the STATs in the nucleus by causing the proteins to aggregate. The net result of these activities is that the genes responsive to activation by IFN-β are not transcribed and IFN activity is aborted (see also Chapter 10).

The V proteins are also involved in the regulation of RNA synthesis after infection. Where studied, they downregulate the production of viral RNA.

Genus *Respirovirus*

The genus *Respirovirus* contains several parainfluenza viruses (abbreviated PIVs) and Sendai virus (from Sendai, Japan, where it was isolated; also called mouse PIV-1) (Table 4.3). The two human respiroviruses, HPIV-1 and HPIV-3, cause a respiratory illness similar to that caused by influenza virus and utilize sialic acid as a receptor, as does influenza. They were once grouped with influenza virus as myxoviruses (*myxo* from mucus because the viruses attach to mucus, which contains sialic acid). When they were separated from influenza virus into a distinct family, they were called parainfluenza viruses and the family was named *Paramyxoviridae*.

The respiratory tract infections caused by HPIV-1 and HPIV-3 may be limited to the upper respiratory tract, causing colds, or may also involve the lower respiratory tract, causing bronchopneumonia, bronchiolitis, or bronchitis. These viruses are widespread around the world and are an important cause of lower respiratory tract disease in young children. Serological studies have shown that most children are infected by HPIV-3 by 2–4 years of age, and that the incidence of infection can be as high as 67 out of 100 children per year during the first 2 years of life (i.e., reinfections

are common). Thus, immunity is incomplete and the viruses continue to reinfect older children and adults. However, subsequent infections are normally less severe and there is a reduction in the incidence of lower respiratory tract disease (which is more serious than infection of the upper respiratory tract). The viruses, as is common for respiratory tract infections, are spread by respiratory droplets.

Attempts to develop vaccines against the HPIVs have not met with success. Because of incomplete immunity produced by natural infections, the primary purpose of a vaccine would be to decrease the severity of natural infection by the virus. Even so, results to date have been disappointing. Inactivated virus vaccines developed for HPIV-1 and -3, as well as for HPIV-2, a rubulavirus, were antigenic but failed to induce resistance to the viruses. This could have resulted from failure to develop IgA following a parenterally administered vaccine (Chapter 10), and attempts to develop effective vaccines are continuing.

Genus *Rubulavirus*

Mumps Virus

The genus *Rubulavirus* gets its name from an old name for mumps, which is the disease produced in humans by mumps virus. The only natural hosts for mumps virus are humans and the virus is transmitted from person to person by contact. The disease has been known (at least) from the fifth century B.C. The incubation period, that is, the period of time between infection by the virus and the development of symptoms, is about 18 days. During the last 7 days of the incubation period, a person sheds virus and is capable of infecting others. Infection of children is usually not serious, but mumps virus infection can cause serious illness, particularly in adults. Infection begins in the upper respiratory tract but becomes systemic with the virus infecting many organs, where it replicates in epithelial cells. It is best known for infection of the parotid salivary glands leading to painful swelling of these glands. More serious disease can result from the replication of the virus in other organs, however. The central nervous system (CNS) is a common target for the virus and 0.5–2.3% cases of mumps encephalitis are fatal. Infection of the pancreas can occur, and it has been suggested that mumps may be associated with sudden onset insulin-dependent diabetes. The heart is sometimes infected, resulting in myocarditis. Infection of the testes in adult males can lead to orchitis and, in rare cases, to sterility. Infection of the fetus can result in spontaneous abortion.

At one time, mumps was one of the common childhood diseases that was contracted by almost everyone. It is now controlled in developed countries by an effective attenuated virus vaccine that was selected by passage of the virus in embryonated eggs. This mumps vaccine is given as part of the MMR (measles–mumps–rubella) combination vaccine.

The dramatic decline in cases of mumps in the United States after introduction of this vaccine is shown in Fig. 4.8. Because mumps is exclusively a human virus that induces effective immunity following infection, and infection of an individual requires direct contact with a person actively shedding the virus, the virus requires a population of at least 200,000 people to sustain it. Such a population density was first attained 4000 or 5000 years ago, before which mumps could not have existed, at least in its current form.

Other Rubulaviruses

Other human rubulaviruses include HPIV-2 and HPIV-4. They are named human parainfluenza viruses because the disease they cause is similar to that caused by HPIV-1 and HPIV-3. However, they are genetically related to the rubulaviruses rather than to the respiroviruses (Fig. 4.5). Other members of the rubulavirus genus infect many mammals and birds. One of the most intensively studied rubulaviruses (studied as a model system for replication of members of the family) has been SV-5 (simian virus 5). During the development of the polio vaccine, rhesus monkey kidney cells were used for replication of poliovirus in culture, and these cultures were often contaminated with monkey viruses. These simian viruses (SVs) were simply numbered as they were isolated, and any particular SV may be totally unrelated to any other. Two of the most widely studied are SV-5 and SV-40, which are not related to one another: SV-5 is an RNA-containing paramyxovirus and SV-40 is a DNA-containing polyomavirus (Chapter 7).

The avian viruses include nine serologically distinct paramyxoviruses. These viruses form a distinct lineage, but clearly group with the rubulaviruses (Fig. 4.5). APMV-1 is also known as Newcastle disease virus (NDV), which causes a highly contagious and fatal disease of birds. NDV has serious economic consequences because it infects chickens, among other avian hosts. When epidemics break out, and they do with some regularity, many birds die, causing economic losses. Quarantines are placed on the movement of birds during epidemics in an effort to curtail the spread of the virus, which has further economic consequences.

Menangle virus is a newly described virus that emerged in Australia in 1997 as the cause of severe reproductive disease in pigs. Menangle virus is a bat virus that was present in flying foxes that formed a large colony near the pig farm where the virus emerged. How the virus was transmitted to pigs is not known with certainty. Two farm workers appeared to have been infected during the outbreak and suffered influenza-like symptoms. Menangle is an example of emerging viral diseases, a topic covered in more detail in Chapter 8.

Genus *Morbillivirus*

The genus *Morbillivirus* contains measles virus as well as a number of nonhuman pathogens that include rinderpest virus, which infects cattle and pigs, and distemper viruses of dogs, dolphins, and porpoises. The relationships among these viruses were illustrated in Fig. 4.5.

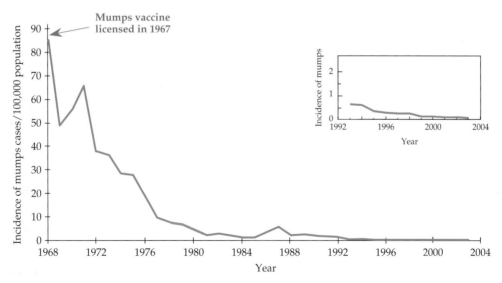

FIGURE 4.8   Incidence of mumps (cases per 100,000 population) in the United States. The minor resurgence of mumps cases in the late 1980s is thought to be due to a pool of susceptible teenagers and young adults who were not aggressively immunized during the first decade after the introduction of the vaccine. In 2003 there were a total of 231 cases of mumps in the United States, the lowest number ever reported for one year (< 0.08 cases per 100,000). The last deaths in the United States from mumps occurred in 2000. However, in 2006 there was a major outbreak of mumps in the Midwest, resulting in 5783 cases. From *MMWR, Summary of Notifiable Diseases-1996*, Vol. 45, p. 45; the comparable Summaries for 2001 and 2003; and *MMWR* (2006) Vol. 55, p. 1152.

Measles

Measles virus causes serious illness in man. Infection begins in the upper respiratory tract but becomes systemic, and many organs become infected. Lymphoid organs and tissues are prominent sites of viral replication, and one consequence of virus infection is immune suppression that lasts for some weeks, apparently due to suppression of T-cell responses. Immune suppression can result in secondary infections that may be serious, even life threatening, and interference with immune function is a major cause of measles mortality. Measles also has uncommon neurological complications, including encephalomyelitis and subacute sclerosing panencephalitis (SSPE). In SSPE, the virus sets up a persistent but modified infection in the brain in which M protein is produced in only low amounts; downregulation of production of M protein appears to be necessary to establish the disease syndrome. Symptoms of SSPE appear several years after measles infection, and the disease progresses slowly but inexorably. Serious complications caused by viral infection of other organs can also occur.

Natural History of Measles Virus

Like mumps, measles is a disease of civilization. The virus is a human virus. Although subhuman primates can be infected by the virus and suffer the same disease as humans, humans are the only reservoir of the virus in nature. Infection requires direct contact with an infected person and recovery from infection results in solid lifelong immunity to the virus. Thus, a minimum size population is required to maintain the virus, in which the continuing birth of new susceptibles occurs at a rate sufficient to maintain continuous virus infection within the community. The requirement for a minimum sized human population to sustain the virus is illustrated in Fig. 4.9. In this figure, data from 1949–1964 (before tourism became as popular as it is today) are plotted that show the duration of measles epidemics on various islands. In Fig. 4.9A, we see that an island must have a population sufficient to produce about 16,000 surviving newborns a year (population about 500,000) in order to maintain the virus continuously in the population. If the population is smaller, the epidemic burns itself out when all susceptibles have been infected. The island is then free of measles until sufficient new susceptibles have been born and measles is once again introduced into the island from outside. The smaller the population, the longer this takes. Note that Guam and Bermuda, with their heavy tourist influx, had measles present more than expected from the curve because the virus is introduced more often, that is, the island population is not truly isolated. Figure 4.9B illustrates that the more densely packed the population, the more readily the virus spreads and therefore the sooner the epidemic burns itself out. The islands shown in this panel all have about the same population, but when the population is compressed into a smaller area, such as in Tonga, person-to-person spread is more efficient and epidemics do not last as long as when the population is dispersed over a larger area, such as in Iceland.

The study of measles epidemics on islands first demonstrated that lifelong immunity arises following infection by measles. After an epidemic of measles in the Faeroe Islands in 1781, the islands were free of measles until the virus was again introduced in 1846 by a Danish visitor. In the 1846 epidemic, 77% of the population of the islands contracted measles, but no one over 65 years of age came down with the disease.

The requirement for a minimum sized population to maintain the virus means that even though measles virus is extraordinarily infectious, the virus could not have existed until perhaps 5000 years ago when human population density became sufficient to support it. At about this time, large population centers arose in the Fertile Crescent, a region of the Middle East encompassing parts of modern Iraq, Syria, Jordan, Israel, Lebanon, and Turkey, which included the upper Tigris and Euphrates rivers and whose climate was conducive to primitive agriculture. These population centers were associated with the cultivation of food plants and the domestication of animals, including bovines. Measles virus is most closely related to rinderpest virus (Fig. 4.5), which infects cattle and swine. An obvious hypothesis is that the close contact between humans and their domesticated animals allowed rinderpest virus, or perhaps another virus of domestic animals, to jump to humans and evolve to become specific for humans. Subsequent coexistence of the virus with its human host led to the present situation where infection results in significant but relatively low morbidity and mortality.

Introduction of Measles into the Americas and Island Populations

Epidemics of measles were undoubtedly widespread in the Old World following the appearance of measles, although it is difficult now to ascertain the causes of epidemics that occurred thousands of years ago. However, it is clear that measles was widespread in Europe at the time the Europeans began their explorations of the Americas and of the many isolated island communities around the world, and Europeans carried measles with them as they traveled. Introduction of measles virus into virgin populations resulted in very high mortality. Mortality was 26% in Fiji islanders when measles was introduced in 1875, for example. It has been estimated that 56 million people in the Americas died of Old World diseases following European exploration of the New World, and measles and smallpox (Chapter 7) were significant contributors to these deaths. The introduction of measles and smallpox by the Spaniards facilitated the conquest of the Americas by them, and the subsequent depopulation of Central and South America allowed the Spaniards to remain dominant.

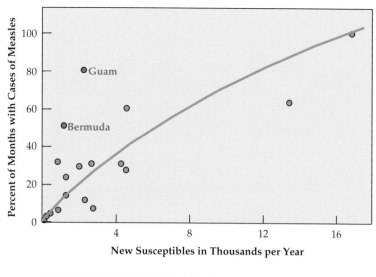

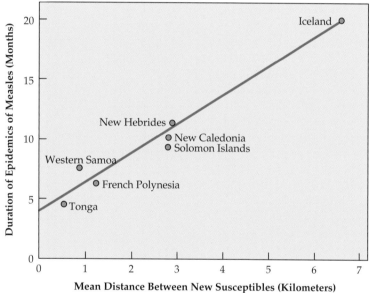

FIGURE 4.9   Effect of population size and density on the epidemiology of measles. Upper panel: percent of months with measles (true endemicity = 100%) in island populations as a function of number of new susceptibles per year. Measles periodically fades out in isolated populations of less that 500,000 (= approximately 15,000 new susceptibles per year). Each dot represents a different island population. Note that Guam (with a transient military population) and Bermuda (with a steady influx of tourists) do not fall on the curve, as they are not truly isolated populations. Vaccination for measles can reduce the number of new susceptibles, even in large urban populations, below the number needed to sustain transmission. Data from Black (1966). Lower panel: relation between the average duration of measles epidemics and the dispersion of populations in isolated islands. All of the islands shown have about the same population, sufficient to introduce 2000–4000 new susceptible children per year. The abscissa is

$$\sqrt{\dfrac{1}{\dfrac{\text{Population input}}{(\text{Land area of the archipelago in kilometers})^2}}}$$

which represents the average distance between infants added to the population each year. Population input is defined as births minus infant mortality.

It has been suggested that the depopulation of the Americas caused by these diseases led the Europeans to introduce Africans as slaves to replace Native Americans being used as slaves.

The very high mortality caused by the virus in naive populations, which contrasts with the low mortality in Europeans, was probably due to two causes. Europeans and other Old World peoples have been continuously exposed to measles for millennia and have been selected for resistance to measles. The people in the Americas had never experienced measles infection, however. A second factor that led to high mortality rates was the introduction of measles into a virgin population, in which not only young children but also all of the adults were susceptible, meaning that the entire population became seriously ill simultaneously. This surely disrupted the ability of the society to maintain itself because there was no one healthy enough to care for the sick.

Vaccination against Measles

At one time, measles virus was epidemic throughout the world and caused one of the childhood illnesses contracted by almost everyone. Because of the extraordinary infectiousness of the virus, very few people escaped infection by it. In the United States, there were about 4 million cases of measles a year, of which about 50,000 required hospitalization and 500 were fatal. There were 4000 cases of measles encephalitis each year, with many patients suffering permanent sequelae. In addition, some fraction of children infected as infants went on to develop SSPE, which is a progressive neurological disease that results in death within about 3 years of the appearance of symptoms. Throughout the world, an estimated 2.5 million children died annually of measles.

Because measles was a widespread and serious disease, attempts to develop a vaccine began at about the same time as attempts to develop a poliovirus vaccine. One vaccine used in the United States from 1963 to 1967 consisted of inactivated measles virus. The vaccine was poorly protective and recipients of this vaccine exposed to measles sometimes developed a more serious form of measles, called "atypical measles," characterized by higher and more prolonged fever, severe skin lesions, and pneumonitis (inflammation of the lungs, from *pneumon* = lung and *itis* = inflammation). The increased severity may have resulted from an unbalanced immune response primed by the formalin inactivated virus or to a lack of local immunity in the respiratory tract (see Chapter 10). Similar problems occurred following vaccination with inactivated respiratory syncytial virus, a paramyxovirus described later.

An attenuated measles virus vaccine, now given as part of the MMR combination vaccine, produced much more satisfactory results. The live virus vaccine gives solid protection from disease caused by the virulent virus and has largely controlled the virus within the United States (Fig. 4.10). Following introduction of the vaccine, the number of cases dropped dramatically. The virus continued to circulate among nonimmunized individuals, however, and thousands of cases per year still occurred, sometimes associated with epidemics of more than 50,000 cases. As vaccine coverage became more effective, cases dropped to new lows, but another epidemic in 1989–1991 caused about 50,000 cases. This epidemic occurred in young immunized adults as well as in young children who had not been immunized. Some of the cases in young adults were due to vaccine failures (about 5% of humans immunized with a single dose of the measles vaccine fail to develop immunity to measles), but other cases appear to have been due to waning immunity. Thus, immunity induced by the vaccine is probably not lifelong, in contrast to natural infection by wild-type measles virus. The guidelines now call for a second immunization on entry to elementary or middle school. This not only boosts immunity in individuals whose immunity is waning, but also usually leads to immunity in those who did not develop immunity after the first dose. In addition, some colleges require immunization on entry. With these changes, the number of cases of measles in the United States was only 100 in 1998.

Molecular genotyping has increased our understanding of the few cases of measles that occur annually in the United States today. In 1988–1992, all the reported isolates of measles virus were subgroup 2, the indigenous North American strain. However, by 1994–1995, all outbreaks were caused by one of four other subgroups that are endemic in other parts of the world. Thus, these outbreaks were initiated by viremic visitors from Asia and Europe. One notable outbreak is thought to have been initiated by a single visitor to Las Vegas and resulted in small epidemics in five states.

After control of measles in the United States and other developed countries, the virus remained epidemic in many parts of the developing world. Control has recently been established throughout most of the Americas, but measles remains a serious pathogen in other parts of the developing world. In fact, measles is the leading cause of vaccine preventable deaths in the world. The World Health Organization has initiated a campaign to eradicate measles but the campaign has met a number of stumbling blocks. Because the virus infects only humans in nature, it should be possible to eradicate it using the same techniques that were used for smallpox and that are being used for poliovirus. A major problem with measles, however, has been the inability to effectively immunize young infants against the disease before they become naturally infected by the virus. Newborns are protected from infection for 6–12 months by maternal antibodies, and a live vaccine does not take while they are thus protected. In many societies, measles is so pervasive that very shortly after the infant becomes susceptible to infection, infection by wild-type virus occurs, thus keeping wild-type virus in circulation. As described in Chapter 10, attempts to overcome maternal immunity by increasing the dose of the vaccine virus have not led to satisfactory results. New guidelines being developed recommend multiple immunizations against measles in such circumstances, starting at an earlier age, in order to catch the child with the vaccine as soon as it becomes susceptible to the virus.

Measles Neuraminidase

The receptor for measles virus is a protein called CD46 that is expressed on the surface of human and monkey cells (Chapter 1), which is bound by the measles H pro-

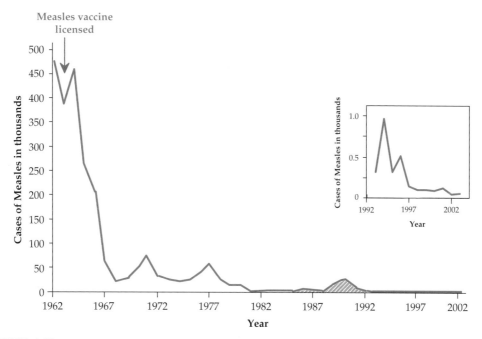

FIGURE 4.10   Cases of measles in the United States by year. Note the difference in the scales of the two graphs. The most recent epidemic of measles, in 1989–1991, with a total of 45,622 cases, was probably due to vaccine failure and waning immunity in children immunized earlier with only a single dose of MMR vaccine. In 2003 a total of 56 cases of measles was reported, of which 24 were internationally imported, and 19 were of persons exposed to the imported infections. In 2005, a total of 66 cases of measles were reported of which 24 were imported and an additional 33 import-linked cases were part of a single outbreak in Indiana initiated by exposure in Romania. From *MMWR, Summary of Notifiable Diseases-1998*; the comparable summary for 2004; and *MMWR* (2006), Vol. 55, p. 1348.

tein. In view of the fact that sialic acid was not the measles receptor, the apparent lack of neuraminidase activity in the H protein was not surprising. Recent studies of the morbillivirus H protein have shown that it is related in structure to the HN protein of paramyxoviruses, however, and that the related rinderpest and peste-des-petits-ruminants viruses possess neuraminidase activity. This neuraminidase activity differs in its specificity from that exhibited by the respiroviruses and the orthomyxoviruses, explaining why it had not been observed previously. Presumably this activity is also found in the measles H protein, and the function of this enzyme in the measles life cycle remains to be determined.

Genus *Henipavirus*

Two species of henipaviruses are currently recognized, Hendra virus and Nipah virus. Hendra virus first emerged in 1994 when an outbreak of severe respiratory illness with a 75% fatality rate occurred in horses near Brisbane, Australia. Two humans also contracted the disease, of whom one died. Hendra virus was quickly isolated and shown to be responsible for the disease. Subsequent studies established that the reservoir of the virus was fruit bats called flying foxes. Small outbreaks of Hendra also occurred in northern Australia in 1994, 1999, and 2004. Hendra virus is widespread and antibodies to it have been found in all four species of fly-

ing foxes in Australia, but from no other animal except the horses infected in these outbreaks.

Nipah virus shares 83% amino acid sequence identity with Hendra virus. It is widely distributed in Southeast Asia and has been isolated from fruit bats in Malaysia, Cambodia, and Bangladesh. It first emerged in 1998–1999 when an outbreak of 258 cases of human encephalitis occurred in Malaysia and Singapore that had a 40% mortality rate. The epidemic was associated with an outbreak of respiratory disease in pigs, and humans infected with the disease were pig farmers or others closely associated with pig farming. There was no evidence for human-to-human transmission in this outbreak.

Recent epidemics of Nipah virus encephalitis have occurred in Bangladesh in 2001, 2003, 2004, and 2005. In these epidemics there was no evidence for the infection of an intermediate animal such as occurred in the Malasian epidemic. Furthermore, in the 2004 epidemic evidence was obtained that person-to-person transmission of the virus had occurred and it is possible that the disease was transmitted directly from bats to humans, possibly by human consumption of partially eaten fruit followed by person-to-person transmission. The fatality rate in these epidemics has been as high as 75%. In nearby India, an epidemic of Nipah occurred in 2001.

Hendra virus and Nipah virus represent emerging pathogens. They are previously unknown viruses that are causing serious disease over widely separated geographic areas.

A more detailed discussion of emerging viral diseases is found in Chapter 8.

Genus *Pneumovirus*

The genus *Pneumovirus*, subfamily *Pneumovirinae*, contains the respiratory syncytial viruses (RSVs). RSVs are known for cattle, mice, sheep, goats, and turkeys, as well as humans. The genome of RSV is more complex than other *Paramyxoviridae*, having more genes (Fig. 4.6). The polymerase gene of RSV is more closely related to those of the filoviruses than to those of the *Paramyxovirinae*, making classification of these viruses problematical.

Human RSV is the most important cause of pneumonia in infants and children worldwide. Half of hospital admissions in the United States in January and February of infants less than 2 years old are due to infection by RSV. Infants are normally infected at 6 weeks to 9 months of age. Infection begins as an upper respiratory tract infection that progresses to the lower respiratory tract in 25–40% of primary infections. Immunity following infection is incomplete and reinfection is common in children and adults, but reinfection tends to produce less severe disease. Symptoms can include bronchitis and pharyngitis (*itis* = inflammation, so inflammation of the mucous membranes of the bronchi or pharynx), rhinorrhea (runny nose), cough, headache, fatigue, and fever. Pneumonia (inflammation of the lungs in which the air sacs become filled with exudate) can result, particularly in infants or the elderly. It is estimated that 17,000 people in the United States die annually from RSV infection, and the great majority of these are people over 65 years old. RSV infection is particularly serious in the immunocompromised. As one example, individuals of any age undergoing bone marrow transplantation have a 90% mortality rate if infected by RSV.

No vaccine is available at the current time for RSV. Because of the widespread prevalence of infection by the virus and the severity of the disease it causes, especially in infants, efforts are ongoing to develop a vaccine that would provide protection against disease or that would at least protect against severe disease. A clinical trial with an inactivated virus vaccine in a group of children some years ago gave disastrous results, however. Not only did the inoculation with the candidate vaccine fail to protect the children against subsequent infection by RSV, but it was found that when infected the vaccinated group suffered a much higher proportion of serious illnesses such as viral pneumonia than did the control group. Thus, immunization potentiated illness, possibly because of an unbalanced immune response. This result has impeded efforts to develop a vaccine and made it clear that a better understanding of the interaction of the virus with the immune system is important.

Many viruses belonging to several different families have now been described that cause respiratory disease, and more viruses belonging to other families will be described later. For comparative purposes, an overview of viruses that cause respiratory disease is shown in Table 4.4. This table is not meant to be comprehensive and includes only a sampling of viruses. Furthermore, some of the viruses in the table, such as measles virus, are better known for disease other than respiratory disease. However, the table makes clear that a large number of diverse viruses can infect the respiratory tract and cause illness.

FAMILY *FILOVIRIDAE*

Table 4.5 lists the known filoviruses, which are classified into two genera, *Marburgvirus* and *Ebolavirus*. Four species of *Ebolavirus* are known, three from Africa (Zaire, Sudan, and Ivory Coast ebolaviruses) and one from the Philippines and/or Southeast Asia (Reston ebolavirus). The filovirus genome is 19 kb in size and contains seven genes, which result in the production of seven or eight proteins following infection (Fig. 4.1). The biology of filoviruses has been very difficult to study. Most filoviruses are severe human pathogens that must be handled under biosafety level 4 conditions (BSL-4), which restricts the number of laboratories that can work with the virus and the number of experiments that can be done. Furthermore, until recently the reservoir of the virus in nature was unknown, limiting studies of the ecology of the virus. A dendrogram showing the relationships among the filoviruses is shown in Fig. 4.11.

The viruses have a genome organization similar to that of other members of the *Mononegavirales*. Their sequences suggest that they are most closely related to the pneumoviruses, and they are assumed to replicate in a manner similar to that for the rhabdoviruses and paramyxoviruses. There are four structural protein genes, encoding the nucleoprotein NP, the glycoprotein GP, and two matrix proteins VP24 (M2) and VP40 (M1). There are three nonstructural protein genes, encoding the viral polymerase and two proteins called VP30 and VP35. The GP gene gives rise to one protein, GP, in Marburg virus. In the Ebola viruses, however, a second glycoprotein, called sGP, is also produced from an edited version of the mRNA for GP. sGP is a soluble, truncated version of GP whose function is unknown, but it is speculated that sGP interferes with the host immune system in some way.

The filovirus virion is enveloped, as is the case for all minus-strand viruses, but rather than being spherical, the virion is long and thread-like (whence the name *filo* as in filament). The infectious virion is 800–1000 nm in length and 80 nm in diameter (Fig. 2.23E), but preparations examined in the electron microscope are pleomorphic and oddly shaped, often appearing as circles or the number 6 but sometimes branched (Figs. 2.23F and G). There is one glycoprotein (GP) in the envelope, present as homotrimers, that is both *N*- and *O*-glycosylated and has a molecular weight of 120–170 kDa.

TABLE 4.4   Viruses Causing Respiratory Disease

Family	Virus[a]	Nucleic acid	Host range	Disease(s)
Orthomyxoviridae	Influenza	(−)RNA	Humans, birds, horses, swine	Rhinitis, pharyngitis, croup, bronchitis, pneumonia
Paramyxoviridae	RSV	(−) RNA	Humans, cattle	Rhinitis, pharyngitis, croup, bronchitis, pneumonia
	Canine distemper		Dogs	Bronchitis, pneumonia
	NDV		Birds	Respiratory distress
	Human parainfluenza		Humans	Rhinitis, pharyngitis, croup, bronchitis, pneumonia
	Measles		Humans	Pneumonia
Bunyaviridae	Sin Nombre	(−) RNA	Humans, rodents	Repiratory distress, pneumonia
Picornaviridae	Rhinoviruses	(+) RNA	Humans	Common cold (rhinitis), pharyngitis
	Coxsackie A		Humans	Rhinitis, pharyngitis
Caliciviridae	Feline calicivirus	(+) RNA	Cats	Rhinitis, tracheitis, pneumonia
Coronaviridae	HCoV	(+) RNA	Humans	Rhinitis
	IBV		Fowl	Bronchitis
	SARS		Humans	Severe acute repiratory disease
Adenoviridae	Human Ad40,41	ds DNA	Humans	Rhinitis, pharyngitis, pneumonia
	CLTV		Dogs	Pharyngitis, tracheitis, bronchitis, and bronchopneumonia
Herpesviridae	Cytomegalovirus	ds DNA	Humans	Pharyngitis, pneumonia
	Herpes simplex, EBV, varicella		Humans	Pharyngitis, pneumonia
	Various alphaherpesvirinae		Cattle, cats, horses, fowl	Rhinotracheitis

[a] Virus name abbreviations: RSV, respiratory syncytial virus; NDV, Newcastle disease virus; HCoV, human coronavirus; IBV, infectious bronchitis virus; CLTV, canine laryngotracheitis; EBV, Epstein-Barr virus.
Adapted from Granoff and Webster (1999), pp. 1493, 1494.

TABLE 4.5   *Filoviridae*

Genus/members	Virus name abbreviation	Usual host(s)[a]	Transmission	Disease	World distribution
Marburgvirus					
Lake Victoria marburgvirus	MARV	Humans	Contact with blood or other body fluids	Severe hemorrhagic disease	Africa
Ebolavirus					
Zaire ebolavirus	ZEBOV				
Sudan ebolavirus	SEBOV	Humans	Contact with blood or other body fluids	Severe hemorrhagic disease	Africa
Cote d'Ivoire ebolavirus	CIEBOV				
Reston ebolavirus	REBOV	Cynomolgus monkeys	?	Severe hemorrhagic disease in monkeys, attenuated in man	Philippines

[a] Natural reservoirs unknown for many years, but were recently found to be bats.

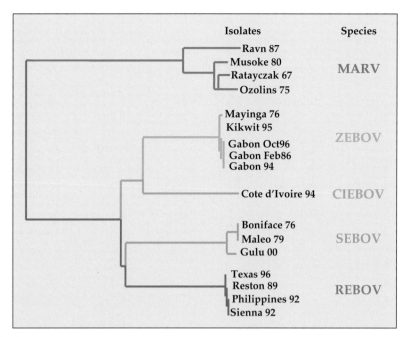

FIGURE 4.11   Phylogenetic tree of the *Filoviridae*, derived from nucleotide sequences for the entire coding region of the glycoprotein genes, using the neighbor-joining method. Adapted from Fauquet *et al.* (2005), Figure 3 on p. 652. Isolates are named by the location and year of isolation. Abbreviations are defined in Table 4.5.

Most of the known filoviruses cause severe hemorrhagic fever in humans with a mortality rate as high as 90%. Because of the dramatic symptoms of the disease involving bleeding from multiple orifices, and the high fatality rate following infection, these viruses, especially Zaire ebolavirus, have been the subject of much discussion in the popular press and have appeared in a number of works of nonfiction as well as fiction. To date, ebolaviruses have caused only a limited number of human cases, but there is always the fear that if the virus were to adapt to humans in a way that allowed for easier transmission, it could become a very big problem.

The filoviruses are examples of emerging viruses and a more detailed discussion of their emergence and the epidemics of disease caused by them are found in Chapter 8. Marburg virus was first isolated in 1967 when it caused an outbreak of hemorrhagic fever in Germany and Yugoslavia, originating from African green monkeys imported from Uganda. The virus is native to central Africa and there have subsequently been epidemics in Kenya, Zaire (the Democratic Republic of Congo), Angola, and Zimbabwe involving from one or a few cases to as many as 374 cases, with an overall mortality rate of 80–90%.

Ebola virus was first isolated during a 1976 epidemic of severe hemorrhagic fever in Zaire and Sudan and named for a river in the region. Subsequent epidemics in these countries and in Gabon and Uganda have occurred at regular intervals (see Chapter 8). The number of cases in an epidemic has been

as high as 600 and the case fatality rate has varied from 50 to 90%. Asymptomatic infection during these epidemics appears to be rare. A milder strain of Ebola virus was isolated from the Ivory Coast in 1994 and three strains or species of African ebolaviruses are now recognized which differ in their virulence (Table 4.5 and Fig. 4.11). Zaire ebolavirus is the most virulent with a case fatality rate approaching 90%, Sudan ebolavirus is less virulent, and Ivory Coast ebolavirus is the least virulent. Human-to-human transmission of the virus requires close contact with the tissues, blood, or other exudates from an infected person and barrier nursing is sufficient to contain the spread of the disease. No vaccine exists for the viruses, but candidate vaccines are in an advanced stage of study.

The natural reservoir of Ebola virus in Africa has recently been shown to be bats. It is clear that monkeys can be infected by the virus and spread it to humans, often when humans use monkeys for food. How the monkeys contract the virus or how human epidemics get started when monkeys are not implicated is not known.

A fourth strain of Ebola virus called Reston ebolavirus first appeared in 1989 in a primate colony in Reston, Virginia, as the causative agent of an epidemic of hemorrhagic fever in monkeys imported from the Philippines. Although highly lethal in monkeys, the virus is not known to cause human disease. Nucleotide sequencing has shown that Reston ebolavirus is closely related to the African Ebola viruses, and the reason it is attenuated in humans is not known.

FAMILY *BORNAVIRIDAE*

Borna disease virus is the sole representative of the family *Bornaviridae* in the order *Mononegavirales* (Table 4.6). It has a nonsegmented, minus-sense genome of 8.9 kb, containing, by one definition, five genes. The general organization of the genome and the suite of genes contained in the genome resembles that of other members of the *Mononegavirales* (Fig. 4.1). However, the virus replicates in the nucleus, not in the cytoplasm. Splicing of mRNAs

occurs to form an incompletely characterized set of alternatively spliced mRNAs from the five genes. Furthermore, alternative start codons are used during translation of some of the mRNAs, a trait shared with other (−)RNA viruses, so that overlapping open reading frames are present in some of the genes. Overall, the number of protein products produced and the complexity of the readout of the genome exceed that of other (−)RNA viruses. The genome organization and transcriptional map of Borna disease virus, as currently understood, is shown in Fig. 4.12.

TABLE 4.6   *Bornaviridae*

Genus/members	Virus name abbreviation	Usual host(s)[a]	Transmission	Disease	World distribution
Bornavirus					
Borna disease	BDV	Horses, sheep other mammals	??	Encephalopathy, fatal paralysis	Europe, possibly worldwide

[a] It was recently reported that the bicolored white-toothed shrew *Crocidura leucodon* is the reservoir host in Switzerland where the disease is endemic. These animals live on the ground and do not climb. Presumably horses are infected when grazing on forage contaminated by excretions from infected shrews; horses fed exclusively from feeding troughs in modern barns are seldom affected (Hilbe *et al.*, 2006).

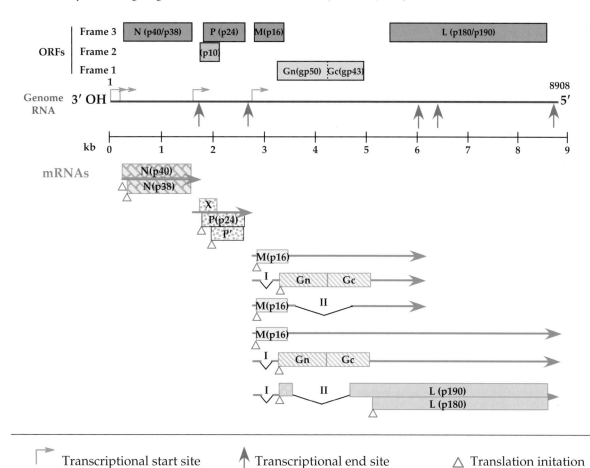

FIGURE 4.12   Genome organization and transcriptional map of an isolate of Borna disease virus. ORFs are represented by the boxes at the top, color coded by reading frame. Messenger RNAs are shown below, with the boxes color coded according to the function of the product as in Figs. 4.1 and 4.6. The positions of two introns, I (nucleotides 1932 to 2025) and II (nucleotides 2410 to 3703), are indicated. Adapted from Fauquet *et al.* (2005), Figure 2 on p. 617.

Borna disease virus is neurotropic and establishes a chronic or persistent infection despite an immune response to viral infection. The chronic infection results at least in part because the virus downregulates its replication, resulting in very low production of infectious virus. Downregulation to establish a persistent infection uses a mechanism different from those described in Chapter 3 for alphaviruses (shut-off of minus-strand RNA) and for pestiviruses (titration of a cellular component required for RNA replication). Borna disease virus, like other (−)RNA viruses, has an inverted terminal repeat at the ends of the genomic RNA that contains promoters for RNA replication. During replication, the four terminal nucleotides at the 5′ ends of both the genomic and antigenomic RNA are often trimmed so that the majority of RNAs are missing these four nucleotides. The truncated RNA can be transcribed to produce mRNA but cannot replicate, thus resulting in downregulation of RNA replication.

Borna disease virus appears to have a very wide host range. It was originally described as a pathogen of sheep and horses in Germany, but is now known to infect a wide variety of warm-blooded vertebrates, birds as well as mammals. The reservoir host in Switzerland has recently been reported to be a shrew (*Crocidura leucodon*). These rodents are exclusively terrestrial and it is thought that horses become infected by grazing on forage contaminated by excretions from infected shrews.

As described, the virus establishes a chronic infection characterized by neurotropism and low production of virus. Infection may be asymptomatic or may result in disease characterized by movement and behavioral abnormalities. Naturally infected horses exhibiting such abnormalities usually recover, but the disease may progress to paralysis and death. Experimentally infected rats and primates also exhibit behavioral abnormalities. Because of these effects on other animals, several recent studies have tried to determine if the virus is associated with neurological disease in man, in particular with schizophrenia. Serological surveys have found that psychiatric patients are more likely to have antibodies to bornavirus than normal controls. Surveys which assay for the presence of viral RNA in peripheral blood mononuclear cells (PMBCs) are even more suggestive: in some surverys up to 66% of psychiatric patients, including schizophrenics, are positive for bornaviral RNA, compared to <5% of normal controls. Furthermore, very small amounts of virus-specific RNA have been isolated from postmortem brain samples from patients suffering from schizophrenia and bipolar disorder, but not from normal individuals or patients suffering from other neurological disorders. Interestingly, a recent study found that two patients hospitalized for severe depression exhibited a rise in bornavirus antigen in PMBCs during the course of the disease, which fell to very low levels on recovery. Whether these different associations are indicative of causality remains to be determined, but it is conceivable that the virus causes recurrent episodes of depression on reactivation of a latent infection.

FAMILY *ORTHOMYXOVIRIDAE*

The family *Orthomyxoviridae* (*ortho* = true or correct) contains three genera of influenza viruses: *Influenzavirus A*, which contains influenza virus A; *Influenzavirus B*, which contains influenza virus B; and *Influenzavirus C*, which contains influenza virus C (Table 4.7). Thogoto virus, a tick-borne virus of mammals, forms a fourth genus, *Thogotovirus* and infectious salmon anemia virus belongs to a fifth genus,

TABLE 4.7   *Orthomyxoviridae*

Genus/members	Virus name abbreviation	Usual host(s)	Transmission	Disease	World distribution
Influenzavirus A					
Influenza A	FLUAV	Humans, birds, swine	Airborne	Respiratory disease	Worldwide
Influenzavirus B					
Influenza B	FLUBV	Humans	Airborne	Respiratory disease	Worldwide
Influenzavirus C					
Influenza C	FLUCV	Humans	Airborne	Respiratory disease	Worldwide
Thogotovirus					
Thogoto virus	THOV	Mammals	Tick-borne	Respiratory disease	Southern Europe, Africa
Isavirus					
Infectious salmon anemia	ISAV	Fish	Waterborne	Anemia, hemorrhagic liver necrosis	North Atlantic, North America

Isavirus. Influenza viruses A and B are closely related, but influenza A infects a wide spectrum of birds and mammals including humans, with birds being the reservoir, whereas influenza B infects primarily humans and humans are the reservoir. Influenza C is more divergent. Eight segments of (−)RNA, totaling about 14 kb, comprise the genomes of influenza A and B viruses (Fig. 4.1) whereas influenza C has only seven segments. Influenza viruses use sialic acid as a receptor, but the form used by influenza A and B viruses differs from that used by influenza C virus, and the enzymes encoded by the viruses to destroy receptors are correspondingly different. All three influenza viruses infect humans and cause disease, but influenza A represents the most serious human pathogen because it causes very large, recurrent epidemics with significant mortality. Influenza A has therefore been the most intensively studied and has been the focus of efforts to control influenza in humans.

Proteins Encoded by the Influenza Viruses

The proteins encoded in the different gene segments of influenza A and influenza C viruses are described in Table 4.8. Influenza A produces 10 proteins from its eight genome segments, and most of these proteins have analogues in other (−)RNA viruses (Fig. 4.1). The matrix protein, M1, and the nucleocapsid protein, NP, perform functions similar to those of M (when present) and N of other (−)RNA viruses, respectively. The three proteins encoded in the three largest segments of influenza, called PB2, PB1, and PA (B or A refers to a basic or acidic pK), possess the RNA polymerase activities encoded in the L protein and the P protein of other (−)RNA viruses. Influenza A and B have two surface glycoproteins, called HA and NA, but influenza C has only one, called HEF. These glycoproteins have the receptor-binding, fusion, and receptor-destroying activities present in surface glycoproteins of (−)RNA viruses.

Two proteins, called NS1 and NS2 (NS for nonstructural), are produced from RNA segment 8. NS1 is produced from the unspliced mRNA (replication occurs in the nucleus). It binds to RNAs in the nucleus, including cellular pre-mRNAs, cellular snRNAs which are involved in splicing, and dsRNA. Its activities inhibit the transport of cellular mRNAs from the nucleus and promote the synthesis of influenza mRNA. NS1 also regulates splicing of influenza mRNAs and their transport from the nucleus to the cytosol. Another function of NS1 is to interfere with the interferon pathway (Chapter 10), in part by binding dsRNA, which is a major inducer of interferon and a cofactor for some proteins in the interferon response to viruses, and in part by interacting with cellular proteins involved in the interferon response. Influenza virus lacking NS1 is very sensitive to interferon and is replication defective in cells or hosts capable of synthesizing interferon,

TABLE 4.8   Genome Segments of Influenza Viruses

Influenza A					Influenza C			
RNA segment	Length (nt)	Encoded Protein		Function[a]	RNA segment	Length (nt)	Encoded Protein	
		Name	(aa)				Name	(aa)
1	2341	PB2	759	Cap recognition, RNA synthesis	1	2365	PB2	774
2	2341	PB1	757	RNA synthesis	2	2363	PB1	754
3	2233	PA	716	RNA synthesis	3	2183	PA	709
4	2073	HA	566	Hemagglutinin, fusion, major surface antigen, sialic acid binding. HEF of FLUCV also has esterase activity	4	2073	HEF	655
5	1565	NP	498	Nucleocapsid protein	5	1809	NP	565
6	1413	NA	454	Neuraminidase				
7	1027	M1	252	Matrix protein	6	Spliced	M1	242
						1180	p42	374
							⇓Signalase	
	Spliced	M2	97	See footnote[b]			M1′(p31)+CM2	259+115
8	934	NS1	230	Nonstructural protein	7	934	NS1	286
	Spliced	NS2	121	See footnote[b]		Spliced	NS2	122

[a] All functions other than those in footnote "b" apply to both influenza A and influenza C.
[b] M2 of Flu A forms an ion channel, and NS2 of FluA is a nuclear export protein; the functions of the comparable moieties of Flu C are unknown.
Source: Adapted from Fields *et al.* (1996) Table 2 on p. 1355 and data in Fauquet *et al.* (2005) p 683.

whereas the wild-type virus is resistant to the interferon pathway. NS2 is produced from a spliced mRNA. It interacts with M1 attached to influenza RNP and promotes the transport of the RNP to the cytoplasm. It is present in small quantities in the virion and so is not truly nonstructural.

Protein M2 is produced from a spliced mRNA from segment 7. It forms ion channels in membranes, probably as a tetramer, that allow passage of H+ ions. During transport of HA to the cell surface, the presence of M2 in the membrane of the transport vesicle causes the pH within the vesicle to equilibrate with that in the cytosol. This prevents low pH activation of the fusion activity of HA during transport, because transport vesicles are otherwise acidic. M2 is also present in virions and is required for the disassembly of the virus and for the activation of the RNA polymerase activity. To become active, the polymerase in the interior of the virus must be exposed to low pH. Influenza virus enters the cell in endosomes, which are progressively acidified. The acidic pH not only triggers a conformational change in HA that results in fusion of the viral membrane with the endosomal membrane, but it also activates the RNA polymerase of the virion through the activity of M2. M2 is the target of the drug amantadine, one of the relatively few drugs that are effective against a viral disease. Amantadine binds M2 of most influenza strains and prevents it from acting as an ion channel, which prevents the activation of the polymerase. When taken early during infection, amantidine ameliorates the symptoms of influenza. A worrisome trend is the appearance of amantidine-resistant variants of influenza, in particular the H5N1 strain referred to as "bird flu."

In most, but not all, influenza A viruses an 11th protein (PB1-F2) is made. This protein is translated from an alternative reading frame from the mRNA of PB1. It is present in mitochrondria in infected cells and may serve to regulate apoptosis by the cell.

Influenza Glycoproteins

Comparison of the glycoproteins of influenza A virus and the paramyxovirus SV-5 is of interest. In both influenza A virus and SV-5, one of the glycoproteins is type 1 (N terminus out) and one is type 2 (C terminus out). In both cases, the type 1 glycoprotein is produced as a precursor that must be cleaved to activate the fusion activity required for entry into cells. The type 1 glycoprotein of influenza A has fusion and receptor-binding (hemagglutinating) activities and is called the hemagglutinin or HA. The precursor is called HA_0 and the cleaved products are called HA_1 and HA_2 (which remain covalently linked by a disulfide bond after cleavage of the peptide bond) (Fig. 1.6). Cleavage is required to activate the fusion activity of the virus and the nature of the cleavage site influences the virulence of the virus. If the cleavage site consists of a single basic amino acid, cleavage is extracellular and influenza replication is restricted to the respiratory tract, and in the case of birds the gut as well, where there are enzymes that can cleave

this site. If the cleavage site consists of multiple basic residues that can be recognized by the intracellular enzyme furin, the virus can replicate systemically in at least some hosts. The SV-5 type 1 glycoprotein has only fusion activity and is called F. As described before, it is produced as a precursor F_0 which is cleaved to F_1 and F_2, and the nature of the cleavage site affects the virulence of the virus (see Viral Glycoproteins under *Paramyxoviridae* earlier in this chapter).

The receptor bound by both influenza A virus and by SV-5 for entry into cells is sialic acid. The type 2 glycoprotein of influenza has neuraminidase activity and is called the neuraminidase or NA. It removes sialic acid from glycoproteins for the same reasons as described for the paramyxoviruses that use sialic acid as a receptor. The type 2 glycoprotein of SV-5 has both neuraminidase activity and receptor-binding (hemagglutinating) activities and is called HN.

Influenza HA is present as a trimer on the surface of the virus (as is F of SV-5). The trimeric spike has a long stalk and a head containing the sialic acid binding sites. As shown in Fig. 1.6, exposure to acid pH in endosomes produces a dramatic rearrangement of the spike in which the fusion peptide, which forms the N terminus of HA_2, is moved over a distance of more than 10nm to the tip of the spike. Here it inserts into the target membrane and promotes fusion of the viral membrane with the target membrane. NA is present as a tetramer (as is HN of SV-5), and forms a spike that is distinguishable in the electron microscope from the HA spike.

There is only one surface glycoprotein in influenza C, the hemagglutinin-esterase-fusion protein (HEF). Influenza C virus has, therefore, one fewer gene segments than influenza A. HEF has receptor-binding (hemagglutination), fusion, and receptor-destroying activities. The receptor is sialic acid, but the activity that destroys the receptor is an esterase activity. The esterase does not remove sialic acid from proteins as does NA of influenza A. Instead it removes the 9-*O*-acetyl group from 9-*O*-acetyl-*N*-acetylneuraminic acid, the receptor used by influenza C, and the virus does not bind to the deacylated sialic acid.

Replication of Influenza RNA and Synthesis of mRNAs

Synthesis of influenza virus RNAs occurs in the nucleus, rather than in the cytoplasm as for most RNA viruses. This makes possible the differential splicing observed for two of the mRNAs. Following infection by the virus, the viral RNPs are transported to the nucleus and mRNA synthesis begins. During synthesis of mRNA, influenza engages in a process called "cap-snatching." Capped cellular pre-mRNAs present in the nucleus are bound by NS1, and the 5'-terminal 10–13 nucleotides, containing the 5' cap, are removed by PB2. This oligonucleotide is used to prime synthesis of mRNA from the influenza genome segments, as illustrated in Fig. 4.13. Once initiated, other aspects of mRNA synthesis resemble

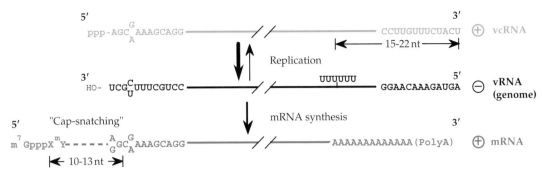

FIGURE 4.13   Relationship between genome RNAs, mRNAs, and vcRNAs of influenza virus. Transcription of mRNAs in the cell nucleus requires a primer of 10–13 nucleotides derived from cellular pre-mRNAs by "cap-snatching," and mRNAs terminate with a poly(A) tail. Those portions of the mRNA which are not complementary to the genome RNA are shown in red. In contrast, vcRNAs are exact complements of the genomic minus strands. Adapted from Strauss and Strauss (1997).

those that occur in rhabdo- and paramyxoviruses. Synthesis continues to near the end of the genome segment, where an oligo(U) stretch is encountered. Here the enzyme stutters to produce a poly(A) tail on the messenger and then releases it. In addition to its role as a primer, using a cap derived from cellular mRNA relieves the virus of the necessity of encoding enzymes required for capping and ensures that the virus mRNA has a cap suitable for the cell in which it is replicating. This mechanism also results in interference with the synthesis and transport of host mRNAs. Furthermore, because the mRNAs have a different 5′ end and lack the 3′ end of the antigenomic RNA, they lack promoters required for replication and packaging and are therefore dedicated mRNAs.

Each genome segment gives rise to one primary mRNA species. However, two of these can be spliced, and both the unspliced and spliced RNAs serve as messengers. Thus, two mRNAs are formed from each of two of the segments, and in total, 10 mRNAs are formed and 10 proteins are produced (11 in the case of viruses that also produce PB1–F2 described earlier). The formation of the two mRNAs from segment 7 and their translation into proteins is illustrated schematically in Fig. 4.14.

When sufficient amounts of viral proteins have been synthesized and transported to the nucleus, viral RNA replication begins. Replication requires encapsidation of progeny genomic and antigenomic RNAs as described for other (−)RNA viruses, and the mechanisms that lead to a switch between synthesis of mRNAs and replication are thought to be similar to those that occur in rhabdoviruses and paramyxoviruses. During replication, the viral genome is copied into a faithful antigenomic RNA (vcRNA) (Fig. 4.13), which is a perfect complement of the genome and serves as a template for production of genomic RNA.

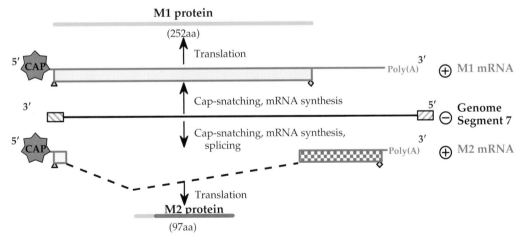

FIGURE 4.14   Synthesis of two mRNAs for the M1 and M2 proteins from gene segment 7 of influenza A. M1 RNA is translated from ORF 1 (open box). M2 RNA starts identically, but after the splice it is translated in ORF2 (checkered box). Both proteins are found in infected cells. The AUG initation codon is shown as a triangle; termination codons are shown as filled diamonds. Patterned boxes at the end of the genome RNA are self-complementary sequences that could form panhandles.

Synthesis of viral RNA, whether plus strand or minus strand, requires that the synthetase interact with both ends of the RNA, whether vRNA or vcRNA; that is, the promoter for synthesis of RNA is composed of elements from both ends of the RNA. This is analogous to what has been found for alphaviruses and flaviviruses, described in Chapter 3, and may be a general mechanism used by many or all RNA viruses. Thirteen nucleotides at the 5' end of the vRNA and 12 nucleotides at the 3' end are highly conserved in influenza A viruses and these seem to contain the entire promoter element. These sequences form an inverted terminal repeat and are capable of forming a panhandle structure (Fig 4.15A), bringing the two ends together where they might interact with the RNA synthetic machinery. An alternative structure, called the corkscrew structure, is thought to be the structure recognized by the synthetase for initiation of RNA synthesis (Fig. 4.15B). Cyclization is also hypothesized to play a role in addition of poly(A) to mRNAs, by causing the polymerase to stutter at the oligo(U) tract located just before the double-strand stem of the circular structure. Figure 4.15C illustrates an experiment to examine the sequence requirements within the panhandle or corkscrew structure.

Assembly of Progeny Virions

Influenza virus matures by budding of nucleocapsids through the cell plasma membrane. Virions are pleomorphic but clinical specimens are primarily filamentous and can be up to a micrometer or more in length. Upon passage in cell culture, most strains eventually give rise to virions that are primarily spherical, averaging 100 nm in diameter. The form that the virions assume is genetically determined. Studies of a strain of influenza A that remained filamentous after passage in cell culture could be induced to form spherical particles by changes in the M1 protein. The significance of filamentous versus spherical particles is unknown, but filamentous forms must have a selective advantage in the infected animal, whereas spherical forms seem to be selected upon passage in cell culture.

During assembly, the eight genome segments are reassorted in progeny virions if the cell is infected with more than one strain of influenza. Reassortment to produce viruses with mixed genomes is efficient—the segments are almost randomly reassorted to give all possible combinations of genome segments in the progeny virions. This

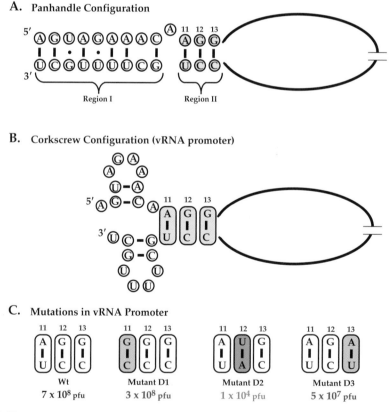

FIGURE 4.15   Models for the influenza A virus promoter. (A) The Panhandle model, with a partially double-stranded structure for the 5' and 3' terminal sequences. (B) The Corkscrew model predicting base pairing within the ends. After Neumann *et al.* (2004). (C) Alternative base pairs introduced into the vRNA promoter and the effect of these changes on viral yield in MDBK cells. Adapted from Catchpole *et al.* (2003), Figure 1.

process is analogous to the reassortment of chromosomes that takes place during sexual reproduction in diploid organisms.

Budding must result in the packaging of the 8 different genomic segments that constitute the viral genome into one virus particle if it is to be infectious. Reoviruses (Chapter 5) have an assembly mechanism whereby the 10–12 different segments are recognized and assorted so that each virus particle has one each of the different segments. The case for influenza virus is not completely clear. Evidence has been presented that the virus appears to package more than 8 segments, possibly about 10, that are randomly chosen from the intracellular pool. Random packaging of 10 segments would result by chance in about 3% of the virions having at least 1 each of the 8 different genome segments. However, more recent data argue that the virions package exactly 8 segments, one each of the 8 different segments. For this to occur, the packaging machinery has to recognize internal sequences in each of the segments and not just a packaging signal in the conserved ends of the viral RNAs.

Influenza A Virus

Natural History of Influenza Virus

Influenza A virus infects a wide variety of birds and mammals. A phylogenic tree that shows the relationships of the NP genes of viruses isolated from humans, pigs, and birds is shown in Fig. 4.16. The human isolates and the pig isolates are closely related; as described later the pig viruses probably originated from a human virus. The human–pig clade is distinct from the avian clade, however.

Influenza A viruses are characterized by their two major surface antigens, HA and NA. There are 16 different HA subtypes (numbered H1 to H16). HAs in different subtypes differ by 30% in sequence and are not immunologically cross protective. There are also 9 different NA subtypes (numbered N1 to N9). The major reservoirs of influenza A in nature are wild ducks and other waterfowl such as gulls, terns, and shearwaters, and viruses containing all 16 subtypes of HA and all 9 subtypes of NA have been isolated from waterfowl. Influenza replicates in the lung and in the gut of birds and the infection is normally asymptomatic (but epidemics of fatal influenza have occurred in turkeys and chickens, and the emerging H5N1 virus has caused fatal infection in a number of different bird species). Ducks can excrete virus in feces for weeks, infecting other ducks via contaminated water, and a significant fraction of ducks may become infected by the virus in this process. Migratory ducks then spread the virus around the world, normally in a north–south direction. The viruses in birds are in stasis. Almost no differences in amino acid sequences of the various proteins are present in viruses separated by many decades, although the nucleic acid sequences encoding these proteins do drift. This together with the fact that the viruses seldom cause disease in their avian reservoirs show that influenza in birds is ancient and the virus has adapted to its primary host.

The gene segments of influenza A virus reassort readily during mixed infection, and viruses with new combinations of genes arise frequently. Newly arising reassortants can cause major epidemics of influenza when introduced into humans, a process called antigenic shift. Not all combinations of genes give rise to viruses that are capable of epidemic spread in humans. Only three subtypes of HA (H1, H2, and H3) and two or three subtypes of NA (N1, N2, and possibly N8) have been found to date in epidemic strains of human influenza virus. The first influenza virus isolated, in 1933, was called H1N1. This virus first appeared as the cause of the great influenza epidemic of 1918 (see later). The virus isolated in the epidemic of 1957 had a different subtype of both HA and NA and was called H2N2. The H2N2 virus replaced the H1N1 virus as the cause of influenza epidemics (Fig. 4.17). The H2N2 virus was itself replaced by H3N2 virus beginning with the epidemic of 1968. Serological surveys suggest that prior to 1918 the virus that circulated was an H3N8 virus that first appeared as the cause of an epidemic in 1890. The reason that only a subset of HAs appear to be capable of causing epidemics in humans is, at least in part, the fact that the receptors for the virus are somewhat different in birds and humans. Sialic acid is linked to galactose predominantly by $\alpha 2,6$ linkages in humans but by $\alpha 2,3$ linkages in birds.

Similarly, only certain types of the other segments are compatible with infection of and epidemic spread in humans. For example, the nucleocapsid gene has diverged into five lineages, but only one of these lineages is present in viruses isolated from humans (see, e.g., Fig. 4.16). NS1 is also at least partially host specific and thus only certain NS1s are compatible with human infection. Other proteins also differ somewhat for optimal replication in birds versus mammals. It is thought that reassortment can result in the introduction of a new HA or NA gene into a human virus, that is, a virus whose other gene segments are optimized for human infection. The HA and NA proteins are the most important antigens of the virus, and change of one or both of these antigens gives rise to a virus for which the majority of the human population has no immunity and which is therefore capable of causing a global pandemic. One possible scenario is that pigs serve as intermediates ("mixing vessels") in the recombination process, because pigs can be infected by both avian and human viruses (they contain sialic acid in both $\alpha 2,3$ and $\alpha 2,6$ linkage) and reassortment could occur in this host.

Influenza A virus is an example of a zoonotic disease in humans. The reservoir of the virus is ducks and other birds, and human infection is irrelevant for the maintenance of the virus in nature.

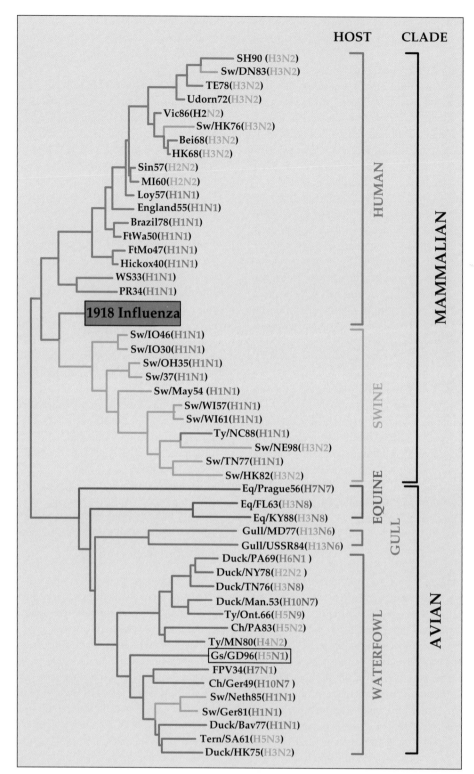

FIGURE 4.16 Phylogenetic tree of the nucleotide sequences of the influenza A virus NP gene sequences, constructed with a neighbor-joining algorithm. Each isolate name includes the location and year of isolation, preceded for non-human viruses by a species designation and a diagonal slash. Species abbreviations: Sw, swine; Ty, turkey; Ch, chicken; Gs, goose. Standard two letter abbreviations for states in the United States are used. Other location abbreviations: SH, Shanghai; DA, Dandong; Vic, Victoria; HK, Hong Kong; Sin, Singapore; Loy, Loygang; Ft Wa, Fort Warren; Ft Mo, Fort Monmouth; Man, Manitoba; Ont, Ontario; GD, Guangdong; Ger, Germany; Neth, Netherlands; Bav, Bavaria; SA, South Africa. The boxed isolate is the probable source of the H5 hemagglutinin in the currently worrisome "bird flu" spreading from China. Adapted from Reid *et al.* (2004), Figure 2.

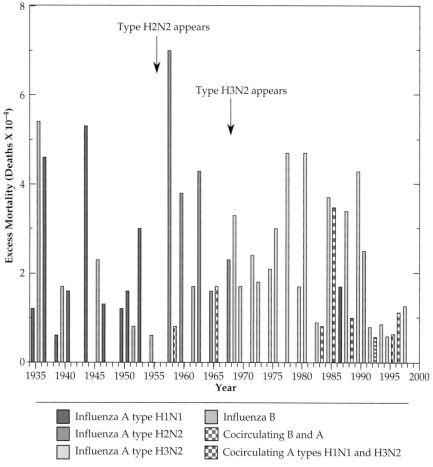

FIGURE 4.17   Excess mortality caused by influenza A and B virus in the United States between 1934 and 1998. "1935" refers to the winter of 1934–1935. Excess mortality due to the three dominant subtypes of influenza A and influenza B are indicated by the colors shown in the key. Green cross-hatched bars are excess mortality in years when both A and B viruses circulated. In 1955 and 1965, type H2N2 circulated with B, in 1983 and 1988 type H1N1 circulated with B, and in 1985, 1992, 1996, and 1998 H3N2 circulated with B. In 1995 H1N1 and H3N2 types of influenza A both circulated (hatched purple bar). Redrawn from Fields *et al.* (1996) p. 1421, with additional data from Thompson *et al.* (2003).

Epidemics of Influenza

Influenza A virus causes a serious human illness, influenza. It is perhaps confusing and unfortunate that the term *flu* is often used to describe any respiratory tract infection (and at times even infections of the gastrointestinal tract), even those that are fairly mild. The symptoms of true influenza are usually more severe than those resulting from other respiratory tract infections and include fever, headache, prostration, and significant muscle aches and pains (myalgias) that last for 3–6 days. Weakness and cough can last 1–2 weeks more. The fever can be high (39–40°C is not uncommon in adults and can be higher, especially in children). The morbidity that accompanies the disease can cause the patient to remain bedridden for a week or longer. In young children, the high fever can result in Reye's syndrome, an encephalopathy that may be fatal. The probability of contracting Reye's syndrome is higher if aspirin is administered to control the fever.

Lower respiratory tract infection can also occur following influenza infection and result in primary viral pneumonia. Invasion of the damaged lungs by pathogenic bacteria may follow and result in secondary bacterial pneumonia. Influenza can be fatal, usually because of pneumonia resulting from viral infection, whether the pneumonia is due to primary viral infection or, more commonly, due to secondary bacterial infection. Fatal infection is more common in the very young (whose immune system is not fully developed) and in the elderly (whose immune system may be waning). Before the advent of antibiotics, bacterial pneumonia killed many following severe bouts of influenza, but even today influenza remains a serious killer. It has been estimated that influenza virus infects 10–20% of the world's population every year causing five million cases of severe illness and 250,000 to 500,000 deaths. In the United States alone the estimated death rate from influenza in an average year is 20,000–30,000 and can be significantly higher in epidemic years. People over 65 are at particular risk from influenza. The annual death rate in the United States from influenza A in people over 65 is 1 per 2200, and in an epidemic year the death rate may be 1 in 300 (i.e., 1 of every 300 people over the age of 65 die of influenza during the epidemic). The excess mortality caused by influenza is illustrated in Fig. 4.17, in which the different strains of influenza A or B responsible for the epidemics are indicated. Although influenza A is usually the most serious cause of mortality, in some years influenza B is more of a problem than influenza A.

The 1918 Influenza Epidemic

A pandemic of influenza erupted in 1918 due to the emergence of a virulent H1N1 strain. This extremely virulent virus swept around the world over a period of about a year and infected an estimated 30% of the world's population, causing 20–100 million deaths. Although the very young and the elderly are normally at the most risk from influenza, this influenza pandemic of 1918–1919 was unusual in that mortality was highest in healthy young adults. The age distributions of people dying of influenza and the related pneumonia are compared for the years 1917 and 1918 in Fig. 4.18. The much higher death rates in the young and the el-

derly in 1917, the normal pattern, is apparent. The dramatic increase in the death rate in the 20- to 29-year-old group in 1918, in which people of this age were more likely to die than the old and the young, is striking. Death rates in young adults 15–34 years of age were more than 20-fold higher in the 1918–1919 pandemic than in the preceding years, and the death toll in young adults in the United States was high enough that overall life expectancy dropped sharply, as illustrated in Fig. 4.19.

The overall mortality was perhaps 2% of the world population but in some regions of the world, for example, regions of Central America and certain islands in the Pacific, 10–20% of the entire population died in the epidemic. In some remote Alaskan villages, more than 70% of all adults died, usually as a result of the simultaneous incapacitation of the entire population so that supportive care was not available. The final death toll can never be known with certainty and estimates vary widely, from 20 to 100 million. The death toll exceeded that produced by World War I, which was ongoing at the time. In fact, 80% of deaths in the U.S. Army during World War I resulted from influenza, and it is thought that the final collapse of the German army in 1918 may have been precipitated by widespread influenza in the troops. The surgeon general of the United States had expressed the hope

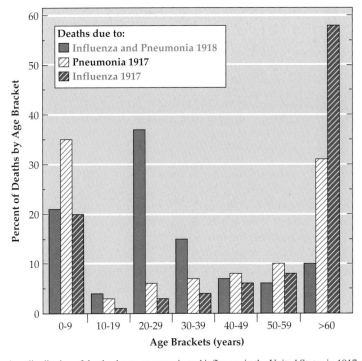

FIGURE 4.18   Age distribution of deaths due to pneumonia and influenza in the United States in 1917 and 1918. Age at death of patients has been divided into 7 intervals of 10 years each. The percent of deaths due to pneumonia in 1917, due to influenza in 1917, and due to the combined effects of pneumonia and influenza during the great epidemic year 1918 which fall into each age bracket are shown. The epidemic shows the atypical preponderance of deaths in the 20–29 and 30–39 year old brackets during the 1918 epidemic. Data from Crosby (1989). For comparison, from 1990 to 1998 only 3.8% of deaths due to influenza and pneumonia occurred in persons <49, 4.75% in persons 50–64, and 91% in persons over 65 years old (updated information from Thompson et al. 2003).

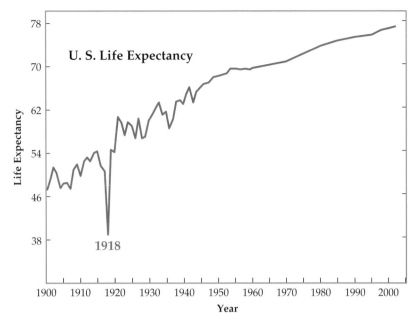

FIGURE 4.19   Life expectancy in the United States, showing the precipitous drop in 1918 because of deaths due to the "Spanish flu." This drop interrupted an otherwise fairly uniform increase in life expectancy that resulted from better health care, sanitation, and living conditions. Note also the leveling off in the late 1980s and 1990s due to AIDS. Adapted from *ASM News*, July 1999, and more recent data from the National Center for Health Statistics.

that WWI would be the first war in which more U.S. soldiers died of war injuries than died of disease, but this hope was shattered by the influenza epidemic. Descriptions of the epidemic with a focus on its effects on U.S. society are found in the books *Flu*, by G. Kolata, *America's Forgotten Pandemic*, by A. W. Crosby, and, quite recently *The Great Influenza*, by John M. Barry.

The reasons for the extreme virulence of the 1918 virus, and why healthy young people were more likely to die, a topic made even more important by the appearance of H5N1 "bird flu" (see Chapter 8) have been addressed recently using the power of modern molecular biology. The pandemic of 1918 occurred before influenza virus could be isolated. However, the sequences of all eight gene segments of the 1918 influenza genes have been obtained starting from a number of tissue isolates. Samples of preserved lung tissue taken at autopsy from two U.S. soldiers who died of influenza in September 1918 in New York and South Carolina were found to contain detectable influenza RNA, albeit in fragmented condition. A third source of influenza RNA came from an Alaskan Inuit victim who died in November 1918 and was buried in permafrost, and whose body was sufficiently well preserved that lung samples containing (fragmented) viral RNA were obtained. Two additional sources of influenza sequences come from two victims of influenza who died of pneumonia in November 1918 and February 1919 at the Royal London Hospital. Reverse transcriptase–polymerase chain reaction technology was used to obtain sequences from influenza

RNA in these tissue samples that could be used to reconstruct the complete sequences of genome segments. The sequences from these five victims are almost identical and showed that the virus belonged to strain H1N1. The HA genes from these five humans differ by only one to three nucleotides despite the fact that they came from five humans whose deaths were separated by over 7500 miles and several months in time. The sequence of this gene places it in the human–swine lineage, not in the avian lineage, and at the root of the tree leading to later isolates of human or swine influenza (Fig. 4.20). Thus, the HA of the virus does not appear to have come directly from an avian source.

It is now possible to use reverse genetics to take a cloned DNA copy of an influenza gene and rescue a virus containing this gene. To do this, cells are transfected with up to 17 plasmids that express the 8 genome segments of influenza as well as the RNA polymerase proteins PB1, PB2, and PA, and the NP protein, and in some cases the other influenza proteins as well. Infectious influenza virus is produced and buds from the cell. Using this system influenza virus has been produced that contains various combinations of the 1918 HA and NA genes with other cloned genes from the 1918 virus or from recent isolates, including virus that contains the complete complement of the 1918 genes and thus is a complete reconstruction of the 1918 virus. Various constructs have been tested in mice. Whereas recent isolates of influenza virus cause only mild disease in mice, the 1918 virus causes severe, often fatal disease. In mice, a virus

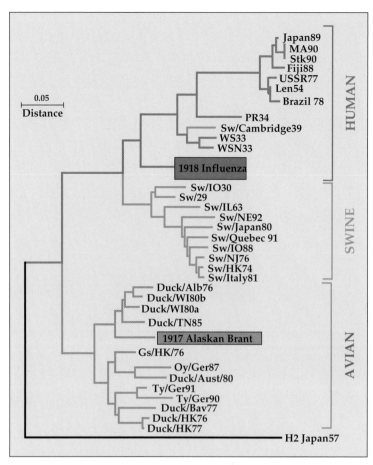

FIGURE 4.20   Phylogeny of the H1 hemagglutinin genes (bases 494–659 aligned to the comparable sequence of
PR34). Viral names include species of isolation followed by location and year of isolation. Species include: Sw, swine;
Gs, goose; Ty, turkey; Oy, oystercatcher. In the United States the standard two letter abbreviation for the state is used;
outside the United States the following abbreviations are used: Len, Leningrad; Ger, Germany; HK, Hong Kong; Bav,
Bavaria; Aust, Australia; Stk, Stockholm; Alb, Alberta. The sequence of the 1918 pandemic strain and the avian strain
most closely related chronologically are boxed. A distance bar, where a distance of 0.05 = 11.2 synonymous differences,
is shown above and beside the tree, and the H2 hemagglutinin of Japan 57 virus is used as an outgroup. Adapted from
Fanning et al. (2002).

containing only the H1 and N1 of the 1918 virus was found
to be highly virulent and caused fatal infection in mice. Virus
grew to high titer in the lungs of the mice and was associated with an influx of neutrophils and macrophages into the
infected lung.

The complete 1918 virus has also been tested, under
BSL-4 high containment conditions, in monkeys as well as
mice. The virus caused severe, usually fatal, disease in monkeys that was marked by much higher replication rates and
more extensive spread in the lungs. It was also marked by an
abnormal innate immune response (see Chapter 10). Certain
elements of the innate response were attenuated, perhaps
because of the activity of the NS1 gene which is known
to interfere with the immune response. In contrast, other
immune responses, in particular inflammatory cytokines,
were enhanced, resulting in a "cytokine storm." The results
are consistent with the hypothesis that in humans the 1918
virus provoked an extreme but unbalanced reaction by the

immune system, and that healthy young people, who have
the strongest immune systems, suffered from more extensive
release of potent cytokines that resulted in more extensive
tissue destruction.

The devastation caused by the 1918 virus raises continuing concern that a strain of influenza of equal virulence might
appear and again cause immense suffering worldwide. New
pandemic strains of influenza appear three or four times a
century. If a pandemic strain emerged from a virus such as
the H5N1 strain of bird flu (see Chapter 10), which has a
very high mortality rate in humans, the resulting epidemic
could indeed be devastating.

Antigenic Shift and Drift

Immunity to influenza A virus following infection is
long lived but may not be complete and is subtype specific
and even strain specific. The continuing appearance of new

strains that arise from antigenic drift and of new subtypes that arise from antigenic shift lead to continuing epidemics. Normally, two or three strains of influenza A circulate in the human population at any one time. Spread from person to person is by respiratory droplets, requiring close proximity, but people travel extensively and new strains of the virus speed around the globe as they arise. Antigenic drift is the process by which mutations accumulate in the virus genome, usually because of immune selection, that result in the development of new strains of the virus. These new strains are partially resistant to the immunity induced by infection with previous strains of virus. After several years of drift, the strain may be sufficiently distinct to cause disease in a person previously infected, but the illness is usually less severe because of partial immunity to the new strain. However, new strains capable of causing serious illness can arise by antigenic shift whereby reassortment results in change of the surface glycoproteins of the virus. The reassortants that cause the biggest problems are those belonging to a new subtype (as illustrated by Fig. 4.17). As described, such a new subtype may cause a pandemic in the human population because there is little immunity to the virus carrying these new surface antigens, as happened in 1918 (H1N1), 1957 (H2N2), and 1968 (H3N2).

H1N1 virus, which had disappeared with the appearance of the H2N2 epidemic strain in 1957, suddenly reappeared in 1977. This H1N1 virus, which first appeared in northern China in May 1977 and was called the Russian flu, was virtually identical to influenza virus isolated from an epidemic in humans in 1950. It circulated in young people who had not been exposed to H1N1 virus. Because it was virtually unchanged despite 27 years having elapsed, it seems unlikely that it arose again *de novo*. Presumably this virus had been preserved in a frozen state, probably in a laboratory freezer. In 1976, in response to reports that investigators outside Western Europe planned to develop and test vaccines against H1N1 influenza, a WHO meeting report urged extreme caution in developing live vaccines from epidemic H1N1 strains because of the possibility of spread of the virus. One year later the virus reappeared.

Vaccination against Influenza A Virus

Because of the seriousness of influenza disease, especially in the elderly, attempts are made each year to vaccinate the population at risk. Because of drift and shift, the vaccine must be reformulated every year to reflect the viruses currently circulating in the human population. There are three strains of virus included in the most common vaccine, an inactivated virus vaccine produced from viruses grown in eggs. These are two influenza A viruses and one influenza B virus. These viruses are chosen from those that are circulating in late spring, because these viruses are usually those that will cause epidemics the following winter. The choice

must be made by late spring in order to allow time for the pharmaceutical companies to prepare the vaccines, and an element of risk is involved that the right choices will not be made. The World Health Organization publishes choices and supplies seed virus based upon the recommendation of an international group of scientists, but the final selections are made by individual health agencies and the choices are usually, but not always, correct. The number of vaccine manufacturers has declined dramatically in the United States over the last 2 decades because of legal liability problems, and what limited capacity that exists for manufacturing flu vaccine is mostly present in Europe. Production problems by one of the manufacturers has resulted in recent shortages of vaccine.

The necessity to grow the virus in fertilized eggs also limits the amount of vaccine that can be produced. There are efforts to develop a cell culture system for virus production for vaccine use, which could then be produced in larger amounts. Efforts are also being made to develop better adjuvents for use with the vaccine, which could reduce the amount of antigen required per inoculation. In addition, obtaining the reassortants required for vaccine production is a time-consuming endeavor using classical methods of coinfecting cells with two different viruses and searching through the progeny for the wanted reassortants. If reverse genetics described earlier can be developed in a way that satisfies the regulatory agencies concerned with vaccine safety, the desired reassortants could be obtained much more quickly, allowing quicker responses to new strains of virus.

In addition to the inactivated virus vaccine that is very widely used, a new live virus vaccine based on a cold attenuated virus has been licensed recently. Reassortment is used to introduce the HA and NA of the predicted epidemic strains into this attenuated virus. Because the attenuation of the virus results from changes in other genome segments, the recombinant strain is also attenuated. The vaccine is administered by nasal spray rather than by injection as is the inactivated virus vaccine. To date this vaccine is only licensed for use in people between the ages of 5 and 49, and thus it cannot be used for the populations most at risk for serious illness, but clinical trials are continuing. It remains to be seen how well accepted this vaccine will be.

The necessity of reformulating the vaccine every year is inconvenient for a number of reasons including the fact that the vaccine cannot be stored for use in the following years. In addition, the vaccine is not always effective because wrong predictions were made about which strains of virus would be the biggest problems. There is an effort being made to develop universal vaccines that would target all strains of influenza A and B, and that would therefore provide protection against all influenza strains and that could be used year after year. One possibility that is being pursued is to use influenza A M2 protein as an antigen. This protein is highly conserved among all A strains but is not normally seen by the immune system for some reason. Preliminary studies have shown that this protein linked to hepatitis B core protein is highly immunogenic in mice and provides protection against influenza A infection in mice, regardless of strain. For influenza B, a subunit vaccine based upon the sequence surrounding the cleavage site of the HA precursor, which includes a highly immunogenic part of the fusion peptide, shows promise in early animal trials.

Swine Flu Virus

Continuing surveillance of influenza strains in nature is required in order to reformulate the vaccines each year. This surveillance also serves to watch for the possible appearance of another killer strain of influenza. An episode that occurred during the Ford administration, however, illustrates the potential difficulties of identifying such a strain and reacting in time. In February of 1976, a young soldier at Fort Dix died of influenza and others became seriously ill. Tests showed that most of the soldiers were suffering from the A/Victoria strain of influenza that was epidemic in the United States at the time or from adenovirus infection. However, the soldier who died and three other soldiers who were ill were infected with an influenza strain that was epidemic in pigs, referred to as swine flu. Serology studies indicated that 200 or more other soldiers had been infected by this virus as well, showing that the virus was being transmitted from person to person. The swine flu virus was closely related to the 1918 pandemic virus, and is thought to have been introduced into pigs in 1918 from humans and to have continued to circulate in pigs after it had died out in humans. Could it be possible that the 1918 virus had reappeared as an epidemic virus in humans? The decision was made by President Ford, in consulation with leading scientists, to begin a crash program to develop a vaccine against swine flu and to begin to immunize the American population. It was thought, with some justification, that to wait for an epidemic to begin before an immunization program was undertaken would mean that it would be too late to be effective, given the speed with which influenza epidemics spread. Further, influenza is usually epidemic in winter, and the early detection of this virus made possible the preparation of a vaccine before the (next) winter flu season set in. Forty million Americans were immunized against swine flu. No epidemic of swine flu developed, however, and litigation began. The pharmaceutical companies had been reluctant to participate in the program, pointing out that at any one time a certain fraction of Americans would develop encephalitis or rheumatoid arthritis or any one of hundreds of other diseases. If disease developed in proximity to receiving a new and relatively untested vaccine, a lawsuit would certainly follow and the potential damages were enormous. The program could only advance when Congress agreed to indemnify the

pharmaceutical houses. The vaccine was never conclusively shown to cause disease, although there seemed to be a slight increase of Guillain-Barré syndrome following inoculation. Litigation went on for years and substantial damages were paid out. In retrospect it is easy to criticize the program as an overreaction, but what would have been the reaction if nothing had been done and an influenza epidemic developed that resulted in 50–100 million Americans becoming seriously ill with 1–2 million deaths? Given the state of knowledge at the time, many leaders felt there was no choice. Further, the decision to vaccinate was not so different from current policy, where strains of influenza A circulating in the spring are incorporated into a vaccine to be given in the fall. A quote from the U.S. Surgeon General at a meeting of the Association of State and Territorial Health Officiers in 1957 is worth thinking about: "I am sure that what any of us do, we will be criticized either for doing too much or for doing too little."

Bird Flu

A recent scare began when 18 people in Hong Kong became seriously ill from influenza in 1997 and 6 died. The culprit was an avian influenza (H5N1) that was epidemic in birds being sold in the markets for food. Avian viruses do not normally infect people, and there was fear that an avian virus had made the jump to humans and might cause an epidemic of lethal influenza. The Hong Kong authorities destroyed 1.6 million domestic birds in order to eradicate the epidemic in birds. No human-to-human transmission took place and the virus disappeared. In 2002, however, H5N1 virus reappeared and by 2006 it has spread throughout Asia and into Africa and Europe. This virus has a mortality rate of about 50% in humans and more than 140 people have died of H5N1 infection as of this date. There is no person-to-person transmission to date, but there is concern that the virus might mutate and cause a wide and devastating pandemic of influenza. This subject is considered at more length in Chapter 8.

Influenza B and C Viruses

Humans are the reservoir of influenza B virus. It causes influenza in humans but there exists only one subtype and antigenic shift does not occur. Antigenic drift does occur, and the virus can cause epidemics of serious illness that result in increased mortality, particularly among the elderly, as shown in Fig. 4.17. For this reason, the current strain of circulating influenza B is included in the annual flu vaccine. However, wide-ranging pandemics do not occur and the virus is therefore not as much of a problem as influenza A. Less attention has accordingly been given to the study of this virus. Influenza C is not a serious human pathogen and has been even less well studied.

Other Orthomyxoviruses

Thogoto virus is present in regions of Africa, southern Europe, and Asia. It is a tick-borne virus that is primarily known from infection of livestock such as cattle, camels, and sheep. There is significant amino acid sequence identity between some of the Thogoto proteins and their counterparts in influenza so these viruses are fairly closely related. It has only six genomic segments, however. Human infection is known to occur in endemic areas.

Infectious salmon anemia virus infects salmonid fish. Atlantic salmon are particularly susceptible to the virus and the virus is a particular problem for fish farming. Farm populations can suffer 100% mortality in outbreaks. The virus has eight gene segments.

FAMILY *BUNYAVIRIDAE*

The family *Bunyaviridae* contains more than 300 viruses grouped into five genera. A representative sampling of these viruses is shown in Table 4.9. Members of four genera, *Orthobunyavirus*, *Nairovirus*, *Phlebovirus*, and *Hantavirus*, infect vertebrates and contain important human pathogens, whereas viruses belonging to the genus *Tospovirus* infect plants. The human pathogens in the family variously cause hemorrhagic fever, hantavirus pulmonary syndrome which can be fatal, encephalitis, or milder febrile illnesses, as shown in the table. Some of these pathogens were listed in Table 3.11, which contains a partial listing of arboviruses that cause disease in humans. All members of the *Bunyaviridae*

TABLE 4.9 *Bunyaviridae*

Genus/members[a]	Virus name abbreviation	Usual host(s)	Transmission/ vector	Disease in humans	World distribution
Orthobunyavirus (~48 viruses)					
Bunyamwera	BUNV	Rodents, rabbits	*Aedes* mosquitoes	Febrile illness	Worldwide
La Crosse	LACV	Humans, rodents	*Aedes triseriatis*	Encephalitis	Midwest United States
Snowshoe hare	SSHV	Lagomorphs	Mosquitoes (*Culiseta* and *Aedes*)	Rarely infects humans	Northern United States
California encephalitis	CEV	Rodents, rabbits	*Aedes melanimon, Ae. dorsalis*	Encephalitis (rare)	Western United States, Canada
Jamestown Canyon	JCV	White-tailed deer	*Aedes* species, *C. inornata*	Increasing	North America
Hantavirus (~22 viruses)					
Hantaan	HTNV	*Apodemus agrarius*	Feces, urine, saliva	Hemorrhagic fever	Eastern Asia, Eastern Europe
Seoul	SEOV	*Rattus* species	Feces, urine, saliva	Hemorrhagic fever	Worldwide
Prospect Hill	PHV	*Microtus pennsylvanicus*	?	None?	United States
Sin Nombre	SNV	*Peromyscus maniculatus*	Feces, urine, saliva	Pulmonary syndrome	Western United States and Canada
Nairovirus (~7 viruses)					
Dugbe	DUGV	Sheep, goats	Tick-borne	?	Africa
Crimean-Congo hemorrhagic fever	C-CHFV	Humans, cattle, sheep, goats	Tick-borne	Hemorrhagic fever	Africa, Eurasia
Phlebovirus (~9 viruses)					
Rift Valley fever	RVFV	Sheep, humans, cattle, goats	Mosquitoes, also contact, aerosols	Hemorrhagic fever	Africa
Sandfly fever Sicilian	SFSV	Humans	Phlebotomous flies	Nonfatal febrile illness	Mediterranean
Uukuniemi	UUKV	Birds	Tick-borne	??	Finland
Tospovirus (~8 viruses)					
Tomato spotted wilt	TSWV	Plants	Thrips	None	Australia, Northern hemisphere

[a] Representative members of each genus are shown; the first virus listed is the type species.

except the hantaviruses are transmitted to their vertebrate or plant hosts by arthropods, and transovarial transmission is important in the maintenance of many of the arboviruses in nature. The hantaviruses, in contrast, are associated with rodents and are transmitted to humans by aerosolized excreta from infected rodents. Thus, their epidemiology resembles that of the arenaviruses considered later, rather than that of other bunyaviruses. In the following discussion, the term bunyavirus refers to any member of the family unless indicated otherwise.

Replication of the *Bunyaviridae*

Genome Organization

The genomes of representative viruses belonging to the five genera of the *Bunyaviridae* are illustrated in Fig. 4.21.

All bunyavirus genomes consist of three segments of RNA, referred to as S(mall), M(edium), and L(arge), that together total from 11 to 19 kb, depending on the virus (Table 4.10). The S segment encodes the nucleocapsid protein, M the two surface glycoproteins, and L the polymerase protein. In addition, viruses belonging to three of the genera encode two nonstructural proteins, NS_s in segment S and NS_m in seg-ment M.

Replication of bunyavirus genomes and the synthesis of mRNAs take place in the cytoplasm. The L protein and N protein are required components of the RNA synthesis machinery. Like influenza viruses, these viruses engage in cap-snatching in order to prime mRNA synthesis. In bunyaviruses, however, the caps are captured from cytoplasmic mRNAs rather than from nuclear pre-mRNAs. The promoter for mRNA synthesis and for RNA replication involves nucleotides located at both ends of the genomic template, which

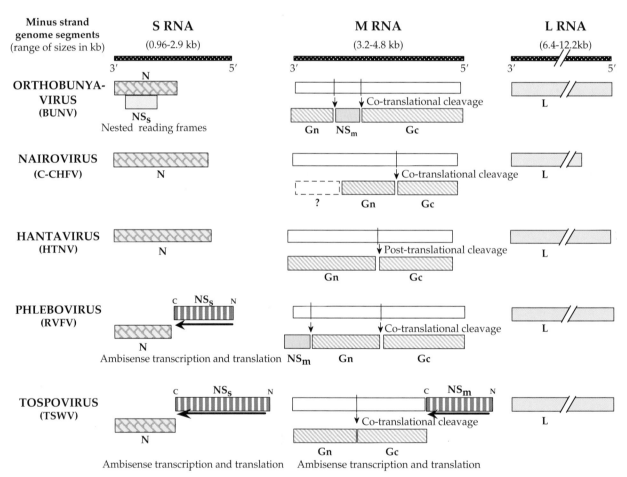

FIGURE 4.21   Genome organization of five genera of *Bunyaviridae*. Protein products encoded in each of the three genome segments and the various strategies used to produce these proteins are shown. Unless otherwise noted the mRNA (not shown) would extend 5' to 3' from left to right, and the protein product is shown N terminal to C terminal in the same direction. The products are illustrated roughly to scale. Structural proteins are N (the nucleocapsid protein) and the two glycoproteins, named Gn and Gc according to their proximity to the N or C termini of the precursor polyprotein; white boxes are precursor proteins. Arrows indicate the direction of synthesis and translation of ambisense mRNAs (magenta hatch). Note that not all phleboviruses encode an NS_m protein on the M segment. Virus abbreviations are as follows: BUNV, Bunyamwera; RVFV, Rift Valley fever; TSWV, tomato spotted wilt; C-CHFV, Crimean-Congo hemorrhagic fever; HTNV, Hantaan.

TABLE 4.10   Deduced Sizes (kD) of Proteins Encoded by *Bunyaviruses*

RNA (nts) Protein (kD)	Genus				
	Orthobunyavirus	**Hantavirus**	**Nairovirus**	**Phlebovirus**	**Tospovirus**
L Segment	6875–6980	6550–6562	12,255	6404–6423	8776–8897
L protein (RNA Polymerase)	259–263	246–247	459	238–241	330–332
M Segment	4458–4526	3616–3696	4888	3231–4215	4821–4972
Glycoprotein Gn	29–41	68–76	30–45	50–72	46–58
Glycoprotein Gc	108–120	52–58	72–84	55–75	72–78
Precursor preG	—	—	78–85, 92–115	—	—
Nonstructural NS_m	15–18	None	None	None or 78	34–37
S Segment	961–980	1696–2059	1712	1690–1869	2916–2992
Nucleoprotein N	10–26	48–54	48–54	24–30	29
Nonstructural NS_s	10–13	None	None	29–32	52

Sizes of precursor proteins are shown in blue, nonstructural proteins are shown in green, and those translated from ambisense transcripts are in red.

are complementary and form hairpin circles, but with some unpaired nucleotides that are thought to be important recognition signals (Table 4.11). Thus, as with many other RNA viruses, the viral RNA polymerase must interact with both ends of the RNA template in order to initiate synthesis, and complementarity between nucleotides at the 5′ and 3′ ends is required for promoter recognition. The terminal complementary sequences are highly conserved within each genus of bunyaviruses but differ between genera (Table 4.11). Perhaps because of this, reassortment occurs only between viruses belonging to the same genus.

During initiation of mRNA, as studied in hantaviruses, the L protein cleaves the 5′-terminal 7–18 nucleotides from a cellular mRNA. Cleavage is after a G residue, which pairs with the C residue at position 3. The primer is elongated by a few residues, and there is then a backward shift of 3 nucle-

otides so that the mRNA attached to the primer begins precisely at the 3′ end of the template. This "prime and realign" strategy (Fig. 4.22) works because of the repeat triplets at the 3′ end of the RNA template (Table 4.11). Transcription continues to near the end of the template, but the termination of an mRNA does not appear to be precise and the exact mechanism used for termination of the mRNAs is not known. No poly(A) is added to the 3′ end of the mRNA upon its release and, thus, the mRNAs are capped but not polyadenylated.

During replication of the genome, an exact complementary copy, called cRNA or vcRNA, is produced. This RNA serves as a template for producing genomic RNA and, in the case of ambisense segments, for producing the mRNA for producing the ambisense-encoded protein. The switch to replication is assumed to use the same mechanisms as used by other (−)RNA viruses.

TABLE 4.11   Terminal Sequences of the Genome Segments of the Five Genera of *Bunyaviridae*

Genus	Nucleotide sequences of the L, M, and S segments
Hantavirus	5′ U A G U A G U A ...
	3′ A U C A U C A U C U G ...
Orthobunyavirus	5′ U C A U C A C A U G A ...
	3′ A G U A G U G U G C U ...
Nairovirus	5′ A G A G U U U C U ...
	3′ U C U C A A A G A ...
Phlebovirus	5′ U G U G U U U C ...
	3′ A C A C A A A G ...
Tospovirus	5′ U C U C G U U A ...
	3′ A G A G C A A U ...

Repeated sequences are underlined.

Expression of Proteins Encoded in S

The S segment of bunyaviruses encodes one or two proteins (Fig. 4.21). In the *Hantavirus* and *Nairovirus* genera, S encodes only N. In the other genera, S encodes both N and NS_s, using one of two different mechanisms. In the genus *Orthobunyavirus*, the two proteins are translated from a single mRNA using two different start codons in different reading frames. The coding region for NS_s is completely contained within that for N. In the phleboviruses and tospoviruses, however, an ambisense coding strategy is used for the two proteins (*ambi* = both). In this strategy, the two genes encoded in a genomic segment are linked tail to tail so that they are in different polarities, as illustrated in Fig. 4.23. The gene for N is present at the 3′ end of the genomic S segment in the minus-sense orientation, and synthesis of the mRNA for N occurs from the genome segment. Expression of this gene occurs early because its mRNA is

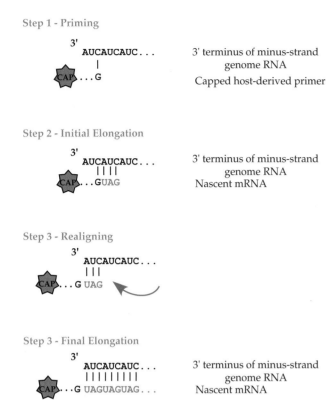

Step 1 - Priming

3'
    AUCAUCAUC...   3' terminus of minus-strand
        | genome RNA
CAP ...G   Capped host-derived primer

Step 2 - Initial Elongation

3'
    AUCAUCAUC...   3' terminus of minus-strand
       ||||| genome RNA
CAP ...GUAG   Nascent mRNA

Step 3 - Realigning

3'
    AUCAUCAUC...
       |||
CAP ...G UAG

Step 3 - Final Elongation

3'
    AUCAUCAUC...   3' terminus of minus-strand
     ||||||||| genome RNA
CAP ...G UAGUAGUAG...   Nascent mRNA

FIGURE 4.22 Steps in the prime–align mechanism of transcription of mRNAs by hantaviruses. This mechanism is made possible by the nucleotide repeats in the 3′ and 5′ termini of bunyaviruses (see Table 4.11). Adapted from Kukkonen *et al.* (2005).

synthesized from the entering genome by the polymerase activity present in viral nucleocapsids. The gene for NS$_s$ is plus sense within the genome, but the genomic RNA does not serve as mRNA. Instead, an mRNA for NS$_s$ is synthesized from the antigenomic RNA. Thus, NS$_s$ is expressed late because its mRNA can only be made after replication of the incoming genomic RNA to produce the antigenomic RNA. Termination of either mRNA occurs at a secondary structure between the genes for N and NS, which appears to cause the polymerase to fall off and release the mRNA.

N has a number of functions in viral infection. It encapsidates the viral RNA, interacts with L to synthesize viral RNAs, and is believed to interact with one of the glycoproteins during virus assembly. In at least some viruses the protein also modifies cellular metabolism, presumably antagonizing antiviral defenses of the cell or otherwise subverting cellular processes to support viral replication. The N protein of hantaviruses, which is larger than those of other bunyaviruses except for that of the nairoviruses, interacts with a number of cellular proteins. The best studied of these are proteins in the small ubiquitin-like protein (SUMO) pathways such as Ubc-9 (which conjugates SUMOs to proteins), SUMO-1 itself, and Daxx (to which SUMOs are conjugated). Sumolation of proteins is an important regulatory process in cellular metabolism.

The NSs protein of Rift Valley fever virus, and presumably of other viruses as well, inhibits host mRNA synthesis, including the mRNA for interferon-α and -β. Thus it suppresses the host immune response and is a major virulence factor.

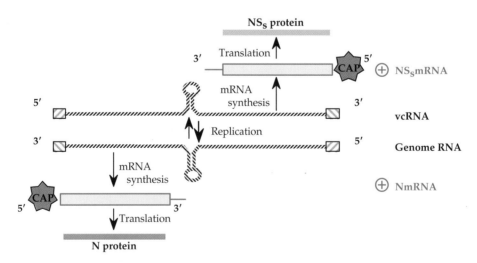

FIGURE 4.23 Ambisense coding strategy of the S RNA of a phlebovirus, family *Bunyaviridae*. The mRNA for the N protein is synthesized from the S genome segment using primers derived by cap-snatching (similar to the mechanism for influenza mRNA priming in Fig. 4.13) from cytoplasmic host mRNAs. The mRNA for the NS$_s$ protein is formed in the same way, but with vcRNA as the template. Diagonally striped boxes are the self-complementary termini. The loops in the middle of the viral genomic and antigenomic RNAs indicate a secondary structure in the RNAs which terminates synthesis of the mRNAs. No poly(A) is added to the 3′ terminus of the mRNAs.

Expression of Proteins Encoded in M

Two glycoproteins, at one time usually called G1 and G2 but now called G_N and G_C, are translated from mRNA made from M (Fig. 4.21). They are produced as a polyprotein that is cleaved by a cellular protease to separate the two glycoproteins, analogous to what happens in some of the (+)RNA viruses that have envelopes (e.g., coronaviruses and flaviviruses). G_N is N terminal in the polyprotein and G_C is C terminal. Where studied, the enzyme responsible for the cleavage is a subtilase, SKI-1/S1P or a related enzyme. G_N and G_C form a heterodimer that is transported to the Golgi apparatus. Virus budding occurs at the Golgi membrane. Heterodimerization recalls the processes that occur in the assembly of alphaviruses and flaviviruses, and like E1 or E of these viruses, respectively, the G_C protein is a class II fusion protein.

The M segments of hantaviruses and nairoviruses encode only the two glycoproteins, but in the other three genera M encodes a third protein called NS_m (Fig. 4.21 and Table 4.10). In phleboviruses and members of the genus *Bunyavirus*, NS_m forms part of the polyprotein translated from the single mRNA produced from M. NS_m is formed during posttranslational processing of the polyprotein. In tospoviruses, an ambisense strategy is used to encode NS_m and the translation strategy is the same as that shown in Fig. 4.23. The function of NS_m is not known.

There is no matrix protein in bunyaviruses. Budding at the Golgi membrane is assumed to involve a direct interaction between the glycoproteins and the nucleocapsid protein. The virion is spherical, 80–120 nm in diameter. The three nucleocapsids are circular when isolated from the virion.

Genus *Orthobunyavirus*

There are about 50 currently recognized species in the genus *Orthobunyavirus*, of which the majority have several distinct strains that are often given separate names. Together, these viruses have a worldwide distribution. Most of these viruses are mosquito-borne, although some are tick-borne and some may be transmitted by culicoid flies or phlebotomines. They are true arboviruses, replicating in the arthropod vector as well as in vertebrates.

Bunyamwera virus, the prototype member of the genus, was first isolated in Uganda in 1943. It causes a febrile illness accompanied by headache, arthralgia, rash, and occasional nervous system involvement. There are 24 named subtypes in the bunyamwera serogroup. Of interest is a reassortant virus isolated during an epidemic of hemorrhagic fever in Kenya and Somalia in 1998. Rift Valley fever virus was responsible for some of the cases, but many were caused by a bunyavirus whose L and S segments were derived from a Bunyamwera virus but whose M segment came from a different bunyavirus. This reassortant virus, named Garissa virus, thus caused a disease different from Bunyamwera virus, perhaps due to the different properties of the M segment.

Viruses belonging to the California encephalitis group, of which La Crosse virus is the best known, are also of medical interest. La Crosse virus was named for the town of La Crosse, Wisconsin, where it was first identified as the causative agent of encephalitis, primarily in children. About 100 cases per year of encephalitis are caused by La Crosse virus, concentrated in the Midwest. Mortality is low (0.3%) but 10% of patients suffer neurological sequelae. No vaccine exists for the virus and control measures have involved control of the mosquito vector. The principal vector of La Crosse is *Aedes triseriatus*. This mosquito breeds in tree holes, but abandoned tires filled with rainwater constitute an important breeding area for it close to human habitation. Such abandoned tires serve as a beautiful incubator for the development of mosquito larvae, and efforts to eliminate this source of mosquitoes, as well as the institution of other mosquito control measures, has resulted in a reduction in the number of cases of disease.

Abandoned tires are important in the transmission of other arboviruses as well. Old tires are abundant in Puerto Rico, for example, and contribute to the endemic transmission of dengue virus, all four serotypes of which are present on the island. Old tires have also been responsible for the introduction into the United States of *Aedes albopictus*, the so-called Asian tiger mosquito that is the vector of dengue virus in Asia. Loads of old tires that were brought from Asia to Houston for recycling contained eggs or larvae of the mosquito. After its introduction into the Houston area, this mosquito spread over large areas of the United States and there is fear that it might become an efficient vector of arboviral disease in this country.

Genus *Phlebovirus*

The ICTV currently recognizes 9 species of phleboviruses but there are an additional 16 tentative species. Many of the species have a number of strains that are given their own names. All are arboviruses transmitted by mosquitoes, phlebotomine flies, or ticks. The most important of these is Rift Valley fever virus, an African virus that was first isolated in 1930 in the Rift Valley of East Africa. The virus is transmitted by mosquitoes and causes hemorrhagic fever in humans. It also causes disease in domestic animals, and many widespread epidemics in cattle, sheep, and humans have occurred over the years in Africa. In 1977–1978, for example, an epizootic in Egypt infected 25–50% of cattle and sheep in some areas, and 200,000 human cases resulted in at least 600 deaths. A more recent large epidemic in East Africa in 1997–1998 was associated with the heaviest rainfall in 35 years, 60–100 times normal in some areas. As described before, the epidemic was caused by Rift Valley fever virus and by Garissa virus. Losses of 70% of sheep and goats and 20–30% of cattle and camels were reported, and there were hundreds of cases of human hemorrhagic fever. Contact with livestock was statistically associated with acute infection with Rift Valley fever virus, indicating that during epidemics contact transmission becomes important as a means of spread to humans. Laboratory-acquired cases contracted through aerosols are also known.

Sand fly fever virus is transmitted by phlebotomine flies and causes an acute, nonfatal influenza-like disease in man. It is found in the Mediterranean area, North Africa, and Southwest Asia. Related viruses are found in South America.

Genus *Nairovirus*

The nairoviruses have a much larger genome than members of the other genera, primarily because the L segment is twice the size of those of the other genera of animal viruses (Table 4.10). They are named for Nairobi sheep disease virus, now considered a strain of Dugbe virus. There are seven species recognized, all of which consist of multiple strains with distinct names. The viruses are tick-borne although a few can also be transmitted by culicoid flies or mosquitoes. Nairobi sheep disease virus causes acute gastroenteritis with hemorrhagic symptoms in sheep and goats, with mortality rates over 90% in some populations. It was first identified as the causative agent of the disease in 1917 and is transmitted by the tick *Rhipicephalus appendiculatus*. Humans can be infected by the virus but suffer only mild illness. There is a close relative of the virus called Ganjam virus present in India which also causes disease in sheep and goats; it is transmitted by the tick *Haemaphysalis intermedia*.

Crimean-Congo hemorrhagic fever virus (CCHF) is the most important nairovirus in terms of human disease. It was first identified in the 1940s in the Crimean region of the former USSR and in the Democratic Republic of Congo. The virus is now known from at least 30 countries. It is found from southern Africa through Eastern Europe and the Middle East to western China. The principal vector is *Hyalomma* ticks, but *Dermacentor* and *Rhipicephalus* ticks can also transmit the virus. Sheep, goats, cattle, ostriches, wild herbivores, and hares become infected by CCHF but most infections result in subclinical disease. In contrast, infection of humans results in severe hemorrhagic fever with a 30% mortality rate. Humans are infected by the bite of a tick or by contact with blood or tissues of infected livestock. Transmission to hospital personnel treating infected patients has occurred.

Genus *Hantavirus*

There are 22 species of hantaviruses currently recognized, and, as with other genera of the bunyaviruses, many of the species have a number of named strains. Many hantaviruses cause serious human disease, including hemorrhagic fevers and hantavirus pulmonary syndrome. Unlike other members of the *Bunyaviridae*, they are not arboviruses. The hantaviruses are associated with rodents, which form their natural reservoir, and are transmitted to humans through contact with aerosolized urine or feces from infected rodents. Each hantavirus establishes persistent infections in one particular species of rodent and is maintained in nature in this way. Humans are not an important host for the virus and do not contribute to its maintenance in nature. Related to this is the fact that the viruses do not cause serious disease in their rodent hosts, but many cause quite serious illness in humans.

An evolutionary tree of hantaviruses is shown in Fig. 4.24. The rodent hosts for the viruses are also indicated. The viruses assort by host rather than by geographical proximity. All of the viruses whose hosts belong to the order *Murinae* group together, as do those that use rodents in the order *Arvicolinae* and those that use rodents in the order *Sigmodontinae*. As one example, consider Prospect Hill virus and New York virus, both found in the northeastern United States. Prospect Hill virus is associated with rodents of the genus *Microtus*, order *Arvicolinae*, and is more closely related to Puumala virus of Europe, which uses *Clethrionomys glareolus*, order *Arvicolinae*, than it is to New York virus. New York virus is associated with rodents in the genus *Peromyscus*, order *Sigmodontinae* and is closely related to Sin Nombre virus of the southwestern United States, which is associated with *Peromyscus maniculatus*. The fact that the evolutionary tree of the hantaviruses resembles that of their rodent hosts rather than being based on geographical proximity is evidence that they have coevolved with their rodent hosts over a very long period of time.

The first of the hantaviruses to be identified was the causative agent of more than 3000 cases of hemorrhagic fever with renal syndrome, now called Korean hemorrhagic fever, that occurred in U.S. troops during the Korean war. The virus was called Hantaan virus after a river in the area where it was isolated. In Korea, Hantaan virus is associated with the field mouse *Apodemus agrarius*. The virus also occurs in Eastern Europe and China, where it is associated with *Apodemus flavicollis* and causes a disease similar to Korean hemorrhagic fever (Fig. 4.25).

Viruses related to Hantaan virus have now been isolated from all over the world, including the Americas. Many Old World viruses cause hemorrhagic fever in humans, and more than 100,000 cases occur worldwide with a case fatality rate between 0.1 and 10%, depending on the virus. Puumula virus occurs in Western Europe (Fig. 4.25) and causes a disease characterized by acute fever with renal involvement. Seoul virus, first identified in Seoul, Korea, is associated with wild urban rats (*Rattus norvegicus*) and has been found all over the world because wild urban rats have been inadvertently

Rodent-borne Hantaviruses   **Rodent hosts**

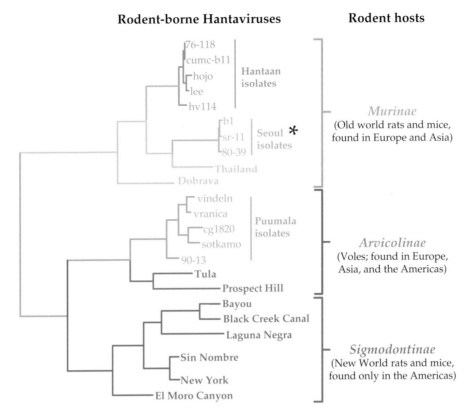

FIGURE 4.24   Phylogenetic tree of rodent-borne hantaviruses derived from the nucleotide sequence of the M RNA segment. This tree illustrates that hantaviruses have coevolved with their rodent hosts for millions of years. However, note (*) that in contrast to other members of this group, Seoul virus, which infects *Rattus norvegicus,* is found worldwide, due to the widespread distribution of these rats. Adapted from Peters (1998a), Figure 2.

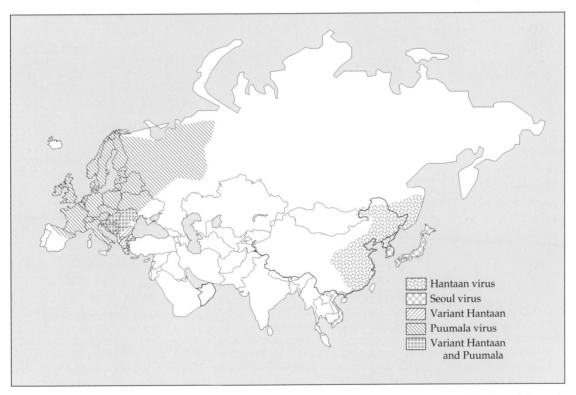

FIGURE 4.25   Map of Eurasia showing the disjunct distribution of different hantaviruses. Adapted from Porterfield (1995) p. 276 and data from Lee (1996).

introduced almost everywhere. It causes a mild form of Korean hemorrhagic fever in Seoul but does not cause apparent illness in most other areas where it has been found. The discovery of Seoul virus led to an intensive study of rats in central Baltimore, where it was found that a high percentage of them were infected with Seoul virus and, furthermore, that a substantial fraction of the people living in the slums of downtown Baltimore showed evidence of infection by hantavirus. No disease is known to be associated with this virus, but statistical studies suggest that infection may lead to high blood pressure and, possibly, renal failure.

The New World hantaviruses that cause disease in humans cause a syndrome called hantavirus pulmonary syndrome or HPS, which has a fatality rate of 20–40%. The first such virus to be identified was Sin Nombre virus, which caused an epidemic of HPS in the Four Corners area of the United States in 1993 that resulted in about 25 deaths. The virus

is associated with the deer mouse *Peromyscus maniculatus*. Sin Nombre virus or related viruses have now been identified in virtually all states within the United States and into Latin America, and fatalities due to infection by the virus have occurred in many states. One of the cases in California is of interest because the person died more than a year before the Four Corners epidemic; retrospective studies of serum collected from the patient at the time of his hospitalization showed that he was infected with a hantavirus. The number of cases of HPS in the Americas from 1993 to 1998, totaled by country, and the names of the viruses responsible in various areas are shown in Fig. 4.26. Of interest is Andes virus, which has the potential for human-to-human transmission.

The mortality rate following infection with Sin Nombre virus or its close relatives is close to 50%. The mortality in the earliest cases was even higher because the pulmonary syndrome results from the rapid extravasation of fluids into

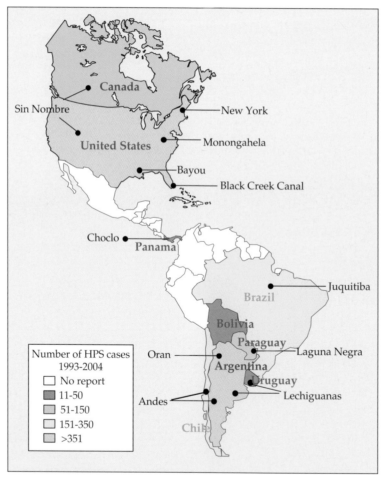

FIGURE 4.26   Cases of hantavirus pulmonary syndrome (HPS) in the Americas, with locations and names of the viruses responsible. Case numbers are cumulative totals from the time that HPS was recognized in 1993 in the Four Corners region of Arizona and New Mexico through 2004. Although several other hantaviruses have been isolated in this region, only those which have been identified as human pathogens are shown. To the current time, cases of HPS have been diagnosed in the United States in 30 states, predominantly in the Western United States, with a few cases as far East as Rhode Island. The total number of cases represented on this map is 1910. Adapted from Peters (1998a) Figure 1 and Table 3, and updated with data from Yates *et al.* (2002), and the Pan American Health Organization at http://www.paho.org/.

the lungs, which can result in respiratory death. This loss of fluids from the intravascular compartment also leads to an increase in the hematocrit (the percentage of blood volume occupied by red blood cells). Early attempts to decrease the hematocrit by supplying fluid intraveneously simply exacerbated the pulmonary edema. Even with the best treatment today, however, the mortality rate is still very high.

It is clear that hantaviruses are widely distributed around the world and have been present in their rodent hosts for a very long time. Although many are capable of causing serious illness in man, the number of human cases is fortunately small. However, there is always the fear that one of these viruses might acquire the ability to spread more readily from human to human and thereby become a more serious problem.

FAMILY *ARENAVIRIDAE*

A listing of the 22 currently recognized arenaviruses is found in Table 4.12. They can be grouped on the basis of sequence alignments and serological cross-reactions into four clades. The Old World viruses form a single clade, whereas the New World viruses group into three different clades, called A, B, C. The genomes consist of two segments of (−)RNA which together total about 11 kb. As for other (−)RNA viruses, the genomic RNA is present in helical nucleocapsids. Budding to acquire the viral envelope is from the plasma membrane (Fig. 2.25B). Virions are spherical but variable in size, with diameters ranging from 50 to 300 nm. It is believed that the number of RNA segments incorporated into a virus particle is not fixed. Multiple copies of the genome segments may be present in virions and this may account, in part if not entirely, for the variation in the size of virions. Also incorporated into the budding virions are variable numbers of ribosomes. The name for the family comes from the Latin word for sand (arena) because the ribosomes in the virions give them a grainy appearance. Why ribosomes are incorporated into virions is not known, as they do not appear to serve a useful function for viral assembly or replication.

The arenaviruses share many features with the hantaviruses. They are associated with rodents and have coevolved with them, as have the hantaviruses. They are transmitted to humans by contact with aerosolized rodent urine or feces; many cause very serious illness, often hemorrhagic fever, with a high mortality rate. Their genome organization and mode of replication has much in common with the hantaviruses, as described later.

Genome Organization and Expression

The genome organization of an arenavirus is illustrated in Fig. 4.27. Arenavirus genomes consist of two segments of RNA, naturally called L(arge) and S(mall). Both genomic RNAs are ambisense in character. The S segment corresponds to the bunyavirus S and M segments linked tail to tail in an ambisense arrangement (Fig. 4.1). The L segment corresponds to the L segments of bunyaviruses but with the addition of a second gene, encoding a protein called Z, in an ambisense orientation. Expression of the encoded genes follows an ambisense strategy as described for some of the bunyaviruses. The mRNA for one gene is synthesized from the genomic RNA and is expressed early, whereas the mRNA for the second gene is synthesized from the antigenomic or vcRNA and is expressed late (Fig. 4.27). As in the bunyaviruses, synthesis of arenavirus mRNA occurs in the cytoplasm using a primer that is snatched from cellular mRNAs, there is a secondary structure in the RNA between the two ambisense genes that causes termination of transcription, and the mRNAs are not polyadenylated.

The genomic S RNA is the template for synthesis of the mRNA for N, and N is therefore expressed early after for the synthesis of infection. Because N is required for the replication of the viral RNA, as is the case for all (−)RNA viruses, this arrangement is necessary if the virus is to replicate. The mRNA for the glycoproteins G1 and G2 is transcribed from the antigenomic copy of S and is therefore expressed late. The glycoproteins are produced as a polyprotein that is cleaved in a process that is similar to what happens in the bunyaviruses. There is an N-terminal signal sequence that leads to the insertion of the precursor called GPC into the endoplasmic reticulum. The signal sequence is removed by cellular signalase. The resulting precursor is cleaved by the cellular subtilase SKI-1/S1P, the same enzyme that processes the hantavirus glycoprotein precursor, into the N-terminal G_N (sometimes called G1 or GP-1) and the C-terminal G_C (sometimes called G2 or GP-2). G_N and G_C remain associated as a heterodimer. Only G_C has a transmembrane anchor, and the process thus resembles what happens in HA of influenza or F of paramyxoviruses where a type I glycoprotein is cleaved into N-terminal and C-terminal subunits that remain associated by noncovalent bonds.

Producing the glycoproteins late has the effect of delaying virus assembly. This allows RNA amplification to proceed for an extended period of time before it is attenuated by the incorporation of nucleocapsids into virions. Attenuation of RNA synthesis is also effected by the Z protein.

In the case of the L segment, the mRNA for protein L is produced early by synthesis from the genomic RNA. Proteins L and N are necessary and sufficient for RNA replication, and this orientation of the genes is necessary for virus replication. The mRNA for protein Z mRNA is transcribed from the antigenomic and thus Z is expressed late, after replication of the RNA begins. Z is a small protein of about 11 kDa that has multiple functions in viral replication. It has a RING finger motif and binds zinc. It downregulates RNA replication and the synthesis of mRNAs. It is also required for budding of virions. In fact, expression of Z in the absence

TABLE 4.12   *Arenaviridae*

Genus/[a] members	Virus name abbreviation	Natural rodent host(s)[b]	Transmission	Disease in humans	World distribution
Old World Arenaviruses					
Lymphocytic choriomeningitis	LCMV	*Mus musculus*	Urine, saliva[c]	Aseptic meningitis	Worldwide
Lassa	LASV	*Mastomys* sp.	Urine, saliva	Hemorrhagic fever (HF)	West Africa
Mopeia	MOPV	*Mastomys natalensis*	Urine, saliva	Nonpathogenic?	Mozambique, Zimbabwe
Mobala	MOBV	*Praomys* sp.	??	??	Central African Republic
Ippy	IPPYV	*Arvicanthis* sp.	??	Nonpathogenic?	Central African Republic
New World Arenaviruses					
Group A[d]					
Tamiami	TAMV	*Sigmodon hispidus*	Urine, saliva	Nonpathogenic?	Florida (U.S.)
Whitewater Arroyo	WWAV	*Neotoma albigula*	Urine, saliva	Three fatal cases of ARDS in California[e]	Western United States
Bear Canyon	BCNV	*Peromyscus californicus*	Urine, saliva	??	Western United States
Paraná	PARV	*Oryzomys buccinatus*	Urine, saliva	Nonpathogenic?	Paraguay
Flexal	FLEV	*Oryzomys* ssp.	Urine, saliva	Nonpathogenic?	Brazil
Pichinde	PICV	*Oryzomys albigularis*	Urine, saliva	Nonpathogenic?	Colombia
Pirital	PIRV	*Sigmodon alstoni*	Urine, saliva	Nonpathogenic?	Venezuela
Allpahuayo	ALLV	*Oecomys bicolor and O. paricola*	Urine, saliva	Nonpathogenic?	Northeastern Peru
Group B					
Guanarito	GTOV	*Zygodontomys brevicauda*	Urine, saliva	Venezuelan HF	Venezuela
Junin	JUNV	*Calomys musculinus*	Urine, saliva	Argentine HF	Argentina
Machupo	MACV	*Calomys callosus*	Urine, saliva	Bolivian HF	Bolivia
Sabiá	SABV	Unknown	???	Isolated from a fatal case, and has caused two severe laboratory infections	Brazil
Amapari	AMAV	*Oryzomys capito, Neacomys guianae*	Urine, saliva	Nonpathogenic	Brazil
Tacaribe	TCRV	*Artibeus* spp. bats[f]	Has been isolated from mosquito	Nonpathogenic	Trinidad
Cupixi	TCRV	*Oryzomys capito*		Nonpathogenic	Brazil
Group C					
Latino	LATV	*Calomys callosus*	??	??	Bolivia
Oliveros	OLVV	*Bolomys obscurus*	??	??	Argentina

[a] LCMV is the type virus of the family.

[b] Most of these viruses cause chronic infections in their natural rodent hosts.

[c] At least one case is known where several recipients contracted LCMV after organ transplants from an asymptomatic donor.

[d] White Water Arroyo, Tamiami, and Bear Canyon viruses have nucleoprotein genes related to those of Pichinde and Pirital in Group A, but glycoprotein genes related to Tacaribe, Junin, and Sabiá in Group B.

[e] ARDS, acute respiratory distress syndrome. Until these cases in 1999–2000, WWAV was not known to cause human illness.

[f] Originally isolated from fruit-eating bats, but subsequent isolation attempts from bats have failed.

Source: Fields *et al.* (1996), Table 1 on p. 1522; Porterfield (1995), Table 11.1 on p. 228; and recent information from Fauquet *et al.* (2005).

Lassa fever virus S RNA (3417 nt)

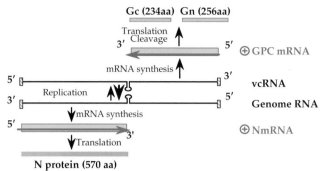

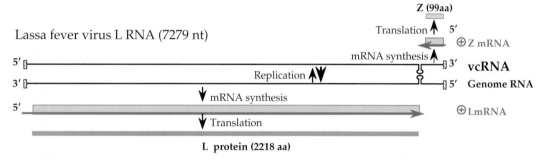

FIGURE 4.27   Genome organization and replication strategy of an arenavirus, Lassa fever virus. The genome consists of two segments of RNA, L and S. Both segments are expressed using an ambisense strategy. The nucleocapsid protein mRNA is synthesized from the 3′ end of the genomic S RNA, while the GPC mRNA is synthesized from the vc S RNA. A similar strategy occurs with the L segment. In this case, however, more than 95% of the coding capacity is used for the L protein, the RNA dependent RNA polymerase. Also note that the 5′ nontranslated region of L mRNA is 157 nt, which is unusually long for an arenavirus. The Z protein is a so-called "ring finger protein" that is involved in regulation of transcription and replication. Drawn from data in Lukashevich *et al.* (1997) and Clegg *et al.* (1990).

of other viral proteins results in the formation of virus-like particles, and Z has a role in budding analogous to the role played by the M proteins of other (−)RNA viruses or the Gag protein of retroviruses. It has been found that Z recruits a cellular protein called Tsg101 to the site of budding. Tsg101 has been shown to be required for budding of (at least) two arenaviruses, of HIV, and of Ebola virus. Tsg is a component of the vacuolar protein sorting machinery of the cell and is therefore active in promoting cellular budding pathways.

Z was originally thought to be a nonstructural protein and was called NS. It is now known to be present in the virion. The stoichiometry of proteins in virions of Lassa virus was found to be 1:160:60:60:20 for L:N:G_N:G_C:Z.

Natural History and Diseases

The natural history of the arenaviruses is very similar to that of the hantaviruses (Table 4.12). They establish a persistent infection in a single rodent host. Many cause hemorrhagic fever in humans following infection by aerosolized virus excreted in urine or feces. They appear to have co-evolved with their hosts: An evolutionary tree of arenaviruses resembles the tree that describes their rodent hosts, as was true of the hantaviruses. The many similarities in genome organization and expression, the association with a single rodent species, and the nature of the disease caused in humans all suggest that the arenaviruses are closely related to the hantaviruses. A reasonable hypothesis is that the arenaviruses arose from the hantaviruses by fusion of the S and M segments to form one segment, which allowed finer control of the virus life cycle.

The cellular receptor for entry of many arenaviruses is α-dystroglycan, which plays a critical role in cell-mediated assembly of basement membranes. This protein is widely distributed in animals and many arenaviruses have a broad tissue tropism. For example, Lassa infection of humans results in high virus titers in spleen, lung, liver, kidney, heart, placenta, and mammary gland. Viruses that have a high binding affinity for α-dystroglycan replicate preferentially in the white pulp of the spleen and infect large numbers of lymphocytes that are important in the immune response to viral infection. The ability of these lymphocytes to act as antigen-presenting cells results in impairment of immune responses resulting in a generalized immunosuppression. Such viruses are more virulent than those that bind less avidly to α-dystroglycan. Immunosuppression may be important for the establishment of persistent infections in the rodent host, in which the virus does not cause disease.

In humans, however, immunosuppression may lead to much more serious illness.

The arenaviruses can be divided into Old World viruses and New World viruses (Table 4.12). Because of their association with a single rodent species, their geographic range is restricted to that of their host, and rodents have a restricted range. The exceptions are rodents that have been distributed widely by humans such as the house mouse and the urban rat. Many arenaviruses cause hemorrhagic fever in man with significant mortality rates (Table 4.12).

Lymphocytic Choriomeningitis Virus

Lymphocytic choriomeningitis virus (LCMV), the prototype virus of the family, is associated with the house mouse *Mus domesticus* and *Mus musculus*. This virus is widespread in Europe, along with its host, and spread to the Americas with the (inadvertent) introduction of the house mouse by European travelers. LCMV has been intensively studied in the laboratory as a model for the arenaviruses, in part because it is less virulent for humans than many arenaviruses, and in part because its natural host is widely used as a laboratory model for animal work. Mice are small, reproduce rapidly, and there is a great deal of experience in maintaining this animal in the laboratory. LCMV is widespread, often being present in colonies of laboratory mice even without overt introduction. It is also present in wild mice and may be present in pets such as hamsters.

LCMV infection of humans usually results in mild or even inapparent illness, although serious illness can result with occasional mortality. In a recent incident, a woman had been infected with LCMV from a pet hamster. She suffered no apparent illness from the viral infection but died of an unrelated cause, a stroke. Her liver, lungs, and kidneys were harvested for transplantation. Transplantation of liver, lungs, and kidney requires immunosuppression so that the transplanted organs are not rejected. Three patients receiving the liver, lungs, and a kidney developed overwhelming infection by LCMV and died. A fourth patient who received a kidney also became quite ill from LCMV infection but survived, aided by reduction in the immunosuppressive drugs being given.

Lassa Virus

The rodent reservoir of Lassa virus is *Mastomys natalensis*. Lassa virus causes outbreaks in West Africa of an often fatal illness in humans called Lassa fever. The mortality rate averages 10–15% but may be as high as 60% in some outbreaks. The virus has a broad tissue tropism and symptoms include fever, myalgia, and severe prostration, often accompanied by hemorrhagic or neurological symptoms. Development of hemorrhagic symptoms indicates a poor prognosis and death often follows. Fatal infection is also characterized by higher viral loads. Survivors of severe infection often suffer nerve damage and may be deaf because of such damage. The full extent of Lassa disease is not known because most Africans infected by the virus do not seek help and there is little monitoring of the disease. However, estimates range from 100,000 to 300,000 cases per year.

Lassa virus was first isolated in 1969 when a nurse in a rural mission hospital in Nigeria became infected. She was transported to Jos, Nigeria where several health care workers became infected. Serum samples were sent to the United States and a well-known virologist at the Yale Arbovirus Research Unit, Dr. Jordi Casals, became infected with the virus while working with it and became very seriously ill. He eventually recovered but later that same year a technician in another laboratory at Yale became infected with Lassa fever virus and died, whereupon Yale ceased to work with the virus. The containment facilities in 1969 were not of the quality of those in current use and virologists in those days literally took their lives in their hands when working with dangerous agents. The study of virology owes a great deal to the courage exhibited by these earlier workers.

Lassa has been imported to the United States on at least one occasion in the form of a viremic individual. A resident of Chicago attended the funerals of relatives in Nigeria who had died of Lassa fever and became infected there. On return to Chicago he began suffering symptoms of Lassa fever but the local hospitals were unable to diagnose the cause of his disease, being unfamiliar with it. He eventually died of Lassa fever, but fortunately there were no secondary cases.

New World Group B Viruses

Several South American arenaviruses belonging to Group B are very important disease agents because they cause large outbreaks of hemorrhagic fever with high mortality rates. The names of a number of these viruses and the places where they are found are shown in Fig. 4.28. They include Junín virus (causative agent of Argentine hemorrhagic fever), Machupo virus (Bolivian hemorrhagic fever), Guanarito virus (Venezuelan hemorrhagic fever), and Sabiá virus (cause of an unnamed disease in Brazil). The diseases caused by these viruses are often referred to as emerging diseases because the number of human cases has increased with development and expanding populations. The increasing number of cases results from development of the pampas or other areas for farming, bringing humans in closer association with the rodent reservoirs. Furthermore, the storage of grain near human habitation results in an increase in the local rodent population, and plowing of the fields leads to the production of aerosols which may transmit the disease to humans. An attenuated virus vaccine against Junín virus has been developed and is widely used in populations at risk. The vaccine is effective and has reduced dramatically the number of cases of Argentine hemorrhagic fever. No vaccines are in use for the other viruses, however.

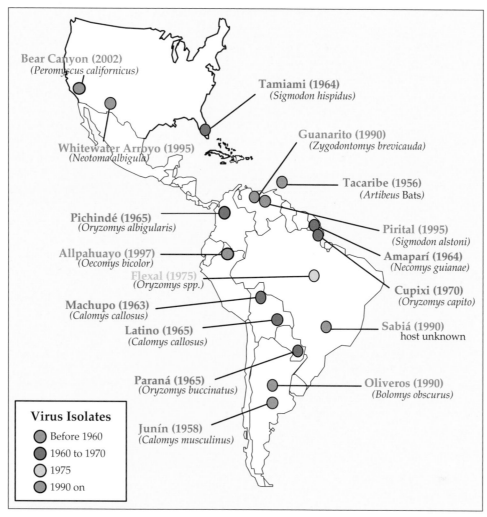

FIGURE 4.28   Arenavirus isolates in the New World. Also shown are the year of first isolation, and the rodent host of each virus where known. Adapted from Peters (1998b), Figure 1.

New World Group A Viruses

Three arenaviruses have been isolated in the United States, Whitewater Arroyo virus, present in the Southwest, Bear Canyon virus in California, and Tamiami virus, present in Florida (Fig. 4.28). None of these viruses, all of which belong to Group A, had been known to cause illness in humans until very recently. In 1999–2000, three Californians died following infection by Whitewater Arroyo virus. The disease these three suffered was ARDS (acute respiratory disease syndrome), although two also had hemorrhagic manifestations. Thus, like the hantaviruses, the U.S. arenaviruses may cause isolated cases of serious illness. There are also a number of Group B viruses in South America (Table 4.12), but these are not known to cause disease in humans.

Agents Causing Hemorrhagic Fevers in Humans

Many viruses, belonging to several different families, have been described that cause hemorrhagic fever in humans. Table 4.13 contains a listing of many of these viruses. These viruses include members of the *Arenaviridae*, *Bunyaviridae*, *Filoviridae*, and *Flaviviridae*. Many cause severe disease with high mortality, but although the disease is severe, with the exception of some arenaviruses, survivors have few sequelae. The dramatic symptom of profuse bleeding has excited the purple prose of many lay authors, best illustrated by recent discussions of Ebola virus, and struck terror in native populations. With the exceptions of yellow fever virus and Junín virus, there are no vaccines, and treatments are primarily supportive, although ribavirin therapy holds some

TABLE 4.13   Viruses That Cause Hemorrhagic Fevers in Humans

Virus	Disease[a]	Geographic range	Vector transmission	%Case mortality[b]	Treatment (prevention)
Arenaviridae					
Junin	Argentine HF	Argentine pampas	Infected field rodents, *Calomys musculinus*	15–30	Antibody effective, ribavirin probably effective; preventive vaccine exists
Machupo	Bolivian HF	Beni province, Bolivia	Infected field rodents, *Calomys callosus*		Ribavirin probably effective
Guanarito	Venezuelan HF	Venezuela	Infected field rodents, *Zygodontomys brevicauda*		No data for humans, ribavirin probably effective
Sabiá	HF	Rural areas near Salo, Brazil	Unidentified infected rodents		Intravenous ribavirin effective in one case
Lassa	Lassa fever	West Africa	Infected *Mastomys* rodents	15	Ribavirin effective
Bunyaviridae					
Rift Valley fever	Rift Valley fever	Sub-Saharan Africa	*Aedes* mosquito	50	Rapid course; ribavirin or antibody might be effective
Crimean-Congo HF	Crimean-Congo HF	Africa, Middle East, Balkans, Russia, W. China	Tick-borne	15–30	Ribavirin used and probably effective
Hantaan, Seoul, Puumala, and others	HFRS	Worldwide (See Fig. 4.25)	Each virus maintained in a single species of infected rodents	Variable[c]	Ribavirin useful; supportive therapy is mainstay
Sin Nombre and others	HPS, also rare HF	Americas (See Fig. 4.26)	As for viruses causing HFRS	40–50	Rapid course makes specific therapy difficult
Filoviridae					
Marburg Ebola	Filovirus HF	Africa	Unknown	Marburg 25 EbolaZ 30–90	No effective therapy, barrier nursing prevents spread in epidemics
Flaviviridae					
Yellow fever	Yellow fever	Africa, South America	*Aedes* mosquito	20	Very effective vaccine
Dengue	DHF, DSS	Tropics and subtropics worldwide	*Aedes* mosquito	<1	Supportive therapy useful; vector control
Kyasanur forest disease	KFD	Mysore State, India	Tick-borne	0.5–9	???
Omsk hemorrhagic fever	OHF	Western Siberia	Poorly understood cycle involves ticks, voles, muskrats??	?	Needs further study

[a] Abbreviations: HF, hemorrhagic fever; HFRS, hemorrhagic fever with renal syndrome; HPS, hantavirus pulmonary syndrome; DHF, dengue hemorrhagic fever; DSS, dengue shock syndrome; KFD, Kyasanur Forest disease; OHF, Omsk hemorrhagic fever.
[b] In humans.
[c] Hantaan is 5–15% fatal, while Puumala is <1% fatal.
Source: This table includes data from Nathanson *et al.* (1996) Table 32.1 on p. 780.

promise for arenavirus disease. Human-to-human transmission is uncommon. Where limited transmission has occurred, it has been by exposure to contaminated blood, or possibly exposure to other bodily fluids, and resulted in limited epidemics for such viruses as Ebola and Machupo virus.

EVOLUTION OF MINUS-STRAND RNA VIRUSES

As has been described, all (−)RNA viruses share a number of features. These include virion structure (enveloped viruses with helical nucleocapsids); mechanisms for replicating the genomic RNA (replication within RNP which requires ongoing protein synthesis; self-complementarity of the ends of the RNA (with its implications for promoter elements involved in replication); mechanisms for synthesis of mRNA (synthesis of leaders or the use of primers for synthesis of mRNA, the presence of intergenic sequences); and the suite of proteins encoded. These similarities make it seem likely that all (−)RNA viruses have diverged from a common ancestor fairly recently on a geological timescale, certainly more recently than the divergence of the extant plus-strand RNA viruses from a common ancestor. The (+)RNA viruses are much more divergent in structure and in the strategies used for replication and expression of the genome, suggesting that they have had a much longer period in which to diverge from one another. Although the suite of proteins encoded is very similar in all (−)RNA viruses, the rate of evolution of RNA viruses is so fast that little sequence identity can be demonstrated between different groups. However, where studies have been performed, evidence for common origin of at least some of these proteins has been shown. As an example, the M proteins of VSV and influenza virus are related and have diverged from a common ancestor. It seems likely that most of the various proteins are related in this way, although it is clear that some viruses have genes that are not represented in all viruses and which presumably arose by recombination events that led to the insertion of new functions, or to deletion events that resulted in a virus with fewer genes, or, probably, to both.

Because the (−)RNA viruses appear to be more recent than the (+)RNA viruses, it is reasonable to postulate that they arose from the (+)RNA viruses. If so, one obvious candidate for the ancestor is a coronavirus. Like the (−)RNA viruses, coronaviruses are enveloped viruses with a helical nucleocapsid that synthesizes RNA leaders and use primers to prime mRNA synthesis, traits in which the coronaviruses differ from other (+)RNA viruses.

If the (−)RNA viruses did arise from the (+)RNA viruses, what traits might account for their success once they arose? One obvious possibility is the ability to synthesize individual mRNAs for each protein needed. This trait also carries with it the necessity to include the RNA synthesis machinery in the virion, but the ability to control the order of synthesis and the translation frequency of the different proteins has obvious advantages for control of the replication cycle. The (−)RNA viruses with segmented genomes also have the ability to undergo ready reassortment, which is clearly advantageous in the *Orthomyxoviridae* and probably important for all viruses with segmented genomes. (+)RNA viruses that infect animals do not have segmented genomes, except for a few insect viruses with bipartite genomes, for reasons that are not clear. (+)RNA viruses of plants with segmented genomes are common, however. It is perhaps suggestive that the (−)RNA viruses have not

been as successful in plants as they have been in animals, and the (−)RNA plant viruses that do exist also replicate in arthropods, which serve as vectors for transmitting the virus to plants.

FURTHER READING

General Reviews

Garoff, H., Hewson, R., and Opstelten, D.-J. (1998). Virus maturation by budding. *Microbiol. Mol. Biol. Rev.* **62**: 1171–1190.

Lamb, R. A. (2006). *Mononegavirales*. Chapter 38 in: *Fields Virology, Fifth Edition* (D. M. Knipe and P. M. Howley, Eds. in chief), Philadelphia, Lippincott Williams & Wilkins, pp. 1357–1362.

Pringle, C. R., and Easton, A. J. (1997). Monopartite negative strand RNA genomes. *Semin. Virol.* **8**: 49–57.

Whelan, S. P. J., Barr, J. N., and Wertz, G. W. (2004). Transcription and replication of nonsegmented negative-strand RNA viruses. *Curr. Top. Microbiol. Immunol.* **283**: 61–119.

Rhabdoviridae

Ball, L. A., Pringle, C. R., Flanagan, B., *et al.* (1999). Phenotypic consequences of rearranging the P, M, and G genes of vesicular stomatitis virus. *J. Virol.* **73**: 4705–4712.

Hanlon, C. A., and Rupprecht, C. E. (1998). The reemergence of rabies. In *Emerging Infections I* (W. M. Scheld, D. Armstrong, and J. M. Hughes, Eds.), Washington, DC, ASM Press, pp. 59–80.

Hooper, P. T., Lunt, R. A., Gould, A. R., *et al.* (1997). A new Lyssavirus—the first endemic rabies-related virus recognized in Australia. *Bull. Inst. Pasteur (Paris)* **95**: 209–218.

Jackson, A. C., and Wunner, W. H. (Eds.) (2002). *Rabies.* Burlington, MA, Elsevier Science, USA.

Jayakar, H. R., Jeetendra, E., and Whitt, M. A. (2004). Rhabdovirus assembly and budding. *Virus Res.* **106**: 117–132.

Lyles, D. S., and Rupprecht, C. E. (2006). *Rhabdoviridae*. Chapter 39 in: *Fields Virology, Fifth Edition* (D. M. Knipe and P. M. Howley, Eds. in chief), Philadelphia, Lippincott Williams & Wilkins, pp. 1363–1408.

Smith, J. S., Orciari, L. A., and Yager, P. A. (1995). Molecular epidemiology of rabies in the United States. *Semin. Virol.* **6**: 387–400.

Springfield, C., Darai, G., and Cattaneo, R. (2005). Characterization of the Tupaia rhabdovirus genome reveals a long open reading frame overlapping with P and a novel gene encoding a small hydrophobic protein. *J. Virol.* **79**: 6781–6790.

Paramyxoviridae

Black, F. L. (1966). Measles endemicity in insular populations: critical community size and its evolutionary implication. *J. Theor. Biol.* **11**: 207–211.

Carbone, K. M., and Rubin, S. A. (2006). Mumps virus. Chapter 43 in: *Fields Virology, Fifth Edition* (D. M. Knipe and P. M. Howley, Eds. in chief), Philadelphia, Lippincott Williams & Wilkins, pp. 1527–1550.

Chua, K. B., Bellini, W. J., Rota, P. A., *et al.* (2000). Nipah virus: A recently emergent deadly paramyxovirus. *Science* **288**: 1432–1435.

Collins, P. L., and Crowe, J. E., Jr. (2006). Respiratory syncytial virus and metapneumoviruses. Chapter 46 in: *Fields Virology, Fifth Edition* (D. M. Knipe, and Howley, Eds. in chief), Philadelphia, Lippincott Williams & Wilkins, pp. 1601–1646.

Curran, J., Latorre, P., and Kolakofsky, D. (1998). Translational gymnastics on the Sendai virus P/C mRNA. *Semin. Virol.* **8**: 351–357.

Daszak, P., Cunningham, A. A., and Hyatt, A. D. (2000). Emerging infectious diseases of wildlife-threats to biodiversity and human health. *Science* **287**: 443–449.

Eaton, B. T., MacKenzie, J. S., and Wang, L.-F (2006). Henipaviruses. Chapter 45 in: *Fields Virology, Fifth Edition* (D. M. Knipe and P. M. Howley, Eds. in chief), Philadelphia, Lippincott Williams & Wilkins, pp. 1587–1600.

Hausmann, S., Garcin, D., Morel, A.-S., *et al.* (1999). Two nucleotides immediately upstream of the essential A6G3 slippery sequence modulate the pattern of G insertions during Sendai Virus mRNA editing. *J. Virol.* **73**: 343–351.

Griffin, D. E. (2006). Measles virus. Chapter 44 in: *Fields Virology, Fifth Edition* (D. M. Knipe and P. M. Howley, Eds. in chief), Philadelphia, Lippincott Williams & Wilkins, pp. 1551–1586.

Karron, R. A., and Collins, P. L. (2006). Parainfluenza viruses. Chapter 42 in: *Fields Virology, Fifth Edition* (D. M. Knipe and P. M. Howley, Eds. in chief), Philadelphia, Lippincott Williams & Wilkins, pp. 1497–1526.

Lamb, R. A., and Parks, G. D. (2006). *Paramyxoviridae*: The viruses and their replication. Chapter 41 in: *Fields Virology, Fifth Edition* (D. M. Knipe and P. M. Howley, Eds. in chief), Philadelphia, Lippincott Williams & Wilkins, pp. 1449–1496 .

Patterson, J. B., Thomas, D., Lewicki, H., Billeter, M. A., and Oldstone, M. B. A. (2000). V and C proteins of measles virus function as virulence factors *in vivo*. *Virology* **267**: 80–89.

Strauss, E. G., and Strauss, J. H. (1991). RNA viruses: Genome structure and evolution. *Curr. Opin. Genet. Dev.* **1**: 485–493.

Vincent, S., Gerlier, D., and Manié, S. N. (2000). Measles virus assembly within membrane rafts. *J. Virol.* **74**: 9911–9915.

Wang, L.-F., Yu, M., Hansson, E., *et al.* (2000). The exceptionally large genome of Hendra virus: Support for creation of a new genus within the family *Paramyxoviridae*. *J. Virol.* **74**: 9972–9979.

Wilson, R. L., Fuentes, S. M., Wang, P., *et al.* (2006). Function of small hydrophobic proteins of parmyxovirus. *J. Virol.* **80**: 1700–1709.

Yanagi, Y., Takeda, M., and Ohno, S. (2006). Measles virus; cellular receptors, tropism, and pathogenesis. *J. Gen. Virol.* **87**: 2767–2779.

Filoviridae

Feldmann, H., and Klenk, H.-D. (1996). Marburg and Ebola viruses. *Adv. Virus Res.* **47**: 1–52.

Peters, C. J., and Kahn, A. S. (1999). Filovirus diseases. *Curr. Top. Microbiol. Immunol.* **235**: 85–95.

Peterson, A. T., Bauer, J. T., and Mills, J. N. (2004). Ecologic and geographic distribution of Filovirus disease. *Emerg. Infect. Dis.* **10**: 40–47.

Pushko, P., Geisbert, J., Parker, M., Jahrling, P., and Smith, J. (2001). Individual and bivalent vaccines based on Alphavirus replicons protect guinea pigs against infection with Lassa and Ebola viruses. *J. Virol.* **75**: 11677–11685.

Sanchez, A., Geisbert, T. W., and Feldmann, H. (2006). *Filoviridae*: Marburg and Ebola viruses. Chapter 40 in: *Fields Virology, Fifth Edition* (D. M. Knipe and P. M. Howley, Eds. in chief), Philadelphia, Lippincott Williams & Wilkins, pp. 1409–1448.

Sullivan, N., Yang, Z.-Y., and Nabel, G. J. (2003). Ebola virus pathogenesis: implications for vaccines and therapies. *J. Virol.* **77**: 9733–9737.

Bornaviridae

Iwata, Y., Takahashi, K., Peng, X., *et al.* (1998). Detection and sequence analysis of borna disease virus p24 RNA from peripheral blood mononuclear cells of patients with mood disorders or schizophrenia and of blood donors. *J. Virol.* **72**: 10044–10049.

Kohno, T., Goto, T., Takasaki, T., *et al.* (1999). Fine structure and morphogenesis of borna disease virus. *J. Virol.* **73**: 760–766.

Lipkin, W. I., and Briese, T. (2006). *Bornaviridae*. Chapter 51 in: *Fields Virology, Fifth Edition* (D. M. Knipe and P. M. Howley, Eds. in chief), Philadelphia, Lippincott Williams & Wilkins, pp. 1829–1852.

Schneider, U., Schwemmle, M., and Staeheli, P. (2005). Genome trimming: a unique strategy for replication control employed by Borna disease virus. *Proc. Natl. Acad. Sci. U.S.A.* **102**: 3441–3446.

Schneemann, A., Schneider, P. A., Lamb, R. A., *et al.* (1995). The remarkable coding strategy of borna disease virus: A new member of the non-segmented negative strand RNA viruses. *Virology* **210**: 1–8.

Orthomyxoviridae

Clouthier, S. C., Rector, T., Brown, N. E. C., and Anderson, E. D. (2002). Genomic organization of infectious salmon anaemia virus. *J. Gen. Virol.* **83**: 421–428.

Crosby, A. W. (1989). *America's Forgotten Pandemic: The Influenza of 1918*. Cambridge, England, Cambridge University Press.

Kolata, G. (1999). *Flu: The Story of the Great Influenza Pandemic of 1918 and the Search for the Virus that Caused It*. New York, Farrar, Straus, & Giroux.

Lipatov, A. S., Govorkova, E. A., Webby, R. J., *et al.* (2004). Influenza: emergence and control. *J. Virol.* **78**: 8951–8959.

Munoz, F. M., Galasso, G. J., Gwaltney, J. M., Jr., *et al.* (2000). Current research on influenza and other respiratory viruses: 2nd international symposium. *Antiviral Res.* **46**: 91–124.

Neumann, G., Watanabe, T., Ito, H., *et al.* (1999). Generation of influenza A viruses entirely from cloned cDNAs. *Proc. Natl. Acad. Sci. U.S.A.* **96**: 9345–9350.

Neumann, G., Brownlee, G. G., Fodor, E., and Kawaoka, Y. (2004). Orthomyxovirus replication, transcription, and polyadenylation. *Curr. Top. Microbiol. Immunol.* **283**: 121–143.

Noda, T., Sagara, H., Yen, A., *et al.* (2006). Architecture of ribonucleoprotein complexes in infleunza A virus particles. *Nature* **439**: 490–492.

Ortín, J. (1998). Multiple levels of posttranscriptional regulation of influenza virus gene expression. *Semin. Virol.* **8**: 335–342.

Palese, P., and Shaw, M. L. (2006). *Orthomyxoviridae*: The viruses and their replication. Chapter 47 in: *Fields Virology, Fifth Edition* (D. M. Knipe and P. M. Howley, Eds. in chief), Philadelphia, Lippincott Williams & Wilkins, pp. 1647–1690.

Reid, A. H., Fanning, T. G., Janczewski, T. A., *et al.* (2000). Characterization of the 1918 "Spanish" influenza virus neuraminidase gene. *Proc. Natl. Acad. Sci. U.S.A.* **97**: 6785–6790.

Reid, A. H., Fanning, T. G., Janczewski, T. A., *et al.* (2004). Novel origin of the 1918 pandemic influenza virus nucleoprotein gene. *J. Virol.* **78**: 12462–12470.

Reid, A. H., and Taubenbeger, J. K. (2003). The origin of the 1918 pandemic influenza virus: a continuing enigma. *J. Gen. Virol.* **84**: 2285–2292,

Taubenberger, J. K., Reid, A. H., and Fanning, T. G. (2000). The 1918 influenza virus: A killer comes into view. *Virology* **274**: 214–245.

Thompson, W. W., Shay, D. K., Weintraub, E., *et al.* (2003). Mortality associated with influenza and respiratory syncytial virus in the United States. *JAMA.* **289**: 179–186.

Bunyaviridae

Barr, J. N., Rodgers, J. W., and Wertz, G. W. (2005). The bunyamwera virus mRNA transcription signal resides within both the 3′ and 5′ terminal regions and allows ambisense transcription from a model RNA segment. *J. Virol.* **79**: 12602–12607.

Galeno, H., Mora, J., Villagra, E., *et al.* (2002). First human isolate of hanta-virus (Andes virus) in the Americas. *Emerg. Infect. Dis.* **8**: 657–661.

Honig, J. E., Osborne, J. C., and Nichol, S. T. (2004). Crimean-Congo hemorrhagic fever virus genome L RNA segment and encoded protein. *Virology* **321**: 29–35.

Ikegami, T., Won, S., Peters, C. J., and Makino, S. (2005). Rift Valley fever virus NSs mRNA is transcribed from an incoming anti-viral-sense S RNA segment. *J. Virol.* **79**: 12106–12111.

Kaukinen, P., Vaheri, A., and Plyusin, A. (2005). Hantavirus nucleocapsid protein: a multifunctional molecule with both housekeeping and ambas-sadorial duties. *Arch. Virol.* **150**: 1693–1713.

Marczinke, B. I., and Nichol, S. T. (2002). Nairobi sheep disease virus, an important tick-borne pathogen of sheep and goats in Africa, is also present in Asia. *Virology* **303**: 146–151.

Schmaljohn, C. S., and Nichol, S. T. (2006). *Bunyaviridae*. Chapter 49 in: *Fields Virology, Fifth Edition* (D. M. Knipe and P. M. Howley, Eds. in chief), Philadelphia, Lippincott Williams & Wilkins, pp. 1741–1790.

Vincent, M. J., Quiroz, E., Gracia, F., *et al.* (2000). Hantavirus pulmonary syndrome in Panama: identification of novel hantaviruses and their likely reservoirs. *Virology* **277**: 14–19.

Yates, T. L., Mills, J. N., Parmenter, *et al.* (2002). The ecology and evolu-tionary history of an emergent disease: hantavirus pulmonary syndrome. *Bioscience* **52**: 989–998.

Arenaviridae

Archer, A. M., and Rico-Hesse, R. (2002). High genetic divergence and recombination in arenaviruses from the Americas. *Virology* **304**: 274–281.

Bowen, M. D., Peters, C. J., and Nichol, S. T. (1997). Phylogenetic analysis of the *Arenaviridae*: Patterns of virus evolution and evidence for cospe-ciation between arenaviruses and their rodent hosts. *Mol. Phylogenet. Evol.* **8**: 301–316.

Buchmeier M. J., de la Torre, J.-C., and Peters, C. J. (2006). *Arenaviridae*: The viruses and their replication. Chapter 50 in: *Fields Virology, Fifth Edition* (D. M. Knipe, P. M. Howley, Eds. in chief), Philadelphia, Lippincott Williams & Wilkins, pp. 1791–1828.

Charrel, R. N., Feldmann, H., Fulhorst, C. F., *et al.* (2002). Phylogeny of New World arenaviruses based on the complete coding sequences of the small genomic segment identified an evolutionary lineage produced by intrasegmental recombination. *Biochem. Biophys. Res. Commun.* **296**: 1118–1124.

Emonet, S., Lemasson, J.-J., Gonzalez, J.-P., *et al.* (2006). Phylogeny and evolution of old world arenaviruses. *Virology* **350**: 251–257.

Gunther, S., and Lenz, O. (2004). Lassa virus. *Crit. Rev. Clin. Lab. Sci.* **41**: 339–390.

Moncayo, A. C., Hice, C. L., Watts, D. M., *et al.* (2001). Allpahuayo virus: a newly recognized arenavirus (*Arenaviridae*) from arboreal rice rats (*Oecomys bicolor* and *Oecomys paricola*) in northeastern Peru. *Virology* **284**: 277–286.

Perez, M., Craven, R. C., and de la Torre, J. C. (2003). The small RING finger protein Z drives arenavirus budding: implications for antiviral strategies. *Proc. Natl. Acad. Sci. U.S.A.* **100**: 12978–12983.

Peters, C. J. (1998). Hemorrhagic fevers: how they wax and wane. In: "Emerging Infections 2" (W. M. Scheld, W. A. Craig, and J. M. Hughes, eds.), Washington, DC, ASM Press, pp. 12–25.

CHAPTER

5

Viruses That Contain Double-Stranded RNA: Family *Reoviridae*

INTRODUCTION

Most of the dsRNA-containing viruses that we know about belong to the family *Reoviridae*. All members of this large family have a genome consisting of 10, 11, or 12 segments of dsRNA totaling 16–27 kb. The family is very successful. Twelve genera are currently recognized and different viruses infect a wide spectrum of vertebrates, invertebrates, plants, and fungi. Several viruses are important pathogens of humans.

Other dsRNA viruses in addition to the family *Reoviridae* include the birnaviruses, icosahedral viruses 60 nm in diameter containing two genome segments of dsRNA totaling about 7 kb, which infect chickens, fish, and arthropods; the totiviruses of fungi and protozoa, which have only one RNA segment; and the partiviruses and hypoviruses of fungi and plants. Much less about these viruses is known and for this reason as well as the fact that no human pathogens are known among these viruses, they will not be considered further.

OVERVIEW OF THE FAMILY *REOVIRIDAE*

Eleven genera of *Reoviridae* are listed in Tables 5.1 and 5.2 together with a partial listing of viruses in each genus, their hosts and modes of transmission, the diseases they cause, and their distributions. A 12th genus, *Mycoreovirus*, contains viruses that infect fungi. As a taxon, *Reoviridae* have a wide host range. Members of the genera *Orthoreovirus*, *Rotavirus*, *Orbivirus*, *Coltivirus*, and *Seadornavirus* infect humans as well as other vertebrates, aquareoviruses infect fish, cypoviruses and entomoreoviruses infect arthropods, and members of three genera infect plants. Unclassified viruses that infect scorpions, crabs, and bed bugs are also known. Characteristics of the five genera that contain human pathogens are compared in Table 5.3.

Many of the *Reoviridae* are transmitted by arthropod vectors (Tables 5.1 and 5.2). The orbiviruses are transmitted by phlebotomine flies, ticks, gnats, and midges of the genus *Culicoides,* and members of this genus are true arboviruses, as are the coltiviruses, transmitted by ticks, and the seadornaviruses, transmitted by mosquitoes. The members of the three genera of plant reoviruses are also transmitted by arthropods and are effectively plant arboviruses (although the term arbovirus is reserved for vertebrate viruses that are transmitted by arthropods). Thus, members of eight genera of *Reoviridae* possess the ability to replicate in arthropods, of which two genera contain viruses that replicate exclusively in insects and six genera contain viruses that have an alternate vertebrate or plant host. The relationships among these various genera are illustrated in the unrooted dendrogram in Fig. 5.1, which is based on the sequence of the RNA polymerase. It is interesting that the various genera do not group in any obvious fashion related to the host or the vector for the virus. As one example, note that the orthoreoviruses are most closely related to the aquareoviruses, not to other viruses of mammals. As a second example, the coltiviruses are more closely related to the plant reoviruses than to the other mammalian reoviruses.

Members of the family *Reoviridae* replicate in the cytoplasm. The virion is icosahedral (T=13), 60–80 nm in diameter, and double or triple shelled (the shells consist of protein). The structures of viruses belonging to three genera of the *Reoviridae* were shown in Chapter 2: reovirus (genus *Orthoreovirus*) in Figs. 2.1 and 2.5; rotavirus (genus *Rotavirus*) in Fig. 2.5; and bluetongue virus (genus *Orbivirus*) in Fig. 2.11 (in this case only the core of the virion is shown). The members of the genus *Orthoreovirus* have been the best studied and have served as a model system for the family, but because of the medical importance of the members of the genus *Rotavirus* and the veterinary

TABLE 5.1 *Reoviridae* of Vertebrates

Genus/members	Virus name abbreviation	Usual host(s)	Transmission or vector	Disease	World distribution
Orthoreoviruses					
Non-fusogenic—Subgroup 1					
Mammalian reoviruses types 1, 2, 3	MRV	Humans, cattle, sheep, swine	Oral–fecal	Gastroenteritis, respiratory disease	Worldwide
Fusogenic—Subgroup 2					
Nelson Bay orthovirus	NBV	Flying foxes (bats)			
Avian orthoreoviruses 8 serotypes	ARV	Birds	Oral–fecal	Wide range of symptoms from inapparent to lethal	Australia, Worldwide
Fusogenic—Subgroup 3					
Baboon orthoreovirus	BRV	Monkeys	Oral–fecal		?
Orbiviruses					
Bluetongue 24 serotypes	BTV	Sheep, cattle	*Culicoides*	Rhinitis, stomatitis	Asia, Americas Africa, Australia,
African horse sickness 10 serotypes	AHSV	Equines	*Culicoides*	Cardiopulmonary disease	Africa
Changuinola	CGLV	Humans	Phlebotamines	Fever	Panama
Kemerovo serogroup:					
Kemerovo	?	Humans			
Great Island	GIV	Seabirds	Ticks	Fever, encephalitis	E. Europe, United States
Chenuda	CNUV				
Wad Medani	WMV	Domestic animals			
Coltiviruses					
Colorado tick fever	CTFV	Humans	Ticks	Fever, encephalitis	North America, Europe
Rotaviruses					
Group A	RV-A	Humans, animals	Oral–fecal	Infant diarrhea	Worldwide
Group B	RV-B	Humans, animals	Oral–fecal	Epidemic adult diarrhea	Primarily China
Group C	RV-C	Humans, animals	Oral–fecal	Clinical significance unknown	Worldwide
Groups D, E, F, G		Birds, mammals	Oral–fecal		
Seadornaviruses					
Kadipiro	KDV	Vertebrates	*Culex* and *Anopheles* mosquitoes	Fever, encephalitis	China, Indonesia
Banna	BAV				
Aquareoviruses					
Six serogroups		Fish	?	?	?

Representative members of each genus are shown, and the first listed is the type species of the genus.

TABLE 5.2 *Reoviridae* of Plants and Insects

Genus/members	Type virus name abbreviation	Usual host(s)	Transmission or vector	Disease	World distribution
Cypoviruses					
Cytoplasmic polyhedrosis viruses (16 species)	BmCPV-1	Arthropods	Ingestion	Diarrhea, starvation due to changes in the gut	Worldwide
Idnoreovirus					
Ten-segmented, insect-derived, non-occluded	DpIRV-1	Arthropods			
Fijivirus					
Fiji disease virus; **five other groups of viruses**	FDV	Plants	Delphacid leafhoppers		Australia, Asia, South America, Northern Europe
Phytoreovirus					
Wound tumor virus	WTV	Plants	Cicadellid leafhoppers		
Rice dwarf virus	RDV	Rice	Cicadellid leafhoppers	Stunting	Southeast Asia, China, Japan, Korea
Rice gall dwarf virus	RGDV				
Oryzavirus					
Rice ragged stunt virus	RRSV	Plants *(Gramineae)*	Planthoppers		
Mycoreovirus					
3 species of 11 or 12 segmented viruses of fungi	CpMYRV-1	Fungi			

Abbreviations: BmCPV-1, *Bombyx mori* cypovirus 1; DpIRV-1, *Diadromus pulchellus* idnoreovirus 1; CpMYRV-1, Mycoreovirus 1.

importance of the members of the genus *Orbivirus*, these viruses have recently come under increased scrutiny.

Genus *Orthoreovirus*

The genus *Orthoreovirus* (the "true" reoviruses, to distinguish the genus from the family) contains three viruses that infect many mammals, including humans, referred to as mammalian orthoreovirus (MRV) types 1, 2, and 3. They were originally named from the first initials of the words *r*espiratory *e*nteric *o*rphan virus—they grow in the respiratory tract and in the enteric tract but were orphans, not known to cause human illness at the time of their discovery. The viruses are widespread and the majority of humans have antibodies against all three serotypes by the time they are adults. Most infections do not result in symptomatic disease or result in only mild symptoms. Studies with human volunteers have shown that some individuals develop a mild disease characterized by headache, pharyngitis, sneezing, rhinorrhea, cough, and malaise.

These three orthoreoviruses, now considered strains of a single species, are virtually ubiquitous viruses of mammals. They have been isolated from many different mammals as well as from sources such as river water and untreated sewage. There is little evidence for host range specificity among these mammalian viruses. However, reovirus infection of lower mammals is sometimes associated with more serious illness than infection of humans.

Other branches of the *Orthoreovirus* genus contain fusogenic reoviruses that infect birds, flying foxes, baboons, or reptiles. In general, the avian viruses do not grow in mammalian cells or must be adapted to mammalian cells before they will grow in them, and thus have a host range distinct from that of the mammalian viruses. The relationships among these viruses are illustrated by the tree in Fig. 5.2. There are five distinct lineages. The three nonfusogenic viruses group closely together, consistent with their classification as a single species, mammalian orthoreovirus (species I). There are several strains of avian orthoreovirus (species II) that form a second lineage. Nelson Bay orthoreovirus of flying foxes

TABLE 5.3   Comparison of *Orthoreovirus, Orbivirus, Rotavirus, Coltivirus,* and *Seadornavirus*

Characteristic	*Orthoreovirus*	*Orbivirus*	*Rotavirus*	*Coltivirus*	*Seadornavirus*
Segments	10	10	11	12	12
Size of genome	23.5 kb	19.2 kb	18.6 kb	28.5 kb	20.0 kb
Type virus	Reovirus type 3	Bluetongue-1	Simian rotavirus SA11	Colorado tick fever	Banna
Portal of entry	Oral	Skin	Oral	Skin	Skin
Tissue tropism	Intestinal tract, upper respiratory tract	Hemopoietic	Intestinal tract	Hemopoietic and muscle	CNS
Vector	None	Culicoid flies, ticks, mosquitoes, phlebotomines	None	Ticks	*Culex* and *Anopheles* mosquitoes
Human disease	Upper respiratory infections, infant enteritis	See Table 5.5	Diarrhea, especially in children <5 yrs old	See Table 5.5	See Table 5.5
Consensus terminal nucleotide sequences					
5′ Terminal	5′-GC(U/A)(U/A)	5′ GU(A/U)AAA	5′GGCUAUUAAA[a] 5′GGC(A/U)NAAAUU[b]	5′(G/C)ACAUUUUGU	5′GUAU(A/U)(A/U)AA
3′ Terminal	UCAUC-3′	AC(U/A)UAC-3′	GAUGUGACC-3′[a] AUAAAAACCC-3′[b]	UGCAGU(G/C)-3′	(A/G)C(C/U)GAC-3′

[a] Rotavirus A.
[b] Rotavirus B.

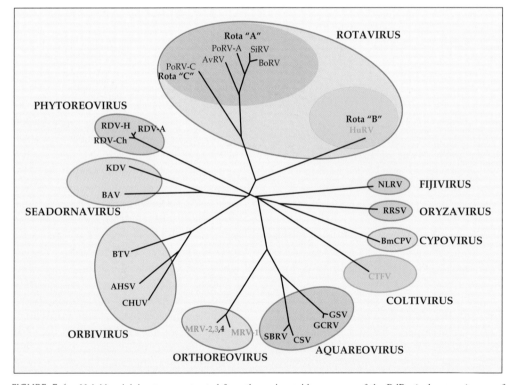

FIGURE 5.1   Neighbor-joining tree constructed from the amino acid sequences of the RdRp (polymerase) gene of representatives of 10 different genera of *Reoviridae*. The size of the oval illustrates the amount of diversion within a genus, and each oval is color coded for the major host of its members (vertebrates, blue; humans, pink; insects, yellow; fish, blue-green; and plants, green). Abbreviations are as follows: PoRV, porcine rotavirus; AvRV, avian rotavirus; SiRV, simian rotavirus; BoVR, bovine rotavirus; HuRV, human rotavirus; NLRV, *Nilaparvata lugens* reovirus; RRSV, rice ragged stunt virus; BAV, Banna virus; KDV, Kadipiro virus; BTV, blue tongue virus; AHSV, African horse sickness virus; CHUV, Palyam virus; MRV-1,2,3,4 mammalian reoviruses types 1–4; CSV, SBRV, group A aquareoviruses; GSV, GCRV, group C aquareoviruses; CTFV, Colorado tick fever virus; BmCPV, *Bombyx mori* cypovirus; RDV-H, RDV-A, RDV-Ch, isolates of rice dwarf virus. Adapted from Fauquet *et al.* (2005) Figure 3 on p. 453.

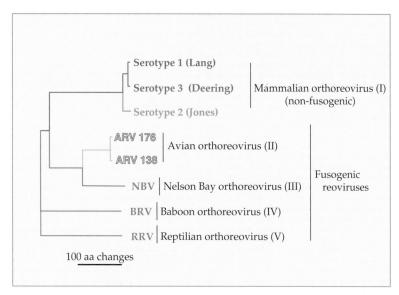

FIGURE 5.2   Phylogenetic tree of the orthoreoviruses, derived from amino acid sequences of the σ2 core proteins, which are encoded in the top three species by the S2 RNA segment, but encoded on a polycistronic S1 segment in the case of BRV and RRV. The orthoreoviruses are now considered to consist of five species (I to V). A virtually indistinguishable tree is given by the σNS sequences, and one with only slightly longer arms, but the same topology, by the outer capsid protein sequences. The scale bar indicates the length of the arm for 100 changes. Adapted from Duncan (1999), and modified according to Fauquet *et al.* (2005).

forms a third lineage (species III). *Baboon orthoreovirus* (species IV) and *reptilian orthoreovirus* (species V) form the remaining two lineages.

The species identified by their alignment in a phylogenetic tree such as Fig. 5.2 are confirmed by other attributes that they share, the most important of which is the ability to undergo reassortment during mixed infection. Members of the same species can reassort their genomes during mixed infection which, as discussed in Chapter 4 for influenza virus, is important for the evolution of these viruses. A second feature, which is certainly related to the ability to undergo recombination, is conservation of sequences at the ends of the genomic RNAs. For MRV, for example, the 5′ end of all RNAs begins GCUA and the 3′ end is UCAUC, whereas the 5′ sequence differs for other species of orthoreoviruses. These conserved sequences are probably important for replication of the RNA and for packaging of the 10 RNA segments. These 5′ and 3′ sequences are compared for different genera of the *Reoviridae* in Table 5.3.

Mammalian orthoreovirus serotype 3 (MRV-3) has been extensively studied as a model for the members of the *Orthoreovirus* genus. In the following discussion, in which aspects of the genome organization, replication, and structure of orthoreoviruses are described, specific details refer to MRV-3. These details are summarized in Table 5.4.

The Genome of Orthoreoviruses

The orthoreovirus virion is a double-shelled icosahedral particle 85 nm in diameter. The genome consists of 10 segments of dsRNA (Fig. 5.3), which range in size from 3.9 to 1.2 kb and sum to 23.5 kb for MRV-3. The 10 segments fall into three size classes called L for large (3 segments called L1, L2, L3), M for medium (3 segments called M1, M2, M3), and S for small (4 segments called S1, S2, S3, S4) (Fig. 5.3 and Table 5.4). Eleven or 12 distinct proteins are produced, described in more detail later. The proteins are named by Greek letters corresponding to the L, M, or S segment that encodes them, but the numbering of the proteins does not reflect the number of the segment encoding it (Table 5.4). Of the 11 proteins, 8 are components of the virion, 4 in the outer shell and 4 in the inner shell. One of the viral structural proteins, μ1, is cleaved during virus assembly to produce two different products called μ1C and μ1N.

Entry of Orthoreoviruses into the Cell

After attachment to receptors, normally molecules that contain sialic acid, orthoreoviruses are internalized into endosomes. In endosomes or in lysosomes, proteolysis of two proteins in the outer shell, σ3 and μ1C, produces what has been termed an ISVP (*i*nfectious *s*ub*v*iral *p*article or *i*ntermediate *s*ub*v*iral *p*article). In this process, μ1C is cleaved to produce two fragments, δ and a small C-terminal φ, whereas σ3 is degraded. These cleavages are illustrated schematically in Fig. 5.4. They can be blocked by agents that prevent acidification of endosomes, which demonstrates the importance of the endosomes in the process. ISVPs can also be produced by treating virions with proteases *in vitro*.

TABLE 5.4   Characteristics of the Proteins Encoded by the Ten Genome Segments
of Mammalian Orthoreovirus Serotype 3

Genome segment	Protein product	Segment (length in nt)	5′ NT (nt)	ORF (aa)	3′ NT (nt)	Function of protein	Location/molecules of protein per virion
L1	λ3	3854	18	1267	35	Poly(C) polymerase, catalytic subunit of polymerase/transcriptase	Core/12
L2	λ2	3916	13	1289	36	Guanylyltransferase	Turrets on core/60
L3	λ1	3896	13	1233	184	NT binding motif, Zn finger	Core/ 120
M1	μ2	2304	13	736	83	Putative transcriptase/polymerase component	Minor core component/12
M2	μ1→μ1C	2203	29	708	50	Major structural protein	Outer capsid shell/600
M3	μNS(μNSC)	2235	18	719	60	Binds ss RNA	Nonstructural
S1	σ1	1416	12	455	39	HA, neut Ag, cell attachment protein	Capsid/36
	σ1NS			120	—	Unknown	Nonstructural
S2	σ2	1331	18	418	59	Structural component	Core/240
S3	σNS	1198	27	366	73	Binds ss RNA	Nonstructural
S4	σ3	1196	32	365	69	Major structural protein	Outer capsid shell/600

Abbreviations: HA, hemagglutinin; neut Ag, contains epitopes recognized by neutralizing antibodies; 5′ NT, nucleotides at the 5′ terminus of the RNA segment that are not translated into protein; 3′ NT, nucleotides at the 3′ terminus of the RNA that are not translated.
Source: Joklik and Roner (1996).

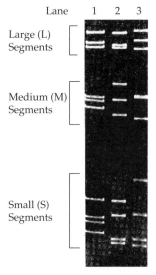

FIGURE 5.3   Gel electrophoresis of the mammalian orthoreovirus RNA genome segments, showing the variation of the segment size with serotype. Lane 1: reovirus serotype 2 (Jones), Lane 2: reovirus serotype 1 (Lang), and Lane 3: serotype 3 (Deering). The segments cluster into three groups: 3L (large), 3M (medium), and 4S (small). From Fields *et al.* (1996), p. 1559.

The ISVP is capable of breeching the endosomal membrane and gaining entry into the cytoplasm. The process of penetration may involve μ1N, whose N terminus is myristoylated and lipophilic. When ISVPs are produced by treatment of virions with proteases *in vitro,* they are capable of penetrating into the cell by way of the plasma membrane,

which confirms the importance of proteolytic processing for the activation of domains required for penetration.

The general scheme of the replication of reoviruses is shown in Fig. 5.5. The various steps are discussed in more detail next.

Synthesis of mRNAs

On penetration of ISVPs into the cytoplasm, they are converted into cores by further loss of δ, φ, and the σ1 fiber, and the rearrangement of the protein λ2 (Fig. 5.4). Core particles are transcriptionally active: Each of the 10 segments of dsRNA within them is used to synthesize an mRNA molecule that is the same length as the plus strand of the genome segment. These mRNAs are capped but not polyadenylated. All enzymatic activities required for the initiation of RNA synthesis, capping, and elongation of the product are present in the core, and occur *in vitro* if cores are supplied with appropriate substrates. Synthesis of mRNA is conservative: The newly synthesized mRNAs are extruded from the core and both strands of the parental dsRNA remain within the core. Extrusion is an active process. Electron microscopic studies have suggested that the mRNAs are extruded from the 12 vertices of the icosahedral structure. The enzymatic activities are organized about these 12 fivefold axes, and it has been suggested that there is an independent transcription unit for each genome segment, consistent with the fact that no member of the family *Reoviridae* has more than 12 genome segments.

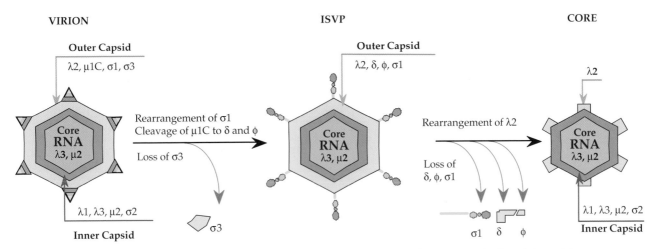

FIGURE 5.4   Structure of a reovirion and subviral particles derived from it. At left, a cross section through the particle shows the two protein shells (in yellow and blue) surrounding the RNA (green). In the middle, the intermediate subviral particle (ISVP) is shown, after the loss of σ3, the cleavage of μ1C to δ and φ, and the extension of σ1. The right illustration shows the core after the loss of σ1, δ, and φ and the rearrangement of λ2. Redrawn from Niebert and Fields (1995).

Translation of Proteins

In MRV, the 10 mRNAs are translated into 12 proteins of which 11 are distinct (Table 5.4). For 8 mRNAs, only one reading frame is used and only one protein is produced. For the mRNA produced from segment M3, only one reading frame is used but two different in-frame AUGs are used to initiate translation. Thus, two proteins (μNS and μNSC) are produced from this segment that differ only in that the longer version has an N-terminal extension. The mRNA from segment S1 is translated using two different, out-of-frame AUGs, however, so that two different proteins (σ1 and σ1NS) are produced. In all orthoreoviruses, the mRNA from the segment corresponding to S1 is translated into two or three different proteins. In MRV the ORF for σ1NS is completely contained within the ORF for σ1, whereas in species other than MRV the reading frames for the two or three proteins overlap only slightly. The various mRNAs are translated with widely different efficiencies so that different amounts of the 11 or 12 proteins are produced.

Of the proteins produced, eight are components of the virion and the rest are nonstructural, present only within the infected cell. Proteins λ1 and σ2, present in 120 and 240 copies, respectively, form the shell of the core. Proteins λ3 and μ2 are present in 12 copies within the core, at the 12 fivefold axes. Protein λ3 is the catalytic subunit of the RNA polymerase, and μ2 is believed to be a component of this enzyme complex.

Pentamers of protein λ2, present in 60 copies, form "turrets" at the 12 fivefold axes of the core through which the mRNAs are extruded. This protein has guanyltransferase activity and is a component of the complex that caps the mRNAs. Protein μ1 and its cleavage products, μ1N and μ1C, together with protein σ3, both present in 600 copies, form the

outer shell of the virion. Protein σ1 is the cell attachment protein on the surface of the virion. Trimers of this protein are located at the 12 fivefold axes. Interestingly, a complete complement of σ1 trimers is not present in all virions. Virions contain from 0 to 12 trimers, with the median number of trimers being 7. Virions devoid of σ1 are not infectious, but virions containing one or more trimers are infectious.

Four proteins of MRV, μNS, μNSC, σ1NS, and σNS, are nonstructural. μNS, μNSC, and σNS are RNA-binding proteins and probably participate in virion assembly. The function of σ1NS is not known; it is found in the nucleus of infected cells.

Assembly of Progeny Virions

The mRNAs serve as intermediates in the replication of reoviruses. Following synthesis and release from the core, mRNAs quickly become associated with proteins μNS, σNS, and σ3 to form complexes containing single-strand RNA. All complexes contain μNS, but only half contain σ3 and one-quarter contain σNS. Complexes containing dsRNA appear later, which contain the three proteins just named but also contain protein λ2 and the RNA polymerase. Significantly, all 10 dsRNA genome segments are present in equimolar quantities in these complexes, suggesting that the selection and assortment of the 10 genome segments into progeny virions is associated with the conversion of (+)RNA into double-stranded RNA. Because the particle-to-infectious virus ratio is almost one, the assembly process is clearly precise.

During maturation, protein μ1 undergoes a cleavage to produce two fragments called μ1N (the N-terminal fragment) and μ1C (the C-terminal fragment). Fragment μ1N is small (4 kDa) and myristoylated, and this process bears a striking

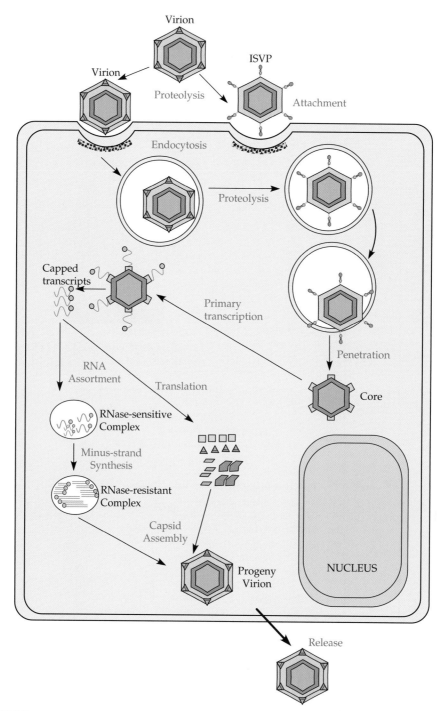

FIGURE 5.5   The reovirus replication cycle. The primary steps are labeled in blue, while the intermediate products are labeled in black. This diagram was adapted from Figure 6 in Fields *et al.* (1996) on page 1706. The color coding of the protein components is the same as in Fig. 5.4.

similarity to the cleavage event that occurs during the maturation of picornaviruses (see Chapters 2 and 3).

It is significant that the dsRNA is never free but always present in particles. As described in Chapter 10, dsRNA is a key molecule in both the induction of interferon and in the antiviral activities induced by interferon. Thus, the sequestering of the dsRNA within particles where it is not available for interaction with cellular sensors that detect the presence of dsRNA or with enzymes that are activated by binding dsRNA is important for avoiding this potent cellular defense against viral infection.

Reoviruses and Cancer

One fascinating recent development is the possibility that orthoreoviruses might be useful in controlling some cancers. They selectively kill mammalian cells whose Ras signaling pathways have been activated. This comes about because orthoreovirus replication is constrained by the interferon stimulated gene RNA-activated protein kinase (PKR) (Chapter 10). However, tumor cells whose Ras pathways are activated lack PKR activity, allowing the virus to replicate vigorously. Since nearly two-thirds of human tumors may have an activated Ras pathway, and since orthoreovirus infection is fairly benign, it is possible that MRV could be used to control or eradicate some tumors.

Genus *Rotavirus*

Rotaviruses are viruses of higher vertebrates and are very widely distributed. They cause gastroenteritis in their various hosts and many different serotypes are known. They have been isolated from monkeys, cattle, dogs, cats, pigs, sheep, horses, chickens, and turkeys, as well as from humans. Viruses isolated from different animals exhibit extensive serological cross-reactivity but limited ability to replicate in other hosts. As one example, rotaviruses isolated from monkeys and cows can infect humans but cause much milder symptoms than human rotaviruses, and have been examined for use as vaccines.

Rotaviruses have been assigned to five species to the current time, *rotavirus A, B, C, D,* and *E*. Two tentative species may also be recognized in the future, *rotavirus F* and *G*. Assignment to species is based on sequence identities and, where known, on the ability of members of a species to undergo reassortment during mixed infection. Human infection is caused by viruses in *rotavirus A, B,* and *C*.

Rotavirus Structure and Replication

Rotaviruses contain 11 genome segments that sum to 18.6 kb (Table 5.3) and which are simply numbered from 1 to 11 in order of decreasing size. As for orthoreoviruses, one segment encodes two proteins in overlapping reading frames and the mRNA transcribed from another segment is translated into two proteins using two in-frame start codons so that they differ only in that one is a truncated form of the other. Of the 12 proteins that differ in primary sequence, 6 are structural and 6 are nonstructural. Proteins are numbered in order of decreasing size as either virion structural proteins (VP1-8, where VP4 is cleaved to produce VP5 and VP8) or nonstructural proteins (NSP1-6).

The replication of the virus and the overall structure of the virion resemble those of the orthoreoviruses, but with some exceptions. Virions are triple shelled rather than double shelled and the virion is distinguishable from that of orthoreoviruses in the electron microscope (Fig. 2.5). The

rotavirus particle resembles a wheel with spokes from which it derives its name (*rota* = wheel in Latin). The assembly of rotaviruses differs in one important detail from that of orthoreoviruses. Subviral particles are assembled in the cytoplasm and bud through the endoplasmic reticulum, acquiring an envelope that is subsequently lost. During this maturation process, the outer capsid layer of the virion is acquired. Perhaps because of this transient association with membranes, two of the rotaviral proteins are N-glycosylated and one is O-glycosylated. In contrast, no orthoreovirus protein is known to be glycosylated.

Activation of infectivity of the virion requires cleavage of one of the major outer capsid proteins, similar to the case for the orthoreoviruses where cleavage of an external protein also occurs. In rotaviruses this process normally occurs in the enteric tract where the virus is exposed to secreted cellular proteases. Uncleaved rotavirus will bind to cells but cannot penetrate and virus produced in cultured cells must be treated with proteases to render it infectious. The protein is cleaved at a hydrophobic sequence that is postulated to possess fusion activity, analogous to the process that occurs in enveloped viruses such as influenza virus where precursor glycoproteins must be cleaved to activate fusion activities.

The Human Rotaviruses

The rotaviruses cause diarrhea, primarily in newborns and the young, and the human rotaviruses are the single most important cause of severe diarrheal diseases of infants and young children. In one study in the United States, nonbacterial infectious gastroenteritis, of which rotaviruses account for about half (Fig. 5.6), was found to be the second most common disease in humans, accounting for 16% of illnesses over a 10-year period and occurring on average 1.5 times per person per year. Severe diarrhea can result in dehydration that can be fatal if fluids are not replaced. In the developing world where hospitalization is not readily available, diarrheal disease, about half of which is caused by rotaviruses (Fig. 5.6), are a major cause of infant mortality. Firm estimates of the extent of diarrheal disease in developing countries are difficult to establish, but in Asia, Africa, and Latin America there may be more than 1 billion cases of diarrhea each year with 2–3 million deaths. The majority of deaths occur in children less than 5 years of age, where 1–4% of diarrheal episodes are fatal. Overall it has been estimated that rotaviral disease causes between 300,000 and 800,000 deaths worldwide each year. The global distribution of deaths caused by rotaviruses is illustrated in Fig. 5.7.

Replication of rotaviruses is normally confined to terminally differentiated enterocytes that line the tips of microvilli in the small intestine. How the deaths of these cells provokes diarrhea has not been definitively established,

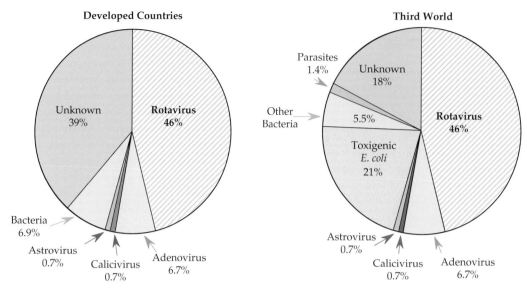

FIGURE 5.6   An estimate of the role of various etiologic agents in severe diarrheal disease requiring hospitalization in infants and young children in developed countries and in the Third World. Adapted from Kapikian (1993).

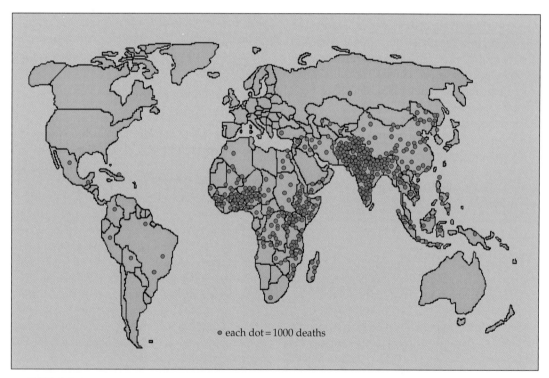

FIGURE 5.7   Deaths (estimated) from rotavirus in 2003 in children under 5 years old. Adapted from Glass (2006). The 10 countries that suffered the greatest losses are Ethiopia (28,905); Nigeria (47,525); the Democratic Republic of Congo (28,905); Tanzania (11,440); India (146,044); Bangladesh (18,986); Indonesia (14,064); China (41,076); Pakistan (36,450); and Afghanistan (17,830). These countries together account for about two-thirds of the worldwide deaths.

although maladsorption and osmotic diarrhea are thought to be involved. Rotaviral diarrhea is watery, suggesting either a secretory process or an osmotic process. It is also known that one of the viral proteins, NSP4, is an enterotoxin that will induce diarrhea when given to mice. Although normally confined to replication in the enteric tract, recent studies have shown that rotavirus infection may result in systemic infection, albeit rarely.

In temperate climates, epidemics of rotaviral disease occur in the winter. This is illustrated in Fig. 5.8 for epidemics in

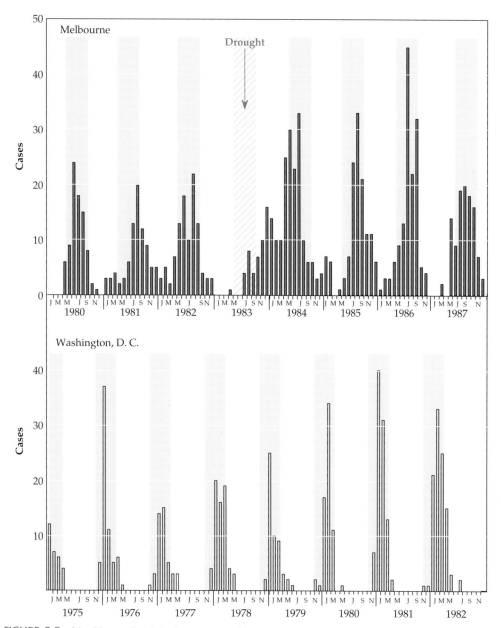

FIGURE 5.8   Monthly rotavirus infections over a multiyear period. Seasonal fluctuations are clear, with the cooler months shaded in blue. Upper panel shows hospitalizations for gastroenteritis in Melbourne, Australia. The lower panel shows hospitalizations in a children's hospital in Washington, D.C. Note the anomalous pattern for Melbourne in 1983, a year in which there was a severe drought. Data from Barnes *et al*. (1998) and Brandt *et al*. (1983).

Melbourne, Australia, and in Washington, D.C. In southern Australia, epidemics peak in the June–September time frame (their winter), while epidemics peak in the January–March time frame in the eastern United States. In the United States, epidemics peak in November and December in the warmer southwestern states and move to progressively later months as they spread to cooler states north and east (Fig. 5.9). As is the case for seasonality in epidemics of disease caused by other viruses, the reasons for the association of rotaviral outbreaks with cool weather are poorly understood.

Rotavirus Vaccines

Because of the seriousness of rotaviral disease on a worldwide basis, there have been ongoing efforts to develop vaccines against rotaviruses. Such vaccines have been directed at newborns and infants, in whom the problem is most acute. Vaccine development and interpretation of results from clinical trials have been complicated by several factors, including the possible presence of maternal antibodies in newborns, the fact that rotaviral infections often do not

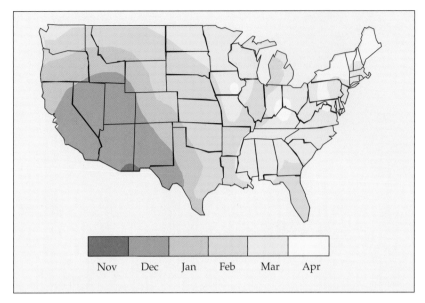

FIGURE 5.9   Average time of peak rotavirus activity in the contiguous 48 states, United States, using cumulative data from July 1991 to June 1997. This contour plot was derived using the median value for time of peak activity reported by each regional or state diagnostic laboratory. The surveillance system and analytic methods used to create this map are described in greater detail in Török *et al.* (1997).

produce absolute and lasting immunity to subsequent reinfection, and the fact that there are multiple serotypes of rotaviruses that infect humans. There are four major serotypes of rotaviruses that cause widespread and serious infections of humans, and thus recent efforts have been directed toward developing a quadrivalent vaccine that would protect against all four serotypes. Furthermore, the objective of vaccination has now been defined as the prevention of severe rotaviral disease in very young children rather than prevention of any rotaviral disease.

With these objectives, a rotavirus vaccine was developed using a virus from another animal species, the rhesus rotavirus, as a human vaccine. The rhesus rotavirus replicates well enough in humans to elicit an immunizing response but does not cause serious illness, and the use of this virus as a human vaccine has been compared to the use of cowpox virus by Jenner to immunize against smallpox (Chapter 7). The rhesus rotavirus will successfully immunize people against only one of the four major rotavirus serotypes, and not against the others. However, rotaviruses contain 11 genome segments and the different segments readily undergo reassortment when tissue culture cells are infected by more than one strain, that is, genome segments can be interchanged in progeny viruses. This property was used to isolate reassorted rhesus rotaviruses in which all of the segments were derived from rhesus rotavirus except for the segments encoding the surface proteins, which were derived from human viruses of the other three major serotypes. Clinical trials showed that a quadrivalent vaccine based on rhesus rotavirus and three reassorted rotaviruses, so that all four serotypes important for humans were present, was successful at preventing

severe rotaviral disease in newborns and the very young. This vaccine (RotaShield) was licensed in 1998 for general use in the United States. It was anticipated that this vaccine would be useful not only in the developing world where rotavirus infection causes many deaths in infants, but also in the developed world where rotaviral infections lead to many cases of diarrhea each year that require hospitilization or visits to physicians. However, on widespread use in the United States, it was found that a small number of infants developed the bowel obstruction called intussusception after immunization. This obstruction sometimes clears spontaneously but can require a fairly benign treatment by medical personnel. In a small minority of cases, surgery is required to correct the defect. The vaccine was subsequently withdrawn.

The withdrawal of the rotavirus vaccine raises interesting legal and moral questions, and illustrates the difficulties associated with developing and introducing vaccines. Roughly 1 out of 2000 infants develop intussusception in the first 2 years of life. It seems clear that vaccination with the rotaviral vaccine triggers intussusception in a small fraction of vaccinees, since immunization with RotaShield increased the risk of intussecption 20- to 30-fold within the first 2 weeks after vaccination, and one in 10,000 vaccinees developed the condition immediately after immunization. However, it is not known whether vaccinated infants are more likely to develop the obstruction in the first 2 years of life than are nonvaccinated infants. It is even possible that the vaccine is actually protective in that fewer vaccinated infants will ultimately develop intussusception than nonvaccinated infants. Furthermore, in the United States, about 55,000 children are hospitalized every year for rotaviral disease and 20 to 40 die. Use of the vaccine in the United States would drastically reduce hospitalizations and the death rate would probably decline since intussusception is almost never fatal with proper treatment. However, because of the legal atmosphere in the United States and the ethical dilemma of giving a problematical vaccine for a disease that is seldom life threatening in developed countries, the vaccine has not been used in the United States since its withdrawal. Since the vaccine was not used in the United States or other developed countries because of concerns about its safety, developing nations did not adopt it, despite the fact that the vaccine would undoubtedly have saved hundreds of thousands of infants from fatal rotaviral infection if it were widely used in such countries, where as many as 0.5% of infected children die. A risk–benefit study in Peru, for example, concluded that the vaccine would prevent 1440 deaths and 23,000 hospitalizations as compared to perhaps 78 cases of intussusception that might be vaccine related.

Now, 8 years and perhaps 5 million deaths later, the problem appears to have been solved with the licensing of two new rotavirus vaccines, Rotateq from Merck, licensed worldwide, and Rotarix from Glaxo Smith Kline, licensed in Latin America, Africa, Asia, and the European Union. No association with intussusception has appeared to date with the use of these vaccines. These vaccines are also live virus vaccines, which is important for use in developing countries because they can be given orally. Furthermore, because the viruses replicate in the enteric tract, such vaccines give rise to IgA that protects the mucosal surfaces infected by rotaviruses.

Genus *Orbivirus*

The genus *Orbivirus* is widely distributed. It contains 21 recognized species, each of which contains multiple serotypes. As examples, *African horse sickness virus* contains 9 serotypes and *bluetongue virus* contains 24 serotypes. Altogether, more than 150 serotypes of orbiviruses are currently recognized. Orbiviruses contain 10 segments of dsRNA totaling 19.2 kb (Table 5.3) in an icosahedral virion 86 nm in diameter. Reminiscent of rotaviruses, orbiviruses may exit the cell by budding through the cell plasma membrane to acquire an unstable envelope that is soon lost. Viruses can also be extruded from cells without acquiring an envelope.

Different orbiviruses infect a wide range of higher vertebrates including ruminants, horses and their relatives, rodents, bats, marsupials, sloths, primates including humans, and birds. Unlike reoviruses and rotaviruses, orbiviruses are arboviruses, transmitted by biting flies, mosquitoes, or ticks, and able to replicate in the vector as well as in the vertebrate host. Among the vectors known for various orbiviruses are insects of the genus *Culicoides* (midges and gnats), phlebotomine flies, culicine and anopheline mosquitoes, and *Ixodes*

ticks. Members of five species, including *African horse sickness virus*, *bluetongue virus*, and *epizootic hemorrhagic disease virus*, cause disease in domestic animals, and members of four other species cause human disease. Human disease caused by naturally acquired orbiviruses may require hospitalization but is normally not life threatening, and no orbivirus is considered to be an important human pathogen. However, the animal pathogens may cause serious illness in domestic or wild animals (Table 5.5).

Veterinary Pathogens

African horse sickness virus (AHSV) has caused many epidemics of fatal illness in horses in sub-Saharan Africa. The distribution of this virus is illustrated in Fig. 5.10C. The first recorded epidemic occurred in the Cape Colony in 1719. Thereafter, disastrous epidemics occurred every 20 years or so up until the twentieth century, when vaccines became available. The virus appears to have been endemic, presumably in the zebra, but wherever horses were introduced, epizootic AHSV was sure to follow. The mortality in introduced horses is close to 90%, and AHSV had a major impact on agriculture, exploration, and conquest in Africa. The military had to operate without cavalry and the early explorers often walked rather than rode. The virus represents an example of an endemic virus that appears to cause little disease in its native host (zebra), but which causes very serious illness when transmitted to a nonnative host (horses).

Another important veterinary pathogen is bluetongue virus, which causes a serious disease in sheep. Bluetongue virus is very widely distributed (Fig. 5.10A) and causes widespread disease, but, interestingly, the strains of bluetongue virus in Australia are not pathogenic. Also widespread are viruses that cause epizootic hemorrhagic disease in animals (Fig. 5.10B), including outbreaks in deer in North America.

Persistence of Orbiviruses on Erythrocytes

At least some orbiviruses have evolved an unusual way of persisting in nature as arboviruses. In tropical areas of the world, arthropods may be continuously active and viruses associated with such arthropods may persist by continuous passage between the vertebrate and the invertebrate host. However, in many areas of the world, including some tropical areas, the vector activity may become low or nonexistent during some periods, such as the dry season or the winter. Arboviruses must be able to survive such periods of low vector activity, and different arboviruses have solved this problem in different ways. Most have evolved ways to persist in the invertebrate host during periods of inactivity. Many can be passed transovarily in the arthropod host, others can survive in diapausing insects. Orbiviruses, in contrast, have evolved a mechanism to persist in the vertebrate host for long periods. Viral infection of vertebrates is normally cleared

TABLE 5.5   Representative Orbiviruses, Coltiviruses, and Seadornaviruses Causing Disease in Humans and Domestic Animals

Genus	Virus	Hosts	Vector	Disease syndromes	Distribution
Orbivirus					
	African horse sickness	Horse, dog, zebra	*Culicoides* (midges)	Cardiopulmonary disease, hemorrhagic fever	Africa, Asia
	Bluetongue	Sheep, cow, goat, deer	*Culicoides* (midges)	Fever, frothing at mouth, shock, coronitis	Africa, Asia, Australia[a], Americas
	Epizootic hemorrhagic disease	Deer	*Culicoides* (midges)	Similar to bluetongue	Americas, Australia, Africa
	Palyam	Cow	*Culicoides* (midges)	Abortion	South Africa, Japan
	Orungo	Humans	Mosquitoes	Febrile illness	Africa
	Changuinola	Humans	Phlebotomines	Febrile illness	Panama
	Kemerovo	Humans	Ticks	Febrile illness, encephalitis	Russian, Eastern Europe
Coltivirus					
	Colorado tick fever	Humans	Ticks	Febrile illness, encephalitis, hemorrhagic fever	North America
	Eyach	Humans	Ticks	Encephalitis ?	Europe
Seadornavirus					
	Banna, Kadipiro	Humans	Mosquito	Meningoencephalitis	Southeast Asia, Indonesia, China

[a] Australian isolates of bluetongue virus are not pathogenic.

rapidly by the immune system so that viremia of sufficient titer to infect a new arthropod taking a blood meal normally lasts only a few days. However, bluetongue virus becomes associated with red blood cells by binding to glycophorins on the surface of the cell, where it persists in indentations in the membrane in a nonreplicating state protected from the immune system. The virus evidently can remain attached and viable for the life of the erythrocyte. When an arthropod takes a blood meal, the virus bound to erythrocytes is able to initiate infection of the arthropod. Because erythrocytes have an average lifetime of 160 days, the virus can persist in a viable state in the vertebrate host for many months.

Genus *Coltivirus*

The coltiviruses possess 12 genome segments summing to 28.5 kb (Table 5.3). The virion is icosahedral and 80 nm in diameter. Like the orbiviruses, the coltiviruses are arboviruses, transmitted by ticks. Two species are recognized, one found in North America and the second in Europe. The prototype virus is Colorado tick fever virus (CTF), from which the genus gets its name (*Col*orado *ti*ck fever). CTF is present in the Rocky Mountain area of North America at elevations from 4000 to 10,000 feet, and two serotypes are known. It has been isolated from a number of mammals including humans and from ticks and mosquitoes that serve as vectors. Transmission to humans is usually by *Dermacentor andersoni* ticks whose range is

shown in Fig. 5.11. The life cycle of the virus is illustrated in Fig. 5.12. Although CTF infects many mammals, including humans, the natural cycle of transmission involves primarily small mammals and larval and nymphal ticks. Adult ticks may transmit the virus to humans and large mammals outside the normal transmission cycle. Transovarial transmission does not occur in ticks and larval ticks are normally infected by feeding on small mammals that are viremic. In humans, CTF causes an illness characterized by fever, myalgia, chills, headache, and malaise, and 20% of cases require hospitalization. The acute illness lasts 5–10 days. Recovery may be uneventful or convalescence may be prolonged for several weeks. CNS infection or hemorrhagic fever may occur, almost always in children. Fatalities are rare (<0.1%).

The second species in the genus is *Eyach virus*. This virus is widely distributed in Europe where the major host is thought to be the European rabbit *Oryctolagus cunniculis*. It is transmitted by *Ixodes* ticks and has been associated with febrile illness and encephalitis in humans. It is hypothesized that the virus was introduced into Europe from North America through migration of lagomorph ancestors from America to Europe about 30 million years ago.

Persistence of CTF in Erythrocytes

CTF has evolved a way to persist in the vertebrate host that resembles that used by *bluetongue virus*. CTF infects

A. Bluetongue virus (BTV)

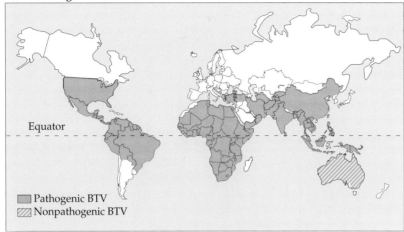

B. Hemorrhagic disease of deer (IHDD), Ibaraki virus, and the Kemerovo group viruses

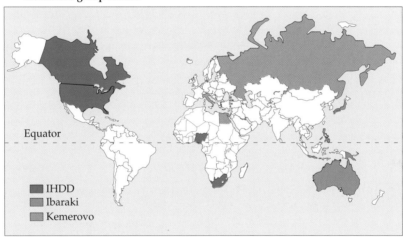

C. African Horse Sickness (AHSV)

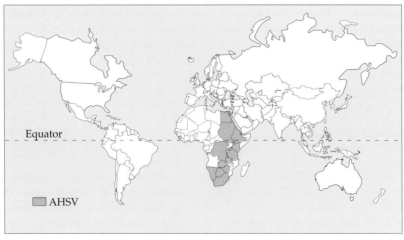

FIGURE 5.10 Geographical distribution of orbiviruses that causes disease in animals and humans. Adapted from Fields *et al.* (1996) p. 1736.

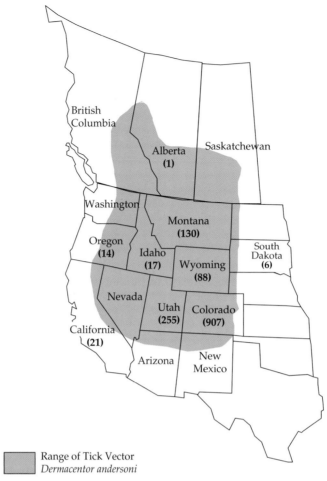

FIGURE 5.11   Distribution of the primary vector of Colorado tick fever, *Dermacentor andersoni*, shown in color, and the number of diagnosed human cases of Colorado tick fever in various states between 1980 and 1988. From Tsai (1991).

bone marrow cells early after infection, including erythrocyte precursors. The virus remains within the erythrocyte after it matures, in a nonreplicating state, and appears to persist for the life of the erythrocyte. When a tick takes a blood meal, it can be infected by the virus within the erythrocyte. The differences in the mechanisms used by CTF and bluetongue virus to persist in the blood of vertebrates are thought to reflect the different vectors used. *Culicoides* flies, the vector of bluetongue, take smaller blood meals and digest the blood meal in a different way than ticks, the vector of CTF.

Genus *Seadornavirus*

Seadornaviruses contain 12 genome segments and are transmitted by mosquitoes. The viruses are found in tropical and subtropical regions of Southeast Asia, primarily China and Indonesia. The name of the genus is derived from **S**outh**e**astern **A**sia **do**deca **RNA viruses**. Three species are

recognized, *Banna virus*, *Kadipiro virus*, and *Liao ning virus*, of which *Banna virus* was the first isolated and the best characterized. Vectors include *Anopheles*, *Culex*, and *Aedes* mosquitoes. Banna has been isolated from humans suffering from encephalitis or febrile illness, but its association with human illness has not been conclusively shown since there are many viruses in this region that cause such diseases.

Comparison of the *Reoviridae* with Other RNA Viruses

Some aspects of the replication cycle of the *Reoviridae* resemble those of single-strand (+)RNA viruses (Chapter 3) and other aspects resemble those of single-strand (−)RNA viruses (Chapter 4). Replication proceeds through a plus-strand intermediate that is a messenger RNA and that exists, at least transiently, free in the cytoplasm, a characteristic of (+)RNA viruses. Reoviruses could have originated from the plus-strand viruses through the acquisition

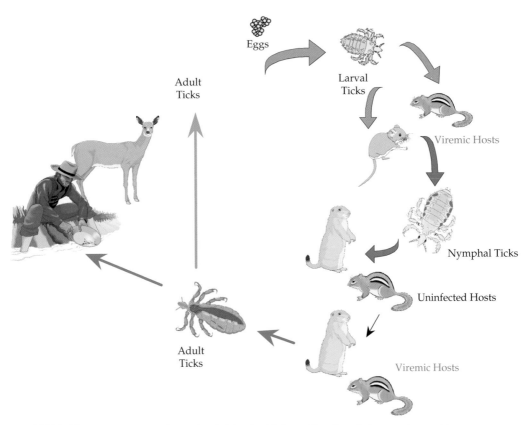

FIGURE 5.12   Natural transmission cycle of Colorado tick fever. Transfers of virus are shown with green arrows. Larval ticks feed on small mammals that can remain viremic for long periods of time and then transmit virus to other small mammals. Adult ticks, while not important for maintaining the virus in nature may then bite nonreservoir hosts such as deer or man. Adult ticks lay eggs to produce the next generation of larval ticks, but no transovarial transmission of virus occurs. Adapted from Bowen (1988) Figure 4.

of an RNA synthetase that is packaged as part of the virion. In contrast, packaging of the RNA synthetase in the virion, which is necessary to begin the infection process, and the retention of the genome in the entering subviral particle, from which it is never released, are features that are shared with the (−)RNA viruses. Packaging of the RNA synthetase makes it feasible for a virus of vertebrates to have its genome in multiple segments, a feature of reoviruses and of many (−)RNA viruses. Segmented genomes allow reassortment to occur during mixed infection, and reassortment is known to occur in nature. The acquisition of novel genome segments from related virus strains has occurred frequently during reovirus evolution and has been advantageous for the survival of the virus in nature.

FURTHER READING

Reoviruses

Chandran, K., Zhang, X., Olson, N. H., *et al.* (2001). Complete in vitro assembly of the reovirus outer capsid produced highly infectious particles suitable for genetic studies of the receptor-binding protein. *J. Virol.* **75**: 5335–5342.

Joklik, W. K., and Roner, M. R. (1996). Molecular recognition in the assembly of the segmented reovirus genome. *Prog. Nucl. Acid Res.* **53**: 249–281.

Lee, P. W. K., and Gilmore, R. (1998). Reovirus cell attachment protein σ1: Structure-function relationships and biogenesis. *Curr. Top. Microbiol. Immunol.* **233**: 137–153.

Saragovi, H. U., Rebai, N., Roux, E., *et al.* (1998). Signal transduction and antiproliferative function of the mammalian receptor for type 3 reovirus. *Curr. Top. Microbiol. Immunol.* **233**: 155–166.

Schiff, L. A., Nibert, M. L., and Tyler, K. L. (2006). Orthoreoviruses and their replication. Chapter 52 in: *Fields Virology, Fifth Edition* (D. M. Knipe and P. M. Howley, Eds. in chief), Philadelphia, Lippincott Williams & Wilkins, pp. 1853–1916.

Shmulevitz, A., Yaameen, Z., Dawe, S., *et al.* (2002). Sequential partially overlapping gene arrangement in the tricistronic S1 segments of avian reovirus and Nelson Bay reovirus: implications for translation initiation. *J. Virol.* **76**: 609–618.

Rotaviruses

Estes, M. K., and Kapikian, A. Z. (2006). Rotaviruses. Chapter 53 in: *Fields Virology, Fifth Edition* (D. M. Knipe and P. M. Howley, Eds. in chief), Philadelphia, Lippincott Williams & Wilkins, pp. 1917–1974.

Glass, R. I., Parashar, U. D., Bresee, J. S., *et al.* (2006). Rotavirus vaccines: current prospects and future challenges. *Lancet* **368**: 323–332.

Matthijnssens, J., Rahman, M., Martella, V., *et al.* (2006) Full genomic analysis of human rotavirus strain B4106 and lapine rotavirus strain 30/96 provides evidence for interspecies transmission. *J. Virol.* **80**: 3801–3810.

Ramig, R. F. (2004). Pathogenesis of intestinal and systemic rotavirus infections. *J. Virol.* **78**: 10213–10220.

Sánchez-San Martín, C., López, T., Arias, C. F., and López, S. (2004). Characterization of rotavirus cell entry. *J. Virol.* **78**: 2310–2318.

Orbiviruses and Coltiviruses

Attoui, H., Jaafar, F. M., de Micco, P., and de Lamballerie, X. (2005). Coltiviruses and Seadornaviruses in North America, Europe and Asia. *Emerg. Infect. Dis.* **11**: 1673–1679.

Roy, P. (2006). Orbiviruses. Chapter 54 in: *Fields Virology, Fifth Edition* (D. M. Knipe and P. M. Howley, Eds. in chief), Philadelphia, Lippincott Williams & Wilkins, pp. 1975–1998.

6

Viruses Whose Life Cycle Uses Reverse Transcriptase

INTRODUCTION

Two families of animal viruses utilize reverse transcriptase (RT) in the replication of their genome, the *Retroviridae* and the *Hepadnaviridae*. Two floating genera of plant viruses also use RT; their life cycles are more similar to the hepadnaviruses than to the retroviruses. The hepadnaviruses and the plant viruses are sometimes called pararetroviruses because their life cycle resembles that of the retroviruses.

For the viruses that use RT, the genetic information in the genome alternates between being present in RNA and present in DNA. RT, which is encoded in the viral genome, converts the RNA genome of retroviruses, or an RNA copy of the DNA genome of hepadnaviruses, into double-stranded (ds) DNA. In the nucleus of the infected cell, cellular RNA polymerase transcribes the DNA genome of hepadnaviruses, or the DNA copy of the retrovirus genome, to produce the RNA to be reverse transcribed. The retroviruses package this RNA in the virion and are allied to retrotransposons that form a prominent feature of eukaryotic genomes. The hepadnaviruses and the plant viruses reverse transcribe the RNA into DNA during packaging, so that the virion contains DNA. Thus, the replication of the genome of retroviruses can be described as RNA→DNA→RNA, whereas the replication of the genome of hepadnaviruses can be described as DNA→ RNA→DNA. Although the two families differ in the timing of when reverse transcription takes place in their life cycles, this difference may not represent a fundamental distinction between them. Recent studies have shown that one genus of retroviruses, the spumaviruses, packages DNA in the virion.

It is an interesting feature of reverse transcription that the RNA template is destroyed in the process of conversion to DNA. RT has associated with it an RNase H activity, which specifically degrades the RNA strand of a DNA–RNA hybrid. This activity is essential for the production of a dsDNA copy of the viral RNA by RT. The destruction of the RNA template makes the process of reverse transcription fundamentally different from other mechanisms used for transcription or copying of nucleic acids, in which the template remains intact.

An essential feature of the infection of cells by retroviruses is that the dsDNA copy of the genome is integrated into the host chromosome, where it is called a provirus. Only integrated DNA is stably and efficiently transcribed by the host machinery. Thus, integration is required for productive infection. During infection by hepadnaviruses, in contrast, the viral DNA does not integrate. Instead, it is maintained in the nucleus as a nonreplicating episome. In contrast to retroviruses, hepadnaviral episomal DNA is stably and efficiently transcribed by the host machinery.

The retroviruses have been intensively studied for years because researchers discovered early that avian retroviruses have the ability to induce leukemias and sarcomas in chickens. The study of these viruses led to the discovery of cellular oncogenes, of RT, and of mechanisms that regulate cycling of the animal cell, and several Nobel Prizes have been awarded for work with the avian retroviruses (Chapter 1). Although clearly important for our understanding of biology, for many years after their discovery retroviruses were in some ways biological curiosities because no human disease was known to be associated with retroviral infection. This changed with the discovery of human T-cell leukemia viruses, now known as primate T-lymphotropic viruses (PTLV), which cause leukemia in humans, and with the appearance of *h*uman *i*mmunodeficiency *v*irus (HIV), which causes *a*cquired *i*mmuno*d*eficiency *s*yndrome (AIDS). These viruses have dramatically altered our understanding of the disease-causing potential of retroviruses.

The most important hepadnavirus is hepatitis B virus, which is a major cause of hepatitis in humans. Like hepatitis C virus, it often establishes a chronic infection that can result in cirrhosis or hepatocellular carcinoma.

FAMILY *RETROVIRIDAE*

The retroviruses are a very large group of viruses that infect invertebrates as well as vertebrates. Most of what we know about this group of viruses comes from studies of viruses that infect birds or mammals. Hundreds have been studied and, although considerable divergences exist, they form a well-defined taxon. All are sufficiently similar to be classified as belonging to a single family, the *Retroviridae*. The family gets its name from the concept that these viruses use retrograde flow of information, from RNA to DNA, whereas the conventional flow of information in living organisms is from DNA to RNA.

The RTs of retroviruses are the most highly conserved elements of these viruses and have been used to study the relationships among them. Figure 6.1 illustrates the relationships among the retroviruses of higher vertebrates based on the sequences of their RTs. Included in the tree is a lineage of fish viruses, now classified as members of the genus *Epsilonretrovirus*. The tree is annotated to show where various new genes entered different virus lineages via recombi-

nation with the host or with other viruses. Based on these sequence relationships, the retroviruses that infect birds and mammals are classified into six genera, as illustrated in Fig. 6.1 and as listed in Table 6.1. Of these, members of three genera are characterized as simple retroviruses, which encode only the genes *gag*, *pro*, *pol*, and *env* (and sometimes *dut*). The other three genera of retroviruses of higher vertebrates, as well as the fish viruses, encode, in addition, regulatory genes that control their life cycle, and they are called complex retroviruses. Notice that these regulatory genes independently entered the four different lineages of complex retroviruses represented by the four different genera (Fig. 6.1). Thus, recombination to acquire new functions has been an ongoing process in the retroviruses. Also notice that the complex retroviruses do not group together. The epsilonretroviruses are more closely related to the gammaretroviruses, which are simple viruses, than they are to other complex retroviruses, and the deltaretroviruses, lentiviruses, and spumaviruses are not particularly closely related.

Members of the different genera differ in their structure as visualized in the electron microscope. The simple viruses were formerly classified on the basis of morphology into groups A, B, C, and D. The nucleocapsids of C-type viruses, now classified as alpharetroviruses and gammaretroviruses, assemble during budding, and the nucleocapsid is centrally located in the mature virion. The nucleocapsids of B-type and D-type viruses, now classified as betaretroviruses,

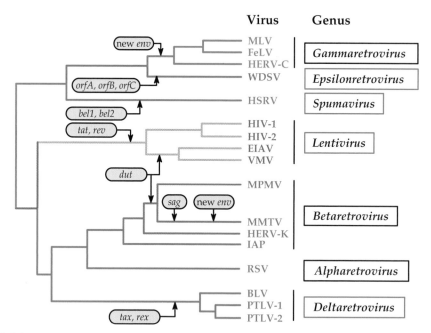

FIGURE 6.1   Phylogenetic tree of the *Retroviridae* drawn from the amino acid sequences of the reverse transcriptases. The lengths of the branches are proportional to the degree of divergence; the names of the "simple" retrovirus genera are boxed in black, the "complex" genera are boxed in red. The greenish ovals indicate the acquisition of new genes during the evolution of current extant viruses. Most of the virus name abbreviations are found in Table 6.1; HERV-C and HERV-K are defective retroviruses in the human genome and IAP is a virus-like element in rodent genomes. Adapted from Coffin *et al.* (1997) Figure 6 on p. 43. and Fields *et al.* (1996) p. 1769.

TABLE 6.1 *Retroviridae*

Genus/members	Virus name abbreviation	Host(s)	Transmission	Disease	World distribution
Orthoretrovirinae					
Alpharetrovirus (simple) formerly avian type C retroviruses					
Avian leukosis	ALV	Birds			Worldwide
Rous sarcoma	RSV	Birds			
Gammaretrovirus (simple) formerly mammalian type C retroviruses					
Moloney murine leukemia	MLV	Mice		T-cell lymphoma	
Feline leukemia	FeLV	Cats		T-cell lymphoma, immunodeficiency	
Betaretrovirus (simple) formerly mammalian type B and type D retroviruses					
Mouse mammary tumor	MMTV	Mice	Vertical, including mothers' milk	Mammary carcinoma, T-cell lymphoma	Worldwide
Mason-Pfizer monkey	MPMV	Monkeys		Unknown	
Deltaretrovirus (complex) formerly the BLV/HTLV group					
Bovine leukemia	BLV	Cows		B-cell lymphoma	
Primate T-lymphotropic[a]	PTLV-1	Humans	Vertical, including mothers' milk, sexual transmission, blood	T-cell lymphoma, neurological disorders	
	PTLV-2	Humans		TSP, HAM[b]	
Epsilonretrovirus (complex)					
Walleye dermal sarcoma	WDSV	Fish		Benign sarcomas	North America
Lentivirus (complex)					
Human immunodeficiency	HIV-1 HIV-2	Humans Humans	Neonatal infection, sexual transmission, blood	AIDS	Worldwide
Simian immunodeficiency	SIV	Monkeys		Simian AIDS	Africa
Visna-maedi	VISNA	Sheep		Neurological disease	No. Europe
Equine infectious anemia	EIAV	Horses		Anemia	Current epidemic in Utah
Spumaretrovirinae					
Spumavirus (complex)					
Chimpanzee foamy	CFV	Monkeys	?	None	
Human spumaretrovirus	HRSV	Humans	?		

[a] Primate T-lymphotropic virus 1 (PTLV-1) was formerly known as T-cell leukemia virus (HTLV).
[b] TSP, tropical spastic pareparesis; HAM, HTLV-associated myelopathy.

assemble before budding and the nucleocapsid is eccentrically located (type B) or bar shaped (type D) in the mature virion (see Figs. 2.1 and 2.21).

Eukaryotic genomes contain a very large number of genetic elements that are related to retroviral genomes. Retrotransposons encode RT and can move around within the genome by a process that uses reverse transcription and insertion, similar to what happens with retroviruses. They are related to retroviruses but have no independent lives as viruses. Other elements in the eukaryotic genome contain additional retrovirus-like genes and appear to have arisen by insertion of retroviral genomes into the germ line at some time in the past. Some of these are still active, capable of giving rise to infectious retroviruses, whereas others are defective. It is clear that this class of elements has been co-evolving with the eukaryotes for a long period of time. The integrated copies of retroviruses in the germ line constitute a form of fossil record that allows us to trace the lineage of at least some retroviruses for 100 million years or longer. For other viruses, whether RNA or DNA, we can only trace ancestry for much shorter periods of time.

Retroviral Genome

The RNA genome in the retrovirion is diploid, consisting of two copies of a 7- to 10-kb single-stranded (ss)RNA that is capped and polyadenylated. The two copies of the genome are normally identical, but during mixed infection hybrid genomes can result. They are joined near their 5′ ends, and perhaps in other regions as well, by hydrogen bonds.

All retroviruses encode the four genes called *gag*, *pro*, *pol*, and *env*, which are always found in this order in the genome. A simplified illustration of a retroviral genome, its relation to the provirus, and the locations of the different gene products in the virion are shown in Fig. 6.2. The name *gag* comes from **g**roup-specific **a**nti**g**en, because the proteins encoded in this gene are more highly conserved, and therefore more widely cross-reactive immunologically, than are the envelope proteins of the virion. The peptides derived from the Gag polyprotein, the precursor polyprotein encoded in the *gag* gene, form the capsid of the retrovirion. The *pro* gene encodes a protease (PR) that is required for the processing of Gag, and *pol* encodes three activities, RT, RNase H, and integrase (IN). RNase H forms a separate domain in the RT-RNase H protein, but functions as an integral component of RT. IN is required for the integration of the dsDNA copy of the retroviral genome into the host chromosome. The fourth gene, *env*, encodes the envelope glycoproteins present at the surface of the enveloped retrovirion. The primary product is Env, which is processed by cleavage to form an N-terminal external protein called SU (for surface) and a C-terminal protein that spans the membrane called TM (for transmembrane). The order of genes in the provirus is the same as in the viral genome.

As described earlier, the genomes of complex retroviruses contain a number of regulatory genes in addition to these four basic genes present in all retroviruses. These genes will be described later.

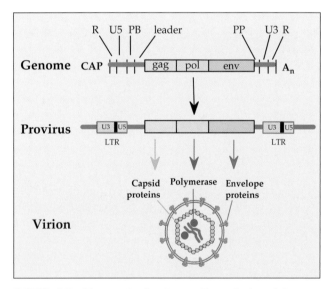

FIGURE 6.2   Diagram showing the overall organization of the genes in the retroviral RNA genome, the comparable organization of the DNA provirus, and the location of the various virus-encoded proteins in the mature virion. The RNA strand is shown in green; the DNA provirus is shown in blue. The ORFs in the genome are color coded to match their products in the virion. The nontranslated regions of the RNA genome and the long terminal repeats (LTRs) are described in detail in the text. The *pro* gene, located between *gag* and *pol,* is not shown here. Adapted from Goff (1997), Figure 3.5 on p. 143.

Reverse Transcription of Viral RNA

A schematic overview of the life cycle of a retrovirus was presented in Fig. 1.13. Most retroviruses penetrate into a cell by fusion with the plasma membrane, but some use the endosomal pathway. Once inside the cell, reverse transcription takes place in a subviral particle to produce a full-length, linear dsDNA copy of the RNA. The composition of this subviral particle is not well defined. It certainly contains RT and its associated activities as well as components derived from Gag, but which Gag components are present and whether cellular proteins form part of this particle are not known.

Reverse transcription begins in the cytoplasm. In most viruses, the full-length dsDNA is produced in the cytoplasm, but in some the finishing touches occur after transfer to the nucleus. After transfer to the nucleus, the viral DNA is integrated into the host chromosome, essentially at random, in a process that requires the activity of IN. In the simple viruses, transfer to the nucleus occurs during cell division, when the nuclear envelope is disassembled. At least some of the complex retroviruses, however, such as HIV, encode proteins that allow the DNA-containing complex to traverse the nuclear membrane. Thus, the simple retroviruses can only productively infect dividing cells, whereas HIV can infect nondividing cells.

Synthesis of First DNA Strand

The process of reverse transcription of the viral genome is illustrated in Fig. 6.3. At the ends of the viral RNA are domains that are essential for production of the dsDNA copy of the RNA genome. These are illustrated schematically in Figs. 6.2 and 6.3, and the sizes of these elements are shown in Table 6.2 for a number of different retroviruses. A direct repeat element called R, 15–230 nucleotides in length depending on the virus, is present at the two ends of the RNA genome. At the 5′ end of the genome, a unique sequence element called U5 (70–220 nt long) is present immediately downstream of R, and at the 3′ end of the genome a unique element called U3 (230–1200 nt) is present immediately upstream of R. On completion of DNA synthesis, these elements give rise to direct repeats present at the ends of the viral DNA, called the long terminal repeats or LTRs, which have the sequence U3–R–U5.

Immediately 3′ of U5 is a primer binding site (PBS), where 18 nucleotides are exactly complementary to the 3′ end of a specific cellular tRNA. Each genomic RNA molecule has bound to it at the PBS one molecule of the appropriate tRNA, which is used as a primer for DNA synthesis. The tRNA used depends on the virus, and several different tRNAs are known to be used by different viruses (Table 6.2).

Reverse transcription begins by extending the primer tRNA through U5 and R, and stops when the 5′ end of the RNA genome is reached (Fig. 6.3, steps 1 and 2). The cDNA

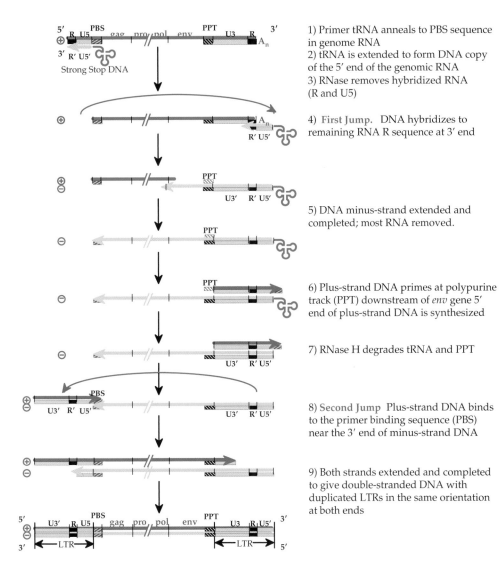

1) Primer tRNA anneals to PBS sequence in genome RNA
2) tRNA is extended to form DNA copy of the 5′ end of the genomic RNA
3) RNase removes hybridized RNA (R and U5)

4) **First Jump.** DNA hybridizes to remaining RNA R sequence at 3′ end

5) DNA minus-strand extended and completed; most RNA removed.

6) Plus-strand DNA primes at polypurine track (PPT) downstream of *env* gene 5′ end of plus-strand DNA is synthesized

7) RNase H degrades tRNA and PPT

8) **Second Jump** Plus-strand DNA binds to the primer binding sequence (PBS) near the 3′ end of minus-strand DNA

9) Both strands extended and completed to give double-stranded DNA with duplicated LTRs in the same orientation at both ends

FIGURE 6.3   Mechanism of retroviral DNA synthesis (reverse transcription). Green lines are RNA, light blue lines are minus-strand DNA, and the dark blue line in the last three steps is plus-strand DNA. Features within the LTRs (U3, R, U5), as well as the PBS and PPT (polypurine tract), are indicated by colored bars beneath the lines designating the nucleic acids, to clarify what is present at each step. Adapted from Fields *et al.* (1996) p. 1792; Goff (1997) Figure 3.6 on p. 145; Coffin *et al.* (1997) Figure 2 on p. 123.

TABLE 6.2   Terminal Regions of Retrovirus Genomes

Genus	Prototype virus	Approximate sizes in bases of terminal elements			Primer tRNA used
		U3	R	U5	
Alpharetrovirus	RSV, ALV	230[a]	20	80	Trp
Betaretrovirus	MMTV	1200[b]	15	120	Lys-3
Gammaretrovirus	MLV	450	70	80	Pro/Gln
Deltaretrovirus	HTLV-1	350	230	220	Pro
Epsilonretrovirus	WDSV	440	80	70	His
Lentivirus	HIV-1	450	100	80	Lys-1,2,3
Spumavirus	HRSV	910	190	160	Lys-1,2

[a] Includes *v-src* gene.
[b] U3 contains *sag* gene.
Source: Adapted from Table 2, page 38, in Coffin *et al.* (1997).

product, called the first strong stop DNA, is then transferred, while still attached to the primer, from the 5′ end of the RNA to the 3′ end ("first jump," Fig. 6.3, steps 3 and 4). It is unknown whether transfer is usually to the 3′ end of the same molecule, to the 3′ end of the second copy of the genome, or randomly to either copy of the RNA in the virus. This transfer uses the repeat element R at the 5′ and 3′ ends—the DNA copy detaches from R at the 5′ end and anneals to R at the 3′ end, so that only one copy of R is present in the DNA transcript. RNase H is presumably important for this. During reverse transcription, the RNA strand is destroyed by RNase H about 18 nt behind the transcription point. The degradation of the RNA strand during transcription of the DNA copy may encourage the jump to the other end. Other components of the particle may also be involved in the jump.

After the jump, reverse transcription resumes until the 5′ end of the RNA template is reached. Note that the RNA strand now ends at PBS because RNase H has degraded the RNA strand of the DNA–RNA hybrid (but not the RNA in PBS because this is an RNA–RNA duplex). This process results in the formation of what is called first-strand DNA or minus-strand DNA, because it is the antimessage sense (step 5).

Synthesis of the Second DNA Strand

The first-strand DNA produced in this way is the template for second-strand (plus-strand) DNA synthesis. The primer for second-strand synthesis is an RNA oligonucleotide, called the polypurine tract or PPT, positioned immediately 5′ of U3. PPT survives RNase H degradation and its 3′ terminus is exact; that is, the cleavages to produce it are precise (Fig. 6.3, step 6). In addition to this precise primer for plus-strand synthesis, which defines the boundary of U3 and thus of the LTR, additional priming sites are used in some retroviruses, such as HIV.

The PPT plus-strand primer is extended through U3, R, U5, and into the region of the primer tRNA, which is still attached to the first-strand DNA. The 18 nucleotides of the tRNA primer that are complementary to the PBS are copied, but further copying is thought to be blocked by a modified nucleotide in the tRNA that cannot be copied. The tRNA primer is then removed by RNase H (step 7) and a second jump occurs. In the second jump, the nascent second-strand DNA is transferred to the other end of the template, using the PBS sequence, which is now present at the 3′ ends of both strands (second jump, step 8). Synthesis of both strands of DNA resumes and it becomes full length and double stranded. The resulting dsDNA is cleaned up, probably by cellular enzymes. In the full-length, linear ds copy of the genome, the LTR sequence U3–R–U5 has been formed at both ends of the DNA genome (Fig. 6.3, step 9).

It is of interest that RT, like DNA polymerases, requires a primer to synthesize DNA, but unlike most DNA polymerases it can copy either DNA or RNA, given the appropriate primers. Thus, the enzyme differs from the RNA replicases of RNA viruses, which use other mechanisms for the initiation of RNA replication

Why Is the Genome Diploid?

Why the retroviral genome is diploid is not clear. In other instances where reverse transcription occurs, such as in the hepadnaviruses and the retrotransposons, the RNA to be copied is not diploid. Furthermore, in vitro studies have shown that retroviral RT can use a single copy of the RNA genome to produce a full-length dsDNA. Thus, two copies of the genome are not essential. However, it is possible that during infection the process is more efficient if the RT can go back and forth between the two copies, and this resulted in selection for a diploid genome in retroviruses. The process of reverse transcription to produce the viral dsDNA is complex with the multiple jumps required, and the diploid genome could be organized in such a way as to make these jumps more efficient. Other possible advantages of a diploid genome that could have resulted in selective pressure for diploidy are the possibility of overcoming at least some damage in the RNA by switching templates, and the fact that switching templates results in recombination. Recombination does occur frequently in retroviruses, and it is clear from many studies that recombination has been important in their evolution.

Integration

After the appearance of the full-length dsDNA genome in the nucleus, the viral integrase catalyzes its insertion into a host cell chromosome by means of a single recombinational event. This process is illustrated in Fig. 6.4. Insertion is essentially random within the host genome. The first event in integration is the removal of two nucleotides from the 3′ end of both strands of the viral DNA. The next two nucleotides are always AC, from 3′ to 5′. The 3′ OH of the now 3′-terminal A residue is used to attack an internucleotide phosphate in the host DNA. Attack is coordinated so that both ends of the viral DNA are inserted at once. The spacing between the two insertion points depends on the viral integrase and is 4, 5, or 6 nucleotides in different viruses. The ends of the inserted structure are then cleaned up, probably by host enzymes. Insertion results in the loss of the two terminal nucleotides of the viral DNA. In addition, a duplication of the 4, 5, or 6 nucleotides of the host that lie between the two insertion points is produced, and these duplicated nucleotides flank the two ends of the viral genome.

The integrated form of the virus, the provirus, is stable. No mechanism for precise excision of the provirus is known, and integration is essentially irreversible. Most retroviruses do not kill the host cell, and the provirus behaves as a simple Mendelian gene that is transmitted to all daughter cells. It is obvious that insertion of such a provirus into the germ

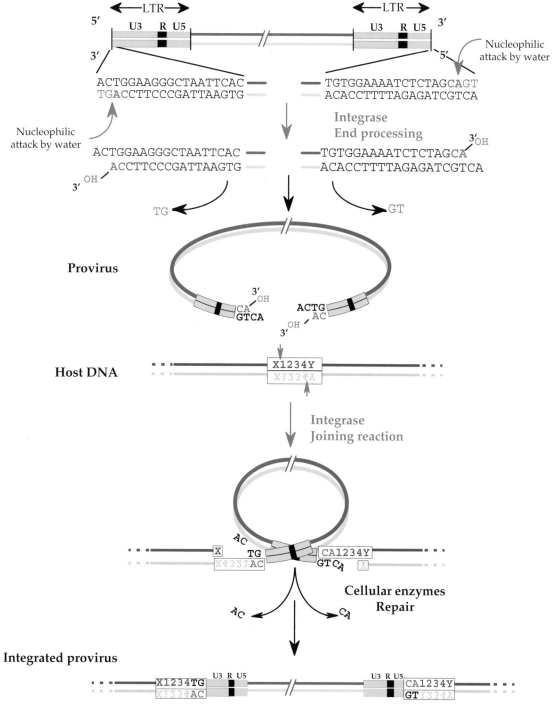

FIGURE 6.4 Steps in the integration of retroviral DNA into the host genome. First, the termini of the blunt-ended viral DNA are attacked by the integrase, and two bases (blue) adjacent to a highly conserved CA dinucleotide (red) are lost by nucleophilic attack by a water molecule, leaving recessed 3′ OH ends. Next, the host DNA is cleaved at the target sequence and the 3′ OH ends of the viral DNA are joined by the integrase to the 5′ phosphates on the host DNA. Overhang removal and gap and nick repair by cellular enzymes complete the integration reaction. Adapted from Hindmarsh and Leis (1999); Goff (1997) Figure 3.10 on p. 145, and Coffin *et al.* (1997) Figure 8 on p. 185.

line would lead to its transfer to progeny organisms, which appears to have occurred many times in the past.

Transcription of RNA in Simple Retroviruses

The LTR of the integrated provirus has all of the signals required for transcription of the provirus by cellular RNA polymerase II. The transcription signals, almost all found in U3, include TATA boxes as well as an array of binding sites for cellular transcription factors that are optimal for the cell in which the particular retrovirus primarily replicates. Fig. 6.5 gives examples of the binding sites for transcription factors in several different retroviruses. This figure illustrates that the constellation of binding sites is complex and is different

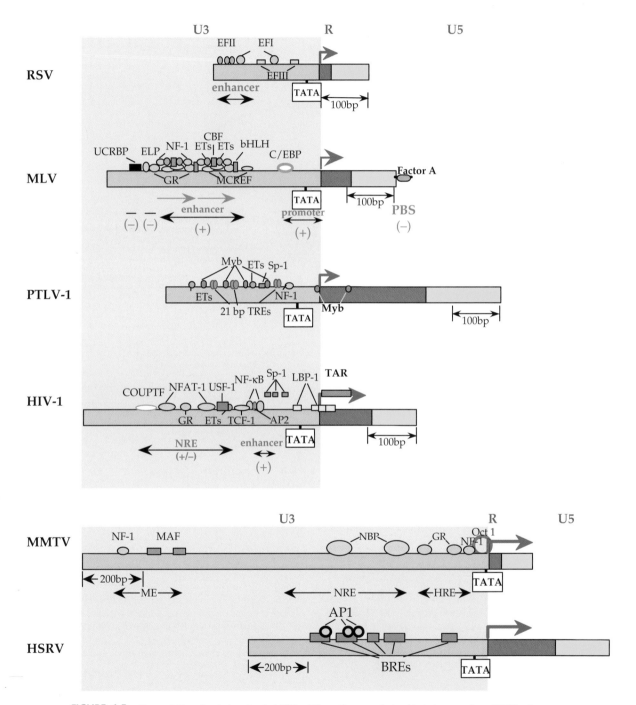

FIGURE 6.5 Transcription signals in retroviral LTRs. Where the same factor binds to a number of LTRs, the same symbol has been used. The TRE elements in PTLV-1 are the *Tax-r*esponsive *e*lements. The blue arrow marks the beginning of transcription. Note that the scale of the lower panel is twofold different from that in the upper panel. Composite of Figures 4, 5, 7, 8, 9, 11 in Chapter 6 of Coffin *et al.* (1997).

in different viruses that replicate in different tissues. Only the upstream (5′) LTR is used to initiate transcription—or at least efficient transcription.

Transcription initiates precisely at the 5′ end of R, as indicated by the arrow in Fig. 6.5, and proceeds through the proviral genome and into the 3′ flanking sequences. The transcript is capped by host cell capping enzymes. There is a poly(A) addition signal (AAUAAA) in the RNA, which is usually present about 20–30 nucleotides upstream of the end of the viral genomic RNA sequence (i.e., upstream of the 3′ end of R). Host cell polyadenylation machinery recognizes the AAUAAA signal, cleaves the RNA transcript precisely at the R–U5 boundary, and polyadenylates the RNA. Thus the processed RNA transcript is identical to the genomic RNA and a round-trip from RNA to DNA to RNA has been made.

In many viruses, the polyadenylation signal is present in U3. Thus, there is only one copy of this signal in the RNA transcript, near the 3′ end. However, in some viruses the signal is present in R and is therefore present in both the 5′ and 3′ regions of the transcript. Interestingly, only the 3′ copy of the signal is active for cleavage and polyadenylation of the transcript.

Some fraction of the genomic RNA is exported to the cytoplasm without further processing. There it serves as mRNA for the synthesis of Gag, Gag–Pro, and/or Gag–Pro–Pol, as described later. Alternatively, it can be packaged into progeny viruses. Genomic RNA that serves as mRNA and genomic RNA that serves as the source of RNA for packaging are maintained in separate pools.

Some fraction of the genomic RNA is spliced before export to the cytoplasm. Only one spliced RNA is made in the simple retroviruses, which serves as mRNA for Env. In most retroviruses, the entire Gag–Pro–Pol region is spliced out and the initiation codon for Env is encoded in *env*. In the avian retroviruses, however, the upstream splice site is located within the Gag coding sequence so that Env begins with the first six codons of Gag.

The need for both spliced and unspliced versions of the viral RNA means that mechanisms must exist to ensure that both are produced and that the ratio of spliced to unspliced RNA is optimal for virus replication. In the simple retroviruses, the splice sites are suboptimal, so that not all RNA is spliced. Experiments have shown that the result is an optimal ratio of spliced to unspliced RNA. Mutations that make splicing more efficient are deleterious for virus growth and revertants quickly arise that restore the proper ratio. The second problem faced by these viruses is the need to export unspliced RNA to the cytoplasm. Eukaryotic cells have control mechanisms to ensure that RNA containing splice sites is not exported from the nucleus. It has been found that sequence elements in the unspliced RNA are required for its export, and it is assumed that these elements interact with cellular proteins that promote export. In the simian ret-

rovirus Mason–Pfizer monkey virus, this element is called the *c*onstitutive *t*ransport *e*lement (CTE). It is 154 nucleotides long and is located in the 3′ nontranslated region of the RNA. In the avian retroviruses, an apparently unrelated element of about the same size, also present in the 3′ nontranslated region, provides the same function for unspliced avian retroviral RNA. Interestingly, the avian sequence does not work in mammalian cells. The monkey virus sequence works in both mammalian and avian cells, but works better in mammalian cells. These findings are consistent with the hypothesis that these transport elements interact with cellular proteins. The inability of unspliced RNA to be exported from the nucleus is one of the reasons that the avian retroviruses will not replicate in mammalian cells.

Translation of Viral Genomic RNA

Retroviral genomic RNA is translated into two or three polyproteins that are eventually processed by the viral protease. The order of genes along the genomic RNA is *gag–pro–pol*, encoding the proteins Gag, Pro, and Pol. Stop codons are present between Gag and Pro, or between Pro and Pol, or in both places. Termination of the polypeptide chain occurs at these stop codons most of the time during translation. These stop codons are suppressed some of the time, however, either by readthrough or by frameshifting (Chapter 1), so that the amount of Pol produced is usually about 5% that of Gag. In viruses with one stop codon, the frequency of suppression is about 5%, but in viruses with two stop codons, the frequency of suppression of each stop codon is higher so that significant amounts of Pol are produced even though suppression of two stop codons is required. This means that the frequency of suppression is variable and can be controlled by changes in the sequence of the viral RNA. Because reinitiation does not occur once the chain is terminated, the polyproteins produced are Gag and/or Gag–Pro and/or Gag–Pro–Pol, depending on the positions of the stop codons (Fig. 6.6).

Pro is produced in three different ways in different retroviruses, as illustrated in Fig. 6.6. In the avian viruses, such as ALV, there is no stop codon between Gag and Pro so that a Gag–Pro polyprotein is produced. Gag and Pro are thus produced in equal amounts. Frameshifting results in the production of a longer polyprotein, Gag–Pro–Pol. Most of the mammalian viruses also only have a single stop codon in the ORF, but it is positioned between Gag and Pro. Readthrough of a UAG stop codon (murine leukemia viruses) or frameshifting (other mammalian viruses with a single stop codon) results in the longer polyprotein. The two polyproteins produced are Gag and Gag–Pro–Pol, and Pro and Pol are produced in the same low amounts. Finally, several mammalian retroviruses, for example, MMTV and PTLV-1, have two stop codons in the ORF, both of which can be suppressed by frameshifting. Thus, three polyproteins

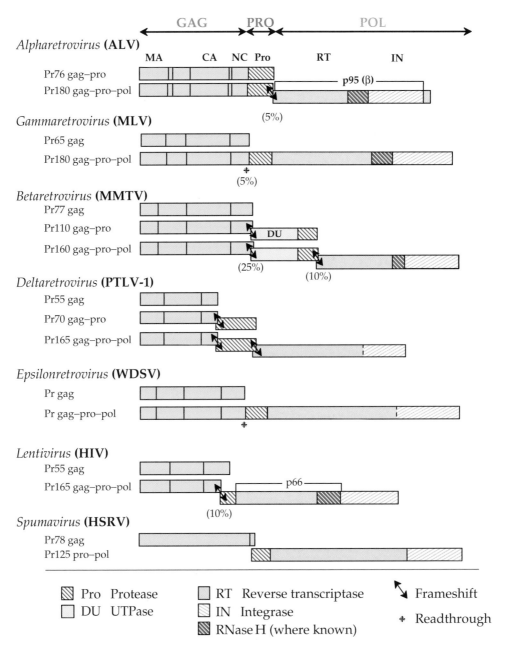

FIGURE 6.6   Organization of the *gag, pro,* and *pol* genes of representative retroviruses belonging to each genus. In some cases two frameshifts are required to generate a complete Gag–Pro–Pol precursor. The *gag* proteins are illustrated in more detail in Fig. 6.7. This figure is a composite of Goff (1997) Figure 3.16 on p. 157; and Coffin *et al.* (1997) pp. 45, 269, 795, and 799. Virus name abbreviations can be found in Table 6.1.

are produced, Gag, Gag–Pro, and Gag–Pro–Pol. In this case Pro is produced at intermediate levels.

Processing of these various polyproteins occurs during assembly of progeny virions. The viral Pro is an aspartate protease whose active site contains two aspartic acid residues (Chapter 1). The protease domain is functional in polypeptides containing the Pro sequence as well as after its release by proteolysis as a small protein of about 100 residues. The enzyme is active only as a homodimer, with each chain in the dimer supplying one of the aspartic acids in the active

site. Because the monomer is not active, there is a delay in processing. The high concentration of viral polyproteins that occurs in viral particles or in previral particles is required to achieve efficient dimerization of the protease and its activation. Experiments have shown that premature activation of the protease, which can be achieved by using genetic tricks, is deleterious for virus assembly. Thus, it is important that processing be delayed until assembly begins or is completed.

During processing of Gag, several different proteins are produced, some of which are quite small, whereas others are larger. The proteins produced from Gag are illustrated schematically in Fig. 6.7 for a number of retroviruses. Gag is cleaved to produce at least three protein products in all retroviruses except the spumaviruses, called MA (membrane-associated or matrix protein), CA (capsid protein), and NC (nucleocapsid protein). These three peptides are always present in that order from the N terminus to the C terminus in the Gag polyprotein. Gag (and thus MA) is myristoylated in most retroviruses and associates with membranes. Myristoylation may serve to recruit Gag to membranes for assembly or budding. NC is a small basic protein that binds RNA and Zn^{2+}. It has a number of functions, including binding to the packaging signals in the viral RNA that lead to its incorporation into virions, the facilitation of binding of the tRNA primer to the genomic RNA, the formation of the genomic RNA dimer, and strand transfer during reverse transcription. CA is a larger protein that is believed to form a shell around the viral RNA and its associated internal proteins (Fig. 2.21). In addition to these three proteins, in most retroviruses other proteins, whose functions are not well understood, are also produced from Gag (Fig. 6.7).

Processing of Pol varies in different viruses. Three patterns of cleavage can be distinguished. In some viruses, RT, consisting of the polymerase domain and the RNase H domain, is released from the upstream Pro and the downstream IN. The active reverse transcriptase is a monomer or a homodimer of RT. In other viruses, cleavage between RT and IN is incomplete, and the active enzyme is a heterodimer of RT and RT-IN. A third pattern occurs in HIV, in which partial cleavage occurs between the polymerase domain and the RNase H domain. In this case, the products of Pol are a truncated RT lacking the RNase H domain (called p51), the full-length RT (called p66), and IN. The active reverse transcriptase is a heterodimer between p51 and p66.

Production and Processing of Env

Translation of the mRNA for Env produces the envelope glycoprotein precursor. Processing of Env and the location of key features are illustrated in Fig. 6.8 for a number of retroviruses. The precursor has an N-terminal signal sequence (labeled L in the figure), which leads to its insertion into the endoplasmic reticulum, a membrane anchor sequence near the C terminus, and a C-terminal cytoplasmic domain (i.e., the protein is a type I integral membrane protein). The signal sequence is removed during translocation into the endoplasmic reticulum, and the protein is glycosylated and transported to the plasma membrane by conventional cellular pathways. As with many other viral glycoproteins, the precursor is cleaved by furin during transport to produce an N-terminal extracellular component called SU (for surface) and a C-terminal membrane-spanning component TM (for

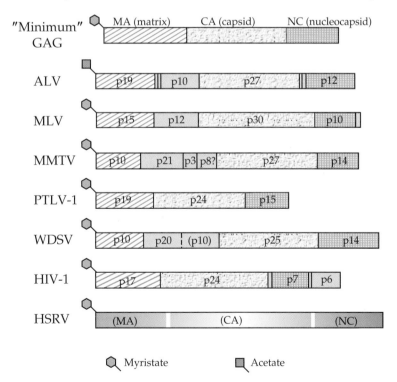

FIGURE 6.7   Organization of the Gag proteins in representatives of each retroviral genus. The viruses are shown in the same order as those in Fig. 6.6. Vertical solid lines mark sites of cleavage by the viral protease. Sequences representing the mature matrix, capsid, and nucleocapsid proteins are indicated with different shadings, and the approximate molecular weights of the processed proteins are shown. Note that the Gag polyprotein of HSRV is not processed. From Coffin *et al.* (1997) pp. 44 and 798.

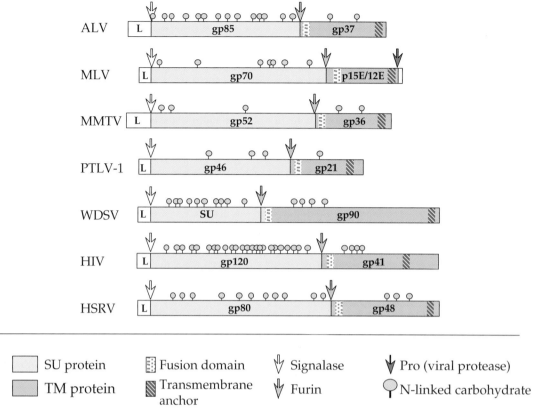

□ SU protein	▤ Fusion domain	⇓ Signalase	⬇ Pro (viral protease)
▦ TM protein	▨ Transmembrane anchor	⇓ Furin	♀ N-linked carbohydrate

FIGURE 6.8   Organization of the envelope proteins of representatives of each genus of retroviruses, shown in the same order as in Figs. 6.6 and 6.7. The domains corresponding to the mature SU and TM proteins are shown in different shades of color. Cleavages by signalase to remove the leader peptide (L) are indicated with white arrows, and those due to furin by yellow arrows. The red arrow marks the site of cleavage in MLV by the viral protease. The fusion domains, the transmembrane domains, and the sites of predicted N-linked carbohydrate addition are shown. Adapted from Coffin *et al.* (1997) Figure 10 on p. 56. Virus name abbreviations can be found in Table 6.1.

transmembrane). SU and TM remain associated. In some cases, the association is stabilized by disulfide bonds, but in other cases there is no covalent linkage. SU is always glycosylated, whereas TM may or may not be glycosylated. Cleavage to produce SU and TM is required for the activation of the fusion activity, which is located near the N terminus of TM. Thus, the production of the envelope glycoproteins of the retroviruses parallels that of the envelope glycoproteins of many enveloped RNA viruses.

Accessory Genes of Complex Retroviruses

In addition to the core genes *gag, pro, pol,* and *env* that are present in all retroviruses, members of the four genera of complex retroviruses possess additional genes that allow them to regulate the development of the infection cycle. Accessory genes are also present in the betaretroviruses, but these are not regulatory in nature. A listing of the accessory genes in different retroviruses is given in Table 6.3, and a complete listing of all of the proteins of HIV is shown in Table 6.4.

The accessory genes of the complex retroviruses are located upstream or downstream of *env* and are translated from spliced mRNAs, with some of the genes requiring multiple splicing for expression. The genome organizations of the different genera are diagrammed in Fig. 6.9 to show the location of these accessory genes. The different accessory genes have been inserted into the retroviral genome at different times during the evolution of these viruses (Fig. 6.1), and insertion of such genes appears to be a dynamic process that is ongoing. As one example, the *vpx* gene of HIV-2 and the *vpr* gene of HIV-1 appear to have been inserted into their respective viruses after the separation of HIV-1 and HIV-2.

The presence of the accessory genes allows more vigorous replication of the retrovirus that possesses them, which can be fatal to the host cell. As described before, the simple retroviruses do not kill the host cell, but instead establish a persistent infection in which the cell survives and produces low levels of virus indefinitely. However, at least some of the complex retroviruses, such as HIV-1, can replicate to

TABLE 6.3   Accessory Genes in Retroviruses

Gene	Functions
Betaretrovirus (MMTV)	
sag	Superantigen
dut	dUTPase
Deltaretrovirus (HTLV/BLV)	
tax	Transcription activator (like tat)
rex	Splicing/RNA transport regulator (like rev)
Lentivirus (HIV-1)	
tat	Transcription activator (like tax)
rev	Splicing/RNA transport regulator (like rex)
vif	
vpr/vpx	
nef	See Table 6.4
vpu	
dut	dUTPase (in nonprimate lentiviruses)
	Facilitates replication in certain cell types.
Spumavirus (HSRV)	
tas (bel 1)	Transcription activator (DNA binding protein)
bel 2	?
bet	? (may be involved in latency)
Epsilonretrovirus (WDSV)	
Orf A	Viral homologue of cyclin D
Orf B	?
Orf C	?

Source: Adapted from Coffin *et al.* (1997) Table 1 on p. 36, and information from Fauquet *et al.* (2005).

high titer in some cell types with the result that the cells die. It is interesting that many endogenous retroviruses are present in the germ line of different vertebrates, as described later. However, none of these are complex retroviruses. It is possible that the regulated lifestyle of the complex viruses would make complex endogenous viruses difficult to silence. The inability to silence such viruses could lead to selection against organisms that contain them in the germ line.

Many of these proteins encoded by the accessory genes are multifunctional and their functions are only partially understood. The intense interest in PTLV and HIV has resulted in extensive study of their accessory genes, *tax* and *rex* in the case of PTLV/BLV and *tat*, *rev*, *vif*, *vpr/vpx*, *nef*, and *vpu* in the case of lentiviruses. The accessory genes of the spumaviruses, *bel*1, *bel*2, and *bet*, as well as those of the epsilonretroviruses, have been less well studied. The functions of these genes can be conveniently grouped into six categories: (1) transport across the nuclear membrane of subviral particles that synthesize viral DNA after infection, allowing the virus to infect quiescent cells;

(2) activation of transcription of the provirus to greatly increase the rate of production of viral RNA; (3) export of unspliced viral RNA to the cytoplasm; (4) arrest of the cell cycle in infected T cells; (5) promotion of virus assembly and release; and (6) protection from cellular defense mechanisms.

Transport of Viral DNA into the Nucleus

HIV-1, and probably other lentiviruses as well, can infect quiescent cells because the viral DNA and associated proteins can cross the nuclear membrane. The product of the *vpr* gene appears to be required for this. MA and IN have also been implicated in the process.

Transactivation of the Transcription of Viral RNA

All complex retroviruses encode a protein that transactivates transcription of viral RNA from the provirus. In PTLV/BLV, the gene for this protein is called *tax*. The Tax protein activates transcription by means of sequence elements in the U3 region of the viral DNA called *T*ax-*r*esponsive *e*lements (TREs), whose locations are shown in Fig. 6.5. Tax interacts with cellular transcription factors that bind to TREs, and this interaction results in increased activity of the transcription factors. One such transcription factor is the *c*AMP *r*esponse *e*lement/*a*ctivating *t*ranscription *f*actor (CREB/ATF). Tax also increases the transcription of certain cellular genes, in some cases through its interactions with CREB/ATF, and in other cases by stimulating the transcription of genes regulated by NFκB. Many of these cellular genes are important in the regulation of T cells, and their stimulation may relate to the pathology of disease caused by the virus.

In the lentiviruses, the gene for the transcriptional activation is called *tat*. The Tat protein works in one of two different ways. Tat of visna virus appears to interact with cellular transcription factors in a manner similar to Tax, using a sequence element in U3, although the cellular factors are different. Tat of HIV and its close relatives, however, stimulates transcription by binding to a sequence element called TAR at the 5′ end of the viral RNA. TAR is composed of the first 60 nucleotides of HIV-1 RNA, which form a stem-loop structure that is essential for the function of TAR. TAR of both HIV-1 and HIV-2 are shown in Fig. 6.10. How the binding of Tat to TAR activates transcription is not yet clear. One model is that Tat interacts not only with the nascent viral RNA but also with cellular transcription factors, and in so doing stabilizes the transcription complex or changes its composition. In this model, the altered transcription complex is more processive, allowing the production of complete RNA genomes rather than truncated transcripts. It may also initiate transcription more frequently.

TABLE 6.4　The HIV Proteins

Protein	mRNA	Size (kD)	Post-translational modifications	Functions
gag	Genomic RNA	p25 (CA)	None	Capsid structural protein
		p17 (MA)	Myristoylated at Gly-2	Matrix protein
		p7	?	RNA-binding protein
		p2	?	RNA-binding protein
pro	Genome RNA frameshifted	p10 (PR)		Viral protease, processes *gag* proteins
pol	Genome RNA frameshifted	p66/p51 RT	Heterodimer, p51 lacks RNase H domain present in p66	Reverse transcriptase
vif	vif mRNA	p23		Viral infectivity factor, essential for spread in macrophages
vpr/vpx	vpr/vpx mRNA[a]	p15	Associates with p7	Augments replication
tat	tat mRNA	p14		Required for replication transactivates RNA synthesis, binds to TAR RNA
rev	rev mRNA	p19		Regulates splicing/RNA transport; binds RRE element and facilitates *env* translation
vpu	vpu/env mRNA	p16	Phosphorylated on Ser	Helps in virion assembly and release, dissociates gp 160/CD4 complex
env	vpu/env mRNA	gp 120 (SU)	24 sites for N-linked glycosylation	Surface glycoprotein, mediates cellular attachment
		gp 41 (TM)	7 sites of N-linked glycosylation	Transmembrane glycoprotein
nef	nef mRNA	p27	Myristoylated at Gly-2, phosphorylated at Tyr-15	Homodimer, causes pleiotropic effects, including downregulation of CD4

[a]vpr is found in HIV-1; vpx is found in HIV-2.
Source: Adapted from Levy (1994) Table 1.4, p. 8.

Export of Unspliced Viral RNA to the Cytoplasm

Complex retroviruses encode proteins that promote the export of unspliced or partially spliced RNA from the nucleus. The proteins are Rex in the case of PTLV/BLV and Rev in the case of lentiviruses. Rex and Rev are translated from multiply spliced mRNAs, as are the transcriptional activators and Nef in the case of the lentiviruses. The multiple splicing events in HIV-1 are illustrated in Fig. 6.11. Early in infection, before Rex (PTLV/BLV) or Rev (HIV-1, -2) are present, only completely spliced mRNAs are exported from the nucleus to the cytoplasm. Thus, the proteins made early are the transcriptional activators, Nef, and the proteins that control the export of unspliced or partially spliced viral RNAs from the nucleus. The transcriptional activators accelerate the production of viral RNAs, and Rex and Rev allow the export of mRNAs for the other viral proteins, which includes the genomic RNA. These processes are illustrated schematically in Fig. 6.12.

Studies with HIV-1 have shown that Rev binds to a sequence element in the viral RNA called the *R*ev *r*esponse *e*lement or RRE. HIV-1 RRE is 234 nucleotides in size and has multiple stem-loop structures that are important for function. It is found in the *env* gene region and is therefore spliced out of the multiply spliced mRNAs (Fig. 6.11). In addition to binding to RRE, Rev also interacts with cel-lular proteins involved with the nuclear export pathway. The end result is that Rev promotes the export of RNAs containing RRE (i.e., unspliced genomic RNA and singly spliced mRNAs) from the nucleus to the cytoplasm (Fig. 6.12). Rev appears to accompany the RNA to the cytoplasm and then to cycle back to the nucleus. Thus, after the appearance of Rev, the infection cycle switches to a late phase with the appearance in the cytoplasm of the mRNAs for Gag–Pro–Pol, Env, Vpu, Vif, and Vpr. Genomic RNA exported to the cytoplasm can also be packaged into progeny virions.

In addition to its function in the export of viral RNAs to the cytoplasm, Rev also has other functions. These functions appear to include regulation of the splicing of HIV RNAs and increasing the efficiency of translation of HIV RNAs (Fig. 6.12).

The PTLV Rex appears to function similarly to Rev. It binds to an RNA element, called the RexRE, that is similar in size to the RRE, and binding of Rex promotes export of viral RNA from the nucleus. However, the RexRE is found in the U3–R region of the genome and is therefore present in all of the viral mRNAs. It is curious that Rex binds HIV-1 RNA, although not in the same place as Rev, and will substitute for Rev in HIV-1. However, Rev does not bind PTLV RNA and will not substitute for Rex.

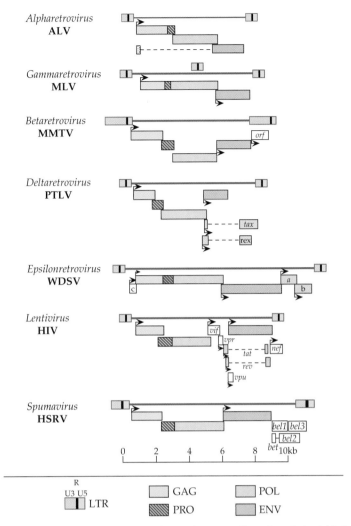

FIGURE 6.9   Coding regions of representatives of each retrovirus genus. Sites of translation initiation are shown with arrows. The locations of the four major genes are shown in different colors. Accessory genes are named. Redrawn from Coffin *et al.* (1997) Figure 5 on p. 37, and from Fields *et al.* (1996) p. 1776.

Cell Cycle Arrest in T Cells

HIV-1 kills infected T cells during active replication by arresting the cell cycle. Cell death results from necrosis, not apoptosis as in the case for many viruses. The result in infected humans is the depletion of T cells and, ultimately, progression to AIDS. The products of the *vif* and *vpr* genes are required for this arrest. Elimination of both genes, but not of either one separately, results in a virus that cannot cause cell cycle arrest.

Assembly Functions

Several accessory proteins of the lentiviruses promote the assembly and release of infectious virus, either directly or indirectly. Vpr, found in most lentiviruses, including HIV-1, and Vpx, found in HIV-2, are structural proteins that interact with Gag. They are found in the virion in amounts similar to that of Gag. Vif is a membrane-associated protein that is also present in the virion, but in amounts similar to Pol rather than Gag. It has an effect on the processing of Gag and is required for the production of infectious virions as described later. Nef, found in the primate lentiviruses, is a multifunctional protein that has a role in the assembly and release of infectious virus. Myristoylated forms of Nef are associated with the virion, where they may be required for full infectivity of the virion. Another function of Nef is the removal of CD4, the receptor for HIV, from the surface of the infected cell. Removal of CD4 prevents the interaction of Env with CD4 and facilitates the assembly and release of viruses. It also prevents superinfection of the cell by released virus. Vpu, a small integral membrane protein found in HIV-1 but not in HIV-2, promotes virus release in two ways. It promotes the degradation of CD4 by

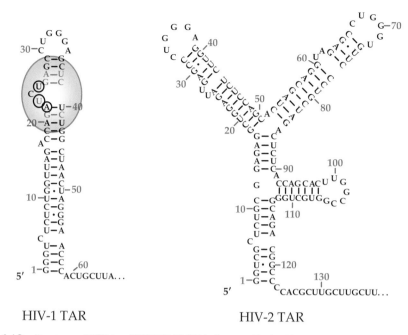

HIV-1 TAR   HIV-2 TAR

FIGURE 6.10   Structures of HIV-1 and HIV-2 TAR RNA elements. Nucleotides are numbered from the 5′ end of the RNA. For HIV-1, positions involved in binding to Tat protein (yellow oval) are circled, and the bases involved in tertiary structure alterations following Tat binding are shown in red. Less is known about the HIV-2 Tat binding. Adapted from Coffin *et al.* (1997) Figure 12 on p. 226.

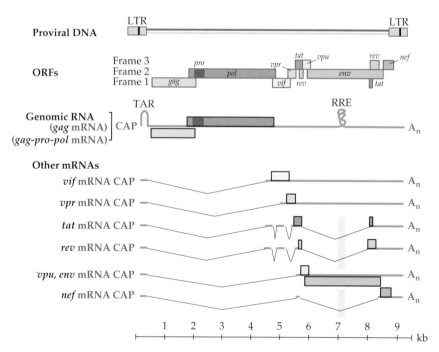

FIGURE 6.11   Genome organization and transcription map of HIV-1, the human immunodeficiency virus. The genome is shown on the top line as the integrated provirus. The LTRs and all open reading frames (ORFs) are indicated. Below this, the unspliced genome RNA is shown, with TAR and RRE (the *R*ev *r*esponse *e*lement in *env*) indicated. The various spliced mRNAs (and the ORFs translated from them) are diagrammed below the RNA genome. The pale blue shading indicates the location of the RRE, which is spliced out of *tat, rev,* and *nef* messages. Redrawn from Coffin *et al.* (1997) p. 803.

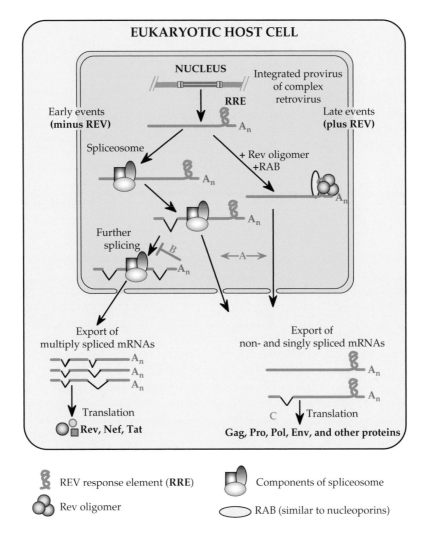

FIGURE 6.12   Model of Rev action. Rev is thought to (A) Mediate export of nonspliced or singly spliced RRE-containing RNAs from the nucleus; (B) Inhibit complete splicing of mRNA; (C) Enhance translation of unspliced and singly spliced mRNAs. Adapted from Coffin *et al*. (1997) p. 244.

a mechanism different from that used by Nef. It also facilitates the release of budding virus from the cell, a function fulfilled by Env in HIV-2.

Vif Abrogates a Cell Defense Mechanism

Vif, so named because it serves as a *v*iral *i*nfectivity *f*actor, is required for the production of infectious virus in certain cell types, such as primary CD4 T lymphocytes. These cells produce a protein called hA3G or APOBEC3G, which deaminates cytosine. This protein is incorporated into progeny virus in the absence of Vif, rendering the virions noninfectious because during first-strand DNA synthesis 10% or more of cytosines can be deaminated. This results in hypermutation of G to A in the plus-sense second-strand DNA, making the provirus nonfunctional. Vif binds to hA3G and recruits ubiquitin ligases such that hA3G is degraded by the proteasome, thereby preventing it from being incorporated into virions.

The dUTPase Gene

The nonprimate lentiviruses and members of the betaretroviruses contain a gene encoding dUTPase, called *dut*. This gene was acquired independently by these two different lineages (Fig. 6.1). The enzyme dephosphorylates dUTP, thereby preventing its incorporation into DNA. A cellular gene usually provides this function for retroviruses (and for DNA viruses). However, possession of this gene allows these two genera of retroviruses to replicate in quiescent cells that do not express adequate levels of the enzyme, such as macrophages.

Assembly of Retroviruses

The retrovirus virion is approximately 100 nm in diameter and acquires its lipid envelope by budding through the plasma membrane of the infected cell. Two forms of budding occur, as illustrated in Fig. 2.21. In the betaretroviruses and the spumaviruses, the capsid is assembled in the cell cytoplasm and then buds through the plasma membrane. In the alpharetroviruses, the gammaretroviruses, and the lentiviruses, the capsid assembles during budding and is not visible as a distinct structure within the cell. Mason–Pfizer monkey virus, a betaretrovirus, changes from budding of preassembled capsids to assembly of capsids during budding as the result of a single amino acid change in MA. Thus, the distinction between the two types of budding does not represent a fundamental difference in budding pathway but simply a matter of the stability of the capsid in the absence of interactions with other components of the virion during budding.

The capsid is formed by the assembly of Gag, Gag–Pro, and/or Gag–Pro–Pol polyproteins into a structure having spherical symmetry. RNA appears to be required for assembly but not the Env protein, even in viruses whose capsids assemble during budding. These polyproteins are incorporated into the capsid in approximately the same ratio as they are produced inside the cell. During or after budding, the viral protease cleaves the polyprotein precursors to the final products. These cleavages are required for the virus to be infectious and result in a change in the structure of the capsid. The final, fully cleaved virion may have the capsid centrally placed within it or it may be eccentrically placed, and the shape of the capsid may be spherical or it may have other shapes, depending on the virus (Fig. 2.21). As described before, it is important that the cleavages be delayed until the virion is partially or completely assembled, or the assembly process will not work properly.

During assembly, the genomic RNA is recruited into the capsid and dimerized. There is a packaging signal in the RNA, usually referred to as ψ, that is often found in the 5′ region downstream of the LTR. The signal is not present in the spliced RNAs of many retroviruses but even in those in which it is present, the spliced RNAs are not packaged. The packaging signal is not absolutely required for packaging, but increases the efficiency of incorporation into the virion by about 100-fold. Because Gag is the only protein required for assembly of capsids, the recognition of the packaging signal must be a property of Gag. During maturation of the virion, the RNA dimer also matures from a less stable form to a more stable form. tRNA is recruited into the capsid by RT, but association with RT is not absolutely required for packaging of the primer tRNA.

The envelope glycoproteins, SU and its associated TM, are incorporated into the virion during budding. The mechanism by which SU/TM is recruited is uncertain. Evidence indicates that MA interacts with TM, but a model in which SU/TM is incorporated nonspecifically, perhaps because it is free to diffuse whereas cellular proteins are not, cannot be excluded. SU/TM is not required for virus assembly. Glycoproteins from other viruses can substitute for SU/TM. More strikingly, capsids can bud to form noninfectious bald particles free of glycoprotein. In a key experiment, it has been found that HIV-1 buds from the basolateral surface of polarized cells, where SU/TM is found. However, when SU/TM is absent, Gag-directed budding occurs from both basolateral and apical surfaces. This result provides evidence that specific interactions between the capsid and the glycoproteins occur during virus budding and are important, but these interactions are not essential for budding to occur.

Alpharetroviruses and Gammaretroviruses

Alpharetroviruses

The alpharetroviruses comprise a large collection of avian leukosis and sarcoma viruses. Dozens of ALVs are known. They are grouped into seven interference groups, A–G, in which members of a group use the same receptor. Infected cells cannot be superinfected by another member of the same interference group, because the receptors for it have been eliminated. Expression of the viral envelope protein is usually responsible for the downregulation of receptors, which is important for the budding and release of virus. It also prevents the cell from becoming infected by hundreds or thousands of progeny viruses. This may be important for the survival of the infected cell and may allow it to produce progeny viruses indefinitely.

Interference groups A, B, C, and D of the ALVs are exogenous viruses that are transmitted as infectious agents from chicken to chicken. Members of group E are endogenous viruses, resident in the germ line of the chicken. The endogenous viruses may be quiescent and not expressed, or parts of the provirus genome or even the entire genome may be expressed. In the latter case, progeny viruses are formed that can infect other cells, leading to widespread expression of the virus in the animal. The endogenous viruses of chickens, as well as those of mammals, tend not to be expressed. Expression often results in disease and a shortened life span, and the animals are selected to not express integrated proviruses. In chickens, the level of expression and the time in the animal's life when expression of an endogenous virus occurs is different for different strains of chickens.

ALVs are commonly present as infectious agents in chicken flocks around the world. The major illness caused by these viruses is a wasting disease characterized by anemia, immunosuppression, and poor growth. Perhaps of more interest to the molecular biologist, these viruses may also cause leukemia or sarcoma, as described later.

Interference groups F and G are viruses of pheasants. These viruses have not been as well studied as the viruses of chickens.

Gammaretroviruses

The gammaretroviruses consist of a large number of leukemia and sarcoma viruses of mice, cats, primates, and other mammals. Also included in the genus is reticuloendotheliosis virus of birds, which causes immunodeficiency. The murine leukemia viruses have been particularly well studied. Both exogenous viruses and endogenous viruses of mice are known. The mouse genome contains 500–1000 endogenous proviruses that are divided into four classes, but only two of these classes are known to encode infectious viruses. One class encodes betaretroviruses, described later, and the other class consists of 50 to 60 copies of gammaretroviruses. The host range of mouse endogenous gammaretroviruses depends on the *env* gene. These viruses may be ecotropic (able to infect only mice), xenotropic (unable to infect mice but able to infect other animals, such as rats), or polytropic (able to infect both mice and other animals). The endogenous viruses are usually transcriptionally silent, but expression does occur in some mouse strains. In an extreme example, AKR mice usually become viremic at an early age and most of the animals eventually die of leukemia.

Feline leukemia virus (FeLV) is an important pathogen of cats. It is an exogenous virus that causes T-cell lymphomas and immunodeficiencies, as well as severe aplastic anemia, in cats. Sarcoma strains of FeLV are also known. A vaccine against FeLV that is given to pet cats is partially effective in preventing FeLV-induced disease.

Gammaretroviruses of other mammals, including primates, are also known, as are sarcoma viruses derived from them. Interestingly, however, no gammaretrovirus is known that infects humans, although the human genome does contain many retroviral-like elements.

Induction of Leukemia by Alpha- and Gammaretroviruses

Alpharetroviruses and gammaretroviruses are important causes of cancer in chickens, mice, cats, and subhuman primates. The cancers are usually forms of leukemia or lymphoma, and arise only after a long latent period. The tumor cells are usually clonal, having developed from a single progenitor tumor cell. Not all infected animals develop cancer. Infection by these retroviruses does not directly produce tumors. Instead, rare insertional or recombinational events must occur that give rise to a tumor cell. Although these events are rare, the very large number of cells infected during the persistent infection established by the virus may render such events probable, and after a sufficiently long latent period the probability that a tumor will arise may be high.

It is only recently that the mechanisms by which unmodified alpha- and gammaretroviruses induce tumors have become at least partially understood. One mechanism involves the insertion of the provirus near to or within a cellular oncogene. This may bring the expression of the gene under the control of the strong viral promoters, and the resulting overexpression of the oncogene may result in a tumor cell. In the case of bursal lymphomas induced by infection of chickens by ALV, for example, it has been found that more than 80% of tumors have ALV provirus inserted near the c-*myc* gene and overexpress the c-*myc* product. In other cases, insertion of the provirus within the oncogene may result in the expression of an mRNA that lacks control sequences (such as sequences that cause the mRNA to be degraded rapidly) or that is translated into an altered protein product that has lost regulatory elements. Such a process is often seen in erythroblastosis induced by ALV in chickens, for example. Integration of the provirus between two exons of c-*erbB* separates the domain of the protein that binds epidermal growth factor and tumor growth factor from the domain that leads to downstream signaling by means of protein tyrosine kinase activity. Unregulated signaling by the modified protein product results in deregulation of growth.

A different mechanism of tumor induction is often seen in mice infected by MLV. Most AKR mice spontaneously express an endogenous virus called AKV1 from birth, and die of thymic lymphomas in their first year of life. For the tumors to develop, multiple recombination events between AKV1 and other endogenous viruses are required to produce an Env protein with altered host range. The altered SU may stimulate T-cell proliferation by binding the interleukin-2 (IL-2) receptor, and other altered interactions with T cells and their precursors may also be important in tumor induction. Alterations in Env are also found in tumors induced by FeLV.

It is obvious from this description that the properties of the virus (host range, the nature of the LTR promoters) as well as the properties of the host cell are important in determining whether a tumor will arise. Thus, different strains of the same virus may have different effects upon infection of the same host, or one strain of virus may affect different strains of its hosts differently.

Alpha- and Gammaretroviruses That Express Oncogenes

Alpha- and gammaretroviruses undergo recombination with cellular oncogenes to produce viruses capable of causing tumors in their hosts. Recombination is thought to occur during reverse transcription of a hybrid genome in which a host mRNA replaces one copy of the viral RNA. Recombinant viruses that express a variety of oncogenes have been repeatedly isolated from spontaneous tumors in animals that are infected by leukosis/leukemia viruses. Such retroviruses will usually transform cells in culture and rapidly cause tumors when inoculated into a new host. Many of these viruses cause sarcomas and the oncogene-containing virus is then referred to as a sarcoma virus. Examples of the genomes of four oncogene-containing retroviruses are shown in Fig. 6.13. Two are derived from avian leukosis viruses and two from murine leukemia viruses. These examples have been chosen to illustrate a range of possibilities for the incorporation and expression of the oncogene in the transforming virus genome.

Most transforming retroviruses are defective, because the oncogene replaces part of the retrovirus genome. However, some isolates of Rous sarcoma virus (RSV) are nondefective. In this case, the v-*src* gene is present in the 3' region of the genome and is translated from an independent, spliced mRNA. Such a nondefective RSV is illustrated in the figure. Also illustrated is an example of avian myeloblastosis virus, which expresses the v-*myb* gene. This virus is defective because the v-*myb* gene replaces most of the *env* gene of the virus. It is translated as a v-myb-env fusion protein from the spliced mRNA that would express the Env protein in ALV.

The Abelson murine leukemia virus expresses v-*abl*. This gene replaces the *pro* and *pol* genes and part of *gag* in the example shown, and the virus is defective. In this case, Abl is produced as a fusion protein linked to the N-terminal domain of Gag. Moloney murine sarcoma virus expresses v-*mos*. In the example shown, v-*mos* replaces *env* and is translated from a spliced mRNA normally used to express Env.

RSV, named for its discoverer Peyton Rous, was one of the earliest such retroviruses discovered. RSV causes sarcomas in chickens and transforms cells in culture. The transforming ability is due to the expression by the virus of the oncogene referred to as *src*. Many isolates of RSV have been made over the years, and by definition all such viruses express *src*. However, the v-*src* expressed by the different isolates differ because the recombination points are different, and the v-*src* genes usually have mutations that distinguish them from the cellular gene. Most RSVs are defective because the recombination event results in the replacement of parts of the viral genome by the cellular oncogene, although nondefective RSVs have also been isolated as shown in Fig. 6.13. In the case of defective RSVs, or of other defective oncogene-containing retroviruses, a replication competent ALV, or its equivalent for other retroviruses, is required to supply the missing functions if progeny virions containing the oncogene are to be produced. However, a defective oncogene-containing virion, once formed, is capable of infecting a cell, making cDNA, and integrating the proviral DNA into the host genome without help. Subsequent expression of the oncogene under the control of the viral LTRs leads to transformation of the infected cell.

Since the discovery of RSV, many different oncogene-expressing retroviruses have been isolated, primarily from chickens, mice, and cats. The identification and study of oncogenes has resulted in the discovery of many proteins

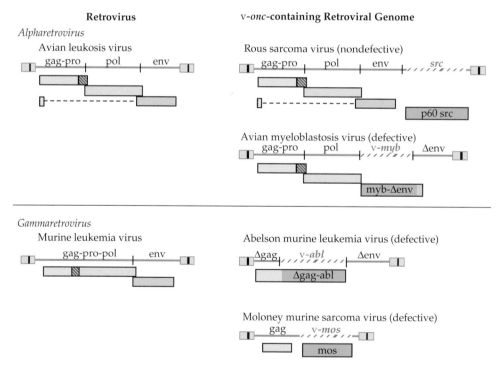

FIGURE 6.13   Representative *v-onc*-containing retroviral genomes and the nondefective retroviruses from which they were derived by capture of cellular oncogenes. In each case the inserted oncogene is shown as a patterned domain in the DNA and its product as a green box. For Rous sarcoma virus the *src* gene is expressed from a new spliced message. For Abelson murine leukemia virus the *abl* gene is present in a fusion protein with a deleted form of *gag*. *Myb* and *mos* are expressed from spliced Env messages, from which most or all of the *env* gene has been replaced by the oncogene. Adapted from Coffin *et al.* (1997) pp. 5 and 504 and from Fields *et al.* (1996), p. 1776.

that are crucial for the regulation of cellular pathways. A sampling of oncogenes that have been found in retroviruses is given in Table 6.5 to illustrate the range of oncogenes that have been captured by retroviruses. Oncogenes encode protein kinases, receptors that respond to growth factors, transcription factors, and other proteins that function at critical points in the cell regulatory cycle. The regulation of cell growth is complex. In broad outline, cells express receptors that bind growth factors. On binding its ligand, the receptor signals the cell to synthesize DNA and divide. This signal is often transmitted by means of a phosphorylation cascade by which transcription factors are activated. The activated transcription factors cause mRNAs to be produced that result in the synthesis of proteins that, in turn, induce DNA synthesis and cell division. These activities are regulated in part by interactions with regulatory proteins called anti-oncogenes, some of which are described in Chapter 7. Thus, oncogenes all have in common that overexpression of the protein, or expression of an altered form of the protein that no longer responds appropriately to regulatory signals, leads to unregulated growth. The regulatory signal could be, for example, the ligand to which a growth receptor normally responds, or domains that interact with inhibitory elements or that normally require phosphorylation for activity, or any one of many other mechanisms that regulate cell growth.

For some oncogenes, overexpression by the strong retrovirus promoter is sufficient for cell transformation. The large amount of oncoprotein overwhelms the cells' regulatory abilities, or the protein is expressed in tissues that normally do not express it and therefore lack the regulatory machinery to control it. For most oncogenes, however, overexpression is not sufficient and the viral oncogene, referred to as v-*onc*, differs from its cellular counterpart, referred to as c-*onc*. The changes in v-*onc* render the oncoprotein nonresponsive to regulatory controls so that the cell is always turned on. One mechanism for the loss of regulatory control is the loss of domains that respond to such control signals. A second mechanism is point mutations that arise during the replication of the recombinant retrovirus. Mutations occur at high frequency during the replication of retroviruses but at a very low frequency during the replication of cells.

The dozens of oncogenes found in retroviruses are too numerous to describe here, but a few examples are cited to illustrate specific points. More than one-third of erythroblastosis in chickens induced by oncogene-containing ALVs arises from viruses that have captured the *erbB* gene, and the viral *erbB* gene has changes in the C-terminal part of the protein (these viruses are called avian erythroblastosis viruses). It was noted before that ALV-induced erythroblastosis often results from the provirus integration into the *erbB* gene. Thus, this gene is very important for the induction of erythroblastosis. As a second example, Ras proteins are G proteins

TABLE 6.5 Selected Retroviral Oncogenes

Oncogene/ functional class	Retrovirus	Viral onco- protein[a]
Growth factors		
sis	Simian sarcoma virus	p28[env-sis]
Hormone receptor (thyroid hormone receptor)		
*erb*A	Avian erythroblastosis virus[b]	p75[gag-erbA]
Tyrosine kinase growth factor receptors		
*erb*B	Avian erythroblastosis virus[b]	gp65[erbB]
sea	S13 Avian erythroblastosis virus	gp160[env-sea]
fms	McDonough feline sarcoma virus	gp180[gag-fms]
kit	Harvey-Zuckerman-4 feline sarcoma virus	gp80[gag-kit]
ros	UR2 avian sarcoma virus	p68[gag-ros]
mpl	Mouse myeloproliferative leukemia virus	p68[gag-mpl]
eyk	Avian retrovirus RPL30	gp37[eyk]
Nonreceptor tyrosine kinases/ signal transduction factors		
src	Rous sarcoma virus	pp60[src]
abl	Abelson murine leukemia virus	p460[gag-abl]
fps[c]	Fujinami avian sarcoma virus	p130[gag-fps]
fes[c]	Snyder-Theilen feline sarcoma virus	p85[gag-fes]
fgr	Gardner-Rasheed feline sarcoma virus	p770[gag-actin-]
yes	Y73 avian sarcoma virus	p90[gag-yes]
Serine-threonine kinases/signal transduction factors		
mos	Moloney murine sarcoma virus	p37[env-mos]
raf[d]	3611 murine sarcoma virus	p75[gag-raf]
mil[d]	MH2 avian myelocytoma virus	p100[gag-mil]
G Proteins (GTPases)		
H-*ras*	Harvey murine sarcoma virus	p21[ras]
K-*ras*	Kirsten murine sarcoma virus	p21[ras]
Transcription factors		
jun	Avian sarcoma virus 17	p65[gag-jun]
fos	Finkel-Biskis-Jenkins murine sarcoma virus	p55[fos]
myc	OK10 avian leukemia virus	p200[gag-pol-myc]
myb	Avian myeloblastosis virus[b]	p45[myb]
ets	Avian myeloblastosis virus[b]	p135[gag-myb-ets]
rel	Avian reticuloendotheliosis virus	p64[rel]
maf	Avian retrovirus AS42	p100[gag-maf]

[a] In these names, p is for protein, gp is for glycoprotein, pp is for phosphoprotein, and the numbers are the approximate apparent molecular weight in kD.
[b] This is a retrovirus with two oncogenes.
[c] *fps* and *fes* are the same oncogene derived from the avian and feline genomes, respectively.
[d] *raf* and *mil* are the same oncogene derived from the murine and avian genomes, respectively.
Source: Fields *et al.* (1996) p. 309.

that are important in signal transduction. Three different *ras* genes have been found in retroviruses and all have changes in codon 12, illustrating the importance of this amino acid for the control pathways that regulate its activity. Mutations in this same codon have also been found in tumors that were not produced by retroviruses.

Two examples of transcription factor oncogenes will be cited. The c-*myc* gene product is tightly regulated and present in low amounts in normal cells. It is active on forming heterodimers with a second factor, present in larger amounts. Overexpression of v-*myc* seems to overwhelm the regulatory pathways and result in high-level expression of genes required for cell proliferation. Even in this case, however, mutations in v-*myc* upregulate its transforming ability. Another transcription factor, the product of *erbA*, is a receptor for thyroid hormone. On binding the hormone, it becomes active as a transcription factor. v-*erbA* expresses a modified protein that no longer needs to bind the hormone in order to be active as a transcription factor, and thus is always on.

Often the expression of a single oncogene, whether modified or not, is not sufficient to achieve full tumorigenic potential. Some transforming retroviruses have been found to express two oncogenes, both of which are required for tumor induction. Furthermore, many oncogene-expressing retroviruses induce tumors only after a lag and in some cases the tumors are clonal. This suggests that for these viruses, the expression of v-*onc* is not in itself sufficient, and additional changes in the infected cell must occur before it becomes fully tumorigenic.

The production of an oncogene-expressing retrovirus that is capable of inducing a tumor is a rare event, but the presence of the virus is signaled by the tumor itself. Most such viruses have been isolated from chickens, mice, or cats because these animals are commonly infected by C-type retroviruses, but also because the occurrence of tumors is more likely to be detected in these animals than in many other animals. Chickens are processed for human food in very large numbers, and tumors, although rare, are often noticed during processing. Mice are used in large numbers in laboratory experiments and receive attentive care from investigators. Pet cats are regularly seen by veterinarians when they become ill.

The study of oncogenes that occur in transforming retroviruses has produced a wealth of information about cell regulatory pathways by which cell growth and the interactions of cells with one another and with various growth hormones are controlled. Many of these pathways were first discovered because of the presence of critical components in retroviruses. Furthermore, some of the oncogenes found in transforming retroviruses were later found to be overexpressed or to be expressed in a mutant form in human cancers, confirming the importance of these products in the production of human tumors. However, although of great importance for our understanding of the regulation of eukaryotic cells, and of importance for the animal in which the tumor produced by the oncogenic retrovirus arose, the transforming retroviruses have little significance for the persistence in nature of their retroviral parent. The production of a tumor shortens the life span of the host, and virtually all transforming retroviruses are defective and require a helper virus for multiplication.

Thus, the probability of transmission of the virus to a new host is low. The appearance of a recombinant retrovirus expressing an oncogene is an accident that soon disappears from the population, along with the unfortunate host.

Betaretroviruses

The best studied member of the betaretroviruses is mouse mammary tumor virus (MMTV). The promoters in the LTR respond to estrogen stimulation, consistent with the growth of the virus in mammary epithelium. MMTV occurs as an endogenous virus, present in 0–4 copies in different strains of mice. It also occurs as an exogenous virus that infects mice by vertical transfer, usually from mother to daughter through the milk. After a latent period, the virus induces mammary tumors with a significant frequency.

MMTV encodes a superantigen (the *sag* gene in Table 6.3). As described in Chapter 10, T cells express a receptor that, when stimulated by binding to a peptide antigen recognized by it, causes the cell to become active and divide. Activated T cells induce B cells to proliferate. Superantigens bind to T-cell receptors and to molecules of the major histocompatibility complex. Binding is not to the domain of the receptor that interacts with peptide antigen, but to a region shared by all T-cell receptors of a given class. A peptide antigen presented by major histocompatibility complex molecules may lead to activation of 10^{-5} of all T cells. However, the Sag protein binds to receptors on as many as 10% of all T cells, leading to their activation. This, in turn, leads to proliferation of B cells, which are the first targets of the virus on infection. In this way, the pool of susceptible cells is increased in size and the infection is amplified. Infected B cells spread the infection to the mammary glands, where mammary epithelium is infected. The virus induces proliferative changes in mammary cells, which eventually lead to tumors.

Mason–Pfizer monkey virus and the so-called simian AIDS virus, or SAIDS virus, are also betaretroviruses. The SAIDS virus causes immunodeficiencies in monkeys but is distinct from simian immunodeficiency virus (SIV). SIV is closely related to HIV, described later.

PTLV Viruses

The two human T-cell leukemia (or lymphotropic) viruses are now usually called primate T-cell lymphotropic viruses (PTLV) (Table 6.1). PTLV-1 causes adult T-cell leukemia (ATL) in humans as well as a neurological disorder that has been termed HTLV-1–associated myelopathy (HAM) (from the former name for the virus). The distributions of PTLV-1 and PTLV-2 are shown in Fig. 6.14. PTLV-1 is prevalent in southern Japan, where transmission appears to be either through sexual intercourse or via breast milk. More than 1 million people in this region are infected by the virus. The virus is also found in other parts of Asia, in the Caribbean

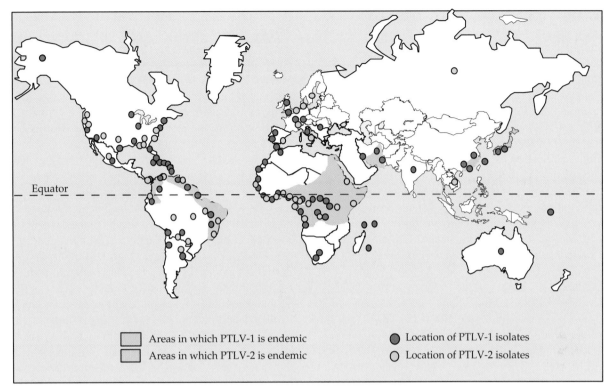

FIGURE 6.14　Global distribution of PTLV-1 and PTLV-2. Shading indicates the areas of endemism, while the dots
are locations of isolates. For PTLV-1 most of the dots refer to subgroup A isolates, but the B subgroup is found in Zaire,
the D subgroup is found in pygmies in Central Africa, and subgroup C is found in Australia and the islands of Melanesia.
Redrawn from Coffin *et al.* (1997) Figure 21 on p. 544.

region, and in central Africa, as well as in the United States
and Western Europe, where it is increasing due to injecting
drug use. Blood transfusions have also been implicated in
the spread of the virus.

ATL has a long latent period, 20–30 years, and only about
1% of infections result in leukemia. Tax is thought to have
a role in induction of leukemia, but the long latent period
implies that other events, probably mutations in the affected
cells, must occur before ATL appears. Patients with acute
ATL live only about 6 months on average.

HAM is a serious neurological disease. It usually develops more rapidly than ATL, and about 1% of PTLV-infected
humans develop the disease. Demyelination in the spinal
cord occurs, as well as a vigorous inflammatory response
involving T cells and other lymphocytes. It is not known
whether the pathology is due to damage to virus-infected
cells by CD8+ T cells (see Chapter 10) or whether the release
of cytokines and chemokines by the inflammatory cells is
responsible for the damage.

The retroviruses are best known for their ability to cause
tumors, including leukemia, or immunodeficiency, but it is
important to note that many retroviruses are associated with
neurological diseases. Some of these diseases are characterized by an inflammatory response, as is HAM, whereas in
others there is no inflammation associated with the disease.
As examples, many strains of MLV cause a spongiform

encephalopathy in mice, in which inflammation is lacking,
and many lentiviruses commonly cause neurological disease,
as will be described later.

PTLV-2 is closely related to PTLV-1. It has been implicated as the causative agent of hairy-cell leukemia, but the
number of cases is small and it is not certain that the virus is
in fact responsible for the disease.

The PTLVs infect not only humans but also a number of
nonhuman primates. In any area, the strains of viruses isolated from infected humans are more closely related to viruses
isolated from monkeys in the region than they are to human
strains isolated from another area. Thus, the viruses seem to
be passed back and forth between monkeys and people.

A virus of cows called bovine leukemia virus (BLV) also
belongs to this genus of complex retroviruses. This virus causes
B-cell lymphomas in a small fraction of infected animals.

Lentiviruses

The lentiviruses are complex retroviruses whose core is
cone shaped in the mature virion. The most important are
the human immunodeficiency viruses, HIV-1 and -2. Other
members of the genus include a number of simian immunodeficiency viruses (SIV), feline immunodeficiency virus
(FIV), bovine immunodeficiency virus (BIV), visna virus
(which causes disease in sheep), equine infectious anemia

virus (EIAV), and caprine arthritis-encephalitis virus (CAEV) (Table 6.1). As described earlier, the regulation of lentivirus replication is complex because lentiviruses encode three to six accessory genes, depending on the virus, to regulate the replication cycle and aid in the assembly of progeny viruses.

All lentiviruses have a specific tropism for macrophages, which comprise the major reservoir of infected cells in an animal. Other cells can also be infected, and this can be important in the disease process (e.g., CD4+ T cells in HIV infection). Lentiviruses establish a lifelong, chronic infection which elicits a vigorous immune response that is unable to clear the infection. Most known lentiviruses cause serious disease in their native host after a long latent period. However, SIV produces an asymptomatic, lifelong infection of its natural host, African monkeys. Intriguingly, SIV causes AIDS when transmitted to Asian monkeys.

Human Immunodeficiency Virus

The two human lentiviruses, HIV-1 and HIV-2, cause the well-known disease called AIDS (*a*cquired *i*mmuno*de*ficiency *s*yndrome). Although both HIV-1 and HIV-2 cause AIDS in people, HIV-2 is not as serious a pathogen. The latent period before disease develops is longer for HIV-2, the virus is not as easily transmissible, and it has not spread as extensively as has HIV-1. It is HIV-1 that is responsible for the vast majority of AIDS in people, and HIV-1 has correspondingly been much more exhaustively studied.

The primary cellular targets of HIV in infected humans are macrophages and CD4+ T cells. These cells express CD4, the primary receptor for all primate lentiviruses, as well as chemokine coreceptors that are also required for entry of the virus (Chapter 1). As described in Chapter 1, strains of HIV are known that preferentially infect macrophages (M-tropic virus) or that preferentially infect T cells (T-tropic), which is a function of the *env* gene of the virus. The infection of macrophages and of CD4+ T cells differs in several important respects. HIV can replicate in fully differentiated, non-dividing macrophages, but infection of macrophages does not lead to extensive cytopathology. Efficient replication of the virus in CD4+ T cells, in contrast, requires that the T cell be activated (Chapter 10) and results in the death of the cell. Most infections of CD4+ T cells do result in the death of the cell, but the virus is also able to establish a latent infection in resting memory T cells, which has important implications for antiviral therapy as described later.

A representative time course of the infection of a human by HIV-1 is shown in Fig. 6.15. Primary infection may be asymptomatic (most infected people do not see a doctor on primary infection) or accompanied by nonspecific symptoms that appear 3–6 weeks after infection. Symptoms may include rash, fever, diarrhea, arthralgia, myalgia, nausea, sore throat, lethargy, headache, or stiff neck. The number of CD4+ T cells declines during this phase. An immune response is mounted that includes both CD8+ cytotoxic T cells (CTLs) and antibody production (Chapter 10). This

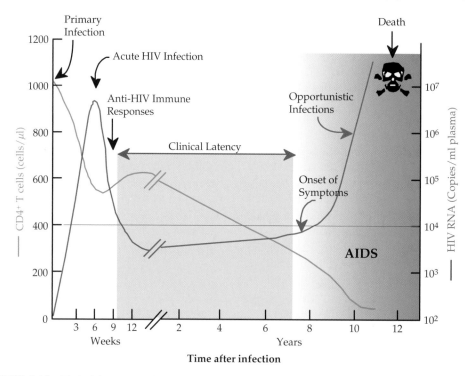

FIGURE 6.15 Typical time course of HIV infection and progression to disease. Patterns of CD4+ T-cell decline and virus load vary greatly among patients. See also Fig. 6.16 for mortality as a function of initial viral load. Some case definitions for AIDS specify a CD4+ count below 400 cells/μl (horizontal line). Adapted from Coffin *et al.* (1997) Figure 5 on p. 600.

reduces the amount of virus in the blood, the number of CD4+ T cells recovers but does not reach preinfection levels, and symptoms of disease ameliorate. However, clearance of the virus is not complete, and a long period of clinical latency ensues. During this period, even though the infected individual is often asymptomatic, the virus continues to replicate actively, especially in lymph nodes, and the immune system continues to fight the infection. It has been estimated that during this clinically quiescent period 10^8 to 10^{10} virus particles per day are released into the peripheral blood supply and cleared. During this time the turnover of CD4+ lymphocytes is also 10^8 to 10^9 cells per day.

In untreated individuals, the continuing replication of virus results in a steady decline in the number of CD4+ T cells (Fig. 6.15) and the continuing appearance of mutations in the virus. In the Env protein, the rate of change in the amino acid sequence averages about 2.5% per year, with a large fraction of changes occurring in certain hypervariable regions. It seems clear that this accumulation of mutations is driven, at least in part, by immune selection.

AIDS, the Disease

With time the immune system is overwhelmed, with the result that the replication of HIV escalates, opportunistic fungal, protozoal, bacterial, and viral infections occur, characteristic cancers appear, and neurological symptoms become apparent. The time to progression to AIDS following infection by HIV is variable but averages about 10 years. Symptoms occur much sooner in infants infected at birth, perhaps because the immune system is not as well developed. In adults infected by the virus, the time to appearance of the symptoms of AIDS is correlated with the level of viral replication during the latent period, as illustrated in Fig. 6.16.

In individuals in which viral replication is high, the CD4+ T-cell count declines faster and the time to appearance of AIDS is shorter. The rate of virus replication is probably a function of the strength of the immune response against the virus, especially the CD8+ CTL response. Individuals with a better immune response control virus replication better and longer. Once the symptoms of AIDS appear, the time to death is usually 1–2 years unless antiviral treatment is administered.

CD4+ T cells, most of which function as T helper cells, are a critical part of the immune system. They are necessary for an immune response to an antigen, whether the response is to produce CD8+ CTLs or to produce circulating antibodies (see Chapter 10). AIDS results when so few CD4+ T cells are present that the body is unable to mount an effective immune response. A partial list of symptoms that appear and opportunistic infections that occur is shown in Table 6.6. The first symptoms usually occur when the CD4+ T-cell concentration falls below 500/μl of blood and may include reactivation of viruses such as herpes zoster, reactivation of bacteria such as *Mycobacteria tuberculosis*, oral lesions caused by fungi, or lymph node pathology. These first symptoms are sometimes referred to as *A*IDS-*r*elated *c*omplex or ARC. Full-blown AIDS is normally signaled by a drop in the CD4+ T-cell concentration below 200/μl. At this point the infected individual becomes susceptible to numerous opportunistic infections, including those of protozoa such as *Pneumocystis carinii*, bacteria such as *Mycobacterium tuberculosis*, and fungi such as *Candida albicans*. Viruses that are normally controlled by the immune system become a problem, such as cytomegalovirus and herpes simplex virus. Several virus-induced cancers become common, such as lymphoma caused by Epstein–Barr virus, Kaposi's sarcoma caused by human herpesvirus 8, or anogenital carcinoma caused by human papilloma

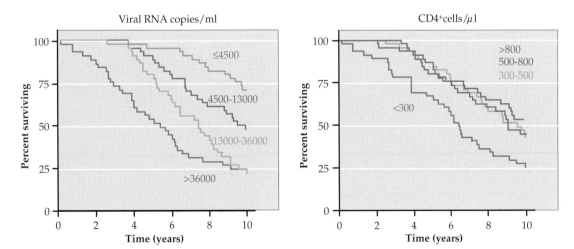

FIGURE 6.16 Relationship between the extent of HIV replication or the CD4+ cell count (two different measures of HIV/AIDS) and clinical progression of disease and time to death. Kaplan–Meier curves are shown for survival versus viral load (left) and concentration of CD4+ T cells (right) at time of diagnosis, plotted by quartiles. Redrawn from Mellors *et al.* (1996).

TABLE 6.6 Pathology and Opportunistic Infections of Patients with HIV/AIDS

Stage of HIV/AIDS	Syndrome/symptoms	Type of infectious agent	Organism
CD4$^+$ T-Cell count 200–500 cells/μl			
	Generalized lymphadenopathy		
	Headache, fever, malaise		
	Generalized weight loss		
	Shingles	Viral	Reactivation of herpes zoster
	Skin lesions	Viral	*Molluscum contagiosum*
	Basal cell carcinomas of skin		
	Oral lesions		
	Thrush	Fungal	*Candida albicans*
	Hairy leukoplakia	Viral	Epstein–Barr virus
	Lung disease (tuberculosis)	Bacterial	Reactivation of *Mycobacterium tuberculosis*
CD4$^+$ T-cell count <200 cells/μl			
	(Syndromes in addition to those mentioned above)		
Microbial Infections	Pneumonia	Protozoan	*Pneumocystis carinii*
	Disseminated toxoplasma	Protozoan	*Toxoplasma gondii*
	Severe diarrhea	Protozoan	*Isospora belli*
	Chronic diarrhea	Protozoan	*Cryptosporidia*
	Tuberculosis	Bacteria	*Mycobacterium tuberculosis*
	Bacterial infections	Bacteria	*Salmonella, Streptococcus, Hemophilus, Cryptococcus*
	CNS disease	Fungal	*Cryptococcus neoformans*
Viral Infections and Malignancies	PML	Viral	JC polyomavirus
	Disseminated disease of lungs, brain, etc.	Viral	Genital herpes, cytomegalovirus
	B-cell lymphoma	Viral	Epstein–Barr virus, HHV-8
	Kaposi's sarcoma	Viral	HHV-8
	Anogenital carcinoma	Viral	Human papilloma virus
	Cervical carcinoma (in women)	Viral	Human papilloma virus
Other Syndromes	Wasting disease	??	??
	Aseptic meningitis		
	AIDS dementia complex	Viral	HIV encephalopathy

Abbreviations: CNS, central nervous system; PML, progressive multifocal leukoencephalopathy; HHV-8, human herpesvirus eight.
Source: Adapted from Coffin *et al.* (1997) Table 1, p. 597.

virus. Many of these opportunistic diseases, such as *P. carinii* pneumonia or Kaposi's sarcoma, are rarely seen in non-HIV-infected people. Others are regularly seen in the general population but HIV-infected individuals have a much higher incidence of the disease, 100-fold higher in the case of lymphoma, for example. Other symptoms of AIDS include a wasting syndrome and neurological abnormalities, described later. The symptoms of AIDS become progressively worse as the CD4$^+$ count continues to drop until virtually no CD4$^+$ cells are present. The decline in numbers of CD4$^+$ cells is accelerated by the increased replication of HIV itself during this terminal phase, when the immune system can no longer control viral infection (Fig. 6.15). Early loss of *g*ut-*a*ssociated *l*ymphoid *t*issue (GALT) is an effect of HIV infection and a harbinger of poor prognosis.

The lymph nodes are organs that trap invading pathogens and present them to immune cells. Large numbers of macrophages and CD4$^+$ T cells are present in lymph nodes, and the nodes are sites of active HIV replication. During the progression of disease following HIV infection, the lymph nodes deteriorate. During the final stages of disease, the architecture of the nodes is completely destroyed and this loss of lymph node function contributes to the loss of immune function.

A wasting syndrome is characteristic of late-stage HIV infection. Bowel involvement results in diarrhea and malabsorption. The wasting syndrome is thought to be a symptom of HIV infection itself, but opportunistic infections of the gut probably exacerbate the symptoms in many individuals.

Most HIV-infected individuals suffer, sooner or later, from neurological disease. Some disease is caused by opportunistic pathogens, such as progressive multifocal leukoencephalopathy caused by JC virus (Chapter 7). However, two-thirds of infected individuals develop an encephalopathy induced by HIV infection itself that produces symptoms including dementia, motor and behavioral abnormalities, and seizures. These symptoms are referred to as AIDS dementia complex. HIV is present in the brains of infected individuals in macrophages and microglia, but does not infect neurons. The cause of the neuronal loss induced by HIV infection is unknown.

Spread of AIDS

HIV is spread sexually, through contaminated blood, and from mother to child. The virus is present in semen, both as free virus and in infected cells, and in vaginal secretions of infected people, and can be transmitted by either homosexual or heterosexual intercourse. The probability of a woman becoming infected during unprotected vaginal intercourse with an infected male is estimated at less than 1/50. The risk of a man becoming infected during heterosexual intercourse is less. The risk is much higher if genital lesions resulting from sexually transmitted diseases are present and in the absence of circumcision. The risk of infection is also much higher during receptive anal intercourse. The use of condoms reduces the risk of transmission by a large factor.

The virus can also be spread by means of contaminated blood, whether through blood transfusions, the use of contaminated products by hemophiliacs, by needle stick in health care workers, the sharing of needles by injecting drug users, or via tattoo needles. Blood transfusion was an important source of infection before the development of tests for the virus and remains a problem in developing countries where blood tests may not be regularly available. Similarly, many hemophiliacs became infected before the development of blood tests. Use of blood tests and screening of donors for risk factors have greatly reduced the risk of transmission of HIV through blood products in developed countries. However, transmission among drug users remains an important source of infection.

Untreated, infected women transmit the virus to a newborn child about one-third of the time. Transmission can occur both during delivery and during breast-feeding. The use of anti-HIV drugs has reduced transmission to infants in countries where the drugs are available.

At the end of 2005, an estimated 40 million people were living with HIV/AIDS, with about 1 million of those in North America (Table 6.7). The cumulative death toll from AIDS since the beginning of the epidemic now exceeds 25 million. Five million new infections occurred worldwide in 2005, and 3 million people died of AIDS that year. In the absence of treatment, virtually all infected people go on to develop AIDS and die within 2 years of the appearance of the symptoms of AIDS.

The focus of HIV infection is sub-Saharan Africa, where 25 million people, or more than 7% of the population, are thought to be infected and where medical care is still primitive. In the course of the epidemic from 1984 to 2005, the number of HIV infections increased dramatically (Fig. 6.17; see also Table 6.7). The extent of the problem is illustrated by the fact that in some large cities in sub-Saharan Africa, an estimated 20–40% of adults may be infected. The spread of HIV has resulted in a marked decrease in life expectancy, which has reversed a long period of increasing life expectancy as health care standards have improved (Fig. 6.18).

In other areas of the world, HIV continues to spread but has not reached the extreme levels found in parts of Africa (Table 6.7). In the Americas and in Southern and Southeast Asia, about 0.6% of the population is infected. Figure 6.19 shows a breakdown of the incidence of AIDS by state in the United States in 2003. Fig. 6.20 shows the explosive increase in AIDS prevalence in Asia between 1991 and 2001. In Western Europe and Australia, estimates are that 0.3% of the population is infected. Table 6.7 documents the spread of the virus between 1996 and 2005.

In North America, HIV has been spread primarily through homosexual contacts and by sharing of needles by drug users who inject drugs, but the frequency of heterosexual transmission is increasing. In developing countries, heterosexual contact is the primary mode of transmission. The result is that in Africa, the male/female infection ratio is approximately 1 to 1, whereas in the United States the male/female ratio is almost 4 to 1 because of the importance of male homosexual contacts in the spread of the virus in this country (Table 6.7). In the United States, the number of newly diagnosed cases of AIDS, especially in white homosexual men, has dropped because awareness of the disease has led to the more frequent use of condoms and a lowered incidence of multiple sex partners, and because more effective drug therapies have been introduced (Fig. 6.21). This trend may be changing for the worse, however, as the development of more effective drugs to treat the disease appears to have resulted in an increase in unprotected sexual activity. Even as the death rate has declined, the total number of persons living with HIV/AIDS continues relentlessly upwards.

The capability of HIV to spread rapidly is illustrated by its recent spread in Thailand. In Thailand, every male is subjected to compulsory military service and is tested for HIV infection at the age of 18 when called up for military

TABLE 6.7 Characteristics of the Global HIV/AIDS Epidemic

Geographical region	Number of people with HIV/AIDS			Deaths from HIV/AIDS			HIV/AIDS in 2005			Comments 2005
	1996	2005	% Change 1996–2005	1996	2005	% Change 1996–2005	% Women	Adult prevalence rate (%)	Primary mode of transmission	
North America	750,000	1.2 million	60	61,300	18,000	−71	26	0.7	1. MSM 2. IDU and hetero	New diagnoses predominantly in African Americans
Caribbean	270,000	300,000	11	14,500	24,000	65	53	2.3	Hetero	Haiti has highest prevalence at 5.6%
Latin America	1.3 million	1.8 million	38	70,900	66,000	−7	27	0.6	1. MSM 2. IDU and hetero	Brazil accounts for >33% of the infected persons in this region
Sub-Saharan Africa	14 million	25.8 million	84	783,700	2.4 million	206	51	7.2	Hetero	Sub-Saharan Africa has just over 10% of the world population and 60% of people living with HIV
North Africa and Middle East	200,000	510,000	155	10,800	58,000	437	37	0.2	1. IDU 2. Hetero	80% of infected persons in this region are in Sudan
W. Europe	510,000	720,000	41	21,000	12,000	−43	28	0.3	1. MSM 2. IDU and hetero	Most infected persons have access to treatment
C. and E. Europe and Central Asia	50,000	1.6 million	3100	1,000	62,000	6100	26	0.9	1. IDU 2. MSM	Most in the Russian Federation, and in persons <30 years old
South and Southeast Asia	5.2 million	7.4 million	42	144,000	480,000	233	30	0.6	Hetero	Epidemic now fueled by new infections in India
East Asia	100,000	870,000	770	1,200	41,000	3320	22	0.1	1. IDU and hetero 2. MSM	Even *low prevalence* translates into large numbers of infected persons, primarily in China
Oceania[a]	13,000	74,000	469	1,000	3600	260	20	0.2	1. MSM 2. IDU and hetero	In New Zealand, 85% of new transmission is in MSM

Abbreviations: IDU, injecting drug use; hetero, heterosexual transmission; MSM, sexual transmission among men who have sex with men.

Percent change was calculated as [(Figure for 2005/Figure for 1996) −1] × 100.

Areas with unusually high prevalence or percent increases are highlighted in red, and reduced numbers are shown in blue.

[a] Includes Papua New Guinea, which has the highest prevalence in the region, and Australia and New Zealand which have two of the lowest.

Source: Data through the end of 2004 is from the website: www.unaids.org/epidemic_update/report/ and updated to 2005 with information from "UNAIDS/WHO AIDS Epidemic Update: December 2005." At the end of 2005 there were 40.3 million people living with HIV/AIDS; during 2005 there were 4.9 million new infections, 3.1 million deaths, 43% of the infected were women, and the overall prevalence rate was 0.63%.

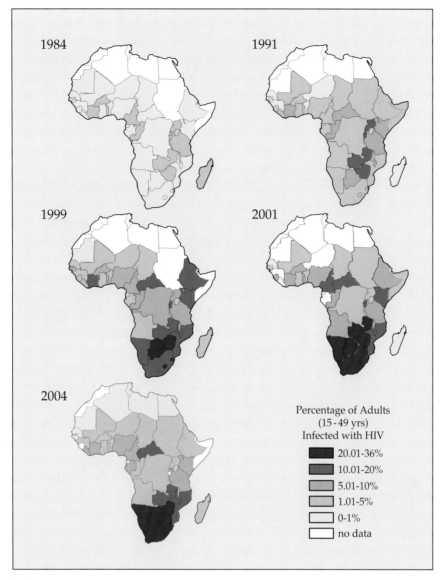

FIGURE 6.17   Explosion of the prevalence of infection with HIV in Africa during the last 20 years. The panels show the percentage of adults (15 to 49 years old) infected with HIV in various countries of Africa in 1984, 1991, 1999, 2001, and in 2004. This map was redrawn from Nations Programmes on HIV/AIDS Conference in Geneva, June, 2000 reported in Schwartlander *et al.* (2000), and updated with information from the *2006 Report on the Global AIDS Epidemic*, from the UNAIDS Programme.

duty. Twenty years ago HIV was almost unknown in Thailand, but within a very few years the disease became prevalent, infecting some 3% of the male population. Thailand is a country that is very tolerant of sexual activity and many young males visit prostitutes. Most of the prostitutes became infected with HIV and began to transfer the infection to many of their male partners, resulting in the observed explosion of infections in the population. In some provinces in northern Thailand, as many as 20% of 18-year-old males were infected with HIV as of 15 years ago. The government began an aggressive education campaign about the disease and how it is acquired and distributed condoms

widely, with the result that the rate of new infections has dropped dramatically (Fig. 6.22).

Prevention and Treatment of AIDS

The most successful method for the prevention of most viral diseases has been the development of vaccines. However, the development of a vaccine against HIV has been a very slow, frustrating, and to date unsuccessful process. This is perhaps not surprising in light of the fact that the virus establishes a persistent lifelong infection in the face of a vigorous immune response on the part of the host that

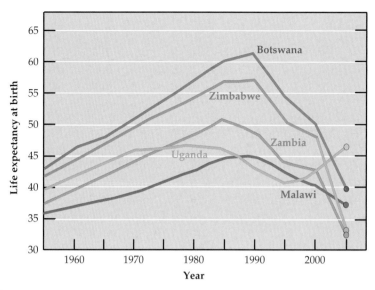

FIGURE 6.18   Changes in life expectancy in selected African countries with high HIV prevalence, 1955 to 2005. Adapted from UNAIDS publications slide 12 of the *AIDS in Africa* series on the Web site www.UNAIDS.org. and updated information from the UN Human Development Report 2004. The points plotted at 2005 are projections.

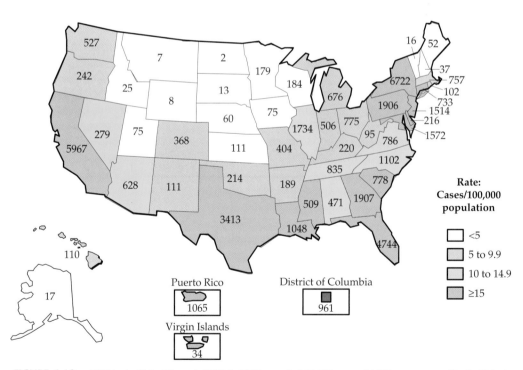

FIGURE 6.19   AIDS in the United States in 2003. In 2003, a total of 44,232 cases of AIDS were reported in the United States. In this map, the rate (cases per 100,000 population) is indicated by the color, and the total number of reported cases is given for each state. The rate for the District of Columbia is so far off scale (169) that a special color is used. Data from *MMWR, Summary of Notifiable Diseases 2003.*

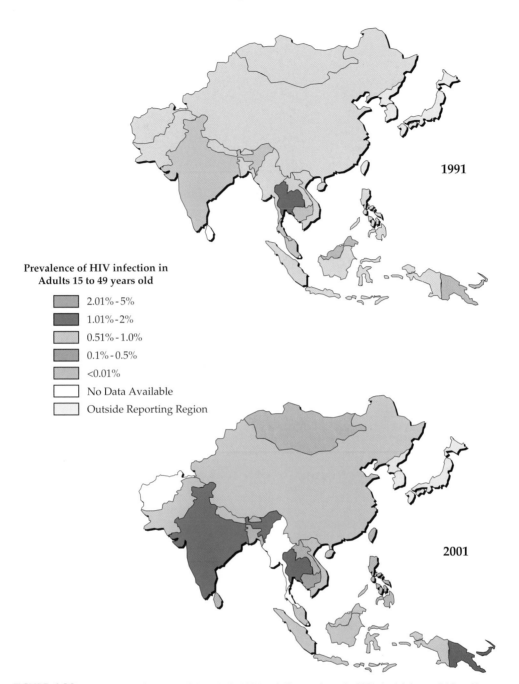

FIGURE 6.20   Prevalence of HIV/AIDS in Asia in 1991 and 10 years later, in 2001, in Adults aged 15 to 49 years. Note that even at a prevalence of less than 2%, due to its enormous population, India now has roughly 6 million persons living with HIV/AIDS. Adapted from *2006 Report on the Global AIDS Epidemic*, from the UNAIDS Programme.

succeeds in controlling the virus for many years. Vaccine development continues to be a priority, but to the current time only public education to slow the spread of the disease and the development of antiviral drugs have had any success in control of the virus.

Much effort has been made to reduce the epidemic spread of HIV by educating the public and thereby altering patterns of behavior, especially sexual behavior. Safe sex practices, such as the use of condoms, have lowered the numbers of new infections, especially for HIV infections among gay white men. Education campaigns in Thailand and Uganda have succeeded in slowing the spread of the disease, and similar campaigns in other areas of the world have met with at least some success. However, although such campaigns

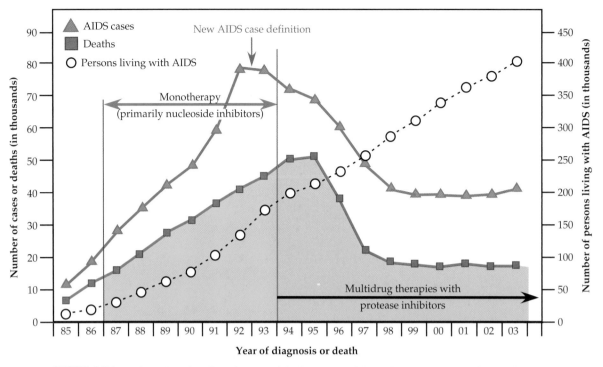

FIGURE 6.21   Estimated number of AIDS cases and deaths per year of diagnosis among persons >13 years old in the United States from 1985 to 2003 and number of Americans living with AIDS. Data from AIDS Surveillance-Trends at: http://www.cdc.gov/hiv/graphics/images/.

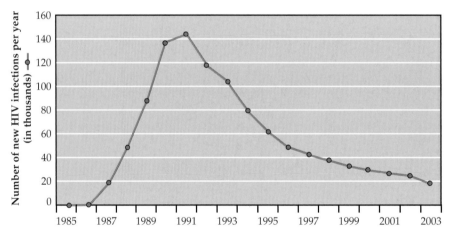

FIGURE 6.22   New HIV infections per year in Thailand from 1985 to 2003. From: *Thailand's Response to HIV/AIDS: Progress and Challenges* (2004) United Nations Development Programme report. Figure 2 on p. 2.

slow the spread of the virus, they do not stop it and will never succeed in eliminating the virus from the population.

The development of drug therapies for control of HIV infection has met with at least partial success. Early drugs for treatment of HIV infection were disappointing at best (see Fig. 6.21). The drug 5′ azidothymidine (AZT) and related nucleoside analogues were first used in monotherapy, and were initially successful in decreasing virus load and allowing partial recovery of immune function. Incorporation of

these analogues into DNA results in premature chain termination, and because the viral RT has a high affinity for these drugs whereas cellular enzymes have a low affinity for them, their primary effect is to disrupt replication of HIV. However, virus mutants that are resistant to the drugs rapidly appear, and control by this agent was found to be of limited duration. Second-generation inhibitors were developed that interfere with the activity of the viral protease, which is required for assembly of infectious virus, or with the viral integrase. The protease inhibitors have been widely used, and variants resistant to them appear much more slowly, probably because multiple changes in the protease are required for resistance. However, resistant virus does eventually emerge when the inhibitor is used in monotherapy.

Third-generation drug therapy, called *h*ighly *a*ctive *a*nti*r*etroviral *t*herapy or HAART, uses a combination of two different nucleoside analogues and a protease inhibitor. This treatment has been quite successful in more than half of HIV patients, in which the viral load declines to undetectable levels with full recovery of immune function. The introduction of combined therapy has led to a marked reduction of the death rate from AIDS in the United States (Fig. 6.21). Figure 6.23 illustrates that by 1995 AIDS had become the leading cause of death in U.S. adults 25–44 years of age. With the advent of HAART, AIDS has fallen to the level of homicide as a killer of young adults.

At first it was hoped that long-term treatment with this regime would result in the elimination of the virus and curing of the infection. However, this does not appear to be possible. Virus production quickly resumes when HAART is discontinued, even in patients that have been in treatment for many years without detectable virus production during that time. Reemergence of the virus may occur because the treatment does not eliminate the virus latently infecting

T cells, although other sources of persistent virus may be present. In any event, it appears that the therapy will have to be continued over the lifetime of the patient.

Although HAART has been successful in many patients, it is clearly not a panacea. First, it is very expensive and suitable for use only in developed countries that can afford a high level of health care. Second, there is significant toxicity that results from the drug treatment, making lifelong treatment problematic. Third, compliance becomes a major issue when any therapy, especially therapy with significant side effects, must be continued indefinitely. It may be for such reasons that the therapy fails in a considerable fraction of HIV patients. Fourth, resistant variants arise, necessitating changes in the regime of drugs being used. However, although this treatment is complex and of limited success, its very existence seemingly has led to a rise in unprotected sexual activity and to an upsurge in the rate of HIV infection among populations at risk.

Attempts to develop a vaccine continue and there is hope that a vaccine may ultimately be produced. It is clear that an immune response that involves both neutralizing antibodies and CD8+ T cells is important in the control of the virus during the clinically latent state. The T-cell response seems to be especially important. Stronger CTL responses result in lower levels of virus production during the latent

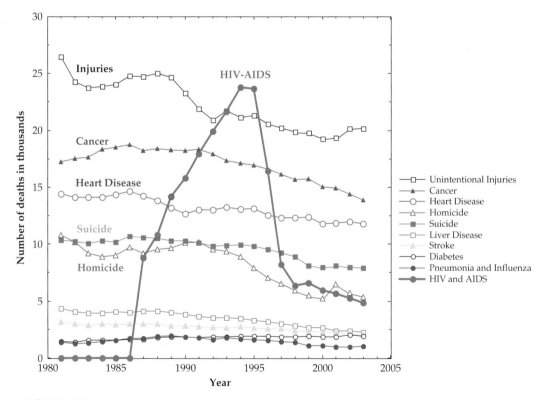

FIGURE 6.23 Annual number of deaths in the United States in adults 25–44 years old for each of the 10 leading causes of death, normalized for population growth. The data shown are the aggregate for all races and both sexes, for the years 1980 to 2003. Data came from CDC, National Center for Injury Prevention and Control and National Vital Statistics System.

state, which results in a slower rate of progression to AIDS. Neutralizing antibodies are also important, but it has been found that fresh isolates of virus from patients are difficult to neutralize, presumably because of the many carbohydrate chains attached to the HIV surface glycoprotein and because of the unusual folding properties of this protein. In addition, variants arise during prolonged infection that are resistant to the mix of antibodies that exist in the patient at the time that the variants arise.

In developing a vaccine, there is the significant question as to whether sterilizing immunity, which is very difficult to achieve, is required, or whether protection from disease can be obtained with nonsterilizing immunity. For virtually all other viruses for which vaccines have been developed, the vaccine is nonsterilizing—subsequent infection with the virulent virus results in a subclinical infection that is quickly damped out. But with a virus that produces a lifelong persistent infection in the face of an immune response, the needs may be different. Recent findings with a very small group of patients have suggested that nonsterilizing immunity may suffice, at least in some cases, but more information is needed before the complete answer is known.

A second, related question is whether is it possible to boost the immune response in people infected with HIV and obtain better control or elimination of the virus. Because of the very long latent period before disease develops, this is an extremely difficult question to approach. Various clinical trials have been conducted to test such possibilities, a topic that is covered in Chapter 11. Clinical trials have also been conducted that test the responses of uninfected people to immunization with the external glycoprotein of HIV. To date, such trials have been disappointing.

The possibility of a live virus vaccine is suggested by the finding that a small fraction of infected people exists who have not progressed to AIDS after many years of infection and whose CD4+ T-cell count has remained fairly stable. In at least some of these individuals, the infecting HIV strain has deletions in *nef*. Initially, experiments with *nef* deletions in SIV in monkeys give comparable results—the monkeys did not develop AIDS and they were protected for some time from infection by wild-type strains of SIV. However, they eventually did succumb to AIDS, illustrating the difficulties in developing a live virus vaccine for a virus that establishes a persistent infection that is ultimately fatal, and for which the symptoms of disease are delayed for many years.

The Origins of HIV

Like measles virus, HIV is a zoonotic disease. HIV-1 derives from SIV that infects chimpanzees and HIV-2 derives from SIV that infects sooty mangabey monkeys. HIV-1 has entered the human population and become established at least three times in the last century, and HIV-2 became established independently. These viruses are now

human viruses and humans are the only reservoir. This topic is covered in more detail in Chapter 8 on Emerging Viral Diseases.

Nonhuman Lentiviruses

SIV, FIV, and BIV cause AIDS or an AIDS-like disease in animals. At least 36 different primate species in sub-Saharan Africa are infected by different SIVs. These viruses cause no disease in their natural hosts but do cause AIDS in Asian monkeys. The disease in Asian monkeys appears very similar to human AIDS, and SIV has been useful in laboratory studies as a surrogate for human AIDS. The primary receptor for SIV is CD4, as is the case for HIV. FIV causes immunodeficiency in cats that appears to result by mechanisms similar to those that produce human AIDS, although the primary receptor for the virus is different. The number of CD4+ T cells declines markedly in the late stages of the disease and the disease is characterized by opportunistic infections as well as by neurological symptoms. BIV infection of cattle results in persistent killing of lymphocytes and lesions in the central nervous system, symptoms that resemble those produced by HIV.

In contrast, equine infectious anemia virus (EIAV) in horses, caprine arthritis-encephalitis virus (CAEV) in goats, and visna virus in sheep produce quite different symptoms. EIAV causes recurrent anemia in horses and is an important veterinary problem. Primary infection by EIAV results in an acute disease that is soon controlled by an immune response. However, variants arise that are resistant to preexisting antibody and cause recurrences of fever and acute anemia. This continues for 6 months to 1 year, after which the infection becomes clinically inapparent. Infection is lifelong, however, as in all lentiviruses. It is thought that the virus binds to, but does not infect, red blood cells. These cells are then destroyed by a complement pathway or by engulfment by macrophages, resulting in the anemia. The virus also appears to suppress the differentiation of precursors to red blood cells. Although anemia is the characteristic disease produced by EIAV, in a small number of horses virus infection results in encephalomyelitis.

CAEV infection of goats produces an inflammatory response that often results in arthritis. The arthritis arises several years after infection and resembles rheumatoid arthritis in humans. Other symptoms produced by infection include encephalomyelitis in 20% of infected animals, which arises within 6 months of infection. Visna virus infection of sheep follows a similar course, but the characteristic disease produced is pneumonia, which arises several years after infection. Encephalomyelitis occurs in less than 5% of infected animals.

Spumaviruses

The spumaviruses comprise a genus of complex retroviruses. There are no known human spumaviruses. A virus previously isolated from human cells in culture, and referred to as human foamy virus or human spumaretrovirus, is now thought to be a virus of chimpanzees or other monkeys. The spumaviruses group with epsilon- and gammaretroviruses (Fig. 6.1), but differ in many properties from other retroviruses. The Gag polyprotein is not cleaved to produce MA, CA, and NC proteins (Fig. 6.7). Spumaviruses appear to bud into the endoplasmic reticulum, rather than from the plasma membrane as do other retroviruses. In addition, Pol is translated from its own mRNA, rather than from the genomic RNA as a fusion protein with Gag (Fig. 6.6). Recent studies have also indicated that the RNA of human spumavirus is converted to DNA during packaging of the virion. It is possible that virions may contain either DNA or RNA, since both are found in virus preparations. The packaging of DNA in the virion rather than RNA, the translation of Pol from its own mRNA, and budding into the endoplasmic reticulum, resemble the events that occur in the pararetroviruses such as the hepadnaviruses, described later. In other respects, the spumaviruses resemble retroviruses rather than pararetroviruses. They use a tRNA primer for reverse transcription, like the retroviruses, and their DNA integrates into the host genome as a provirus. Thus, they are retroviruses that are intermediate in their properties between the classical retroviruses and the pararetroviruses.

Retroelements in the Genomes of Living Organisms

Reverse transcriptase is an ancient enzyme. Its origins probably go back to the RNA world at the time of the invention of DNA, when such an enzyme would have been useful for converting informational RNA molecules into DNA, the form in which genetic information is now stored. It appears to resemble the RNA polymerases of RNA viruses more than DNA polymerases or DNA-dependent RNA polymerases, and may have evolved from a primordial RNA polymerase of the RNA world. The enzyme still exists in modern eukaryotes and plays a role in the replication of eukaryotic cells. Telomerase is a eukaryotic enzyme that repairs the ends of the linear chromosomes, which are progressively shortened during replication. Cells lacking telomerase cease to divide after the ends of the chromosomes become too short to support replication. Telomerase is a ribonucleoprotein enzyme that synthesizes repeat sequence elements that are attached to the end of the chromosomes. These elements are specific for a given species. This process involves reverse transcription of an RNA template that is a component of the telomerase, and the ends of the chromosomes serve as primers. The composition of telomerase has not been completely elucidated, but it does have a reverse transcriptase activity that is related to the RT present in a variety of retroelements.

There has been a long period of time in which elements that utilize reverse transcription for insertion into eukaryotic

and prokaryotic chromosomes have been able to evolve. The number of different kinds of retroelements is consequently very large and encompasses not only the retroviruses described before but also endogenous retroviruses, retrotransposons of various types, retrointrons, retroplasmids, and retrons, among others. A partial listing of these elements is given in Table 6.8, and the phylogenetic relationships among the RTs of those elements that possess RT are shown in the tree in Fig. 6.24.

As a class, retroelements are found in all organisms from prokaryotes to mammals. Many of these elements are capable of amplification and are mobile, able to move within the genome. Some encode RT, whereas others utilize RT of other elements for their propagation. These elements can be thought of as selfish DNA. They replicate to fill up the host genome to what is, in essence, its carrying capacity. These elements must be benign and have only limited effects on the replication potential of the organisms in which they reside. Otherwise their host would be selected against during evolution. The number of retroelements in eukaryotic genomes is often very large. In mammals, 5–10% of the genome appears to consist of endogeneous retroviruses, and all retroelements together comprise about half of the genome. The endogeneous retroviruses are provirus-like, containing LTRs and primer binding sites flanking regions with detectable relationships to *gag* and *pol*. Some of these are active endogenous viruses that have inserted into the

TABLE 6.8   Retroelements and Their Distribution in Nature

Class	Distribution
Eukaryotic retroelements	
Retroviruses (*Retroviridae* Table 6.1)	Vertebrates
Pararetroviruses	
Hepadnaviruses (*Hepadnaviridae*, Table 6.10)	Mammals, birds
Caulimoviruses	Plants
Retrotransposons	
LTR retrotransposons[a]	Animals, plants, fungi, protozoa
Non-LTR Retrotransposons	Animals, plants, fungi, protozoa
Mitochondrial elements	
Group II introns (retrointrons)	Mitochondria of fungi and plants, plastids of algae
Mauriceville plasmid	Mitochondria of *Neurospora*
RTL gene	Mitochondria of *Chlamydomonas*
Prokaryotic retroelements	
msDNA-associated RT[b]	*Myxococcus xanthus, E. coli*, other bacteria

[a] Includes the two families *Pseudoviridae* and *Metaviridae*; see Table 6.9.
[b] msDNA, multicopy single-stranded DNA.
Source: Adapted from Eickbush (1994), Table 1, p. 122.

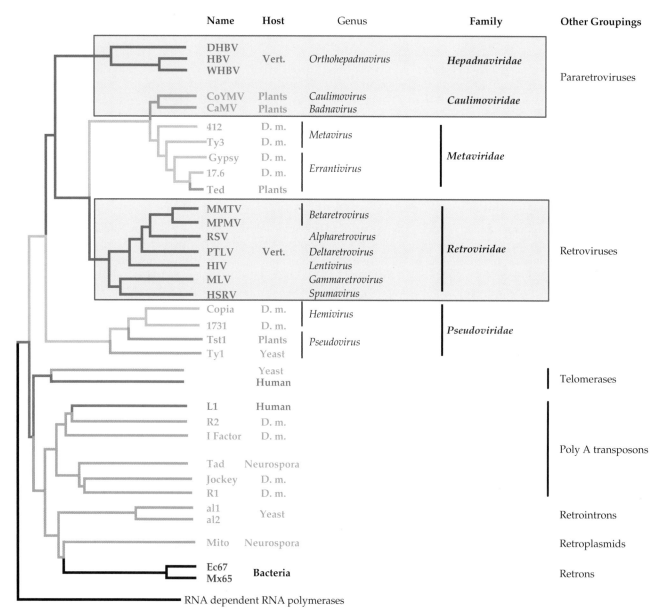

FIGURE 6.24   Phylogenetic tree of the retroelements, based upon the sequences of the reverse transcriptases. The blue boxes enclose the retroviruses and pararetroviruses. Plant hosts are shown in green, insects in brown, fungi in purple, vertebrates in blue, and humans in red. The vertical distances are arbitrary; the length of the horizontal branches is proportional to the number of changes. The tree has been rooted with the RNA-dependent RNA polymerase sequences. Abbreviations of the retroviruses were given in Table 6.1; abbreviations of orthohepadnaviruses are found in Table 6.10. Names of representative *Pseudoviridae* and *Metaviridae* are given in Table 6.9; the Sirevirus and Semotivirus genera, as well as four more genera of *Caulimoviridae*, which are exclusively plant viruses, are not shown. CoYMV is commelina yellow mottle and CaMV is cauliflower mosaic virus; D. m. is *Drosophila melanogaster* (fruit fly). Redrawn from Coffin *et al.* (1997) Figure 17 on p. 411, plus information from Fauquet *et al.* (2005), p. 418.

genome recently, whereas others are inert. The time at which any element entered the germ line can often be estimated from comparative studies of these elements in different animals, and these proviruses or other retroelements constitute a kind of fossil record for these elements. As one example, chickens have one to four endogenous proviruses closely related to the ALVs, but there are no such endogenous viruses in turkey or quail. Thus, insertion of these

ALVs into the chicken genome occurred after chicken and turkey separated.

Next, four classes of eukaryotic elements, the endogenous retroviruses, the LTR-containing retrotransposons, the poly(A)-containing retrotransposons, and the group II retrointrons, are described in more detail. The relationships among these elements and their relationships to the infectious retroviruses described earlier, and to the pararetroviruses described later, are illustrated by the phylogenetic tree in Fig. 6.24. The genetic organizations of these elements are compared in Fig. 6.25.

Endogenous Retroviruses

Endogenous retroviruses are proviruses that have become established in the germ line of many different organisms at various times in the past. They align with the retroviruses in the tree in Fig. 6.24, and their genetic organization is identical to that of the simple retroviruses (see Fig. 6.25). Most are defective, having deletions or mutations in them that prevent them from undergoing a full round of replication. Some are not defective, however, and can, upon activation, give rise to progeny virus that can infect other cells. Most endogenous retroviruses are silent—the genes that they encode are not expressed or are expressed only under restricted conditions,

although in some animals one or more endogenous retroviruses are normally expressed during the lifetime of an animal.

The best studied endogenous retroviruses are those of chickens and of mice. Mice contain more than 1000 endogenous viruses or elements that are thought to have originated from endogenous viruses. Most of these are defective and unable to undergo a complete cycle to give rise to infectious virus, but up to 100 copies of nondefective endogenous retroviruses are present in some strains of mice. As described before, these viruses can be ecotropic, xenotropic, polytropic, or modified polytropic, depending on the receptor recognized by their envelope gene. In most strains of mice, these viruses are not expressed, although in many cases they can be induced to replicate by treatment of the cell with mutagenic agents or agents that lead to demethylation of DNA. In some strains of mice, the endogenous viruses

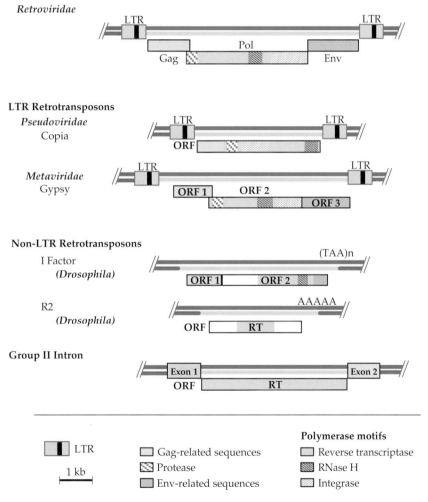

FIGURE 6.25 Comparative genome organizations of retroviruses, retrotransposons, and retrons. In each case the integrated form of the element is shown, flanked with dark blue bars representing the host DNA. ORFs for each element are shown below the DNA, and boxes on different lines represent different reading frames. Note that Gypsy has now been classified as an *Errantivirus* in the family *Metaviridae,* and Copia is a *Hemivirus* in the family *Pseudoviridae.* Various motifs related to retroviruses are indicated with different colors and shadings, as shown in the key. At the 3' end of many non-LTR retrotransposons there are a variable number of TAA repeats (I factor) or adenylate residues (R2). Adapted from Eickbush (1994).

are transcriptionally active and expressed at some stage during the lifetime of the mouse without having to be induced. Early expression and vigorous replication in strains of mice such as AKR leads to the development of leukemia in most of these mice.

An interesting question deals with how the xenotropic viruses, which cannot infect mice, entered the mouse germ line. The simplest explanation would be that these viruses entered the germ line of an ancestor of the modern mouse that did express receptors for the virus. In this model, during the evolution of mice these receptors were silenced or mutated to a form that was no longer usable by the xenotropic viruses. It is possible that modern mice were selected to not express a functional receptor for these viruses, in order to minimize the virus load from endogenous viruses.

LTR-Containing Retrotransposons

The LTR-containing retrotransposons are essentially endogenous retroviruses that lack the envelope gene (Fig. 6.25B). They contain LTRs and encode Gag-related proteins, protease, and RT with its associated activities. RNA transcribed from the elements is translated to produce Gag and Pol, and the RNA is assembled into intracellular capsid-like particles that reverse transcribe it back into DNA. Subsequent integration into the genome allows the elements to spread within the genome. Further, the LTR-containing

retrotransposons are obviously related to retroviruses. The LTRs that they possess are closely related to retroviral LTRs, including the conserved two terminal nucleotides of the provirus. They have tRNA primer binding sites and polypurine tracts, so that reverse transcription of the RNA appears to follow the same pathways as in retroviruses.

Thus, the LTR-containing retrotransposons are effectively intracellular viruses that replicate only within an individual cell. As such they have recently been classified into two families of virus-like agents, the *Pseudoviridae* and the *Metaviridae* (Table 6.9). These agents are ancient, being found in plants, insects, and fungi as well as in animals including vertebrates. The argument has been made that the *Metaviridae* gave rise to the current retroviruses, that they represent an ancestral form of retroviruses in which an RT-containing element became mobile by associating with *gag*-like genes that allowed it to spread within the cell. Later acquisition of an envelope gene in vertebrate *Metaviridae* gave rise to a complete retrovirus. The *Metaviridae* lineage is a sister lineage to the *Retroviridae* as shown in Fig. 6.24, consistent with this hypothesis.

The two families of LTR retrotransposons, which represent two independent evolutionary lines (Fig. 6.24), each contain three genera (Table 6.9). In the *Metaviridae* are Ty3 of yeast and Gypsy of *Drosophila*, among other elements. They are closely related to the pararetroviruses of plants (Fig. 6.24). The *Pseudoviridae* contain Ty1 of yeast and

TABLE 6.9   LTR Retrotransposons

Family/genera/ type virus	Type virus name abbreviation	Transmission	Distribution	Comments
***Pseudoviridae* (LTR retrotransposons of the Ty1-copia family)**				
Pseudovirus		Exclusively vertical	Worldwide	Genus contains several other viruses of plants and fungi
Saccharomyces cerevisiae Ty1 virus	SceTy1V			
Hemivirus		Unknown	Worldwide	Found in insect, fungal, and viral genomes
Drosophila melanogaster copia virus	DmeCopV			
Serivirus		Unknown		This genus found only in plants
Glycine max SIRE1 virus	GmaSIRV			
***Metaviridae* (LTR retrotransposons of the Ty3-gypsy family)**				
Metavirus		Unknown		Genus contains several other viruses of plants, insects, and fungi
Saccharomyces cerevisiae Ty3 virus	SceTy3V			
Errantivirus		Unknown		Most infect insects
Drosophila melanogaster gypsy virus[a]	DmeGypV			
Semotivirus[a]		Unknown		Primarily in nematodes, but also insects and vertebrates
Ascaris lumbricoides Tas virus	AluTasV			

[a] All members of these genera have a env-like ORFs, but the ORFs of errantiviruses have no sequence similarity to those of semotiviruses.
Source: Fauquet *et al.* (2005).

Copia of *Drosophila*. As noted, the Ty3/Gypsy lineage is more closely related to retroviruses than it is to the Copia lineage (Fig. 6.24). There are also elements in mammals that resemble LTR retrotransposons, but these may belong to a distinct lineage that arose more recently.

Poly(A) Retrotransposons

Non-LTR-containing retrotransposons, often called poly(A)-containing retrotransposons because many have a poly(A) tract or an A-rich tract at the 3' end, are a very large family of elements found in virtually all eukaryotes. Many of these elements encode RT, and of these many encode Gag-like proteins. The structures of two elements from *Drosophila melanogaster* are illustrated in Fig. 6.25C. Both encode RT. The I factor element encodes Gag-like proteins, whereas R2 encodes proteins that bind nucleic acid but are not obviously related to Gag. The elements that encode RT use it to move by reverse transcription, but because they lack LTRs, the mechanism of reverse transcription is different.

The human genome contains many non-LTR-containing retrotransposons. About 20% of the genome, in fact, consists of repeated elements called LINEs (*l*ong *i*nterspersed *n*uclear *e*lements, often abbreviated L1). L1 belongs to the same lineage as the *Drosophila* elements R2 and I factor (Fig. 6.24). L1 elements encode Gag-like proteins and RT, but lack LTRs as well as an envelope gene. It is thought that reverse transcription is coincident with integration of L1 into a new location. In this model, the cell chromosomal DNA is nicked by an endonuclease encoded in the retrotransposon, and the nick site is used as a primer for reverse transcription of the L1 RNA.

In addition to the L1-like retrotransposons, a simpler class of poly(A) transposons exists that do not encode RT. This class includes elements in the human genome called SINEs, for *s*hort *i*nterspersed *n*uclear *e*lements. SINEs include the human ALU sequences present in large numbers in the genome. It is thought that these elements borrow the transcription machinery of the L1-type elements during retrotransposition.

Group II Retrointrons

Group II introns are self-splicing introns that encode RT. The structure of such an intron is shown in Fig. 6.25D. These introns are able to move and may have been the source of introns in nuclear genes. They are mostly found in prokaryotes and in organelles. The RT is translated as a fusion protein from unspliced RNA. Thus, the amount of RT produced, which determines the ability of the element to move, is regulated by the efficiency of splicing. The retrointrons group with the retroplasmids and retrons in the RT tree to form a distinct lineage (Fig. 6.24).

Effects of Retroelements on the Host

In many ways, retroelements are not that different from extracellular viruses that have learned to infect organisms and pass from organism to organism. These elements multiply to fill an ecological niche, but are limited to intracellular spread. However, these retroelements, as noted before, must be benign in order to avoid being selected against. The ability of retroelements to cause disease appears to be quite limited, and in many organisms they may cause no disease. In at least a few cases, they may even serve a useful function.

The ability of endogenous viruses to cause leukemia in AKR mice or mammary tumors in some mice seems to contradict this observation. However, these diseases are largely characteristic of inbred laboratory mice. Wild mice control their endogenous viruses much more successfully.

There is no evidence that endogenous viruses ever cause disease in humans. Humans appear to have no endogenous proviruses that are ever expressed, although humans do have defective proviruses that have been present in the germ line for a long time. The absence of endogenous viruses may be a function of the long life span of humans. Long-lived animals have a longer time span to express endogenous viruses, which would lead to selection against animals that contain such viruses.

FAMILY *HEPADNAVIRIDAE*

The hepadnaviruses (*hepa* from **hepa**totropic, dna from their **DNA** genome) share with retroviruses the property of encoding RT and replicating via an RNA-to-DNA step. They package DNA in the virion, however. The process of reverse transcription shares features with that described for the retroviruses but also differs in many important details, as described later. Because of the similarities in their mode of replication to that used by retroviruses, the hepadnaviruses and the plant viruses that replicate via RT are referred to as pararetroviruses. The hepadnaviruses form a distinct taxon in phylogenetic trees, however, and are not particularly closely related to the plant pararetroviruses or to the retroviruses (Fig. 6.24). In fact, the plant pararetroviruses, even though they package DNA in the virion, appear to be more retrovirus-like in their replication than are the hepadnaviruses, consistent with their position in the tree.

The hepadnaviruses consist of two genera (Table 6.10). The genus *Orthohepadnavirus* contains mammalian viruses and the genus *Avihepadnavirus* contains viruses of birds. The mammalian viruses include hepatitis B virus of humans (HBV), woodchuck hepatitis virus (WHV), ground squirrel hepatitis virus (GSHV), and viruses of both Old World and New World primates including woolly monkeys, orangutans, gorillas, gibbons, and chimpanzees. The primate viruses are all closely related and are usually treated as strains of a single

TABLE 6.10   *Hepadnaviridae*[a]

Genus/members	Virus name abbreviation	Natural host(s)	Transmission	Disease
Orthohepadnavirus				
Hepatitis B virus	HBV	Humans, chimpanzees, gibbons,	Horizontal, vertical, IDU, sexual, blood	ACS, hepatitis, cirrhosis, HCC[b]
Ground squirrel hepatitis B	GSHV	Ground squirrels, woodchucks, chipmunks	Horizontal, sexual, blood	ACS, hepatitis, HCC
Woodchuck hepatitis B	WHBV	Woodchucks	Horizontal, sexual, blood	ACS, hepatitis, HCC
Woolly monkey hepatitis B	WMHBV	Woolly monkeys	Horizontal, sexual, blood	Hepatitis
Avihepadnavirus				
Duck hepatitis B virus	DHBV	Ducks, geese	Predominantly vertical	ACS, hepatitis
Heron hepatitis B virus	HHBV	Herons	Predominantly vertical	

[a] Hepatitis B virus has a worldwide distribution in humans, as shown in Fig. 6.29.
[b] Abbreviations: ACS, asymptomatic carrier state; HCC, hepatocellular carcinoma; IDU, injecting drug users.
Source: Adapted from Fields *et al.* (1996) Table 1 on p. 2708, with taxonomy according to Fauquet *et al.* (2005).

virus species with the exception of the woolly monkey virus, which differs by 20% in nucleotide sequence from the other primate viruses and is considered a distinct species. It is reasonable to assume that the human virus, which has diverged into a number of strains, arose from one or more of the non-human primate viruses. The rodent viruses are distinct from the primate viruses and diverge by 40% from HBV.

The avian viruses include duck hepatitis B virus (DHBV), heron hepatitis B virus (HHBV), Ross' Goose hepatitis B virus, snow goose hepatitis B virus, and stork hepatitis B virus. These viruses appear to be widespread and common—it is estimated, for example, that up to 50% of free-living herons in North America may be naturally infected with HHBV. The bird viruses form a distinct lineage. They are closely related to one another but more distantly related to the mammalian viruses.

As their names imply, all of the known hepadnaviruses are hepatotropic, infecting liver cells, and all can cause hepatitis in their native host. All have a very narrow host range that may be determined at least in part by the identity of the receptors used for entry.

The hepadnaviral genome is circular and approximately 3.2 kb in size, as illustrated in Fig. 6.26A. It consists of DNA that is mostly, but not completely, double stranded. One DNA strand, the minus strand, is unit length and has a protein covalently attached to the 5′ end, as described later. The other strand, the plus strand, is variable in length, but less than unit length, and has an RNA oligonucleotide at its 5′ end. Thus neither DNA strand is closed and circularity is maintained by cohesive ends.

Hepadnavirus virions are enveloped and about 42 nm in diameter. The nucleocapsid or core of the virion contains a major core protein called the HBV core antigen, abbreviated HBcAg. The external glycoproteins are called HBV surface antigens or HBsAg. Budding is through internal membranes.

In the description here, the focus is on HBV, but DHBV has been important for working out the mechanisms of replication. None of the established cell lines support the complete infection cycle of any hepadnavirus, making study of virus replication difficult. Much of what we know comes from studies of infected liver in experimental animals or studies in explanted primary hepatocytes. DHBV replicates well in primary duck hepatocytes, but the mammalian viruses replicate poorly in explanted hepatocytes. Interestingly, a number of hepatoma cell lines will support viral replication if transfected with viral DNA. However, attempts to infect them with virus do not result in replication, for reasons that are not clear.

Transcription of the Viral DNA

A schematic of the life cycle of a hepadnavirus was shown in Fig. 1.15. After infection of a cell, the viral nucleocapsid is transported to the nucleus. Transport depends on the phosphorylation of the capsid protein and is mediated by cellular transport receptors importing α and β. The size of the capsid, about 35 nm, is at the upper limit for transport through the nuclear pore. The viral genome is uncoated in the nucleus and converted to a covalently closed, circular, dsDNA molecule, called cccDNA. In this process, the protein attached to the minus strand is removed, as is the RNA oligonucleotide at the 5′ end of the plus strand, gaps are filled in, and the ends of the DNA strands are closed. Host repair enzymes are assumed to carry out this process. The resulting cccDNA does not integrate into the host genome nor does it replicate as an episome; rather it is maintained as a single copy of circular DNA. Note that a primary site of replication of the virus, and the cells in which most of the studies of replication have been conducted, are terminally differentiated hepatocytes which divide only rarely and in which there is no ongoing DNA synthesis. Thus, the virus has evolved other means for amplification of its genome.

The cccDNA is transcribed by cellular RNA polymerase II to produce several mRNAs (Fig. 6.26B). Only one strand is transcribed. Four different promoters in the DNA of the mammalian viruses lead to the production of unspliced transcripts of lengths 3.5 kb (i.e., slightly greater than unit length), 2.4, 2.1, and 0.7 kb, all of which terminate at the same poly(A) addition site (purple arrow in the figure). More than one start point is used in the case of two of the promoters, and from these two promoters, RNAs with two different 5′ ends are transcribed that serve different functions, as illustrated in Fig. 6.26B. The RNA transcripts are capped and polyadenylated. The polyadenylation signal in the mammalian hepadnaviruses is TATAAA rather than AATAAA, and the use of this suboptimal signal appears to require viral sequences upstream of this site.

Transcription of viral RNA is most efficient in hepatocytes. At least some of the promoters require transcription factors such as hepatocyte nuclear factor 1, present primarily in hepatocytes, for optimal activity. Furthermore, at least two enhancer sequences are known to be present in the DNA that function most efficiently in hepatocytes. In addition to the various cellular factors, the X gene product upregulates transcription of viral DNA.

The four classes of mRNAs are exported to the cytoplasm for translation and assembly of virions. Export is facilitated by a sequence element of about 500 nucleotides called the posttranscriptional regulatory element (PRE). This element is required because the major hepadnaviral RNAs are not spliced. Thus, PRE is functionally analogous to RRE of HIV or CTE of the simple retroviruses, but it is not known if the mechanisms by which these elements effect export are the same.

Synthesis of Viral Proteins

Synthesis of the viral proteins is complex. Four genes are usually recognized. The core gene gives rise to two products called precore (preC) and core (also referred to as HBcAg or simply C). The polymerase gene gives rise to RT-RNase H (usually called the polymerase or P). The surface protein gene gives rise to three proteins in mammals called preS1, preS2, and S (also called HBsAg), but to two proteins in birds corresponding to preS1 and S. The X gene gives rise to a protein called X, so called because its function was originally unknown and its complete range of functions is still obscure. The X protein has been long considered to be produced only by the mammalian viruses, but a region corresponding to X is present in the avian viruses and an X

protein may be produced using a non-AUG start codon. As described before, four classes of mRNAs are produced by initiation at the four promoters which form an overlapping set that lead to the production of seven proteins in the mammalian viruses and to five or six proteins in the avian viruses (Fig. 6.26B).

Both C and P are translated from the largest (3.5 kb) mRNAs, which are slightly longer than unit length. Two mRNAs are produced starting at this promoter. One mRNA is slightly longer and is translated to produce the protein called preC. The shorter form of the 3.5-kb mRNA, which is also called pgRNA, lacks the AUG used to initiate translation of preC. A downstream AUG in this mRNA, which is in the same reading frame as preC, is used to initiate translation of the protein called C, which is the major capsid protein of the virion. PreC has a different fate. It is inserted into the endoplasmic reticulum during synthesis and transported through secretory vesicles, undergoing cleavages to remove an N-terminal signal sequence and some C-terminal residues. It is secreted from the cell as a 17-kDa protein called HBeAg. HBeAg may be important for the establishment of a chronic infection in infants (see later). HBcAg, HBsAg, and HBeAg are all used as clinical markers of infection and virus replication. In general, HBeAg is associated with more aggressive clinical hepatitis.

The gene for P is downstream of C (Fig. 6.26B). It is in a different reading frame than C and partially overlaps C. The mechanism by which translation of P is initiated is not yet resolved. It is translated from the same mRNA as is C (i.e., pgRNA), but initiation is internal, using the start AUG of P, rather than being produced by some form of frameshifting and cleavage. Internal initiation does not appear to use an IRES but appears to be cap-dependent, and some form of ribosome scanning has been invoked in order to position the ribosome at this start site. This process is inefficient, and about 200 copies of C are produced for each copy of P.

Three forms of S are produced, a long version called L or preS1, a medium size version called M or preS2, and a short version S (Fig. 6.26B). These differ only at their N termini and are produced by using different in-frame AUG initiation codons. PreS1 is translated from the 2.4-kb mRNA, whereas preS2 and S are translated from two forms of the 2.1-kb mRNA in a manner similar to preC and C. As stated, only the proteins corresponding to L and S are produced by the avian viruses.

X is translated from the 0.7-kb mRNA. As described, it is not known if an X protein is produced by the avian hepadnaviruses.

The viral genome is very compact (Fig. 6.26). Over half of it is translated in two reading frames. The P gene requires about three-quarters of the coding capacity of the genome and this gene overlaps each of the other three genes. The S gene is completely contained within the P gene.

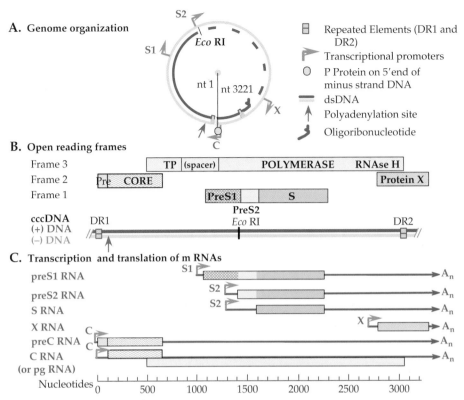

FIGURE 6.26 Genome organization of hepatitis B virus. (A) Circular map of the genome showing the locations of the repeated elements (DR1 and DR2), the four known promoters (blue-green arrows), and the gap of variable length in the plus strand of the DNA. This map is numbered from the beginning of the pgRNA. Some authors number the nucleotides from the unique Eco RI site (here at nt 1407). (B) Linearized map of cccDNA showing the open reading frames. Below this are shown the six mRNAs and the proteins (colored blocks) translated from them. Note that the pC RNA and C RNA (pg RNA) are of more than genome length, and that all mRNAs end at the polyadenylation site (purple arrow) at nt 121 (but have been shown extended to the right for clarity). (C) Transcription and translation of mRNAs. All transcripts are made from the (−)DNA (light blue) as template, in a clockwise direction on the circular map, or left to right in the linearized map. Data for constructing this diagram come from Yen (1998), Hu and Seeger (1997), and Fields *et al.* (1996) p. 2706, and details provided by Dr. James Ou.

Replication of the Viral Genome

The 3.5-kb pgRNA serves not only as a messenger but also as an intermediate in viral genome replication. This process is illustrated in Fig. 6.27. Protein P (which has RT activity, including RNase H) uses this RNA as a template to make the (partially) dsDNA copy found in the virion. As is the case for retroviruses, transfer of initiated complexes from one end of the genome to the other occurs twice during reverse transcription. Unlike retroviruses, however, the primer for first-strand synthesis is not a tRNA but protein P itself, which remains covalently attached to the 5′ end of the first strand.

DNA synthesis takes place in capsids and the first step is therefore the encapsidation of the pgRNA. Only pgRNA is packaged. This RNA is the messenger for both C and P, and both C and P are required for encapsidation. P binds to a specific sequence called epsilon present in the 5′ region of the RNA (Fig. 6.27, step 1). The signal is found within a stem-loop structure present within direct repeats of about 200 nt at the two ends of the RNA, which is illustrated in Fig. 6.28. Interestingly, although ε is present at

both the 5′ and 3′ ends of the RNA, only the 5′ signal functions for encapsidation. Once P binds to ε, C is recruited and the capsid assembles. In the absence of P, C assembles into capsids that package RNAs randomly. Thus, the specificity in packaging of viral RNA lies in the interaction of P with the RNA, unlike most viruses, including the retroviruses, in which it is the capsid protein that recognizes a packaging signal in the viral genome.

In the nucleocapsid, first-strand synthesis is initiated by using the −OH group of a specific tyrosine in P as the primer. This tyrosine is present in an N-terminal domain (labeled TP in Fig. 6.26) that is distinct from the domain that constitutes the RT. Four nucleotides are added, copied from ε (Fig. 6.27, step 1; see also Fig. 6.28). P with its covalently attached chain is then transferred to the DR1 acceptor site at the 3′ end of the RNA (which has a sequence complementary to the four nucleotides used to start DNA synthesis, which are indicated by the red arrow in step (1). DNA synthesis of first strand then continues until the 5′ end of the RNA is reached (Fig. 6.27, steps 2, 3, and 4).

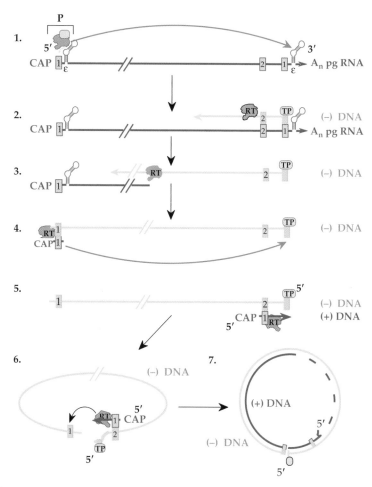

FIGURE 6.27   Mechanism of HBV DNA synthesis. Pregenomic RNA (pgRNA) is capped, polyadenylated, and greater than genome length (green line). It contains two copies each of DR1 and ε, the encapsidation sequence shown in Figure 6.28. Step (1) Priming of reverse transcription occurs when P protein (consisting of TP, RT, and RNaseH domains) makes a tetranucleotide copy of the bulge in the ε structure. This tetranucleotide is covalently linked to P. The nascent DNA strand is then translocated to DR1 at the 3′ end of pgRNA. Step (2) TP (terminal protein) is cleaved from P (but see text for an alternative hypothesis) and remains attached to the 5′ end of the minus strand, while the minus-strand DNA (light blue) is extended right to left by RT. Step (3) During minus-strand DNA synthesis, RNase H activity degrades pgRNA until P (RT) reaches the 5′ end of the template RNA. Step (4) A short RNA oligomer is left annealed to a short terminal duplication. Step (5) The RNA oligomer is translocated to DR2 where it primes plus-strand DNA synthesis (medium blue) left to right. Step (6) During plus-strand elongation, a second template transfer circularizes the genome and in step (7) the plus strand is extended a variable length to give mature progeny viral DNA. Modified from Locarnini *et al.* (1996).

The RNase H activity of P degrades the RNA strand during synthesis of first-strand DNA, but the extreme 5′ end of the RNA is not degraded. This 5′ piece, which is capped and about 18 nucleotides long, is transferred to the DR2 acceptor site near the 5′ end of the first-strand cDNA (step 4), where it serves as a primer for second-strand DNA synthesis (step 5). After reaching the 5′ end of the first-strand cDNA, continued synthesis of second-strand cDNA requires translocation to the 3′ end of the first strand, as is the case for retroviruses (step 6). In the case of hepadnaviruses, cyclization of the DNA occurs during this translocation, promoted by terminal redundancies. Second-strand synthesis is usually not complete so that the genomic double-strand DNA has a single-strand gap, of variable length, in it (step 7).

In the model for DNA synthesis shown in Fig. 6.27, the terminal protein (TP) domain is cleaved from the polymerase domain after initiation of DNA synthesis. However, reports conflict as to whether cleavage does or does not occur. If no cleavage occurs, the entire P protein remains covalently attached to the 5′ end of the DNA, rather than just TP. If P remains attached to the 5′ end of the minus strand, it will keep the end of the first-strand DNA with it at all times. This could simplify the second jump and the cyclization of the DNA, since DR2 would be close by.

Some mature capsids thus produced, which now contain copies of the viral DNA genome, are transported back into the nucleus and release their DNA to amplify the replication cycle. This results in the accumulation in the nucleus

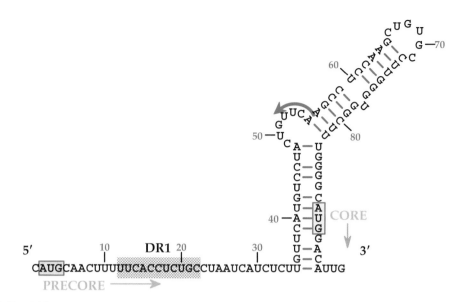

FIGURE 6.28   Two-dimensional structure of the RNA element that forms the packaging signal in hepatitis B RNA. Numbering is that for the sequence as presented in GenBank. The initiation codons for the precore and core proteins are marked, the direct repeat sequence (DR1) is shaded, and the blue arrow shows the initiation sequence for DNA replication (see Figure 6.27). Redrawn from Buckwold and Ou (1999), Figure 2.

of about 20 copies of viral DNA to serve as templates for mRNA synthesis. This is thought to occur primarily early in infection, before accumulation of large amounts of S protein.

Assembly of the Virion

Cores, with their partially dsDNA, bud through intracellular membranes to produce mature virions called Dane particles. The 42-nm virions contain both L (preS1) and S in their envelopes, with at least one-fourth as much L as S, and both L and S are required for virion assembly. In the virion, both L and S have both their N and C termini outside. Thus, they must span the membrane at least twice, and some models propose that they span the membrane four times. L is myristoylated at the N terminus. Myristoylation is required for the infectivity of the virus but not for assembly, suggesting that myristoylation serves a function in entry. Virions also contain M in amounts equivalent to L, but M does not appear to be required for assembly or for infectivity of virions.

In addition to the 42-nm Dane particles, 20-nm particles are also produced in abundance (10^4- to 10^6-fold excess over Dane particles). These particles contain S and M but little or no L, and lack the core. These particles form when S alone is expressed in cells, and thus S has the ability to produce a bud in the absence of other viral components. The virus may produce such vast quantities of the 20-nm particles, which can result in concentrations as high as 10^{13}/ml in serum, in order to tolerize the immune system, since the concentration

of HBsAg contained in these particles can reach more than 100 µg/ml in serum.

HBV and Hepatitis in Humans

HBV has a pronounced tropism for hepatocytes, as do all hepadnaviruses, and causes hepatitis in humans. Infection of neonates or very young children is usually asymptomatic, but infection nevertheless has serious consequences because chronic infection often results. Infection of adults results in serious disease characterized by liver dysfunction accompanied by jaundice in about one-third of infections, although death from fulminant hepatitis is uncommon (the fatality rate is around 1% of acute cases). The incubation period is long, 45–120 days, and convalesence is usually extended (more than 2 months), but more than 90% of adults infected by HBV recover completely. In the United States, almost 500,000 cases of hepatitis are reported annually, of which about 40% are due to HBV (Table 3.6). It is reasonable to assume that the total number of new infections with HBV is perhaps 10-fold the number of reported cases of HBV-induced hepatitis.

In a small number of adult cases, less than 10%, the infection is not cleared and becomes chronic. Infection of neonates or young children results in high levels of chronicity, however. Up to 90% of neonates infected with HBV become chronically infected, and infection of 3-year-old children may result in 30% chronicity. It is thought that the immature state of the immune system in the very young is important in the development of a chronic infection. Chronic infection may remain asymptomatic and may even eventually clear in a small fraction of cases, especially if infection occurred as an adult. However, other patients develop chronic active hepatitis that may progress to cirrhosis and death.

HBV is spread primarily by contact with contaminated blood, by sexual intercourse, and from mother to child during delivery or breast-feeding. Persistently infected individuals can have very high titers of virus in the blood, up to 10^{10}/ml, and the virus resists drying for up to 1 week. Thus, contact with infected blood need not be extensive to transmit the virus. It has been suggested that household contact leads to spread via sharing of razors, for example. Medical personnel are at risk of contracting the virus from their patients, not only by needle stick, which is responsible for many cases in unvaccinated individuals, but through other contact with contaminated blood. The virus also spreads readily among institutionalized individuals. At one time, blood transfusion was a source of spread of virus, but with the development of sensitive assays for the presence of the virus, the risk of infection following blood transfusion in developed countries is now 1/200,000 per unit of blood.

Chronic infection acquired at birth is thought to be the major mechanism by which the virus persists in nature. Up to 90% of babies born to mothers who are acutely or chronically infected with HBV and positive for HBeAg will be infected by HBV, and most of these will become chronically infected. In the United States, there are an estimated 1.2 million carriers of HBV, and worldwide there are an estimated 350 million. The fraction of the population chronically infected with HBV varies from 0.1 to 0.5% in developed countries to 5 to 15% in Southeast Asia and sub-Saharan Africa. A map that illustrates the prevalence of HBV in different regions of the world is shown in Fig. 6.29.

HBV and Hepatocellular Carcinoma

Liver cancer causes more than 500,000 deaths a year worldwide, and about 90% of primary malignant tumors of the liver are hepatocellular carcinoma (HCC). HCC is more common in men than women, by 4 to 1, and is in the top 10 in frequency of cancers in humans. HCC is more common in regions that exhibit high chronicity for HBV (compare Fig. 6.30 with Fig. 6.29). The association of HBV with HCC is also shown by data such as the finding that in areas in which chronic infection occurs in 5–10% of the population, 50–80% of HCC patients are chronically infected with HBV.

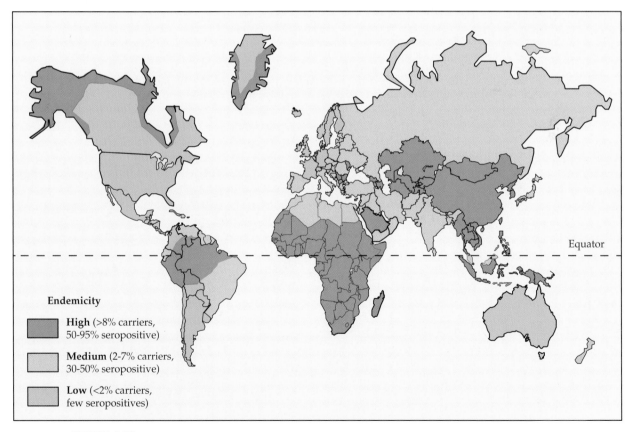

Endemicity

High (>8% carriers, 50-95% seropositive)

Medium (2-7% carriers, 30-50% seropositive)

Low (<2% carriers, few seropositives)

Equator

FIGURE 6.29   Worldwide hepatitis B prevalence as of 2005. Each country is designated as having high, medium, or low levels of endemicity, based on estimates of carrier frequency. Correlated with this is a range of persons who show serological evidence of past infection. From CDC Web page: http://www.cdc.gov/ncidod/diseases/hepatitis/slides.

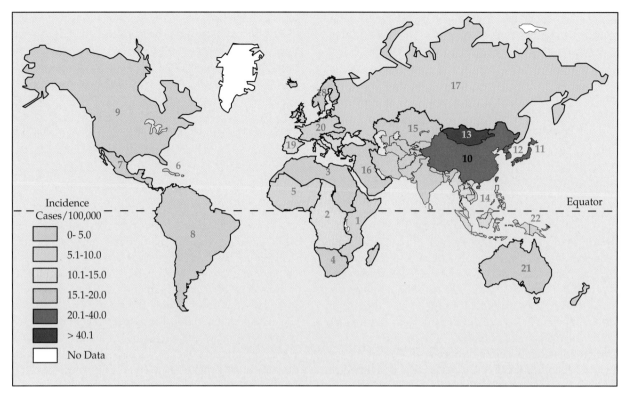

FIGURE 6.30 Average annual incidence (cases per 100,000) of liver cancer in various regions of the world in 2002. Of these cases, 67% are due to chronic hepatitis B and 26% are due to chronic hepatitis C infection. The names of the regions and the incidence for men/women are (**1**) Eastern Africa 10.7/4.8; (**2**) Middle Africa 15.4/9.0; (**3**) Northern Africa 2.6/1.6; (**4**) Southern Africa 4.3/1.8; (**5**) Western Africa 9.0/3.5; (**6**) Caribbean 7.5/4.7; (**7**) Central America 3.2/3.6; (**8**) South America 2.9/2.6; (**9**) North America 7.0/3.2; (**10**) China 37.9/15.1; (**11**) Japan 43.8/19.2; (**12**) North and South Korea 45.7/13.0; (**13**) Mongolia 53.1/34.2; (**14**) Southeast Asia 13.2/4.6; (**15**) South Central Asia 1.8/1.1; (**16**) Western Asia 3.0/1.5; (**17**) Eastern Europe 6.8/4.3; (**18**) Northern Europe 5.41/3.4; (**19**) Southern Europe 19.8/8.8; (**20**) Western Europe 10.1/3.6; (**21**) Australia/New Zealand 5.4/2.0; (**22**) Melanesia 10.2/5.5. Data from Globocan 2002 at: http://www-dep.iarc.fr/.

The lifetime risk of HCC in chronically infected persons is estimated to be 10–25%. It is clear that chronic HBV infection contributes to a large fraction of HCC cases.

The mechanism by which chronic infection by HBV leads to HCC is not altogether clear, and may not be the same in all cases. One possibility is that long-term infection, characterized by continuing destruction of liver cells followed by regrowth, results eventually in the appearance and selection of tumor cells. Up to 90% of patients with HCC associated with HBV infection have cirrhosis, implying extensive liver damage. It also appears that HCC may result from other causes of liver disease such as alcohol-induced cirrhosis or chronic infection by HCV (see Chapter 3). Furthermore, HCC often appears only after 30–40 years of chronic infection by HBV. Thus, there is an association between HCC and continuing liver damage and regeneration over very long periods.

Chronic infection of woodchucks by WHV results in HCC, and in 40% of HCC cases in this system there is integration of the WHV DNA genome near *N-myc*2. There is no evidence in humans that insertional mutagenesis of HBV DNA is responsible for HCC in humans, but there is some evidence that the X protein might be responsible, at least in part, for HCC caused by HBV. The HBV X gene can induce HCC in transgenic mice. This protein binds to p53, a known anti-oncogene (Chapter 7) that regulates signaling pathways and modifies the activities of transcription factors. Although it is probable that the X gene product is responsible for induction of HCC in some fraction of cases, it cannot be the whole story because chronic infection by hepatitis C virus (Chapter 3) also leads to HCC, and HCV lacks the X gene.

The Immune System and HBV

HBV infection of itself does not lead to the death of infected hepatocytes. Whether *in vivo* or in cell culture, a persistent, noncytolytic infection is established by the virus. Liver damage during HBV infection results instead from the activities of cytotoxic T lymphocytes (CTLs) (see Chapter 10), which attempt to clear the infection by killing infected cells. It appears that the strength of the CTL response determines the course of infection. A vigorous response results in clearance and recovery, although often after frank hepatitis with jaundice. A weak response results in chronic infection with little symptomology. An intermediate response results in chronic infection characterized by chronic hepatitis.

Because of the potential seriousness of chronic infection by HBV, including the potential to infect others, continuing efforts are being made to develop methods of controlling or clearing the infection in chronically infected people. HBeAg was used as a marker of severity of infection until recently. This has now been replaced clinically by direct measurement of viral titers, which is used to assess the response to therapy.

The first treatment that showed at least partial success was use of high doses of interferon for extended periods of time. This succeeds in clearing the viral infection in a small minority of patients. Interferon is a cytokine that boosts the immune response (see Chapter 10), and these doses of interferon appear to enable the immune system to eradicate the virus in those patients that respond. However, the drug is poorly tolerated, with significant side effects. Nucleoside analogues have also been tested. Most are not effective but lamivudine treatment resulted in improvement in more than 50% of patients in a large trial, and the apparent clearing of infection in 16% of patients. Resistant viruses appeared in more than 10% of cases, however, limiting the effectiveness

of continuing treatment. Adefovir dipivoxil, an analogue of adenosine, is also effective in controlling virus replication in about half of patients tested but most patients relapse when therapy is terminated. Thus, there is as yet no therapy that is effective in clearing virus infection in even half of patients.

Liver transplantation is offered to some patients with HBV infection. However, circulating virus invariably reinfects the graft.

Vaccination against HBV

Several vaccines have been developed to prevent infection by HBV. The first vaccine, which was licensed in 1981, was prepared from blood plasma from chronically infected individuals. It consisted of highly purified preparations of 20-nm particles that were treated to inactivate any residual virus infectivity (whether HBV or any other virus). This vaccine was effective and safe, but obvious difficulties accompany the production of large amounts of such a vaccine. Recombinant vaccines in which the S gene is expressed in yeast or in Chinese hamster ovary cells have now replaced this early vaccine. These vaccines are cheaper and can be produced in large quantities.

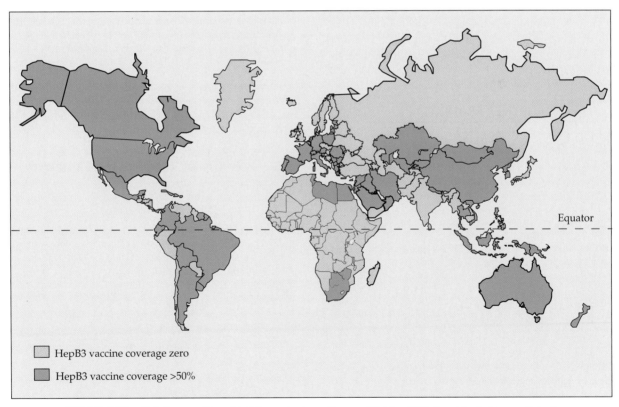

☐ HepB3 vaccine coverage zero

▦ HepB3 vaccine coverage >50%

FIGURE 6.31   Global status of hepatitis B immunization policy, as of October 2005. Countries that have policies in place for routine immunization of infants (as measured by estimates of coverage for the third dose of hepatitis B vaccine: HepB3). Countries whose reported HepB3 coverage is zero are presumed not to have such programs in place. From WHO Program on Immunization, Surveillance, Assessment, and Monitoring at: http://www.who.int/immunization_monitoring/en/globalsummary/timeseries/tscoveragehepb3.htm.

The current vaccination schedule calls for three injections of S protein over a period of 6 months. Further boosting, assessed by anti-HBV titers in serum, may be necessary to achieve adequate immunity. These vaccines are quite effective if received prior to exposure or shortly after infection (the incubation period of HBV is quite long, as noted earlier). They are unable to eliminate the infection in a chronically infected individual, however. Adults in high-risk groups, such as medical personnel, laboratory personnel who handle potentially infected blood samples, sexually active individuals, drug users who inject drugs, or individuals in ethnic groups that have a high incidence of HBV such as Alaskan natives, are encouraged to become vaccinated. Vaccination of all infants at birth is also recommended by the American Academy of Pediatricians. One reason for vaccination of infants is to prevent the establishment of chronic infection in babies born to infected mothers. Immunization of infants starting within 12 hours of birth and going through three or four shots has proved to be effective in preventing chronic HBV infection even in infants that would otherwise have become chronically infected (born to mothers positive for HBsAg). Vaccination of infants is also recommended in an effort to eventually eliminate the virus from the human population. Because chronic infection of neonates is a major reservoir of the virus in nature, elimination of this source of the virus could lead to the eradication of the virus in the human population. Many countries have now initiated programs to immunize infants, and the current status of vaccination against HBV on a worldwide scale is shown in Fig. 6.31.

FURTHER READING

Biology of Retroviruses

Coffin, J. M., Hughes, S. H., and Varmus, H. E. (Eds.) (1997). *Retroviruses.* Cold Spring Harbor, NY, Cold Spring Harbor Laboratory Press.

Goff, S. P. (2006). *Retroviridae*: The retroviruses and their replication. Chapter 55 in: *Fields Virology, Fifth Edition* (D. M. Knipe and P. M. Howley, Eds. in chief), Philadelphia, Lippincott Williams & Wilkins, pp. 1999–2070.

Herniou, E., Martin, J., Miller, K., *et al*. (1998). Retroviral diversity and distribution in vertebrates. *J. Virol.* **72**: 5955–5966.

Hindmarsh, P., and Leis, J. (1999). Retroviral DNA integration. *Microbiol. Mol. Biol. Rev.* **63**: 836–843.

Holzschu, D. L., Martineau, D., Fodor, S. K., *et al*. (1995). Nucleotide sequence and protein analysis of a complex piscine retrovirus: walleye dermal sarcoma virus. *J. Virol.* **69**: 5320–5331.

Lairmore, M. D., and Franchini, F. (2006). Human T-Cell leukemia virus Types 1 and 2. Chapter 56 in: *Fields Virology, Fifth Edition* (D. M. Knipe and P. M. Howley, Eds. in chief), Philadelphia, Lippincott Williams & Wilkins, pp. 2071–2106.

Lentiviruses and HIV/AIDS

Berson, J. F., and Doms, R. W. (1998). Structure-function studies of the HIV-1 coreceptors. *Semin. Immunol.* **10**: 237–248.

Burton, D. R., Stanfield, R. L., and Wilson, I. A. (2005). Antibody vs. HIV in a clash of evolutionary titans. *Proc. Natl. Acad. Sci. U.S.A.* **102**: 14943–14948.

Coskun, A. K., van Maanen, M., Nguyen, V., and Sutton, R. E. (2006). Human chromosome 2 carries a gene required for production of infectious human immunodeficiency virus type I. *J. Virol.* **80**: 3406–3415.

Cullen, B. R. (1998). Posttranscriptional regulation by the HIV-1 Rev protein. *Semin. Virol.* **8**: 327–334.

Desrosiers, R. C. (2006). Nonhuman lentiviruses. Chapter 59 in: *Fields Virology, Fifth Edition* (D. M. Knipe and P. M. Howley, Eds. in chief), Philadelphia, Lippincott Williams & Wilkins, pp. 2215–2244.

Freed, E. O. (1998). HIV-1 Gag proteins: diverse functions in the virus life cycle—minireview. *Virology* **251**: 1–15.

Heeney, J. L., Dalgleish, A. G., and Weiss, R. A. (2006). Origin of HIV and the evolution of resistance to AIDS. *Science* **313**: 462–466.

Huthoff, H., and Malim, M. H. (2005). Cytidine deamination and resistance to retroviral infection: towards a structural understanding of the APOBEC proteins. *Virology* **334**: 147–153.

Kuritzkes, D. R., and Walker, B. D. (2006). HIV-1 pathogenesis, clinical manifestations and treatment. Chapter 58 in: *Fields Virology, Fifth Edition* (D. M. Knipe and P. M. Howley, Eds. in chief), Philadelphia, Lippincott Williams & Wilkins, pp. 2187–2214.

Leitner, T., and Albert, J. (1999). The molecular clock of HIV-1 unveiled through analysis of a known transmission history. *Proc. Natl. Acad. Sci. U.S.A.* **96**: 10752–10757.

Persaud, D., Zhou, Y., Siliciano, J. M., and Siliciano, R. F. (2003). Latency in human immunodeficiency virus type I infection: no easy answers. *J. Virol.* **77**: 1659–1665.

Riddler, S. A., and Mellors, J. W. (1997). HIV-1 viral load and clinical outcome: review of recent studies. *AIDS* **11**: S141–S148.

Saad, J. S., Miller, J., Tai, J., Kim, A., Ghanam, R. H., and Summers, M. F. (2006). Structural basis for targeting HIV-1 gag proteins to the plasma membrane for virus assembly. *Proc. Natl. Acad. Sci. U.S.A.* **103**: 11364–11369.

Si, Z., Vandegraaff, N., O'hUigin, C., *et al*. (2006). Evolution of a cytoplasmic tripartite motif (TRIM) protein in cows that restricts retroviral infection. *Proc. Natl. Acad. Sci. U.S.A.* **103**: 7454–7459.

United Nations Development Programme (2004). *Thailand's Response to HIV/AIDS: Progress and Challenges.*

Whetter, L. E., Ojukwu, I. C., Novembre, F. J., and Dewhurst, S. (1999). Pathogenesis of simian immunodeficiency virus infection. *J. Gen. Virol.* **80**: 1557–1568.

Zhu, P., Liu, J., Bess, J., Jr., *et al*. (2006). Distribution and three-dimensional structure of AIDS virus envelope spikes. *Nature* **441**: 847–852.

Foamy Viruses

Lecellier, C.-H., Neves, M., Giron, M.-L., Tobaly-Tapiero, J., and Saïb, A. (2002). Further characterization of equine foamy virus reveals unusual features among the foamy viruses. *J. Virol.* **76**: 7220–7227.

Linial, M. (2006). Foamy viruses. Chapter 60 in: *Fields Virology, Fifth Edition* (D. M. Knipe and P. M. Howley, Eds. in chief), Philadelphia, Lippincott Williams & Wilkins, pp. 2245–2262.

Retroelements and Endogenous Retrovirus Sequences

Eickbush, T. H. (1994). Origin and evolutionary relationships of retroelements. In *The Evolutionary Biology of Viruses* (S. S. Morse, Ed.), New York, Raven Press, pp. 121–157.

Jern, P., Sperber, G. O., and Blomberg, J. (2006). Divergent patterns of recent retroviral integrations in the human and chimpanzee genomes: probable transmissions between other primates and chimpanzees. *J. Virol.* **80**: 1367–1375.

Medstrand, P., van de Lagemaat, L. N., Dunn, C. A., *et al.* (2005). Impact of transposable elements on the evolution of mammalian gene regulation. *Cytogenet. Genome Res.* **110**: 342–352.

Hepadnaviruses

Buckwold, V. E., and Ou, J.-H. (1999). Hepatitis B C-gene expression and function: the lessons learned from viral mutants. *Curr. Top. Virol.* **1**: 71–81.

Feitelson, M. A. (1999). Hepatitis B in hepatocarcinogenesis. *J. Cell. Physiol.* **181**: 188–202.

Lin, L., Prassolov, A., Funk, A., *et al.* (2005). Evidence from nature: interspecies spread of heron hepatitis B viruses. *J. Gen. Virol.* **86**: 1335–1342.

Ou, J.-H. J. (Ed.) (2001) *Hepatitis Viruses,* The Netherlands, Kluwer Academic Publishers.

Rabe, B., Vlachou, A., Panté, N., Helenius, A., and Kann, M. (2003). Nuclear import of hepatitis B virus capsids and release of the viral genome. *Proc. Natl. Acad. Sci. U.S.A.* **100**: 9849–9854.

Seeger, C., Zoulim, F., and Mason, W. S. (2006). Hepadnaviruses. Chapter 76 in: *Fields Virology, Fifth Edition* (D. M. Knipe and P. M. Howley, Eds. in chief), Philadelphia, Lippincott Williams & Wilkins, pp. 2977–3030.

Tavis, J. E. (1996). The replication strategy of the hepadnaviruses. *Viral Hepatitis Rev.* **2**: 205–218.

Xu, Z., Yen, T. S. B., Wu, L., *et al.* (2002). Enhancement of hepatitis B virus replication by its X protein in transgenic mice. *J. Virol.* **76**: 2579–2584.

Yen, T. S. B. (1998). Posttranscriptional regulation of gene expression in hepadnaviruses. *Semin. Virol.* **8**: 319–326.

CHAPTER

7

DNA-Containing Viruses

INTRODUCTION

Twenty-two families of viruses that contain double-stranded DNA (dsDNA) as their genome are currently recognized by the International Committee on Taxonomy of Viruses (ICTV). The families that contain viruses of vertebrates are listed in Table 7.1 together with their hosts, and those that contain viruses of nonvertebrates including bacteria are listed in Table 7.2. Unassigned genera of dsDNA viruses also exist and new families of these viruses will continue to be recognized in the future. There are dsDNA viruses that infect bacteria, archaea, amoebae, mycoplasma, algae, fungi, invertebrates, and vertebrates, but interestingly there are no known dsDNA viruses of plants. Members of seven of these families infect vertebrates of which members of five families, the *Poxviridae*, the *Herpesviridae*, the *Adenoviridae*, the *Polyomaviridae*, and the *Papillomaviridae*, infect humans and cause disease. Only these five families are considered further. The genomes of these viruses vary in size from about 5 to 375 kb.

There are also five families of DNA viruses that contain single-stranded DNA (ssDNA) as their genome, as listed in Table 7.3. These viruses infect bacteria, mycoplasma, spiroplasma, plants, invertebrates, and vertebrates. Members of only one family, the *Parvoviridae*, infect humans and cause disease, and only this family is considered further. The parvoviruses have a small genome (4–6 kb), from which they receive their name (Latin *parvus* = small). Note that in the classification used here, viruses with a DNA genome that replicate through an RNA intermediate, such as the hepadnaviruses and the caulimoviruses, are not referred to as conventional DNA viruses and were considered in Chapter 6.

During infection, most vertebrate DNA viruses stimulate host-cell DNA replication, or at least the early stages of DNA replication, in order to prepare a suitable environment for their own DNA replication. Such a favorable environment

includes the presence of cellular factors required for DNA replication as well as an increase in the amount of substrates required for making DNA. Further, some viruses cause cells to proliferate, at least early in the infection cycle. For this reason, most DNA viruses are at least potentially transforming and many DNA viruses are known to cause tumors in humans or in other animals.

In contrast, the parvoviruses do not encode proteins that stimulate cellular DNA replication. For this reason, they can only replicate in cells that are actively dividing and their target tissues in the host are organs that undergo continual renewal. Members of one genus of the *Parvoviridae*, *Dependovirus*, require a helper virus for replication, and replication will occur only in cells coinfected with the helper.

Where studied, almost all viruses interfere with the defenses of the vertebrate host against viruses and the large dsDNA viruses are particularly remarkable in this regard. With their large genomes they can afford to dedicate dozens of genes to the control of the host immune response. Components of the host immune system that are targeted include the interferon system, cytotoxic T-cell responses, other immune effector functions, and the complement system. In part because of this ability to interfere with multiple host defense systems, infection with some DNA viruses results in a latent or a persistent infection that can endure for the life of the infected individual. Some aspects of these interference pathways are described in this chapter, where needed to understand the replication cycle and epidemiology of the virus, but a more detailed description of defense mechanisms is presented in Chapter 10.

Some DNA viruses, such as the poxviruses and the adenoviruses, cause epidemics of symptomatic disease in vertebrates from which recovery is complete (if the infection is not fatal) and immunity is established. Other vertebrate DNA viruses, such as the herpesviruses and the polyomaviruses, establish long-term infections that persist despite

TABLE 7.1 Double-Stranded DNA Viruses That Infect Humans and Other Vertebrates

Family/subfamily	Genera	Genome size (kb)[a]	Type species
Poxviridae[b]		130–375	
Chordopoxvirinae	See Table 7.4		
Entomopoxvirinae	*Alphaentomopoxvirus*		*Melolontha melolontha* entomopoxvirus
	Betaentomopoxvirus		*Amsacta moorei* entomopoxvirus 'L'
	Gammaentomopoxvirus		*Chironomus luridus* entomopoxvirus
Iridoviridae		170–200	
	Iridovirus		Invertebrate iridescent 6
	Chloriridovirus		Invertebrate iridescent 3
	Ranavirus		Frog 3
	Lymphocystivirus		Lymphocystis disease 1
	Megalocytivirus		Infectious spleen and kidney necrosis
Herpesviridae		~125–235	
Alphaherpesvirinae			
Betaherpesvirinae	See Table 7.7		
Gammaherpesvirinae			
	Ictalurivirus		Channel catfish herpesvirus
Adenoviridae	See Table 7.10	20–25	
Polyomaviridae	See Table 7.12	5	
Papillomaviridae	See Table 7.14	8	
Asfarviridae	*Asfivirus*	170	African swine fever

[a] kb, kilobase pairs.

[b] Viruses in families/genera in blue type all infect vertebrate hosts other than humans; families and subfamilies in red have members that infect humans.

TABLE 7.2 Families of Double-Stranded DNA Viruses That Infect Nonvertebrate Hosts

Family	Genera	Genome size[a] (kb)	Type species	Host
Myoviridae	Six genera of phages	~170	Enterobacteria phage T4	Bacteria
Siphoviridae	Seven genera of phages	48.5	Enterobacteria phage λ	Bacteria
Podoviridae	Four genera of phages	40	Enterobacteria phage T7	Bacteria
Tectiviridae	*Tectivirus*	147–157	Enterobacteria phage PRD1	Bacteria
Corticoviridae	*Corticovirus*	9	Alteromonas phage PM2	Bacteria
Plasmaviridae	*Plasmavirus*	12	Acholeplasma phage L2	Mycoplasma
Lipothrixviridae	Three genera of phages	15.9	Thermoproteus 1	Archaea
Rudiviridae	*Rudivirus*	?	Sulfolobus SIRV 1	Archaea
Fuselloviridae	*Fusellovirus*	15.5	Sulfolobus SSV 1	Archaea
Guttaviridae	*Guttavirus*	?	Sulfolobus SNDV[b]	Archaea
Phycodnaviridae	Six genera of algal viruses	>300	*Paramecium bursaria* Chlorella 1	Algae
Polydnaviridae	*Ichnovirus*	2–28	*Campoletus sonorensis* ichnovirus	Invertebrates
	Bracovirus	2–28	*Cotesia melanoscela* bracovirus	Invertebrates
Ascoviridae	*Ascovirus*	120–180	*Spodoptera frugiperda* ascovirus	Invertebrates
Baculoviridae	*Nucleopolyhedrovirus*	80–180	*Autographa californica* nucleopolyhedrovirus	Invertebrates
	Granulovirus	80–180	*Cydia pomonella* granulovirus	Invertebrates
Nimaviridae	*Whispovirus*	300	White spot syndrome virus 1	Invertebrates

[a] kb, kilobase pairs.

[b] Now called sulfolobus neozealandicus droplet-shaped virus.

TABLE 7.3   Single-Stranded DNA Viruses

Family	Genera	Genome size (kb)	Type species	Host[a]
Inoviridae		4.4–8.5		
	Inovirus		Enterobacteriaphage *M13*	Bacteria
	Plectrovirus		*Acholeplasma* phage *MV-L51*	Mycoplasma
Microviridae		~4.4–6.0		
	Microvirus		Enterobacteria phage ΦΧ*174*	Bacteria
	Spiromicrovirus		*Spiroplasma* phage *4*	Spiroplasma
	Bdellomicrovirus		*Bdellovibrio* phage *MAC1*	Bacteria
	Chlamydiamicrovirus		*Chlamydia* phage *1*	Bacteria
Geminiviridae		2.5–3.0		
	Mastrevirus		Maize streak	Plants
	Curtovirus		Beet curly top	Plants
	Begovirus		Bean golden mosaic-Puerto Rico	Plants
	Topocuvirus		Tomato pseudo-curly top virus	Plants
Circoviridae				
	Circovirus	1.7–2.3	Porcine circovirus	Vertebrates
	Gyrovirus	2.3	Chicken anemia	Vertebrates
[Unassigned genus] [formerly *Circoviridae*]	*Anellovirus*	3.8	Torque teno	Vertebrates
Parvoviridae		4.0–6.0		
Parvovirinae	See Table 7.16			Vertebrates
Densovirinae				
	Densovirus		*Junonia coenia* densovirus	Invertebrates
	Iteravirus		*Bombyx mori* densovirus	Silkworms
	Brevidensovirus		*Aedes aegypti* densovirus	Mosquitos

Vertebrates in red indicate humans are among the vertebrates infected. Vertebrates in blue indicate nonhuman hosts only.

a vigorous immune response. For such a strategy of long-term persistence to be successful, infection in the majority of hosts must be inapparent or cause only moderate symptoms that are not unduly deleterious. Spread may be epidemic and accompanied by symptoms during primary infection, but for some herpesviruses, vertical transmission to infant progeny occurs without producing symptoms and persists for the life of the animal. For such viruses, transmission needs to occur only once per generation for the virus to persist and a minimal population size is not required to maintain the virus in nature.

FAMILY *POXVIRIDAE*

The poxviruses are a very large family of dsDNA-containing viruses that infect mammals, birds, and insects. Eleven genera are recognized, eight of which are classified as members of the subfamily *Chordopoxvirinae* and infect vertebrates (Table 7.4), and three of which are classified as members of the subfamily *Entomopoxvirinae* and infect invertebrates (Table 7.1). Of the eight genera of the *Chordopoxvirinae*, seven contain viruses that infect mammals and one, *Avipoxvirus*, contains viruses that infect birds. Only two human poxviruses (viruses for which humans are the reservoir) are known, variola or smallpox virus, a member of the genus *Orthopoxvirus*, and molluscum contagiosum virus, the only member of the genus *Molluscipoxvirus*.

The host range of any particular poxvirus is usually narrow. The two human poxviruses infect only humans in nature and other poxviruses are similarly limited in their natural host range. However, a number of mammalian poxviruses whose primary host is not humans can cause natural, albeit usually limited, infections of humans, and still other poxviruses, including the avian poxviruses, can infect humans under experimental conditions. Such mammalian and avian poxviruses have been used as agents for vaccination against virulent human viruses (smallpox, described in detail later) or as vectors to express foreign antigens for the purposes of immunization (Chapter 11), because they normally cause only a limited or an abortive infection of humans and can be engineered to express foreign antigens.

TABLE 7.4 *Poxviridae* (*Chordopoxvirinae*: Poxviruses infecting vertebrates)

Genus/ members	Virus name abbreviation	Usual host(s)[a]	Transmission	Disease	World distribution
Orthopoxvirus					
Vaccinia	VACV	Unknown/humans, bovines	Contact	Localized lesions	Worldwide
Variola virus	VARV	Humans/none	Contact	Smallpox (now extinct)	Worldwide
Monkeypox	MPXV	Squirrels/humans, monkeys	Contact	Smallpox-like	West and Central Africa
Cowpox	CPXV	Rodents/humans, cats, bovines, zoo animals	Contact	Localized lesions	Europe, W. Asia
Camelpox	CMLV	Camels/none	Contact, aerosols	Localized lesions	Africa, Asia
Ectromelia	ECTV	Unknown/laboratory mouse colonies, foxes, mink	Contact, aerosols	Lesions plus disseminated disease	Europe
Volepox	VPXV	Voles/none	Contact	?	Western United States
Parapoxvirus					
Orf	ORFV	Sheep/humans, ruminants	Contact	Localized lesions	Worldwide
Bovine papular stomatitis	BPSV	Cattle/humans	Contact	Localized lesions	Worldwide
Pseudocowpox	PCPV	Cattle/humans	Contact	Localized lesions	Worldwide
Parapox of red deer	PVNZ	Red deer/none	Contact	?	New Zealand
Yatapoxvirus					
Yaba monkey tumor	YMTV	Primates/human laboratory infections	MTBA[b]	Localized lesions Many nodular	East and Central Africa
Tanapox	TANV	?Rodents/primates, humans	MTBA[b]		
Molluscipoxvirus					
Molluscum contagiosum	MOCV	Humans/none	Contact, including sexual transmission lesions		Worldwide
Capripoxvirus					
Sheeppox	SPPV	Sheep/none	Contact, fomites, MTBA		Asia, Africa
Lumpy skin disease	LSDV	Cattle/none			
Suipoxvirus					
Swinepox	SWPV	Swine/none	MTBA, primarily by lice	Generalized skin disease	Worldwide
Leporipoxvirus					
Myxoma	MYXV	*Sylvilagus* rabbits	MTBA, primarily mosquitos	Benign tumors in natural hosts, severe disease in European rabbits	South America, Western United States, introduced into Australia
Rabbit fibroma	SFV	*Sylvilagus* rabbits	MTBA	Benign tumors in natural hosts	Eastern United States
Hare fibroma	FIBV	European hare			Europe
Squirrel fibroma	SQFV	*Sciurus* squirrels			Eastern and Western United States
Avipoxvirus					
Many species including fowlpox		Birds	Contact, MTBA	Lesions of skin and digestive tract	Worldwide

[a] Hosts are listed as "reservoir host/other naturally infected hosts."

[b] MTBA, mechanical transmission by arthropods.

The poxviruses are exceptional among eukaryotic DNA viruses because they replicate in the cytoplasm. The structure of the virion is also unusual, being shaped like a brick rather than round or filamentous like most viruses. It has been argued that the *Poxviridae*, *Iridoviridae*, *Asfarviridae*, and *Phycodnaviridae* evolved from a common ancestor because they share a number of genes that distinguish them from other DNA viruses. The first three of these families contain viruses of vertebrates that replicate in the cytoplasm, which as noted is an unusual trait for DNA viruses, whereas the last family contains algal viruses that replicate in the nucleus.

Structure of the Virion

Poxvirions are large and enveloped. They have a central core containing the DNA complexed with multiple proteins. The core is biconcave in mammalian viruses (Figs. 2.1 and 2.24). One or two lateral bodies (normally two in vertebrate viruses and one in invertebrate viruses) flank the core. These contain proteins required for the initiation of viral replication. One or two viral envelopes, which contain a number of virus-encoded glycoproteins, surround the core and lateral bodies. Altogether, virions contain 30 or more structural proteins.

Vaccinia virus, which serves as a model for the family, is shaped like a brick with rounded edges and has dimensions of approximately $360 \times 270 \times 250$ nm. Two forms of infectious vaccinia virions are known, an intracellular form and an extracellular form. The intracellular form, referred to as the intracellular mature virus or IMV, has a lipid-containing surface membrane and can be released by disruption of the infected cell. The extracellular form, called the extracellular enveloped virus or EEV, has a second, lipid-containing envelope around it, which contains glycoproteins that are not present in the intracellular form.

Vaccinia virus forms in viral factories in the cytoplasm of the infected cell. A spherical immature virus 280 nm in diameter first forms in the cytoplasm whose surface membrane has been acquired from modified membranes called crescents. Maturation of the immature virus to form the IMV involves proteolytic cleavage by at least two viral proteases that lead to the reorganization of the core and the lateral bodies and the formation of the brick-shaped structure. Some IMV become further enveloped by a double membrane from the trans-Golgi or early tubular endosomes, the outer of which is lost as the virus exits the cell to become the EEV. Both IMV and EEV are infectious but attach to cells differently. The extra membrane in EEV helps to protect the virus from immune surveillance.

The cellular receptors used by poxviruses to enter cells have not been characterized. The virus enters by fusion of the IMV membrane with the cell plasma membrane. In the case of EEV, the outer membrane is first disrupted upon binding to cellular receptors, allowing the IMV membrane to fuse with the plasma membrane.

Replication of Poxviruses

An overview of the poxvirus replication cycle is shown in Fig. 7.1. Following infection, an RNA polymerase within the core is activated by the release of the core into the cytoplasm. This polymerase, together with accessory enzymes such as the capping enzyme and poly(A) polymerase, synthesizes mRNAs that are capped and polyadenylated. These are extruded from the core and translated by the host-cell machinery. Translation of early mRNAs leads to further uncoating of the virus and the development of a regulatory pathway by which mid-cycle genes and finally late genes are expressed. Early and mid-cycle functions include interference with host defense mechanisms and the replication of the viral genome, whereas late genes are primarily involved with formation of progeny virions.

Poxviruses have large genomes, from 130 to 380 kb, and encode hundreds of proteins. Because virus replication, including DNA replication, occurs in the cytoplasm, poxviruses must encode all enzymes required for DNA replication and production of mRNAs. Thus, the encoded proteins include DNA and RNA polymerases, a poly(A) polymerase to polyadenylate mRNAs, a capping enzyme, several enzymes with functions in nucleotide metabolism, protein kinases, DNA topoisomerases, as well as the proteins that form components of the virion and the proteins that interfere with host defense mechanisms. Table 7.5 lists many enzymatic proteins encoded by vaccinia virus and Table 7.6 lists structural proteins and proteins without enzymatic activity. Poxvirus-encoded molecules that interfere with host defenses are discussed in Chapter 10.

A schematic of the vaccinia virus genome is shown in Fig. 7.2. It is double stranded and linear but the ends of the genome are covalently closed so that the genome consists, in essence, of a very large single-stranded circular molecule that is self-complementary. The ends of the genome possess inverted terminal repeats that are involved in the initiation of DNA replication. A model for vaccinia DNA replication is shown in Fig. 7.3. The mechanisms by which DNA replication is initiated have not been completely worked out, but it is thought that a viral enzyme specifically nicks the DNA in or near the terminal repetitions, and the 3′ end of the nicked DNA forms a primer for DNA synthesis. Continued elongation of the DNA chain leads to production of concatenated progeny DNA. These concatamers must subsequently be resolved into genome-length segments whose ends are then covalently closed.

Replication of DNA and assembly of progeny virions occurs in what have been called viral factories in the cytoplasm. These are electron-dense areas that contain viral DNA and membranes. Progeny DNA is assembled into cores and assembly of virions occurs by condensation of membranes around the viral core as described before.

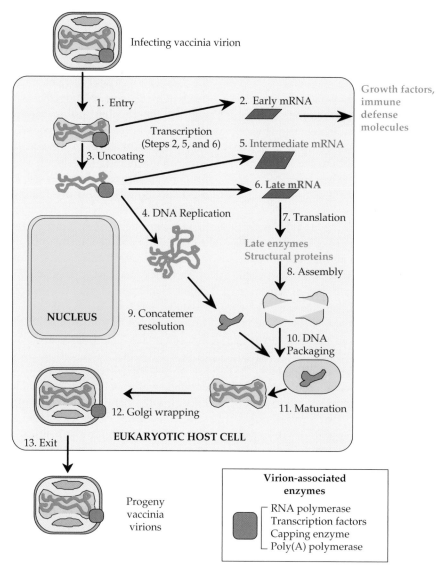

FIGURE 7.1   The replication cycle of vaccinia virus. Sequential steps are numbered in order. Note that despite the fact that vaccinia is a DNA virus, the entire replication cycle takes place in the cytoplasm. Adapted from Fields *et al.* (1996) p. 2638.

Interactions with the Host

Poxviruses produce proteins that stimulate cell development and proliferation. Many poxviruses produce homologues of epidermal growth factor, which are excreted from the infected cell. Binding of this factor to its receptor induces the cell to enter pathways leading to proliferation and differentiation. At least one poxvirus, orf virus, produces a homologue of vascular endothelial growth factor, which again induces cell proliferation upon binding to its receptor.

Poxviruses also produce a large number of proteins whose function is to interfere with the host immune system. Vaccinia virus, variola virus, and other poxviruses encode proteins that interfere with the complement system, with the interferon system, with the activity of TNF (tumor necrosis factor), with the cytokine system so as to prevent an inflammatory response to the virus, or with the induction of apoptosis (programmed cell death), among others. The discovery of proteins that interfere with host defenses is fairly recent, and intensive studies are ongoing to understand the full extent of viral inhibition of host defenses. The complexity of the interference by the virus with host antiviral defenses is illustrated by the recent discovery that a viral protein called CrmB encoded by variola virus, which was known to be a homologue of the receptor for TNF and to inhibit the activity of TNF by binding to it, has a separate domain that binds a number of chemokines and inhibits their activity as well. Thus, this single protein inhibits a number of host defense molecules. This discovery led to the further discovery that

TABLE 7.5   Vaccinia-Encoded Enzymes

Functional group name of enzyme	ORF[a]	kD	Properties
DNA Replication			
DNA polymerase	E9L	110	
Protein kinase	B1R	34	Phosphorylates H5R
Unknown	D5R	90	Replication fork, ATP/GTP binding motif A
Uracil DNA glycosylase	D4R	25	
Nicking-joining enzyme	??	50	Concatemer resolution
DNA toposiomerase	H6R	32	
ssDNA binding protein	I3L	30	
DNA ligase	A50R	63	Nonessential
DNA helicase	A18R	57	DNA-dependent ATPase
Early DNA-related metabolism			
Thymidine kinase	J2R	20	
Thymidylate kinase	A4w8R	23	
Ribonucleotide reductase			Provide dNTPs
M1	I4L	87	Large subunit
M2	F4L	37	Small subunit
dUTPase	F2L	15	dUTP→ dUMP, downregulate dUTP
DNA repair?	D9R	25	hydrolyze 8-oxo-GTP
	D10R	29	
DNA dependent NTPase	D11L	72	NPH I[b]
RNA Transcription			
RNA polymerase			Multisubunit enzyme
RPO147	J6R	147	
RPO132	A24R	133	
RPO35	A29L	35	
RPO30	E4L	30	Transcription factor
RPO22	J4R	22	
RPO19	A5R	19	
RPO18	D7R	18	
RNA polymerase-associated protein	H4L	94	RAP94, early promoter-specificity factor
Early transcription factor			DNA-dependent ATPase
	A7L	82	ETR subunit 1
	D6R	74	ETR subunit II
Poly(A) polymerase	E1L	55	Catalytic subunit
	J3R	39	Stimulatory subunit, methyltransferase
Capping enzyme			RNA triphosphatase, guanyltransferase
	D1R	97	Large subunit, catalytic activities
	D12L	33	Small subunit, stimulates transferase
RNA/DNA dependent NTPase	I8R	77	NPH II[b]
Protein kinase 2	F10L	52	Phosphorylates serines and threonines
Glutaredoxin	O2L	12	Thioltransferase, dehydroascorbate reductase

[a] ORFs are named and color coded according to the restriction map shown in Fig. 7.2.
[b] NPH, nucleoside triphosphate phosphohydrolase.
Source: Adapted from Fields *et al.* (1996) Table 3 on p. 2645 and data from Goebel *et al.* (1990).

TABLE 7.6   Vaccinia-Encoded Nonenzymatic Components

Location in virion	ORF[a]	kD	Properties
Membrane of intracellular mature virus	I5L	8.7	Hydrophobic
	L1R	27.3	Myristylated, hydrophobic
	H3L	37.5	Hydrophobic
	H5R	22.3	Phosphorylated by B1R to give 34–36 kD
	D8L	35.3	Cell-surface binding, virulence
	D13L	61.9	Rifampicin resistance
	A13L	7.7	Oligomeric
	A14L	10.0	Oligomeric
	A17L	23.0	Dimer, neutralizing epitope
	A27L	12.6	Fusion protein, neutralizing epitope, required for EEV, nonessential
Core of intracellular mature virus	F17R	11.3	Phosphoprotein, DNA binding
	I7L	49.0	Homology to toposiomerase II
	G7L	41.9	Processed
	L4R	28.5	Structural protein VP8
	D2L	16.9	
	D3R	28.0	
	A3L	72.6	Major core protein P4b
	A4L	30.8	
	A10L	102.3	Major core protein P4a
	A12L	20.5	Processed
Specific to enveloped extracellular virus (EEV)	F13L	41.8	Envelope antigen
	A34R	19.5	N-glycosylated, homologous to lectin, EEV release
	A36R	25.1	
	A56R	34.8	N- and O-glycosylated, hemagglutinin, nonessential
	B5R	35.1	Complement control protein, required for EEV, homologue to C3L

[a] ORFs are named and color coded according to the restriction map shown in Fig. 7.2.
Source: Adapted from Fields *et al.* (1996) Table 2 on p. 2643 with additional information from Goebel *et al.* (1990).

variola and other poxviruses produce a family of chemokine inhibitors. It is clear that these various viral functions are required for successful viral infection of their hosts in nature, and the existence of these viral activities has been very useful for our understanding of host defenses against viral infection. We will return to this topic in Chapter 10.

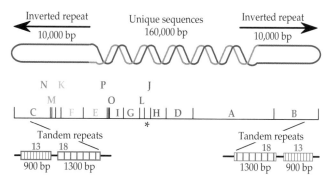

FIGURE 7.2   Schematic representation of the DNA of vaccinia virus. Upper part: linear double-stranded DNA with terminal hairpins and inverted repeats (not to scale). Center line is the HindIII restriction map (* indicates the TK gene). Color coding of the HindIII fragments is the same as that used in Tables 7.5 and 7.6. Lower line diagrams the internal structure of the terminal repeats. Adapted from Fenner *et al.* (1988).

Genus *Orthopoxvirus*

The best known poxviruses are the orthopoxviruses. Vaccinia virus has been widely studied in the laboratory as a model for the replication of poxviruses and has been used to immunize hundreds of millions of people against smallpox virus, also known as variola (from the Latin word for "spotted"). The extensive knowledge of vaccinia virus gained from laboratory and clinical studies has also led to its use as a vector to express foreign antigens in cultured cells or in animals (Chapter 11). Other members of the genus infect a variety of domestic and wild animals (Table 7.4). Two similarity trees are shown in Fig. 7.4. One illustrates the relationships among the genera of the *Chordopoxvirinae*, and the second illustrates the relationships among the orthopoxviruses. All orthopoxviruses are sufficiently closely related that they are cross protective.

Smallpox Disease

Smallpox once caused vast epidemics in human populations. It was already endemic in India 2000 years ago and had spread to China, Japan, Europe, and northern Africa by 700 A.D. It was introduced into the New World by the Europeans during their explorations and settlement. Infection resulted in a fatality rate of 20–30% in most populations and at one time the virus infected virtually the entire population of Europe. Thus, the virus was responsible for a significant fraction of all human deaths on the Continent. Data from London in the 1700s show that, depending on the year, between 4 and 18% of all deaths in the city were due to smallpox. When introduced into virgin populations in the New World, the mortality rate was much higher, probably due in part to a lack of previous selection for resistance to smallpox and in part due to the breakdown of the social system caused

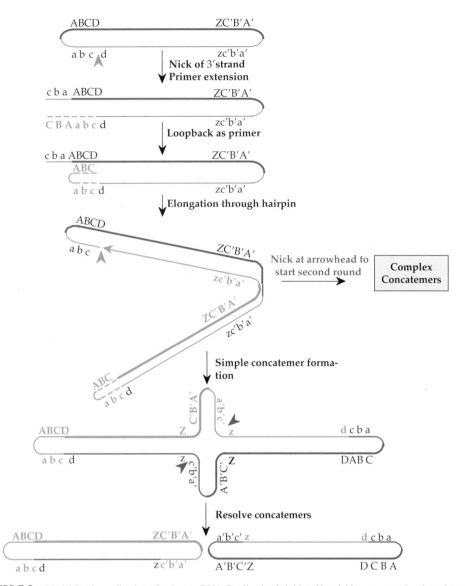

FIGURE 7.3   Model for the replication of orthopox DNA. Replication is initiated by nicking one strand at the red arrow near the left end of the DNA. This is followed by primer extension and loopback to form an internal primer. Extension then occurs through the hairpin at the right end of the molecule to form a concatemer. Concatemer resolution occurs by nicking at the blue arrows. Parental DNA is shown in blue, new strands in red. Redrawn from data in Moyer and Graves (1981) and Traktman (1990).

by the simultaneous infection of most of the population. As one example, smallpox coming up from Mexico swept across the high plains of the western United States in the late 1700s, reducing the population of the Mandan, Ojibwa, Pawnee, Arikara, and other tribes by as much as two-thirds. Then in 1837 an epidemic of smallpox in Native American populations along the upper Missouri River began with its introduction by passengers on a steamboat that came upriver from St. Louis. In this epidemic it is estimated that 99% of the Mandan died and half of the Hidatsa and Arikara. Other tribes also suffered enormous losses leading to the depopulation of the region. The importance of smallpox, measles, and other Old World plagues in the conquest and settlement

of the Americas by the Europeans was discussed in Chapter 4. A detailed description of the effect of smallpox virus on human civilization through the ages can be found in the book *Princes and Peasants: Smallpox in History* by D.R. Hopkins. The history of nations has often been changed by the early death of rulers from smallpox or from the appearance of smallpox in armies. As one example of interest to Americans, an invasion of Canada by American troops during the Revolutionary War, with the idea of adding Canada to the American colonies, failed in part because an epidemic of smallpox swept through the American troops.

Smallpox virus was exclusively a human virus, maintained by person-to-person contact, and recovery from the disease

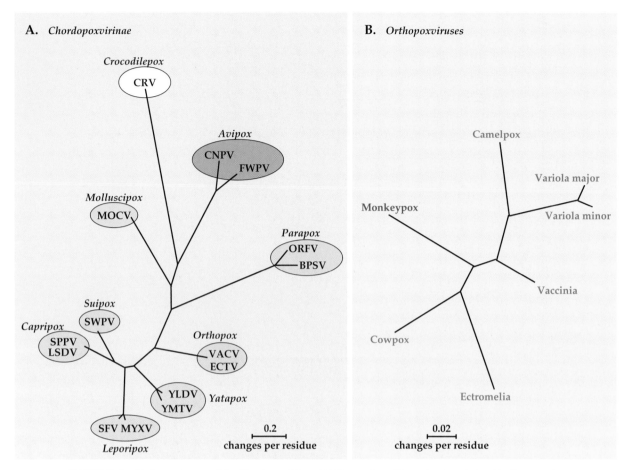

FIGURE 7.4   Phylogenetic trees of the *Chordopoxvirinae*. (A) Unrooted tree illustrating the relationships between the various genera. Eighty-three proteins from crocodilepox were aligned with similar data sets from other *Chordopoxvirinae* using the program MUSCLE. Adapted from Afonso *et al.* (2006) Figure 5. (B) Unrooted tree of the *Orthopox* genus from aligned coding sequences of the DNA polymerase gene. Adapted from Chen *et al.* (2003) Figure 3. Virus abbreviations: CRV, crocodilepox; CNPV, canarypox; FWPV, fowlpox; ORFV, Orf; BPSV, bovine papular stomatitis; VACV, vaccinia; ECTV, ectromelia; YLDV, Yaba-like disease; YMTV, Yaba monkey tumor; MYXV, myxoma; SFV, Shope fibroma; LSDV, lumpy skin disease; SPPV, sheeppox; SWPV, swinepox; MOCV, molluscum contagiosum. Domains in (A) and virus names in (B) are color coded by host, blue for vertebrates; pink and red for primates including humans; and purple for birds. Scales indicate approximate number of changes per amino acid residue.

resulted in lifelong immunity to the virus. Thus, it was a disease of civilization. As was described for measles virus, it must have originated from a nonhuman poxvirus, possibly a poxvirus of a domesticated animal, at the time that the development of human civilization led to the presence of large population centers and the domestication of animals.

Although infection by smallpox begins as an upper respiratory disease, spread by aerosolization of virus shed from skin poxes, infection becomes systemic and many organs are infected. The virus replicates extensively in the skin, leading to pocks that can cover the entire body. These pocks leave scars, particularly in the facial area, which mark a person for life, and at one time the fear of scarring on recovery from smallpox appears to have been at least as terrifying as the fear of death caused by the disease. A photograph of an infected child in Fig. 7.5A, taken on the eighth day of rash, illustrates the ubiquitous nature of pox formation that can occur. Figure

7.5B is a photograph of a man who was blinded by smallpox infection and illustrates the extent of facial scarring that can occur as a result of smallpox.

Immunization against Smallpox

So great was the fear caused by smallpox that a technique of immunization was developed by the tenth century, called variolation, in which people were deliberately infected by smallpox using dried skin from pox lesions. The infecting virus was delivered by blowing the inoculum up the nose or by inoculation into the skin. Infection via either method led to relatively low mortality, about 1–2% compared with 20–30% following natural infection, and the disease was milder with less scarring.

Jenner developed the modern concept of vaccination when he introduced and popularized the use of cowpox virus as a

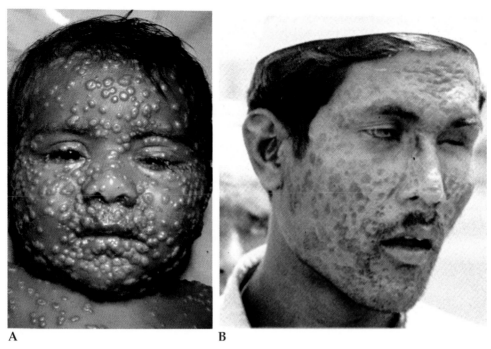

FIGURE 7.5   Effects of smallpox. (A) Child with smallpox, on the eighth day of the rash. The rash began to develop one day after the onset of fever. From Fenner *et al.* (1988) p. 15. (B) Adult after recovery, who is now blind and has deeply pigmented pocks. From Fenner *et al.* (1988) p. 57.

vaccine against smallpox. It was known that milkmaids had beautiful complexions because they never contracted smallpox, and there was some evidence that their resistance to smallpox resulted from infection with cowpox virus, which was occupationally acquired. In milkmaids, cowpox virus caused localized lesions, usually on the hands where they came in direct contact with the virus during milking, but the lesions did not spread. The virus is antigenically related to smallpox virus and will protect against infection with smallpox. Jenner's investigation in the 1790s included inoculation of a young boy with cowpox virus and then challenging him with virulent smallpox virus, to which the boy was found to be resistant, an experiment that could not be performed today because of ethical considerations.

Jenner then introduced cowpox virus as a vaccine against smallpox in the general population. This concept was controversial at first, but gradually won widespread acceptance. The virus now used for immunization is referred to as vaccinia, derived from the Latin name for cow. However, the relationship between modern vaccinia virus and cowpox virus is a mystery. At some time in the last century, the source of the virus was changed and no one knows where vaccinia virus came from. Modern vaccinia virus exhibits a fairly wide experimental host range. It has been suggested that it was derived from a domestic animal other than the cow, perhaps a horse.

Jenner "vaccinated" people by placing a drop of vaccinia-containing solution on the skin and then scarifying the skin in some way, allowing the virus to replicate in this region.

Vaccinia virus infection is localized to the area inoculated in the vast majority of people, leading to a localized lesion that results in a pock. This vaccination procedure gave solid immunity against smallpox but because of the limited replication of the virus in humans, the immunity was not considered to be lifelong and periodic reimmunization was practiced. Vaccinia virus has been used to immunize hundreds of millions of people over the years. It is generally safe, but infection of a small fraction of people results in a more serious disease. The most serious reaction in people with apparently normal immune systems is encephalitis, which is usually fatal but very rare (1 in a million vaccinees). Skin rash occurs more often but is not life threatening. In people whose immune system is impaired, progressive vaccinia may develop because the individual cannot control the replication of the virus, which results in a fatal illness.

Eradication of Smallpox

Following the introduction of vaccination, the incidence of smallpox declined and the virus was largely eliminated from developed countries. Interestingly, there also appeared a variant of smallpox that produced only a 1% mortality rate, called variola minor or alastrim (from the Portuguese word meaning something that "burns like tinder, scatters, and spreads from place to place"). Variola minor was endemic in Africa and the Americas and coexisted with variola major. Beginning in the 1960s, the World Health Organization

(WHO) began an intensive campaign to eradicate smallpox virus from human populations. Such a campaign was possible because smallpox was exclusively a human virus, there was only a single serotype, vaccination with vaccinia virus was safe and effective, and inapparent infection was virtually unknown. Every case of smallpox was tracked down, infected patients were quarantined, and all contacts were immunized against smallpox. Over a period of about 10 years, smallpox epidemics became less frequent and more completely contained until finally the virus was, in fact, eliminated (Fig. 7.6). The last case of natural smallpox occurred in 1977, and vaccination against the virus is no longer practiced because of the possible side effects of the vaccine. Smallpox is the first virus to be deliberately exterminated, and the effort and dedication required to accomplish this were remarkable.

What Do We Do Now?

The remaining, known stocks of smallpox virus are now stored in two laboratories, the Centers for Disease Control and Prevention in Atlanta and the Vector Laboratories in Novosibirsk. These viruses are still being worked with and there is an effort being made to sequence many different strains of the virus and to obtain cDNA clones of them. The WHO has adopted the principle that all remaining stocks of smallpox should at some time be destroyed,

the rationale being that smallpox virus represents a great threat to the human population, which is now largely non-vaccinated and within another generation will be completely susceptible to infection by smallpox. If the virus were to be released, whether accidentally or deliberately, it could again cause devastating epidemics. However, the original date for destruction of viral stocks has already passed and the deadline has been extended more than once. This issue has been controversial because many scientists believe that further useful information can be obtained from the study of smallpox virus. The genes that block human antiviral defenses have only recently been described, for example, and there are surely secrets of viral virulence that we do not understand. Furthermore, new forms of smallpox virus might arise. For example, monkeypox virus causes a disease in humans that is similar to smallpox but with greatly reduced transmissibility. Variants or recombinants able to spread more readily could well arise. (Could this have been the origin of smallpox in the first place?) Thirdly, there is no guarantee that if the known smallpox stocks in the United States or in Russia were destroyed, smallpox virus would cease to exist. There may be other stocks in laboratories around the world (the virus was very widespread at one time and extensively studied). It is known that the Russians weaponized smallpox at one time, and the virus would make a good agent for bioterrorists, an all-too-real concern in today's world.

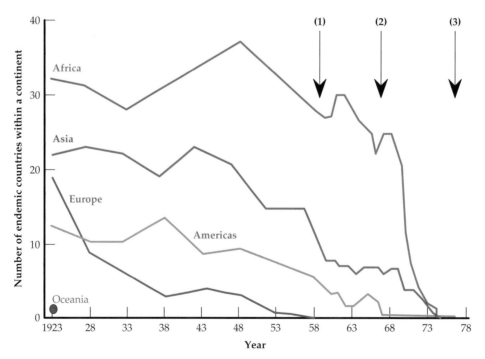

FIGURE 7.6 Number of countries reporting the occurrence of endemic smallpox between 1923 and 1978, grouped by continents. (1) marks the WHO resolution on global smallpox eradication in 1959; (2) shows the start of the intensified eradication program in 1967; (3) marks the last case of smallpox on October 22, 1977 in Somalia. From Fenner (1983).

Because of concerns that smallpox might reappear, there is a need to develop new vaccines against it. Although smallpox was eradicated using vaccinia virus as a vaccine, the virus is more reactogenic than is tolerated in modern vaccines, and the production of the vaccine uses outmoded technology. Further, people with an impaired immune system, an increasing fraction of the population, cannot be given vaccinia virus. To develop a new vaccine that is safe and effective, an animal model is important. A model of smallpox disease in cynomolgus macaques has been developed, using infection by variola virus itself, another reason to keep stocks of the virus. Approaches to producing an effective vaccine that is safer than vaccinia include using highly attenuated virus. One approach is to use replication incompetent virus as a possible vaccine. Replication incompetent viruses can be derived from vaccinia virus by deleting essential genes, and viruses such as canarypox virus are inherently replication incompetent in mammals. Such vaccine candidates can now be tested in a monkey model for their effectiveness in preventing disease following smallpox infection.

Monkeypox Virus

Monkeypox virus was thought to be a virus of monkeys but is now known to be associated with squirrels and rodents.

It infects humans under natural conditions but was thought to be only a rare zoonosis. The disease in humans caused by monkeypox virus is clinically very similar to smallpox, and it was not recognized as a distinct human pathogen until the eradication of smallpox in Africa. There are two strains of monkeypox virus, a West African strain found in Sierra Leone, Nigeria, Liberia, Ivory Coast, and Gambia and a central African strain found in Zaire (now the Democratic Republic of Congo) and the Central African Republic. The West African virus causes many fewer cases and the cases are less severe with no deaths reported to date. The central African strain has caused almost a thousand cases of severe disease since 1970 with a fatality rate of about 10% in individuals who had not been vaccinated for smallpox. Epidemics of monkeypox in the Democratic Republic of Congo in 1996–1998 resulted in about 400 cases of disease. The first wave of these epidemics lasted from February to August 1996 and involved 89 cases of clinical disease with six deaths. A follow-up investigation of monkeypox in the area in February 1997, which included a hut-by-hut search for active cases in 12 villages, gave evidence that up to 73% of monkeypox cases resulted from secondary human-to-human transmission (Fig. 7.7). Furthermore, three of the deaths were in children less than 3 years old and a large proportion of cases were in persons <15 years old. Thus, it appears

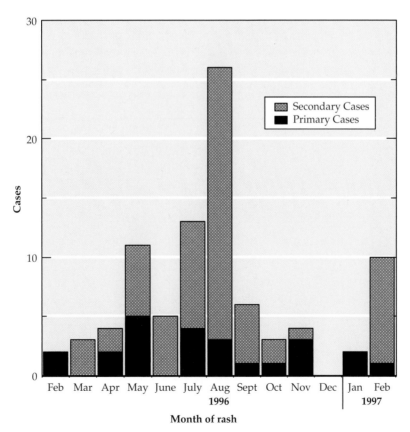

FIGURE 7.7   Number of monkeypox cases by date of rash onset in 12 villages in the Katako-Kombe health zone, Kasai-Oriental, Zaire, February 1996–February 1997. Note that most of the cases are secondary cases resulting from human-to-human transmission. From *MMWR* (1997).

that the outbreak is the result of the cessation of vaccination against smallpox in the late 1970s, since smallpox vaccination is also protective against monkeypox. Monkeypox virus is therefore a human pathogen that can cause fatal illness, spread from person to person, and cause outbreaks of disease in susceptible populations. To date, spread has been limited, but the potential for wider spread exits if the virus mutates so that it is more readily transmissible from person to person.

Monkeypox in the United States

An outbreak of monkeypox virus occurred in the Midwestern United States in 2003. The virus was imported with a shipment from Ghana of more than 700 squirrels and rodents that included Gambian giant rats, rope squirrels, brushtail porcupines, tree squirrels, striped mice, and dormice. These animals, at least some of which were infected with monkeypox, were intended as pets and were distributed to many states. One such transfer of some Gambian giant rats and dormice went to a facility in Illinois where they were housed in the vicinity of U.S. prairie dogs that were also intended as pets. The prairie dogs became infected by the virus and were subsequently distributed in seven states including Illinois. A total of 71 people in six states became ill with monkeypox contracted from the prairie dogs, 18 of whom were hospitalized. Half of the cases were confirmed by laboratory testing (Fig. 7.8). There were no deaths from this West African strain of monkeypox, which as described causes a milder illness in humans than does the Central African strain. To prevent the continuing spread of monkey-

pox in this epidemic, 30 persons were immunized with the smallpox vaccine. These vaccines included veterinarians, health care workers, laboratory workers, and household contacts of patients. One vaccinee, who reported a rash, was confirmed as having monkeypox.

Molluscum Contagiosum

Molluscum contagiosum virus is the sole representative of the genus *Molluscipoxvirus*. It is a widely distributed human virus that is spread by contact, including sexual contact. The virus causes a skin disease characterized by raised lesions. The disease is chronic but usually resolves within a few months, and the illness is considered to be trivial in immunocompetent individuals. In HIV-infected patients, however, the disease can be more troublesome. It has not been possible to grow the virus in cultured cells, and all virus for laboratory study has been derived from lesions of infected individuals, which usually contain large amounts of virus. Despite this limitation, the genome of the virus has been completely sequenced and molecular biological studies are under way.

Rabbit Myxoma Virus

Rabbit myxoma virus, a member of the genus *Leporipoxvirus*, has been widely used in Australia to control populations of the European rabbit, which was introduced with disastrous results into Australia by European settlers. The history of the virus in Australia represents a facinating

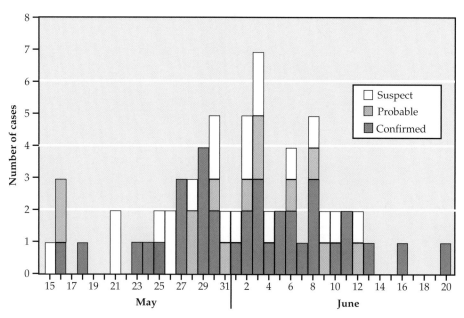

FIGURE 7.8   Number of monkeypox cases by date of illness onset during May and June of 2003 in Illinois, Indiana, Kansas, Missouri, Ohio, and Wisconsin, reported as of July 8, 2003. Date of onset was unknown for two additional cases. From *MMWR* (2003) Vol. 52, p. 642.

story of viral evolution in the field and is one of the best studied examples of coevolution of both a virus and its host. The story has important implications for our understanding of virus–host interactions.

Rabbit myxoma virus is native to the Americas, where it causes a localized skin fibroma in American rabbits. The virus is mechanically transmitted by mosquitoes. It does not replicate in the mosquito (it is not an arbovirus), but the mosquito can transmit the virus when mouthparts become contaminated by feeding on an infected rabbit with skin lesions. It was discovered early that, although the virus causes a trivial illness in American rabbits, it causes a systemic infection in European rabbits that is fatal more than 99% of the time, a disease called myxomatosis.

In Australia, a land of marsupials, the European rabbit was introduced for hunting or as potential food for foxes, which in turn had been introduced for the pleasures of fox hunting. The rabbits multiplied as rabbits proverbially do and became a plague. The enormous populations of these introduced rabbits depleted agricultural crops and also threatened to overwhelm a number of Australian marsupials by competing with them for the food supply. A significant percentage of Australian native mammals have gone extinct and rabbits have played an important role in the extinction of a number of them. Rabbits have also caused problems from erosion by depleting large areas of ground cover. Campaigns

in which people went out in large groups and killed as many rabbits as possible were undertaken in order to reduce the rabbit population, but it was a losing game. To make matters worse, the introduced foxes preferred the Australian fauna to the introduced rabbits anyway, and constituted a second plague on the native fauna, which have not evolved to deal with mammalian predators.

In an effort to eliminate or control the rabbit population, rabbit myxoma virus was introduced. At first this strategy was very successful and the rabbit population was reduced by an estimated 95%. It was believed that the virus would have to be broadcast every year in order to continue the campaign against the rabbits, but there was hope that the rabbit plague would end. However, soon after its release, new strains of virus arose that were less virulent, as illustrated in Fig. 7.9. These strains prevailed because they persisted more successfully in the rabbit population—less virulent viruses remained within a host longer and were transmitted to new rabbits more successfully than virus that rapidly killed its host. Furthermore, rabbits that survived infection with the less virulent strains were now immune to the virulent virus, making it more difficult to control the rabbits by reintroduction of a virulent strain (although rabbits have a short life span and herd immunity is not very important in resistance to disease). With time the virus became progressively less virulent but, interestingly, strains of very low virulence never

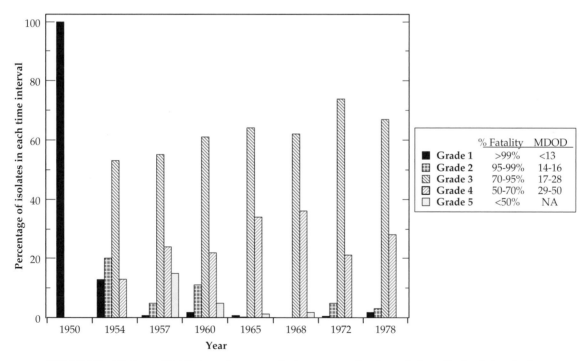

FIGURE 7.9   Virulence of field isolates of myxomatosis virus over a number of years in Australia after the introduction of a grade 1 virus for biological control in 1950. Field isolates were classed as belonging to one of the five grades of virulence on the basis of average survival times (MDOD = mean day of death) of six laboratory rabbits inoculated with each isolate. This measure closely correlates with case fatality rates (which defined the original grades of virulence). Isolates were collected over intervals of 3 to 5 years, and the collective data for each interval are plotted at a year in the midpoint of the interval. Data from Fenner (1983).

became dominant components of the virus population. Very low virulence strains did arise (Fig. 7.9, grade 5) but did not persist, presumably because they were transmitted less efficiently than were strains of moderate virulence. After a few years, the dominant strains were grades 3 and 4, which kill 50–95% of nonimmune, wild-type (i.e., not selected for resistance to myxomatosis) European rabbits.

At the same time that less virulent virus strains were being selected, rabbits that were more resistant to myxomatosis were also being selected. The enormously high death rate caused by viral infection coupled with the short generation time of rabbits rapidly led to the selection of rabbits that exhibited increased resistance to the disease caused by the virus, as illustrated in Fig. 7.10. Perhaps this is the reason that strains of low virulence first arose and then faded from the virus population. Notice that after seven epidemics of myxomatosis, virus strains of grade 3, the dominant strain in the virus population, now caused severe disease in less than 60% of selected rabbits, compared with 95% of unselected (wild type) rabbits. With continuing passage of time, the virus and rabbit populations evolved such that rabbit myxoma virus was approximately as serious for the rabbit population as smallpox was for man, that is, about 40% mortality following viral infection. Nevertheless, the rabbit population, about 600 million before myxoma virus was introduced, has never completely recovered and is about half that today. Efforts are being made to select more virulent strains of virus that would kill a larger percentage of the rabbits. Rabbit calicivirus (Chapter 3) is also being used for control of rabbits.

The history of myxomatosis in Australia makes clear that a situation in which a virus rapidly kills the vast majority of its hosts is inherently unstable. There is selective pressure on the virus to attenuate its virulence and on the host to become resistant to the virus. Because both rabbits and rabbit myxoma virus multiply rapidly, the accommodation of the virus and host occurred rapidly.

FAMILY *HERPESVIRIDAE*

There are more than 100 known herpesviruses which are currently classified into three subfamilies called alpha, beta, and gamma. All but one of the known viruses infect vertebrates. A partial listing of these viruses is given in Table 7.7, together with their hosts and the diseases they cause. Most known herpesviruses infect mammals or birds, but reptilian, amphibian, and fish herpesviruses also exist. One invertebrate virus, of oysters, has been characterized. The viral genome is large, 120–230 kb, and the viruses encode

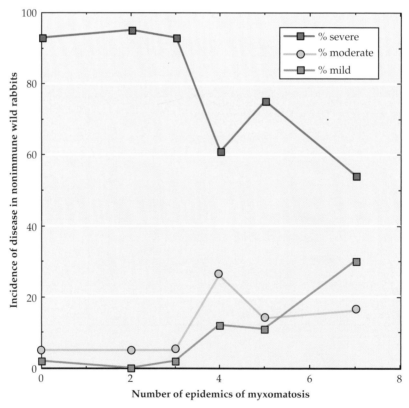

FIGURE 7.10   Incidence and severity of disease in nonimmune wild rabbits experimentally inoculated with a strain of myxomatosis of virulence grade 3 (which induces 70–95% mortality in laboratory rabbits) following a given number of epidemics of myxomatosis. Data from Fenner (1983).

TABLE 7.7   *Herpesviridae*[a]

Subfamily/genus/ members	Virus name abbreviation	Usual host(s)	Transmission	Disease
Alphaherpesvirinae				
Simplexvirus				
Herpes simplex 1	HHV-1 or HSV-1[b]	Humans	Infected cells	Cold sores on face and lips
Herpes simplex 2	HHV-2 or HSV-2	Humans	Infected cells	Genital ulcers
Monkey virus B[c]	CeHV-1 or HBV	Monkeys	Saliva	Cold-sore like lesions in macaques, fatal infection in man
Simian agent 8	CeHV-2 or SA8	Vervet monkeys		
Saimiriine herpesvirus 1	SaHV-1	Marmosets		
Ateline herpesvirus 1	AtHV-1	Spider monkeys		
Bovine herpesvirus 2	BoHV-2	Cattle		
Several less well known members that infect monkeys and wallabies are not listed here.				
Varicellovirus				
Varicella-zoster	HHV-3 or VZV	Humans	Aerosols	Chickenpox, shingles
Pseudorabies[d]	SuHV-1 o r PRV	Swine		
Equid herpesvirus 1,2,4,8,9	EHV-1,-4 etc.	Horses	Aerosols, contact	Respiratory disease, abortigenic disease
Bovine herpesvirus 1	BoHV-1	Cattle		
Felid herpesvirus 1	FeHV-1	Cats		
Several less well known members infecting deer, horses, dogs, goats, and cats are not listed separately here.				
Mardivirus				
Marek's disease[e]	GAHV-2,-3	Chickens	Aerosols, contact	T-cell lymphoma
Meleagrid herpesvirus 1	MEHV-1	Turkeys		
Iltovirus				
ILTV[f]	GaHV-1 or ILTV	Chickens		
Betaherpesvirinae				
Cytomegalovirus				
Cytomegalovirus	HHV-5 or CMV	Humans	Saliva, urogenital excretions	Disseminated disease in neonates or immunocompromised hosts leading to CNS involvement, hearing loss, and fatal pneumonitis
Cercopithecine herpesvirus -5, -8	CeHV-5,-8	Primates		
Muromegalovirus				
Murid herpesvirus-1,-2	MuHV-1,-2	Mice,rats	Saliva	
Roseolovirus				
Human herpesvirus 6	HHV-6	Humans	Contact, saliva	Exanthum subitum or sixth disease, may be associated with chronic fatigue syndrome and/or multiple sclerosis
Human herpesvirus 7	HHV-7	Humans	Saliva, urogenital excretions	Unknown
Gammaherpesvirinae				
Lymphocryptovirus				
Epstein-Barr	HHV4 or EBV	Humans	Saliva, contact	Infectious mononucleosis, Hodgkin's lymphoma, Burkitt's lymphoma
Numerous other members that infect marmosets, monkeys, orangutans, and apes are not listed separately here.				

(Continues)

TABLE 7.7   (*Continued*)

Subfamily/genus/ members	Virus name abbreviation	Usual host(s)	Transmission	Disease
Rhadinovirus				
Saimiriine herpesvirus 2	SaHV-2	Squirrel monkeys		
Human herpesvirus 8	HHV-8	Humans	Contact	Kaposi's Sarcoma
Ateline herpesvirus 2	AtHV-2	Spider monkeys		

Numerous other members that infect equids, mice, sheep, and monkeys are not listed separately here.

[a] Herpesviruses are generally worldwide in distribution.
[b] Although the newer nomenclature lists an adjective describing the species, followed by "herpesvirus" and a number, for example, human herpesvirus 3 or HHV-3, many authors still use the former names, "varicella-zoster" or VZV, so both forms are shown for the commoner viruses.
[c] Monkey virus B is cercopithecine herpesvirus 1; Simian agent 8 is cercopithecine herpesvirus 2.
[d] Suid herpesvirus 1.
[e] Gallid herpesviruses 2 and 3.
[f] ILTV, infectious laryngotracheitis virus or Gallid herpesvirus 1.

many dozens of proteins, which allows them to finely regulate their life cycle. Virions are enveloped, 100–300 nm in size, with an icosahedral nucleocapsid (Figs. 2.1, 2.5, and 2.20).

Classification of Herpesviruses

A phylogenetic tree of 32 herpesviruses belonging to nine genera is shown in Fig. 7.11. This figure illustrates the divi-

sion of the viruses into three major lineages classified as distinct subfamilies as well as the division into the nine genera illustrated. All of the viruses in this figure infect mammals or birds. The mammalian viruses, including the human viruses, are scattered among all three subfamilies, whereas the known bird viruses belong to two genera in the subfamily *Alphaherpesvirinae*. Characterized viruses of reptiles also belong to the subfamily *Alphaherpesvirinae*. However, viruses of amphibians and

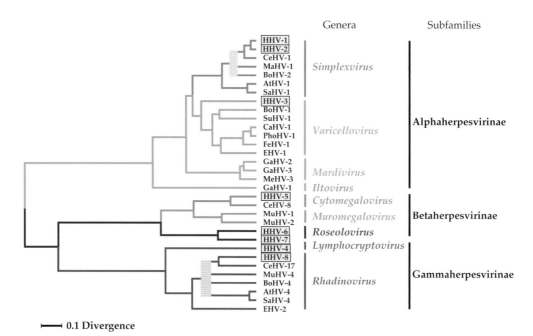

FIGURE 7.11   Composite phylogenetic tree for herpes viruses based on amino acid sequence alignments of eight sets of homologous genes. Most abbreviations are found in Table 7.7. MaHV-1, macropod (wallaby) herpesvirus 1; CaHV-1, canid (dog) herpesvirus 1; PhoHV-1, phocid (seal) herpesvirus 1. Human herpesviruses are boxed, to emphasize their distribution throughout the genera. The tree was generated by maximum likelihood; uncertain branches are shown in heavy patterned lines. Adapted from Fauquet *et al.* (2005), Figure 5, p. 211.

fish are only distantly related to these three subfamilies and will probably be classified into a new subfamily in the future. Furthermore, the oyster virus is distinct from both the mammalian viruses and the fish and amphibian viruses and probably represents yet another subfamily.

Herpesviruses are ancient viruses that have coevolved with their hosts. Figure 7.12 illustrates this for several viruses in two genera of the *Alphaherpesvirinae*. Figure 7.12A compares the tree for four simplexviruses with the tree of their host species. The trees are congruent. Similarly, Fig. 7.12B compares the tree of four varicelloviruses with that of their hosts, and again the trees are congruent. Thus, it is clear that the evolution of these herpesviruses has gone hand-in-hand with that of their hosts.

The tree in Fig. 7.11 also illustrates another interesting feature. All of the simplexviruses are primate viruses, infecting humans and various species of monkeys, except for bovine herpesvirus-2 (BoHV-2). BoHV-2 fits well in the virus tree, but cattle obviously do not fit in the primate tree of Fig. 7.12A. Thus, this bovine virus has not coevolved with cattle but appears to have been obtained from a primate. An obvious hypothesis is that this represents the spread of a virus from humans to their domestic animals, the other side of the coin from the transfer of viruses like measles to humans from domestic animals. Intriguingly, BoHV-2 causes lesions largely confined to the udders of dairy cattle, which could have come into contact with a human virus during milking.

Epidemiology of Herpesviruses

The herpesviruses have a narrow host range and any particular herpesvirus is adapted to use only a single vertebrate host in nature. All herpesviruses are capable of establishing a latent infection in their natural host whereby they persist for the life of the animal. Latent infection is established in one specific set of cells that are nonpermissive or semipermissive for virus growth and which differ from virus to virus. A different set of cells is lytically infected so as to produce a progeny virus that is capable of spreading to new hosts. Lytic infection is invariably fatal for the infected cell. Reactivation of latent virus and lytic infection of permissive cells allow the virus to reemerge unchanged and infect new hosts. For some herpesviruses reactivation occurs only sporadically, sometimes only at very long intervals, whereas for other herpesviruses reactivation occurs more or less continuously and infectious virus is usually present.

The ability of the viruses to latently infect their hosts for life presents a set of constraints and opportunities for the spread of the virus in nature different from those affecting most other viruses. The disease caused by the virus must be relatively innocuous, or at least not life threatening, in an immunocompetent native host if lifelong latent infection is to be a viable strategy for the virus. However, the ability of the virus to remain latent and reemerge after long intervals means that the virus can persist even if the human population is small. Thus, these viruses could have been present in human populations since humans arose, being passed on from their nonhuman ancestors, as suggested by Fig. 7.12. The interplay between the virus and the host required to establish lifelong infection in the face of a vigorous immune response, described later and in Chapter 10, is further evidence that the herpesviruses have coevolved with their hosts. Further support for this idea is the fact that most herpesviruses are worldwide in distribution. In the case of the human herpesviruses, most are present in all populations of people on earth, including the most isolated and remote tribes of people that have been examined.

Biology of Herpesviruses

The classification of herpesviruses was originally based on biological properties, which differ among the three subfamilies (Table 7.8). Although sequence data is now the preferred means of classification, the subdivision into three subfamilies continues unchanged. Alphaherpesviruses have a broad host range in the laboratory and will infect a wide variety of cultured cells or experimental animals. They spread rapidly in cultured cells with a short reproductive cycle and efficiently destroy the infected cells. In their natural host, latent infections are usually established in sensory neurons and lytic infection often occurs in epidermal cells. The human alphaherpesviruses belong to two genera, *Simplexvirus* and *Varicellovirus* (Table 7.7 and Fig. 7.11). Betaherpesviruses have a restricted host range and a long infection cycle in culture, and infected cells often become enlarged (cytomegaly). In the natural host, the virus is maintained in latent form in secretory glands, lymphoreticular cells, kidneys, and other tissues. The human viruses belong to two genera,

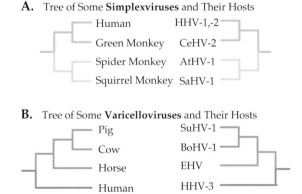

A. Tree of Some **Simplexviruses** and Their Hosts

B. Tree of Some **Varicelloviruses** and Their Hosts

FIGURE 7.12   Evolutionary relationships among the alphaherpesviruses and their hosts. (A) Comparisons of the host and viral trees for several simplexviruses. (B) Comparison of host and viral trees for several members of the *Varicellovirus* genus. Virus abbreviations can be found in Table 7.7. Adapted from McGeoch *et al.* (2000).

TABLE 7.8   Biological Characteristics of the Three Subfamilies of Herpesviruses

Characteristic	Alphaherpesvirinae	Betaherpesvirinae	Gammaherpesvirinae
Host range	Variable, often broad	Restricted	Limited to family of natural host
Reproductive cycle	Short	Long	Relatively long
Infection in cell culture	Spreads rapidly Infects many cell types	Progresses slowly	Infects primarily lymphoblastoid cells
Cytotoxicity	Much cell destruction	Enlarged cells form	Some lytic infections of epithelial and fibroblastic cells
Latency	Primarily in sensory ganglia	Maintained in many cells including secretory glands, lymphoreticular cells, kidneys, and others	Specific for either B or T lymphocytes
Characteristic genes	Genes in the U_s sequence and its flanking repeats	Genes corresponding to the HHV-5 US22 family	Genes correspoonding to BNRF-1, BTRF-1, and BRLF-1 of HHV-4
Genera	*Simplexvirus*	*Cytomegalovirus*	*Lymphocryptovirus*
Human viruses	HHV-1 (HSV-1) HHV-2 (HSV-2)	HHV-5 (CMV)	HHV-4 (EBV)
	Varicellovirus	*Roseolovirus*	*Rhadinovirus*
	HHV-3 (VZV)	HHV-6 HHV-7	HHV-8

Cytomegalovirus and *Roseolovirus*. Gammaherpesviruses have the narrowest host range and experimentally infect only members of the family or order to which the natural host belongs. They replicate in lymphoblastoid cells, and some can lytically infect epithelium and fibroblasts. They are specific for B or T cells and infection is frequently latent. There are two human gammaherpesviruses, HHV-4 belonging to genus *Lymphocryptovirus* and HHV-8 belonging to the genus *Rhadinovirus*.

The eight known human herpesviruses can be referred to as human herpesviruses (HHV) 1 through 8, but the older names for HHV-1 to 5 are still in common use and are used later in this chapter (see Table 7.7). Although they generally cause inapparent or innocuous disease, serious illness can result, particularly in neonates or in immunocompromised people. Some cancers are also associated with certain of these herpesviruses. In addition to the human herpesviruses, at least one herpesvirus of monkeys, *Cercopithecine herpesvirus-1* or B virus, causes a serious, usually fatal illness of humans that is of concern to animal handlers. These nine viruses and the diseases they cause are described later.

Structure of the Viral Genome

The genomes of herpesviruses are linear dsDNA. Repeated sequence elements are present in most and herpesvirus genomes can grouped into six classes on the basis of the location of reiterated domains, as illustrated in Fig. 7.13A. In some viruses, including herpes simplex, inverted repeats in the DNA lead to inversions of parts of the DNA sequences relative to one another during the process of replication. These different orientations are called isomers. In others, the linear sequence of the DNA is fixed (only one isomer exists). The repeated domains in the viruses in Class E of this figure give rise to four isomers; those in Class D give rise to two isomers. The genome organizations are important for the expression and replication of the genome, but grouping by genome organization does not correlate with the taxonomy of herpesviruses based on biological or sequence criteria (Table 7.7). Thus, the development of the constellation of repeated sequences may be a more recent occurrence.

Comparison of the genomes of herpesviruses makes clear that multiple rearrangements have occurred during the evolution of these viruses. This is illustrated in Fig. 7.13B where the genomes of three human herpesviruses belonging to the three different subfamilies are compared. Even though all share a substantial number of genes, the positions of these genes differ in the various genomes.

Structure of the Virion

Herpesviruses are enveloped and approximately spherical, with a diameter of 100–300 nm (Fig. 2.1). They possess a 100-nm icosahedral nucleocapsid ($T=16$) that contains at least six proteins and the viral DNA. The nucleocapsid is surrounded by or embedded within a structure known as the tegument (Figs. 2.5 and 2.20). The tegument is composed of about 20 different virus-encoded proteins and its thickness can vary, even within a single virion. Outside the tegument is the envelope containing a dozen or more virus glycoproteins. Many, perhaps all, of the glycoproteins are present in 600 or more spikes of which several different morphological types can be distinguished. One type of spike is 20 nm long and has a globule at its terminus.

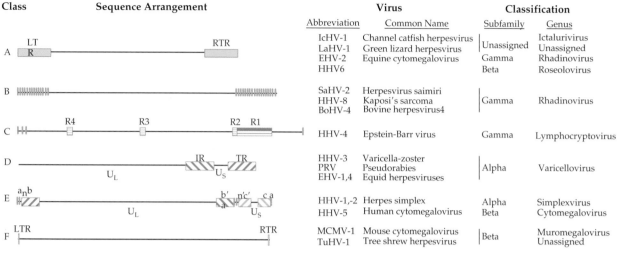

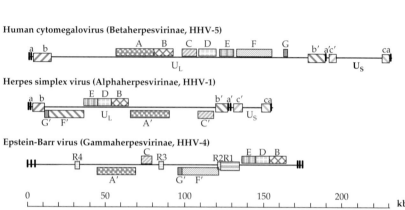

0   50   100   150   200   kbp

FIGURE 7.13   Upper panel: Grouping of herpesviruses by their genome organization. The narrow lines are the unique regions of the genomes and the rectangles are repeated domains. These are designated the left and right terminal repeats (LTR and RTR) in Group A, internal repeats R1 to R4 in Group C, and the internal and terminal repeats (IR and TR) in Group D. In Group E, both the long and short unique regions are flanked by inverted terminal repeats (shown as **ab** and **b′a′**). In contrast to the LTR and RTR in group A which are almost 10 kbp long, the LTR and RTR in Group F are only 30 bp. Note that this grouping does not correspond exactly with the taxonomy of these viruses as shown in Table 7.7, which is based on a number of biological characters. Redrawn from Fauquet *et al.* (2005) p. 195. Lower panel: A more detailed view of human cytomegalovirus, herpes simplex type 1, and Epstein–Barr virus showing conserved sequence blocks, which are distinguished by color and pattern. Blocks shown below the midline in HSV and EBV are those in which the orientation is reversed from that in HHV-5. Redrawn from Gompels *et al.* (1995). Note that in HHV-5 and HHV-1 the entire U_L region can exist in either of two orientations relative to U_S.

The nucleocapsid assembles in the nucleus. The mechanism by which it is enveloped and released from the cell is controversial, and more than one mechanism may be used by various herpesviruses. One popular model is that the capsid buds through the nuclear membrane into the cytoplasm (Fig. 2.25A). The nuclear envelope is a double membrane structure and the nucleocapsid acquires an envelope upon budding through the inner leaflet of the double membrane, then loses it upon fusion with the outer leaflet, resulting in the naked nucleocapsid being deposited in the cytoplasm. In the cytoplasm the nucleocapsid acquires the tegument by means of a complex series of protein–protein interactions. The tegument interacts with both the nucleocapsid and, during budding, with the viral glycoproteins, and serves the same function as do the matrix proteins of enveloped RNA viruses. The tegumented nucleocapsid buds into trans-Golgi vesicles, acquiring an envelope, and the completed virion is released from the cell when the vesicle fuses with the plasma membrane.

Infection by Herpesviruses

Many, perhaps all, herpesviruses utilize accessory receptors to accelerate virus binding and entry into cells, as described in Chapter 1, and many or all appear to be able to utilize more than one high-affinity or entry receptor. In the case of HSV, glycosaminoglycans serve as accessory receptors and three classes of entry receptors can be used. These are a member of the tumor necrosis receptor family called herpesvirus entry mediator (HVEM), two cell adhesion molecules that are members of the Ig superfamily, called nectin-1 and nectin-2, and specific sites in certain isoforms of heparan sulfate. In the case of Epstein–Barr virus, a protein called complement receptor 2 (also referred to as CD21) serves as an accessory receptor and entry receptors include HLA class II molecules. Entry of herpesviruses may occur by fusion with the plasma membrane or fusion with an endosomal membrane, depending on the virus and the cell type. Fusion requires the presence of cholesterol in the viral membrane in at least some herpesviruses, a requirement also shown by alphaviruses, influenza virus, and HIV-1.

Herpesviruses contain multiple glycoproteins in their envelope, and fusion requires the activity of several of these glycoproteins. Thus, for HSV, the activity of 4 of the 12 or more glycoproteins in the envelope are required for fusion.

REPLICATION OF HERPESVIRUSES

The best studied herpesvirus is herpes simplex virus type 1, which has been used as a model for the entire family, and a more detailed description of its replication will be presented in the section that discusses this virus. Here some generalities of the replication cycle of herpesviruses are described.

It has long been thought that herpesvirus DNA replicates by a rolling circle mechanism as illustrated in Fig. 1.9 Although the DNA is linear in the virus, there is evidence that it has no ends in the infected cell, indicative of circularization. As described in Chapter 1, a circular genome obviates the need for a special mechanism to repair the ends of the DNA during replication. Recent data suggesting that during productive infection, as opposed to latent infection, the DNA remains linear have not been confirmed and the favored hypothesis remains that replication of the DNA is by a rolling circle mechanism.

Replication of herpes DNA occurs in the nucleus, and all herpesviruses encode a large number of enzymes that are involved in nucleic acid metabolism. Among these proteins are a DNA polymerase and a protein that binds to the viral origins of replication in order to initiate DNA replication.

Herpesviruses encode more than 70 proteins. The promoters used to transcribe mRNAs fall into different classes such that there is a temporal program for the expression of genes during the lytic cycle. The proteins encoded by the first genes to be expressed, the immediate-early genes, have regulatory functions. Expression of these genes permits the expression of the early genes, most of which are involved in DNA replication. Expression of the early genes, in turn, permits the expression of the late genes, most of which encode structural proteins required for assembly of virions. The mRNAs are transcribed by host RNA polymerase II and the promoters are in general 40–120 nucleotides in length.

Most promoters reside outside the open reading frames, but some are found within them. The genome organization is complex. Some genes lie within other genes, and for a few genes both strands of the DNA are transcribed into RNA.

The use of host RNA polymerases to transcribe mRNAs limits the host range of a herpesvirus, since cells that do not express the factors required for recognition of the viral promoters are not susceptible to a complete replication cycle. The mRNAs are polyadenylated following the cellular AATAAA consensus poly(A) signal and exported to the cytoplasm for translation. From most transcripts only one protein is translated. Most mRNAs are not spliced, but some transcripts are spliced or multiply spliced and partially overlapping transcripts may exist that form nested sets, using a common polyadenylation signal. Many of the proteins synthesized are dispensable in cell culture and function to extend the host range and tissue tropism of the virus or to subvert host antiviral defenses.

Herpes Simplex Viruses (HHV-1 and -2)

Herpes simplex virus (HSV) causes fever blisters or cold sores around the lips or in the genital area. The disease caused by the virus has been known for thousands of years. The characteristic vesicles were described by the ancient Greeks and have been noted in writings through the ages. The name of the virus comes from the Greek word *herpes*, to creep or crawl, referring to the lesions caused by the virus. HSV exists as two serotypes, HSV-1 and HSV-2, also known as HHV-1 and HHV-2. These alphaherpesviruses share about 50% sequence identity and produce a similar disease, but usually infect different parts of the body. HSV-1 infects the facial area, whereas HSV-2 infects the genital area. In humans, the viruses lytically infect epidermal and mucosal cells; they latently infect neurons. In the laboratory they lytically infect cells of many different origins and will infect many experimental animals. Because of the relative ease of experimental manipulation, HSV has been intensively studied as a model for the entire group of herpesviruses. The large size of the genome, 150 kb, and the large number of encoded genes have made a detailed understanding of the virus genome organization very complicated. However, due to the efforts of large numbers of workers, maps such as that shown in Fig. 7.14 can now be constructed. This map illustrates the different genes of HSV, their functions, and their classification into three temporal groups described later. The complexity of the genome and the existence of very many genes are clear from this diagram.

Lytic Infection by the Virus

A schematic representation of the lytic cycle of herpesvirus infection is shown in Fig. 7.15. Infection is normally initiated by fusion of the viral membrane with the cell plasma membrane. The nucleocapsid is then transported to nuclear pores

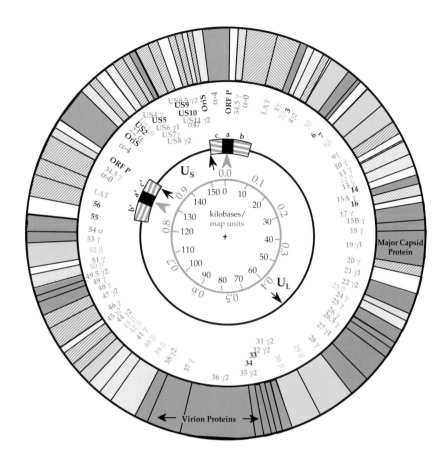

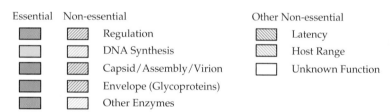

FIGURE 7.14   Functional organization of the HSV-1 genome. Circles are described from the center outwards. Circle 1 shows kilobase pairs in black and map units in green. Circle 2 shows the general organization of the gemone: U$_L$, the long unique region; U$_S$, the short unique region; and a, b, c, the inverted repeats. The black arrows are the three origins of DNA replication, and the green arrowheads show the sites of cleavage/linearization of concatameric or circular DNA. Circle 3 identifies the ORFs, color coded according to their kinetic class (α in **red**, β in ochre, γ in **blue**). Black numbers are ORFs not belonging to a particular class. The outer circle indicates the functions of the ORFs, where known, by color coding as indicated in the key below. Solid colors indicate ORFs required for replication in tissue culture cells, while the corresponding patterned ORFs can be deleted without affecting replication in culture. Only one of the two copies of α-0 (diagonal red stripes) is required. Adapted from Fields *et al.* (1996) p. 2245.

where the viral DNA is released and enters the nucleus (stages 1–3). There the viral DNA is transcribed by RNA polymerase II to produce mRNAs. Of the 80 or so mRNAs produced, only 4 appear to be spliced. Transcription is regulated and can be divided into three phases, α, β, and γ. Transcription of α genes, also called immediate-early genes, occurs immediately on entry of the DNA into the nucleus. Transcription of these genes is transactivated by TIF, a virus-encoded protein that is present in the tegument of the virion. TIF interacts with a cellular transcription factor called Oct-1, which recognizes octomer sequences, and, to simplify somewhat, a complex of Oct-1, TIF, and another cellular factor called C1 binds to the consensus sequence TAATGARAT in the HSV genome (R=A or G) (stage 3). This binding results in transcription of the five α genes (stage 4) that are translated into proteins called ICP0 (ICP = intracellular protein, to distinguish them from virion proteins), ICP4, ICP22, ICP27, and ICP47 (stage 5). These proteins have regulatory roles in viral replication and are required to activate the β genes (stage 6). ICP4 transactivates HSV genes and functions together with ICP0. ICP27

A. Early Events

B. Late Events

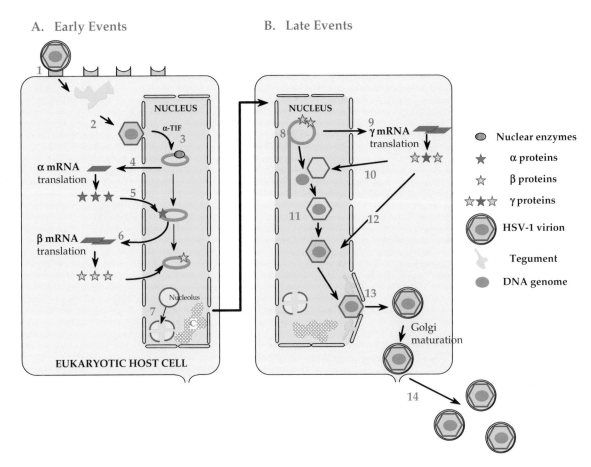

FIGURE 7.15   Replication of a herpesvirus. (A) Early events: Stage (1) attachment to cellular receptor and entry by pH-independent fusion with plasmalemma; stage (2) release of tegument proteins causing shutoff of host protein synthesis, nucleocapsid goes to a nuclear pore; stage (3) α-TIF (VP16) is transported to the nucleus; viral DNA enters nucleus and circularizes; stage (4) transcription of early α genes by nuclear enzymes, and export of α mRNA; stage (5) immediate early α proteins are transported to the nucleus; stage (6) α proteins are involved in β mRNA synthesis; stage (7) chromatin (C) is degraded and nucleoli disaggregated. (B) Late events: Stage (8) β proteins replicate DNA by rolling circle to give head-to-tail concatemers; stage (9) β proteins lead to transcription of late γ mRNAs that are translated into structural proteins; stage (10) formation of empty capsids; stage (11) packaging of unit-length DNA into capsids; stage (12) addition of further structural proteins; stage (13) particles receive envelope (and tegument?) at nuclear membrane; stage (14) particles mature in Golgi and exit by exocytosis. Redrawn from data in Roizman and Sears (1990).

activates expression of β genes and blocks splicing of cellular pre-mRNAs, thereby interfering with host protein synthesis. It shuttles between the nucleus and the cytoplasm at late times. ICP22 is not required in cultured cells and its function is unknown. ICP47 blocks presentation of antigens to cytotoxic T cells (CTLs) (see Chapter 10).

Most of the β genes encode proteins required for DNA replication, which include a DNA polymerase, a primase/helicase, a DNase, both double-strand and single-strand DNA-binding proteins, thymidine kinase, ribonucleotide reductase, dUTPase, uracil DNA glycolase, and a protein kinase. The activities of these proteins result in morphological changes in the nucleus, which include fragmentation of the nucleolus and degradation of the host-cell chromosomes (stage 7), and allow DNA replication (stage 8) to begin. There are three origins of replication, which are 800–1000

nucleotides in length and have a 100- to 200-nt critical core, but only one origin appears to be required. During DNA replication, up to 50% of the DNA in the cell becomes viral.

DNA replication is required for transcription of the γ genes, most of which encode the proteins that form progeny virions, of which there are more than 30 (stages 9 and 10). Assembly and release of virions (stages 11, 12, and 13) (see also Fig. 2.25A) was described before. Cleavages in capsid proteins occur during assembly, associated with uptake of DNA, and may serve a maturation function.

More than half of the 80 or so genes in HSV are not required for replication of the virus in cultured cells. However, all appear to be required for infection of humans and maintenance of the virus in nature.

Host-cell macromolecular synthesis is shut off after lytic infection. A component of the virion tegument called vhs (for *v*irion *h*ost *s*hutoff) is an RNase that inhibits host protein synthesis by degrading cellular mRNA, so that inhibition begins very early. The enzyme appears to show some specificity in the mRNAs degraded, as some mRNAs are degraded rapidly, some more slowly, and some not at all. This enzyme also accelerates the turnover of viral mRNAs, helping the transition from one stage of the infection cycle to the next. At late times after infection, the enzyme is rendered inactive by another viral protein. Later synthesis of new viral proteins leads to a more profound inhibition of host expression, due in part to the fragmentation of the nucleolus and degradation of the host cell chromosomes. These events invariably result in the death of the host cell.

Latent Infection

After the infection of epithelial tissues, HSV infects sensory nerves that serve these tissues and establishes a latent infection. Virus, probably as nucleocapsids, is transported up axonal processes to sensory ganglia, where it is estimated that 5–10 copies of viral DNA, in the form of circular episomes, take up residence. Less than 1% of the neurons within a ganglion appear to be latently infected. Much of what we know about latent infection comes from studies of animal models. HSV will infect and establish a latent infection in mice, guinea pigs, and rabbits, but it is not known how faithfully these animal models reflect the situation in humans. Only one viral transcript is detected in latently infected neurons, called latency-associated transcript or LAT. This transcript promotes neuronal survival by interfering with apoptosis of infected neurons by means of an RNA silencing pathway (Chapter 10). The detailed mechanisms by which latency is established and maintained, and of how latency is abolished upon reactivation of the virus, are not understood. It is assumed that part of the answer is that neurons are nonpermissive, or at best semipermissive, for virus replication.

Reactivation of virus occurs sporadically, in response to stressful stimuli such as fever, exposure to UV light, menstruation, or emotional stress. The frequency of recurrence varies in different people from monthly to less than once per year. Stresses that are well known to induce reactivation include high fever resulting from infection with influenza virus and prolonged exposure to sunlight. On induction, limited replication of virus occurs in the neuron and it travels down the axon, where it infects epithelial cells served by that neuron. Thus, fever blisters erupt in the same tissues as were originally infected, and these lesions contain infectious virus. Classically, these lesions occur around the lips (HSV-1) or in the genital region (HSV-2). Virus replication in epithelial cells is quickly controlled by the immune system and the lesions heal within 2 weeks or so. Although not yet resolved, it appears that the limited replication of the virus in the neuron is probably fatal for that neuron.

Primary Infection and Maintenance of the Virus

Primary infection is established when a seronegative individual comes in contact with the virus, usually from a person who is secreting the virus at the time. Thus, a person with a reactivated infection can serve as the source of infection, but virus may be actively secreted even when no lesions are present. Acute infection is accompanied by the formation of vesicles that are sites of virus replication and contain infectious virus. In the case of HSV-1 these vesicles are normally found in the facial skin and in the oral mucosa, and latent infection is established in the trigeminal ganglion. HSV-2 is sexually transmitted and infects the genital area. Latent infection is established in neurons in the sciatic ganglion. At one time, HSV-2 was thought to be a causative agent of cervical carcinoma, but this disease is now known to be associated with papillomavirus infections, which are also sexually transmitted and produce warts in the genital region.

HSV-1 is extremely common in human populations. It has been found in every population examined, with from 50 to more than 90% of adults having been infected. Primary infection often occurs early in life, but may be delayed in some fraction of the population, especially in developed countries. Because of its mode of transmission, HSV-2 is less common and primary infection occurs later in life. Estimates of its prevalence in human populations center near 10%, but the only population found to be completely free of antibodies to HSV-2 was a group of Roman Catholic nuns. The acquisition of seropositivity to HSV-1 and HSV-2 in the United States in different populations is shown in Fig. 7.16.

Serious Disease Caused by HSV

Primary infection with HSV is usually inapparent or produces only minor illness. In one study of children, for example, 70% of infections were found to be asymptomatic. However, HSV infection of neonates is almost always symptomatic and frequently fatal. Infection *in utero*, during delivery, or shortly after birth usually leads to a disseminated infection often accompanied by encephalitis. The neurotropism of the virus also leads to serious illness on occasion in postneonatal individuals. The Centers for Disease Control and Prevention estimates that about 50 cases of herpes encephalitis occur yearly in the United States that are usually fatal if untreated, and this incidence may be underestimated. Of these cases, about half are due to primary infection and half to reactivated infection. HSV keratoconjunctivitis also occurs and can lead to impairment of vision. Herpetic whitlow is an occupational hazard of dentists and other health care workers, characterized by painful herpetic lesions on the fingers. And as is true of many viral infections, HSV infection or reactivation can be very serious in individuals whose immune function is compromised by suppressive therapy for organ transplant or by infection with HIV.

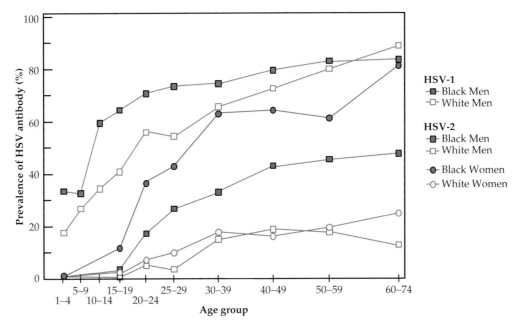

FIGURE 7.16   Seropositivity to herpes simplex virus types 1 and 2 as a function of age, sex, and race in the United States in 1978. Data for this figure came from Johnson *et al.* (1989), and Nahmias *et al.* (1990).

HSV infections can be treated with acyclovir, a guanine analogue that is incorporated into DNA and results in chain termination. It has little toxicity for host cells but inhibits the replication of HSV-1, HSV-2, and VZV DNA. It is less effective against other herpesviruses.

Interference with Host Antiviral Defenses

HSV interferes with several host defense mechanisms. These activities will be described in more detail in Chapter 10, in which the host defense mechanisms themselves are described, but a brief summary of HSV activities is presented here. The virus interferes with the interferon system, with the lysis of infected cells by cytotoxic T lymphocytes (CTLs), with complement-mediated lysis of infected cells, with antibody-dependent lysis of infected cells, and with the lysis of infected cells by a cell suicide pathway called apoptosis. The antiviral effect of interferon is blocked by viral protein 34.5, which causes a cellular phosphatase to remove the inactivating phosphate put on eukaryotic translation initiation factor eIF-2 (Chapter 10). CTL lysis of infected cells is blocked by protein 47, which inhibits the presentation of peptide antigens to the T cells by major histocompatibility complex class I molecules. Complement-mediated lysis is blocked by HSV proteins that interact with complement. Antibody-dependent cellular toxicity is blocked by HSV-1 IgG Fc receptors, composed of gE and gI, that are expressed on the surface of the infected cells. Finally, HSV gene products suppress apoptosis by the infected cell. The net result is to allow the production of

virus by the infected cell for an extended period before the host antiviral defenses can shut it down.

Establishment of latency in neurons is also an important part of the strategy evolved by HSV to avoid host defenses. Neurons are immunologically privileged, and elaborate mechanisms to protect the infected neuron for long periods of time are not necessary. Furthermore, because neurons are nonrenewing and long lived, the establishment of latency in these cells allows the virus to persist indefinitely even in the absence of reactivation.

Varicella-Zoster Virus (HHV-3)

Varicella-zoster virus (VZV, also known as HHV-3) is an alphaherpesvirus that is the prototype member of the genus *Varicellovirus*. Other members of this genus infect monkeys, horses, and pigs (see Table 7.7 and Figs. 7.11 and 7.12). The VZV genome is 125 kb in size and contains at least 69 different genes, of which all but 5 are homologous to genes in HSV. The homologous genes are almost all colinear with the corresponding genes in HSV, and thus the gene map of VZV is essentially the same as that for HSV. Molecular studies of VZV have been hampered by the inability to produce high titered virus stocks in cultured cells, because the virus remains cell associated. Where known, however, the VZV life cycle closely resembles that of HSV, as would be expected from their close relationship. This close relationship also exhibits itself biologically: VZV, like HSV, lytically infects a number of different cells but most characteristically epidermal cells resulting in skin lesions, and VZV, like HSV, establishes a lifelong latent infection in sensory ganglia. The vesicle fluid present in VZV skin lesions contains large amounts of free virus and can spread the disease to susceptible persons.

Chickenpox

VZV causes two different diseases known as chickenpox (varicella) and shingles (zoster). Chickenpox is a highly contagious childhood disease contracted by contact with other children with chickenpox or with an adult with shingles. The virus is transmitted by the aerosols and virus replication begins in the upper respiratory tract. It later disseminates through the bloodstream to other areas of the body. The characteristic feature of the disease is a rash of vesicular lesions in the skin that are often quite itchy. Up to 2000 occur in some patients, but fewer than 300 is the norm. Other symptoms include fever, malaise, and loss of appetite.

The disease, which has an incubation period of 10–21 days, is normally self-limited, lasting a week or less, but serious complications can occur. The most common complication in otherwise healthy children is bacterial infection of the skin lesions, which can become serious. Rare complications include viral pneumonia, central nervous system involvement leading to encephalitis or cerebellar ataxia (loss of muscle coordination during voluntary movements), involvement of the liver leading to hepatitis, or involvement of other organs. Primary varicella infection of adults is a more serious illness than primary infection of children and complications are more frequent. Viral pneumonia is not uncommon in adults, and adult infection can result in male sterility or acute liver failure, albeit rarely, as well as other complications.

Shingles

Following primary infection, which results in chickenpox, VZV sets up a latent infection in dorsal root ganglia in the spinal cord, where it may reactivate later in life. Reactivation is less common than for HSV-1 and is age related, occurring more frequently in older people, presumably as a result of waning immunity. Reactivation may occur without symptoms, but most commonly reactivation produces the disease known as shingles. Shingles is characterized by painful eruptions of vesicular lesions in skin, usually in the upper back, served by a single sensory ganglion (and therefore the lesions do not cross the midline). The disease normally resolves within a few weeks, but neuralgia (nerve pain, from *neuro* = nerve and *algia* = pain) can continue for up to a year or more and be quite debilitating. More extensive dissemination of the virus occurs in a significant percentage of patients, most of whom have some underlying immunologic defect or are immunosuppressed. Dissemination can result in serious complications, as described for primary varicella infection. The

vesicular lesions of shingles contain live virus that can infect children and give them chickenpox. Thus, the virus is able to remain latent for decades and then erupt in an essentially unchanged form to start a new epidemic of chickenpox.

Episodes of zoster are more frequent in older people but a second episode of zoster in a person is rare. It appears that the reactivation of the virus leads to a boost in immunity to the virus that prevents further episodes. Immunity to VZV is primarily a function of CTLs. Children that suffer from agammaglobulinemia, who are unable to make antibodies, have a normal course of infection by VZV, but children that are deficient in CTL production often die. Furthermore, it has been found that an accelerated CTL response is associated with asymptomatic infection or a mild disease. Finally, reactivation of VZV to produce zoster is correlated with decreased CTL responsiveness against VZV, but not with decreased titer of IgG antibodies against the virus.

VZV in At-Risk Populations

Varicella or zoster is a serious illness in people with compromised immune systems, and zoster is a frequent complication in patients undergoing immune suppression or who have AIDS or leukemia. Before the introduction of antiviral drugs, in particular acyclovir, which is fairly effective for treatment of VZV infections, children who contracted varicella while undergoing immunosuppressive therapy for leukemia suffered a very high rate of visceral dissemination and pneumonia, with a fatality rate of about 10%.

Primary infection with VZV is also serious in pregnant women, leading to significant mortality in both the mother and the infant. Congenital varicella syndrome may occur when infection is in the first trimester of pregnancy, during active fetal organogenesis. Varicella infection of the neonate is serious as well, with a high mortality rate in the absence of treatment.

Epidemiology of VZV

The geographical pattern of infection by VZV is peculiar. The virus has a worldwide distribution but infection is much more common in temperate regions. In temperate regions, infection by VZV is almost universal and occurs mostly in early childhood, in association with epidemics that have a peak frequency in winter and spring. In tropical regions, however, only about half of the population contracts chickenpox. The difference in attack rate is not due to differences in susceptibility to the virus, because the attack rate is very high when uninfected adults move to temperate climates. Why this difference in attack rates occurs remains a mystery.

It is interesting to consider the differences in the epidemiology of VZV and HSV-1 and the rationale for such differences. HSV reactivates fairly often and the virus is typically spread to young children by adults with reactivated HSV-1 when they fondle or kiss the child. There is no requirement that the virus spread from child to child in order for it to spread within a population. Because of this, virus perpetuation does not require large-scale production of virus during primary infection and primary infection is usually asymptomatic. In contrast, VZV reactivates very infrequently. Epidemics may begin with the exposure of a susceptible child to an adult with shingles, but the major mechanism for dispersal of the virus in a population is by epidemic spread among young children. The requirement for child-to-child spread for perpetuation requires that large-scale virus production occur during primary infection, and this results in the primary infection being symptomatic.

There is now a live attenuated virus vaccine for chickenpox, licensed in 1995, that was originally developed because of the severe complications of chickenpox in children undergoing chemotherapy for cancer. Although the disease is not normally serious and the vaccine is not mandated by the authorities, it has nonetheless been well received in the United States. Before 1995, there were 4 million cases of varicella annually in the United States, with approximately 100 deaths and 10,000 hospitalizations each year. Because chickenpox ceased to be a notifiable disease in 1997, complete statistics on the current incidence of chickenpox in the United States are not available. However, 20 states still report

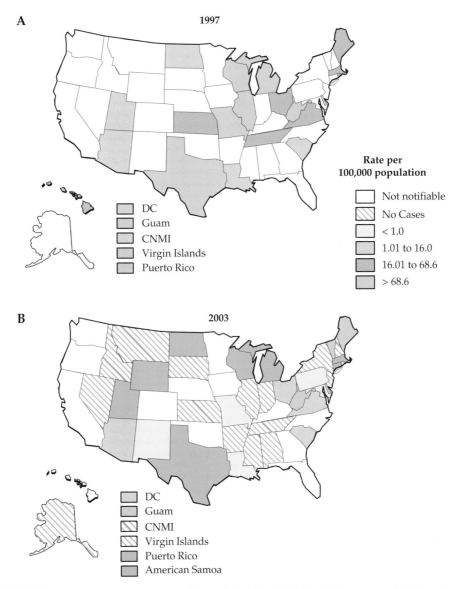

FIGURE 7.17   (A) and (B) Incidence of varicella in the United States in 1997 and in 2003, respectively. In neither year was varicella a nationally notifiable disease, but some states have active surveillance programs. The color indicates the incidence rate per 100,000 population. The overall rate for the entire United States was 7.27 in 2003. Data from *MMWR, Summary of Notifiable Diseases-United States for 1998* and *2003*.

cases to the Centers for Disease Control and Prevention. The incidence of chickenpox by state for those states that still report is shown in Fig. 7.17 for 1997 and 2003 and illustrate the overall decline in the incidence of chickenpox. The decline is illustrated by year for four states that still report (Fig. 7.18). In those states, the incidence of chickenpox has declined by about 80% over a 10-year period. A different formulation of the VZV vaccine has been licensed for adults over 60 years old, designed to boost the immune response to VZV and thus prevent shingles.

Epstein–Barr Virus (HHV-4)

Epstein–Barr virus (EBV or HHV-4) is a gammaherpesvirus. It is named after Tony Epstein and Yvonne Barr, who first described the virus in tumor cells from patients with Burkitt's lymphoma. EBV is classified as a member of the genus *Lymphocryptovirus*. About 20 viruses infecting Old World primates and about 10 viruses that infect New World primates have been identified that also belong to this genus. Other members of the gammaherpesvirus subfamily belong to the genus *Rhadinovirus* and include herpesvirus saimiri (a virus of squirrel monkeys) and HHV-8, the virus responsible for Kaposi's sarcoma. Equine herpesviruses 2 and 5, originally

classified as betaherpesviruses, are also now considered to be gammaherpesviruses. Gammaherpesviruses establish latent infection in B lymphocytes and cause these cells to proliferate. This ability to induce proliferation can result in cancers in their native host or in related species.

Primary Infection with EBV

EBV infection is virtually universal. More than 90% of the world's adult population is persistently infected by the virus. Primary infection occurs by transmission of virus present in saliva. In most human societies, the majority of the population is infected by age 3. In developed countries, however, infection is often delayed until the teens. Primary infection in infants is normally asymptomatic, but in young adults primary infection often results in the disease known as infectious mononucleosis or the "kissing disease." This disease is characterized by fatigue, fever, rash, and swelling of lymph nodes, the spleen, and, in a minority of patients, the liver, for extended periods of time.

Only B cells can be infected by free virus and humans who cannot produce B cells, a syndrome called X-linked agammaglobulinemia, cannot be infected by EBV. The receptor used to enter B cells is a protein called CD21, a member of the Ig

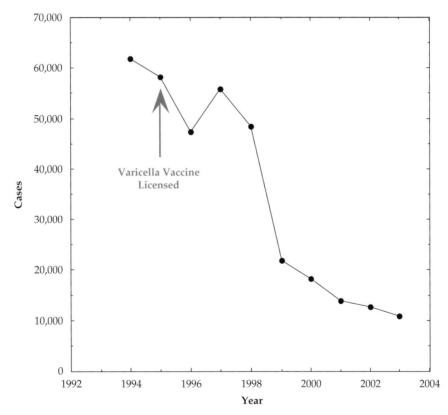

FIGURE 7.18   Number of reported cases of varicella in Michigan, Rhode Island, Texas, and West Virginia from 1994 to 2003. These states maintained adequate surveillance and in each of the years 1990 through 1995 reported cases consisting of >5% of their birth cohorts. The number of cases in 2003 represents an 81% decline compared with cases reported in the 3 years before the vaccine was licensed in 1995. From *MMWR, Summary of Notifiable Diseases-United States, 2003.*

superfamily expressed at high levels in B cells. Infection of B cells does not normally result in the production of infectious virus, although B cells that differentiate to plasma cells may undergo a productive infection cycle. Infection of B cells does result in the stimulation of the cells to proliferate, thus expanding the number of infected cells. This proliferation leads to a potent T-cell response that controls the number of B cells, and it is these B-cell–T-cell proliferative cycles that can result in the symptoms of infectious mononucleosis. As a result of T-cell killing of infected B cells, the virus life cycle changes to establish a latent infection in which a limited set of viral genes is expressed. Latent infection of memory B cells results in lifelong persistence of the virus.

Newly infected B cells are able to transfer the virus to epithelial cells in which the virus undergoes a complete replication cycle, and this appears to be the primary mechanism by which free, infectious virus is produced. Infection of cells in the oral mucosa results in virus being present in saliva, by which means the virus can be transmitted to uninfected individuals. Lytic replication of virus in the oropharyngeal epithelium persists indefinitely, as persistently infected B cells continue to seed it, although the extent of shedding of virus into the saliva declines with time.

Replication of EBV

EBV will readily infect B cells in culture and establish a latent infection in which a limited set of viral genes is expressed but production of progeny virus is limited. Although it has been shown recently that epithelial cells, in which the virus undergoes a complete replication cycle, can be infected by coculture with freshly infected B cells, most studies of EBV infection have used B cells.

The infection of B cells involves complicated interactions of the virus with the host cell. Three different forms of latent infection have been distinguished in B cells. These were first described in cells isolated from tumors but can now be reproduced in cultured cells. These different forms of latency, referred to as latency (Lat) I, II, and III, differ in the extent to which the EBV genome is expressed, as illustrated in Fig. 7.19. Lat I cells express only one protein, EBNA1 (*Epstein–Barr nuclear antigen*). They also express RNAs that are not translated, among them RNAs called EBERs (*Epstein–Barr early RNA*) and BamAs (these last are transcribed from a region of the genome found in a BamHI restriction fragment called the A fragment). Lat II cells express additional proteins called LMPs (*latent membrane protein*). Lat III cells express still more proteins

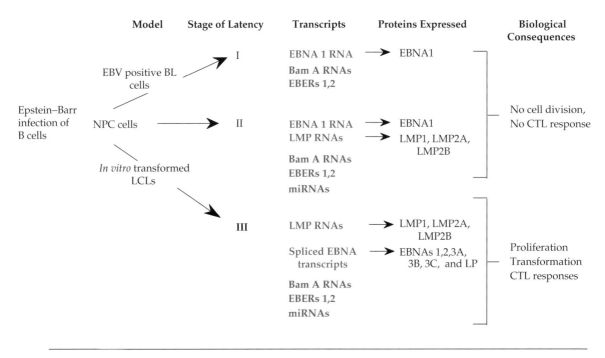

LCLs	Lymphoblastoid cell lines	EBNA-1 RNA	Translated RNAs
BL	Burkitt's lymphoma cell lines	Bam A RNAs	Nontranslated RNAs
NPC	Nasopharyngeal carcinoma cell lines		

FIGURE 7.19   Stages of latency after Epstein–Barr infection. Latency I was first described in BL cell lines, but can be reproduced by fusion of *in vitro* transformed LCLs with EBV-negative hemopoietic cell lines. Similarly, latency II was first described in NCP cells but can be reproduced by fusing LCLs with certain human epithelial, fibroblast, or hemopoietic lines. Data from Fields *et al.* (1996), pp. 2399–2402.

called EBNAs 2, 3A, 3B, 3C, and LP. One model for the functions of these different types of latency is that Lat III is first established and serves to amplify the pool of infected B lymphocytes, since Lat III cells are stimulated to divide. Lat III cells are targets of CTLs, however, which kill these cells in an attempt to control the viral infection. In contrast, Lat I and II cells are resting cells. The more limited set of proteins produced does not stimulate B cells to proliferate nor make the cells targets of CTLs, and it is these cells that maintain the latent infection in the host. In this model, during the lifelong infection by the virus, Lat I and II cells may sporadically become permissive for lytic growth and produce virus, or may sporadically switch to the Lat III state, which results in cell division and the stimulation of CTLs that control these cells. Colonization of B cells is limited, and it is estimated that only about one in 10^5 B cells is infected by virus in asymptomatic carriers.

The RNAs and proteins expressed in latently infected cells are important for maintaining the latent state and inducing cellular proliferation. EBV DNA has three high-affinity binding sites for EBNA1 and this protein enables the circular EBV DNA to be maintained as an episome in the infected B cells. LMP1 is an integral membrane protein with six membrane-spanning domains. It has transforming activity when expressed in certain rodent cell lines and is presumably important for stimulation of B-cell division. It also induces the production of cellular bcl-2, which protects the infected B cells from undergoing apoptosis. LMP2 is also an integral membrane protein. It appears to prevent complete activation of the B cell so that the infection remains latent. EBNA LP, 2, 3A, and 3C are required for continuing growth of the infected B cell.

The EBER RNAs are abundantly produced in infected cells. They are small, nonpolyadenylated RNAs that are located mainly in the nucleus. They appear to be similar to the VA RNAs of adenoviruses described later. The various small RNAs produced upon EBV infection appear to be involved in countering host antiviral defenses and in regulating aspects of cellular metabolism, functions that are important for the maintenance of latency and for the maintenance of lifelong persistent infection. For example, microRNAs produced by EBV may target regulators of cell proliferation and apoptosis, chemokines and cytokines, transcriptional regulators, and components of signal transduction pathways. The topic of microRNAs is covered in Chapter 10.

Latently infected B cells sporadically become permissive for virus replication, perhaps as a result of differentiating into plasma cells. In the infected human, this results in production of virus and continued seeding of the oral mucosa. In the laboratory, treatments of latently infected B lymphocytes have been developed that result in the conversion of a substantial fraction of them to permissivity. This has allowed studies of lytic infection and production of progeny virus in cultured cells. Replication of EBV under these conditions appears to involve pathways similar to those of HSV, with which it shares many genes (Fig. 7.13B).

The control of latently infected B lymphocytes by CTLs is obviously to the advantage of the virus as well as of the host. If the initial unlimited proliferation of infected B cells continued indefinitely, it would be lethal for the host. This in fact happens in people whose immune system is compromised, as described later.

Burkitt's Lymphoma

The ability of EBV to latently infect B cells and stimulate continuous cell division leads to an association with a number of human cancers (Table 7.9). The first to be described was Burkitt's lymphoma (BL) by Denis Burkitt in the 1960s in Africa. He attributed the disease to viral infection and it is now known to be associated with EBV. BL is a childhood malignancy that is worldwide but occurs predominantly in regions of Africa and New Guinea with a high incidence of malaria. The tumors arise in lymph nodes, frequently in submandibular nodes. The disease is fatal if not treated, but treatment with chemotherapeutic agents is effective and the majority of patients can be cured. The association with malaria was originally proposed to result from the suppression of CTLs induced by the microbe. Such suppression could result in an enlarged pool of EBV-transformed cells, which are the progenitor cells that become malignant. However, a more recent hypothesis proposes that it is the expansion of germinal centers that occurs in malaria that is responsible for the increased incidence of lymphomas. Germinal centers serve as sites in which somatic mutation occurs in the V gene of an antibody in order to increase the affinity of the antibody for its cognate antigen (see Chapter 10). BL cells are characterized by deregulated expression of the cellular oncogene c-myc, and the c-myc expressed consistently carries mutations. Chromosomal translocations have been found in all BL tumors that result in placing the c-myc oncogene upstream of the immunoglobulin (Ig) genes. Ig genes are expressed in B cells to high levels, and the translocation of c-myc to near the Ig locus leads to its deregulated expression in B cells. Perhaps its association with the Ig locus also leads to hypermutation of c-myc and the development of a mutant form of the gene that results in the development of a tumor. In any event, it appears that overexpression of a mutated c-myc is essential for the development of BL.

Changes in genes other than c-myc also appear to be required for development of BL. The cellular tumor suppressor gene p53 is often altered in BL, and changes in other cellular oncogenes or in chromosome architecture occur in some cases. These alterations may be important for the development of the full malignant phenotype. In addition,

TABLE 7.9   Tumors Caused by Epstein–Barr Infection

Tumor/subtype	Latent period	EBV positivity (%)	EB antigens	World distribution
Burkitt's lymphoma				
Endemic	3–8 years	100		Infection primarily of children
		90		<15 years of age in regions endemic
			EBNA1	for malaria (*P. falciparum*)
Sporadic	3–8 years	15–85		
AIDS-associated	3–8 years post HIV	30–40		
Nasopharyngeal carcinoma	>30 years	100	EBNA 1, LMP1, LMP2	Mostly in SE Asia
Hodgkin's disease				
Mixed cell	>30 year	80–90	EBNA 1, LMP1, LMP2	Worldwide, but more common in Western hemisphere
Nodular sclerosing	>10 years	30		
T-cell lymphoma				
Fatal IM	<6 months	?100	?	Chinese and Caucasians
Nasal	>30 years	100	EBNA 1, LMP1, LMP2	
AILD pleomorphic	>30 years	?40	?	
Immunoblastic lymphoma				
Fatal IM	<6 months	100		
Transplant-associated	<6 months after transplant	100	EBNA 1,2,3A, 3B, 3C	
Transplant-associated	>1 year	100	LMP1 and LMP2	
AIDS-associated	5–10 years post HIV	70–80		

Post EBV infection if not otherwise noted. IM, infectious mononucleosis; AILD, angioimmunoblastic-lymphadenopathy-like.
Source: Adapted from Fields *et al.* (1996) Table 2, p. 133.

BL cells downregulate several functions that are required for recognition and lysis by CTLs. In these cells the expression of class I major histocompatibility complex proteins (MHC) and of the transporter proteins (TAPs) that are required to transfer antigenic peptides across the endoplasmic reticulum (see Chapter 10) are reduced, effectively downregulating the presentation of antigens to CTLs by class I MHC. BL cells also reduce or eliminate expression of cofactors required for efficient interaction with T lymphocytes. Thus, a BL cell resists lysis by CTLs, which are active in immune surveillance.

Thus, the transition from an infected B cell to a malignant cell that can form a fatal tumor in an immunologically competent person is a multistep process. Establishment of a latent infection in B lymphocytes by EBV is only the first step. Many other events must follow, most of which are rare.

Hodgkin's disease is another form of malignant lymphoma associated with EBV. The disease often strikes young adults (hockey fans will remember that Mario Lemieux, a star of the Pittsburgh Penguins, underwent treatment for Hodgkin's disease). There is a second peak in incidence after age 45. The disease is worldwide, although more common in developed countries. It is estimated that perhaps 50% of all Hodgkin's disease is due to EBV. It is assumed that other events must occur in order for infected cells to become malignant, as is the case for BL.

T-Cell Lymphomas

The primary target of EBV in humans is B cells. However, EBV has also been associated with some T-cell lymphomas (Table 7.9). Nothing is known about the process by which the virus infects T cells and causes tumors.

Nasopharyngeal Carcinoma

EBV is also associated with nasopharyngeal carcinoma (NPC). This disease is worldwide but has a much higher incidence in Southeast Asia, in Eskimos, and among some populations in northern and eastern Africa. The available evidence suggests that both genetic and environmental factors are important for the higher incidence in these populations. Studies have found that a particular MHC haplotype (Chapter 10) is correlated with the relative risk of developing NPC. However, dietary factors are also important because immigrant Chinese in the United States have a lowered frequency of NPC than people in China, although the rate is still higher than that in Caucasians.

NPC is a carcinoma rather than a lymphoma, arising in epithelial cells of the nasopharynx. Little is known about how the carcinoma arises, but it presumably requires a non-lytic infection by EBV in which transforming genes are expressed. As for other tumors, several transforming events are probably required for the carcinoma to develop.

EBV Infection in People with Compromised Immune Systems

People who are immunodeficient because of infection with HIV or are pharmacologically immunosuppressed following organ transplant are at greatly increased risk for the development of lymphomas caused by EBV, as shown in Table 7.9. These lymphomas may develop after a very short latent period, less than 6 months. Of interest is the finding that AIDS patients develop two forms of lymphoma. The first arises early, while the immune system is relatively intact, and is a form of Burkitt's lymphoma, having the same c-myc chromosomal translocation as described earlier. BL may arise at higher frequency in AIDS patients in comparison to people with normal immune systems because of expansion of the infected B-cell population, giving rise to an expanded pool of potential precursor cells. The second form arises late, when the immune system is highly compromised. The late form appears to result from failure of CTLs to control the EBV-infected B-cell population.

A fatal, infectious mononucleosis-like illness in young males has been described that is X linked (i.e., the susceptibility gene is carried on the X chromosome, of which males have only one copy). The disease is apparently due to a defect in the immune system that allows EBV-infected B cells to proliferate out of control. The disease is fatal 75% of the time, and death usually results from uncontrolled immunoblastic lymphoma.

Cytomegalovirus (HHV-5)

The cytomegaloviruses (CMVs) are betaherpesviruses that, like all herpesviruses, are species specific in their natural host range. They replicate slowly in cultured cells and have a restricted host range in the laboratory. Human CMV (HCMV) will infect cultured human skin or lung fibroblasts as well as some peripheral blood monocytes. It will also infect chimpanzee cells. Lytic replication in cultured cells resembles that of HSV. There is regulated transcription of α, β, and γ genes, and many of the genes are shared with HSV (Fig. 7.13B). However, the CMV replication cycle differs from HSV in one important aspect. CMV infection leads to the stimulation of host-cell DNA, RNA, and protein synthesis throughout infection, whereas infection with HSV results in the immediate shutoff of host-cell macromolecular synthesis.

Infection of Humans with CMV

Transmission of HCMV requires close contact between a susceptible person and a person shedding virus. Virus present in oropharyngeal secretions, breast milk, or other bodily secretions is probably responsible for transmission. It can also be transmitted by blood transfusion. The virus is ubiquitous, present in all human populations, and most humans become infected as infants. In different populations, 40–100% of persons become infected before the age of puberty. HCMV infections are usually asymptomatic, but primary infection of adults can result in infectious mononucleosis, and primary infection or reactivation of viral replication in neonates or in the immunocompromised can have serious consequences.

HCMV infects epithelial cells in many different tissues, in contrast to its restricted host range in cultured cells. Infection characteristically results in cell enlargement, from which the virus gets its name, and the presence of intranuclear inclusions. Shedding of infectious virus may persist for an extended period of time following primary infection, in fact, for years if infection is congenital or occurs very early in life. Following control of infection by CTLs, HCMV becomes latent, as do all herpesviruses. Latency is probably established in leukocytes. Infection is lifelong and, as for other herpesviruses, reactivation of viral infection can occur and result in renewed shedding of virus.

Infection in Populations at Risk for Disease

Congenital infection by HCMV can be very serious if the infant is not protected by maternal antibodies. About 1% of infants born in the United States are infected *in utero*, either as the result of reactivation of a latent infection in a seropositive mother or as the result of primary infection in a seronegative mother. In the case of mothers who are seropositive, maternal antibodies against HCMV, which are protective against disease, are transferred to the fetus. Congenital infection then occurs with a frequency of only 0.2–2% and symptomatic disease does not occur in the infected fetus. However, primary HCMV infection during pregnancy of women who were previously seronegative results in infection of the fetus up to 50% of the time, and about 10% of infections result in symptomatic infection in the newborn. Infection may be fatal or may result in long-term neurological sequelae, which may include defects in hearing or vision, seizures, microcephaly, or lethargy. Up to 80% of symptomatic infants suffer severe neurological problems, and neurological impairment may occur even in the absence of symptomatic infection. Hearing loss is the most common neurological sequela and congenital HCMV infection is the most common cause of hearing loss in the United States other than that caused by genetic factors.

Like many other herpesvirus infections, HCMV infection, whether primary or resulting from reactivation of latent infection, is extremely serious in patients with compromised immune systems. It is often the most common infection following organ transplant and can result in life-threatening systemic disease. It is also a major life-threatening disease in AIDS patients. Latent HCMV is present in most humans and systemic spread occurs when the CD4$^+$ lymphocyte count falls to very low levels. Systemic disease affects virtually every organ in the body, but infection of the lungs, central nervous system, and the gastrointestinal tract are the most common and most serious. Infection of the lungs can lead to fatal pneumonitis. Infection of the central nervous system commonly results in retinitis, which develops in 20% of long-lived AIDS patients. Infection of almost any region of the gastrointestinal tract can occur and result in severe ulcerations that can lead to perforation of the gut.

The serious nature of disease in the immunocompromised shows that HCMV is an invasive virus that will infect many organs if not controlled by a vigorous immune response. Disease in transplant patients and in patients with AIDS is exacerbated by the expression of genes in HCMV that interfere with many aspects of the immune response. In immunocompetent people, these immunity-defeating mechanisms allow the virus to live in harmony with the host, establishing a lifelong infection that is associated with little or no disease. However, in the immunocompromised, the thwarting of an immune response that is at best weak leads to uncontrolled virus growth and serious illness. The mechanisms by which HCMV interferes with the immune response are described in Chapter 10. They include the synthesis of several proteins that block the presentation of antigens to CTLs by class I MHC, of a protein that interferes with the interferon response, and of a homologue to cellular interleukin-10, which suppresses inflammatory responses, among others.

Human Herpesviruses 6, 7, and 8

Three newly described human herpesviruses have come to light in the last 2 decades and have simply been given the sequential numbers HHV-6, -7, and -8. They all appear to be typical herpesviruses that establish latent infections worldwide. These infections are normally accompanied by no disease or only mild disease symptoms. The silence of their infections caused them to be overlooked until recently, when the ability of HHV-6 and HHV-8 to cause disease in immunocompromised populations, especially in AIDS patients, led to their discovery.

HHV-6 and -7 have a tropism for lymphocytes, especially CD4$^+$ T cells. On the basis of this tropism they were first considered to be gammaherpesviruses. However, they are genetically related to betaherpesviruses like CMV and are now classified as betaherpesviruses, in the genus *Roseolovirus*.

HHV-6 occurs as two major types, called A and B. The virus is probably transmitted by oral secretions. In one study, 90% of adults were reported to have infectious virus in their saliva, although other studies have given lower numbers. About half of children infected by HHV-6 suffer a disease called roseola infantum, exanthem subitum, or sixth disease, a mild disease of childhood that is characterized by fever and rash lasting 3–5 days. In one study, acute infection with HHV-6 accounted for 20% of visits to emergency rooms for febrile illness in 6- to 8-month-old infants. More severe symptoms or neurological complications occur but infrequently. Primary infection of adults is rare because most people are infected as infants, but symptoms are more serious when it does occur. The virus establishes latency in monocytes and macrophages and a persistent infection in salivary glands and respiratory secretions.

As for other herpesviruses, primary infection or recurrence of infection in immunosuppressed people or people with AIDS can be life threatening. HHV-6 was first described in 1986 because of its association with lymphoproliferative disorders in AIDS patients. The virus can also cause serious problems in immunosuppressed populations, in particular in patients undergoing bone marrow, kidney, or liver transplants, where infection or reactivation can result in bone marrow suppression, pneumonitis, encephalitis, hepatitis, fever, or rejection of the transplanted organ.

Conflicting evidence has also suggested that HHV-6 might play a role in multiple sclerosis (MS). This chronic disease is characterized by inflammation and demyelination of neurons and has long been thought to have a viral etiology. A number of different viruses have been suggested to be implicated in MS and HHV-6 is now one of them. Whether HHV-6 is in fact involved in MS remains to be seen.

HHV-7 was found during studies of HHV-6 in peripheral T cells. At present there is no clear evidence for the involvement of HHV-7 in human disease. Of possible clinical importance is the fact that HHV-7 may use the same receptor to infect CD4$^+$ T cells as does HIV, which may allow HHV-7 to be used as a vector to express anti-HIV genes in the specific target population infected by HIV.

HHV-8 is the most recently described human herpesvirus. It establishes a latent infection in B lymphocytes and is classified as a gammaherpesvirus, genus *Rhadinovirus*. It has a prevalence of 5% in the United States and is sexually transmitted. It was discovered through its association with Kaposi's sarcoma, the most common tumor found in patients with AIDS. Tumor cells are of endothelial origin and are multifocal. AIDS patients are much more likely to develop Kaposi's sarcoma than are immunosuppressed patients, and there must be a synergism between the infections of HIV and HHV-8. Intriguingly, Kaposi's sarcoma is 15-fold more common in homosexual male AIDS patients than in patients who acquired HIV by a nonsexual route. HHV-8 is also associated with body cavity–based lymphoma, a lymphoid tumor in some AIDS patients, and with multifocal Castleman's disease, a rare lymphoproliferative disorder. In common with a number of other herpesviruses, HHV-8 encodes factors that stimulate cells to divide, that interfere with the host immune response, and that block apoptosis. It also produces a protein in latently infected cells that tethers the viral DNA to mitotic chromosomes.

Monkey B Virus

Many herpesviruses infect vertebrates other than humans, but only one nonhuman herpesvirus is known to be highly pathogenic for humans. Cercopithecine herpesvirus 1 or monkey virus B is indigenous to Old World monkeys in the genus *Macaca*. In its native host it causes a disease that is similar to that caused by HSV-1 in humans. A latent, lifelong infection is established in the monkey that seldom leads to serious illness. Sporadic reactivation of the virus occurs that results in the formation of vesicular lesions, particularly on the tongue and cheeks. However, infection of humans or of a number of monkeys other than macaques results in a very serious neurological disease that has a high fatality rate. As described earlier, HSV-1 is also neurotropic and occasionally causes fatal encephalitis.

B virus has usually been transmitted to humans through the bite of a monkey in which infectious virus was present in the saliva. However, transmission has also occurred by other means. In at least one case transmission resulted from contact of infectious material with the eye and two cases are thought to have resulted from exposure to aerosols. One case of transmission of virus from an animal worker to his wife has also been documented. The majority of human infections have resulted in fatal neurological disease, but some infections resulted in only mild disease and the establishment of a latent infection. Acyclovir has been used to treat persons infected by B virus with apparent success, but the number of cases is small and no controlled trials of efficacy have been conducted.

Recurrence of herpes infections is often associated with stress. Newly captured or shipped animals are subject to a great deal of stress, so that active infection in these animals is not uncommon. Animal handlers or researchers using these animals are at risk for the disease and must take proper precautions when handling macaques. Because of the dangers from B virus infection, most laboratories in the United States use only monkeys that lack antibodies to B virus, an indication that they are not infected. Although this greatly reduces the risk of handling the animals, it does not eliminate all risk because occasionally monkeys that are infected do not have detectable antibodies.

FAMILY *ADENOVIRIDAE*

Adenoviruses are widespread viruses of mammals and birds, but a few of the known viruses are able to infect reptiles or frogs. The virions are a $T=25$ icosahedron, 70–90 nm in diameter, with fibers 9 to 77 nm in length projecting from the 12 fivefold axes of the icosahedron (Figs. 2.1 and 2.12). Virions contain about a dozen proteins, of which 4 are present in the core. The major structural proteins are a hexon protein called II, three copies of which form a hexon, of which there are 240 in the virion, and a penton protein called III, five copies of which form a penton base, of which there are 12 in the virion. The genome of adenoviruses is a linear dsDNA of size 26 to 45 kb. A terminal protein that served as a primer during DNA replication is covalently attached to the 5′ end of both strands.

Adenoviruses are named after adenoids, a glandlike collection of lymphoid tissue in the nasopharynx. Many human adenoviruses establish a long-term infection in this tissue and adenoviruses were first isolated from human adenoids. Four genera are currently recognized. The genus *Mastadenovirus* contains viruses that infect only mammals and the genus *Aviadenovirus* contains viruses that infect only birds, whereas the genera *Atadenovirus* and *Siadenovirus* contain viruses that infect a wide variety of vertebrates (Table 7.10). Most viruses are species specific and in general will only undergo a complete replication cycle in cells isolated from their native host.

Fifty-one human adenoviruses have been distinguished on the basis of serological reactivity—an adenovirus is considered distinct if it resists neutralization by antisera against the other known adenoviruses. All belong to the genus *Mastadenovirus* and have a genome size of 30–36 kb. The 51 viruses are simply numbered in order of their isolation and are often referred to as Ad1, Ad2, etc., or more formally as HAdV-1, etc., to distinguish them from adenoviruses that infect other species. The human viruses were originally divided into six subgroups on the basis of serological cross-reactions in a hemagglutination-inhibition assay. In this assay, the ability of an antiserum to bind to the virus and prevent it from agglutinating red blood cells is examined. An antiserum against one of the viruses of subgroup A, for example, inhibits hemagglutination by all members of that subgroup but not by members of other subgroups. This grouping correlated with a number of other properties of the viruses as well, such as their ability to form tumors in rodents. These original subgroups are now considered to be different adenovirus species, *human adenovirus A* through *F*, with grouping relying on sequence identities where possible. Two viruses are considered to belong to the same species if they differ by less than 10% in their sequence, and by this criterion the human adenovirus species may contain adenoviruses of other animals.

Because they replicate to high titer in cultured human cells, several human adenoviruses have been intensively studied,

TABLE 7.10 *Adenoviridae*

Genus/members	Serotypes	Usual host(s)	Disease in natural host
Mastadenovirus			
Human adenovirus A	Types 12,18,31	Humans	Enteritis
Human adenovirus B	Types 3,7,11,14,16,21,34,35	Humans	Enteritis; military recruits' disease (3, 7, 14, 21); type 35 causes pneumonia in elderly and immunocompromised humans
Human adenovirus C	Types 1,2,5,6	Humans	Respiratory infection in children
Human adenovirus D	Types 8,9,10,13,15,17, 19,20, 22-30,32,33,36-39,42-47	Humans	Enteritis
Human adenovirus E	Human type 4, simian types 22,23,24,25	Humans	Enteritis, pneumonia and upper respiratory disease in military recruits
Human adenovirus F	Humantypes 40,41, simian type 19	Humans	Infant diarrhea
Murine adenovirus A		Mice	Asymptomatic or mild respiratory disease
Bovine adenovirus A, B, C		Cattle	
Ovine adenovirus A, B		Sheep	
Porcine adenovirus A, B, C		Swine	
Canine adenovirus	Types 1 and 2	Dogs	Hepatitis (type 1)
			Respiratory disease (type 2)
Aviadenovirus			
Fowl adenovirus A, B, C, D, E.		Chickens, ducks	Hepatitis, bronchitis, duck hepatitis (rare)
Goose adenovirus		Geese	
Atadenovirus			
Ovine adenovirus D		Sheep	Basis of some vectors for human gene therapy
Bovine adenovirus D		Cattle	
Duck adenovirus A		Ducks	Egg drop syndrome
Siadenovirus			
Frog adenovirus		Amphibians	Nonpathogenic
Turkey adenovirus 3		Turkey, pheasants, chickens	Hemorrhagic enteritis in turkeys, marble spleen disease in pheasants, splenomegaly in chickens

[a] In general adenoviruses are transmitted by both aerosols and fomites and by the oral–fecal route.

in particular Ad2, Ad5, and Ad12. Further interest has been generated by the fact that, although they will not undergo a complete replication cycle in rodent cells, they will infect and transform rodent cells in culture. Members of HAdV-A will also cause tumors in rodents at a high rate and members of HAdV-B at a moderate rate. Other human adenoviruses cause tumors in rodents at a low or undetectable rate. There is no evidence that adenoviruses are associated with tumors in humans, however. Because of the extensive background of information about them and the fact that they usually cause only minor illness, attempts are being made to use them as expression vectors for gene therapy (Chapter 11).

Transcription of Adenovirus mRNAs

Adenovirus replication takes place in the nucleus. After the entry of the infecting genome into the nucleus, the infecting genome is transcribed by host RNA polymerase II to produce a set of early RNAs. Later, a set of late RNAs is produced. A transcription map of Ad2 is shown in Fig. 7.20. Transcription of early RNAs occurs from five promoters, three on the so-called R strand and two from the L strand. The R strand is transcribed rightward on the chromosome as conventionally drawn, and the L strand is transcribed leftward. There are two other promoters for transcription of delayed early mRNAs. Multiple splicing of these transcripts leads to the production of about 30 mRNAs. The proteins translated from these early mRNAs are required for replication of the viral genome. The E1A and E1B gene products are oncogenes that stimulate the cell to enter S phase and thereby induce an ideal environment for the replication of the viral DNA. Their mode of action is described later. Proteins from the E2 region are directly involved in replicating adenovirus DNA and include a DNA polymerase, an ssDNA-binding protein, and a precursor to the terminal protein, which is involved in initiation of DNA replication. E3

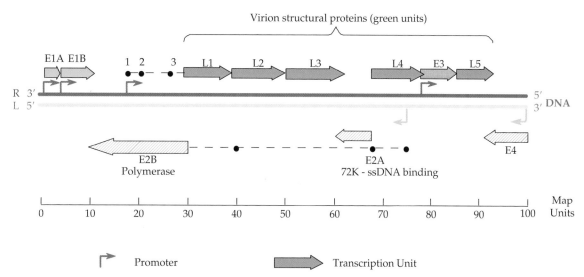

FIGURE 7.20   Genome organization of human adenovirus type 2. The double blue lines represent the two linear DNA strands that make up the genome of 36 kbp. The genes have been mapped by superimposing an arbitrary scale of 100 map units. Each arrow represents a transcription unit composed of a nested set of spliced messages, transcribed in the direction of the arrow. Black dots indicate short spliced sequences that form leaders. General functions of the various transcription units are shown. Proteins in the E3 cluster interact with the host immune system, and E4 genes are involved in DNA replication. The major late transcription unit includes the leaders "1," "2," and "3" and the L1, L2, L3, L4, and L5 families of genes. Adapted from Wold and Golding (1991).

proteins modulate the host response to adenovirus infection, and this region is nonessential in cultured cells. The functions of the E3 proteins are described in Chapter 10. Region E4 encodes proteins involved in transcription and transport of viral mRNAs and in DNA replication.

Late genes are all transcribed beginning from a single promoter on the R strand, and transcription leads to an RNA product that is about 80% the length of the adenovirus genome. Multiple splicing occurs to produce at least 18 different mRNAs that fall into five families, based on the use of five different polyadenylation sites. Each late mRNA has the same tripartite leader, formed by splicing. The late mRNAs are translated into the proteins required for the assembly of progeny virions.

The multiple splicing events that occur during processing of adenoviral mRNA, especially the late mRNA, which is made in abundance, led to the discovery of RNA splicing by Phillip Sharp. Upon examination of adenoviral mRNA–DNA hybrids with an electron microscope, he observed that regions of the genome were missing from the mRNA transcripts. For his discovery of RNA splicing, he was awarded a Nobel Prize, along with Richard Roberts, in 1993 (Table 1.1).

In addition to the many genes transcribed by RNA polymerase II, one or two adenovirus genes, called VA, are transcribed by host RNA polymerase III. Short VA RNA molecules are produced that are not translated. They function to inhibit the host interferon system and the host RNAi system and will be described in Chapter 10, after the description of these host systems.

Replication of the Viral DNA

Inverted terminal repeats that contain the origins of replication are present at the ends of the adenovirus genome. DNA synthesis is initiated at one of the two ends and proceeds to the other end. There is no lagging strand synthesis and the partner of the strand being copied is displaced as an ssDNA (illustrated schematically in Fig. 1.9C). The precursor of the terminal protein serves as a primer during initiation. It forms a complex in solution with the adenovirus DNA polymerase, and it is assumed that these two proteins bind to the origin of replication as a complex during initiation of DNA replication. There are also binding sites within the terminal repeats for several cellular proteins that stimulate the initiation of DNA synthesis. The first step in initiation is the covalent linkage of dCMP, the first nucleotide in adenovirus DNA, to the preterminal protein. Subsequent chain elongation requires the activity of the adenovirus DNA polymerase and of the ssDNA binding protein, as well as of a cellular topoisomerase. The use of a protein primer eliminates the need for a primase to initiate DNA synthesis with an RNA primer, and thus solves the problem of how to maintain the ends of the linear adenoviral DNA molecule during replication.

The products of the first round of replication are a double-strand progeny genome and a single-strand copy of one of the two strands of the genome. Initiation of DNA synthesis can also occur on this ssDNA. It is proposed that a panhandle structure is formed by the terminal repeats so that an origin of replication is present that is identical to that in the dsDNA. Copying this ssDNA renders it double stranded and completes the production of two copies of the dsDNA genome from the parental genome.

Assembly and Release of Progeny Virions

Progeny viruses are assembled in the nucleus from pre-assembled hexons and pentons (see Fig. 2.12). Viral DNA is required for assembly. A packaging signal of about 260 nucleotides at the left end of the viral DNA leads to polarized encapsidation starting from this end. During assembly, the viral protease cleaves at least four viral products, and these cleavages are required to stabilize the particle and make it infectious. Release of virions from the cell is associated with the disruption of intermediate filaments. Vimentin is cleaved early after infection by an unknown protease, and cytokeratin K18 is cleaved late by the viral protease. A schematic of the relative timing of the major events in the adenovirus life cycle is presented in Fig. 7.21. Virus infection is lytic and the cell eventually dies.

Adenovirus Oncogenes

Two early genes of adenovirus, E1A and E1B, encode proteins that induce the cell to enter S phase, in which cellular DNA is replicated. E1A targets a number of cell proteins that are involved in cell cycling, forming complexes with Rb, p107, p130, p300, and several other cellular proteins. These proteins are listed in Table 7.11 for two forms of E1A that result from differential splicing, called 12S E1A (because it is translated

from an mRNA that sediments at 12S) and 13S E1A (because its mRNA sediments at 13S). Rb or retinoblastoma susceptibility protein is a tumor suppressor. It was first identified because it is absent in patients suffering from retinoblastoma. In its hypophosphorylated form, Rb binds a cellular transcription factor called E2F and causes the cell cycle to arrest in G1. Hyperphosphorylation of Rb causes it to dissociate from E2F. Free E2F activates the transcription of genes that cause the cell to enter S phase. The binding of Rb by E1A prevents it from complexing with E2F, and E2F is thus free to induce the cell to enter S phase. p107 and p130 are other members of the Rb family that also interact with E2F, as well as with cyclins and cyclin-dependent kinases, whose activities are disrupted by binding to E1A. Binding of E1A to p300 appears to be an independent method of disrupting the cell cycle; p300 is thought to bind to DNA and activate transcription of factors involved in cell cycle progression.

The E1B 55-kDa protein also targets a tumor suppressor protein called p53. p53 is another cellular protein that regulates cell cycle progression. It is the most commonly mutated gene associated with human tumors. It is both a transcriptional activator and repressor. It activates the transcription of genes whose function is to arrest cell cycle progression. E1B blocks the activity of p53, and when p53 is inactive or absent, cell cycle progression continues. Ad5 E1B binds p53 and sequesters it outside the nucleus, whereas Ad12 E1B does not appear to bind p53 but inhibits its activity in some indirect way.

The net result of the expression of E1A and E1B is the continued cycling of the cell. The expression of these two genes alone will transform cells in culture. Rat cells transformed by Ad12 (HAdV-A) will produce tumors in syngeneic newborn

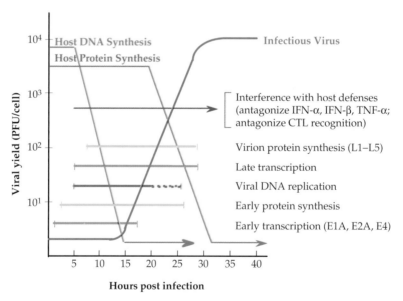

FIGURE 7.21 Relative timing of major events during the adenovirus life cycle. Example shown is for HeLa cells infected with adenovirus at a multiplicity of infection (MOI) of 10. L1–L5 are the major late transcription units mapped in Fig. 7.20. Adapted from Fields *et al.* (1996), p. 2119 and Ginzberg (1984), Fig. 1 on p. 2.

TABLE 7.11 Cellular Proteins That Bind Directly
to Adenovirus E1A Proteins[a]

E1A binding protein	Binds to[b]		Other proteins in complex
	12S E1A	13S E1A	
pRB	+	+	None known
p107	+	+	Cyclin A, cdk2/cyclin E, cdk2
p130	+	+	Cyclin A, cdk2/cyclin E, cdk2
p300	+	+	None known
TATA binding proteins (TBP)	−	+	TBP-associated factors
ATF-2	−	+	None known
YY1	−	+	None known

[a] E1A is the first viral transcription unit to be expressed. In the early phase of infection, differential splicing results in two mRNAs from this gene, which sediment at 12S and 13S respectively. The translation products of these two mRNAs are referred to as 12S E1A and 13S E1A.

[b] RB is the retinoblastoma susceptibility protein; cdk2 is cyclin-dependent kinase 2; ATF-2 is a member of the ATF family of transcription factors; YY1 is a human transcriptional repressor. These cellular proteins have been shown to bind directly to the E1A protein in a biologically relevant way. See also Fig. 7.22.

Source: Adapted from Fields *et al.* (1996), p. 2122.

A. Adenovirus

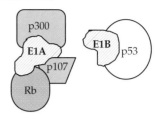

B. SV40

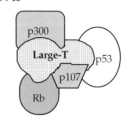

C. Papillomavirus

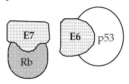

FIGURE 7.22 Known interactions between the oncogenic proteins (shaded with pink patterns) of an adenovirus, a polyomavirus, and a papillomavirus and cellular proteins that are regulators of cell cycle progression. (A) Protein E1A of adenoviruses binds to the Rb family, promoting entry into S phase. The 19-kD form of E1B also binds to p53, blocking apoptosis. (B) The large T antigen of the polyomavirus, SV40, interacts with the Rb family of proteins as well as with, p53 (see also Fig. 7.28). (C) The human papillomavirus proteins E6 and E7 bind Rb and p53, respectively, the latter promoting the destruction of p53. Adapted from Berg and Singer (1997), Fig. 1.35 on p. 61.

rats but cells transformed by Ad2 or Ad5 (HAdV-C) are not tumorigenic. The differences in the ability to induce tumors appear to be in the immune responses of the host. In particular, Ad12 appears to interfere more effectively with host CTL responses, as described in Chapter 10. Although some adenoviruses cause tumors in rats, there is no evidence that they do so in humans. The regulation of cell cycling may be different in rodents and humans, or the adenovirus proteins may interact with rodent and human regulatory proteins in different ways.

The proteins targeted by E1A and E1B are key regulatory elements of the cell. The polyomaviruses and the papillomaviruses target many of these same proteins in order to induce cell cycling, emphasizing their importance in the control of the cell cycle. The oncogenes of these three families of viruses and their interactions with these key cellular proteins are illustrated schematically in Fig. 7.22.

Interference with Host Defenses

Adenoviruses interfere with host antiviral defenses in multiple ways. This topic is covered in detail in Chapter 10, so only a summary is presented here. Adenoviruses have two independent mechanisms to suppress the interferon system. First, E1A inhibits the transcription of interferon response genes by inhibiting the activity of ISGF3, a cellular transcription factor. Second, VA RNA prevents the activation of a protein kinase called PKR, which is one of the major

effector products of the interferon pathway. VA RNAs are highly structured RNAs that are abundantly produced and act as dsRNA decoys. VA RNA binds to PKR, which requires bound dsRNA for activity, but does not activate the enzyme. VA RNA also binds to several molecules in the RNA interference (RNAi) pathway, including Exportin 5 required for the export of pre-microRNAs from the nucleus, and DICER required for processing of these RNAs, and VA RNA processed by DICER binds to RISC (see Chapter 10). The result is to saturate this pathway, thereby inactivating it.

Adenoviruses also inhibit the lysis of infected cells by CTLs. Ad12 E1A protein blocks the transcription of the genes for class I MHC molecules, whereas the Ad2 or Ad5 E3 19-kDa protein prevents the export of class I MHC molecules to the cell surface. In either case, presentation of peptide antigens to CTLs is blocked.

Apoptosis is a cell suicide pathway by which cells die after infection by many viruses. Adenoviruses encode proteins that delay the advent of apoptosis in order to give the virus more time to replicate. First, E3 region proteins interfere with the action of TNF-α, blocking TNF-α-induced apoptosis. Second, E1B blocks, or at least delays, apoptosis otherwise induced by E1A.

Because adenoviruses block the antiviral defenses of the host, they have the ability to persist in the infected host for considerable periods of time. Virus may be present in tonsils and adenoids and may be shed in the stools for a year or more following primary infection (Fig. 7.23).

Adenoviruses and Human Disease

The human adenoviruses replicate primarily in the upper respiratory tract or in the gastrointestinal tract. Some replicate well in both while others express a tropism for one or the other. Spread of the viruses is by a respiratory route or by an oral–fecal route. Many infections by adenoviruses appear to be asymptomatic or to result in only mild illness, but about 5% of acute respiratory disease in children under 5 years old is due to adenovirus infection. Some serotypes can also cause gastroenteritis, but the overall importance of these viruses as causative agents of gastroenteritis is not resolved. Ad1, 2, and 5 are the most common viruses found in human populations, and antibodies to these viruses are present in about one-half of all children. Ad7, and

to some extent Ad3 and Ad4, are the adenoviruses most often associated with severe disease, and Ad7 accounts for about 20% of adenoviruses reported to the World Health Organization.

Virus can be shed for months following primary infection, especially in stools (Fig. 7.23). Children who are shedding virus are infectious and spread is particularly efficient between close family members. Figure 7.24 diagrams the spread of infection within four families in which older siblings were excreting adenovirus at the time of the birth of the youngest child. In all cases the new baby was infected within a year. Day care centers are also important in the spread of these viruses.

Adenoviruses also cause respiratory disease in adults and probably account for about 3% of such illnesses. The disease is usually mild, but Ad4 and Ad7 have caused epidemics of more serious respiratory illness in military recruits. Such epidemics of acute respiratory disease have resulted in the infection of 80% of the recruits in a unit and 20–40% of these have required hospitalization. The stress, crowding, and bringing together of young men from different backgrounds and from all over the country seems to potentiate the illness. Illness may be exacerbated if infection begins by deep inhalation of aerosolized virus, which results during the vigorous exercise that is required of recruits. Such epidemics result in protective immunity because seasoned troops do not suffer recurrent epidemics

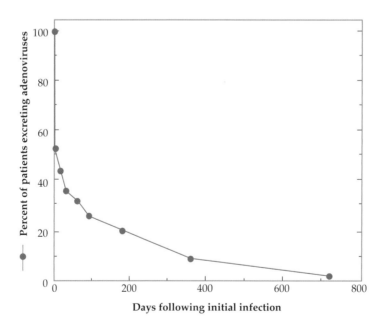

FIGURE 7.23 Adenovirus shedding by patients. The percent of 133 patients who shed virus in stools for at least the number of days indicated after an initial adenovirus infection is plotted. Note that the last point is almost 2 years. Data from Strauss (1984).

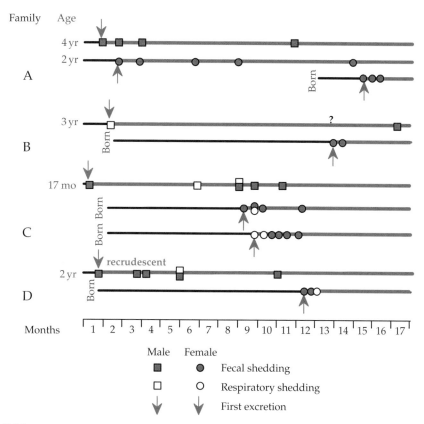

FIGURE 7.24   Adenovirus transmission between siblings in four families. Schematic representation of infection with adenovirus of newborn babies in families where one or more older siblings was already infected. The newborns became infected between 3 and 12 months after birth. It is believed that shedding of virus was continual (red lines) after the first virus-positive stool sample from a given child. From Fox *et al.* (1969).

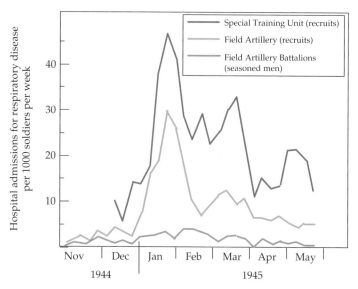

FIGURE 7.25   Admission to the base hospital for treatment of respiratory disease, presumably primarily adenovirus related, of two recruit groups and one group of seasoned troops. From Dingle and Langmuir (1968).

(Fig. 7.25). Starting in 1971, the military used a vaccine to prevent the disease. This vaccine consisted of live nonattenuated virus that was encapsulated in coated capsules and taken orally. Infection of the enteric tract is asymptomatic for these viruses under these conditions and results in a good antibody response, presumably including mucosal antibodies to protect the respiratory tract. Use of this vaccine reduced adenovirus-specific disease rates in trainees by 95 to 99% and reduced the total respiratory disease rates by 50 to 60%, illustrating the importance of Ad4 and Ad7 as causative agents of severe respiratory disease. The vaccine is no longer available, however. It was made by a single manufacturer whose facilities for making the vaccine became outdated and needed upgrading. The army was unwilling or unable to find a mechanism to help with the cost of the upgrade, and since the military was the sole purchaser of the vaccine, production was discontinued in 1995. The last of the vaccine was exhausted in 1999 and epidemics of severe adenovirus respiratory disease in recruits are once again a serious problem.

Adenoviral epidemics such as have occurred in military recruits are rare in the civilian sector but do occur. As one example, there was an epidemic of Ad11 in 1997 in a South Dakota job-training facility. Like a military camp, this facility had young adults housed in crowded conditions with regular multiple introductions of new susceptible persons. Illness with severe morbidity also occurs in pediatric populations, school environments, and health care settings.

Adenoviruses are being developed as vectors to immunize people against other viruses and for gene therapy. Two approaches are being used. One is to use the military experience with the Ad4 and Ad7 oral vaccines to develop vaccines that could be delivered safely using this route. A second approach is to disable the virus by deleting some of the early genes. Such an attenuated virus has been used in clinical trials for treatment of cystic fibrosis. In these trials, virus that expresses the gene that is defective in such patients is delivered directly to the lungs. Trials that use adenoviruses in an attempt to treat other diseases are also ongoing. The use of adenovirus for gene therapy has not been very successful to date, and the death of a patient in a gene therapy trial from adenovirus led to retrenchment in this area. This topic is discussed in more detail in Chapter 11.

FAMILY *POLYOMAVIRIDAE*

In the past the polyomaviruses and the papillomaviruses were considered to be two subfamilies within the family *Papovaviridae*. The name *papova* came from *pa*pilloma virus/*po*lyoma virus/simian *va*cuolating virus (= SV40), three characteristic members of the enlarged family. However, the mode of replication, genome organization, and evolutionary history of polyomaviruses and papillomaviruses are distinct and they are now classified into two distinct families.

To date, 14 different polyomaviruses have been described and others probably exist. These viruses infect warm-blooded vertebrates. Mammalian viruses include viruses infecting humans, monkeys, cattle, rabbits, rats, mice, hamsters, while bird viruses infecting parakeets, geese, and other birds are known. The majority of infections of their native host by the mammalian viruses are asymptomatic and the viruses establish a lifelong infection. Many can cause tumors in nonnative hosts, however, and may be associated with tumors in their native hosts, and have been intensively studied as models for tumor induction. In contrast, the bird viruses cause severe disease with a high mortality rate in their native host. A partial listing of the members of the family *Polyomaviridae* is given in Table 7.12. The two best studied viruses are mouse polyomavirus (MPyV), so called because it causes many (*poly*) different kinds of tumors (*omas*) in mice, and simian virus 40 (SV40), which infects monkeys. These viruses share about 60% nucleic acid sequence identity. SV40 was first recognized as a contaminant in monkey kidney cultures used for the production of poliovirus vaccine. Two polyomaviruses for which humans are the natural host are known, BK virus and JC virus. SV40 also circulates in humans, possibly as a result of its introduction with the contaminated poliovirus vaccine. These three viruses have received increasing attention as disease agents.

The structures of the two polyomaviruses SV40 and MPyV have been solved to atomic resolution (Fig. 2.10). They are icosahedral viruses 45 nm in diameter with pseudo-$T=7$ symmetry (Figs. 2.1 and 2.5). The capsid consists of 72 pentamers of structural protein L1, and each pentamer contains one molecule of either structural protein L2 or L3. The polyomavirus genome is a circular dsDNA molecule that is 5 kb in size. In the virus it is complexed with cellular histones to form a supercoiled minichromosome.

The Early Genes

Transcription maps of SV40 and MPyV are shown in Fig. 7.26, in which the circular genomes are represented as linear structures for ease of presentation. The genomes are divided into two domains, an early domain that is transcribed from one DNA strand and a late domain transcribed from the other strand. An origin (Ori) is present at the junction of the two domains. Ori serves as an origin of replication for DNA replication and a transcription start site for transcription by RNA polymerase II. It also contains enhancer elements.

Following infection, the viral DNA is transported to the nucleus, whether as part of a virion or subviral particle or as free DNA is not clear, and early mRNA is transcribed by cellular RNA polymerase. The MPyV or SV40 promoters

TABLE 7.12   *Polyomaviridae*[a]

Genus/members	Virus name abbreviation	Usual host(s)	Transmission	Disease
Polyomavirus				
Simian virus 40	SV-40	Monkeys	Reactivation of persistent infection, virus shedding in urine, contact, aerosols, sexual transmission	PML[b]-like disease in rhesus monkeys and immunocompromised macaques; may cause childhood brain tumors in humans
Murine polyoma	MPyV	Mice		Tumors when inoculated into newborn mice, virus persists in the kidney
Bovine polyomavirus	BPyV	Cattle		Common infection, persists in the kidney
BK polyomavirus	BKPyV	Humans	Transmission as above plus tissue transplantation in humans	Common infection of early childhood; virus persists; tumors in immuncompromised humans
JC polyomavirus	JCPyV	Humans		Common infection of late childhood; causes PML[b] in immunocompromised hosts
Budgerigar fledgling polyomavirus	BFPyV	Parakeets		Fatal illness in fledgling parakeets
Plus other viruses infecting monkeys, hamsters, and rabbits				

[a] Most polyomaviruses are worldwide in distribution.
[b] PML, progressive multifocal leukoencephalopathy.

are strong promoters, active in many cells. Because of this, the SV40 promoter has been used to drive expression of foreign genes in many different expression systems. Early mRNA transcribed from the viral genome terminates about halfway around the genome and is differentially spliced to yield several mRNAs (Fig. 7.26). The translation products of these mRNAs are called T antigens (tumor antigens) because their expression in the absence of productive viral infection leads to cell transformation and the formation of tumors in animals. The two major SV40 proteins are called the small t antigen and the large T antigen, whereas the three major MPyV antigens are called small t, middle T, and large T antigens. In addition, a 17k T antigen has been reported in SV40 and a short-lived tiny t antigen has been reported for MPyV.

The large T antigens (size: about 700 amino acids) are multifunctional proteins that interact with viral promoters and several cellular proteins. A schematic representation of the large T antigen of SV40 that illustrates the locations of many of its activities within the protein sequence is shown in Fig. 7.27. These activities are differentially regulated by phosphorylation of serine and threonine residues, as shown. The large T antigen possesses a nuclear localization signal that causes it to be transported to the nucleus, a domain that binds to sequence elements within Ori, phosphorylation sites in two widely separated regions that serve to regulate its activities, domains that interact with DNA polymerase α–primase, a domain that controls the host range of the virus, a domain that has DNA helicase activity, and domains that bind cellular proteins that play regulatory roles in cell cycling, among others.

Binding of large T antigen to the Ori regulates its own production. Binding is also required for DNA replication. Binding of SV40 large T antigen to Rb and p53, as well as to p107 and p300, is illustrated in Fig. 7.22. Binding to these tumor-suppressor proteins results in continued cell cycling and transformation of the cell. Binding to Rb and its relatives p107 and p300 releases transcription factors required for cell cycling, of which the best studied and perhaps most important are members of the E2F family. p53 is a transcription factor whose activity can prevent DNA synthesis, cause G2 and G1 growth arrest, and induce apoptosis. As described earlier, cell cycling induces a state suitable for viral DNA replication. Expression of large T in the absence of other SV40 proteins is sufficient to transform a cell, although such a transformed cell does not express the full complement of transformed attributes unless small t is also expressed.

The activities in MPyV virus T antigens are distributed somewhat differently. Polyoma large T antigen binds Rb but does not bind p53. Binding to Rb induces cellular DNA synthesis. Cell transformation is a function of middle T antigen. Middle T antigen binds to *src*, a known proto-oncogene located at the cell surface, and activates its kinase activity (Fig. 7.28). The phosphorylated middle T antigen interacts with other cellular factors and induces cell transformation, and it has been shown that expression of middle T antigen alone is sufficient to transform a cell.

The oncoproteins of these small DNA viruses are excellent examples of how a viral protein may be multifunctional and mediate a variety of functions important for the outcome of infection.

A. SV40 (5243 bp)

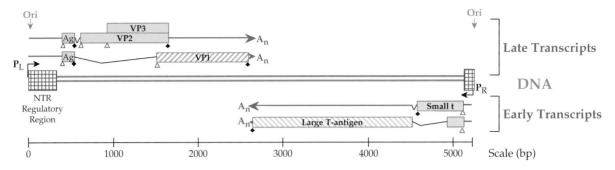

B. Mouse Polyoma (5297 bp)

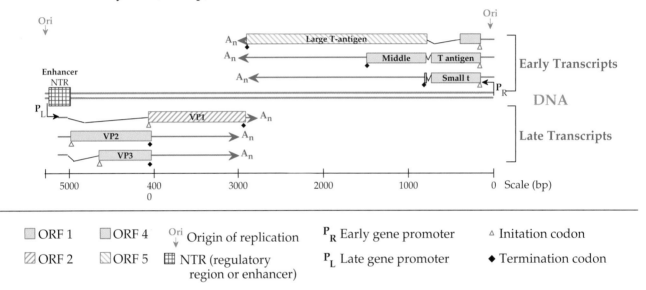

FIGURE 7.26   Comparisons of the genome organizations of two polyomaviruses: SV40 and mouse polyoma. Both are circular genomes which have been linearized at the origin of replication (Ori) for ease of presentation. The genome of mouse polyoma virus is shown here in the opposite sense (right to left) to that of SV40. Ag is the agnoprotein. Redrawn after Brady and Salzman (1986).

DNA Replication

Large T antigen binds to specific sites within the Ori to promote replication of the viral genome. Binding first unwinds the DNA. Then T antigen associates with replication protein A followed by DNA polymerase α-primase to form an initiation complex. Association with primase is species specific. As a result, SV40 productively infects only monkey cells and mouse polyomavirus infects only mouse cells. After initiation by primase, DNA polymerase takes over and replication proceeds. DNA synthesis is bidirectional and when the replication forks meet about halfway round the molecule, the daughter genomes separate, aided by topoisomerase II (Fig. 1.9A).

The Late Genes

Large T antigen also regulates the transcription of late mRNAs, which are transcribed from the opposite strand as the early mRNAs (Fig. 7.26). Differential splicing leads to two mRNAs in SV40 and three mRNAs in polyoma. One mRNA is translated into VP1, the major virion structural protein. In SV40 the second mRNA is translated into both VP2 and VP3, whereas in polyoma virus VP2 and VP3 are translated from different mRNAs. VP3 is a truncated form of VP2, consisting of the C-terminal 60% or so of VP2. Both are minor components of the virion. VP2 is myristylated and may serve an entry function. The three structural proteins are transported to the nucleus and assembly of virions takes place there.

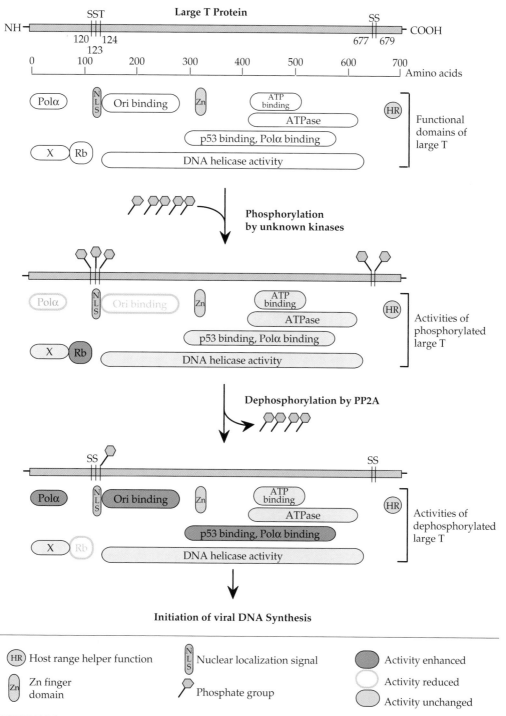

FIGURE 7.27 Functional domains of SV40 large T antigen. The top panel illustrates the location of various functional domains and of the serine and threonine residues that are phosphorylated. The second panel shows the functions of fully phosphorylated large T antigen. The bottom panel shows the activities of large T singly phosphorylated on threonine 124, which can bind to the origin of replication to initiate DNA synthesis. In the middle and lower panels, blue domains are unchanged in activity, gray domains are reduced in activity, and red functions are strongly increased. Adapted from Berg and Singer (1997), Figures 1.23 and 1.24 and Fields *et al.* (1996), Figure 6 on p. 2011.

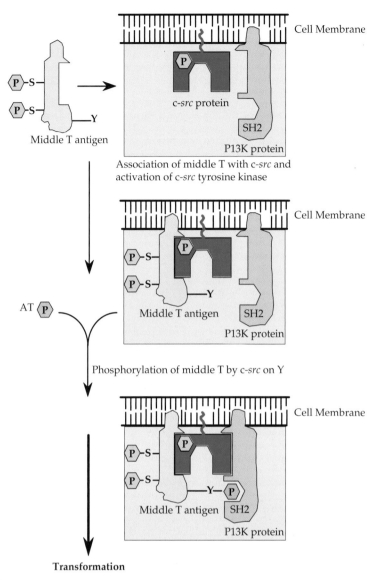

FIGURE 7.28   Interactions of polyoma middle T antigen with membrane proteins to initiate cellular transformation. Phosphatidylinositol-3-kinase (Pl3K) and c-*src* are localized in the plasma membrane. The *src* kinase phosphorylates middle T on a tyrosine residue (Y). This generates binding sites for a variety of other cellular signaling proteins such as Pl3K. The resultant complex closely resembles an activated growth factor receptor bound to signal transduction proteins. Serine residues (S) on middle T that are phosphorylated by other protein kinases are also indicated. Drawn from data in DiMaio *et al.* (1998).

VP1, when expressed alone, assembles into virus-like particles. If VP2 and VP3 are coexpressed with VP1, they assemble into virus-like particles whose composition is the same as that of the virion.

Release of virions is an active process. Membrane vesicles form and transport virions to the cell surface, where they are released. Infection by polyomaviruses is lytic. Expression of the late genes and assembly of progeny virions results in the death of the cell.

Polyomavirus Infection of Humans

The two well-known human polyomaviruses are BK virus and JC virus. These viruses were first isolated in 1971, JC from the brain of a patient with progressive multifocal leukoencephalopathy (PML) and BK from the urine of an immunosuppressed renal transplant patient. They were named after the initials of the patients from whom they were isolated. These two viruses share 75%

TABLE 7.13   Virus-Coded Proteins of Primate Polyomaviruses

Protein	Number of amino acids			% Amino acid identity			Function
	JC	BK	SV40	JC/BK	JC/SV40	BK/SV40	
Late proteins							
VP1	356	362	364	77.9	76.4	82.4	Major capsid protein, attaches to cellular receptors, hemagglutination, HI and NT epitopes
VP2	344	351	352	78.8	73.4	78.5	Minor capsid protein
VP3	225	232	234	74.5	67.2	73.6	Minor capsid protein
Agnoprotein	71	66	62	59.1	45.0	53.2	Facilitates capsid assembly
Early proteins							
Large T	688	695	708	86.6	72.0	73.9	Initiation of replication, stimulates host DNA synthesis; modulates transcription, transformation
Small T	172	172	174	79.6	67.8	69.5	Necessary for efficient viral DNA replication

Source: Adapted from Fields *et al.* (1996), p. 2030 and additional data from Walker and Frisque (1986).

nucleotide sequence identity. A comparison of the proteins encoded by JC, BK, and SV40 viruses is given in Table 7.13. Shown are the number of amino acids in each protein and the amino acid identity between any two of these viruses. This table makes obvious the close relationship among these viruses.

Most humans in the United States are infected with BK virus before the age of 10 (Fig. 7.29). Infection with JC virus usually occurs somewhat later, but by the age of 14 the majority of the population has been infected. Primary infection with BK virus has been associated with mild respiratory disease or cystitis (bladder infection) in young children, but most infections with either BK or JC virus are not associated with illness. The viruses establish a latent infection that persists indefinitely, and the virus may reactivate after many years. The latent infection appears to be established in the

kidneys, in B lymphocytes, and, for JC virus, perhaps in the brain. Reactivation may be brought about by immunosuppression or by factors such as pregnancy or diabetes, and results in the excretion of virus in the urine. Reactivation of JC virus resulting from immune suppression is serious because it can replicate in oligodendrocytes, which produce myelin in the brain, and the death of these cells results in PML. PML is a rare, subacute, demyelinating disease of the central nervous system that has a worldwide distribution. It is an infrequent complication of a wide variety of conditions, including Hodgkin's disease, chronic diseases such as tuberculosis, primary acquired immunodeficiency diseases such as AIDS, or immunosuppression following organ transplant. The frequency of PML, once considered a rare disease, has increased with the AIDS epidemic and PML is now recognized as one of the AIDS-defining illnesses—it occurs in about 5% of all AIDS patients. PML may be more likely to occur when immunosuppression is due to infection by HIV because HIV-1 transactivates the JC late promoter. The disease progresses rapidly and can lead to mental deterioration and death within 3–6 months after onset.

Reactivation of BK virus can cause kidney disease. Such reactivation as a result of immune suppression accompanying kidney transplantation is a leading cause of allograft failure.

Polyomaviruses cause tumors in laboratory animals and many attempts to associate BK or JC virus with human cancer have been made. Association of these viruses with many human malignancies has been reported, but proof that they cause these malignancies is still lacking. Both viruses persistently infect most humans, and the presence of one of these viruses in association with a tumor cell does not prove causality, as the tumor cell could provide a suitable

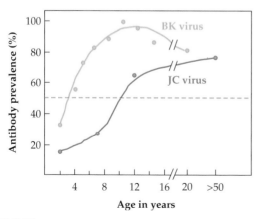

FIGURE 7.29   Prevalence of antibodies to BK and JC viruses in humans in the United States as a function of age. From Fields *et al.* (1996), p. 2039.

environment for reactivation of the virus rather than to result from infection by the virus. Nonetheless, these viruses are known to be capable of transforming cells and of causing tumors in experimental animals, giving rise to the possibility, if not probability, that they cause tumors in humans. Transformation of cells by JC virus appears to affect primarily cells of neuronal origin, and there is suggestive evidence that in humans the virus may cause tumors in the brain and perhaps in other organs as well. JC virus causes solid tumors in nonhuman primates as well as in rodents, giving further reason to believe it might be responsible for some human cancer. For BK virus there also exists suggestive evidence that the virus may cause a number of human tumors.

Early lots of poliovirus vaccine, both the inactivated Salk vaccine and the live Sabin vaccine, were contaminated with live SV40 virus, which was a contaminant in the monkey kidneys used to produce the virus for the vaccines. A number of other live virus vaccines produced around this time were similarly contaminated with SV40. As a result, many millions of people were infected with the virus. SV40 shares 69% sequence identity with JC and BK viruses (Table 7.13), and causes tumors in experimental animals, giving rise to concern that it might cause tumors in humans. Extensive study of the cohort of people infected as a result of contaminated vaccines, both in the United States and in Europe, has not revealed any convincing evidence for tumors associated with SV40 infection, although suggestive evidence exists that implicates SV40 in a number of tumors, especially brain tumors and mesothelioma tumors of the lungs. The problem is complicated by the fact that SV40 now circulates in the human population, whether as a result of its introduction with the poliovirus vaccine or from another source is not known. It is also complicated by the fact that although the virus is often found associated with certain tumors, it is not found in all such tumors. Is the association with such tumors, therefore, adventitious or causative? The case of mesothelioma illustrates this point. These tumors are clearly associated with exposure to asbestos but SV40 is often associated with them. Is the virus a cofactor in the development of the tumor or simply a freeloader? It should be noted that in animal models polyomaviruses are most likely to cause tumors in animals that are nonpermissive for virus replication. SV40 is primarily a monkey virus and human cells are only semipermissive for virus replication, suggesting that this virus may be more likely to cause tumors in humans than the human viruses BK and JC.

FAMILY *PAPILLOMAVIRIDAE*

Papillomaviruses resemble polyomaviruses in structure but are larger (Fig. 2.5). The virion is 55 nm in diameter, and the circular dsDNA genome is 8 kb in size. Papillomaviruses are highly diverse viruses that have been found in all mammals and birds that have been carefully examined, with the exception of the laboratory mouse, and probably occur in most mammals and birds. The viruses are highly host specific. To date 118 virus types have been completely described, most of which, for obvious reasons, are human viruses (HPVs). Cell culture systems for the study of these viruses are very limited because they undergo a complete replication cycle only in terminally differentiated cells, and most virus isolates are characterized on the basis of nucleotide sequences of virus DNA obtained directly from patients or infected animals. Only when the complete genome has been characterized is the virus given a name, which consists of a sequential number together with a designation for its host. Thus, although there are 96 recognized types of HPV, more exist and continue to be characterized.

The classification of papillomaviruses has recently undergone major revisions. Classification is now based on the sequence of the most highly conserved protein in the family, the major capsid protein called L1. If a virus that differs by more than 10% in the nucleotide sequence encoding this protein from that of the virus to which it is most closely related, it is recognized as a different type and given a new number. Isolates that differ by 2 to 10% in the sequence of this gene are considered to be different subtypes of the same virus type, and if they differ by less than 2% they are classified as different strains of the same virus type or subtype. Grouping into genera and species is also based on the sequence of the L1 gene. Viruses that share less than 60% nucleotide sequence identity in the L1 gene are classified into different genera, and species within a genus share between 60 and 70% sequence identity. Thus, virus types within a species share 71–90% identity. This definition of a species is biologically relevant, because it groups virus types that share important biological traits.

Sixteen currently recognized genera of papillomaviruses are listed in Table 7.14. Genera are simply named by Greek letters. The HPVs fall into 5 genera at present, only one of which, *Alphapapillomavirus*, is known to contain a nonhuman virus as well as human viruses, but even in this case all of the known viruses in this genus are primate viruses. The 96 HPV types are grouped into 27 species using the rules outlined earlier. Each species is named after the prototype HPV type in that species. Thus, for example, 14 species are recognized in the genus *Alphapapillomavirus*, most of which consist of multiple virus types as defined before. As an example of one of the species within the genus, *Alphapapillomavirus* species 16, this species contains HPV types 16, 31, 33, 35, 52, 58, and 67. *Beta-*, *Gamma-*, *Mu-*, and *Nupapillomavirus* genera contain 5, 5, 2, and 1 species, respectively.

Infection by Papillomaviruses

Papillomaviruses infect epithelial cells, either mucosal or cutaneous. Each virus is usually more or less restricted

TABLE 7.14   *Papillomaviridae*

Genus/members	Virus name abbreviation	Usual host(s)	Disease/genome characteristics
***Alphapapillomavirus*[a]**			
Human papillomaviruses 2, 6, 7, 10, 16, 18, 26, 32, 34, 53, 54, 61, 71, cand 90	HPV-2, etc.	Humans	Oral and anogenital mucosal lesions E5 ORF conserved between early and late regions
Rhesus monkey papillomavirus **1**	RhPV-1	Primates	
Betapapillomavirus			
Human papillomaviruses 5, 9, 49, cand 92, cand 96	HPV-5 etc.	Humans	Epidermodysplasia verruciformis activated by immunosuppression
Gammapapillomavirus			
Human papillomaviruses 4, 48, 50, 60, 88	HPV-4 etc.	Humans	Cutaneous lesions E5 ORF is absent
Deltapapillomavirus			
European elk papillomavirus	EEPV	Elk	Fibropapillomas
Other ovine and bovine papillomaviruses		Cattle, sheep and deer	
Epsilonpapillomavirus			
Bovine papillomavirus 5	BPV-5	Cattle	Cutaneous papillomas
Zetapapillomavirus			
Equine papillomavirus 1	EcPV	Horses	Cutaneous papillomas
Etapapillomavirus			
Chaffinch papillomavirus	FcPV	Birds	Cutaneous lesions E6 ORF absent
Thetapapillomavirus			
Timneh African gray parrot papillomavirus	PePV	Birds	Cutaneous lesions E4, E5, E6 ORFs are absent
Iotapapillomavirus			
Mastomys natalensis papillomavirus	MnPV	African soft-furred rat	Cutaneous lesions E2 ORF larger and E5 ORF absent
Kappapapillomavirus			
Cottontail rabbit papillomavirus	CRPV	Cottontail rabbits	Cutaneous and mucosal lestions E6 larger, and extra E8 ORF
Lambdapapillomavirus			
Canine oral papillomavirus	COPV	Dogs	Cutaneous and mucosal lesions
Feline papillomavirus	FdPV	Cats	Region between early and late genes very long
Mupapillomavirus			
Human papillomaviruses **1, 63**	HPV-6, -63	Humans	Cutaneous lesions
Nupapillomavirus			
Human papillomavirus 41	HPV-41	Humans	Benign and malignant cutaneous lesions
Xipapillomavirus			
Bovine papillomavirus 3	BPV-3	Cattle	True papillomas ORF 6 absent
Omikronpapillomavirus			
Phocoena spinipinnis papillomavirus	PsPV	Cetaceans	Genital warts ORF 7 absent
Pipapillomavirus			
Hamster oral papillomavirus	HaOPV	Hamsters	Mucosal lesions

[a] Transmission in most cases is by close contact, including sexual contact and the viruses are found worldwide. For human viruses, viruses with red numbers cause malignancies, while those in blue are primarily benign. Nonhuman virus names are colorcoded according to their hosts.

to specific sites in the host. The receptors for the virus are unknown, but they enter the cell by receptor-mediated endocytosis. In order to enter the cytoplasm from the endosomal compartment, cleavage of the viral minor capsid protein L2 by furin is required. This is the same protease that cleaves the glycoproteins of many enveloped viruses to render them infectious, but here it is used by the virus upon entry rather than during assembly of the virion. Although furin is a type I membrane protein that is known to be present in the trans-Golgi network, it is also present on the cell surface, and it is probably here that it cleaves L2.

The virus first infects cells of the basal proliferative layer of epithelium, probably at the site of a cut or abrasion that gives it access to this layer. The viral DNA enters the nucleus where it is maintained as a low-copy number plasmid and only the early genes are expressed. When the cells divide, the viral DNA is transmitted to both daughter cells by means of tethering the DNA to mitotic chromosomes by the E2 protein. This mechanism of insuring distribution of viral DNA to progeny cells is also used by two herpesviruses, EBV and HHV8. Only when the cells become terminally differentiated does the amplification of the viral DNA begin in earnest, the expression of late genes commences, and the progeny virions are assembled and shed.

Our knowledge of the replication of the papillomaviruses is limited because none will undergo a full replication cycle in any simple tissue culture system. Bovine papillomavirus (BPV-1) has been the most extensively characterized because it proliferates in dermal cells as well as in terminal epithelial cells. It readily infects and transforms rodent cells, in which the early proteins are expressed and viral DNA replication occurs. In transformed cells, the BPV genome is maintained as a stable plasmid and this feature has permitted the use of BPV-1 as an expression vector. Much effort has also been put into the study of the human papillomaviruses (HPVs) because of their association with human cancer, but these studies have been hampered because human papillomaviruses will only grow in humans and human cells and will only undergo a full lytic cycle in terminally differentiated cells. Two recent developments have helped in the study of HPVs. A tissue culture systems has been developed in which epidermal cells will differentiate, permitting at least limited studies of a full growth cycle. This method is laborious, however. Another approach has been the development of packaging systems in which viral DNA and proteins are expressed from expression plasmids, and large numbers of virus particles are formed. Other studies have simply used expression of various viral proteins from expression vectors to study the properties of these proteins, or have used grafts of infected human tissue in immunocompromised mice to study a full replication cycle.

Transcription of mRNAs

The genome organizations of two papillomaviruses, BPV-1 and HPV-11, are shown in Fig. 7.30A and B and a detailed transcription map of HPV-11 showing the exons, transcriptional promoters, and polyadenylation sites is shown below in Fig. 7.30C. Unlike the polyomaviruses, all papillomavirus mRNAs are transcribed in the same direction from only one of the two strands. The genomes of papillomaviruses are circular, but the genomes have been linearized in the figure for ease of presentation. In the absence of cell culture systems for the virus, most studies of RNA transcription have used RNA extracted from papillomas or from carcinoma cell lines. Because of this, the maps are probably not complete.

The transcription of papillomavirus RNAs is complex, as illustrated in the figure. There are multiple promoters and splice sites and differential use of these in different cells. Furthermore, there is extensive overlap of genes in the genome. Different reading frames encoding different peptide sequences may be linked in different ways by alternative splicing.

In BPV-1 there are at least seven promoters for transcription of RNA and more than 20 different mRNAs have been identified. Six promoters are used for the transcription of early mRNAs, all of which terminate at a poly(A) addition site at position 4180 (A_E). The seventh promoter is used for the transcription of the late mRNAs, which terminate at a poly(A) addition site at 7156 (A_L). The early genes of BPV-1 include E1 and E2, required for DNA replication, and E5, E6, and E7, required for cell transformation. The late genes, which are expressed only in terminally differentiated epithelial cells, include L1 and L2, which encode proteins present in the virion. L1 is the major capsid protein and when expressed alone assembles into virus-like particles that appear identical to virions. If L2 is coexpressed with L1, it is also incorporated into the virus-like particles. Another late gene is E4, which although located in the early region is expressed from the late promoter.

The pattern of transcription in HPV-11, illustrated in Fig. 7.30C, is slightly different. Only three promoters are known, two for the early genes and one for the late genes. There are poly(A) addition sites for early and late transcripts. Proteins corresponding to those of BPV-1 are produced during HPV-11 infection, but the complexity of the pattern of proteins produced and the difficulties in studying HPV replication make exact comparisons difficult. However, three regions of the genome are recognized, the early region encoding nonstructural proteins (about 4 kb of DNA), the late region encoding the structural proteins (about 3 kb of DNA), and the noncoding long control region or upstream regulatory region of about 1 kb that contains cis-acting elements that regulate viral replication and gene expression.

A. Bovine Papillomavirus 1 (7945 bp)

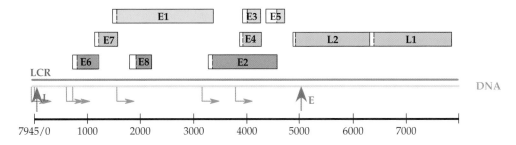

B. Human Papillomavirus 11 (7933 bp)

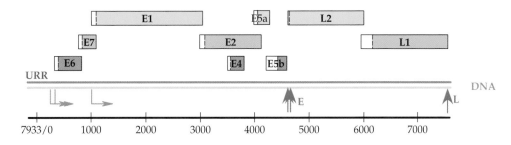

C. Transcription Map of HPV 11

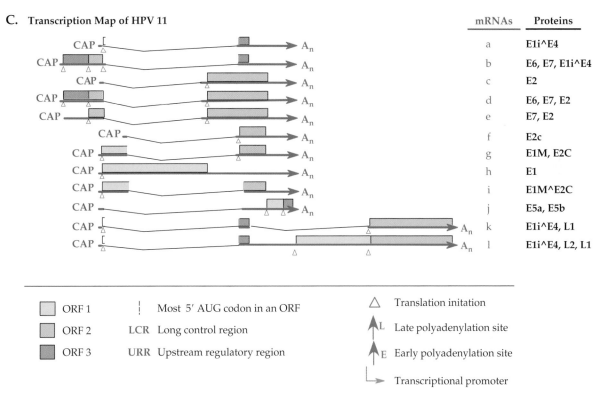

FIGURE 7.30 Genome organization and transcription map of papillomaviruses. (A) Genome organization of bovine papillomavirus 1. (B) Genome organization of human papillomavirus 11. (C) Transcription map of human papillomavirus 11. Genomes have been linearized at the upstream regulatory region (LCR or URR) for ease of presentation. All ORFs are transcribed from left to right from one DNA strand. Promoters are shown as turquoise arrows, sites of poly(A) addition with purple arrows labeled late (L) or early (E), initiation codons are shown as open red triangles. The symbol "^" joins two ORFs that are translated together as a fusion protein from a spliced RNA. Adapted from Nathanson *et al.* (1996), p. 27 and from Fields *et al.* (1996), pp. 2051 and 2052.

DNA Replication

DNA replication requires the activities of E1 and E2. The E1 protein binds to the origin of replication, has helicase activity, binds α-primase, and is presumed to promote the initiation of DNA replication. Thus, it has many of the functions of the polyomavirus T antigens. E2 is a regulatory protein that is produced in multiple forms (Fig. 7.31). It can either transactivate or repress genes, depending on the location of binding sites for it within a gene. It plays an important role in the regulation of transcription, and interacts with E1 to promote recognition of the promoter for efficient replication of DNA. The interaction with E1 may help recruit host replication factors to the origin of replication or promote the assembly of the preinitiation complex.

DNA replication occurs in two phases. In cells that the virus infects nonproductively (cells transformed by BPV-1 or cells in the dermal layer of the epithelium), the DNA is maintained as a multiple copy plasmid (50–400 copies per cell). After replicating sufficiently to reach this number of copies, further DNA replication is limited to that required to maintain the copy number as cells divide. A complete replication cycle occurs only in terminally differentiated cells, where large numbers of DNA genomes are produced for incorporation into progeny virions. Since these terminally differentiated cells do not divide, they are intrinsically incompetent in supporting DNA synthesis. Thus, large-scale production of viral DNA genomes in these cells requires the activity of viral transforming genes.

The Transforming Genes

Three genes of papillomaviruses, E5, E6, and E7, have been shown to be involved in transforming cells. E5 of BPV-1 is a small polypeptide (44 amino acids) that is believed to

have a short N-terminal cytoplasmic domain, a transmembrane region, and a more extensive C-terminal extracellular domain. It activates the receptor for platelet-derived growth factor, perhaps by binding to it and causing it to dimerize. Dimerization of many growth factor receptors present at the surface of cells results in activation of a protein kinase and phosphorylation of tyrosines, which leads to the activation of transcription factors whose activities stimulate cell proliferation. BPV-1 E5 also binds other cellular proteins that may be involved in transformation. Transformation of established rodent cells as defined by a number of criteria can be obtained by expression of E5 alone, but the expression of both E6 and E7 is required for the fully transformed phenotype.

The E5 encoded by HPVs has also been shown to induce some transforming alterations, but whether this protein plays an important role in transformation is uncertain. More is known about HPV E6 and E7. Expression of E6 and E7 from high-risk strains of HPV (strains that are often associated with human cancer) are capable of facilitating the immortalization of primary human keratinocytes. E7 is a small zinc-binding protein that is phosphorylated and is capable of transforming cells when expressed alone. It binds the cellular tumor suppressor protein Rb as well as p107 and p130 (Fig. 7.22). Rb undergoes changes in phosphorylation induced by cyclin-dependent kinases at the G–S1 boundary. In its hypophosphorylated form it inhibits cell cycle progression. The viral oncogene preferentially binds the hypophosphorylated form, thus preventing its inhibitory activity and inducing cycling of the cell (and therefore DNA synthesis). Genetic studies have shown that binding to Rb is required in order for E7 to transform cells. It is of considerable interest that E7 from low-risk HPVs binds Rb only one-tenth as efficiently as E7 from high-risk strains, and that the E7 from low-risk strains is inefficient in transformation assays.

E6 from high-risk HPVs, but not from low-risk HPVs, can complex with the tumor suppressor protein p53 (Fig. 7.22). SV40 and adenovirus oncoproteins also bind p53, but simply sequester it. In contrast, binding of HPV E6 leads to the degradation of p53 by the ubiquitin-mediated degradation pathway. Removal of p53 has the effect of inducing DNA synthesis, as described earlier. The importance of Rb and p53 in regulating cell cycling is made clear by the fact that three different viruses—high-risk HPVs, adenoviruses, and SV40—all target these proteins in order to provide an atmosphere conducive for DNA replication by the virus. HPV E6 can also activate transcription of hTERT, the catalytic subunit of telomerase. Full transformation of human epithelial cells requires telomerase expression as well as the expression of viral and cellular oncogenes.

Papillomaviral Disease

On infection, papillomaviruses induce cellular proliferation that leads to the production of warts or papillomas.

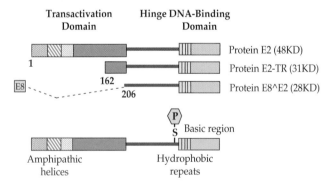

FIGURE 7.31 Structure of BPV-1 virus E2 protein. Full-length E2 contains a transactivation domain at the N terminus, linked by a hinge to a DNA-binding domain at the C terminus. There are 3 forms of E2, all of which contain the 85–amino acid long DNA-binding domain. The bottom diagram identifies other functional features of E2 such as the amphipathic helices, hydrophobic repeats, a basic region, and a phosphorylation site. Data from McBride *et al.* (1989).

The viruses are usually restricted to a specific area of the body such as the skin, the mouth, the throat, or the genital tract. In most infections these eventually resolve, but in some cases tumors can result. Papillomaviruses of humans, cattle, sheep, and cottontail rabbit have been shown to be associated with cancers in their natural hosts.

Human Papillomaviruses

More than 200 different strains of HPV have been identified by analyses of full or partial sequences of viral DNAs isolated from individual lesions and classified into 96 different types to date. These 96 types, numbered from 1 to 96, are classified in turn into 27 species found in 5 different genera. Species are named after the type within that species with the lowest number, which is therefore the first described type within that species, and species numbering is thus not sequential. A listing of 19 of these species that shows the tissues infected and the probability that infection leads to cancer is shown in Table 7.15. HPVs cause warts in the skin, genital tract, mouth, or respiratory tract. The warts are normally self-limited proliferative lesions that regress after some time because of an immune response. Cytotoxic T lymphocytes are thought to play an important role in regression, and warts are often more numerous under conditions where the immune system is suppressed. HPVs are spread by direct contact and infection begins at the site of an abrasion in which the virus can contact the deeper epithelial layers.

HPVs are not only specific for humans but also for the tissues infected. HPVs cause either cutaneous lesions or mucosal lesions. Eleven of the HPV species shown in Table 7.15 cause skin warts. Only a few of these are responsible for most skin warts, which are common in school children. The remainder have been isolated only from patients who suffer from a rare disease called epidermodysplasia verruciformis. These patients are unable to resolve their warts, probably because of an inherited immunologic defect, and wartlike lesions appear all over the body. These warts often become malignant after many years, especially on areas of the skin exposed to sunlight, but these tumors are generally slow growing and do not metastasize.

TABLE 7.15 Common Clinical Lesions Associated with Human Papillomaviruses

| Type/isolates | Anatomical site | Disease | | Risk | Cancers |
		Common name	Medical term		
Cutaneous HPVs					
HPV 1, 4	Sole, palm	Plantar warts	Verruca plantaris	None	None
HPV 2, 4, 10, 26	Cutaneous	Common warts	Verruca vulgaris	None	None
HPV 2, 5, 10	Cutaneous	Flat warts	Verruca plana	None	None
HPV 5, 50	Face, trunk, esophagus	Benign warts	EV	High	Skin carcinomas[a]
HPV 5, 9	Cutaneous	Flat warts	Verruca plana		In ISP
HPV 5, 9	Cutaneous	Macular lesions	EV (benign)	Some	In ISP
HPV 5, 9, 49	Cutaneous		EV	Some	Some SCC
HPV 9	Cutaneous		Malignant melanoma	High	Skin carcinomas[a]
HPV 41, 48	Cutaneous		SCC	Some	In ISP
HPV 49	Cutaneous	Plantar warts	Verruca plantaris	?	In ISP
Mucosal-Associated HPVs					
HPV 6	Anogenital, larynx	Genital warts	CD	Low	Rare[b]
HPV 6, 7, 32, 34	Anogenital	Anogenital warts	IN	Low	Rare[b]
HPV 6, 7, 26, 61	Genital	Genital warts	CIN	Low	
HPV 61, 34	Larynx	Oral papillomas		Low	In ISP
HPV 26	Genital mucosa	Genital warts	CIN	Some	Some malignant progression
HPV 16, 18, 53	Genital mucosa	Genital warts	CIN	High	1–3% progress to cervical carcinomas, cofactors unknown

[a] 30–40% undergo neoplastic conversion in sun-exposed area.
[b] On rare occasions these HPV types have also been found associated with carcinoma.
Abbreviations: SCC, squamous cell carcinoma; EV, epidermodysplasia verruciformis; CD, condyloma acuminatum; CIN, cervical intraepithelial neoplasia; IN, intraepithelial neoplasia; ISP, immunosuppressed patients.
Source: Adapted from Fields *et al.* (1996), Table 1, pp. 2048–2049, Tables 3 and 4; p. 2085; and Alani and Münger (1998).

Nine of the species shown in Table 7.15 cause genital warts (species 26 infects both mucosal and cutaneous surfaces). Genital warts are among the most common sexually transmitted diseases. One study found that 46% of college women examined were positive for HPV DNA in the genital tract. Older women have a lower incidence of HPV, either because they have had fewer recent partners or because they have acquired some immunity. In many cases the infection is cleared completely after some months, but the virus may remain in a latent or persistent form in apparently normal tissue adjacent to the wart and the lesions may recur. Immunosuppression results in an increased incidence of warts.

Genital HPVs are clearly associated with cervical cancer. Cancer is a rare complication of HPV infection that may take decades to develop and it requires additional genetic mutations. Because of the prevalence of HPVs, however, there are about 500,000 new cases of cervical carcinoma diagnosed annually worldwide and most, perhaps all, are associated with HPV. In developed countries with a high standard of health care, cervical cancer accounts for about 7% of cancer in women, but in developing countries cervical cancer accounts for 24% of all cancer in women. In the United States about 4000 deaths occur annually from cervical cancer.

Genital HPVs can be divided into low risk (rarely or never associated with cancer), intermediate risk, and high risk (often associated with cancer). HPV-16 accounts for about half of the cancers, and HPV-16 and HPV-18 together account for more than 70% of the cancers. As described earlier, papillomaviruses encode oncoproteins that interfere with the functions of tumor suppressor proteins, and the high-risk HPV-16 and HPV-18 interfere most strongly. The high-risk viruses also induce the production of telomerase. These activities are almost certainly the basis of papillomavirus-induced cancer, but development of tumors requires more than just the expression of these genes. Integration of the HPV genome into the host chromosome, or at least that part which encodes E6 and E7, occurs and this integration results in higher expression of the viral transforming genes. In addition, these viruses destabilize host chromosomes so that chromosomal abnormalities occur, including aneuploidy. Thus progression to a cancer is a long-term process that requires many changes for the transformed cell to become immortalized and invasive.

HPV types 6 and 11 (two types belonging to species 6), which cause genital warts, may also infect the mouth, nasal cavity, larynx, or lungs. Infection of the larynx may be problematic because of the resulting obstruction of the airways or because of hoarseness caused by infection of the vocal cords. Surgical removal of the papillomas may be required. These papillomas tend to recur, requiring further operations. There are in addition two HPVs known that infect only the oral cavity.

Because of the association with cervical cancer, efforts are being made to develop vaccines. A vaccine directed against HPV-16 and HPV-18 has been shown in clinical trials to be 100% effective in preventing infection by these two viruses, and also 92% effective against low-risk HPVs. On June 8, 2006 the Merck vaccine "Gardasil" was approved by the Federal Drug Administration for 9- to 26-year-old females. This quadrivalent vaccine contains antigens from HPV types 16, 18, 6, and 11. Widespread use of such a vaccine will result in preventing the majority of cervical carcinomas as well as vocal cord warts caused by HPV, although how long the immunity will last is as yet to be determined.

A second possible approach to the control of papillomaviruses is the use of human α-defensins to block infection by the virus. These are small peptides secreted by humans that have bacteriocidal and antiviral activity (Chapter 10). Certain of the defensins block infection by papillomaviruses at high concentrations and, interestingly, in some women such concentrations of these peptides occur naturally in the genital tract. Application of these peptides could ablate virus infection, a possibility that is perhaps even more significant because the use of condoms affords very little protection against papillomavirus infection.

A phylogenetic tree of selected human papilloma viruses constructed using sequences in the E2 gene is shown in Fig. 7.32. There is no simple relationship between the relatedness of the different viruses as illustrated by this tree and the target tissue infected by the viruses or the risk of neoplastic transformation following infection by them. Note, for example, that HPV-16 and HPV-18, which cause the majority of cervical carcinoma, are widely separated in the tree, although both are alphapapillomaviruses.

FAMILY *PARVOVIRIDAE*

The parvoviruses are small icosahedral viruses that are 18–26 nm in diameter (Figs. 2.1 and 2.5). They contain ssDNA of about 5 kb as their genome. Different viruses variously package the minus strand (the strand complementary to the messenger sense) or a mixture of plus and minus strands. Two subfamilies are recognized. The *Densovirinae* are viruses of insects and consist of three recognized genera. The *Parvovirinae* are viruses of birds and mammals and five genera are recognized at present. A partial listing of the members of the *Parvovirinae* is shown in Table 7.16. These viruses are species specific and also specific for the spectrum of tissues that can be infected. Unlike other DNA viruses, the parvoviruses do not encode genes that induce the cell to enter S phase, and they can only replicate in cells that are actively replicating. The members of the *Dependovirus* genus are further limited in their replication in that normally they can only replicate in cells that are infected by an adenovirus, a herpesvirus, or a papillomavirus.

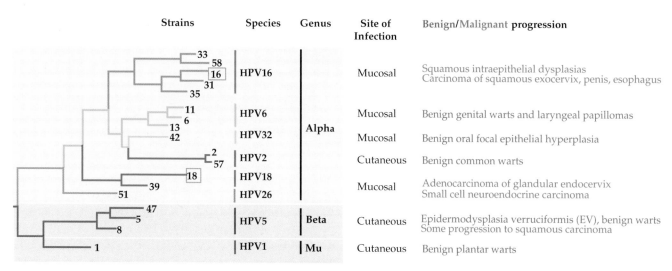

Strains	Species	Genus	Site of Infection	Benign/Malignant progression

FIGURE 7.32 Phylogenetic tree of the human papillomaviruses based on the nucleotide sequence of the amino-terminal half of the E2 gene. The sites of infection and the potential for neoplastic progression are shown. The numbers of the strains are shown (compare Table 7.15), the strains are grouped into species according to new taxonomy, and the two high-risk viruses that cause most cases of cervical carcinoma are boxed. Note that there is no simple relationship between position on the phylogenetic tree and either the site of infection or probability of progression to malignancy, although the two viruses causing the most malignancies are both in the *Alphapapillomavirus* genus. Adapted from Nathanson *et al.* (1996) p. 273 and taxonomic data from Fauquet *et al.* (2005).

TABLE 7.16 *Parvovirinae*

Genus/members	Host(s)	Transmission	Disease
Parvovirus			
Minute virus of mice	Mice	Contact, fomites	?
Feline panleukopenia	Dogs, cats	Contact, fomites	Enteritis in adults, myocarditis in pups
Kilham rat parvovirus	Rats		Stillbirth, abortion, fetal death, mummification
Porcine parvovirus	Swine		
Erythrovirus			
B19 Several primate parvoviruses	Humans		Fifth disease, aplastic anemia, hydrops fetalis, arthritis immunodeficiency
Dependovirus			
Adeno-associated virus 1–5	Humans	Transplacental (AAV-1), vertical (AAV-2)	None
Adeno-associated viruses of other species	Cattle, dogs, sheep		None
Goose parvovirus	Geese	Vertical transmission	Hepatitis
Amdovirus			
Aleutian mink disease	Mink	Contact, fomites	Chronic immune complex disease
Bocavirus			
Bovine parvovirus	Cattle		Enteritis
Human bocavirus	Humans		Respiratory infections

Transcription of the Viral Genome

The genome organizations and transcription maps for two human parvoviruses belonging to different genera are shown in Fig. 7.33. The parvovirus genome contains two genes, each of which is transcribed into multiple mRNAs. The non-structural or replication gene, referred to as NS or REP, is located at the 5′ end of the plus-sense copy of the genome and the gene for the capsid proteins, referred to as VP or CAP, is located at the 3′ end. The two genes are present in the same orientation so that only one strand is transcribed.

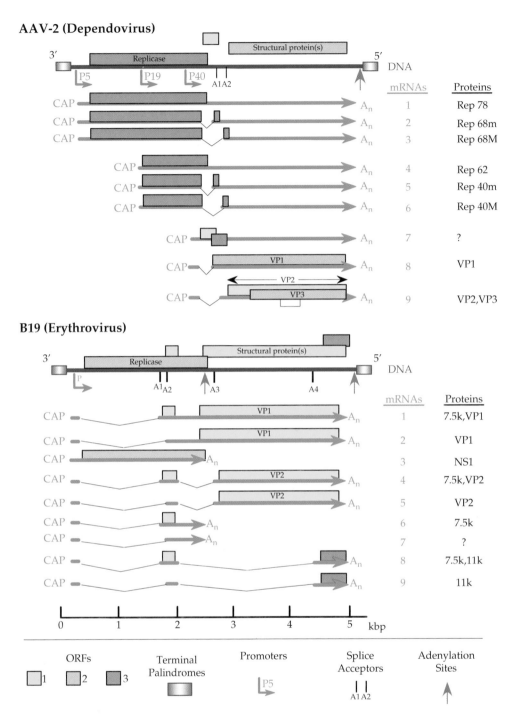

FIGURE 7.33 Genome organizations and transcription/translation schemes for two human parvoviruses belonging to two different genera. Adeno-associated virus (AAV) is a dependovirus and B19 virus is an erythrovirus. These viruses use 3 and 1 promoters, respectively, to make a set of spliced and unspliced messages, all transcribed from one DNA strand, from which the various virus proteins are translated. Terminal palindromes are shown as shaded boxes. Adapted from Heegaard and Brown (2002), Figure 2; Mouw and Pintel (2000), Figure 1.

B19 has only one promoter for initiation of transcription but two poly(A) addition sites, whereas AAV has three promoters for transcription but only one poly(A) addition site. The use of multiple promoters or poly(A) sites is combined with alternative splicing events to give rise to a number of mRNAs. The best understood translation products are a nonstructural protein of about 80 kDa and the two or three capsid proteins.

Replication of the Viral DNA

Replication of parvoviral DNA occurs in the nucleus. Replication of the DNA and transcription of mRNAs are effected by host DNA and RNA polymerases but the NS or REP protein of the virus is required. The activities of this protein include a site-specific DNA-binding activity, a site-specific DNA nuclease activity, and helicase activity.

Parvoviral DNA is linear and possesses palindromic sequences at the two ends. In some viruses the palindromic sequences at the two ends are the same, whereas in other viruses they are different. These palindromic sequences are 100–300 nucleotides long, depending on the virus, and can fold back to form a very stable hairpin structure, as illustrated in Fig. 7.34 for two different parvoviruses. The hairpin primes DNA replication, with the 3' end serving as a primer that is elongated to form a double-stranded intermediate, as illustrated in Fig. 7.35. How this intermediate is used to continue DNA replication and how it is resolved to give plus and minus DNA genomes is not clear, although it is known that the viral nonstructural protein is involved. Models have been proposed that involve either continued rolling of the hairpin or the formation of cruciform structures that might be resolved by cellular recombination enzymes. In the favored model, the rolling hairpin is resolved by an endonuclease activity in NS/REP, as illustrated schematically in the figure, an activity known to be present in the protein. Resolution in this way results in the flipping back and forth of the terminal sequence. Such flipping is known to occur for at least some viruses, which results in the palindromic sequence being present in two orientations that are distinguishable. In some viruses, only one end of the genome flips, whereas in others both ends flip.

Assembly of the Virion

The parvovirus virion is a *T*=1 icosahedron that is constructed of 60 molecules of capsid protein (see Fig. 2.5). The major capsid protein is called VP2 and is about 60 kDa in size. Smaller amounts of a larger capsid protein called VP1 (about 80 kDa) are present in all parvoviruses, and some contain in addition a third capsid protein called VP3. The two or three structural proteins share significant sequence overlap, being in essence variously truncated forms of the largest protein (Fig. 7.33). The structure of the major capsid protein of a parvovirus has been solved to atomic resolution and it possesses the same eight-stranded antiparallel β sandwich present in many RNA and DNA viruses (Chapter 2), which suggests that these various capsid proteins share a common ancestry.

The parvoviral genome is packaged into preassembled capsids starting from the 3' end. Packaging requires the helicase activity of NS/REP and the expenditure of ATP. In viruses for which the two ends of the genome are the same, equal numbers of DNA genomes of both plus and minus orientation are made and packaged. However, in viruses whose genomes have different palindromic sequences at the two ends, only the minus sense genome is packaged. Thus, a packaging signal at the 3' end is specifically recognized for packaging. In experiments in which a virus that normally packages only the minus-sense genome is induced to package both plus- and minus-sense strands by changing the 3' end sequence of the plus-sense strand, it was found that the plus-sense genome was packaged poorly and got hung up after about half of the genome was packaged. Thus, the minus-sense DNA has evolved so that secondary structures that form in single-strand nucleic acids do not interfere with packaging, whereas the plus-sense genome, not being subject to such selective pressure, has secondary structures within it that preclude efficient packaging.

Genus *Erythrovirus*

B19 virus, the only known human virus in the genus *Erythrovirus*, was until recently the only recognized human pathogen among the parvoviruses. Infection of humans with B19 is accompanied by nonspecific flulike symptoms followed by symptoms of erythema infectiosum (fifth disease), which presents as a generalized erythematous rash with a "slapped cheek" appearance and inflammation of joints. Children infected by the virus are usually not very ill. However, illness in adults can be more serious because the joint inflammation may mimic rheumatoid arthritis and can persist for months or years. In addition, virus infection of people with some forms of anemia can be quite serious.

B19 has a tropism for human erythroid progenitor cells, which are rapidly dividing cells capable of supporting virus replication. The virus is cytolytic and the infected cell dies. Thus, infection of erythroid progenitor cells results in a suppression of erythropoiesis for 5–7 days following infection by the virus. In healthy humans, whose red blood cells last for 160 days, this is a not a serious event. However, in people suffering from chronic anemias the inability to synthesize red blood cells for a week may be serious and occasionally fatal. In particular, patients with hemolytic anemia have a low hemoglobin concentration in the blood because their red blood cells have a short life span, only 15–20 days, so that arrest of erythropoiesis in the bone marrow leads to a sharp fall in hemoglobin concentration and worsening symptoms of anemia. Other populations at increased risk following B19 infection include patients with compromised immune

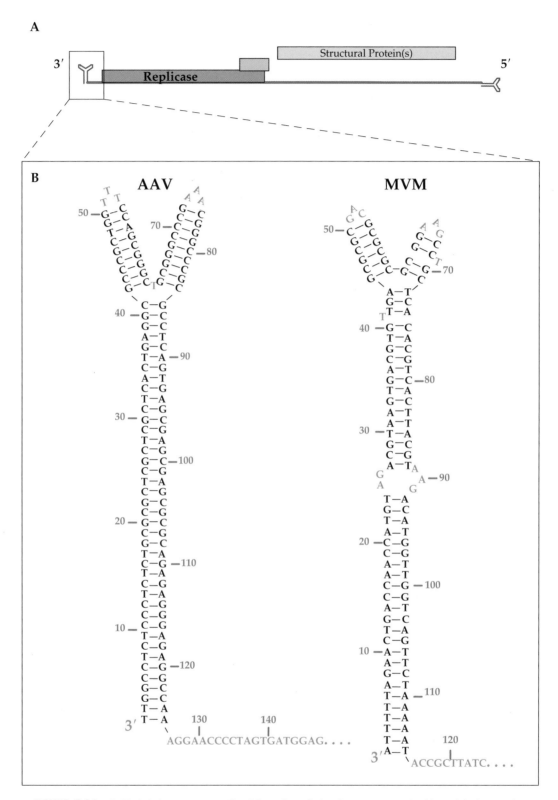

FIGURE 7.34 Stable hairpin structures predicted from the palindromic sequences at the 3′ termini of parvovirus virion DNAs. (A) Diagram of the genome, showing the location of the hairpin. (B) The most stable secondary structure predicted by the 3′ terminal nucleotide sequences of MVM DNA and AAV DNA. Adapted from Fields *et al.* (1996) p. 2175.

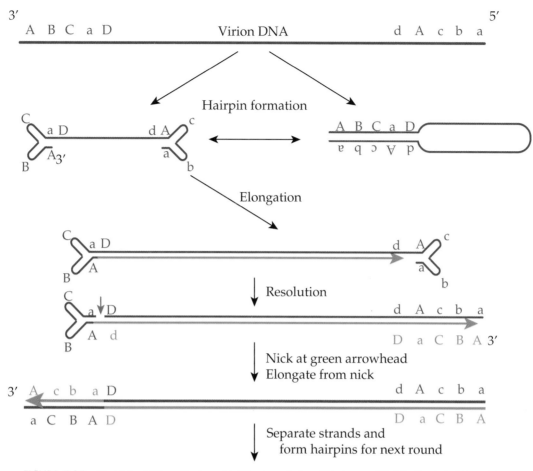

FIGURE 7.35   Model for DNA replication of AAV, a dependovirus. This is a modified "rolling hairpin" model, and results in inversion of both of the repeated sequences at the termini. Models for replication of the autonomous parvoviruses like MMV are more complex, since in that case the 5′ terminal sequence of the virion DNA is inverted during replication, while the 3′ terminal sequence is not. Adapted from Brister and Muzyczka (2000).

systems (in which infection by B19 can result in persistent anemia) and pregnant women. Congenital infection with B19 can be serious and can lead to fetal abnormalities or death, due to arrest of red blood cell formation and consequent anemia at critical times during development.

The receptor for the B19 virus is erythrocyte P antigen. This antigen is expressed on cells of the erythroid lineage, but only precursor cells can be productively infected. Mature erythrocytes are terminally differentiated, lack a nucleus, do not divide, and cannot support virus replication. In addition to possessing a receptor that allows the virus to enter, other factors required for replication are also furnished by erythroid precursor cells. Transfection of viral DNA into other cells does not lead to a complete replication cycle and the pattern of RNA transcription differs from that in permissive cells shown in Fig. 7.33.

B19 is a common virus. About 50% of adults have antibodies against the virus, which show that they have been previously infected, and in the elderly this rises to more than 90%. Infection of healthy people with normal immune systems leads to a solid immunologic response and subsequent immunity to the virus.

Genus *Dependovirus*

The genus *Dependovirus* contains viruses that can replicate without a helper under certain conditions and in certain tissues, but these viruses normally depend upon a helper virus for replication. Known helper viruses include adenoviruses, herpesviruses, and papillomaviruses, but because the dependoviruses were first found associated with adenoviruses, they are called adeno-associated viruses or AAVs. AAVs of humans and of numerous other vertebrates are known. More than 90% of human adults have antibodies to AAV, which shows that the virus is widely distributed and common.

On infection of a nonpermissive cell (one that has not been infected by a helper and is otherwise nonpermissive), AAV establishes a latent infection in which its DNA integrates into the host chromosome. In human cells integration is specific—it occurs on chromosome 19 within a defined region. There appears to be a binding site for the nonstructural protein at this location that directs integration to a site within a nearby region of several hundred nucleotides. On infection of the cell by an adenovirus or other helper, the AAV DNA is excised and replicates. Latent infection in humans appears to be common.

Adenoviruses are the best studied helpers, and several regions of the adenovirus genome are involved in helper function, including E1A, E1B, E2A, E4, and VA. At least some papillomaviruses can also serve as helpers, and in the case of HPV-16 the E1, E2, and E6 genes contribute to AAV replication. Replication of AAV downregulates the infection by HPV-16, which lowers the risk of cervical carcinoma induced by the papillomavirus. Retrospective studies have found that patients with cervical carcinoma are markedly deficient in antibodies to AAV as compared with normal controls. It is also known that independent replication of AAV in cultured cells can be obtained by treatment of the cells with toxic agents such as UV or one of several chemical carcinogens. This treatment must induce the production of cellular factors that are required for a full virus replication cycle. The helper function supplied by the helper virus could then be to independently supply these factors or to induce the production of these cellular factors. Thus, protection from the development of cervical carcinoma by AAV could be due not only to its effect upon the replication of papillomaviruses but also to increased replication in cancer cells, resulting in the death of the cell. AAV infection might therefore protect against the formation of tumors and be a virus that actually has a beneficial function in humans.

Genus *Bocavirus*

A recent isolate of a bocavirus from human respiratory tract samples has been tentatively called human bocavirus. The virus was subsequently detected in 17 humans and appears to be associated with lower-respiratory-tract infection in children. This may be a second parvovirus that causes disease in humans.

Genus *Parvovirus*

The genus *Parvovirus* contains viruses that infect a number of different mammals and birds. Several of these cause important diseases in domestic animals or in wildlife, including feline panleukopenia virus (FPV), porcine parvovirus, and canine parvovirus (CPV).

CPV is an example of an emerging nonhuman pathogen. It appeared suddenly in 1978 and spread around the globe in less than 6 months. The virus causes gastroenteritis in adult dogs and myocarditis in pups, which often leads to death from acute heart failure. It was called CPV-2 to distinguish it from a previously known canine parvovirus called minute virus of canines, which is sometimes referred to as CPV-1. CPV-2 is closely related to FPV and to parvoviruses that circulate in foxes and other wild animals, all of which share 98% sequence identity. Molecular genetic studies have suggested that FPV, which does not replicate in dog cells, could have jumped the species barrier if as few as two amino acid changes occurred in the major coat protein VP2. Once the ancestor to CPV-2 arose from FPV or a related virus, further selection for growth in dogs could have produced the virus CPV-2, which is much better adapted to dogs as a natural host. Consistent with this hypothesis is the finding that antibodies to a CPV-2-like virus were first detected in serum from European dogs taken in the early to mid 1970s, but only in 1978 did explosive spread of the virus occur. Also of interest for this hypothesis is the fact that a variant of CPV-2 called CPV-2a replaced CPV-2 between 1979 and 1981, and a newer variant 2b appeared in 1984. Since this time CPV-2a and -2b have changed little and appear to be in worldwide equilibrium.

A dendrogram of these various parvoviruses is shown in Fig. 7.36. Although CPV-2 is closely related to FPV and its relatives, it clearly belongs to its own clade, which shows that the lineage of the canine virus is distinct. This is different from the pattern shown by the lineages of FPV and the viruses of foxes, mink, and other wildlife. Here the clades do not assort with species, which suggests that the various strains are freely transmissible among these various animals and circulate within this expanded group of species. Thus it is of interest that CPV-2a and CPV-2b replicate well in cats, although CPV-2 does not. Also shown in the figure is a timeline for the emergence of CPV and its divergence into the different lines.

TORQUE TENO VIRUS: A NEWLY DESCRIBED HUMAN VIRUS

A new DNA-containing virus was isolated recently from the blood of a Japanese patient and first called TT virus, after the initials of the patient, but subsequently renamed Torque teno virus. The viral genome was circular single-strand DNA, of size 3739 nucleotides. Studies of sera from other people around the world demonstrated that the virus was widespread and common. Recent estimates are that 12% of Japanese, 1% of Americans, and 2% of English are infected, but that a much larger fraction of the population in tropical countries such as Papua New Guinea (74% of people) and Gambia (83%) are infected. The nucleotide sequences of different isolates from around the world exhibit up to 50% sequence diversity, and in dendrograms of these sequences the viruses fall into five different groups. The viruses have been classified as belonging to the genus *Anellovirus*. This genus was initially placed in the family *Circoviridae* but is now unclassified as to family.

Phylogenetic tree of the canine and feline parvoviruses

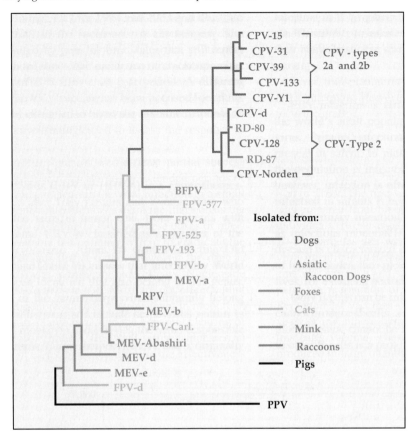

Evolution of the canine parvoviruses

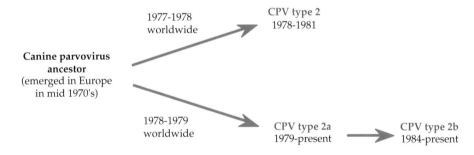

FIGURE 7.36 Phylogeny and evolution of feline and canine parvoviruses. Upper panel: phylogenetic tree of parvoviruses from dogs (CPV), Asiatic raccoon dogs (RD), cats (FPV), mink (MEV), raccoons (RPV), and foxes (BFPV), using porcine parvovirus (PPV) as an outgroup. The tree was constructed using the sequences of VP1 and VP2 and illustrates that all of the canine isolates form a distinct clade. Lower panel: diagram of the evolution of the canine parvoviruses into the two types circulating today. From Parrish (1997).

TT virus was isolated from a patient with hepatitis. Thus, a major interest in the virus is a possible association with hepatitis in humans, but there is no evidence at present that it does cause hepatitis in humans.

The isolation of a new virus that is widespread and common illustrates that there may yet be other viruses that infect humans but of which we are unaware. Whether Torque teno virus or other viruses that are as yet unknown are associated with human disease remains to be determined.

FURTHER READING

Poxviridae

Chen, N., Danila, M. I., Feng, Z., et al. (2003). The genomic sequence of ectromelia virus, the causative agent of mousepox. *Virology*, **317**: 165–186.

Damon, I. K. (2006). Poxviruses. Chapter 75 in: *Fields Virology, Fifth Edition* (D. M. Knipe and P. M. Howley, Eds. in chief), Philadelphia, Lippincott Williams & Wilkins, pp. 2947–2976.

Goebel, S. J., Johnson, G. P., Perkus, M. E., et al. (1990). The complete DNA sequence of vaccinia virus. *Virology* **179**: 247–266.

Griffiths, G., Roos, N., Schleich, S., and Locker, J. K. (2001). Structure and assembly of intracellular mature vaccinia virus: thin section analyses. *J. Virol.* **75**: 11056–11070.

Iyer, L. M., Aravind, L., and Koonin, E. V. (2001). Common origin of four diverse families of large eukaryotic DNA viruses. *J. Virol.* **75**: 11720–11734.

Jones, L. Y. (2005). Tribal fever. *Smithsonian* **May**: 91–97.

Kerr, P. J., Perkins, H. D., Inglis, B., et al. (2004). Expression of rabbit IL-4 by recombinant myxoma viruses enhances virulence and overcomes genetic resistance to myxomatosis. *Virology* **324**: 117–128.

Law, M., Carter, G. C., Roberts, K. L., Hollinshead, M., and Smith, G. L. (2006). Ligand-induced and nonfusogenic dissolution of a viral membrane. *Proc. Natl. Acad. Sci. U.S.A.* **103**: 5989–5994.

Lefkowitz, E. J., Wang, C., and Upton, C. (2006). Poxviruses: past, present, and future. *Virus Res.* **117**: 105–118.

Likos, A. M., Sammons, S. A., Olson, V. A., et al. (2005). A tale of two clades: monkeypox viruses. *J. Gen. Virol.* **86**: 2661–2672.

McFadden, G. (2005). Poxvirus tropism. *Nature Rev. Microbiol.* **3**: 201–213.

Moss, B. (2006). *Poxviridae*: The viruses and their replication. Chapter 74 in: *Fields Virology, Fifth Edition* (D. M. Knipe and P. M. Howley, Eds. in chief), Philadelphia, Lippincott Williams & Wilkins, pp. 2905–2946.

Herpesviridae

Barbera, A. J., Chodaparambit, J. V., Kelley-Clark, B., et al. (2006). The nucleosomal surface as a docking station for Kaposi's sarcoma herpesvirus LANA. *Science* **311**: 856–861.

Boehmer, P. E., and Lehman, I. R. (1997). Herpes simplex virus DNA replication. *Annu. Rev. Biochem.* **66**: 347–384.

Bornkamm, G. W., Behrends, U., and Mautner, J. (2006). The infectious kiss: newly infected B cells deliver Epstein-Barr virus to epithelial cells. *Proc. Natl. Acad. Sci. U.S.A.* **103**: 7201–7202.

Cohen, J. L., Straus, S. E., and Arvin, A. M. (2006). Varicella-Zoster virus. Chapter 70 in: *Fields Virology, Fifth Edition* (D. M. Knipe and P. M. Howley, Eds. in chief), Philadelphia, Lippincott Williams & Wilkins, pp. 2773–2818.

Enbom, M. (2001). Human herpesvirus 6 in the pathogenesis of multiple sclerosis. *APMIS* **109**: 410–411.

Fleming, D. T., McQuillan, G. M., Johnson, R. E., et al. (1997). Herpes simplex virus type 2 in the United States, 1976–1994. *N. Engl. J. Med.* **337**: 1105–1111.

Ganem, D. (2006). Kaposi's sarcoma–associated herpesvirus. Chapter 72 in: *Fields Virology, Fifth Edition* (D. M. Knipe and P. M. Howley, Eds. in chief), Philadelphia, Lippincott Williams & Wilkins, pp. 2847–2888.

Gaspar, M., and Shenk, T. (2006). Human cytomegalovirus inhibits a DNA damage response by mislocating checkpoint proteins. *Proc. Natl. Acad. Sci. U.S.A.* **103**: 2821–2826.

Grünewald, K., Desai, P., Winkler, D. C., et al. (2003). Three-dimensional structure of herpes simplex virus from cryo-electron tomography. *Science* **302**: 1396–1398.

Kieff, E. D., and Rickinson, A. B. (2006). Epstein-Barr virus and its replication. Chapter 68A in: *Fields Virology, Fifth Edition* (D. M. Knipe

and P. M. Howley, Eds. in chief), Philadelphia, Lippincott Williams & Wilkins, pp. 2603–2654.

Koch, W. C. (2001). Fifth (human parvovirus) and sixth (herpesvirus 6) diseases. *Curr. Opin. Infect Dis.* **14**: 343–356.

Kotenko, S. V., Saccani, S., Izotova, L. S., Mirochnitchenko, O. V., and Pestka, S. (2000). Human cytomegalovirus harbors its own unique IL-10 homolog (cmvIL-10). *Proc. Natl. Acad. Sci. U.S.A.* **97**: 1695–1700.

Leuzinger, H., Ziegler, U., Schraner, E. M., et al. (2005). Herpes simplex virus 1 envelopment follows two diverse pathways. *J. Virol.* **79**: 13047–13059.

McGeoch, D. J., Rixon, F. J., and Davison, A. J. (2006). Topics in *Herpesvirus* genomics and evolution. *Virus Res.* **117**: 90–104.

Mocarski, E. S., Jr., Shenk, T., and Pass, R. F. (2006). Cytomegaloviruses. Chapter 69 in: *Fields Virology, Fifth Edition* (D. M. Knipe and P. M. Howley, Eds. in chief), Philadelphia, Lippincott Williams & Wilkins, pp. 2701–2772.

Pellett, P. E., and Roizman, B. (2006). The family *Herpesviridae*, a brief introduction. Chapter 66 in: *Fields Virology, Fifth Edition* (D. M. Knipe and P. M. Howley, Eds. in chief), Philadelphia, Lippincott Williams & Wilkins, pp. 2479–2500.

Phelan, A., and Clements, J. B. (1998). Posttranscriptional regulation in herpes simplex virus. *Semin. Virol.* **8**: 309–318.

Rickinson, A. B., and Kieff, E. D. (2006). Epstein-Barr virus. Chapter 68B in: *Fields Virology, Fifth Edition* (D. M. Knipe and P. M. Howley, Eds. in chief), Philadelphia, Lippincott Williams & Wilkins, pp. 2655–2700.

Roizman, B., Knipe, D. M., and Whitley, R. J. (2006). Herpes Simplex viruses. Chapter 67 in: *Fields Virology, Fifth Edition* (D. M. Knipe and P. M. Howley, Eds. in chief), Philadelphia, Lippincott Williams & Wilkins, pp. 2501–2602.

Schlieker, C., Korbel, G. A., Kattenhorn, L. M., and Ploegh, H. L. (2005). A deubiquitinating activity is conserved in the large tegument protein of the *Herpesviridae*. *J. Virol.* **79**: 15582–15585.

Schulz, T. (1998). Kaposi's sarcoma-associated herpesvirus (human herpesvirus-8) (Review Article). *J. Gen. Virol.* **79**: 1573–1591.

Spear, P. G., and Longnecker, R. (2003). Herpesvirus entry: an update. *J. Virol.* **77**: 10179–10185.

Taddeo, B., Zhang, W., and Roizman, B. (2006). The U$_L$41 protein of herpes simplex virus 1 degrades RNA by endonucleolytic cleavage in absence of other cellular or viral proteins. *Proc. Natl. Acad. Sci. U.S.A.* **103**: 2827–2832.

Whitley, R. J., and Hilliard, J. (2006). Cercopithecine herpes virus 1 (B Virus). Chapter 73 in: *Fields Virology, Fifth Edition* (D. M. Knipe and P. M. Howley, Eds. in chief), Philadelphia, Lippincott Williams & Wilkins, pp. 2889–2904.

Yamanishi, K., Mori, Y., and Pellett, P. E. (2006). Human herpesviruses 6 and 7. Chapter 71 in: *Fields Virology, Fifth Edition* (D. M. Knipe and P. M. Howley, Eds. in chief), Philadelphia, Lippincott Williams & Wilkins, pp. 2819–2846.

Adenoviridae

Berk, A. J. (2006). *Adenoviridae*: The viruses and their replication. Chapter 63 in: *Fields Virology, Fifth Edition* (D. M. Knipe and P. M. Howley, Eds. in chief), Philadelphia, Lippincott Williams & Wilkins, pp. 2355–2394.

Ebner, K., Pinsker, W., and Lion, T. (2005). Complete sequence analysis of the hexon gene in the entire spectrum of human adenovirus serotypes: Phylogenetic, taxonomic, and clinical implications. *J. Virol.* **79**: 12635–12642.

Erdman, D. D., Xu, W., Gerber, S. I., et al. (2002). Molecular epidemiology of adenovirus type 7 in the United States, 1966–2000. *Emerg. Infect. Dis.* **8**: 269–276.

Gray, G. C., Callahan, J. D., Hawksworth, A. W., et al. (1999). Respiratory diseases among U. S. military personnel: countering emerging threats. *Emerg. Infect. Dis.* **5**: 379–387.

Gray, G. C., Goswami, P. R., Malasig, M. D., *et al.* (2000). Adult adenovirus infections: loss of orphaned vaccines precipitates military respiratory disease epidemics. *Clin. Infect. Dis.* **31**: 663–670.

Hay, R. T., Freeman, A., Leith, I., *et al.* (1995). Molecular interactions during adenovirus DNA replication. *Curr. Top. Microbiol. Immunol.* **199**: 31–48.

Leppard, K. N. (1998). Regulated RNA processing and RNA transport during adenovirus infection. *Semin. Virol.* **8**: 301–308.

Polyomaviridae

Agostini, H. T., Ryschkewitsch, C. F., and Stoner, G. L. (1998). JC virus type 1 has multiple subtypes: three new complete genomes. *J. Gen. Virol.* **79**: 801–805.

Benjamin, T. L. (2005). Polyoma viruses. In: *The Mouse in Biomedical Research, 2nd Edition,* Vol. 2, pp. 1–33.

Garcea, R. L., and Imperiale, M. J. (2003). Simian virus 40 infection of humans. *J. Virol.* **77**: 5039–5045.

Imperiale, M. J., and Major, E. O. (2006). Polyomaviruses. Chapter 61 in: *Fields Virology, Fifth Edition* (D. M. Knipe and P. M. Howley, Eds. in chief), Philadelphia, Lippincott Williams & Wilkins, pp. 2263–2298.

Moens, U., and Van Ghelue, M. (2005). Polymorphism in the genome of non-passaged human polyomavirus BK: implications for cell tropism and the pathological role of the virus. *Virology* **331**: 209–231.

Shackelton, L. A., Rambaut, A., Pybus, O. G., and Holmes, E. C. (2006). JC virus evolution and its association with human populations. *J. Virol.* **80**: 9928–9933.

Sullivan, C. S., and Pipas, J. M. (2002). T antigens of simian virus 40: Molecular chaperones for viral replication and tumorigenesis. *Microbiol. Mol. Biol. Rev.* **66**: 179–202.

Papillomaviridae

Alani, R. M., and Münger, K. (1998). Human papillomaviruses and associated malignancies. *J. Clin. Oncol.* **16**: 330–337.

Howley, P. M., and Lowy, D. R. (2006). Papillomaviruses. Chapter 62 in: *Fields Virology, Fifth Edition* (D. M. Knipe and P. M. Howley, Eds. in chief), Philadelphia, Lippincott Williams & Wilkins, pp. 2299–2354.

Oliveira, J. G., Colf, L. A., and McBride, A. A. (2006). Variations in the association of papillomavirus E2 proteins with mitotic chromosomes. *Proc. Natl. Acad. Sci. U.S.A.* **103:** 1047–1050.

Pyeon, D., Lambert, P. F., and Ahlquist, P. (2005). Production of infectious human papillomavirus independently of viral replication and epithelial cell differentiation. *Proc. Natl. Acad. Sci. U.S.A.* **102**: 9311–9316.

Richards, R. M., Lowy, D. R., Schiller, J. T., and Day, P. M. (2006). Cleavage of the papillomavirus minor capsid protein, L2, at a furin consensus site is necessary for infection. *Proc. Natl. Acad. Sci. U.S.A.* **103**: 1522–1527.

deVilliers, E.-M., Fauquet, C., Broker, T. R., Bernard, H.-U., and zur Hausen, H. (2004). Classification of papillomaviruses. *Virology* **324**: 17–27.

Parvoviridae

Allender, T., Tammi, M. T., Eriksson, M., *et al.* (2005). Cloning of a human parvovirus by molecular screening of respiratory tract samples. *Proc. Natl. Acad. Sci. U.S.A.* **102**: 12891–12896.

Berns, K., and Parrish, C. R. (2006). *Parvoviridae*. Chapter 65 in: *Fields Virology, Fifth Edition* (D. M. Knipe and Howley, Eds. in chief), Philadelphia, Lippincott Williams & Wilkins, pp. 2437–2478.

Cotmore, S. F., and Tattersall, P. (2005). Encapsidation of minute virus of mice DNA: Aspects of the translocation mechanism revealed by the structure of partially packaged genomes. *Virology* **336**: 100–112.

Norja, P., Hokynar, K., Aaltonen, L.-M., *et al.* (2006). Bioportfolio: Lifelong persistence of variant and prototypic erythrovirus DNA genomes in human tissue. *Proc. Natl. Acad. Sci. U.S.A.* **103**: 7450–7453.

Shackelton, L. A., and Holmes, E. C. (2006). Phylogenetic evidence for the rapid evolution of human B19 erythrovirus. *J. Virol.* **80**: 3666–3669.

Shackelton, L. A., Parrish, C. R., Truyen, U., and Holmes, E. C. (2005). High rate of viral evolution associated with the emergence of carnivore parvovirus. *Proc. Natl. Acad. Sci. U.S.A.* **102**: 3779–3784.

Thom, K., Morrison, C., Lewis, J. C. M., and Simmonds, P. (2003). Distribution of TT virus (TTV), TTV-like minivirus, and related viruses in humans and nonhuman primates. *Virology* **306**: 324–333.

Vihinen-Ranta, M., Suikkanen, S., and Parrish, C. R. (2004). Pathways of cell infection by parvoviruses and adeno-associated viruses. *J. Virol.* **78**: 6709–6714.

You, H., Liu, Y., Prasad, C. K., *et al.* (2005). Multiple human papillomavirus genes affect the adeno-associated virus life cycle. *Virology* **344**: 532–540.

Oncogenesis

Friedlander, A., and Patarca, R. (1999). DNA viruses and oncogenesis. *Crit. Rev. Oncogen.* **10**: 161–238.

Levine, A. J. (1997). p53, the cellular gatekeeper for growth and division. *Cell* **88**: 323–331.

Münger, K., Baldwin, A., Edwards, K. M., *et al.* (2004). Mechanisms of human papillomavirus-induced oncogenesis. *J. Virol.* **78**: 11451–11460.

Nevins, J. R. (2006). Cellular transformation by viruses. Chapter 8 in: *Fields Virology, Fifth Edition* (D. M. Knipe and P. M. Howley, Eds. in chief), Philadelphia, Lippincott Williams & Wilkins, pp. 209–248.

Schiffman, M., Herrero, R., DeSalle, R., *et al.* (2005). The carcinogenicity of human papillomavirus types reflects viral evolution. *Virology* **337**: 76–84.

C H A P T E R

8

Emerging and Reemerging Viral Diseases

The litany of the viruses described in the previous chapters of this volume makes clear that humans have been subjected to a large number of viral diseases throughout our history. Some of these viruses evolved along with humans and have been present since the earliest human walked the earth. Such viruses include the various herpesviruses, for example, which were present as human pathogens at the time that humans first appeared. Others have been acquired from zoonotic sources. These are animal viruses that have acquired the ability to infect humans. Upon jumping from their animal sources to humans, some of these viruses became human viruses that infect only humans, and humans became the vertebrate reservoir of this new virus. Such viruses include measles, described in Chapter 4, and the dengue viruses, described in Chapter 3 and in this chapter. Many of these viruses entered the human population long ago. Arguments were presented in Chapter 4 that measles virus could not have existed as a human virus until perhaps 5000 years ago when the human population first reached the numbers required to sustain the virus in the population, and that this virus probably jumped from cattle to humans after humans domesticated these animals. Others have entered the human population more recently. The four serotypes of dengue virus, for example, appeared to have jumped independently from monkeys to humans between 200 and 1000 years ago, and HIV was established as a human virus within the last 50–100 years. Other zoonotic viruses that infect humans do so only peripherally and humans do not serve as the vertebrate reservoir of these viruses. Examples are West Nile virus, Eastern equine encephalitis virus, Ebola virus, and rabies virus.

As the human population expands it impinges on wildlife more and more, and changes in habitat caused by humans lead to closer interactions between humans and wildlife, with the result that an increasing number of zoonotic viruses are causing epidemics of serious human disease. Some, like

HIV, became human viruses while others, like influenza virus, remain zoonotic viruses. There is growing concern that viruses to date have caused small epidemics in humans but may acquire the ability to cause very large epidemics. In this chapter we will consider a number of viruses that are known to have caused epidemics in humans only within the last century, or that have the potential to cause wide-ranging epidemics in the future, or that are undergoing dramatic range expansion at present.

BAT-ASSOCIATED VIRUSES

A number of emerging viruses are bat viruses that have recently entered the human population and caused small or large epidemics of disease. Although these viruses can cause serious illnesses in humans, they usually cause little or no illness in bats. The recent emergence of bat viruses as human pathogens may seem strange because so many viruses are now known to come from bats, but in fact bats form a sizable proportion of the diversity of mammals. More than 900 species are currently recognized and these constitute more than 20% of all mammalian species. Furthermore, bats are intensely social creatures that are ideally suited to pass viruses back and forth among large populations. Humans impinge more and more into the habitats of bats, and this, as well as disruptions of bat colonies caused by humans, has led to more contact between bats and humans or their domestic animals. Furthermore, in many areas of the world bats are used as food or for medicinal purposes, resulting in human–bat contacts.

Almost all bats are nocturnal. They are classified in the order *Chiroptera*, which has two major divisions or suborders. *Megachiroptera* are mostly large, fruit-eating bats that are classified in a single family, *Pteropodidae*. There are about 170 species distributed throughout the tropics of the Old World. They find their food, consisting of fruits, flowers, and pollen, using eyesight and an excellent sense of smell. Of these, more than a third, 65, belong to the genus *Pteropus* and are called flying foxes. The *Pteropus* flying foxes are found from Australia across southern Asia and India to Madagascar (Fig. 8.1). They weigh from 300 grams to more than a kilogram and have a wingspan of 0.6 to 1.7 meters. *Microchiroptera* are in general smaller and most eat insects. They are virtually worldwide in distribution. They have evolved echolocation to navigate and find their prey in the dark. Bats play an important role in the ecology of the planet, dispersing seeds, pollinating plants, and reducing the number of night flying insects such as mosquitoes.

Rabies Virus

Rabies virus is an example of a virus for which bats are an important reservoir. Although we think of rabies as being primarily associated with canines such as dogs and other mammals such as skunks and raccoons, and these animals do serve as important reservoirs for rabies that enter the human population, bats are also an important reservoir. In fact, in the United States over the last few years the majority of human cases of rabies have been bat-associated rabies. In some of these cases of bat rabies in humans, exposure to bats is documented and the bite of infected bats is known to have transmitted the virus, but in other cases there is no known contact with bats, leading to the suggestion that inhaling aspirated droplets containing rabies may be the cause of the infection.

In South America, vampire bats, which feed on the blood of mammals after biting them with their sharp incisors, have been an important vector in the spread of rabies to livestock and humans. This has resulted in campaigns to indiscriminately slaughter bats using a variety of methods, including poison and the destruction of roosts and caves with explosives. Although these campaigns have resulted in enormous numbers of bats being killed, these campaigns have had no effect on the spread of rabies. Thus, reduction of bat numbers is not effective in the control of rabies but does destroy ecologically important animals.

The various bat lyssaviruses, which can also cause rabies in humans, were described in Chapter 4. Although only a few human cases are known that arose from infection by these viruses, they have the potential to spread more widely. Further, bites or scratches from bats need to be treated as potential

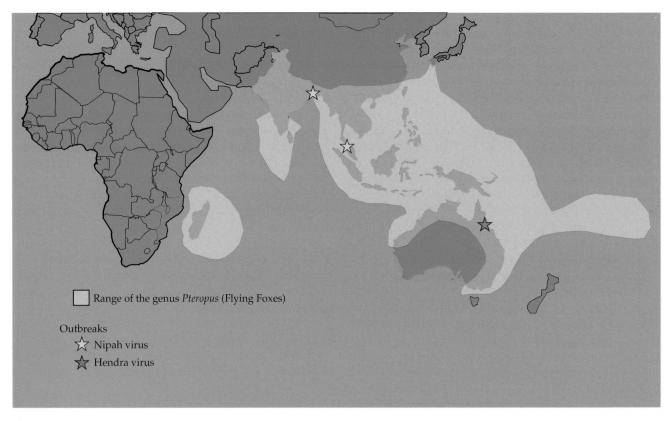

FIGURE 8.1   Illustration of the distribution of the genus *Pteropus* in the Old World. These large bats, called flying foxes, are present from Madagascar across the Indian subcontinent and throughout the tropical and subtropical regions of Indonesia, Australia, and the Philippines, and as far east as the Cook Islands. Adapted from Figure 1B in Eaton *et al.* (2006).

exposure to rabies and treated accordingly. This is expensive and in effect only available in developed countries.

Henipaviruses

In September of 1994, a number of cases of severe respiratory illness occurred in racehorses near Brisbane, Australia. The first horse to become ill was a pregnant mare that was pastured in a field, which was then moved into a stable with 23 other thoroughbreds. The disease spread among the horses in this stable and to an adjoining stable and ultimately 17 horses became ill, of which 13 died. Of the four horses that survived, two were left with mild neurological sequelae. Three other horses were infected but did not suffer symptoms. Two humans that nursed the horses became ill with a severe respiratory disease, of whom one died of respiratory and kidney failure. Spread of the disease required close contact, and imposition of quarantine measures contained the outbreak. A previously unknown virus was isolated from the sick animals that was found to be a paramyxovirus and it was initially named equine morbillivirus. Sequencing of the genome showed that it was not closely related to the morbilliviruses; however, and the virus was assigned to a new genus called *Megamyxovirus* because of the large size of the genome. Subsequent studies established that the virus was a bat virus and widespread in flying foxes in eastern Australia. The virus was renamed Hendra virus after the Brisbane suburb where the outbreak occurred and the genus was renamed *Henipavirus*. Analysis of sera from healthy humans and horses in the area failed to detect the presence of antibody, and analysis of more than 5000 sera from a variety of wild animals trapped in the areas also failed to detect antibody in any animal other than flying foxes. Flying foxes can be readily infected experimentally with the virus but do not suffer illness upon infection.

A second outbreak of Hendra virus began in August 1994 about 1000 km north of Brisbane. Two horses died and the owner of the horses became mildly ill with neurological symptoms from which he appeared to recover. However, in October of 1995 the owner suffered a relapse and died of encephalitis. At this point an investigation showed that Hendra virus was to blame for the illness of the horses and the death of the owner. In January 1999, a fatal case of Hendra infection occurred in a horse near Cairns, Australia, and in 2004 a horse died of Hendra infection in Townsville. In the 2004 incident the veterinarian who attended the horse was infected but recovered after a mild illness.

Extensive studies of flying foxes have shown that Hendra virus is present in all four species of flying fox that occur in Australia. Almost half of flying foxes have been found to have antibodies to the virus, so it is widespread and common. It is only rarely transmitted to other animals, however, at least to date, as shown by the extensive serological studies and the limited occurrence of clinical illness caused by the virus in humans and their horses.

A related virus, 83% identical to Hendra virus at the amino acid level, emerged in 1998 that represents a much more serious threat to human health. From September 1998 to April 1999, an outbreak of 258 cases of human encephalitis occurred in Malaysia and Singapore that had a 40% mortality rate. Clinical symptoms included fever, headache, myalgia (muscle aches), drowsiness, and disorientation that sometimes progressed to coma within 48 hours. The disease was associated with an outbreak of respiratory disease in pigs with or without neurological symptoms, and humans infected with the disease were pig farmers or others closely associated with pig farming. It was first thought that the outbreak was due to infection by Japanese encephalitis (JE) virus (see Chapter 3 on the importance of pigs as amplifying hosts for this virus) and the Malaysian government vaccinated 2.4 million pigs against JE virus. When this did not slow the epidemic, 1.1 million pigs were culled in an attempt to reduce the incidence of disease. In March 1999, with the assistance of the Centers for Disease Control and Prevention, the virus responsible for the epidemic was identified as a Hendra-like virus, a virus related to but distinct from Hendra virus. Retrospective studies suggested that the virus had been responsible for disease in pigs in Malaysia for several years and it seems clear that the human cases were contracted from pigs. There is no evidence for human-to-human transmission in this outbreak. The virus responsible has been called Nipah virus, after the village in Malaysia where the disease first appeared, and it is classified as a second member of the genus *Henipavirus* (which gets its name from Hendra and Nipah viruses).

Like Hendra virus, the reservoir of Nipah virus is flying foxes and the virus has been isolated from flying foxes in the area. It has been suggested that the outbreak occurred in part because the destruction of the natural habitat of the flying foxes caused by deforestation and consequent food shortage led the bats to forage in nearby orchards located very near piggeries. There, half-eaten fruit or regurgitated fruit that was contaminated with virus-containing saliva from the bats could be eaten by pigs, causing them to become infected.

More recent epidemics of Nipah virus encephalitis have occurred in southern Asia. Epidemics in Bangladesh occurred in 2001, 2003, 2004, and 2005. No evidence for the intermediate infection of an animal, as occurred in the Malaysian epidemic, has been seen in these epidemics. Furthermore, in the 2004 epidemic evidence was obtained that person-to-person transmission of the virus had occurred. It is likely that the disease was transmitted directly from bats to humans, possibly by human consumption of partially eaten fruit that was contaminated with bat saliva containing the virus, followed by person-to-person transmission. It is known in Bangladesh, for example, that during the fruiting season young boys climb trees to pick fruit. If this fruit was partially eaten by bats, the fruit could be contaminated with the virus from the bat. The fatality rate in these epidemics was as high as 75%. In nearby India, an epidemic of Nipah occurred in 2001. Flying foxes are widely distributed throughout this area (Fig. 8.1) and Nipah virus has been isolated from them in both Malaysia and Bangladash. The virus has also been isolated from flying foxes in Cambodia although human infection has not been documented to date.

The very wide distribution of Hendra and Nipah viruses, the possibility of person-to-person transmission, and the increasing contacts between humans and their domestic animals with fruit bats carrying the virus, suggests that epidemics will continue to occur. As indicated for rabies, eradication of the bats is neither desirable nor feasible. However, simple solutions exist to reduce the contacts of humans and their animals with the bats, such as not locating fruit orchards near piggeries.

SARS Coronavirus

SARS virus (*s*evere *a*cute *r*espiratory *s*yndrome virus) occurs in a number of cave-dwelling species of horseshoe bats in China belonging to the genus *Rhinolophus*. Field studies have found that 30–70% of bats belonging to this genus

have been infected by the virus. In the autumn of 2002 an epidemic of SARS in humans began in Guangdong Province in China. The disease is an atypical pneumonia characterized by high fever, myalgia, and lymphopenia (smaller numbers of lymphocytes). By February of 2003 there were 305 cases with five deaths. The infection was then spread to other areas by a Chinese doctor who had been treating patients in Guangdong. He traveled to Hong Kong on 21 February 2003 and while staying in a hotel he developed symptoms of SARS and died shortly thereafter. Ten guests at the hotel who were housed on the same floor or nearby floors became infected and before developing symptoms traveled to Singapore, Vietnam, Canada, and the United States, spreading the epidemic. The epidemic also spread independently to Beijing in April of 2003. The epidemic finally waned in the summer of 2003 when the World Health Organization reported the cumulative total of 8098 probable cases of SARS with 774 deaths worldwide in 29 countries (Fig. 8.2). The death rate from the disease was thus about 10% although it is age related. Children either do not contract the virus or show little reaction to it, whereas the death rate in people over 65 can be as high as 50%.

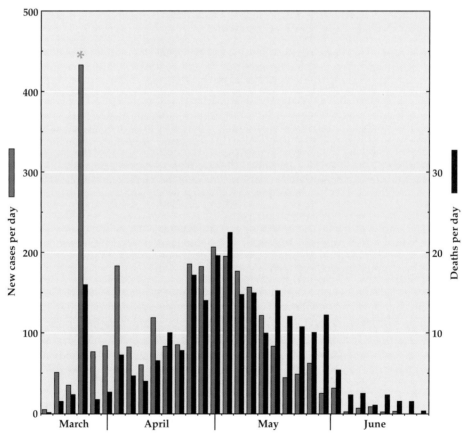

FIGURE 8.2   SARS epidemic of 2003. Cases and deaths reported between March 17 and June 23 are plotted. The data for this graph were reported in the outbreak updates from the World Health Organization, and can be found at: http://www.who.int/csr/don/archive/disease/severe_acute_respiratory_syndrome/en/. Cases and deaths have been normalized for the length of the reporting interval, which varied from 2 to 4 days. The asterisk marks the date on which cumulative totals were first released by China.

The disease probably started in markets in China in which a number of exotic animals including bats, masked palm civets (*Paguma larvata*), and raccoon dogs (*Nyctereuctes procyonoides*) are sold for food. Civets and raccoon dogs from the markets were found to be infected by the virus and it is believed that either these animals or bats being consumed as food spread the disease to humans, followed by human-to-human spread of the virus. It is almost certain that the civets in the markets contracted the virus there from bats because civets on farms were largely free from the SARS virus. In addition, 13% of tested merchants in the markets in Guangdong had SARS antibodies (showing they had been infected by the virus).

Adaptation of SARS Virus to a Human Receptor

Recent studies have shown that the SARS virus, a coronavirus, had to adapt to human receptors in order to cause severe illness. Infection by the bat virus or the civet virus appears to cause only mild illness. As stated before, merchants who were infected did not develop illness, and some persons who work with wildlife were found to be seropositive for SARS but suffered no illness. However, several changes are present in the virulent SARS virus isolated from humans. There is a deletion of 29 nucleotides upstream of the start codon for the N protein and there are four amino acid changes in the spike protein. It is believed that the crucial changes are two amino acid changes in the spike protein that allow the virus to bind to the human receptor called ACE2 (angiotensin-converting enzyme 2) 1000-fold more avidly than does the civet strain or the bat strain. This is perhaps the reason why the virus has not to date reappeared in the human population, together with extensive culling of animals in the food markets in China. If occasional human cases occur they are likely to be mild unless the virus has the opportunity to mutate in humans to form the virulent strain of the virus. However, this did happen in 2002 and may happen again in the future. There is need to develop vaccines or antiviral treatments for the virus, as well as to maintain the Chinese food markets in a way that does not encourage the spread of the virus.

The ACE2 protein is highly conserved among mammals and it is perhaps surprising that one of the few amino acid differences in the human form of this protein occurs in the virus-binding site and causes such a change in the ability of SARS to utilize ACE2 as a receptor. In view of the fact that to become virulent the virus must mutate to bind more strongly to the human form of the ACE2 protein, it is interesting that there is a second receptor for SARS virus, the protein called CD209L or L-SIGN. Why this second receptor cannot compensate for the failure of unmodified SARS to infect humans efficiently is unknown. It may be significant that another human coronavirus, NL63, only recently discovered, also uses ACE2 as its receptor. Other coronaviruses use different receptors, including aminopeptidase N (also called CD13) by the human coronavirus 229E, transmissible gastroenteritus virus of swine, and feline infectious peritonitis virus, and carcinoembryonic antigens by mouse hepatitis virus.

A Second Bat Coronavirus

It is noteworthy that at least one other coronavirus, as yet unnamed, circulates in bats belonging to the genus *Miniopterus*. In *Miniopterus pusillus* more than 60% of the bats were found to be positive for this virus. This virus is distinct from the SARS virus. It belongs to group 1 coronaviruses whereas SARS belongs to group 2 coronaviruses. This new virus is not known to infect humans or to cause disease.

The Zoonotic Origin of a Human Coronavirus

SARS is a zoonotic disease of humans caused by a coronavirus. It is of interest that human coronavirus HCoV OC43 also appears to have a zoonotic source. It is very similar to a virus of cattle, bovine coronavirus (BCoV). From studies of the rate that mutations have been fixed in these viruses, it has been estimated that the virus entered the human population around 1890.

Filoviruses

Marburg Virus

The filoviruses first came to the attention of science in 1967 when outbreaks of hemorrhagic fever occurred in Marburg and Frankfurt, Germany, and in Belgrade, Yugoslavia. The cause was a virus subsequently named Marburg that was present in African green monkeys imported from Uganda whose kidneys were being processed for cell culture production (for use in preparing poliovirus vaccine). Twenty-five laboratory workers were infected and six secondary cases resulted; of these 31 infected people, 7 died. The monkeys in the shipment, which originated in Uganda, also died. Subsequent studies with the virus isolated during the outbreak showed that it caused lethal illness in African green monkeys following experimental infection. There were 3 cases of Marburg in South Africa in 1975 (the source of infection was probably Zimbabwe) with one death, 2 cases in Kenya in 1980 (infection probably in Uganda), 1 case in Kenya in 1987, an outbreak of 149 cases with 123 deaths in Zaire (now the Democratic Republic of Congo) in 1998–2000, and an outbreak of 374 cases with 329 deaths in northern Angola in 2005. The number of cases is surely underreported since many people in remote areas do not seek medical assistance when ill, and counting of new graves in such locations indicates that the death toll is higher than officially reported. The locations of these outbreaks are shown on the map in Fig. 8.3. The reported fatality rate in the larger outbreaks was 80–90%.

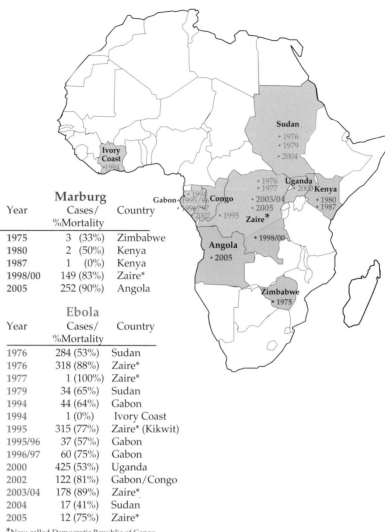

Marburg

Year	Cases/ %Mortality	Country
1975	3 (33%)	Zimbabwe
1980	2 (50%)	Kenya
1987	1 (0%)	Kenya
1998/00	149 (83%)	Zaire*
2005	252 (90%)	Angola

Ebola

Year	Cases/ %Mortality	Country
1976	284 (53%)	Sudan
1976	318 (88%)	Zaire*
1977	1 (100%)	Zaire*
1979	34 (65%)	Sudan
1994	44 (64%)	Gabon
1994	1 (0%)	Ivory Coast
1995	315 (77%)	Zaire* (Kikwit)
1995/96	37 (57%)	Gabon
1996/97	60 (75%)	Gabon
2000	425 (53%)	Uganda
2002	122 (81%)	Gabon/Congo
2003/04	178 (89%)	Zaire*
2004	17 (41%)	Sudan
2005	12 (75%)	Zaire*

*Now called Democratic Republic of Congo

FIGURE 8.3   Map of Africa showing the different filovirus outbreaks. Data from Porterfield (1995) p. 320, and later data from Georges-Courbot *et al.* (1997); Peters and Khan (1999), and news bulletins from the World Health Organization (2005) at: http://www.who.int/disease-outbreak-news/. Note that in recent years, outbreaks of Ebola disease have occurred almost annually in the center of the range, particularly in Gabon and the Democratic Republic of Congo. On the contrary, the recent epidemic of Marburg in Angola was the first in 5 years.

African Ebola Virus

Ebola virus was first isolated during a 1976 epidemic of severe hemorrhagic fever in Zaire and Sudan and named for a river in the region. During this epidemic, the more than 600 cases resulted in 430 deaths and asymptomatic infection appeared to be rare. One case of Ebola occurred in 1977, and in 1979 there were 34 cases with 22 deaths in the Sudan. In this latter epidemic, an index case was brought to the hospital and the virus spread to four people there, who then spread it to their families. After this, Ebola disease in Africa disappeared until 1994. In late 1994, a Swiss etholo-gist working in the Ivory Coast performed necropsies on chimps. She contracted Ebola but survived, and a new strain of Ebola was isolated from her blood. Then, in May 1995,

there was an epidemic in Kikwit, Zaire, that resulted in at least 315 cases with >75% mortality. This was followed by several deaths in western Africa that resulted from consump-tion of a monkey that had died of Ebola. Then there was a prolonged series of smaller outbreaks in Gabon from 1995 through 1997. In 2000, Ebola appeared in Uganda for the first time and caused an epidemic of more than 425 cases. There have been further outbreaks in 2002, 2003, 2004, and 2005 in various countries including Gabon, the Democratic Republic of Congo, and Sudan. A map showing these vari-ous filoviral outbreaks is shown in Fig. 8.3. Three strains or species of African Ebola viruses are now recognized which differ in their virulence. Zaire ebolavirus is the most virulent with a case fatality rate approaching 90%, Sudan ebolavirus is less virulent, and Ivory Coast ebolavirus is the least virulent.

The natural reservoir of Ebola virus in Africa has recently been shown to be bats. Three species of fruit bats collected in Gabon and the Democratic Republic of Congo, close to areas where an epidemic of Ebola had devastated local gorilla and chimpanzee populations, showed evidence of infection. Significant numbers of *Hypsignathus monstrosus* (4 of 17 tested), *Epomops franqueti* (8 of 117), and *Myonycteris torquata* (4 of 58) were found to have antibodies to Ebola virus, and viral nucleic acid was detected in liver or spleen of other bats of these three species (4 of 21, 5 of 117, and 4 of 41, respectively). No viral RNA was found in any other animal species tested, which included 222 birds and 129 small vertebrates, among others. The infection of the bats appears to be asymptomatic, consistent with the hypothesis that the fruit bat is the reservoir of Ebola virus. It is probable that bats are also the reservoir of Marburg virus.

It is clear that monkeys can be infected by the virus and spread it to humans, but how the monkeys contract it is not known. Perhaps the monkeys eat fruit that has been partially eaten by the bats. There have been serious die-offs of gorillas and chimpanzees in the last decade or so that have severely impacted the populations of these animals in some areas. Some human epidemics get started when monkeys dying of the disease are butchered for food, but in other epidemics monkeys are not implicated and how the epidemic starts is not known. Again, perhaps consumption of partially eaten fruit is to blame. The recent large die-offs of apes and chimpanzees in some areas together with the increasing frequency of human infection indicates that the virus is spreading more widely. Whether this is due to a new strain of virus that is more easily spread to humans and nonhuman primates, or to the recent introduction of the virus from a source outside the areas now experiencing epidemics, or to human alterations of the environment leading to more contact between humans and nonhuman primates with the bats that carry the virus is not known.

Reston Ebola Virus

A fourth strain of Ebola virus originating from the Philippines (or, perhaps, another region of Asia) first appeared as the causative agent of an epidemic of hemorrhagic fever in monkeys imported from the Philippines. This epidemic occurred in Reston, Virginia, near Washington, D.C., in 1989. The deaths were at first attributed to simian hemorrhagic fever virus (SHFV), but investigation by the U.S. Army Medical Research Institute for Infectious Diseases and the Centers for Disease Control and Prevention found that both SHFV and Ebola virus were present in the monkeys. Believing that the community was at risk for Ebola, made even the more alarming because the epidemic was occurring in the neighborhood of the central government of

the United States, the army team quickly decided to euthanize the monkeys and decontaminate the facility. Follow-up studies showed that four animal handlers at the facility had been infected by the virus but had suffered no illness. Thus the strain of Ebola present in the Reston monkeys, called Reston ebolavirus, seems to be nonpathogenic for humans although it remains pathogenic for monkeys. The story of the Reston incident was recounted in a book called *The Hot Zone* by Richard Preston. Since this first epidemic of Reston ebolavirus, a new outbreak has occurred in Reston, two outbreaks have occurred in an animal facility in Alice, Texas, and an outbreak took place in Sienna, Italy. All outbreaks occurred in monkeys imported from the Philippines and it seems probable that the virus is native to the Philippines. No serious illness in humans has occurred in any of these outbreaks. Nucleotide sequencing has shown that Reston ebolavirus is closely related to the African ebolaviruses. The reason it is attenuated in humans is still unexplained, and high containment is used for studies of Reston ebolavirus in the laboratory.

Are Filoviruses a Major Threat?

To date, the filoviruses have caused only a limited number of human infections. However, if a filovirus were to adapt to humans such that human-to-human transmission occurred readily, it could become a major problem. The probability of such an event is unknown.

VIRUSES ASSOCIATED WITH BIRDS

Many viruses are known for which birds are the vertebrate reservoir but which can also infect humans, causing outbreaks of serious illness. Because many species of birds travel large distances during their annual migration, viruses associated with such species are spread rapidly over a wide geographic area. Here we consider the rapid spread of West Nile virus in the Americas and the alarming appearance of new strains of influenza virus that have the potential to cause very large epidemics of very serious influenza in human populations.

West Nile Virus in the Americas

Appearance and Spread of West Nile in the New World

In the summer of 1999, West Nile virus, a mosquito-borne flavivirus (Chapter 3), appeared in North America for the first time. There were 62 human cases of West Nile disease in the New York City area, of whom 7 died of encephalitis. Numerous birds also died, including exotics in zoos as well as native birds. With the end of the mosquito season the epidemic died out, but the virus had become established. Over the next 6 years the virus rapidly spread across the United States and north into Canada as well as south into Central America, northern South America, and the Caribbean. The march of the virus across the United States is illustrated in Figs. 8.4 and 8.5. The rapid spread of this virus into a new ecological area requiring adaptation to new mosquito vectors as well as new vertebrate hosts is extraordinary. In the process, more than 20,000 Americans became ill from WN virus infection and more than 800 died of neurological complications (Fig. 8.6). Figures 8.5 and 8.6 illustrate an interesting and not well understood phenomenon. After the front passes through an area, there are many fewer human cases of disease in the following years. Yet only a minority of humans in any area had been infected and are therefore immune to the virus, and the virus continues to be present. In some way the transmission cycle has been interrupted, perhaps by the die-off of the birds that serve as the primary amplifying hosts.

WN virus strains from the United States are 99.7% identical to strains from Israel, and the U.S. strain certainly originated in the Middle East. It presumably arrived in New York on a jet aircraft, and there are three possible vectors that could have carried the virus. It is conceivable that a viremic human introduced the virus, although this seems unlikely because humans are poorly able to pass on the virus as described in Chapter 3. A second possibility is that the virus arrived in a viremic bird being imported legally or illegally, although there is no evidence for such an event. It seems highly unlikely that a migratory bird that was off course introduced the virus because West Nile first appeared near a major international airport. The favored hypothesis is that the virus arrived in an infected mosquito that came along for the ride in a jet aircraft. Introduction of the virus was almost certainly a singular event.

Effects of West Nile on Wildlife

West Nile virus has had profound and well documented effects on horses and wildlife as it moved across North America. Many horses died of WN disease, and a vaccine has now been introduced in order to protect horses. The effect upon bird populations has been particularly dramatic. Crows, jays, and raptors (hawks and owls) are particularly sensitive to the virus. Almost all crows die after infection, for example, and in many parts of the country the crow population crashed with the arrival of the virus, although there are recent signs that the population is at least partially recovering. Many raptors prey upon small rodents and are

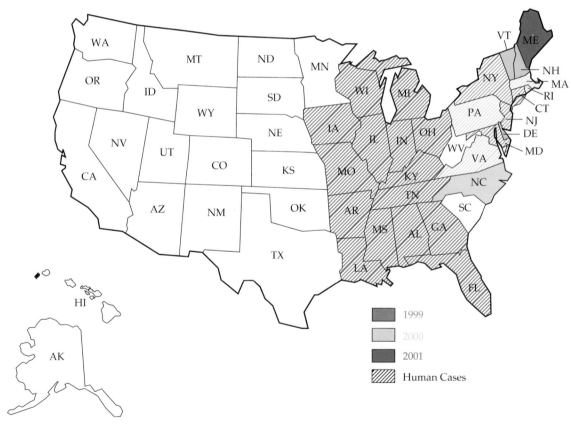

FIGURE 8.4   The early spread of West Nile virus across the United States from 1999 to 2001. Data are from the archives of the West Nile Virus Surveillance site from the Centers for Disease Control and Prevention at: http://www.cdc.gov/ncidod/dvbid/westnile/surv&control.htm.

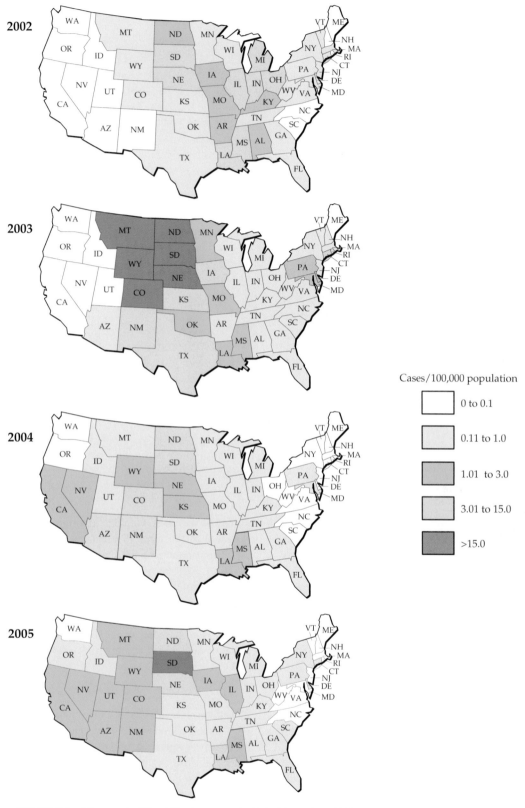

FIGURE 8.5   The spread of West Nile virus across the United States from 2002 to 2005. Data are from the archives of the West Nile Virus Surveillance site from the Centers for Disease Control and Prevention at: http://www.cdc.gov/ncidod/dvbid/westnile/surv&control.htm.

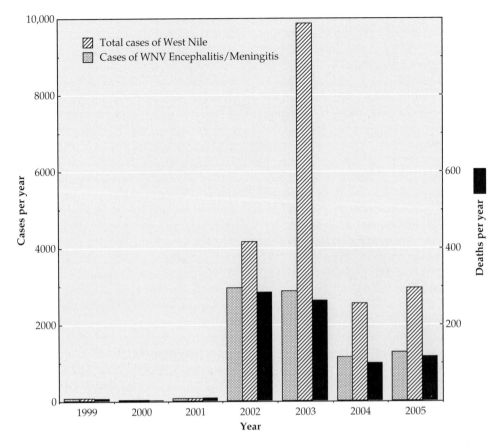

FIGURE 8.6   Total number of cases of West Nile fever, cases of encephalitis and meningitis due to WN virus, and deaths from WN virus are shown for 1999 through 2005. Data are from the archives of the West Nile Virus Surveillance site from the Centers for Disease Control and Prevention at: http://www.cdc.gov/ncidod/dvbid/westnile/surv&control.htm.

important in controlling their numbers, and a crash in their numbers results in an increase in rodent populations. As an example of the sensitivity of some owls to WN virus, there was an outbreak of WN disease in North American owls in a rehabilitation facility in Ontario in 2002, associated with the die-off of corvids in the area. Large owls with a northern breeding range were particularly susceptible. Snowy owls, northern hawk owls, great gray owls, and boreal owls experienced 100% mortality, and northern saw-whet owls experienced 92% mortality.

The Future of West Nile in the Americas

WN virus is now well established in the Americas and will surely continue to cause sporadic cases of human illness, if not sporadic epidemics, and sporadic die-offs of birds. It is possible that the interruption of its transmission cycle described above is due to extensive die-off of susceptible species of birds that serve as its major amplifying vectors. If so, the resurgence of the crow population and the populations of other amplifying species could lead to renewed outbreaks similar to those that occurred with the first appearance and

spread of the virus. One encouraging finding relating to the potential of the virus to cause future epidemics of disease is the appearance of an attenuated variant of WN virus in Texas in 2003. In a mouse model, this variant is attenuated in neurovirulence. The emergence of WN virus was associated with the appearance in Europe and the Middle East of a more virulent strain of virus, perhaps enabling it to spread more easily. It seems possible that once the virus becomes endemic, attenuated strains may become dominant.

Avian Influenza

Influenza A virus caused three pandemics of severe influenza in the twentieth century. The pandemic of 1918 was especially severe, infecting perhaps 30% of the world's population and causing up to 100 million deaths. These pandemics were associated with the appearance of new subtypes of HA and NA, the surface glycoproteins of the virus, in human adapted viruses. The reservoir of influenza A is birds and the bird viruses must ordinarily undergo adaptation in some way to humans in order to infect and cause epidemic spread. In part this adaptation involves reassortment of flu segments to incorporate human adapted segments. It had been thought that virus circulating in birds will not infect humans and cause disease without prior adaptation, and that only HA subtypes H1, H2, and H3 were compatible with spread in humans. Recently, however, there have been numerous cases of direct human infection by several strains of avian influenza resulting in serious, even fatal, illness involving new subtypes of HA. Three such viruses, H5N1, H7N7, and H9N2 will be considered here.

H5N1 Influenza

H5N1 virus was first detected in China in 1996 where it caused the death of a number of geese. After undergoing reassortment to obtain new genes, it surfaced in Hong Kong in 1997 where it became widespread in live poultry markets. Eighteen people were infected by the virus and six died. The virus was eradicated by culling all domestic chickens in Hong Kong. Different reassortants of H5N1 continued to arise and in 2002 an epidemic of influenza killed most birds, domestic and wild, in Hong Kong nature parks. Two people were infected and one died. The virus then spread widely over eastern Asia. The virus killed wild as well as domestic waterfowl and chickens, and together with repeated efforts to stop the spread of the virus by culling of birds, the death of more than 140 million birds occurred. The virus has a high mortality rate in humans, about 50%, and as of early 2007 more than 150 people have died as a result of H5N1 infection. There is no evidence for human-to-human transmission at the current time. All human infections appear to have originated from close contact with infected birds.

The virus has continued to spread westward. It has now reached many countries in Europe and established a beachhead in Africa. It is thought that the virus will soon reach North America, brought by migratory birds. The mechanism of its rapid spread is not completely clear. It is thought that transport of domestic birds may be responsible in part for the spread, but there is reason to believe that wild waterfowl, ducks, geese, and swans, are also spreading the virus. These birds undergo seasonal migrations over large distances, and episodes of dead and dying migratory birds have been associated with the appearance of the virus in new areas. The virus is continuing to evolve. Unlike most avian influenza viruses, it is highly pathogenic in chickens and many strains are also pathogenic in ducks, both domestic and wild. Virulence in wild ducks seems to be moderating, as some recent isolates do not kill wild ducks, but virulence in chickens continues. All isolates contain the multibasic cleavage site in HA that is recognized by furin, which is correlated with increased virulence.

Studies of H5N1 virus in mice and ferrets have shown that many virus isolates are highly virulent for these mammals. In one study with ferrets, isolates from humans caused a fatal infection characterized by broad tissue tropism, including infection of the brain. In this same study, isolates from birds caused nonlethal infection with virus replication restricted to the upper respiratory tract.

There have been three or four pandemics of influenza every century for as long as can be ascertained from historical studies and there is every reason to believe that pandemics will come again in the twenty-first century. If H5N1 virus should acquire the ability to spread readily from person to person, and if its high lethality for humans should continue unabated, it could cause a devastating pandemic. Efforts are being made to prepare defenses against this virus, including vaccines and antiviral drugs, supported by the U.S. government and other governments. Promising vaccines are being developed and tested in clinical trials, and plans are to stockpile such a vaccine for possible use if an epidemic of this virus arises. This effort is complicated by the fact that H5N1 influenza, like all influenzas, continues to evolve, and stockpiled vaccine might not be completely effective against the strain that ultimately might emerge as a pandemic strain. Nevertheless, there is hope that at least partial immunity might be effected by such a vaccine that could ameliorate the symptoms of the disease and protect against the extreme virulence of the virus. As described in Chapter 4, new approaches to flu vaccine development are being tried, including development of a universal flu vaccine that would work against any influenza A strain, but such an approach will clearly require many more years of research before a licensed vaccine for general use could be produced. Efforts are also being directed toward producing and stockpiling antiviral compounds directed against influenza. Unfortunately, most H5N1 isolates tested are resistant to amantadine and related compounds, antiflu agents that have been in use for years and that are readily available. It is thought that this is the result of the wide use of amantadine by Chinese farmers to protect their chicken flocks from influenza, thereby selecting for amantadine-resistant variants of the viruses. This effect may be moderating, however, as some recent isolates are sensitive to amantadine. Inhibitors of the influenza neuraminidase such as oseltamivir (Tamiflu) and zanamivir appear to be reasonably effective against H5N1 virus, but these drugs must be used early in infection if they are to be effective. Production of these compounds is limiting at present, but production is being accelerated in order to stockpile them for possible use if or when a pandemic erupts.

H7N7 Influenza

An epidemic of H7N7 influenza A erupted in The Netherlands in 2003. A total of 89 human infections was recorded, of whom 86 were directly involved in handling poultry. Human-to-human transmission occurred and three family members were also affected. The primary disease syndrome was conjunctivitis but there were two cases characterized by mild influenza-like symptoms only and one fatal case characterized by pneumonia followed by respiratory distress syndrome. The virus was eradicated by culling of poultry in the country.

H9N2 Influenza

In 1999, two cases of human infection by H9N2 influenza A occurred in Hong Kong and five cases in Guangdong Province, China. Then in 2003 another case occurred in Hong Kong. The disease was characterized by mild influenza-like symptoms and recovery was uneventful.

VIRUSES ASSOCIATED WITH PRIMATES

Dengue Virus

The Origin of Dengue Viruses

Dengue viruses are mosquito-borne flaviviruses that cause widespread epidemics in humans (Chapter 3). Forest cycles of dengue have been documented in Africa and Southeast Asia in which the vertebrate reservoir is monkeys and various species of *Aedes* mosquitoes maintain the virus. Recent sequencing studies have now shown that the sylvatic monkey viruses evolved in monkeys into four serotypes. This evolution from a common ancestor occurred in the African/Southeast Asian region at some time in the distant past. These different dengue viruses then jumped independently into humans to become human dengue viruses. A dendrogram of dengue viruses of monkeys (sylvatic strains) and humans (endemic/epidemic strains) is shown in Fig. 8.7. Notice, for example, that the monkey dengue-4 virus groups with human dengue-4 but forms a distinct lineage from that of the other dengue viruses. Similarly, monkey dengue-1 and dengue-2 group with the respective human viruses but form distinct lineages. Sequencing of sylvatic dengue-3 has not yet been done but it will presumably group in the same way. From the extent of the divergence in sequences between the monkey viruses and the human viruses, it is estimated that the jump to humans occurred on the order of 200 years ago for DEN-1, 600 years ago for DEN-4, and 1000 years ago for DEN-2. It is further estimated that the African and Malaysian sylvatic viruses diverged about 800 years ago. Such estimates are subject to considerable uncertainty but are probably valid to within a factor of two. Since human dengue virus is an exclusively human virus that is epidemic in nature and induces lifelong immunity, it could not have existed until human populations were large enough to support the continued existence of such a virus. This topic is covered in more detail in Chapter 4 when discussing measles virus, but sufficiently large human populations arose only within the last thousand years or so in the regions in which dengue viruses first arose and flourished.

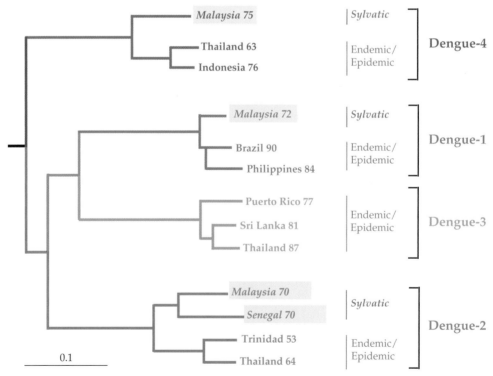

FIGURE 8.7   Phylogenetic tree of the four dengue virus types derived from E protein gene nucleotide sequences of sylvatic (in monkeys) and representative endemic/epidemic (in humans) DEN strains using maximum parsimony. The scale shows a genetic distance of 0.1 or 10% nucleotide sequence divergence. Adapted from Wang *et al.* (2000).

Dengue in the Americas

Dengue viruses have been continuously active over large areas of Asia and the Pacific region for many years. They have recently dramatically expanded their range in the Americas. The viruses may have caused large epidemics in the Americas, including the United States, in the 1800s and into the early 1900s. However, it is impossible to determine with certainty from descriptions of the disease written at the time whether dengue was the causative agent of these epidemics or whether other viruses that cause similar illnesses might have been responsible. In any event, dengue almost died out in the Americas following World War II because of efforts to control *Aedes aegypti*, the urban vector of the virus. With the discovery of DDT in the mid 1900s, a serious effort was made in the Americas to eradicate *Ae.*

aegypti from large regions, in order to control viral diseases spread by these mosquitoes, which include not only dengue but also yellow fever and other arboviruses. These efforts succeeded in eliminating the mosquito from large areas of Central and South America, as illustrated in Fig. 8.8A. However, in 1970 these efforts were abandoned because of the expense involved and the detrimental effects of DDT on the environment, and by 2000 the mosquito had reestablished itself over most of the region (Fig. 8.8A). The reintroduction of multiple dengue strains into the Americas from foci in Asia after the reestablishment of *Ae. aegypti* resulted in the outbreak of dengue hemorrhagic fever associated with huge epidemics of dengue fever (Fig. 8.8B). The history of the increasing infection rate is illustrated in Fig. 8.9 by data from Brazil. Before 1993, epidemics of dengue were sporadic, occurring

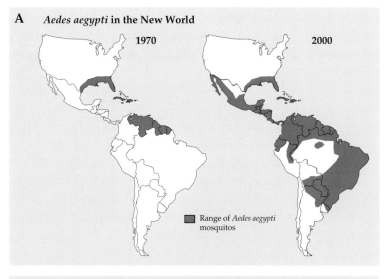

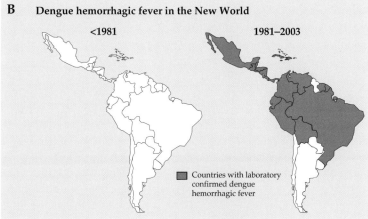

FIGURE 8.8   Changing distribution of dengue hemorrhagic fever (DHF), and the vector for dengue virus in the New World. (A) Distribution of the vector mosquito *Aedes aegypti* in the Americas in 1970 and 2000. *Aedes aegypti* spread rapidly during this period due to the collapse of mosquito control programs and urbanization. (B) Increase and spread of dengue hemorrhagic fever, from 1981 to 2003. Data for these graphs came from the dengue fever information sheets from the Centers for Disease Control and Prevention Web site at: html://www.cdc.gov.

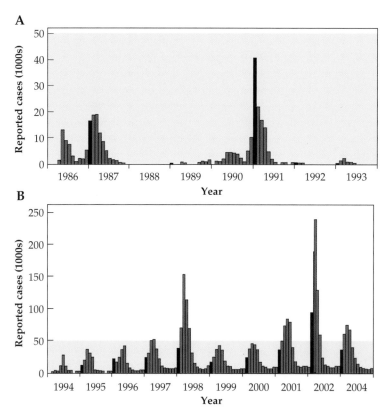

FIGURE 8.9   Number of dengue fever cases reported per month in Brazil (A) between 1986 and 1993 and (B) between 1994 and 2004. Note that the areas shaded in pink in the two graphs represent the same number of cases (50,000). Bars for January cases are filled in black. Adapted from Siquiera *et al.* (2005).

every few years and then dying out. After this, however, epidemics have occurred every year and the total number of cases has increased dramatically (notice the difference in scales).

Prior to the 1980s, a "native American" strain of DEN-2 circulated and there was very little dengue hemorrhagic fever (DHF) in the Americas. A strain of DEN-3 circulated in the 1960s and 1970s but it then disappeared. DEN-1 was introduced into the Americas in 1977 and DEN-4 in 1981 and these viruses then radiated throughout large regions of the Caribbean and northern South America. The first epidemic of DHF occurred in 1981, but interestingly, it was due to the introduction of a new strain of DEN-2 from Asia. This DEN-2 strain grows more vigorously than the native American strain, which is not associated with DHF, and led to the DHF epidemic. Then in 1994 the Southeast Asian strain of DEN-3 responsible for the DHF epidemic in Sri Lanka described in Chapter 3 reached the Americas. The result of all these introductions has been a dramatic increase in the incidence of DHF and dengue shock syndrome (DSS) in the Americas. Whereas in the 1970s there were very few cases of DHF in the Americas, there were more than 10,000 cases in the 1980s and more than 60,000 cases in the 1990s. Cases have continued to increase in number. In the last 5 years (2001–2005 inclusive), more than 50,000 cases have been reported that resulted in about 700 deaths.

The evolution of DEN-1 in the Americas after its introduction in 1977 is illustrated in Fig. 8.10. Sequencing of strains isolated in various years after 1977 show that silent nucleotide substitutions (i.e., synonymous substitutions that do not result in a coding change) have been fixed at the rate of 0.2% per year. However, essentially no coding changes have occurred. Thus, coding changes are not acceptable and viruses containing such changes do not persist.

Human Immunodeficiency Virus

HIVs are human viruses that have become established in the human population within the last century. The two human viruses HIV-1 and HIV-2 derive from different simian immunodeficiency viruses (SIVs), HIV-1 from SIVcpz and HIV-2 from SIVsmm. Further, HIV-1 has become established at least three times by independent entry of SIVcpz into humans. A phylogenetic tree of the primate lentiviruses is shown in Fig. 8.11. There are three lineages of HIV-1, all of which are related to SIV isolated from chimpanzees (SIVcpz). Of these three lineages, the M lineage is found worldwide and is responsible for the majority of human infections. The O lineage is found only in western Africa and in France. The N lineage represents a third introduction of SIVcpz into humans. The structure of the dendrogram makes clear that these three viruses independently entered the human population.

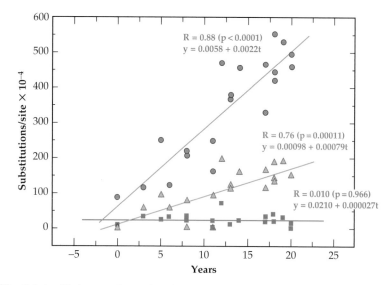

FIGURE 8.10   Relationship between the number of total (▲), synonymous (●), and nonsynonymous (■) nucleotide substitutions per site and the number of years since the introduction of dengue 1 into the Americas in 1977. R is the regression constant. Adapted from Figure 2 in Goncalvez *et al.* (2002).

Epidemiology of SIVcpz

Until recently, the extent of SIVcpz infection of chimpanzees was not known. The chimp is an endangered species, limited in numbers, difficult to study, and only a few isolations of SIV from chimps had been made. In recent studies to examine the extent of SIV infection of wild chimps, 1300 stool samples from wild chimps were collected in the field and laboriously tested for the presence of anti-SIV antibodies and for SIV RNA by RT–PCR. The individual responsible for the stools was identified by examining the host DNA in the sample using highly polymorphic microsatellite loci. Several different chimp populations were included in these studies.

There are four different subspecies of chimpanzee. The type subspecies, *Pan troglodytes troglodytes*, is found in West Africa in southern Cameroon, Gabon, and Congo (Fig. 8.12). *Pan t. schweinfurthii* is further east, primarily in the Democratic Republic of Congo but penetrating northward into the Central African Republic and eastward into a swath from southern Sudan down to Tanzania. *Pan t. vellerosus* is north of the range of *troglodytes*, in northern and western Cameroon. Finally, *Pan t. verus* is west of the range of *vellerosus*, in a broad zone from Senegal to Ghana. Subspeciation has resulted in part because chimps do not swim and large rivers fragment the various populations. Of these four subspecies only two, *troglodytes* and *schweinfurthii*, are naturally infected by SIV. Rates of infection vary in different populations but average about 20%, with some populations exhibiting almost 50% infected individuals. Thus, SIVcpz is a naturally occurring, widespread virus for which two subspecies of chimps are the reservoir. The virus does not

appear to cause disease in chimps, similar to the case for other SIVs that infect African monkeys.

Surprisingly, SIVcpz is itself a recombinant virus. The 5′ half of the genome is derived from SIV infecting red-capped mangabeys, whereas the 3′ half is derived from SIV infecting greater spot-nosed or mustached or mona monkeys (Fig. 8.13). The recombination probably occurred in a chimp that had been infected by the two SIVs. Chimps eat other monkeys and could have become infected in this process in the same way that humans probably became infected with SIVcpz upon slaughtering and eating chimps. The fact that only two of the four subspecies of chimps are infected with SIVcpz argues that this virus arose after subspeciation of the chimps had taken place. The spread of this virus in chimps might in fact be a fairly recent occurrence.

Establishment and Spread of HIV

After infection of humans by SIVcpz, the virus had to adapt to humans and become a human virus in order to be transmitted from person to person and to spread widely. It seems probable that humans have become infected with SIVcpz repeatedly but in most cases the virus failed to adapt to humans or failed to become epidemic because of the low transmissibility of the virus. Following a human infection, it may have smoldered in a small number of people but then died out. When the viruses crossed the species barrier and became firmly established in the human population as HIV-1 is not clear. HIV-1 has been isolated from serum collected in 1959 in Zaire and antibodies to HIV have been found in serum collected in 1963 in Burkina Faso, so

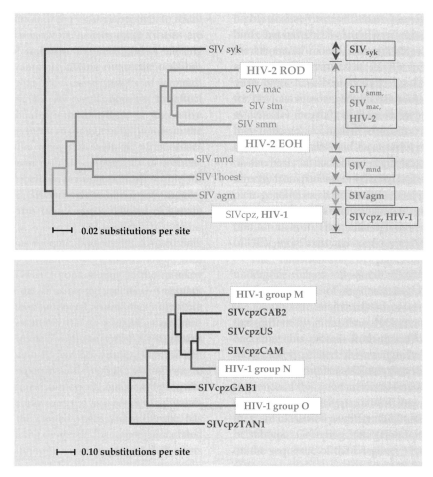

FIGURE 8.11   Phylogenetic trees of the primate lentiviruses. Upper panel: a tree constructed using the neighbor-joining method on selected SIV and HIV *pol* sequences. Horizontal branch lengths are to the scale shown below. HIV strain names are arbitrary. SIV names include the name of the host from which they were obtained: syk, Sykes monkey; smm, sooty mangabey; mac, rhesus macaque; mnd, mandrill; l'hoest, l'hoest monkey; agm, African green monkey; cpz, chimpanzee. The boxes at the right give the names of the five major lineages of primate lentiviruses identified to date. Redrawn from Whetter *et al.* (1999), Figure 1. Lower panel: a tree constructed from maximum-likelihood analysis of full-length *env* sequences from HIV-1 isolates of groups M, N, and O and a number of SIV strains from chimpanzees, corresponding to the box on the lowest branch in panel (A). Human viruses are in red (within a white box), viruses from *Pan troglodytes troglodytes* are in dark red, and the strain from *P. t. schweinfurthii* is in purple. Note the difference in scale. Redrawn from Sharp *et al.* (2005).

HIV-1 has been in the human population at least that long. From sequencing studies of the glycoprotein gene and examination of the rate of divergence, one estimate is that the virus might have entered the human population about 70 years ago, although estimates of divergence rates are controversial. Recent changes in human behavior, including more extensive travel by truck, bus, and plane, changes in sexual practices, and the use of injectable drugs, as well as the increase in the human population, could have allowed the virus to reach major population centers and spread more extensively than in the past, becoming epidemic worldwide. The spread of the virus could also have been aided by the appearance of mutants that were more easily transmissible from person to person. The large increase in population during the last century has certainly resulted in more opportunities for the introduction and spread of the

virus in humans, and therefore for the selection of such transmissible mutants.

HIV-2 represents a distinct lineage that is closely related to SIV of sooty mangabey monkeys (smm) and of macaques (mac). SIV of African green monkeys (agm) and of mandrills (mnd) form other lineages that are more closely related to HIV-2 than to HIV-1. It is clear that SIVsmm and SIVagm are naturally occurring infectious agents that are widespread in Africa and have coevolved with their monkey hosts. Sequence comparisons have shown that different isolates of SIV group with their hosts rather than by geography, and they are therefore adapted to their hosts. They cause no disease in their natural host, but SIVsmm does cause AIDS when transferred to Asian macaques in captivity. HIV-2 is found primarily in western central Africa, where its distribution is almost coincident with that of mangabey monkeys. It

FIGURE 8.12   Natural ranges of four chimpanzee subspecies. Note that only chimpanzees belonging to the subspecies *trogodytes* and *schweinfurthii* have been found naturally infected with SIVcpz. The strains of SIVcpz that have been found cluster into two divergent lineages corresponding to the chimp subspecies lineages, and the strain of SIVcpz found in *troglodytes* was the source of HIV-1. Figure has been adapted from Figure 1 in Sharp *et al.* (2005).

seems clear that HIV-2 represents a separate introduction of SIVsmm into humans.

Repetitive Introductions of Monkey Retroviruses into Humans

It has been found recently that the multiple introductions of SIV into humans that became HIV-1 and -2 do not represent the only human infections by monkey retroviruses. Studies of bush meat hunters in Cameroon showed that at least six retroviruses have crossed from monkeys into humans who were exposed to fresh bush meat. These include two previously unknown retroviruses, HTLV-3 (PTLV-3) and HTLV-4 (PTLV-4). Thus, infection of humans by simian retroviruses has not been a rare event. A book called *The River* by Edward Hooper claimed that the HIV epidemic started because an early version of polio vaccine was contaminated by HIV (or its progenitor). No evidence for such

contamination exists, and recent analysis of lots of this vaccine that had been stored for 40 years by three different laboratories failed to find any trace of SIV/HIV in the vaccine. Further, given that monkey viruses have repeatedly entered the human population, it is unnecessary to postulate human vaccine activity for the origin of HIV. Rumors based upon this claim, however, have led to suspension of vaccination for polio in some African nations, resulting in a resurgence of poliomyelitis in Africa that has spread to other areas previously free of polio (see Chapter 3).

VIRUSES ASSOCIATED WITH RODENTS

Hantaviruses

Hantaviruses are associated with rodents and infect humans through aerosols containing virus from rodent

SIVrcm (red-capped mangabey)

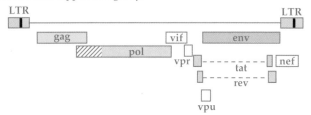

SIVcpz (chimpanzee/HIV-1)

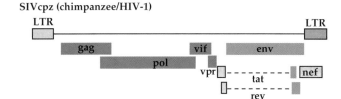

SIVgsn/SIVmus/SIVmon

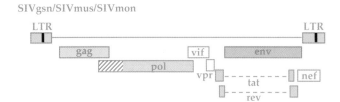

FIGURE 8.13 Diagram of the recombinant origin of SIVcpz. The various genes of SIV in red-capped mangabeys are shown at the top, outlined in magenta. Similarly the genome of SIVgsn (greater spot-nosed)/SIVmus (mustached)/SIV mon (mona) monkeys is shown at the bottom, outlined in blue. In the recombinant genome of chimpanzee/HIV-1 (center), SIVcpz genes derived from SIVrcm are magenta, genes from SIVgsn are in blue, and genes of unknown origin are in gray. Adapted from Figure 4 in Sharp *et al.* (2005).

excreta. The first hantavirus to come to medical attention was Hantaan virus which caused more than 3000 cases of hemorrhagic fever with renal syndrome in U.S. troops during the Korean War. Since then, many hantaviruses have been identified in both the Old World and in the Americas that cause serious human illness. They are examples of emerging viruses because as the number of humans increases and as they invade more habitat occupied by rodents carrying hantaviruses, the incidence of infection in humans has risen. Very interesting in this regard was the isolation, in May 1993, of a new hantavirus that causes acute respiratory distress in humans that can lead to rapid death, a syndrome now called *h*antavirus *p*ulmonary *s*yndrome (HPS) and originally called *a*cute *r*espiratory *d*isease *s*yndrome (ARDS). The virus was isolated by the CDC in collaboration with local health authorities following an epidemic in the Four Corners area of the southwestern United States that resulted in approximately 25 deaths. The virus is associated with the deer mouse *Peromyscus maniculatus*. It is thought that the epidemic may

have resulted from an abundance of pine nuts in the area during a good growing year, leading the local people to harvest larger amounts of these than usual and store them in their homes when their normal storage areas became full. With abundant food available, the rodent population exploded and invaded homes to get to the pine nuts, and it is thought that this more intimate contact between humans and rodents may have led to the epidemic. The hantavirus responsible for this epidemic is now called Sin Nombre virus, which is Spanish for "without a name." Early suggestions that it be called Four Corners virus or Muerto Canyon virus (after a geographical feature in the area) drew objections from local residents who did not want this major tourist area identified with a fatal disease. Eventually the CDC simply named it Sin Nombre (there is a small creek in the area called the Sin Nombre River that serves as justification for the choice of name).

With the discovery of Sin Nombre virus, searches for viruses in other regions of North America resulted in the isolation of many viruses related to Sin Nombre. These viruses are associated with other rodents in the order *Sigmodontinae* and have been given names of local features in order to distinguish them. These include New York, Monongahela, Bayou, and Black Creek Canal viruses, all of which have caused HPS in the United States (see Fig. 4.26). Related viruses are also found in Latin America. In fact, studies have now shown that hantaviruses are present in virtually all states within the United States and into Latin America, and that fatalities due to infection by the virus have occurred in most states. Retrospective studies of stored sera collected from patients who died of ARDS in the past have identified earlier cases of HPS. Thus these viruses are widespread and have caused many fatal cases of human disease over the years.

As noted in Chapter 4, the epidemiology of arenaviruses is similar to that of the hantaviruses. Several South American arenaviruses have caused increasing numbers of cases of human hemorrhagic fever because of increased contact between humans and the rodent carriers of the viruses. The development of the Pampas of Argentina, in particular, led to increased incidence of human arenavirus disease.

FURTHER READING

Emerging Viral Zoonoses

Anishchenko, M., Bowen, R. A., Paessler, S., *et al.* (2006). Venezuelan encephalitis emergence mediated by a phylogenetically predicted viral mutation. *Proc. Natl. Acad. Sci. U.S.A.* **103**: 4994–4999.

Barclay, W. (2006). Influenza vaccines. *Microbiol. Today* **33**(1): 17–19.

Davis, C. T., Ebel, G. D., Lanciotti, R. S., *et al.* (2005). Phylogenetic analysis of North American West Nile virus isolates, 2001–2004: evidence for the emergence of a dominant genotype. *Virology* **342**: 252–265.

Eaton, B. T., Broder, C. C., Middleton, D., and Wang, L.-F. (2006). Hendra and Nipah viruses: different and dangerous. *Nature Rev. Microbiol.* **4**: 23–35.

Johnson, R. T. (2003). Emerging viral infections of the nervous system. *J. Neurovirol.* **9**: 140–147.

Kallio-Kokko, H., Uzcategui, N., Vapalahti, I., and Vaheri, A. (2005). Viral zoonoses in Europe. *FEMS Microbiol. Rev.* **29**: 1051–1077.

Kobasa, D., and Kawaoka, Y. (2005). Emerging influenza viruses: past and present. *Curr. Mol. Med.* **5**: 791–803.

Olsen, B., Munster, V. J., Wallensten, A., *et al.* (2006). The global patterns of influenza A virus in wild birds. *Science* **312**: 384–388.

Peters, C. J. (2006). Emerging viral diseases. Chapter 18 in: *Fields Virology, Fifth Edition* (D. M. Knipe and P. M. Howley, Eds. in chief), Philadelphia, Lippincott Williams & Wilkins, pp. 605–626.

Sharp, P. M., Shaw, G. M., and Hahn, B. H. (2005). Simian immunodeficiency virus infection of chimpanzees. *J. Virol.* **79**: 3892–3902.

Siquiera, J. B., Jr., Martelli, C. M. T., Coelho, G. E., da Rocha Simplício, A. C., and Hatch, D. L. (2005). Dengue and dengue hemorrhagic fever, Brazil, 1981–2002. *Emerg. Infect. Dis.* **11**: 48–53.

Wang, E., Ni, H., Xu, R., *et al.* (2000). Evolutionary relationships of endemic/epidemic and sylvatic dengue viruses. *J. Virol.* **74**: 3227–3234.

Webster, R. G., Peiris, M., Chen, H. L., and Guan, Y. (2006). H5N1 outbreaks and enzootic influenza. *Emerg. Infect. Dis.* **12**: 3–8.

Bats and Viruses

Dobson, A. P. (2005). What links bats to emerging infectious diseases? *Science* **310**: 628–629.

Fooks, A. R., Brookes, S. M., Johnson, N., McElhinney, L. M., and Huston, A. M. (2003). European bat lyssaviruses: an emerging zoonosis. *Epidemiol. Infect.* **131**: 1029–1039.

Jia, G., Zhang, Y., Wu, T., Zhang, S., and Wang, Y. (2003). Fruit bats as a natural reservoir of zoonotic viruses. *Chin. Sci. Bull.* **48**: 1179–1182.

Leroy, E. M., Kumulungui, B., Pourrut, X., *et al.* (2005). Fruit bats as reservoirs of Ebola virus. *Nature* **438**: 575–576.

Mackenzie, J. S., Field, H. E., and Guyatt, K. J. (2003). Managing emerging diseases borne by fruit bats (flying foxes), with particular reference to henipaviruses and Australian bat lyssavirus. *J. Appl. Microbiol.* **94**: 59–69S.

Mayen, F. (2003). Haematophagous bats in Brazil, their role in rabies transmission, impact on public health, livestock industry and alternatives to an indiscriminate reduction of bat population. *J. Vet. Med.* **50**: 469–472.

Osborne, J. C., Rupprecht, C. E., Olson, J. G., *et al.* (2003). Isolation of Kaeng Khoi virus from dead *Chaerephon plicata* bats in Cambodia. *J. Gen. Virol.* **84**: 2685–2689.

Reynes, J.-M., Counor, D., Ong, S., *et al.* (2005). Nipah virus in Lyle's flying foxes, Cambodia. *Emerg. Infect. Dis.* **12**: 1041–1047.

Warrilow, D. (2005). Australian bat lyssavirus: a recently discovered new rhabdovirus. *Curr. Top. Microbiol. Immunol.* **292**: 25–44.

Woo, P. C. Y., Lau, S. K. P., Li, K. S. M., *et al.* (2006). Molecular diversity of coronaviruses in bats. *Virology* **351**: 180–187.

SARS

Vijayanand, P., Wilkins, E., and Woodhead, M. (2004). Severe acute respiratory syndrome (SARS): a review. *Clin. Med.* **4**: 152–160.

Stadler, K., Masignani, V., Eickmann, M., *et al.* (2003). SARS—Beginning to understand a new virus. *Nature Rev. Microbiol.* **1**: 209–218.

Poon, L. L. M., Chu, D. K. W., Chan, K. H., *et al.* (2005). Identification of a novel coronavirus in bats. *J. Virol.* **79**: 2001–2009.

Li, W., Shi, Z., Yu, M., *et al.* (2005). Bats are natural reservoirs of SARS-like coronaviruses. *Science* **310**: 676–679.

Snijder, E. J., Bredenbeek, P. J., Dobbe, J. C., *et al.* (2003). Unique and conserved features of genome and proteome of SARS-coronavirus, an early split-off from the coronavirus group 2 lineage. *J. Mol. Biol.* **331**: 991–1004.

9

Subviral Agents

INTRODUCTION

This chapter considers a number of infectious agents that are subcellular, but that are not viruses in the strict sense of the term. Some of these are not capable of independent replication but require a helper virus, in which case the agent is effectively a parasite of a parasite. Others replicate independently but use unconventional means to achieve their replication and spread. Many of the agents discussed here cause important diseases in plants or animals, including humans.

The agents to be considered include *d*efective *i*nterfering (DI) viruses that arise by deletions and rearrangements in the genome of a virus. DIs require coinfection by a helper virus to replicate. They may play an important role in modulation of viral disease or they may simply be artifacts that arise in laboratory studies. Related to DIs, at least conceptually, are satellites of viruses, which can replicate only in the presence of a helper. Satellites are known for many plant viruses and are known to influence the virulence of the helper viruses. Completely different are agents called viroids. Viroids consist of small, naked RNA molecules that are capable of directing their own replication. They do not encode protein, but instead contain promoter elements that cause cellular enzymes to replicate them. Many are important plant pathogens. Related to viroids are virusoids, which are satellites of viruses that resemble packaged viroids. There are also agents that combine the attributes of both satellites and viroids, such as hepatitis delta virus, which is an important human pathogen. Finally, prion diseases, caused by infectious agents whose identity is controversial but which may consist only of protein, are discussed.

DEFECTIVE INTERFERING VIRUSES

Defective interfering viruses are a special class of defective viruses that arise by recombination and rearrangement of viral genomes during replication. DIs are *defective* because they have lost essential functions required for replication. Thus, they require the simultaneous infection of a cell by a helper virus, which is normally the parental wild-type virus from which the DI arose. They *interfere* with the replication of the parental virus by competition for resources within the cell. These resources include the machinery that replicates the viral nucleic acid, which is in part encoded by the helper virus, and the proteins that encapsidate the viral genome to form virions.

DIs of many RNA viruses have been the best studied. Because DI RNAs must retain all *cis*-acting sequences required for the replication of the RNA and its encapsidation into progeny particles, sequencing of such DI RNAs can provide clues as to the identity of these sequences. Identification of *cis*-acting sequences is important for the construction of virus vectors used to express a particular gene of interest, whether in a laboratory experiment or for gene therapy.

The most highly evolved DI RNAs are often not translated and consist of deleted and rearranged versions of the parental genome. In the case of alphaviruses, whose genome is about 12 kb (Chapter 3), DI RNAs have been described that are about 2 kb in length. However, they have a sequence complexity of only 600 nucleotides, because sequences are repeated one or more times. The sequences of two such DI RNAs of Semliki Forest virus (SFV) are illustrated schematically in Fig. 9.1. From the sequences of these DIs as well as DIs of other alphaviruses, specific functions for the elements found in these DIs have been proposed. Other approaches

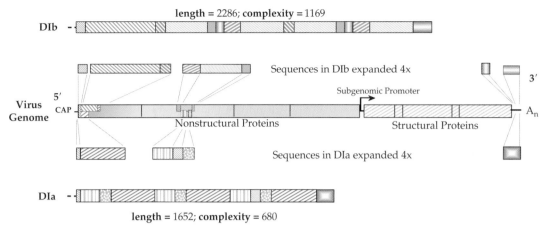

FIGURE 9.1   Schematic representation of DIs (defective interfering particles) found after high multiplicity infection of Semliki Forest virus. The central block shows the genome of the nondefective virus, with vertical lines demarking the four nonstructural and five structural polypeptides. The blocks of sequence found in two different DIs are expanded fourfold below and above. Their location in the DI genome is illustrated with blocks of identical shading. Note that some blocks of unique sequence are repeated three times in DIa and one block is repeated four times in DIb. Adapted from Strauss and Strauss (1997), Figure 1.

have then been used to confirm the hypotheses derived from such sequence studies. Thus, the 3′ end of the parental RNA, which is retained in all alphavirus DI RNAs, forms a promoter for the initiation of minus-strand RNA synthesis from the plus-strand genome. The 5′ end of the RNA is also preserved in many DI RNAs, such as those illustrated in Fig. 9.1. Surprisingly, however, it has been replaced by a cellular tRNA in some DI RNAs. The complement of this sequence is present at the 3′ end of the minus strand, where it forms a promoter for initiation of genomic RNA synthesis. The finding that the DI RNAs with the tRNA as the 5′ terminus have a selective advantage over the parental genome during RNA replication suggests that this promoter is a structural element recognized by the viral replicase. It also suggests that the element present in the genomic RNA is suboptimal, perhaps because the genomic RNA must be translated as well as replicated. Finally, repeated sequences from two regions of the genome are present in all alphavirus DI RNAs. It is thought that one sequence (shown as red patterned blocks in Fig. 9.1) is an enhancer element for RNA replication and the second (shown as yellow and green patterned blocks) is a packaging signal. Repetition of these elements may increase the efficiency of replication and packaging of the DI RNA.

Vesicular stomatitis virus (VSV) (Chapter 4) DI RNAs vary in size from a third to half the length of the virion RNA. Some DI RNAs are simply deleted RNA genomes, but others have rearrangements at the ends of the RNAs. Representative examples are illustrated in Fig. 9.2A. During replication of the RNA, the sequences at the ends must contain promoter elements for initiation of RNA synthesis. More genomic RNA (minus strand that is packaged in virions) is made than antigenomic RNA (which functions only as a template for genomic RNA synthesis) and therefore the promoter at the

3′ end of the antigenomic RNA is stronger than the promoter at the 3′ end of the genomic RNA. Thus, it is not surprising that some DI RNAs have the stronger promoter at the 3′ ends of both (+) and (−)RNA (as in Class II DIs), ensuring more rapid replication of the DI RNAs. The DI RNAs may have the luxury of doing this because they are not translated nor do they serve as templates for the synthesis of mRNAs.

The well-studied alphavirus DI RNAs and the VSV DI RNAs are not translated. For many DI RNAs, however, translation is required for efficient DI RNA replication. The best studied examples of this are DIs of poliovirus and of coronaviruses (these viruses are described in Chapter 3). DI RNAs of poliovirus are uncommon and contain deletions in the structural protein region. It has been suggested that in this case it is the translation product that is required for efficient replication of the RNA (the replicase translated from the RNA may preferentially use as a template for replication the RNA from which it was translated). In contrast, for at least one well-studied DI of a coronavirus, translation of the RNA is required for efficient replication, but the translation product is not important. In this case, translation may stabilize the DI RNA, since there appears to be a cellular pathway to rid the cell of mRNAs that are not translatable. If so, it is uncertain how DI RNAs that are not translated avoid this pathway. Some representative naturally occurring DIs of mouse hepatitis virus, a murine coronavirus, are illustrated in Fig. 9.2B.

Because DI RNAs are replicated by the helper virus machinery and encapsidated by the capsid proteins of the helper virus, they interfere with the parental virus by diverting these resources to the production of DI particles rather than to the production of infectious virus particles. It was the first noted by von Magnus in the early 1950s that

Vesicular stomatitis virus (VSV) and DIs derived from it

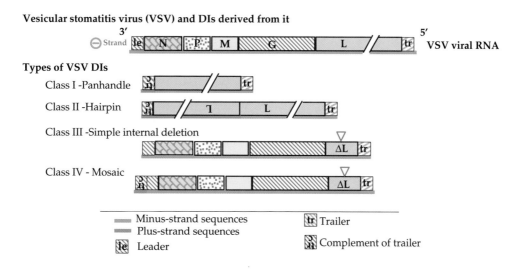

Types of VSV DIs

Class I - Panhandle

Class II - Hairpin

Class III - Simple internal deletion

Class IV - Mosaic

Minus-strand sequences
Plus-strand sequences
le Leader
tr Trailer
Complement of trailer

Murine hepatitis virus (MHV) and DIs derived from it

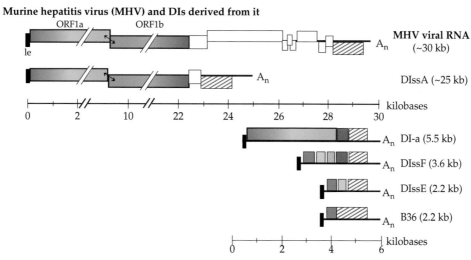

FIGURE 9.2   Types of DIs generated from a rhabdovirus and a coronavirus. Upper panel: diagrammatic representation of the VSV genome and members of the four classes of DI particles. The leader and trailer are shown as patterned blocks. The genome is shown 3′ to 5′ for the minus strand (ochre underline). The parts of the DIs corresponding to the complement of the minus strand are underlined in green. A red triangle marks the internal deletion in the L gene, which is found in Class III and Class IV DIs. Adapted from Whelan and Wertz (1997). Lower panel: structures of naturally occurring DI RNAs of MHV (a murine coronavirus). DIssA, DI-a, etc. were isolated from MHV-infected cells. The bottom line shows a synthetic DI replicon called B36. Sequences in the DIs are color coded by their region of origin in the parental virus genome. Adapted from Brian and Spaan (1997) Figure 1.

influenza virus, passed at high multiplicity for many passages, produced yields that cycled between high and low. This effect is illustrated schematically in Fig. 9.3A. We now know that this is due to the presence of DI particles. In early passages virus yields are high. When DIs arise, they depress the yield of virus. Because high multiplicities of infection are required to maintain DI replication, so that cells are infected with both the helper and the DI, low yields of virus lead to a reduction in DI replication in the next passage or two. Reduced DI replication leads to higher

yields of virus. Thus, the yield of infectious virus continues to fluctuate.

In a laboratory setting, at least, DIs can drive the evolution of the wild-type virus. This is shown schematically in Fig. 9.3B. When virus is passed at high multiplicity for very many generations, mutants often arise that have altered promoters that are recognized by mutant replication proteins. Such mutants are resistant to the DIs that are in the population at the time, because the mutant replication proteins do not recognize the promoters in the DIs. The mutant virus

A.

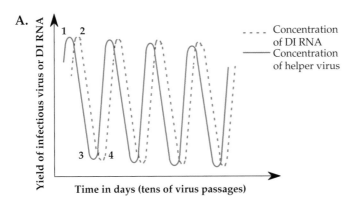

1) Early in an undiluted passage series, standard virus replicates well, but DI's are beginning to accumulate.

2) DIs replicate at high efficiency, and interfere with standard virus.

3) So little standard virus is produced that there is little helper function, and DI replication drops.

4) With little DI replication, interference is reduced and standard virus titers rise again.

B.

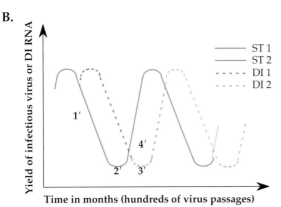

1') DI 1 interferes strongly with replication of standard virus (ST1).

2') A new variant of standard virus (ST2) emerges that is resistant to interference by DI 1, and does not serve as a helper for DI 1.

3') Without helper assistance, DI 1 disappears, and ST 2 replicates vigorously.

4') New DIs of ST 2 (DI 2) appear and begin to depress ST 2 replication.

FIGURE 9.3   Stylized illustration of the influence of defective interfering particles on viral evolution. (A) Short-term generation of DI particles during undiluted passage and the cyclical fluctuations in infectious virus yield and concentration of DIs. This effect was first described by von Magnus in 1946 during repeated passage of influenza virus at high multiplicity. (B) Role of DI particles in driving long-term evolution of viruses. The net result of hundreds of passages is that variant 2 and its DIs completely replace the original wild type or standard virus 1. Adapted from *Encyclopedia of Viruses,* Figure 2 on p. 373.

rapidly takes over the population of virus because of its selective advantage. However, new DIs then arise that will interfere with the mutant virus, and the cycle repeats.

It is unclear whether DI particles serve a biological role in nature or whether they are artifacts of abortive recombination that appear in the laboratory because of the high multiplicities of infection that are often used. It has been argued that DIs may arise late in the infection of an animal by a virulent virus and lead to attenuation of symptoms by reducing the yield of infectious virulent virus. As described in earlier chapters, such attenuation could be important for the persistence of a virus in nature. Hepatitis delta virus, described later, is a satellite that replicates only in a cell infected by HBV. Thus, it is clear that it is possible to achieve the multiplicities required to maintain a defective virus in at least some circumstances. Reconstruction experiments in mice have shown that it is possible to attenuate the virulence of lymphocytic choriomeningitis virus by injecting DIs along with the virus. However, it is not clear that DIs will arise in an acute infection in time to ameliorate symptoms. Thus, it has not been possible to provide firm evidence that DIs

actually modulate the virulence of their parents in nature, and the question remains open.

SATELLITES AND SATELLITE VIRUSES

A number of satellites and satellite viruses are listed in Table 9.1. The dependoviruses, a genus in the family *Parvoviridae*, were considered in Chapter 7. Although the dependoviruses require a helper, the helper is only needed in order to stimulate the cell to enter a stage in which the dependovirus can replicate. The helper does not provide any function directly related to dependovirus replication or packaging. Satellites and satellite viruses, however, are usually more intimately dependent on the presence of a helper virus to furnish functions directly required for the replication of the satellite genome or for its encapsidation into progeny particles.

Many plant viruses have satellites associated with them. When a satellite encodes its own coat protein, it is sometimes referred to as a satellite virus. Otherwise it is simply called a

TABLE 9.1   Satellites and Satellite Viruses

Group	Genome size	Helper virus	Host(s)	Comments
dsDNA satellite				
Bacteriophage P4	11.5 kb (10–15 genes)	P2 bacteriophage	Bacteria	All structural proteins from P2
ssDNA satellite ciruses[a]				
Dependovirus (AAV)	4.7 kb	Adenovirus herpesvirus	Vertebrates	See Table 7.16
dsRNA satellites				
M satellites of yeast	1 to 1.8 kb	*Totiviridae*	Yeast	Encode "killer" proteins; encapsidated in helper coat protein
ssRNA satellite viruses				
Chronic bee-paralysis virus associated satellite	3 RNAs, each 1 kb	Chronic bee-paralysis virus	Bees	
Tobacco necrosis virus satellite	1239 nt	Tobacco necrosis virus	Plants	
ssRNA satellites				
Hepatitis delta virus	1.7 kb	Hepatitis B virus	Humans	Encode two forms of δ antigen encapsidated by helper proteins
B-type mRNA satellites	0.8 to 1.5 kb	Various plant viruses	Plants	Encode nonstructural proteins, rarely modify disease syndrome
C-type linear RNA satellites	<0.7 kb	Various plant viruses	Plants	Commonly modify disease caused by helper
D-type circular RNA satellites "virusoids"	~350 nt	Various plant viruses	Plants	Self-cleaving molecules

[a] When a satellite encodes its own coat protein, it is known as a satellite virus.

satellite. One of the best studied satellite systems is tobacco necrosis virus (TNV) and its satellite, tobacco necrosis virus satellite (TNVS). TNV has a plus-strand RNA genome of about 3.8 kb. The TNV virion is icosahedral with $T=3$, and contains 180 copies of a single-coat protein species of about 30 kDa. Associated with many isolates of TNV in nature is TNVS. TNVS has an RNA genome of 1239 nucleotides. The TNVS virion is a $T=1$ icosahedral structure formed by 60 molecules of a single species of capsid protein encoded by the satellite RNA. The satellite RNA encodes only this single protein that encapsidates its own RNA. All of the functions required to replicate the RNA are provided by the helper TNV.

Some satellites of RNA plant viruses encode only a nonstructural protein required for RNA replication, and the RNA is encapsidated by the capsid protein of the helper virus. In other cases, the satellite is not translated into protein and depends on the helper for all of its functions, in which case it is functionally analogous to DI RNAs. A distinct class of satellite RNAs, called virusoids, consist of viroid-like RNAs that are encapsidated in the capsid protein of the helper virus. These are discussed in the next section.

Although satellites are quite common among plant viruses, they are almost unknown among animal viruses. It is very common among plant viruses to have the genome divided among two or more segments that are separately encapsidated into different particles, a situation that does not occur among animal viruses. Evidently, the mechanisms by which plant viruses are transmitted allow the infection of a plant, and of individual cells within a plant, by multiple particles that together constitute a virus or that constitute a virus and its satellites. Transmission of animal viruses between hosts or among the cells of a host does not appear to allow multiple infections with sufficient frequency to maintain virus systems that are constituted by multiple particles, with the exceptions of hepatitis δ virus, described later, and dependoviruses, which have evolved ways to persist within a cell until the helper virus comes along (Chapter 7). The defense mechanisms of the animal host may also play a role in this restriction.

VIROIDS AND VIRUSOIDS

Viroids are small, circular RNA molecules that do not encode any protein and that are infectious as naked RNA molecules. Sequenced viroids range from 246 to 375 nucleotides and possess extensive internal base pairing that results in the RNA being rodlike and about 15 nm long. A partial listing of viroids is given in Table 9.2. All known viroids

TABLE 9.2   Viroids

Family/genera	Type species	Genome size	Host(s)	Comments
Popsiviroidae[a]				
Popsiviroid	PSTVd	356 to 375 nt	Plants	Presence of central conserved region and lack of self-cleavage mediated by hammerhead ribozyme; replicate by an asymmetric rolling circle strategy in nucleus of infected cells, probably using RNA polymerase II
Hostuviroid	HpSVd	295 to 303 nt		
Cocaviroid	CCCVd	246 to 301 nt		
Apscaviroid	ASSVd	306 to 369 nt		
Coleviroid	CbVd-1	248 to 361 nt		
Avsunviroidae[b]				
Avsunviroid	ASBVd	246 to 250 nt	Plants	Lack central conserved region; replicate by a symmetric strategy in chloroplasts of infected plants using chloroplastic RNA polymerase; can form self-cleaving hammerhead ribozymes in both plus and minus strands
Pelamoviroid	PLMVd	337 to 399 nt		

[a] Formerly known as the Group B viroids.
[b] Formerly known as the Group A viroids.

infect plants. However, hepatitis δ, which infects humans, has many viroid-like properties and may be related to viroids. Many viroids are important agricultural pathogens, whereas others replicate without causing symptoms. Viroids are often transmitted through vegetative propagation of plants, but can also be transmitted during agricultural or horticultural practices in which contaminated instruments are used. Some viroids can be transmitted through seeds and at least one viroid is transmitted by an aphid.

On infection of a plant cell, viroid RNA is transported to the nucleus. The circular RNA appears to be copied by host-cell RNA polymerase II, using a rolling circle mechanism in which multimeric antigenomic sense RNA molecules are produced. The multimeric antigenome sense RNA can then be used as a template to make multimeric genomic sense RNA. This synthesis may also be performed by RNA polymerase II. The concatenated RNAs are cleaved and cyclized to produce the progeny viroid RNA molecules. In an infected cell, as many as 10^4 viroid RNAs can accumulate, most of them in the nucleus.

Some viroids are capable of self-cleavage by the concatenated RNAs to produce genome-length RNAs, followed by self-ligation to cyclize the unit-length molecule. Other viroids are not capable of self-cleavage and ligation. There are five groups of non-self-cleaving viroids, classified by the sequences in the central conserved region. The structures of these five groups are shown in Fig. 9.4. The conserved domains highlighted in the figure are thought to be important for the replication of the viroid (i.e., to form promoters recognized by RNA polymerase II) and for its cleavage to produce unit-length molecules. A pathogenesis domain is also highlighted. Changes in this domain affect the

virulence of the viroid on infection of its plant host. For non-self-cleaving viroids, it is assumed that the concatenated RNAs are cleaved and ligated by host-cell enzymes.

The self-cleaving viroids possess a hammerhead ribozyme structure, illustrated in Fig. 9.5. The ribozyme activity cleaves the concatenated RNA at the points indicated by the arrows and ligates the ends to form circular molecules. The viroid RNA is very compact in its structure, with extensive secondary structure, including pseudoknots.

There also exist a large number of satellites called virusoids. Virusoid RNAs are about 350 nt in length, and the RNA is a single-stranded, covalently closed circle. The mechanisms by which virusoid RNA is replicated have not been precisely determined, but they appear to be viroid-like and may replicate by the same mechanisms as viroids. At least some virusoid RNAs are capable of self-cleavage. Virusoid RNAs are encapsidated by the capsid protein of the helper virus of which the virus is a satellite. Thus, transmission occurs by conventional virus-like means, and virusoids may have arisen from viroids that evolved a mechanism for packaging using a helper virus.

HEPATITIS δ

The hepatitis delta (δ) agent or virus (HDV) is a satellite of hepatitis B virus (HBV). It has a worldwide distribution, although strains isolated from different regions of the world differ by up to 40% in their nucleotide sequence. The distribution of HDV is not uniform around the world. Regions of particularly high prevalence include the Mediterranean basin, the Middle East, Central Asia, West Africa, the Amazon

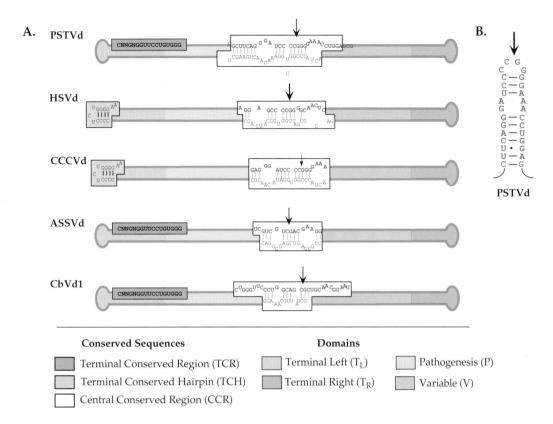

FIGURE 9.4 Models for the genomes of the type species of the five genera of non-self-cleaving viroids. They are as follows: Popsiviroid: PSTVd, potato spindle tuber viroid; Hostuviroid: HSVd, hop stunt viroid; Cocadviroid: CCCVd, coconut cadang-cadang viroid; Apscaviroid: ASSVd, apple scar skin viroid; Coleviroid: CbVd-1, Coleus blumei viroid-1. (A) The RNA strand is shown as a green closed loop. Four functional domains (T_L, T_R, P, and V) are indicated with different colors of shading. Three conserved sequences are boxed. The *c*entral *c*onserved *r*egion (CCR) is a white box, the *t*erminal *c*onserved *r*egion (TCR) is a lavender box, and the *t*erminal *c*onserved *h*airpin (TCH) is a green box. The nucleotides in the upper strand of the CCR (dark green) can in each case form a stable stem and loop structure with the top of the loop at the black arrow, as shown for PSTVd in (B). In this alternative configuration, the nucleotides that are invariant within all five groups are shown in red. Adapted from Flores *et al.* (1997).

Basin, and certain islands in the South Pacific. HDV will only replicate in cells that are simultaneously infected with HBV and its distribution is thus dependent in part upon the distribution of HBV, which was shown in Fig. 6.29. However, as shown in Fig. 9.6, HDV is not uniformly distributed throughout the range of HBV. The percentage of hepatitis B patients that are also infected by infection with HDV ranges from 5% to more than 60% in different geographic areas.

Infection of humans by HDV can either occur by simultaneous infection with both HBV and HDV, or by superinfection with HDV of a person who is chronically infected with HBV. In the case of coinfection, a chronic infection by HDV, which requires that HBV also establish a chronic infection, is established only 1–3% of the time. Most often the infection is completely resolved and recovery occurs. In contrast, superinfection of chronically infected HBV patients with HDV leads to chronic infection by HDV in 70–80% of

patients. The different outcomes following infection with HDV are illustrated schematically in Fig. 9.7, in which the symptomology at different times after infection is indicated.

The illness caused by HDV is usually more serious than that caused by HBV alone. The mortality rate from HDV infection is 2–20%, 10-fold higher than the rate for HBV infection alone, which is the next most severe form of viral hepatitis. Most cases of HDV infection are probably clinically important. It is estimated that 460 million people in the world are chronically infected with HBV of whom perhaps 20 million are also chronically infected by HDV. All persons chronically infected with HBV that are not already infected by HDV are at risk for contracting HDV and suffering a more severe form of hepatitis.

The mechanisms by which HDV is transmitted are not understood. It is conjectured that poor hygiene together with intimate contact among people who are infected with the

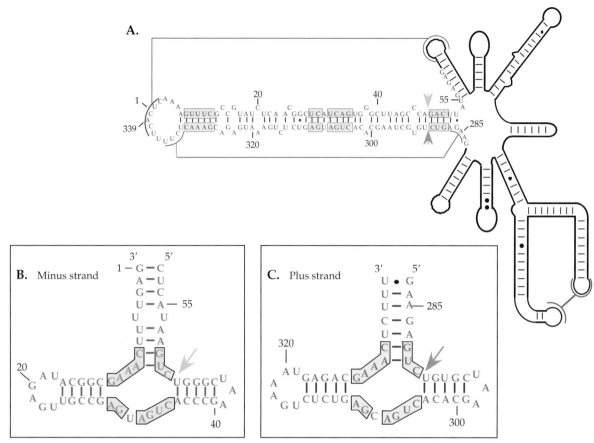

FIGURE 9.5   Predicted secondary structure of peach latent mosaic viroid (PLMV) (Family *Avsunviroidae*, Genus *Pelamoviroid*) RNA in solution. (A) The entire viroid. For most of the molecule, only the backbone is shown, with hydrogen bonds indicated by bars and G–U pairs indicated with dots. The numbering of the nucleotides is arbitrary. The structure is predicted to be even more compact due to pseudoknots formed by the regions joined by purple lines. The nucleotides making up the minus-strand and plus-strand hammerhead ribozymes are shown in green and blue letters, respectively. The nucleotides conserved in all hammerheads in *Avsunviroidae* are boxed and shaded, and the sites of cleavage are marked by arrows. (B) and (C) The structures of the two hammerhead ribozymes, using the same color conventions. The minus-strand hammerhead is made up of the complement of the sequence shown in green in (A). Adapted from Flores *et al.* (1997) and Pelchat *et al.* (2000).

virus may be an important source of transmission in many parts of the world. In developed countries, contaminated blood products and sharing of needles by drug users are important in the spread of the virus, but these are not important modes of transmission worldwide. Sexual transmission may occur, but again this does not appear to be an important component of the transmission of the virus on a worldwide basis.

Since HDV depends on HBV for its propagation, control of HDV is dependent on control of HBV. Current HBV vaccines are highly effective at preventing HBV infection and the increasing levels of vaccination against HBV, together with increased screening for the presence of HBV in blood products, has resulted in a dramatic reduction in recent years of new infections by HDV.

Replication of the HDV Genome and Synthesis of mRNA

The HDV genome is a single-stranded, covalently closed circular RNA molecule of 1.7 kb. The HDV genome can be thought of as a viroid into which has been inserted a gene encoding a single polypeptide, the hepatitis δ antigen (HDAg). As is the case for viroid RNA, HDV RNA has a high degree of secondary structure, with about 70% of the molecule being base paired internally so that it forms a rod-like structure. The HDV genome has minus-sense polarity, that is, it is complementary to the sequence that is translated into the HDAg. The structures of the genomic RNA, the mRNA for the HDAg, and of the antigenomic RNA are illustrated in Fig. 9.8A. The 0.8-kb mRNA is capped and

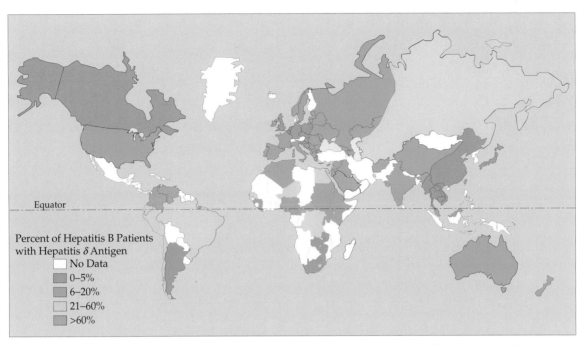

FIGURE 9.6   Worldwide distribution of hepatitis δ infection as measured by the presence of hepatitis δ antigen in the serum of hepatitis B infected patients with hepatitis. Adapted from Fields *et al.* (1996) p. 2826.

polyadenylated as is common for mRNAs. The 1.7-kb antigenome is an exact complement of the genomic RNA and, like genomic RNA, is also circular.

Following infection by the agent, the RNA is transferred to the nucleus. The HDAg, of which there are about 70 copies in the virion, is required for this. In the nucleus, the RNA is replicated by mechanisms related to those used by viroid RNA. However, there is the added complication that an mRNA for HDAg must also be produced. Thus, there are three elements to the replication of RNA: the production of an antigenomic RNA template from genomic RNA, the production of mRNA for HDAg from genomic RNA, and the production of genomic RNA from antigenomic RNA templates. It is believed that synthesis of genomes from antigenomic templates and synthesis of the mRNA from genomic templates are carried out by RNA polymerase II. However, synthesis of antigenomes from genomic templates may utilize another polymerase, perhaps RNA polymerase I. Other, currently unknown host factors also participate, and the HDAg, of which there are two kinds, S-HDAg and L-HDAg, as described later, both of which can be modified in various ways, is absolutely required. These replication steps are illustrated schematically in Fig. 9.9A and B.

Replication of the RNA, whether production of genomes or antigenomes, is thought to utilize a rolling circle mechanism in which concatenated RNAs are produced that are cleaved to unit length by the self-cleavage activity present in both genomic and antigenomic molecules. The resulting unit-length molecules are then cyclized. Although HDV RNA is capable of self-ligation, this process appears to be inefficient and it is thought that a cellular ligase is responsible for most cyclization of HDV RNA monomers.

There are differences in the synthesis of genomic and antigenomic RNA, including differences in their rates of synthesis, different sensitivities of their synthesis to drugs, in the requirement for different forms of HDAg (described later) for genomic and antigenomic RNA synthesis, and in the transport of the RNAs to different places within the cell after synthesis. Furthermore, synthesis of genomic and antigenomic RNA may occur in different places in the nucleus. Thus, synthesis of genomes is sensitive to inhibition by amanitin, requires S-HDAg that has been phosphorylated, methylated, and acetylated, and the RNA product is immediately exported to the cytoplasm. Synthesis may occur in the nucleoplasm. In contrast, synthesis of antigenomes is resistant to amanitin, requires L-HDAg but does not require its phosphorylation or acetylation, and the RNA product is retained in the nucleus. Synthesis perhaps occurs in the nucleolus. Approximately 10 times as much genomic RNA is produced as antigenomic RNA.

The genomic RNA is used as a template to produce the mRNA for the HDAg. The mechanism by which this RNA is produced is not fully understood. It may resemble the process for production of antigenomic RNA but occurs in a different place in the nucleus utilizing a different RNA polymerase and different forms of HDAg. Synthesis of mRNA is also

A.   Simultaneous Coinfection with Hepatitis B and Hepatitis δ

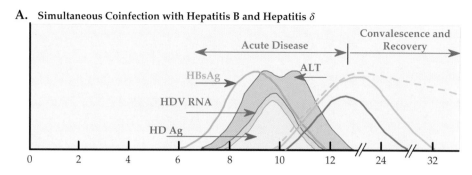

B.   Acute Hepatitis δ after Superinfection of a Chronic Hepatitis B Patient

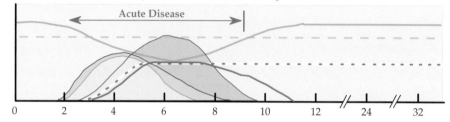

C.   Chronic Hepatitis δ after Superinfection of a Chronic Hepatitis B Patient

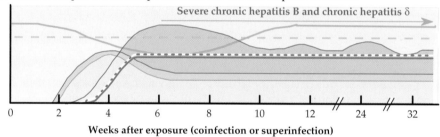

Weeks after exposure (coinfection or superinfection)

Serology

Antigens, RNA, and liver function markers	Antibody levels
ALT level (alanine aminotransferase)	IgM / IgG Anti HBc (anti-hepatitis B core)
HD RNA (hepatitis δ RNA)	
HDAg (hepatitis δ antigen)	IgM / IgG Anti HD (anti-hepatitis δ antigen)
HBsAg (hepatitis B surface antigen)	

FIGURE 9.7   Patterns of anti-hepatitis B core antigen antibody, hepatitis B antigen, hepatitis δ RNA, hepatitis δ antigen, and ALT in patient serum during different types of coinfection with hepatitis δ and B. (A) Simultaneous infection by both types. (B) and (C) Superinfection by hepatitis δ of a patient with chronic hepatitis B infection. Many infections start with acute hepatitis δ as in (B). Some proportion of superinfections progress to chronic hepatitis with elevated liver enzymes and sustained production of hepatitis δ RNA and protein as in (C). Adapted from Fields *et al.* (1996) p. 2825.

sensitive to amanitin, suggesting that RNA Pol II is involved and synthesis may occur in the nucleoplasm. There is a polyadenylation site following the open reading frame (ORF) for HDAg, and cellular enzymes are assumed to cut the pre-mRNA and polyadenylate it similar to what happens with cellular mRNAs. The fact that the mRNA is also capped suggests that the origin of synthesis may differ from that used for RNA replication so that the process resembles cellular production of mRNA rather than the replication of HDV RNA.

HDV Delta Antigen

The mRNA for HDAg is exported to the cytoplasm and translated into a polypeptide of 195 amino acids, referred to as the small δ antigen or S-HDAg. This protein is required for RNA replication. Thus, for example, *in vitro* systems to study the replication of HDV RNA must be supplemented with S-HDAg for replication to occur. S-HDAg is a component of the infecting particle and is therefore present in the infecting

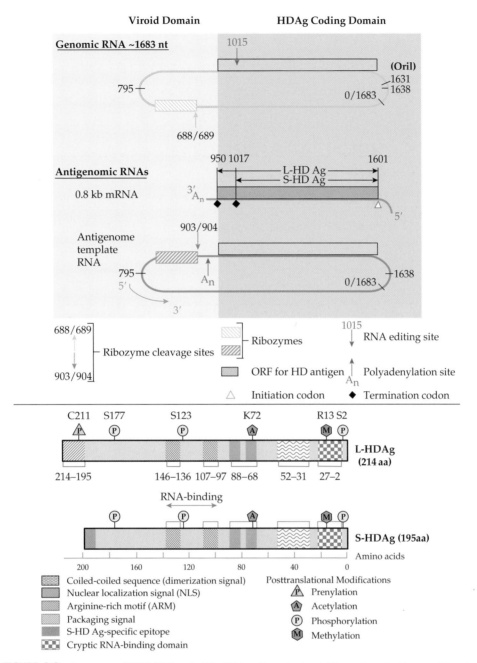

FIGURE 9.8   Structures of HDV RNA and of the HDAg. Above horizontal line: schematic diagram of the structure of HDV RNA. Nucleotides are numbered from the unique *Hind* III site in the cDNA clone of the prototype HDV. Numbering is 5′ to 3′ in the genomic RNA. Nucleotides 795 and 1638 represent the ends of the rodlike structure. Below line: schematic diagram of the structural and functional domains of the hepatitis delta antigen (HDAg). The protein is shown in the same orientation as the mRNA in part (A), with the amino acids numbered from right to left. Other features are described in the key. Adapted from Modahl and Lai (2000), and Lai (2005).

cell when RNA replication first begins. Production of new protein after infection enables RNA replication to accelerate.

A second form of δ antigen is also produced during infection. An RNA-editing event occurs in about one-third of the antigenomic templates, in which the termination codon UAG, at position 196 of the ORF for the δ antigen is changed to UGG, encoding tryptophan (see Fig. 9.8). This change is thought to be effected by deamination, in the antigenome, of the adenosine in the UAG codon to produce inosine. A cellular adenosine deaminase has been described that probably performs this function. Inosine pairs as guanosine, and continued replication of the RNA will lead to the substitution of G for A. This RNA editing site is specific and requires specific sequences within the antigenomic RNA for it to occur.

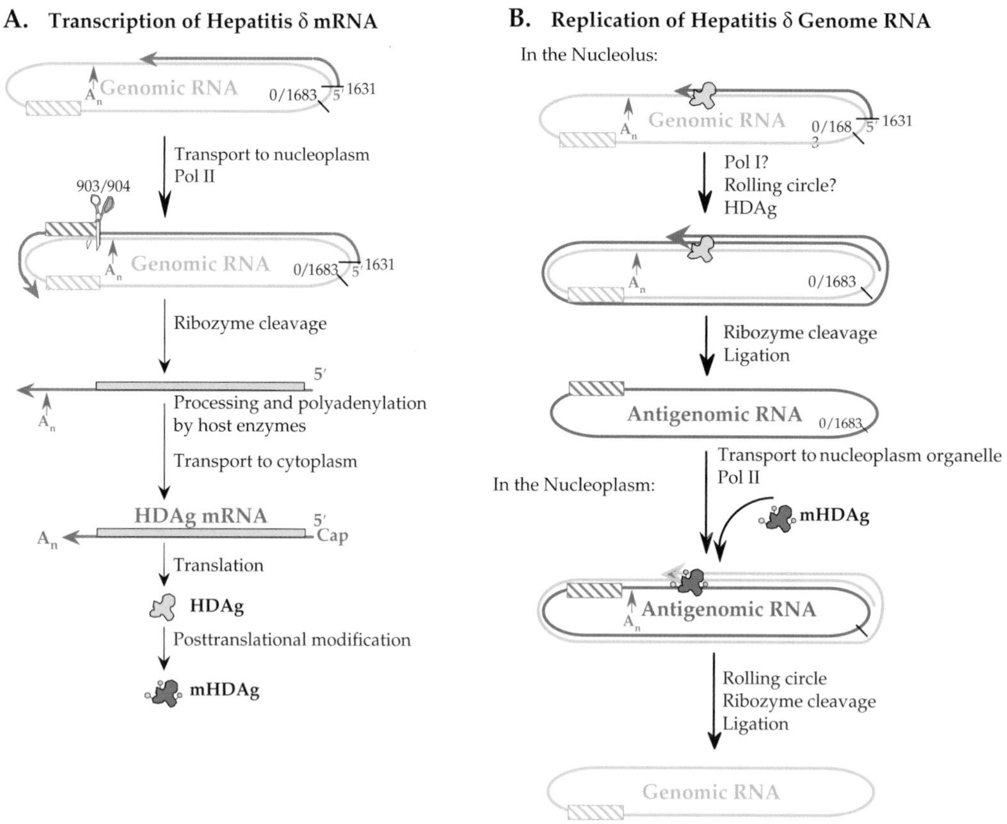

A. Transcription of Hepatitis δ mRNA

B. Replication of Hepatitis δ Genome RNA

FIGURE 9.9   Transcription and replication of hepatitis δ RNA. (A) Transcription of hepatitis δ mRNA. RNA synthesis by Pol II begins about 30 nucleotides upstream of the HDAg ORF (at nt 1631). Plus-strand synthesis proceeds through the ORF. The nascent strand is cleaved by the ribozyme at nt 903/904 and the mRNA is subsequently processed and polyadenylated. HDAg mRNA is translated in the cytoplasm and some HDAg is posttranslationally modified (mHDAg). (B) Replication of hepatitis δ genome RNA. Genome RNA is transported to the nucleolus where it is replicated by Pol I in the presence of HDAg as a rolling circle. After ribozyme cleavage and ligation the antigenome RNA is transported out of the nucleolus with mHDAg and genome-sense RNA is synthesized by Pol II in a second rolling circle, then cleaved and ligated as before. Adapted from Lai (2005).

Change of the termination codon to a tryptophan codon leads to the production of a polypeptide that is 19 residues longer, for a total length of 214 amino acids, referred to as the large δ antigen or L-HDAg. Because editing is required, it is only produced later in the infection cycle. The extent of editing is controlled, perhaps by S-HDAg. Obviously, only genomes that are not edited can give rise to infectious virions. S-HDAg is required for replication, and only nonedited genomes encode it.

HDAg can be phosphorylated on Ser-2, Ser-177, and Ser-123, methylated on Arg-13, and acetylated on Lys-72. These modifications change the activities of the protein as well as its subcellular localization. As stated, S-HDAg is required for RNA replication. L-HDAg can be isoprenylated on a cysteine four residues from the C terminus that is therefore not present in S-HDAg. L-HDAg suppresses RNA replication and leads to a shift from replication of RNA to encapsidation of RNA into progeny virions. It is specifically required for virus assembly, and isoprenylation is required for this activity. A map of functional domains of the L- and S-HDAgs is shown in Fig. 9.8B.

Assembly of Virus

Assembly of HDV virions begins with the formation of a nucleocapsid or core that contains the HDV genome and both the L and S forms of the δ antigen. The core is 19 nm in diameter and matures by budding, using the HBV surface antigens. Budding appears to be the same as for HBV (Chapter 6), and the three surface antigens of HBV form the protein component of the outer envelope surrounding the HDV capsid. Thus, although the RNA of HDV can replicate independently of HBV, assembly of progeny virions requires the simultaneous infection of the cell by HBV to supply the surface glycoproteins needed to produce infectious particles.

HDV is extremely prolific. The serum of an infected individual can contain up to 10^{12} RNA-containing HDV particles per milliliter.

Host Range of HDV

The only known natural hosts for HDV are humans, but HDV can be experimentally transmitted to chimpanzees and to woodchucks. Infection of chimps requires coinfection with HBV, and this provides a useful primate model for the study of the agent. A second model system is furnished by woodchucks. Woodchucks can be chronically infected with woodchuck hepatitis virus (WHV, Chapter 6), a relative of HBV, and WHV can provide helper activity for HDV. Chronically infected animals can be infected with HDV, and in this case the surface properties of the HDV virion are determined by the helper WHV rather than by HBV.

PRIONS AND PRION DISEASES

Transmissible spongiform encephalopathies (TSEs), now often referred to as prion diseases, are progressive, fatal diseases of humans and of other animals. A listing of TSEs is given in Table 9.2. TSEs of humans include kuru, Creutzfeldt-Jakob disease (CJD), Gerstmann-Sträussler-Scheinker syndrome (GSS), and fatal familial insomnia (FFI). TSEs are characterized by neuronal loss that appears as a spongiform degeneration in sections of brain tissue, often accompanied by amyloid plaques or fibrils. The most prominent symptoms of disease are usually dementia (loss of intellectual abilities) or ataxia (loss of muscle control during voluntary movement) that results from the progressive loss of brain function. The disease always has a fatal outcome. In humans, death usually occurs within 6 months to 1 year of the first appearance of symptoms.

TSEs can be contracted by inoculation with or ingestion of brain tissue or other tissues containing the infectious agent, and thus they can be transmitted as an infectious disease. Kuru first came to light as an infectious disease and many cases of CJD in humans have been acquired by infection. However, TSEs can also occur as sporadic diseases for which there is no evidence of infection by an outside agent. In humans, CJD occurs sporadically with a frequency of about 10^{-6}. Finally, TSEs can appear as inherited diseases. GSS, most FFI, and some cases of CJD occur as dominant inherited diseases, associated with mutations in the gene for the prion protein. Inheritance of the mutant gene dramatically increases the probability of developing TSE, such that the probability of acquiring the disease over a lifetime may approach 100%. In most cases of sporadic or inherited TSE, the disease is transmissible as an infectious disease once it occurs.

There is now considerable evidence that all TSEs are related and result from defects in the metabolism of the prion protein. The pattern of symptoms associated with a particular TSE may vary, however, depending in part on how the disease was contracted; on the source of the infecting agent; and on the nature of mutations in the prion protein. Thus, although the prion protein is central to disease in every case, symptomology can differ, in part because the particular area of the brain most affected can vary.

There is no immune response associated with any TSE. No antibodies are formed and no inflammation marked by the infiltration of mononuclear cells is present. As stated, it is relentlessly progressive and always results in death.

Kuru

Kuru was a disease of epidemic proportions among the Fore people of New Guinea that reached a prevalence of about 1% of the population. The disease was characterized by progressive ataxia that led to total incapacitation and death, normally in 12–18 months after the appearance of symptoms in adults or 3–12 months in children. The demonstration that kuru was transmissible to primates by inoculation of brain tissue from people dying of the disease was the first demonstration of the transmissibility of a TSE in humans. These transmission studies and other studies of kuru resulted in a Nobel prize for Carleton Gajdusek in 1976, the first of two prizes for work with TSEs (Chapter 1).

Kuru is believed to have been spread among the Fore people by cannibalism in which the bodies of relatives who had died were eaten in a ritualistic feast. Women and children were more often affected than men, and it is thought this was because they prepared the body for the feast and they ate the brains of deceased relatives. Men were less often affected, it is conjectured, because they ate primarily other body parts. It has been postulated that the epidemic began when a member of the tribe died of a sporadic case of CJD, and the disease was then spread to others through cannibalism. Through the efforts of missionaries, cannibalism ceased many years ago and the disease has become progressively rarer. Now only older people who contracted the infectious agent during the time of cannibalism continue to develop the illness. From studies of the continuing development of kuru in older Fore people, it is known that the disease can appear as long as 40 years after the event that resulted in infection with the agent.

Sporadic and Iatrogenic CJD

CJD in man is usually a sporadic illness that occurs with a frequency of about 10^{-6} that is uniform around the world. However, once the disease has arisen it is transmissible by inoculation of infected material into experimental animals such as primates and transgenic mice. CJD has also been transmitted iatrogenically to humans. Iatrogenic cases have occurred in recipients of pituitary-derived human growth hormone obtained from cadavers, some of whom died of CJD; in recipients of homographs of dura mater derived from cadavers; through implantation into epilepsy patients of contaminated silver electrodes that had been incompletely sterilized; and through corneal transplants. The infectious agents of TSEs are extremely difficult to inactivate and require extraordinary sterilization techniques in order to destroy their infectivity. Better methods of sterilization have been introduced, and human growth hormone is now produced in bacteria from recombinant DNA plasmids, so that the iatrogenic spread of CJD has been greatly reduced.

Sporadic FFI has also been described. No case of sporadic GSS is known, however.

Inherited Forms of Human TSE

About 5% of CJD cases arise in a familial, autosomal dominant fashion and are associated with mutations in the gene for the prion protein. GSS and most cases of FFI are also inherited forms of TSE associated with mutations in the prion protein. Many of the responsible mutations are illustrated in Fig. 9.10, in which a schematic diagram of the human prion protein is presented.

A dozen single amino acid substitutions in the prion protein have been found to be associated with inherited CJD, GSS, or FFI. Additionally, an element normally containing five repeats of a 24-nucleotide sequence (encoding an 8-amino-acid repeat, P-Q/H-G-G-G-W-C-Q) has been found to contain one to nine extra repeats, probably originating from unequal crossing over, in some cases of inherited CJD or GSS. These three diseases are distinguished on the basis of symptomology, which is overlapping. CJD is characterized by ataxia, dementia, and behavioral disturbances. GSS is usually characterized by cerebellar disorders accompanied by a decline in cognitive ability. FFI, as its name suggests, is characterized by abnormal sleep patterns, including intractable insomnia.

The penetrance of the different mutations varies but is usually very high. For example, CJD caused by the change from glutamic acid-200 to lysine (E200K), when residue 129 is homozygous for methionine, has been estimated to have a penetrance of 0.45 by age 60 and a penetrance of more than 0.96 above age 80. Thus, a person with this mutation is almost certain to develop CJD if he or she lives long enough.

Attempts have been made in many cases of inherited TSEs to transmit the disease to subhuman primates or to mice. Transmission has been achieved in most cases tested. Thus once the disease arises, it is transmissible to animals that do not contain the mutation.

In addition to mutations associated with inherited TSEs, several polymorphisms in the prion protein are known that are not associated with disease (Fig. 9.10). The polymorphism at residue 129 is of particular importance. Homozygosity at this position affects the probability of contracting TSE.

TSEs in Other Animals

Naturally occurring TSEs of a number of other mammals are known. The oldest known TSE, in fact, is that of sheep, and is called scrapie. Scrapie has been known for more than 200

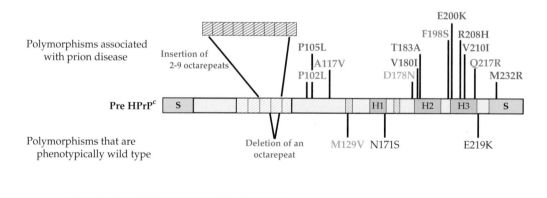

FIGURE 9.10   Mutations found in the human prion protein gene. Polymorphisms that are phenotypically wild type are shown below the schematic of the gene; mutations that segregate with inherited prion diseases are shown above the gene. GSS, FFI, and CJD are defined in Table 9.3. Adapted from Prusiner (1998) and Riek *et al.* (1996).

TABLE 9.3 Prion Diseases

Disease (abbreviation)	Natural host	Experimental hosts	Cause of disease
Scrapie	Sheep and goats	Mice, hamsters, rats	Infection in genetically susceptible sheep
Transmissible mink encephalopathy (TME)	Mink	Hamsters, ferrets	Infection with prions from sheep or cattle
Chronic wasting disease	Mule deer, white tail deer, and elk	Ferrets, mice	Unknown
Bovine spongiform encephalopathy (BSE)	Cattle	Mice	Infection with prion-contaminated meat and bonemeal
Feline spongiform encephalopathy (FSE)	Cats	Mice	Infection with prion-contaminated beef
Exotic ungulate encephalopathy (EUE)	Nyala, oryx, and greater kudu	Mice	Infection with prion-contaminated meat and bonemeal
Kuru	Humans	Primates, mice	Infection through ritual cannibalism
Creutzfeldt-Jakob disease	Humans	Primates, mice	
iCJD (iatrogenic)	Humans		Infection from prion-contaminated human growth hormone, dura mater grafts, etc.
sCJD (sporadic)	Humans		Somatic mutation or spontaneous conversion of PrP^c to PrP^{Sc}
nvCJD (new variant)	Humans		Ingestion of bovine prions?
fCJD (familial)	Humans		Germ line mutation in PrP gene
Gerstmann-Sträussler-Scheinker syndrome (GSS)	Humans		Germ line mutation in PrP gene
Fatal familia insomnia (FFI)	Humans	Primates, mice	Germ line mutation in PrP gene (D178N, M129)
Fatal sporadic insomnia (FSI)	Humans		Somatic mutation or spontaneous conversion of PrP^c to PrP^{Sc}

Source: Adapted from Granoff and Webster (1999), p. 1389.

years and is widely distributed in Europe, Asia, and America. The name comes from the tendency of animals to rub themselves against upright posts, apparently because of intense itching that arises from this neurological disease. Scrapie appears to be transmitted horizontally in sheep flocks, but the mechanism by which it is transmitted is not understood. The infectious agent is very resistant to inactivation and may persist in pastures for a long time. It may be ingested, but other mechanisms for persistence have also been proposed.

Scrapie appears to have been transmitted to a number of other mammals. In some cases the spread has been to animals that share pasturage with infected sheep, such as white-tailed deer, mule deer, and elk (where the disease is called chronic wasting disease). In these cases, it is thought that infection occurs by the same mechanisms that maintain scrapie in sheep flocks. In other cases, spread has occurred via food derived from infected sheep that was fed to mink (transmissible mink encephalopathy), domestic cats or exotic cats in zoos (feline spongiform encephalopathy), ungulates in zoos (exotic ungulate encephalopathy), and perhaps to cattle. However, there is no evidence that scrapie has ever spread to humans, despite the long history of human consumption of scrapie-infected sheep.

Bovine spongiform encephalopathy (BSE), also called mad cow disease, is a TSE of cattle that was recently an epidemic

in Britain. The epidemic was maintained by feeding to cattle the processed offal from cattle and other animals, that is, by a form of animal cannibalism as happened with kuru in humans. Although this practice was of long standing, it did not cause trouble until recently, when a change in the rendering process was introduced. It is believed that this change allowed the BSE agent to survive the processing steps, whereas formerly it had been killed during rendering. The result was an epidemic of BSE that spread across all of Britain (Fig. 9.11). At the height of the epidemic, there were more than 35,000 cases of BSE per year in Britain (Fig. 9.12).

The original source of the BSE that led to the epidemic is uncertain. It may have arisen from a spontaneous case of BSE, similar to spontaneous CJD in humans, although spontaneous BSE in cattle appears to be rare or nonexistent. A second possibility is that it may have arisen from infection with scrapie from infected sheep, since sheep offal was included in the rendered offal.

Once the epidemic of BSE in cattle in Britain was recognized, legislation was introduced that banned the feeding of any ruminant-derived protein to ruminants. Also introduced was legislation to make BSE a notifiable disease and to prohibit the use of brain, spinal cord, and certain other offals from any bovine animal in human food. These initial bans were subsequently enlarged and extended in various ways

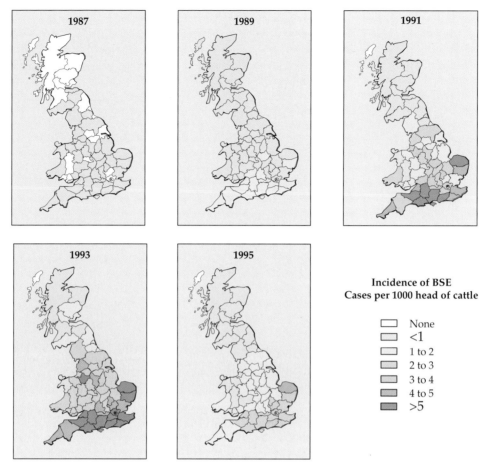

FIGURE 9.11   Spread of the BSE epidemic in the British Isles. Geographic distribution of the incidence of BSE per head of cattle by county from 1989 to 1995. Adapted from Anderson *et al.* (1996).

(see Fig. 9.12 and its legend). The ruminant feed ban resulted in the waning of the epidemic in cattle in Britain, but new cases continued to arise, whether the result of a long latent period of the infectious agent, or from contaminated ruminant feed that continued to enter the system, or from alternative modes of transmission, such as passing the infection from mother to calf. With the recognition of new variant CJD in people, the issue of eradicating BSE became more pressing and culling of cattle was undertaken. This culling of herds containing BSE-infected cattle together with the subsequent culling of herds infected with foot-and-mouth disease virus, as well as the continued enforcement of the ruminant feed bans, have resulted in a marked reduction in the incidence of BSE, although not to its total eradication (Fig. 9.12).

New Variant CJD in Humans

At the beginning of the BSE epidemic, public health officials in Britain had little fear that the epidemic might pose a threat to human health. There is a species barrier to the transmission of the TSE from any particular animal to another animal. Even in cases where transmission does occur in experimental systems, there is a requirement for an adaptation event before the agent can be readily transmitted. Humans were thought not to be sensitive to animal TSE agents because of this species barrier. In particular, no evidence for the transmission of scrapie to humans has ever been found despite the fact that people all over the world, but especially in Britain, have eaten sheep infected with scrapie for 2 centuries.

In 1995 and 1996, however, 12 cases of a variant form of human CJD occurred in Britain. These new variant CJD cases (nvCJD) were characterized by an unusually early age of onset, with some cases in their teens, and by a different symptomology. A comparison of the ages at which people in Britain contracted sporadic CJD during the last 25 years with that of the ages of the first 21 cases of nvCJD is shown in Fig. 9.13A. Sporadic CJD is primarily a disease of people in their 50s, 60s, and 70s, with a peak of occurrence in the early 60s. Cases in people under 40 are rare. Variant CJD to date has been a disease of young people, primarily people in their teens, 20s, and early 30s. Symptomology also differs. Sporadic CJD is characterized by dementia as an early symptom, whereas variant CJD

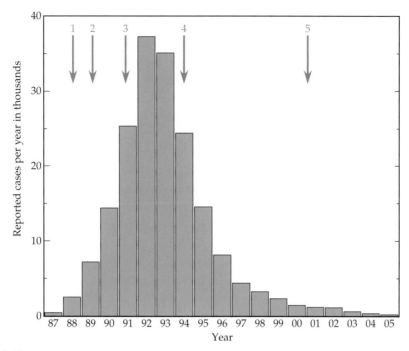

FIGURE 9.12 Confirmed cases of BSE (bovine spongiform encephalopathy) in British cattle per year between 1987 and 2005. Arrows indicate (**1**) ruminant feed ban (1988); (**2**) specified offals ban, to prevent offals proteins from entering the human food chain (1989); (**3**) extended specified offals ban [prohibiting feeding of offal proteins to pigs and poultry (1991)]; and (**4**) offals ban futher extended to include offals from bovines < 6 months old (1994). (**5**) Prohibit the use of any animal protein (excluding milk and fish meal) from feed for any farmed animal species (2001). Note that data after 2002 could be biased by the large number of cattle slaughtered during the foot-and-mouth-disease virus (FMDV) epidemic in 2001, which must have contained some infected animals. Adapted from Anderson *et al.* (1996), the "2004 Institute of Food Science and Technology Information Statement on BSE," and data from http://www.oie.int/fr/info/fr_esbru.htm.

is characterized by psychiatric symptoms, usually depression, and the patient is often first seen by a psychiatrist. Third, time to death averages somewhat longer in variant CJD than in sporadic CJD. The number of cases of nvCJD rose for several years, plateaued in the year 2000, and then declined, as shown in Fig. 9.13B. Also shown in this figure for comparison are the number of cases of sporadic CJD each year in Britain; the rise in the number of cases of sporadic CJD reported over this time frame is probably due to increased recognition of CJD disease, catalyzed in part by the nvCJD epidemic. Through 2005 there had been a total of about 150 cases of nvCJD.

There is now a considerable body of evidence that nvCJD is caused by infection with BSE and results from eating BSE-contaminated meat. For one, the BSE prion and the human nvCJD prion are closely related and differ from other CJD prions (see later). For another, the nvCJD epidemic closely parallels the BSE epidemic with an 8-year lag. It appears, therefore, that the incubation period of nvCJD, at least to date, averages about 8 years. It is not clear how many cases may ultimately arise. As described

earlier, kuru has a long latent period, with disease developing as long as 40 years after infection. The decline in the incidence of nvCJD following control of BSE, however, suggests that the dynamics of nvCJD disease are different and that only small numbers of disease will continue to arise, perhaps the result of the species barrier that exists for the transmission of BSE to humans. Further, it is not understood why the young are so much more sensitive to nvCJD than are the old. A sensationalized and gripping account of kuru, CJD, and BSE is found in the book *Deadly Feasts* by Richard Rhodes.

Prion Protein

The nature of the infectious agents responsible for scrapie and other TSEs has been controversial, in part because the study of these agents has presented enormous technical difficulties. The kuru agent was shown to be transmissible to other primates many years ago, but the incubation period is very long in these animals (more than 10 years in some cases) and they are expen-

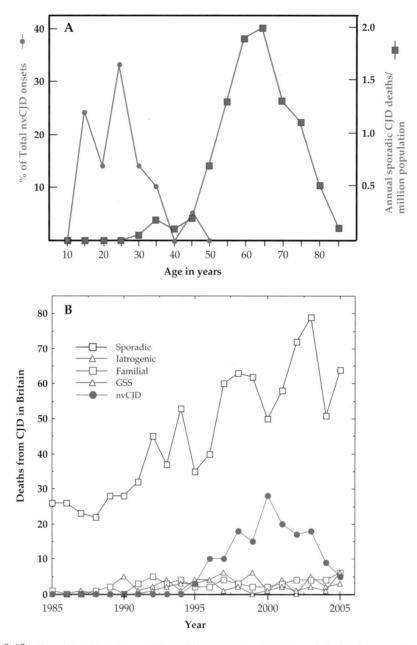

FIGURE 9.13 Creutzfeldt-Jakob disease (CJD) in Britain. (A) Age distribution of the first 21 cases of new variant CJD (nvCJD) in 1995 and 1996, compared with the annual age-specific death rates for sporadic CJD (573 cases) in Britain between 1970 and 1994. The scales have been chosen to optimize comparison of the age distributions. Adapted from Nathanson (1999) Figure 7 on p. 449. (B) Cases of CJD of various etiologies and Gerstmann-Straüssler-Scheinker syndrome (GSS) in Britain from 1985 to 2005. The rise in the number of cases of sporadic CJD reported is probably due to increased recognition of the disease. Data are Monthly CJD Statistics, from the Department of Health of the United Kingdom.

sive to maintain, which limited early progress in the study of the molecular biology of the agents. The subsequent discovery that many TSEs could be transmitted to mice and hamsters, in which the incubation period was much shorter, as short as 60 days in some instances, speeded up progress. Transgenic mice, in particular,

have been very useful because the genetic background can be controlled. However, such studies remain slow and tedious because an infectivity assay often takes more than 1 year.

Studies in mice and other animals, as well as the finding that mutations in the prion protein are associated with inherited TSEs in humans, have made clear that the prion protein, abbreviated PrP, is intimately involved in the transmission of TSE and in the disease process. The normal cellular protein is referred to as PrP^c. The structure in solution of the C-terminal half of the mouse version of this protein (residues 121–231) is illustrated in Fig. 9.14. The protein has a high content of α helix. In this half of the protein, there are three α-helical domains of 11, 15, and 18 residues, and only a short (four residues in each strand) two-stranded antiparallel β sheet. The N-terminal 98 residues of this protein form a flexible random coil in solution, as determined by nuclear magnetic resonance imaging.

The prion protein is synthesized as a larger precursor of 254 amino acids that contains both N-terminal and C-terminal extensions (Fig. 9.15). The N-terminal extension is a signal sequence that leads to the translocation of PrP into the lumen of the endoplasmic reticulum. It is removed by signal peptidase, as are most N-terminal signal sequences. The C-terminal extension is removed by another cellular protease and the protein is attached to a phosphoinositol glycolipid anchor that anchors the protein in the membrane. The protein is N-glycosylated on two sites. The processed protein is transported to the plasma membrane and transiently

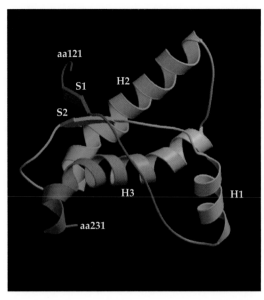

FIGURE 9.14 The structure of the prion protein. The structure of residues 121–231 of the mouse prion protein in solution as determined by NMR is shown. The protein is color coded from blue at residue 121 to red at residue 231, with β sheets shown as flat arrows and α helices as coils. The second and third helices are linked with a disulfide bond (not shown). (Compare with Figures 9.10 and 9.15.) Adapted from Riek *et al.* (1996).

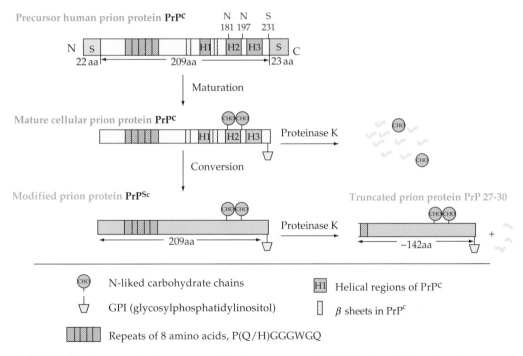

FIGURE 9.15 Isoforms of the human prion protein. The precursor protein is 254 amino acids long. Maturation involves removal of the N-terminal signal sequence and the C-terminal 23 amino acids (two boxes marked S) N-linked glycosylation at Asn181 and Asn197, and linkage of GPI near the new C terminus. After exposure to scrapie prions, the protein is converted to PrP^Sc, which is partially resistant to proteinase K. This conversion involves loss of some helical regions (H's) in the cellular form, and formation of new β sheets. Adapted from Weissmann (1996), Riek *et al.* (1996), and Prusiner (1998).

displayed on the surface of the cell with a half-life of about 5 hours. It is then recycled into endosomal compartments and eventually into lysosomes, where it is degraded.

The function of PrP^c is unknown. It is expressed by a number of different cells, including neurons, hematopoietic stem cells, and follicular dendritic cells. Many knockout mice that lack the gene for PrP have been constructed and most appear normal. The conservation of this protein indicates that it must perform some important function, but apparently its functions can be replaced by other proteins through the redundancy of many mammalian pathways. Recent studies have suggested that the protein is important for the renewal of hematopoietic stem cells and for the development of neurons, and that when an organism is stressed this developmental function becomes critical.

The brains of most humans or experimental animals exhibiting TSEs contain a conformational variant of PrP^c called PrP^{Sc} (Sc for scrapie) or PrP^{res} (res for resistant to protease). PrP^{Sc} is found in aggregates that are largely resistant to digestion with proteases. Treatment of PrP^{Sc} with proteases and subsequent disaggregation of the proteolyzed PrP^{Sc} give rise to a molecule that is truncated by about 80 amino acids at its amino terminus (Fig. 9.15). In contrast, PrP^c is completely destroyed by such protease treatment, and the normal PrP is also referred to as PrP^{sen} (for sensitive to protease). Circular dichroism and infrared spectroscopy indicate that PrP^{Sc} has a much higher content of β sheet than does PrP^c, 43 versus 3%, and a lower content of α helix, 30 versus 42%, suggesting a profound conformational rearrangement of the prion protein in the process of conversion from PrP^c to PrP^{Sc}.

Studies with Mice

Transgenic mice have been useful in the study of TSEs. Mice have been made that lack the gene for the prion protein, or that express wild-type or mutant prion proteins at levels from less than normal to several times normal. Most mice that make no PrP^c are normal, as described. However, such mice are resistant to scrapie infection. They do not become ill, and no infectious material is produced in the brain after inoculation of scrapie. In contrast, mice that overexpress PrP are more sensitive to infection with scrapie. The incubation period is shorter, and the animals die more quickly after inoculation with scrapie.

Thus, the presence of PrP is essential for the development of TSE in mice. It has also been found that individual neurons must be able to produce PrP if they are to be sensitive to scrapie-induced death. Neurografts from a donor mouse that expresses PrP have been implanted into mice that lack the PrP gene. Upon inoculation of scrapie into the brain, the neurons in the graft develop a typical scrapie-induced disease pathology. However, neurons outside the graft remain healthy.

Ex Vivo and in Vitro Studies

Cell lines have been established that are persistently infected with scrapie. These cells continuously produce PrP^{Sc}, which allows biochemical studies to be performed over a shorter time frame. The infected cells produce infectious material that causes scrapie when inoculated into mice. Of great interest has been the development of an *in vitro* system for the conversion of PrP^c to PrP^{Sc}. In this system, radioactive PrP^c is mixed with unlabeled PrP^{Sc}, and the conversion of the labeled PrP^c to PrP^{Sc} is followed by its becoming resistant to protease. These studies make clear that PrP^c can be converted to PrP^{Sc} by exposure to PrP^{Sc} in a process that does not require the activities of intact cells. However, so much infectivity is associated with the PrP^{Sc} added to the reaction mixture that no increase in infectivity can be demonstrated. Thus, these studies do not address the question of the nature of the infectious agent. These studies have also been useful in the study of the species barrier, which can be quantitatively examined in such reactions.

Protein-Only Hypothesis

It is clear that PrP is important in the development of TSEs. There are two unresolved questions about PrP and the disease process, however. First, is the infectious agent that leads to TSE PrP^{Sc} itself or is it another entity, such as a virus? Second, does PrP^{Sc}, or some other modified form of PrP, cause the symptoms of the disease, or is it simply a side effect of the disease process?

Preparations of the infectious agent purified from scrapie-infected mouse brain consist largely of PrP^{Sc}. There is very little nucleic acid in infectious preparations of PrP^{Sc}. In particular, there is no homogeneous DNA or RNA molecule that might arise from a virus, for example. This has led to the hypothesis that PrP^{Sc} is itself the infectious agent. In this model, "infectious" PrP^{Sc} induces PrP^c to assume the PrP^{Sc} conformation, and the accumulation of PrP^{Sc} in the brain leads to the pathology associated with TSEs. Most of the experimental data are compatible with such a model. PrP^{Sc} does induce PrP^c to assume the PrP^{Sc} conformation, as described earlier. Mutations in PrP^c could make it easier for the protein to assume the PrP^{Sc} conformation, compatible with the observation that some mutations result in inheritance of TSEs. The species barrier could result from lowered interaction affinities between proteins of different sequence. However, it has not been possible to prove this hypothesis. PrP^{Sc} preparations have a very low specific infectivity, with at least 10^5 molecules of PrP^{Sc} required for infection. Thus it remains possible that contaminants in the preparation might be required for infectivity. It has not been possible to demonstrate an increase in infectivity associated with the conversion of PrP^c to PrP^{Sc}, as described before, which would provide solid evidence that PrP^{Sc} is infectious.

In addition to the inability to prove the protein-only hypothesis, which could be due to the technical difficulties associated with this system, there are specific conceptual difficulties with PrP^{Sc} as the infectious agent. One of the major criticisms of the protein-only hypothesis is the fact that as many as 20 different strains of scrapie exist as assayed in mice. These strains of scrapie differ in properties such as the length of the incubation period following infection before disease becomes apparent, the areas of the brain affected, and the symptoms of the disease, but these properties do not vary within a strain. Such properties are expected for an infectious entity with a nucleic acid genome, but are difficult to reconcile with the properties of an infectious protein. If the protein-only hypothesis is true, these differences in properties could only result from differences in the conformation of PrP^{Sc} in the different strains. How is it that a single, fairly small protein can take up so many different conformations and that each can induce the production of more protein having the same conformation?

Supporters of the protein-only hypothesis suggest that a limited number of conformational states of the prion protein would be sufficient to explain the multiple strains of scrapie that exist. They point to experimental data that show that at least two demonstrably different conformational states of the prion proteins of two different mammals exist that "breed true." Two different strains of transmissible mink encephalopathy that produced different disease characteristics in mink were passaged in hamsters. The PrP^{Sc} from the two strains, isolated from infected brain, are differently truncated at the amino terminus on treatment with proteases *in vitro*. Thus the conformations of the PrP^{Sc} in these two strains, both derived from hamster PrP, must be different. Furthermore, this difference can be reproduced in an *in vitro* reaction in which PrP^{c} is mixed with the two different types of PrP^{Sc}. Each type of PrP^{Sc} induces PrP^{c} to assume its own distinct conformation, as shown by the protease resistant fragment that is produced from the PrP^{c} on it conversion to PrP^{Sc}.

In a second example of demonstrably different prion conformations, human prions isolated from two different cases of TSE, one FFI and the second CJD, were found to be differently truncated after protease treatment. Passage of these TSEs in transgenic mice that expressed a chimeric mouse–human prion protein gave rise to prions in infected brain that reproduced the differences in truncation. Thus, these conformational differences breed true when passed in mice.

These studies demonstrate that PrP^{Sc} can exist in at least two conformational forms, that the different conformational forms can produce different symptoms, and that the different forms are capable of propagation by inducing PrP^{c} to take up their own particular conformation. Thus, the experimental data are consistent with the protein-only hypothesis, although it has not been proven conclusively and many still doubt its validity. The hypothesis received a vote of confidence when its most outspoken and passionate advocate, Dr. Stanley Prusiner, was awarded the 1997 Nobel Prize for his "discovery of prions."

Transport of Infectivity to the Brain

Related to the conceptual problem of an infectious protein is how it might be transported to the brain after ingestion with food. This problem has been addressed in studies that ascertain in which tissues PrP^{Sc} is present following ingestion of PrP^{Sc}, and studies with transgenic mice that express PrP only in certain tissues. These various studies are compatible with a model in which infectivity is transported via axons following direct neuroinvasion of peripheral nerves. In the case of infection with only low doses of infectious material, amplification in follicular dendritic cells in lymphoid tissue may be required before neuroinvasion occurs. Thus, in terms of the protein-only model, PrP^{Sc} might induce the conversion of PrP^{c} to PrP^{Sc} in cells in Peyer's patches, which then spreads via lymphatic tissue to peripheral nerves by sequential conversion of PrP.

Formation of the PrP^{Sc} Seed

If PrP^{Sc} can transmit the disease to a new susceptible host, how is it formed in the first place? Current models propose that the conformational change resulting in PrP^{Sc} occurs rarely, but that once PrP^{Sc} is formed, it acts as a seed to induce the formation of more PrP^{Sc}. Two models to explain the conversion of PrP^{c} to PrP^{Sc} by PrP^{Sc} have been proposed. In one, PrP^{Sc} (which may be present in an aggregate) and PrP^{c} form a complex, and the PrP^{Sc} induces the conformational change in PrP^{c}. In the second model, PrP^{c} undergoes spontaneous transitions to different conformational states that are short lived and revert quickly to the native PrP^{c} conformational state. These conformational variants, however, can be locked into place by interaction with PrP^{Sc}. In either case, the altered PrP joins the aggregated PrP^{Sc} to form a larger aggregate. Since the aggregated PrP^{Sc} is insoluble, the reaction is essentially irreversible. Such a process could also explain the species barrier. The PrP proteins of different animals differ slightly in sequence. PrP^{c} that is identical in sequence to the PrP^{Sc} seed could interact with such a seed more readily than with a PrP^{Sc} seed that differs in sequence.

The protein-only hypothesis still requires a seed of PrP^{Sc} to begin the reaction. One possibility is that it can form spontaneously with a very low probability. Perhaps spontaneous changes in the conformation of PrP^{c} to the PrP^{Sc} conformation might be fixed if this change occurred simultaneously in a number of adjacent or interacting molecules. The effect of mutations in PrP^{c} might be to increase the probability of change to the PrP^{Sc} conformation, with the result that disease occurs more frequently. Such a model is compatible with data for human TSEs, where sporadic CJD occurs, albeit infrequently. However, sporadic disease has not been seen in shorter-lived animals. No sporadic BSE has been described, and in countries where scrapie in sheep has been eradicated, such as New Zealand and Australia, no recurrence of disease has been observed.

Does PrP^Sc Cause the Disease?

If PrP^Sc is responsible for the pathology of TSE disease, and not simply a by-product of disease, the mechanism by which it causes disease is uncertain. An early model suggested that PrP^Sc itself is neurotoxic. However, it has been shown that a neuron must be able to express PrP^c before it

can be killed by exposure to PrP^Sc. Thus simple neurotoxicity of PrP^Sc is not the cause of neuronal death. However, it is possible that conversion of PrP^c to PrP^Sc at the surface of the cell, which is known to occur, followed by accumulation of PrP^Sc in lysosomes as the neuron attempts to recycle it, could be toxic. In this model, it is the resistance of PrP^Sc to proteases in the lysosome that results in toxicity.

Recent findings have suggested another possibility. PrP can be expressed as a membrane-spanning protein as well as a protein anchored by a glycolipid anchor (Fig. 9.16). One membrane-spanning form, called ^CtmPrP, has its C terminus

A. Conformations of the human prion protein translated *in vitro*

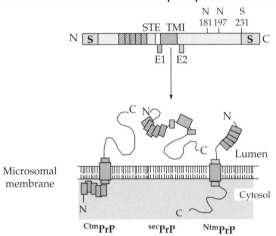

B. Maturation of ^secPrP in cells

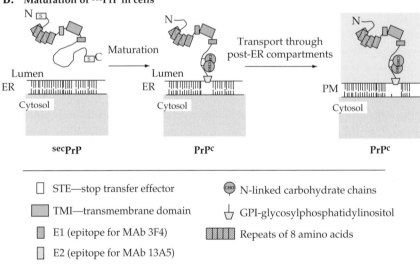

FIGURE 9.16   Postulated topology of PrP proteins in membranes. (A) Topology of PrP proteins in membranes after translation in a cell-free system supplemented with microsomes. The topology was determined by a combination of protease digestion from the cytosolic compartment and identification of the domains protected within the lumen using the two MAbs, 3F4 and 13A5. Mutations have been shown to affect the ratio of the three forms shown, and greater concentrations of ^CtmPrP are associated with neurodegenerative disease in mice. (B) Model for maturation and association with membranes of ^SecPrP in cells. ER, endoplasmic reticulum; PM, plasma membrane. Adapted from Hegde *et al.* (1998).

outside the cell and the N terminus inside, with a transmembrane domain near the middle of the molecule. Preliminary data suggest that this form of PrP is neurotoxic: ^{Ctm}PrP has been found in brains of animals, including humans, suffering from TSE but not in normal brains. This has led to a model in which ^{Ctm}PrP is regularly produced at some frequency, but the normal cell has a mechanism to eliminate it. Overproduction of ^{Ctm}PrP, either by mutation or by a failure to eliminate it, leads to the symptoms of TSE. In this model, production of PrP^{Sc} might somehow result in the accumulation of ^{Ctm}PrP, perhaps by overwhelming the ability of the cell to eliminate it.

PRIONS OF YEAST

Prions, defined as agents that possess two (or more) conformational forms, a soluble "normal" form and an aggregated form that can induce the conversion of the normal form to more of itself, have also been found in fungi. Two prions have been found in yeast (*Saccharomyces cerevisiae*) and a third in *Podospora* spp. The yeast prions have the characteristics of disease but the *Podospora* prion performs a normal cell function (controlling heterokaryon compatibility). The yeast prions are called [URE3], which affects nitrogen catabolism, and [PSI], which affects the termination of polypeptide chains during translation. A diagram of these proteins is shown in Fig. 9.17. [PSI] is a prion form of Sup35p, which is a translation release factor.

In the [PSI] state, Sup35p assumes an altered conformation and aggregates, like PrP^{Sc}. The [PSI] state is dominant and can be transmitted to other yeast cells by transfer of cytoplasm containing [PSI]. Thus, the prion state induces the normal cell protein to assume the prion state, as with the model for PrP^{Sc}. The effect of the [PSI] state on the cell is to render Sup35p nonfunctional, and thus has the same effect as deletion of the gene encoding the protein. Loss of Sup35p activity leads to increased readthrough of stop codons during translation, and renders nonsense suppressor tRNAs much more active. [URE3] is the prion state of Ure2p, a protein involved in nitrogen catabolism. Like [PSI], [URE3] is an aggregated form of a conformational variant of Ure2p, and is dominant and transmissible. Loss of Ure2p by the cell affects the metabolism of nitrogen. Normal cells can assume the prion state with a low frequency, but once assumed the prion state is retained. Cells in the prion state can be cured by certain treatments that break up the protein aggregates and cause the protein to assume a normal conformation. Studies of yeast prions have shown that Sup35p produced in bacterial cells can be converted to [PSI] *in vitro* by introduction of a small seed of [PSI]. The [PSI] produced *in vitro* can be used in turn to convert more Sup35p to [PSI]. The [PSI] produced *in vitro* is infectious—when introduced into yeast it induces the assumption of the prion state. Thus, these studies clearly show that yeast prions are infectious and that only protein is required for infectivity, providing further support for the protein-only hypothesis of mammalian TSEs.

Yeast Prion Protein Ure2p

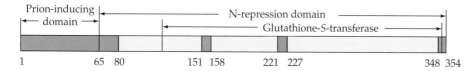

Yeast Prion Protein Sup35p

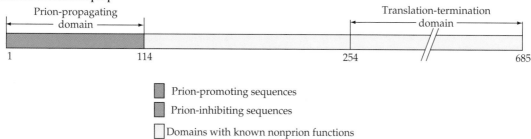

FIGURE 9.17   Comparison of two yeast prion proteins. The prion domains (red) of Ure2p and Sup35p are rich in Asn and Gln residues, which are important for prion generation and propagation. Adapted from Figure 10 of Wickner *et al.* (1999).

FURTHER READING

Defective Interfering Particles

Roux, L. (1999). Defective interfering viruses. In *Encyclopedia of Virology* (A. Granoff and R. G. Webster, Eds.), San Diego, Academic Press, Vol. 1, pp. 371–375.

Viroids

Flores, R., Di Serio, F., and Hernández, C. (1997). Viroids: the noncoding genomes. *Semin. Virol.* **8**: 65–73.

Pelchat, M., Lévesque, D., Ouellet, J., *et al.* (2000). Sequencing of peach latent mosaic viroid variants from nine North American peach cultivars shows that this RNA folds into a complex secondary structure. *Virology* **271**: 37–45.

Hepatitis Delta

Casey, J. L. (2002). RNA editing in hepatitis delta virus genotype III requires a branched double-hairpin RNA structure. *J. Virol.* **76**: 7385–7397.

Macnaughton, T. B., Shi, S. T., Modahl, L. E., and Lai, M. M. C. (2002). Rolling circle replication of hepatitis delta virus RNA is carried out by two different cellular RNA polymerases. *J. Virol.* **76**: 3920–3927.

Modahl, L. E., and Lai, M. M. C. (2000). Hepatitis delta virus: the molecular basis of laboratory diagnosis. *Crit. Rev. Clin. Lab. Sci.* **37**: 45–92.

Polson, A. G., Ley, H. J. III, Bass, B. L., *et al.* (1998). Hepatitis delta virus RNA editing is highly specific for amber/W site and is suppressed by hepatitis delta antigen. *Mol. Cell. Biol.* **18**: 1919–1926.

Taylor, J. M., Farci, P., and Purcell, R. H. (2006). Hepatitis Delta Virus Chapter 77 in: *Fields Virology, Fifth Edition* (D. M. Knipe and P. M. Howley, Eds. in chief), Philadelphia, Lippincott Williams & Wilkins, pp. 3031–3046.

Prions and Prion Diseases

Anderson, R. M., Donnelly, C. A., Ferguson, N. M., *et al.* (1996). Transmission dynamics and epidemiology of BSE in British cattle. *Nature* **382**: 779–788.

Angers, R. C., Browning, S. R., Seward, T. S., *et al.* (2006). Prions in skeletal muscles of deer with chronic wasting disease. *Science* **311**: 1117.

Baker, H. F., and Ridley, R. M. (1996). What went wrong in BSE? From prion disease to public disaster. *Brain Res. Bull.* **40**: 237–244.

Baron, G. S., Magalhães, A. C., Prado, M. A. M., and Caughey, B. (2006). Mouse-adapted scrapie infection of SN56 cells: greater efficiency with microsome-associated versus purified PrP-res. *J. Virol.* **80**: 2106–2117.

Bons, N., Mestre-Frances, N., Belli, P., *et al.* (1999). Natural and experimental oral infection of nonhuman primates by bovine spongiform encephalopathy agents. *Proc. Natl. Acad. Sci. U.S.A.* **96**: 4046–4051.

Brown, P. (2004). Mad-cow disease in cattle and human beings. *American Scientist* **92**: 334–341.

Calzolai, L., Lysek, D. A., Pérez, D. R., Güntert, P., and Wüthrich, K. (2005). Prion protein NMR structures of chickens, turtles, and frogs. *Proc. Natl. Acad. Sci. U.S.A.* **102**: 651–655.

Chabry, J., Priola, S. A., Wehrly, K., *et al.* (1999). Species-independent inhibition of abnormal prion protein (PrP) formation by a peptide containing a conserved PrP sequence. *J. Virol.* **73**: 6245–6250.

Chesebro, B. (1999). Prion protein and the transmissible spongiform encephalopathy diseases. *Neuron* **24**: 503–506.

Gossert, A. D., Bonjour, S., Lysek, D. A., Fiorito, F., and Wüthrich, K. (2005). Prion protein NMR structures of elk and of mouse/elk hybrids. *Proc. Natl. Acad. Sci. U.S.A.* **102**: 646–650.

Haywood, A. M. (2004). Transmissible spongiform encephalopathies. In *Infections of the Central Nervous System, 3rd edition* (W. M. Scheld, R. J. Whitley, and C. M. Marra, Eds.), Philadelphia, Lippincott, Williams & Wilkins, pp. 261–272.

Hegde, R. S., Mastrianni, J. A., Scott, M. R., *et al.* (1998). A transmembrane form of the prion protein in neurodegenerative disease. *Science* **279**: 827–834.

Lysek, A. A., Schorn, C., Nivon, L. G., *et al.* (2005). Prion protein NMR structures of cats, dogs, pigs, and sheep. *Proc. Natl. Acad. Sci. U.S.A.* **102**: 640–645.

Peretz, D., Supattapone, S., Giles, K., *et al.* (2006). Inactivation of prions by acidic sodium dodecyl sulfate. *J. Virol.* **80**: 322–331.

Prusiner, S. B. (Ed.) (1999). *Prion Biology and Diseases.* Cold Spring Harbor, NY, Cold Spring Harbor Laboratory Press.

Prusiner, S. (2006). Prions. Chapter 79 in: *Fields Virology, Fifth Edition* (D. M. Knipe and P. M. Howley, Eds. in chief), Philadelphia, Lippincott Williams & Wilkins, pp. 3059–3091.

Rhodes, R. (1997). *Deadly Feasts.* New York, Simon & Schuster.

Riek, R., Hornemann, S., Wider, G., *et al.* (1996). NMR Structure of the mouse prion protein domain PrP(121–231). *Nature* **382**: 180–183.

Ross, E. D., Edskes, H. K., Terry, M. J., and Wickner, R. B. (2005). Primary sequence independence for prion formation. *Proc. Natl. Acad. Sci. U.S.A.* **102**: 12825–12830.

Safar, J. G., Geschwind, M. D., Deering, C., *et al.* (2005). Diagnosis of human prion disease. *Proc. Natl. Acad. Sci. U.S.A.* **102**: 3501–3506.

Schonberger, L. B. (1998). New-variant Creutzfeldt-Jakob disease and bovine spongiform encephalopathy: the strengthening etiologic link between two emerging diseases. In *Emerging Infections 2* (W. M. Scheld, W. A. Craig, and J. M. Hughes, Eds.), Washington, DC, ASM Press, pp. 1–15.

Weissmann, C. (1996). Molecular biology of transmissible spongiform encephalopathies. *FEBS Lett.* **389**: 3–11.

Windl, O., Buchholz, M., Neubauer, A., *et al.* (2005). Breaking an absolute species barrier: transgenic mice expressing the mink PrP gene are susceptible to transmissible mink encephalopathy. *J. Virol.* **79**: 14971–14975.

Zhang, C. C., Steele, A. D., Lindquist, S., and Lodish, H. F. (2006). Prion protein is expressed on long-term repopulating hematopoietic stem cells and is important for their self-renewal. *Proc. Natl. Acad. Sci. U.S.A.* **103**: 2184–2189.

Yeast Prions

King, C.-Y., and Diaz-Avalos, R. (2004). Protein-only transmission of three yeast prion strains. *Nature* **428**: 319–328.

Roberts, B. T., and Wickner, R. B. (2003). Heritable activity: a prion that propagates by covalent autoactivation. *Genes Dev.* **17**: 2083–2087.

Wickner, R. B., Taylor, K. L., Edskes, H. K., *et al.* (1999). Prions in *Saccharomyces* and *Podospora* spp.: protein-based inheritance. *Microbiol. Mol. Biol. Rev.* **63**: 844–861.

10

Host Defenses against Viral Infection and Viral Counterdefenses

INTRODUCTION

On infection of a mammal by a microorganism, a complex response is generated that attempts to eliminate the infectious agent. These responses can be conveniently grouped into two categories, innate immune responses and adaptive immune responses. The adaptive or acquired immune response requires time to develop, is specific to the invading pathogen, and is followed by immunologic memory that usually renders the host immune, or at least less susceptible, to subsequent infections by the same organism. The two most important components of this response are the production of cytotoxic T lymphocytes (CTLs) and the production of humoral antibodies. Innate responses, in contrast, generate early responses to infection, are not specific to the pathogen, and do not render the organism resistant to subsequent infection by the same pathogen. Cytokines such as interferon are among the most effective innate responses. Innate and adaptive immune responses do not function independently of one another: The proper functioning of the immune system requires the activities of both, and, in fact, the activation of the adaptive response requires the prior activation of the innate system. The various activities of the two systems constitute a multifaceted, interactive, and complex series of responses to infection by a pathogen. Figure 10.1 illustrates some of the cells and other participants involved in innate and acquired immunity.

ADAPTIVE IMMUNE SYSTEM

The adaptive immune system contains two major arms. The cellular arm leads to the production of CTLs, also called killer T cells. The humoral arm leads to the production of antibodies that are secreted by B cells. T-helper cells are important players in orchestrating both of these arms.

Major Histocompatibility Complex

Both cellular immunity and humoral immunity require the activation of a class of lymphocytes called T cells (T from thymus, where these cells mature). T cells recognize peptide antigens 8–20 amino acids in length that are presented to them by cell surface proteins encoded in the *m*ajor *h*istocompatibility *c*omplex (MHC) locus (in humans, the MHC is called HLA, from *h*uman *l*ymphocyte *a*ntigen, and is encoded in chromosome 6). The two types of MHC molecules that present antigenic peptides are called class I and class II. Both class I and class II MHC molecules are integral membrane proteins that are composed of two polypeptide chains.

Class I and Class II MHC

The MHC class I molecule is a heterodimer composed of a heavy chain of about 350 amino acids, which is encoded within the MHC locus, and a light chain of about 100 amino acids, $\beta 2$ microglobulin, which is encoded elsewhere. The structure of an MHC class I molecule is shown schematically in Fig. 10.2A, and as determined by X-ray crystallography in Fig. 10.2B. The MHC class I heavy chain consists of three extracellular domains called $\alpha 1$, $\alpha 2$, and $\alpha 3$, a transmembrane domain, and a cytoplasmic domain. $\beta 2$ microglobulin forms a fourth extracellular domain and is held in the complex by noncovalent interactions. The $\alpha 1$ and $\alpha 2$ domains, which are structurally related to one another, form a platform with helical walls. The walls form a groove in which the antigenic peptide, consisting usually of 8–10 amino acids, is anchored. The $\alpha 3$ domain and $\beta 2$ microglobulin are homologous (derived from a common ancestral polypeptide by gene duplication) and are members of the immunoglobulin (Ig) superfamily. They share some sequence identity and have common structural features.

Innate Immunity

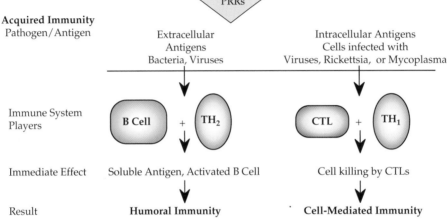

FIGURE 10.1   The mechanisms of innate and acquired immunity are integrated to provide the basis for humoral and cellular immunity. PRRs are *p*attern *r*ecognition *r*eceptors. Adapted from Mims *et al.* (1993), p. 5.8.

The structure of MHC class II molecules is similar to that of class I molecules, as illustrated in Fig. 10.2A. Class II molecules are composed of a heterodimer of two proteins encoded within the MHC locus. These two proteins, designated α and β, each contain two extracellular domains (and thus the assembled molecule contains four extracellular domains as does class I MHC). Both proteins are anchored in the plasma membrane by membrane-spanning anchors and have cytoplasmic domains. The distal $\alpha 1$ and $\beta 1$ domains form a platform with a groove that binds an antigenic peptide for presentation to T cells, but in this case the peptide is longer, usually 14–18 amino acids in length. The proximal domains, $\alpha 2$ and $\beta 2$, are members of the Ig superfamily.

The number of peptides that can be presented by an individual class I or class II molecule is large. Only certain residues in the peptide, called anchor residues, interact specifically with the MHC molecule. The remainder of the peptide can vary in sequence. Furthermore, the MHC is

exceedingly polymorphic and there are hundreds of different alleles within the human population. The haploid number of genes in humans that encode heavy chains used in class I MHC molecules is three (called HLA-A, -B, -C), and a diploid individual can make up to six different class I MHC molecules with differing requirements for anchor residues. In the case of class II MHC, six α and seven β genes have been identified in humans (of which the most important loci are called DR, DQ, and DP). Thus, any individual can present very many different peptides to T cells. Note that since the MHC is polymorphic, different individuals present different peptides to T cells.

T-Cell Recognition of Peptide Antigens

T cells express a T-cell receptor on their surface, which is able to recognize a specific peptide presented in the context of class I or class II MHC molecules. The T-cell receptor is a heterodimer formed by one α and one β chain, or by

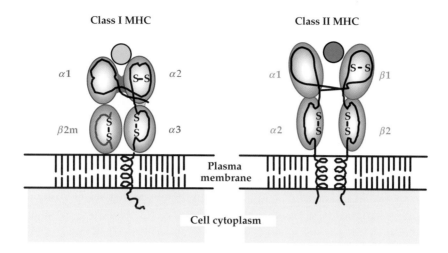

Class I MHC   **Class II MHC**

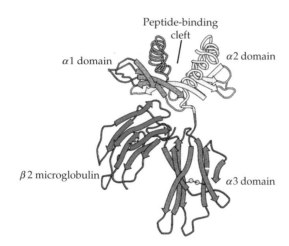

Peptide-binding cleft

α1 domain   α2 domain

β2 microglobulin   α3 domain

FIGURE 10.2   Upper panels: schematic representations of MHC class I and MHC class II molecules. The orientation of these molecules at the cell surface is indicated, as are the domain structures of the extracellular portions of the proteins, the membrane-spanning domains and the cytoplasmic domains, and how they function in the cell. The yellow and green spheres represent bound peptide antigens. Lower panel: three-dimensional ribbon diagram of the structure of MHC-1 (HLA-A2), as determined by X-ray crystallography. From Bjorkman *et al.* (1987) as reprinted in Kuby (1997).

one γ and one δ chain, as illustrated in Fig. 10.3. The great majority of circulating T cells possess receptors formed by αβ dimers.

During maturation of T cells, three or four separate regions of the genes for α or β, respectively, are brought together by deletion of the intervening sequences, with the rearrangements to form the β chain occurring first. This process is illustrated in Fig. 10.4. The regions are the V (variable), D (diversity), J (joining), and C (constant) regions for β, or the V, J, and C regions for α. The V and C regions belong to the Ig superfamily, whereas D and J are shorter, unrelated

domains. V and J, or V, D, and J, are first joined at the DNA level by a process that deletes the intervening DNA. The combined VJ or VDJ is then joined to C by splicing of the pre-mRNA transcript. Multiple copies of each of the four segments exist in the germ line and combinatorial joining of these results in a very large number of possible α or β subunits. Furthermore, the joining of VJ and VDJ is imprecise, and additional diversification results from repair of the joining regions. Combinatorial joining of α and β subunits results in the production of an even larger repertoire of T-cell receptors. An estimate of the possible diversity of

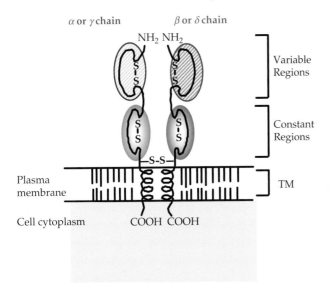

FIGURE 10.3 Structure of the T-cell receptor (TCR). Receptors are heterodimers of α/β or γ/δ TCR chains. Each chain has a transmembrane anchor (TM). The C-terminal domains and the TM are relatively constant in sequence, but the N-terminal domains are variable. Adapted from Chen and Alt (1997) Figures 7.6 and 7.7 on pp. 344 and 345.

T-cell receptors, sometimes referred to as the theoretical repertoire, is shown in Table 10.1. The theoretical repertoire, how many T-cell receptors could possibly be formed, is perhaps 10^{18}, which is an exceedingly large number.

Once any individual T cell develops and expresses a T-cell receptor, no further recombination occurs and the receptor does not change thereafter. The function of T cells, simply stated, is to discriminate self from nonself, and T cells that express an appropriate receptor that serves this function are selected in the thymus (see later) and released into the circulation, where they are long lived. Any individual has many, many T cells in the body and these continue to develop throughout life. The repertoire present in any human at any one time certainly exceeds 10^8 different T-cell receptors, expressed on different T cells.

Each different T-cell receptor is potentially able to recognize a different peptide epitope (the epitope is the surface of the peptide that interacts with the T-cell receptor). The T-cell receptor recognizes the peptide epitope in the context of class I or class II MHC; that is, the T-cell

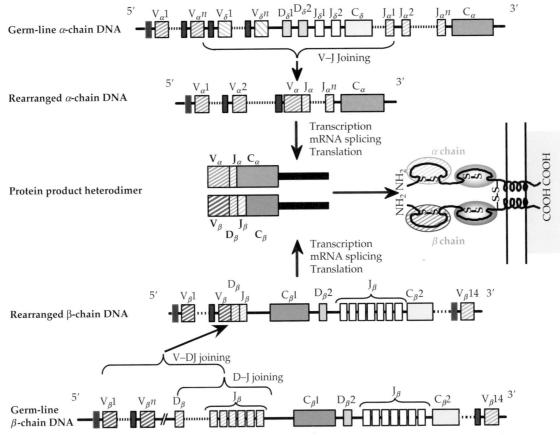

FIGURE 10.4 Gene rearrangements to generate T-cell receptor diversity. The example shown generates an αβ T-cell receptor. The α chain undergoes a V_{α}–J_{α} joining, whereas the β chain undergoes two joins: D_{β} to J_{β} followed by V_{β} to $D_{\beta}J_{\beta}$. Primary RNA transcripts are spliced to give mRNAs in which RNA sequences encoding VJ or VDJ are joined to RNA sequences encoding constant domains, C_{α} or C_{β}. These spliced mRNAs are then translated into the α and β chains of the TCR. Adapted from Kuby (1997) Figure 11.6 on p. 269.

TABLE 10.1 Comparison of Diversity in Human Immunoglobulin and T-Cell Receptor Genes

Mechanism of diversity	Immunogobulins			αβ T-cell receptors		γδ T-cell receptors	
	Heavy chains	Light chains		α chain	β chain	γ chain	δ chain
		κ chain	λ chain				
Multiple germ-line gene segments							
V	65	40	30	~70	52	12	>4
D	27	0	0	0	2	0	3
J	6	5	4	61	13	3,2	3
Combinatorial joining							
Combinatorial V–J–D combinations	$65 \times 27 \times 6 = 1.0 \times 10^4$	$40 \times 5 = 200$	$30 \times 4 = 120$	$70 \times 61 = ~430$	$52 \times 2 \times 13 = 1.3 \times 10^3$	$12 \times 3 = 36$	$4 \times 3 \times 3 = 36$
D segments read in 3 frames	Rarely	—	—	—	Often	—	—
Joints with N and P nucleotides	2		(1)	2	1	??	??
V gene pairs	$1.0 \times 10^4 \times 320 = \mathbf{3.2 \times 10^6}$			$430 \times 1.2 \times 10^3 = \mathbf{5.8 \times 10^6}$		$36 \times 36 = \mathbf{1.3 \times 10^3}$	
Junctional diversity	$~3 \times 10^7$			$~2 \times 10^{11}$		???	
Total diversity	$\mathbf{~10^{14}}$			$\mathbf{~10^{18}}$		???	

Source: Data from Janeway *et al.* (1999) pp. 62, 93, 158.

receptor interacts with both the peptide and the MHC molecule (Fig. 10.5). The T cell cannot recognize the peptide alone or even recognize the peptide presented by the wrong MHC molecule. Similarly, the T cell cannot recognize a class I or class II MHC molecule with the wrong peptide in its cleft. The discovery of this requirement for dual recognition, which was made using a viral system, resulted in a Nobel Prize for Doherty and Zinkernagel (Table 1.1). Such a requirement for multiple interactions is a recurring theme in the immune system, and it has evolved because this potent and potentially harmful system must be carefully regulated.

The T-cell receptor is part of a complex containing accessory molecules that are required for the function of the receptor. Two such molecules are CD4 and CD8, and mature T cells possess either CD4 or CD8 (but not both). CD8 contains one Ig domain attached to a stalk region, whereas CD4 contains four Ig domains (Fig. 10.5). CD8+ T cells recognize peptides presented by class I MHC (Fig. 10.5A). CD4+ T cells, in contrast, recognize peptide epitopes presented in the context of class II MHC (Fig. 10.5B). CD8 or CD4 interacts with constant regions of class I or class II MHC molecules, respectively, and increases the binding affinity of the T cell for its cognate MHC–peptide complex by about 100-fold. Class I and class II MHC molecules acquire the peptides that they present in fundamentally different ways and are components of two different responses to infection by microorganisms

Cytotoxic T Cells

The majority of CD8+ T cells are or become CTLs, although a minority of CD4+ T cells are also CTLs. CD8+ T cells are class I MHC restricted, as described. Because class I MHC molecules are expressed on most mammalian cells, the major exception being neurons and, in humans, red blood cells, which express little or no class I MHC, most cells in an individual are capable of presenting peptides to T cells in a class I MHC context.

The peptides presented by class I MHC are derived from intracellular proteins and represent a sampling of all proteins being synthesized within the cell. The pathway involved is illustrated in Fig. 10.6. The peptides are generated by proteolysis of intracellular proteins by an enzyme system referred to as the proteasome. The proteasome is a large complex, possessing many subunits, that is present in the cytoplasm. It possesses ATP-dependent proteolytic activity and is the major cellular proteolytic site other than the lysosome. In addition to its function in the immune response, the proteasome is important for turnover of many proteins within the cell and for degradation of misfolded proteins. Peptides resulting from degradation of intracellular proteins are actively secreted, in a process that requires hydrolysis of ATP, into the lumen of the endoplasmic reticulum (ER) by a transporter called TAP (*t*ransporter associated with *a*ntigen *p*resentation). TAP is encoded in the MHC and consists of a heterodimer anchored in membranes of the ER.

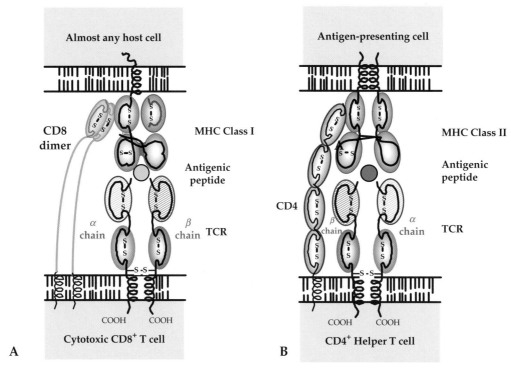

FIGURE 10.5   (A) Interaction of the TCR on a cytotoxic CD8+ T cell with an MHC class I molecule complexed with an antigenic peptide on almost any cell. The TCR interacts with both the peptide and with the MHC molecules. The CD8 homodimer interacts with a conserved region of the MHC α3 domain. (B) Interaction of the TCR on a CD4+ helper T cell with an MHC class II molecule complexed with an antigenic peptide on the antigen-presenting cell. The TCR interacts with both the peptide and with the MHC molecule. The membrane distal domain of CD4 recognizes a conserved region of the MHC β2 domain. Adapted from Chen and Alt (1997) pp. 344 and 345; and data from Kuby (1997) pp. 275–277.

Proteins secreted into the lumen of the ER during synthesis are also sampled. There is a pathway that recycles lumenal proteins back to the cytoplasm. This pathway may serve to rid the ER of misfolded proteins as well as enabling the sampling of proteins destined for the plasma membrane or other intracellular organelles. On reentry into the cytoplasm, a cellular glyconase removes carbohydrates from glycoproteins, and the protein backbone is degraded by the proteasome. Thus, viral proteins that are inserted into the lumen of the ER during synthesis, such as glycoproteins used to assemble progeny virions, are also sampled by the proteasome–TAP pathway.

A peptide delivered to the lumen of the ER by TAP can be bound by a class I MHC molecule if it has the right anchor residues. Delivery of the peptide to the MHC is a complex process that ensures that the MHC receives a high-affinity peptide. The peptide can be shortened to the proper length (8–10 residues) by a protease called ERAAP if it is too long. The MHC heavy chain is bound by the membrane-associated chaperone calnexin upon synthesis and folds into its proper conformation. Calnexin then dissociates, β2 microglobulin associates with the heavy chain, and the MHC is delivered to the peptide-loading complex, which consists of several

proteins including the two subunits of TAP (TAP1 and TAP2), a transmembrane glycoprotein called tapasin, the soluble chaperone calreticulin, and thiol oxidoreductase Erp57. Binding of peptide stabilizes the class I molecule and facilitates its release and transport to the cell surface. Class I MHC that is transported without a peptide is unstable at 37°C and is degraded.

The end result is that class I MHC presents a random sampling of peptides derived from proteins being synthesized within the cell for inspection at the cell surface by any T cell that may be in the vicinity. The peptides bound to a single isoform of class I MHC molecules present on the surface of cells in culture have been analyzed by very sensitive techniques. More than 10,000 different peptides, present at 2 to 4000 copies per cell, were identified. This great diversity of peptides consists mostly of self-peptides, but peptides derived from intracellular viruses or other intracellular pathogens will be represented if the cell is infected. If a patrolling T cell has a receptor that binds specifically to a peptide being presented by the class I MHC on another cell, the T cell may become activated and may proliferate. Once activated, CTLs kill cells that present the epitope they recognize. In different assays, the number of MHC–peptide complexes required for

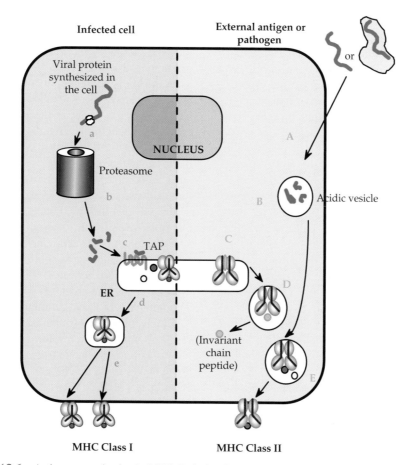

Infected cell

External antigen or pathogen

Viral protein synthesized in the cell

a

NUCLEUS

Proteasome

b

c TAP

ER

d

e

MHC Class I

A

B Acidic vesicle

C

D

(Invariant chain peptide)

E

MHC Class II

FIGURE 10.6   Antigen processing by the MHC. Both class I and class II pathways are shown. On the left, a viral protein is synthesized in the cell (**a**) degraded (**b**) and viral peptides are bound by TAP in the ER and transported into the ER lumen, where they are bound by MHC class I molecules (**c**), MHC containing bound peptide is transported by cellular vesicles through the Golgi apparatus (**d**) to the cell surface (**e**) where it is expressed for inspection by T cells. On the right, an external antigen or pathogen is endocytosed into the cell (**A**) and is degraded in an acidic vesicle (**B**). Class II molecules are synthesized and imported into the ER (**C**) with a trimer of an invariant chain peptide in the binding site, and transported to the trans-Golgi (**D**) where the invariant chain is lost. The Golgi vesicle fuses with the endosome containing the antigenic peptide (**E**), and the peptide–MHC class II complex is transported to the cell surface. Adapted from Fields *et al.* (1996) p. 351.

recognition by a CTL has been estimated to be between one and several hundred, and may depend on the affinity of binding of the MHC–peptide target by the T-cell receptor as well as the state of activation of the T cell.

The activation of a T cell is a multistage process. Naïve T cells are activated to become effector cells by interaction with professional antigen presenting cells (APCs), usually macrophages or dendritic cells, that present antigen in the context of class I or class II MHC. APCs can present antigens in a class I context that are synthesized within the cell and processed as described before, but they can also present antigens in a class I context that they acquire from the external environment (called cross-presentation). Cross-presentation allows APCs to respond to a viral antigen, for example, even if the virus does not replicate within the APC. These antigens may arise from the dissolution of an infected cell that dies, and enter the APC by phagocytosis or pinocytosis.

Inside the APC, the viral antigen may enter the cytoplasm and be processed by the methods described earlier, or may be processed and bound by MHC class I within the endosomal compartment. Activation of a T cell by an APC requires not only the recognition of the cognate antigen in a class I context, but also that additional immunostimulatory signals be present. For example, if Toll-like receptors, described later, have been activated in the APC, it expresses additional proteins that enable it to activate the T cell. In the absence of such prior activation of the APC, it will tolerize a T cell rather than activating it, illustrating another layer of control to avoid possible autoimmune responses.

Proliferation of activated CTLs requires further stimulation by cytokines such as interleukin-2 (IL-2). The source of IL-2 is usually a class of T-helper (T_H) cells (most T_H cells are CD4+, as described later). Activated T_H-1 cells secrete IL-2 as well as tumor necrosis factor β (TNF-β), interferon γ (IFN-γ), and other cytokines. Proliferation of T cells that recognize a specific peptide derived from a viral protein means that a vigorous CTL response against an invading pathogen ensues.

CTLs kill target cells by inducing apoptosis, a cell suicide pathway described in a later section of this chapter. One of three different mechanisms is used to induce apoptosis. These mechanisms are listed in outline form here but are described in more detail when the apoptotic pathway is considered. In one mechanism, the T cell releases the contents of granules, which contain perforins and proteases among other components, into the target cell. Perforins form pore structures in the plasma membrane of the target cell that allow ions to leach out of the cell, and the proteases participate in the activation of cell pathways leading to apoptosis. In a second mechanism, apoptosis is induced by triggering the Fas death receptor on the surface of the target cell. These two mechanisms lead to cell death within 4–6 hours. A third mechanism utilizes the TNF-α death pathway and is a slower process, leading to cell death in 18–24 hours. Killing of target cells is a drastic response and the CTL pathway is directed toward eliminating internal pathogens, usually viruses. Because early proteins encoded by the virus can be sampled by the MHC–T-cell receptor pathway as well as late proteins, it is possible for a T cell to kill a virus-infected cell before it has time to synthesize much progeny virus. Many viruses counter this pathway by interfering with the ability of an infected cell to express class I MHC at its surface, as described later in this chapter.

Although three killing mechanisms are used by different CTLs, they are not redundant. Mice that lack the perforin gene are unable to control infection by lymphocytic choriomeningitis virus and half die within a month of infection. This virus is not pathogenic in normal mice or even in immunocompromised mice. Thus death must result from immunopathology caused by an unbalanced or incomplete T-cell response.

Since the class I pathway presents peptides derived from self as well as from viruses that may have infected the cell, how then do the CTLs know not to kill cells expressing self-antigens? The answer lies in part in the selection of an appropriate repertoire of T cells. T cells bearing T-cell receptors recognizing many different possible peptide antigens arise by random combinatorial joining and diversity-inducing processes. T cells undergo their early differentiation in the thymus, where selection occurs. Only T cells that express a T-cell receptor capable of recognizing class I MHC (or class II MHC) bearing a peptide are selected (called positive selection); T cells that do not express an appropriate receptor die. However, if such a T-cell receptor has a high affinity for self-peptides present in the thymus, then that T cell also dies (called negative selection). In this process of selection only about 2% of T cells survive and most of these encode receptors that recognize nonself antigens. These T cells

are released into the circulation to patrol for cells that are infected by viruses or other intracellular pathogens.

A second level of control that reduces the incidence of reaction against self lies at the level of cytokine induction. Virus infection or infection by other parasites activate innate response elements, of which Toll-like receptors are the best known, that result in the production of inflammatory cytokines. Cytokines such as IL-2 or IFN are required as a second signal for CTL activation, and IFNs also upregulate the presentation of antigens by MHC molecules as well as other aspects of the immune response. Thus, the inflammatory response makes it more likely that T cells will respond to antigens that they recognize. As described before, APCs may tolerize a T cell to an antigen rather than activating it if the appropriate immunostimulatory factors are not present, another layer of control to prevent autoimmunity. Furthermore, once activated, CTLs undergo apoptosis if the cytokine signals are no longer present, thus damping out any autoimmune responses that might occur.

Killing of virus-infected cells by CTLs in an effort to eradicate viral infection relies on the ability of the cells of most organs to regenerate from progenitor cells. In this context it makes sense that neuronal cells express only low levels of class I MHC. These cells are terminally differentiated and nondividing, and if killed by a CTL they are unable to regenerate.

Although the ability of CTLs to kill virus-infected cells is well established, recent findings indicate that CTLs, as well as other activated cells of the immune system, may also use noncytolytic means to control and clear many virus infections. This control is thought to be achieved by the secretion of cytokines such as IFN-γ and TNF-α. Dengue virus infection in the brains of mice is one example in which noncytolytic clearance appears to be important. During dengue virus infection of neurons, CTLs are actively recruited into the brain and are essential for the clearance of virus in immunized mice, at least under some conditions. Neurons are immunologically privileged, as noted before, and noncytolytic mechanisms of control are important. Hepatitis B virus infection of hepatocytes is another system in which there is evidence that noncytolytic control is important, even though hepatocytes do regenerate and killing of infected hepatocytes does occur, causing the symptoms of hepatitis. Other examples are also known. Noncytolytic clearance does not appear to be universal, however. It appears to be possible only in some tissues and for some viruses.

T-Helper Cells

Most CD4+ T cells are helper cells. Some CD8+ T cells are also helper cells, but most helper cells are CD4+. CD4+ T cells recognize peptides presented in the context of class II MHC molecules—they are referred to as class II–restricted cells (see Fig. 10.5). Class II MHC molecules, unlike class I, are present on only a restricted set of cells within the organism and are most abundant on B cells, macrophages, dendritic cells, and, in humans, activated T cells, that is, cells of the immune system itself.

The peptides presented by class II MHC molecules are derived from extracellular proteins, and thus these MHC molecules sample the extracellular environment. Proteins, whole viruses, or other microorganisms are taken up by antigen-presenting cells and degraded within intracellular organelles. Peptides derived from these sources can be bound by class II MHC molecules being transported to the cell surface. The process of producing a peptide and transferring it to a class II MHC molecule is complicated, as is the case for class I MHC, but will not be described here since viruses are not known to interfere with the class II pathway, whereas they do interfere with the class I pathway. The peptides derived from this pathway are kept separate from peptides generated through the proteasome–TAP pathway, and the end result is that the class II pathway presents peptides derived from the external environment, whereas the class I pathway normally presents peptides derived from the intracellular environment (Fig. 10.6) (but see the previous discussion of cross-presentation).

Professional APCs, which may be macrophages, B cells, or dendritic cells, present peptides to T cells. They also express accessory proteins that deliver a second signal to the T cell that is required for its activation. When a T_H cell is activated by the interaction of its receptor with its target peptide presented by class II MHC, and the appropriate second signals are present, as described before, the cell proliferates and secretes cytokines that are important for eliciting an immune response. Activated T_H cells are essential for any immune response, whether CTLs to kill infected cells or B cells to make antibody (see later). It is the disappearance of T_H cells during HIV infection that leads to the symptoms of AIDS.

T_H cells are not homogeneous. Different T_H cells secret different panels of cytokines and have different functions in the immune response. Two main types have been recognized, which perhaps represent extremes in function. T_H-1 cells are highly effective for CTL activation and function in the cellular immune pathway. T_H-2 cells are optimal for the activation of B cells and function in the humoral immune pathway. They secrete IL-4, IL-5, IL-6, and IL-10. T_H-1 and T_H-2 cells can be mutually antagonistic. The cytokines secreted by one suppress the other, and a balanced immune response often requires a balanced activation of these two classes of helpers. T_H-2 cells usually deliver their cytokine signals directly to the B cells that they help, following cell–cell contact. T_H-1 cells, in contrast, do not deliver their signals directly to the T cells that they help.

B Cells and Secretion of Antibodies

B cells, so named because they are derived from the bone marrow in mammals, secrete antibodies, which are members of the Ig superfamily. The essential subunit of an antibody is a heterodimer of a heavy (H) chain (which has four or five Ig domains) and a light (L) chain (which has two Ig domains). This subunit is always present as a dimer in which two H-L heterodimers are linked through the H chains. The structure of a light chain, as determined by X-ray crystallography, is shown in Fig. 10.7. This structure illustrates the Ig fold that is common to all Ig domains.

The five classes of antibodies are illustrated in Fig. 10.8. An IgG molecule consists of a dimer of H-L heterodimers in which the H chain is of the γ class. IgD and IgE antibodies are also dimers of H-L heterodimers, but in this case the H

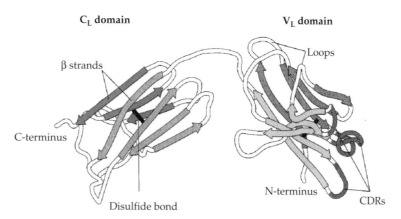

FIGURE 10.7   Ribbon diagram of an immunoglobulin light chain depicting the immunoglobulin-fold structure of its variable and constant domains. Two β-pleated sheets in each domain (colored in red and pink in the constant domain and yellow and brown in the variable domain) are held together by hydrophobic interactions and a single disulfide bond (dark bar). The hypervariable regions also known as *c*omplementarity *d*etermining *r*egions (CDRs), shown in blue, form part of the antigen-binding site. Adapted from Kuby (1997) p. 113.

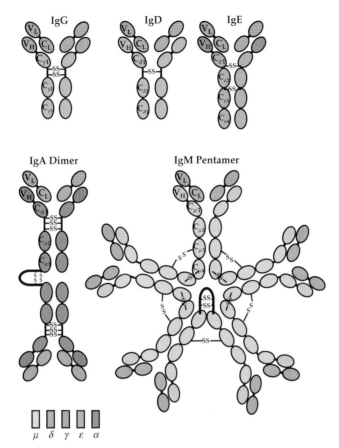

FIGURE 10.8 Structure of the different classes of secreted immunoglobulin molecules. Note that IgM and IgA are secreted as pentamers and dimers, respectively, linked by a J-chain. Adapted from J. Gally (1973). IgG, IgA, and IgD heavy chains have four domains and a hinge region; IgE and IgM lack the hinge. Not shown are intrachain disulfide bonds and bonds linking the light and heavy chains.

chains are of the δ or ε class, respectively. IgA antibodies contain four H-L heterodimers (it consists of a dimer of the dimeric unit, as shown). In this case, the H chain is of the α class. IgM contains 10 H-L heterodimers (it is a pentamer of dimeric units) formed with μ H chains.

A more detailed representation of an IgG molecule is shown in Fig. 10.9. The terminal domains of both H and L chains are variable. Within the variable domains, there are regions that are more variable than other regions, called hypervariable regions, also known as *c*omplementarity-*d*etermining *r*egions (CDRs) (see Fig. 10.7). The combining site of the antibody, the region that specifically binds to an antigen recognized by the antibody, is formed by the variable regions of both the H and L chains.

Formation of Light and Heavy Chains

The L and H chains of antibody molecules are formed in a process that is similar to that used to form T-cell receptors.

The two Ig domains of the L chain are called V (for variable) and C (for constant) and between these two domains is a J (for joining) domain. All three domains are encoded separately in the genome. There are multiple copies of V, J, and C in the genome (Table 10.1), and these gene segments are polymorphic in the population. The light chain genes fall into two different families, called κ and λ (Table 10.1), which are encoded on two different chromosomes in humans. During maturation of a B cell, a V-gene segment, a J-gene segment, and a C region of the light chain are brought into juxtaposition to one another, as illustrated in the top panel of Fig. 10.10. In this process, a V-gene segment is fused to a J-gene segment by deleting the intervening DNA. The C region is brought into play by RNA splicing: transcription of the VJ region continues through the C region, and splicing of the pre-mRNA joins the J region to the C region.

The heavy chain, which is the first to rearrange, is formed by a similar sequence of events, but in this case there is an additional gene segment D (for diversity) that introduces additional diversity in the recombination process. As for the light chain, there are multiple copies of V, D, and J in the genome (Table 10.1), and the population is polymorphic for these gene segments. During B-cell maturation, a D-gene segment is first joined to a J segment, and the DJ segment is then joined to a V segment (bottom panel of Fig. 10.10).

Recombinational rearrangements to form the antigen-binding site occur only during the maturation of the B cell. Once a B cell is mature, no further rearrangements occur in this region, although during an immune response the antigen-binding site is subjected to hypermutation in germinal centers, as described later.

The many combinatorial possibilities of V and J light-chain gene segments, and of V, D, and J heavy-chain gene segments, lead to the possible production of a very large number of light chains and heavy chains (Table 10.1). In addition, joining V with J, or joining V, D, and J, is imprecise, leading to additional diversity. Finally, joining an L chain with an H chain to form the heterodimer generates still more possible antigen recognition sites, since the antibody recognition site is formed by the V regions of both the H and the L chains. The total number of possible combinations is very large (Table 10.1). It is important to note that any individual B cell usually produces only one H chain and one L chain, and thus each B cell produces only one antibody recognition site.

Activation of B Cells

During maturation of B cells, a large population of cells results, each of which has one antibody-combining site. The theoretical repertoire, how many different types of B cells could conceivably be produced using the known mechanisms that are active during B-cell development, is thought to be much larger than the minimal estimate of 10^{14} shown

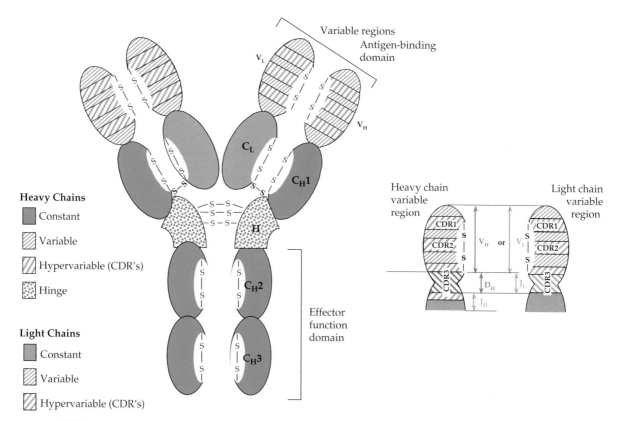

FIGURE 10.9 Diagrammatic representation of an IgG molecule. Each antibody molecule is composed of two heavy chains (which contain four domains, each consisting of an Ig fold like that shown in Fig. 10.7) and two light chains, which each have two Ig domains. The distal domains of both the light and heavy chains are the variable regions, composed of interspersed framework regions and hypervariable regions (diagonal shading) known as complementarity-determining regions or CDRs. The right part of the figure illustrates the role of the V, D, and J gene segments in encoding CDR1, 2, and 3 regions of Ig variable region genes. Adapted from Chen and Alt (1997) pp. 340 and 345.

in Table 10.1. In humans there may be 10^{10} B cells with differing specificities circulating at any time, and it is estimated that about 1 in 10^5 cells produces an antibody that will bind, with differing affinities, to any particular antigen that is being examined.

The antibody molecule is first produced as an integral membrane protein that is displayed on the surface of the B cell, anchored through a membrane-spanning region and containing an intracellular cytoplasmic domain. If the antibody displayed on the surface of the cell binds antigen, the B cell is activated. If the cell receives a second signal from a T_H-2 cell, it proliferates to form cells that secrete antibody. Memory cells also arise that serve to protect the organism from future infection by the same pathogen.

The T_H-2 cell signal may be delivered to the B cell either through a specific pathway or through a nonspecific pathway. Antigen, which could be in the form of a whole virus or in the form of a protein, that is bound to antibody present on the surface of the B cell can be internalized by the B cell and degraded by the MHC class II antigen-processing pathway. Peptides derived from the degraded virus or protein can then be presented on the surface of the B cell in the context of class II MHC molecules. Class II–restricted T-helper cells

that recognize this peptide will secrete cytokines that stimulate the B cell to proliferate and secrete antibodies. Note that the peptide displayed by the class II molecule does not have to be related to the epitope recognized by the antibody displayed on the B-cell surface. It may, in fact, be derived from an entirely different protein. Thus, while T cells respond to peptide epitopes, the antibody molecules can recognize much more complex antigens, such as whole proteins or viruses or even nonprotein antigens like carbohydrates. The protein epitopes recognized by antibodies are most often what are called conformational or nonlinear epitopes, which are formed by residues physically located at different places in the linear sequence of a protein but which form a contiguous surface in the protein after it folds into its three-dimensional conformation. Such discontinuous epitopes are destroyed if the protein is denatured and the different components of the epitope separated from one another. A certain fraction of antibodies, however, recognize continuous epitopes, which are formed by a linear sequence of amino acids present in the protein.

In addition to specific activation of B cells by T-helper cells, B cells can also be activated through area stimulation. If a B cell is in the vicinity of T-helper cells that

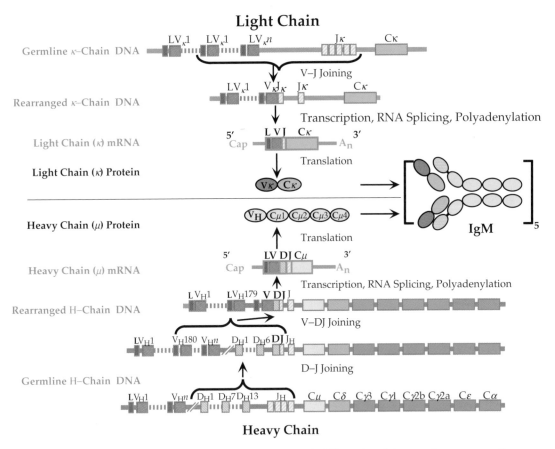

FIGURE 10.10   Formation of the IgM heavy and light chains. (Top) Illustration of the germ-line genes encoding κ immunoglobulin light chains encoded on human chromosome 2. The first line shows the germ-line genes, and the next line illustrates the B-cell genes after V–J recombination. The next line down shows the RNA transcript after splicing to join the VJ region to the C region. This mRNA is translated by cytoplasmic ribosomes into the light chain, which then combines with a heavy chain. The leader, L, is translated into a signal sequence that is removed posttranslationally. (A similar series of rearrangements occurs among the gene segments for the λ light chains encoded on human chromosome 22.) (Bottom) A comparable illustration of the heavy chain genes in the germ line and in the B cell, after two rounds of rearrangement known as "D–J joining," and "V–DJ joining." The final IgM molecule is a pentamer held together with disulfide bonds and a "J chain" which links the Fc regions. Adapted from Kuby (1997) Figures 7.4 and 7.5 on pp. 172 and 173.

are releasing cytokines to activate B cells, it may also be stimulated. The importance of area stimulation, and the frequency with which it occurs, in the context of fighting off a viral infection is not clear. In many cases of viral infection, a generalized and active inflammatory response occurs that involves the release of many cytokines, and in which many different antigens are being presented. Area stimulation could be important in developing a rapid response during such events. However, in such a process antibodies against self might also be produced, and such processes must be controlled.

Secretion of Antibodies

A B cell stimulated by exposure to its cognate antigen and by help from a T_H-2 cell proliferates and begins to secrete

antibodies. The first antibodies secreted are IgM, whose structure is illustrated in Fig. 10.8. IgM antibodies, which circulate in the blood, can be detected as early as a few days after virus infection. Their production quickly wanes and over a period of weeks or months the concentration of IgM in the blood decreases to very low or undetectable levels, as illustrated schematically in Fig. 10.11. Thus, the presence of IgM antibodies specific for a virus is usually a sign of acute, or at least very recent, infection.

During further maturation of the B cell, class switching occurs, as illustrated in Fig. 10.12. Recombinational events in the heavy-chain region lead to the substitution of the IgG, IgE, or IgA heavy chain for that of IgM. Homologous recombination within the intron just downstream of the J gene results in deletion of the intervening DNA such that the active VDJ gene is brought into contact with the C region

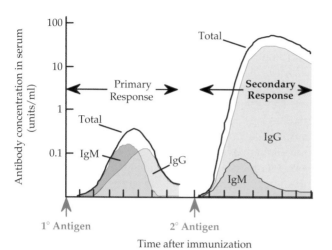

FIGURE 10.11   Time course of development of circulating antibodies after primary and secondary immunizations. No timescale has been shown, since the actual results vary with antigen, adjuvant, site of injection, and animal species. From Kuby (1997) Figure 16.19 on p. 398.

IgA) or directly (e.g., IgM to IgE without passing through an IgG phase). The situation in the human chromosome is more complicated (Fig. 10.12). There are two α genes, the first of which lies between two of the four γ genes. Thus, there are more ways to switch from one class of antibody to another.

Production of IgG (or of IgE or IgA) thus occurs later after infection (illustrated schematically in Fig. 10.11). At least 2 weeks are required before there is production of large amounts of IgG. Once a B cell begins to make IgG, it is no longer able to make IgM because the gene encoding the M heavy chain (Cμ in Fig. 10.12) has been deleted. Note that the antigen-combining site of an antibody is not changed by class switching because the V region of the H chain is not involved. However, hypermutation in the combining site to increase the affinity of the antibody for the antigen, which occurs in germinal bodies, is associated with class switching because a deaminase that is induced upon activation of the B cell is involved in both class switching and hypermutation.

IgG circulates in the blood and therefore very many cells exposed to blood and blood products are exposed to IgG. IgG lasts only 20–30 days in the blood, but B cells continue

of a different class of heavy chain. The order of heavy-chain genes in the mouse chromosome is μδγεα, and class switching can happen either sequentially (IgM to IgG to IgE to

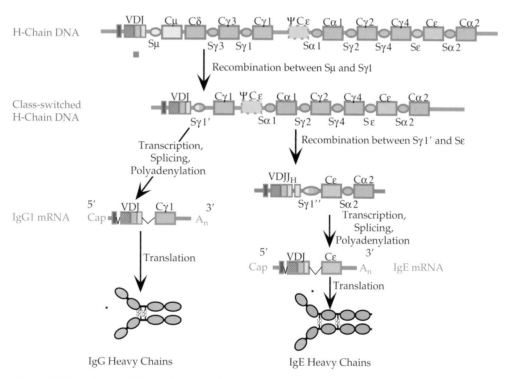

FIGURE 10.12   Immunoglobulin class switching to produce heavy chains for IgG and IgE, following initial rearrangements to produce IgM. Switch sites, designated "S", are located upstream of each CH gene except Cδ. In humans there is only one allele for Cμ, Cδ, and Cε, but there are two alleles for Cα, designated α1 and α2 and four alleles for Cγ, designated γ1, γ2, γ3, and γ4, which differ somewhat in sequence. In addition there is a pseudogene related to Cε. Because there is no switch site for Cδ, IgD is produced only in conjunction with IgM, by alternative mRNA splicing. Adapted from Kuby (1997) Figure 7.17 on p. 185; and Janeway et al. (2004) Figure 4.18.

to produce antibody for long periods of time, and although production wanes with time, IgG remains circulating in the blood for years or decades. Because of this, and the ease of obtaining blood samples, IgG has been extensively used to monitor past exposure to different pathogens and the immune status of individuals for any particular pathogen. As illustrated in Fig. 10.11, on a second exposure to an antigen, IgG concentrations rise dramatically and remain in the circulation at much higher levels. IgM concentrations do not show this anamnestic effect.

Because of its widespread circulation within the body, its high concentrations, and the anamnestic effect that occurs on secondary exposure, IgG is important in the control of viral diseases. It is also important in protecting the fetus and the very young. IgG crosses the placenta during pregnancy and is present in fetal blood. This transfers maternal immunity to the fetus, and this immunity lasts for the first few months of postnatal life.

IgA is also present in the blood, but its importance lies in the fact that it is secreted. It is present on mucosal surfaces where it helps prevent viral diseases such as those caused by rhinoviruses or influenza viruses in the respiratory tract or by rotaviruses or enteroviruses in the intestinal tract. It is present in secretions such as tears, saliva, and genital tract secretions, where it plays an antiparasite role. Because it is present in milk, it also serves to transfer maternal immunity to the gut of the infant.

IgE is important for control of infection by multicellular parasites. It can bind to mast cells via specific receptors and cause an inflammatory response, leading to the destruction of parasites. It is IgE that produces allergic symptoms that occur when pollen granules or mites or other comparatively large particles are recognized by IgE, producing an inflammatory response.

IgD is present in a membrane-bound form at the surface of immature B cells, along with IgM, where it helps in the activation of the cell on exposure to antigen as described before. It is also present in very small amounts in the blood. There is no class switching mechanism to express IgD and it is only produced in combination with IgM by means of differential splicing of mRNAs.

Although the B-cell repertoire first arises by combinatorial joining of the V and J light-chain gene segments and of V, D, and J heavy-chain gene segments, the antibody response is fine-tuned once a B cell has been activated. There is an error-inducing mechanism during B-cell replication in which the genes encoding the V segments of both heavy and light chains undergo hypermutation. This activity takes place in germinal centers of the lymph node and there is concurrent selection for B cells that bind more tightly to the antigen. Over a period of time B cells are selected that bind to the antigen with higher and higher affinity. Hypermutation in germinal centers has been postulated to play a role in the development of Burkitt's lymphoma, as described in Chapter 7.

Antibodies may bind to a virus and neutralize its infectivity, and such neutralizing antibodies are thought to be of critical importance in controlling virus infection. However, other antibodies may bind to a virus without inactivating it and such nonneutralizing antibodies can also be protective. There are several mechanisms by which nonneutralizing antibodies may protect. Aggregation of virions by antibodies leads to a reduction in the total number of infectious viruses. The maturation of enveloped viruses by budding through the cell plasma membrane can be inhibited by the binding of antibodies to viral proteins present on the surface of the cell. Antibody-dependent lysis of an infected cell can occur by a complement-mediated pathway or by a natural killer cell pathway called antibody-dependent cell-mediated cytotoxicity. These pathways are triggered by the binding of antibody to viral protein expressed on the cell surface. Binding of antibody or of certain components of complement to a virus can also increase the rate of phagocytosis by macrophages, which results in the destruction of the antigen, a process called opsonization. Complement and natural killer cells, which are components of innate immunity as well as of acquired immunity, are described later. Other mechanisms that result in protection by antibodies also exist.

Immunologic Memory

In the course of the B-cell response, memory B cells are formed. Memory B cells persist after the antigen disappears from the body, and are primed to react quickly and vigorously on renewed stimulation by the cognate antigen. Renewed activity of memory B cells still requires T-helper cell stimulation, but the secondary response leads to immediate production of antibody of high affinity, because there is no need for a long maturation process, and is so vigorous that it results in the production of much larger amounts of antibody than are produced during a primary response (Fig. 10.11).

Memory T cells are also generated in the course of an immune response. After expansion of T-cell clones following stimulation by the cognate antigen, most activated T cells die by apoptosis when no longer stimulated by the presence of antigen and cytokines. Immunologic memory remains, however. It had been widely thought that such memory requires a continuous supply of antigen with which T cells interact. This antigen might be present because the virus establishes a chronic infection, or antigen might be sequestered in regions of the body and made available over an extended period of time. It was thought even possible that T cells continued to be stimulated because of cross-reactivity with self antigens. Data have now accumulated, however, that memory T cells exist in a quiescent state for long periods of time and are capable of rapid reactivation on renewed exposure to antigen. Thus, memory in T cells is now thought to be similar to memory in B cells.

It is the existence of memory cells that renders an animal immune to the virus or other pathogen that first evoked the immune response. Memory cells are primed for such a rapid and vigorous response that the invading organism is stopped early during the infection process, before disease is established. Residual antibodies circulating in the blood or present on mucosal surfaces may even prevent infection altogether (sterilizing immunity).

The immune status of an individual is often tested by examining the blood for the presence of antibodies. Such an assay is imperfect, however. The presence of antibodies in the blood is usually a good indication that a person is immune. However, although such antibodies fade with time, in many cases a person remains immune despite the absence of detectable antibodies in the blood, because memory cells that are primed to react quickly are still present.

Complement System

The complement system is composed in part of more than 20 soluble proteins that circulate in the blood. These proteins are activated through a proteolytic cascade to produce effector molecules that aid in the control of viral infection or infection by other pathogens. Complement forms part of both the adaptive immune system and the innate immune system. Its activities turn antibodies into effective killers of viruses or of virus-infected cells, as well as of other pathogens, and in this role it is a component of the adaptive response. Complement can also be activated by interaction with parasites in the absence of antibody, however, and in this role it is a component of the innate responses.

The classical pathway of complement activation (an adaptive response) involves interaction of a complex of complement molecules called C1 with IgG or IgM. This could be IgG or IgM bound to antigen present at the surface of an infected cell or a bacterium, for example. The alternative pathway of complement activation (innate response) does not involve interaction with antibody, but rather requires the deposition of a molecule called C3b on the surface of a particle, such as a parasite. Once bound, C1 or C3b interacts with other components of the complement system. The result is the activation of a cascade of proteases whose cleavage activities result in the formation of effector molecules. One group of effector molecules forms a complex that inserts into the lipid bilayers of cell membranes and results in the lysis of the cell. Many enveloped viruses can also be killed by this lytic mechanism. The system must be finely regulated so that activated components of complement are produced only in response to pathogens and so that cell killing is confined to infected cells or parasites. Control of complement activation and action is therefore suitably complicated.

Other effector molecules that result from activation of complement have activities that aid in the control of viral infection or infection by other parasites by mechanisms other than cell lysis. Some products enhance the neutralization of viruses by antibody. By binding to virus that is coated with antibody, they render the virus less capable of binding to its receptor; cause aggregation of the virus, resulting in fewer infectious units; and increase the uptake of viruses by phagocytic cells (opsonization). Other molecules induce an inflammatory response, in part by inducing the release of agents by mast cells and basophils, or take part in the activation of B cells, or help in the clearing of immune complexes.

Numerous studies have shown that complement is required for an effective immune response. As one example, mice lacking a receptor for an early product of the complement cascade, called C3d, are unable to mount an antibody response. As a second example, depletion of complement in mice infected with Sindbis virus leads to a prolonged viremia and a more severe central nervous system disease. As a third example, people are known who are genetically deficient for components of the complement pathway. Many suffer from immune-complex diseases as a consequence, because they are unable to effectively clear immune complexes. They also suffer from an increased incidence of bacterial infections.

Adaptive Immunity in the Control of Virus Infection

It has been conjectured that the two arms of the adaptive immune system evolved to fight off different pathogens. The CTL response seems well adapted for the control of viral infections because these pathogens replicate intracellularly, but less well adapted for controlling extracellular pathogens such as bacteria or protozoa. Conversely, the humoral response and the associated complement system seem better adapted for the control of extracellular pathogens. Consistent with this model, children who are deficient in the production of antibodies do not in general show an increased susceptibility to viral diseases but do show a marked increase in susceptibility to bacterial infection. Children unable to make gammaglobulin, for example, recover normally from infection by measles virus and are immune to reinfection, demonstrating the importance of T-cell immunity in this disease. Conversely, impairment of CTL function in children often leads to increased frequency and severity of virus infections. Such findings suggest that CTLs evolved primarily to deal with viral infections and remain of prime importance in dealing with viral infections, whereas the humoral system evolved to deal with infections by free-living organisms such as bacteria, protozoa, and yeast.

Although this model may well be correct, it is clear that humoral antibodies are also important in the control of viral disease. Many experiments have shown that passively transferred antibodies alone can protect against viral infection. Further, whether reinfection by a virus results in disease or asymptomatic infection is often correlated with the level of antibodies against the virus in the blood. These observations are particularly relevant to the protection of an unborn or newborn child from viral disease. Maternal antibodies are actively introduced into the fetal bloodstream during intra-uterine development. Such antibodies are critical for the protection of the fetus and the newborn against viral infections early in life before its own immune system develops. It is also clear that neutralizing antibodies are of prime importance in preventing reinfection by at least some viruses, such as influenza virus. Previous infection leading to a vigorous T-cell response directed against many of the viral proteins does not protect against subsequent reinfection by variants which are altered only in their surface glycoproteins. As another example of the importance of humoral antibodies in combating viral infections, CTL-induced cytolysis is not very effective in the control of viral infection of the brain. Neurons are terminally differentiated and cannot be replaced, and express only low levels of MHC class I molecules. However, CTLs do appear to be important in control of viral infections in the brain, perhaps because they secret IFN-γ when activated. Other mechanisms involving humoral antibodies are also probably important. It has been shown in a mouse model that humoral antibodies can cure persistently infected neurons of viral infection.

Thus, a broad and varied immune response to viral infection is important for both the suppression of the original virus infection and in evoking a status of immunity to subsequent reinfection by the virus. Experiences with vaccines used in humans support this idea. Some vaccines have been found to provoke an unbalanced response that renders subsequent infection by the virulent virus more serious, such as early vaccines against measles virus and respiratory syncytial virus, discussed elsewhere.

Vaccination against Viruses

For most viruses, once a person has been infected and recovered, he or she is immune to subsequent reinfection by the same virus. This is the concept behind immunization, also called vaccination, in which a person is exposed to a virus, either live attenuated virus or inactivated virus, or to components of the virus, in order to establish the immune state.

Live Virus Vaccines

Immunization has been practiced for centuries, having been introduced a millennium ago for smallpox. In a process called variolation, less virulent strains of smallpox virus were introduced into humans by intranasal inoculation. The disease induced by this procedure had a lower fatality rate than that caused by the epidemic disease (although the fatality rate was still significant), and the extent of pocking or scarring was less (which was of importance to

people concerned with their appearance). This technique was greatly refined about 200 years ago by Jenner, who immunized people against smallpox with a nonhuman virus derived from cows. Cowpox virus is antigenically related to smallpox virus and induces immunity to smallpox, but does not cause such severe disease as smallpox. This immunization procedure was very successful and smallpox has now been eradicated using modern versions of Jenner's original vaccine. The process of using a nonhuman virus to induce immunity in humans against a related human virus has been referred to as "the Jennerian approach." Jenner's use of cowpox virus to immunize against smallpox gave us the name "vaccination" and "vaccine," from the Latin word vacca meaning cow.

Since Jenner's time, other approaches to vaccination have been developed. Rather than using a nonhuman virus as a vaccine, it is more common to use an attenuated strain of a virulent human virus. Attenuation has classically been achieved by passing the virus in animals or in cultured cells from animals. Passage selects for viruses better adapted to grow in the nonhuman host and often results in a virus that is attenuated in humans. One of the earliest vaccines to be developed in this way was a rabies vaccine developed by Pasteur by passing the virus in rabbits. A more modern method for producing a vaccine strain, which is still often used today, was introduced by Theiler and Smith, who passaged virulent yellow fever virus in chicken tissue and chicken cells in culture. After 100 passages, a marked change in virulence of the virus occurred. Although the passaged virus retained its ability to infect humans, it no longer caused disease. Passage of virus in tissue culture cells and selection of attenuated variants has been used to produce vaccines for measles, mumps, and rubella, among others. With modern technology, it is now possible to introduce mutations into a viral genome that might be expected to attenuate the virus and to test the effects of such mutations in model systems (see Chapter 11). Although no currently licensed human vaccines have been produced in this way, it is expected that this approach will be useful for future vaccines.

In a few cases it is possible to infect humans with a virulent virus in a way that does not lead to disease. Oral vaccines have been developed for adenoviruses 4 and 7 in which the virus is encapsulated in a protective coating that does not dissolve until the virus reaches the intestine. The viruses replicate in the intestine but do not produce disease, although they do induce immunity against adenovirus respiratory disease.

Live virus vaccines in general induce a more protective and longer lasting immunity than do inactivated virus vaccines. They replicate and therefore produce large amounts of antigen over a period of days or weeks that continues to stimulate the immune system. Furthermore, the viral antigens are presented in the context of the normal viral infection and these vaccines induce the full range of immune responses, which includes production of CTLs as well as antibody. Live virus vaccines may be more effective than inactivated vaccines in eradicating the wild-type virus from a society. As described in Chapter 3, an inactivated poliovirus vaccine protects the individual from disease but it does not prevent the wild-type virus from circulating. Finally, live virus vaccines are cheaper to produce and administer than inactivated virus vaccines because a single dose containing smaller amounts of (live) virus is usually sufficient to induce immunity.

Although they have many advantages, live virus vaccines also suffer from a number of potential problems. Attenuating the virus sufficiently so that it does not cause disease while retaining its potency for inducing immunity can be difficult to achieve. The human population is outbred and individuals differ greatly in health status, immune competence, and ability to fight off viral infections, yet a single vaccine must be useful for all, or at least most, individuals in the population. Furthermore, many viruses quickly become overattenuated on passage, losing their ability to induce immunity on infection of humans. Another problem is the potential for reversion of the virus to virulence and the possible virulence of the attenuated virus in normal or, especially, immunocompromised people. In the case of the Sabin polio vaccine, about 10 vaccine-related cases of paralytic polio occurred every year in the United States due to reversion to virulence of the virus (Fig. 3.4) before it was replaced with an inactivated vaccine. Certain lots of yellow fever virus vaccine have also been found to contain partial revertants that can cause encephalitis, especially in infants. Because of this, each lot of yellow fever vaccine must be carefully monitored for virulence and vaccination of infants under 6 months of age is not recommended. Despite this care, a few deaths from vaccine-induced yellow fever have occurred very recently, which appear to be due to residual virulence of the virus in persons with immune systems that are inadequate to handle the infection with this virus because of age or other reasons.

Interference caused by activation of the innate immune system can also cause problems. The effectiveness of a live virus vaccine may be diminished because of interference from a preexisting infection. Interference also makes it difficult (although not impossible) to immunize simultaneously against multiple viruses when using live virus vaccines. This becomes a particular problem with vaccines that contain multiple components, such as vaccines for the four serotypes of dengue virus or for four serotypes of rotaviruses. However, the mumps–measles–rubella vaccine contains three live, attenuated viruses and this vaccine has been very successful in controlling these three viruses with a single inoculation (albeit that a booster is recommended after 10 years or so). Problems can also arise from the presence of adventitious infectious agents in the vaccine, since it has not been treated with inactivating agents. Early lots of the live Sabin poliovirus vaccine were con-

taminated with SV40, for example, which was present in the monkey kidney cells being used to prepare the vaccine. Millions of people were unknowingly infected with SV40, which fortunately appears to cause no disease in humans, or at least to cause disease (some rare brain tumors) only very rarely. Finally, living viruses are often unstable, making it difficult to transport vaccines over long distances and to store them so that they maintain their potency, a problem that is more acute in developing tropical countries.

Although there are potential difficulties, live virus vaccines have many advantages and have been extremely successful in the control of viral diseases. Smallpox virus has been eradicated, measles virus and poliovirus are on the brink of eradication, and many other serious diseases have been controlled by live virus vaccines. A partial list of currently licensed virus vaccines, many of which use live viruses, is given in Table 10.2.

TABLE 10.2 Available Viral Vaccines

Live attenuated virus	Killed virus	Subunit vaccines
Poliovirus (Sabin)	Polio (Salk)	Hepatitis B
Measles	Rabies	West Nile[a]
Mumps	Influenza	Human papillomavirus[b]
Rubella	Hepatitis A	
Yellow fever	Japanese encephalitis	
Vaccinia	Western equine encephalitis	
Varicella[c]		
Rotavirus[d]		
Adenovirus (in military recruits)		
Junin (Argentine hemorrhagic fever)		
Influenza[e]		

[a] Recombinant vaccine on yellow fever 17D backbone in clinical trials; already in use as veterinary vaccine.
[b] Recombinant vaccine (Gardasil) based on virus-like-particles composed of proteins L1 and L2 approved for girls 9–26 years old by the FDA on June 8, 2006. Quadrivalent vaccine contains antigens from HPV16, HPV18, and types 6 and 11 of HPV 6.
[c] Live attenuated vaccine (Zostavax) to protect adults over 60 who have previously had varicella from contracting shingles, approved May 25, 2006.
[d] Second generation pentavalent vaccine (Rota Teq) approved by the FDA on Feb. 3, 2006.
[e] New generation live vaccines composed of reassortants of previous vaccine strains with HA and NA of current epidemic strain being developed.
Source: Data for this table came from Granoff and Webster (1999) p. 1862; from Fields et al. (1996), p. 371; and from Arvin and Greenberg (2006).

Inactivated Virus and Subunit Vaccines

The second general approach to vaccination is to use inactivated virus or subunits of the virus, such as the surface proteins of the virus. The original Salk poliovirus vaccine and current vaccines against influenza and rabies viruses use inactivated virus. For these vaccines, virus is prepared and purified, and virus infectivity is destroyed by treatment with formalin or other inactivating agents. These inactivated viruses are injected, often intramuscularly, and induce an immune response. Subunit vaccines, on the other hand, are usually produced by expressing the surface proteins of a virus in a cell culture system. The proteins are purified and injected. Licensed subunit vaccines include the modern vaccine against hepatitis B virus.

Inactivated virus vaccines or subunit vaccines suffer from a different set of problems from live virus vaccines, but in turn have a number of advantages. Their advantages include the fact that they are usually stable; that interference by infection with other viruses does not occur; and, with proper monitoring, viral virulence is not a problem. Their stability makes transport and storage of the vaccines easier and more reliable. Their insensitivity to interference makes it possible to immunize against many viruses simultaneously. The fact that the viruses are inactivated, or that no live virus was ever present in a subunit vaccine, means that no virus infection occurs with its potential for disease. Early lots of the Salk inactivated poliovirus vaccine were contaminated with live poliovirus, which resulted in a number of cases of polio caused by the vaccine, but better methods of inactivation and of monitoring residual infectivity have solved this problem.

Difficulties with these vaccines include the expense of preparing the large amounts of material required, the necessity for multiple inoculations in the case of most such vaccines, and the failure to induce a full range of immune responses. Because no virus replication occurs, large amounts of material must be injected to induce an adequate immune response. Further, lack of virus replication means that an inflammatory response that is required for an efficient immune response must be obtained by using adjuvents that are incorporated into the vaccine. In practice, inactivated virus or subunit vaccines are designed to provoke only a very limited inflammatory response, and multiple immunizations are usually required in order to achieve effective immunity. Even so, a full range of immunity is not achieved. Worse, in two cases immunization with inactivated virus resulted in more serious illness on subsequent infection with epidemic virus. Inactivated virus vaccines against measles virus and respiratory syncytial virus did not protect against infection by the respective viruses, and led to the development of atypical disease that was more severe than that caused by the viruses in nonimmune people. The measles vaccine was subsequently replaced with a live virus vaccine, which has been very successful, but no successful vaccine has yet been developed for respiratory syncytial virus.

Despite these problems, many successful vaccines have been introduced that use inactivated viruses or subunits of viruses, and some of these are listed in Table 10.2. Characteristics of live virus vaccines versus nonliving vaccines are compared in Table 10.3

Modern biotechnology makes possible another approach to subunit vaccines, the use of a nonpathogenic virus as a vector to express proteins from a virulent virus. This approach potentially overcomes many of the disadvantages of inactivated virus vaccines, or of subunit vaccines based on purified components, because the antigens are presented in the context of a viral infection and in a native conformation. To date, none of the licensed vaccines use this procedure, but clinical trials are in progress or are planned to begin soon for a number of such vaccines. Two examples will be cited. In one series of trials, vaccinia virus is being used to express the HIV surface glycoproteins. Antibodies to the HIV glycoproteins are induced, but whether they are protective remains to be determined. In another series of trials, the vaccine strain of yellow fever virus has been engineered to express the surface glycoproteins of Japanese encephalitis virus, of dengue virus, or of West Nile virus. Based on the successful use of yellow fever vaccine for more than 50 years, the prognosis is good that these new vaccines will be successful. The use of viruses as vectors is covered in more detail in Chapter 11.

DNA Vaccines

A new and potentially exciting approach to vaccination is to inject DNA that encodes genes whose expression leads to immunity to a virus. This approach has been tested in model systems with promising results and clinical trials in humans have recently begun. Plasmid DNA containing the gene for,

TABLE 10.3   Characteristics of Live and Killed Vaccines

Activity	Live attenuated virus vaccine	Killed virus and subunit
Antibody induction	+++	+++
MHC class I (CD8 cells)	+++	–
Cytotoxic T-cells	+++	–
MHC class II (CD4 cells)	+++	+++
Humoral antibody	+++	+++
All viral antigens	Usually	Seldom
Longevity of immunity	Months/years	Months
Cross-reactivity among viral strains	+++	+
Risk of viral disease	+	–

Source: Data for this table came from Granoff and Webster (1999), p. 1862 and from Fields *et al.* (1996), p. 371.

say, a surface glycoprotein of influenza virus under the control of a general mammalian promoter is injected intramuscularly. Muscle cells take up DNA and import it into the nucleus, a surprising finding. Expression of the influenza gene may elicit an immune reaction against the encoded influenza protein. Immunization with DNA might produce protective immunity in the best of cases, or DNA immunization might serve to prime an individual for later inoculation with a conventional vaccine so as to obtain a stronger response.

Immunization with DNA has many potential advantages. The encoded protein is made intracellularly and therefore a CTL response is generated, because peptides are presented by class I MHC molecules. In addition, the proteins are made and are present in a native conformation, which may make possible the production of antibodies against a wider range of conformational epitopes. Other advantages include the fact that multiple proteins can be encoded in plasmid DNA, which allows immunization against multiple products of one virus or against several different viruses, and that cytokine genes can also be present in the plasmid, which can serve as adjuvents to obtain not only a stronger immune response but to direct the immune response toward desired pathways. Another important feature is that the expression of the foreign protein continues for some time, potentially leading to a better immune response. Finally, plasmid DNA is easy to modify to express different genes, cheap to make in quantity, and easy to purify. Once safety trials have been conducted for a DNA vaccine, it should be possible to modify it to express antigens from other viruses with minimal safety concerns.

Human trials with DNA vaccines are just beginning. A Phase 1 safety trial of an HIV vaccine has just been completed using 86 volunteers, and the results were sufficiently encouraging to begin Phase 2 efficacy trials. The DNA vaccine uses multiple DNAs expressing HIV-1 genes from three subtypes of the virus. Three intramuscular injections to prime the individuals were given using a needle-free device. This was followed by a boost using adenovirus type 5 that expressed HIV proteins. At the highest doses used, almost all individuals produced antibodies against HIV as well as both CD4+ and CD8+ T cells that were reactive against HIV antigens.

Plant-Based Vaccines

Another possible approach to the production of vaccines is to use modern biotechnology to introduce genes for viral proteins into a food plant. If eating this plant induces immunity, it would provide a cheap and easy alternative to vaccines that require extensive purification followed by injection. No human vaccines have been developed to date using this technology, but the concept has been tested in animals. Piglets fed transgenic corn that expresses the spike protein of swine transmissible gastroenteritis virus (TGEV) were better protected from TGEV infection than were animals vaccinated with a commercial live TGEV vaccine. Mice immunized with a single dose of DNA expressing the measles H protein and then multiply boosted by being fed transgenic tobacco plants expressing measles H protein developed significantly higher neutralizing antibody titers to measles virus than did mice immunized with either the DNA or the plant food alone. Thus, in principle plant-based vaccines are capable of inducing immunity. Whether practical vaccines can be developed using this approach remains a question for the future.

Vaccine Development: The Example of Measles Vaccines

Vaccines against viruses have been enormously successful in controlling viral diseases that have plagued mankind for millennia. Smallpox virus has been eradicated worldwide, poliovirus has been eradicated from the Americas and may be eradicated worldwide within the next few years, and many epidemic diseases have been controlled, at least in regions that have access to the vaccines. As an example of potential difficulties in vaccine development, however, the measles virus vaccines will be discussed. Measles virus is ubiquitous. Not so long ago, virtually the entire population of the world was infected by the virus as children. Very few people escaped infection by the virus. Measles can be a serious illness, with significant mortality and debilitating sequelae (Chapter 4). Concerted efforts to produce a vaccine began many years ago. One early measles virus vaccine, used from 1963 to 1967, was made with inactivated virus. Vaccination with this vaccine potentiated a more serious illness when a person was infected with the epidemic virus, however, characterized by a more severe rash and called "atypical measles." The atypical disease is thought to be produced by an atypical or unbalanced T-cell response to the virus following immunization. This response controls virus infection only poorly and leads to collateral damage from T-cell attacks on virus-infected cells. An attenuated virus vaccine was more successful. The first live virus vaccine used was protective but insufficiently attenuated and thus more reactogenic than would be ideal. It was subsequently replaced with a more highly attenuated vaccine.

The attenuated virus vaccine currently in use has been highly successful in controlling measles in many developing countries as well as in the developed countries. The distribution of the virus as of 1998 is shown in Fig. 10.13A, illustrating that by this time the virus had been eradicated in many developed countries and had substantially reduced the incidence of measles over most of the world. By 2004, the virus had been essentially eradicated from the Americas and most other areas, but pockets of endemicity remain in Africa (Fig. 10.13B) where the virus continues to be an important cause of infant mortality and morbidity. The failure

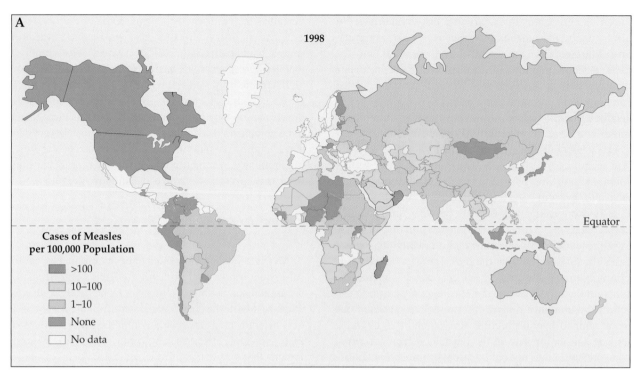

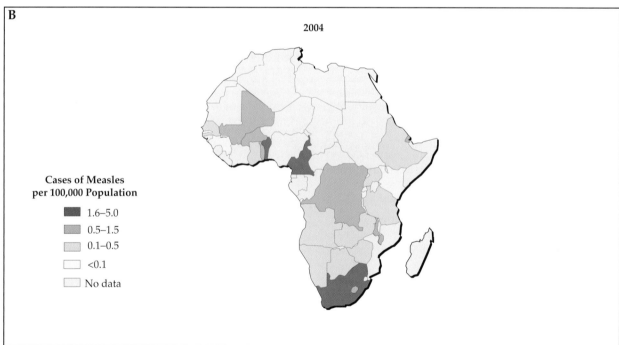

FIGURE 10.13   Incidence of measles. (A) Reported incidence rate of measles per country as of 14 August 1998. Rates are shown as cases per 100,000 population. From Web site of World Health Organization. (B) The incidence of measles in Africa in 2004. Note the drastically different scales. In 2004 the Americas were virtually free of measles, although Southeast Asia (including India) still records upwards of 200,000 deaths from measles annually.

to completely eradicate the virus in these areas is due to the difficulty of immunizing infants with a live virus vaccine while maternal antibodies are still protective. In regions of Africa, measles is so widely epidemic that many infants contract the disease as soon as the protection due to maternal antibodies wears off. Multiple, spaced immunizations for each infant would be required to successfully immunize the population of new susceptibles before the epidemic virus gets there, which is difficult to achieve in countries with limited health care facilities. In an effort to get around this problem, very young infants were immunized with higher doses of the live virus vaccine in a trial in Senegal, on the theory that using a larger dose would overcome the effects of maternally derived immunity. This trial was a failure, however, because for unknown reasons the cumulative mortality in infants receiving the high titer vaccine was greater than that in controls (Fig. 10.14). Thus at present it is very difficult to devise methods that will effectively immunize young infants in these countries against measles, but programs are being developed to multiply immunize infants at different ages using national immunization days in order to overcome these problems.

INNATE IMMUNE SYSTEM

The innate immune system is composed of a large number of elements that attempt to control infection by pathogens, but this system is independent of the identity of the particular pathogen. Complement is a component of the innate system

as well as the adaptive system and has been described. Other components of the innate system include the cytokines, a complicated set of small proteins that are powerful regulators of the immune system. Many cytokines are critical for the function of the adaptive immune system, as described in part before. Cytokines are also important components of the innate system. The innate system also includes natural killer cells, cytotoxic lymphocytes that kill cells that do not express adequate levels of class I MHC molecules. Apoptosis is another innate defense mechanism, because most cells are programmed to die if they sense that they are infected.

Toll-Like Receptors

Initiation of an immune response requires the recognition of "*p*athogen-*a*ssociated *m*olecular *p*atterns" or PAMPs for short, which are conserved molecular motifs present in or produced by invading pathogens but which are not present in their hosts. The sensors that recognize PAMPs are called *p*attern *r*ecognition *r*eceptors or PRRs. The best known PRRs are *T*oll-*l*ike *r*eceptors (TLRs) but other PRRs exist, such as the protein called RIG-1 described later.

TLRs are components of ancient pathways of innate immunity to invading microbes. They recognize a number of conserved structures in microbes and upon activation lead to the production of antimicrobial peptides or to activation of components of the innate and the adaptive immune systems. The Toll pathway was first discovered in the fruit fly, *Drosophila melanogaster*. Activation of this pathway in *Drosophila* results in the production of peptides that

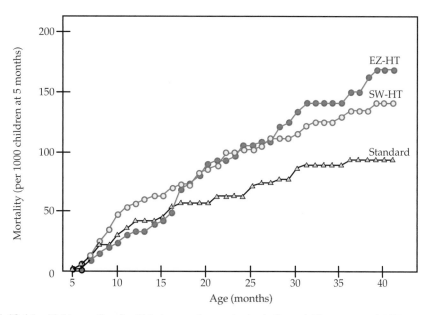

FIGURE 10.14   Child mortality after high-titer measles vaccination in Senegal. Three groups of children were given either $10^{5.4}$ PFU of Edmonston-Zagreb vaccine at 5 months of age (EZ-HT), or $10^{5.4}$ PFU Schwartz vaccine (SW-HT) at 5 months of age or the standard schedule (placebo at 5 months and $10^{3.7}$ PFU vaccine at 10 months of age.) Adapted from Garenne *et al.* (1991).

render the fly resistant to fungal infection and to infection by Gram-positive bacteria. Flies that lack this pathway are very sensitive to infection by fungi or Gram-positive bacteria. Resistance of flies to Gram-negative bacteria is mediated by another pathway called the IMD pathway. Innate defenses against microorganisms are also known to exist in starfish, nematodes, plants, and virtually all other organisms that have been examined, and many of these defense mechanisms are known to rely on Toll-like proteins.

Mammalian homologues of Toll have been described only recently. There are at least 11 TLRs in humans that can be grouped into five subfamilies. They are integral membrane proteins, some of which are expressed at the cell surface and some of which are present in intracellular vesicles. They possess extracytoplasmic leucine-rich domains that are important for recognition of the molecular stimuli and intracytoplasmic domains that signal, via various intermediates, to start production of effector proteins. A partial list of the ligands that stimulate various TLRs is shown in Table 10.4. Some TLRs recognize nonself molecules as is the case for TLR1, 2, 4, 5, and 6, whereas others recognize nucleic acids in unfamiliar contexts as in the case for TLR3, 7, 8, and 9. In general, TLRs in the first class are expressed at the cell surface whereas TLRs of the second class are expressed in

intracellular compartments, usually thought to be endosomal compartments. Examples of TLRs activated by nonself molecules include TLR4, activated by exposure to lipopolysaccharide (LPS), a component of the cell wall of gram-negative bacteria, as well as by other molecules, including the envelope proteins of some viruses; TLR5, activated by exposure to bacterial flagellin; and TLR2, activated by components from Gram-negative bacteria, mycoplasma, spirochetes, and trypanosomes, among others. TLRs activated by exposure to nucleic acids include TLR9, activated by exposure to nonmethylated CpG in DNA, which is rare in mammalian DNA (CpG itself is underrepresented in mammalian DNA and it is usually methylated) but common in bacteria and in DNA viruses; TLR3, activated by double-stranded RNA, which is produced by most viruses after infection; and TLR7 and TLR8, which recognize single-stranded RNA. Overall, then, a wide variety of molecules produced by invading viruses, bacteria, fungi, and other pathogens is recognized by various TLRs and an immune response is generated. Of interest also is the fact that TLR4 responds to heat shock proteins (HSP60 and HSP70), which are produced by the host in response to stress, which can include the stress induced by infection.

TLRs that are expressed on the cell surface respond to extracellular stimuli, whether present there because of cell

TABLE 10.4   Characterization of Toll-Like Receptors

Toll-like receptor	Major cell types	Ligands	Source of ligand
TLRs found predominantly in the cell plasmalemma			
TLR1/TLR2 heterodimer	Monocytes, mDCs	Tri-acyl lipopeptides	Bacteria/mycobacteria
		GP-1 anchored proteins	Parasites
TLR2/TLR6 heterodimer	Monocytes, mDCs, pDCs	Di-acyl lipopeptides, lipoteichoic acid,	Bacteria/mycobacteria, gram-positive bacteria
		Zymosan	Fungi
TLR4 (homodimer)	Monocytes, differentiated DCs	LPS	Gram-negative bacteria
		Taxol	Plant
		HRSV fusion protein	*Paramyxoviridae*
		Heat shock proteins	Mammalian host
		Fibrinogen	Mammalian host
		Envelope proteins	MMTV (*Retroviridae*)
TLR 5	Monocytes	Flagellin	Motile bacteria
TLRs found predominantly in intracellular compartments			
TLR3	mDCs (also surface of fibroblasts)	dsRNA	Many viruses; *Reoviridae*
TLR7	???	ssRNA, some siRNAs	Many viruses including: HIV (*Retroviridae*), VSV (*Rhabdoviridae*), Influenza
TLR8	Monocytes, mDCs	ssRNA	Many viruses; *Rhabdoviridae*
TLR9	pDCs	CpG DNA	Bacteria
		CpG DNA	DNA viruses, esp. *Herpesviridae*

Abbreviations: mDCs, myeloid dendritic cells; pDCs, plasmacytoid dendritic cells; HRSV, human respiratory syncytial virus; MMTV, mouse mammary tumor virus. Interactions most important for controlling viral infection are shown in red.
Source: Data from Takeda *et al.* (2003); O'Neill (2005); Iwasaki and Medzhitov (2004); Kawai and Akira (2006).

death that releases these components or because extracellular microorganisms are present. TLRs in intracellular compartments may also be responding to extracellular molecules that enter the endosomal compartment by endocytosis, or they may be intracellular molecules that enter the intracellular compartments where TLRs are present during infection or replication. Many (most?) TLRs function as dimers. Heterodimers of TLR1 and TLR2 and of TLR6 and TLR2 have been found, and these two heterodimers have different specificities. TLR4 is commonly found as a homodimer.

Various TLRs are expressed in most tissues of the body. Particularly noteworthy is the expression of multiple TLRs by monocytes, macrophages, dendritic cells, and mast cells, key players in the immune system. Activation of TLRs leads to activation of transcription factors such as NFκB that results in the production of a number of end products. In some cases, antimicrobial peptides are produced, including in humans various defensins described later. Of equal or greater importance, however, is the expression of proinflammatory genes such as TNF-α, IL-6, IL-1β, and IL-12 in response to activation of several TLRs. Of particular importance, type I interferons are produced in response to activation of TLR3, 4, 7, 8, and 9. The inflammatory response that results from the activation of TLRs is crucial for the activation of other components of the innate system as well as for the activation of the adaptive immune system. In particular, type I IFNs are key players in starting and organizing an immune response. TLR-signaling also leads to the maturation of dendritic cells, key players in the adaptive immune system. Activation of dendritic cells may be direct, by infection of these cells by viruses, for example, or indirect, as the result of signaling by other infected cells. There appear to be multiple kinds of dendritic cells that express different combinations of TLRs. Upon activation of a dendritic cell, it migrates to lymph nodes where it presents antigens to T cells and stimulates T cells that recognize the antigen presented by it to become active. Thus, the maturation of dendritic cells induced by activation of TLRs is critical for the activation of the adaptive immune response. Overall, therefore, TLRs are not only important as the first line of defense against invading microorganisms via the production of antimicrobial peptides or the induction of cytokine expression that slow their growth and in some cases eradicate the infection, but also for activation of an adaptive immune response to eliminate the invading pathogen and cause the organism to become immune.

Defensins

Organisms from plants to mammals produce antimicrobial peptides that are important in defending the organism from infection. Mammals produce two classes of such peptides, called cathelicidins and defensins. Defensins are peptides of 18 to 45 amino acids that possess three disulfide bonds, have a net positive charge, and are produced by cleavage of propeptides. They fall into three subfamilies called α-, β-, and θ-defensins. α- and β-defensins are linear polypeptides whereas θ-defensins are cyclic 18mers. Only α- and β-defensins are produced in humans. They are mainly produced by leukocytes and epithelial cells.

α- and β-defensins have diverged from a common β-defensin ancestral gene that is expressed as far back in evolution as snakes. There are multiple genes for them. Six human α-defensins have been identified, and gene-based searches indicate that there are about 30 β-defensin genes in humans and more than this in mice. These polypeptides will kill a number of microbes *in vitro*, and model studies in mice have shown that they are important in protecting mice from bacteria that infect the lungs or the gut. Production of α-defensins is constituitive for the most part, whereas production of β-defensins is usually inducible. Induction can be through the activation of Toll-like receptors or through the activity of cytokines. Defensins appear to work by permeabilizing the microbial membrane, but they also appear to recruit other elements of the immune system to the site of infection. They may have evolved to fight off bacteria, fungi, and protozoa, but studies have shown that they are also effective against at least some viruses. Their antiviral activity is of two types. They can directly inactivate some enveloped viruses, probably in the same way that they inactivate bacteria. They can also interact with potential target cells to interfere with virus replication by mechanisms that are poorly understood, and these mechanisms are potentially effective against both enveloped and nonenveloped viruses. It is noteworthy that both α- and β-defensins have been found in breast milk, suggesting that they are important for protecting the infant from infection.

Natural Killer Cells

Natural killer (NK) cells are cytolytic cells that kill by an antigen-independent mechanism. They are important for the control of many virus infections, as shown by the fact that ablation of NK cells leads to more serious disease. NK cell activity increases within the first 2–3 days after infection by a virus, stimulated by the presence of interferon and perhaps by other cytokines, and thereafter the number of cells declines. One of the functions of these cells is to kill cells that do not express class I MHC or that express it in only low amounts. Such cells, of course, do not present antigen to CTLs and therefore escape normal immune surveillance. NK cells express two sets of receptors on their surface. One set of receptors interacts with MHC class I molecules on the surface of the target cell. This interaction inhibits killing by the NK cell. The second set of NK receptors interacts with activating molecules on the surface of cells. Interaction with activating molecules will stimulate the NK cell to kill the target cell if the NK cell is not sufficiently inhibited by its interaction with class I molecules.

Many different viruses downregulate the production of MHC in infected cells in order to escape immune surveillance. The elimination of these infected cells by NK cells is of considerable importance to the host. It seems clear that NK cells evolved to rid the body of cells that are not subject to normal immune surveillance, which could include tumor cells as well as virus-infected cells that no longer express adequate levels of MHC. In the one case known of a human with a deficiency in NK cells, as well as in experimental studies with mice in which NK cells are depleted, infection by a number of viruses that downregulate production of MHC results in a much more severe disease than that produced in individuals able to mount a normal NK response.

Apoptosis

Apoptosis or programmed cell death is a defense of last resort by an infected cell. It is a cell suicide pathway in which mitochondria cease to function, chromatin in the nucleus condenses, nuclear DNA is degraded, and the cell fragments into smaller membrane-bound vesicles that are phagocytosed by neighboring cells. Apoptosis serves to eliminate cells that are no longer needed, cells that are damaged in some way (such as by exposure to UV light or to free radicals), or cells that are dangerous to the host because they are infected or because their cell cycle is deregulated (and they are therefore potential tumor cells). Cells that are no longer needed include excess cells that are produced during development and must be eliminated (a normal occurrence during development) and T cells responsible for the control of an infectious agent after the infection has been eliminated (and which may even be dangerous if left around). Thus, apoptosis serves many roles in an organism and the basic machinery has been conserved throughout evolution. In mammals it serves as a component of both the adaptive immune response (ridding the body of unneeded lymphocytes and killing of infected cells by CTLs) and the innate immune response (suicide by infected cells).

Several different mechanisms for the induction of apoptosis exist, as would be expected from the many functions served by apoptosis. Furthermore, the control of apoptosis is very complicated, which is necessary to control an event as drastic as cell death. Some of the pathways used to induce apoptosis are illustrated in Fig. 10.15. One pathway involves

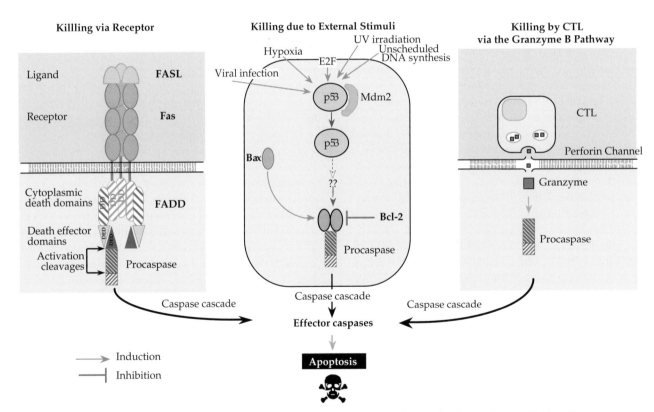

FIGURE 10.15   Three pathways that lead to cell death by apoptosis. In each case a signal comes from outside the cell. This signal may be the binding of a ligand to its receptor (left panel); stress upon the cell caused by viral infection, DNA damage caused by exposure to UV light, deregulation of the cell cycle, or other stresses (center panel); or from a cytotoxic T cell (right panel). The signal leads to the activation of a procaspase by proteolytic cleavage. The activated caspase begins a cascade that leads to apoptosis. Data for this figure came from Ashkenazi and Dixit (1998) and Raff (1998).

death receptors on the cell surface, which initiate apoptotic pathways when their ligands are present and bind to them. One such receptor is the Fas receptor (also known as CD95 or Apo-1), which can be stimulated by Fas ligand present on the surface of CTLs. Fas contains domains known as death domains in the cytoplasmic region, and trimerization of Fas induced by ligand binding results in the recruitment of other cellular proteins by these death domains. This in turn results in the activation, by cleavage, of a protease known as caspase-8. A popular model for activation is that two molecules of caspase-8 must be brought into proximity so that they cleave one another. Caspase-8 then cleaves other caspases, thereby activating them in turn. At least 10 caspases are known, and they are normally present as inactive procaspases. A cascade of caspase cleavages leads, ultimately, to the activation of effector caspases. The effector caspases cause cell death by cleaving many cellular proteins, including, for example, lamin in the nuclear membrane and a repair enzyme called poly(ADP-ribose) polymerase.

Another death receptor that may be present at the cell surface is a receptor for TNF (tumor necrosis factor). TNF can be secreted by CTLs, as well as by other lymphocytes, in response to inflammatory signals. TNF receptor oligomerization also results in cleavage of caspase-8 and the induction of the caspase cascade. The sensitivity of cells to the induction of apoptosis by these pathways thus depends on the concentration of Fas and TNF receptor in the surface, and this concentration can be regulated by other events.

Fas and TNF receptors are ubiquitous, present on most cells. Many other receptors that can lead to apoptosis are also known. Some of these are also ubiquitous, others are present on only a limited set of cells.

The caspase cascade can also be initiated by other mechanisms. Withdrawal of growth factors, such as occurs during development or occurs when activated T cells are no longer stimulated by interaction with their antigens, can lead to activation of caspases. Events that damage DNA or the protein synthetic machinery can lead to apoptosis. At least some of these apoptotic signals are delivered through changes in p53 concentrations. p53 is a key player in the regulation of the cell cycle (Chapter 7), and the apoptotic pathway is induced if its concentration is not tightly regulated. These mechanisms involve Bax and related molecules, as well as Bcl-2 and related molecules, as key players in the activation of the caspase cascade.

In addition to being able to initiate the caspase cascade indirectly by signaling through the Fas receptor or the TNF receptor, T cells can also initiate the caspase cascade directly. They do this by introducing granzymes into the cell by way of channels formed by perforin secreted by the T cell. Granzyme B cleaves procaspases downstream of caspase-8 to begin the cascade.

Control of apoptosis is complex. Many cells make proteins that render them less susceptible to apoptosis. One such anti-apoptotic protein is called Bcl-2, which blocks the action of Bax and interferes with the caspase cascade, among other activities. Bcl-2 is made by mature neurons, for example, in response to signals received when they form connections with the target cells that they enervate. Immature neurons do not make adequate amounts of Bcl-2 and die when nerve growth factor is removed. There is a complicated interplay between Bcl-2 and related proteins and between Bax and related proteins that we are only beginning to understand.

Mitochondria are also important players in apoptosis. A decrease in the potential across the mitochondrial membrane makes the cell more sensitive to apoptosis. Decreased potential is accompanied by synthesis of reactive oxygen species and the release of cytochrome C into the cytoplasm, both of which are proapoptotic. Cytochrome C forms complexes with other cellular proteins that promote the caspase cascade. Bax, a proapoptotic protein, may exert its effect by lowering the mitochondrial membrane potential, and Bcl-2 may prevent this.

The importance of apoptosis for the regulation of the organism is clear. Individuals with deficits in apoptotic pathways may suffer from many diseases, including developmental abnormalities, neurodegenerative disorders, or autoimmune diseases. In normal individuals, tumor cells can only thrive if they avoid apoptosis despite the fact that their cell cycle is deregulated, and mutations that suppress apoptosis are important for the development of tumors.

Interferons and Other Cytokines

The cytokines constitute a large family of proteins, defined by their structure, that have important regulatory roles in an animal. More than 30 cytokines have been described, and for some of these there is more than one gene. Most have a molecular mass of about 30 kDa. Many, but not all, are glycoproteins. Some function as monomers but others act as homodimers or homotrimers and at least one acts as a heterodimer. Cytokines are inducible agents and constitutive production is normally low or absent. On induction, their production is short lived. Cytokines effect their action by binding to receptors on the surface of target cells. These receptors bind cytokines with high affinity—the dissociation constants range from 10^{-9} to 10^{-12} M. Binding to the receptor induces a cascade of events in the target cell that leads to changes in gene expression within the cell. These changes in gene expression may lead to many different effects. They are important in orchestrating the immune response by inducing cell proliferation in cells like T cells and B cells and causing changes in the state of differentiation of cells like dendritic cells. Hematopoietic cells are important targets of all cytokines, although most cytokines have a diverse range of actions. The cytokine system is redundant in that many cytokines evoke a similar spectrum of action, and it is pleiotrophic in that one cytokine may have many different target cells. This system is also complex because many different cytokines are usually induced at the same time and they may act synergistically or antagonistically to achieve a result. Cytokines are important players in both the adaptive and innate immune systems. The importance of cytokines in the maturation of T cells and B cells has been described. Fig. 10.16 presents an overview of cytokine networks that illustrates their complexity, and a partial listing of cytokines is given in Table 10.5. Because of the complexity of their activities, we are only beginning to understand their many functions, but as these functions begin to be understood, attempts are being made to use different cytokines therapeutically as indicated in the table.

A second family of proteins that play an important role in the regulation of the immune response is the chemokines. Chemokines are small proteins, 70–80 amino acids in size. More than 30 are known in humans. Some serve housekeeping functions and are produced constitutively; others are proinflammatory and usually inducible. Chemokines serve to attract leukocytes. The housekeeping chemokines are important for the development and homeostasis of the

hematopoietic system (e.g., maintenance of lymphoid organs), whereas the proinflammatory chemokines recruit immune cells to sites of infection, inflammation, or tissue damage. The receptors for chemokines are distinct from those of cytokines. An example of a cytokine receptor is described later in this chapter for interferon. Chemokine receptors belong to the family of seven-transmembrane-domain, G-protein-coupled receptors, and one was illustrated schematically in Fig. 1.4B as the receptor for HIV. In some treatments, chemokines are considered a class of cytokines. Here, because these two classes of molecules differ in structure and mechanism of action, the term cytokine refers only to nonchemokine cytokines.

Importance of Interferons in the Defense against Viruses

Among the best known cytokines that function in the defense against viruses are the interferons (IFNs). There are two kinds of IFN, called type I and type II. The IFNs

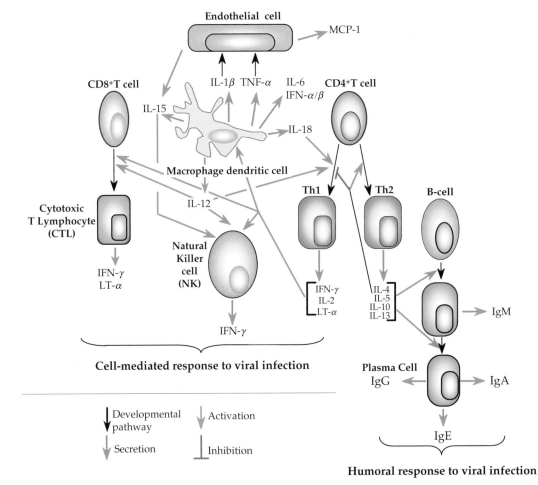

FIGURE 10.16   Overview of the cytokine networks important for innate and acquired antiviral immune responses. IFN, interferon; IL, interleukin; Ig, immunoglobulin; TNF, tumor necrosis factor; LT, lymphotoxin; MCP, monocyte chemotactic protein. Adapted from Griffin (1999) Figure 1 on p. 340.

TABLE 10.5   Some Cytokines and Their Therapeutic Uses

Functional group	Name (abbreviation)	Normal biological function	Therapeutic targets	Side effects of therapy
Antiviral Cytokines	Type I interferon (IFN-α,β)	Inhibits viral replication	Chronic hepatitis B, hepatitis C, herpes zoster, papilloma viruses, rhinovirus, HIV(?), warts	Fever, malaise, fatigue, muscle pain; toxic to kidney, liver, heart, bone marrow
	Type II interferon (IFN-γ)	Inhibits viral replication, upregulates expression of class I and class II MHC, enhances activity of macrophages	Lepromatous leprosy, leishmaniasis, toxoplasmosis	As above for Type I interferons
Inflammatory Cytokines	Tumor necrosis factor (TNF)	Cytotoxic for tumor cells, induces cytokine secretion by inflammatory cells	Anti-TNF in septic shock	Shock with marked hypotension
	Interleukin 1 (IL-1)	Costimulates T-helper cells, promotes maturation of B cells, enhances activity of NK cells, attracts macrophages and neutrophils	Receptor antagonist in septic shock	???
	Interleukin 6 (IL-6)	Promotes differentiation of B cells, stimulates Ab secretion by plasma cells		
Regulators of lymphocyte functions	Interleukin 2 (IL-2)	Induces proliferation of T cells, B cells, and CTLs, stimulates NK cells	Leprosy, local treatment of skin lesions	Vascular leak syndrome, hypotension, edema, ascites, renal failure, hepatic failure, mental changes, and coma
	Interleukin 4 (IL-4)	Stimulates activity of B cells, and proliferation of activated B cells, induces class switch to IgG and IgE		
	Interleukin 5 (IL-5)	Stimulates activity of B cells, and proliferation of activated B cells, induces class switch to IgA		
	Interleukin 7 (IL-7)	Induces differentiation of stem cells, increases IL-2 in resting cells		
	Interleukin 9 (IL-9)	Mitogenic activity		
	Interleukin 10 (IL-10)	Suppresses cytokines in macrophages	Septic shock	
	Interleukin 12 (IL-12)	Induces differentiation of T cells into CTLs		
	Interleukin 13 (IL-13)	Regulates inflammatory response in macrophages		
	Transforming growth factor (TGF-β)	Chemotactically attracts macrophages, limits inflammatory response, promotes wound healing	Septic shock	Symptoms similar to those for IL-2, especially shock and hypotension

Source: Data for this table came from Mims *et al*. (1993) p. 37.3, and Figure 7.13, and Kuby (1997) pp. 318, 319.

induce cells to become resistant to viral infection, an innate defense against viruses, but also play important roles in the adaptive immune response. The importance of IFNs in the defense of mammals against viral infection has been shown by experiments in which an IFN response is ablated. Early experiments in mice used injection of antibodies against IFN to block its activity. More stringent ablation of IFN activity has been accomplished by using transgenic mice in which the receptor for either type I or type II IFN has been abolished. In general, mice that lack a type I IFN response are extremely sensitive to infection by viruses. Viruses grow to much higher titer in such animals and, in the more dramatic examples, virus infection may be lethal although infection by the same virus in animals able to mount a normal IFN response may be asymptomatic. Many viruses have evolved mechanisms to ablate the activity of IFNs, and in many cases it has been shown that viral mutants that have lost the ability to resist IFN are severely crippled, again demonstrating their importance in controlling viral infection. Interestingly, mice lacking a type I IFN response are able to handle bacterial infections reasonably well. Conversely, mice lacking a type II IFN response are extremely sensitive to bacterial infection but handle viral infections well. One known exception is vaccinia virus, which is lethal in mice lacking either IFN response. Although the relative importance of the two IFNs in the defense against viruses versus bacteria may differ, both are important in the defense against viruses.

Types of IFNs

Several characteristics of type I and type II IFNs are shown in Table 10.6. The type I IFNs are 165–172 amino acids in length and have been classified into four different subfamilies, called α, β, ω, and τ, of which α and β are the best studied. In humans there are 14 IFN-α genes and 1 IFN-β gene, none of which contain introns. IFN-β shares 25–30% amino acid sequence identity with any particular IFN-α. In humans, IFN-β is glycosylated whereas IFN-α is not, but in mice both are glycosylated. Thus, glycosylation is not a fundamental property distinguishing α from β IFNs. IFN-α and IFN-β use the same receptors and therefore evoke the same responses in target cells.

Type II IFN contains only one member, IFN-γ. In humans there is one IFN-γ gene, which contains three introns. The receptor used by IFN-γ is distinct from that used by type I IFNs, but the phosphorylation cascade induced by IFN-γ contains some elements that are shared with the type I cascade. Thus the responses to the two IFNs are partially overlapping.

Induction of IFN

Type I IFNs are induced upon the activation of a number of PRRs, as described earlier. Induction may be direct (e.g., upon activation of TLR3, 4, 7, 8, 9 or RIG-1) or the result of production of other cytokines (e.g., TNF-α). The best studied activating agent that results in type I IFN production, and perhaps the most important agent, is double-stranded RNA (dsRNA), whether infection is by a DNA- or RNA-containing virus. Two dsRNA sensors are known whose activation results in production of type I IFN, illustrated schematically in Fig. 10.17. One sensor is TLR3 and the second sensor is a protein called RIG-1. TLR3 is present in intracellular compartments in many cells but present at the cell surface of fibroblasts, where it senses external dsRNA. As noted earlier, it is possible that intracellular TLR3 also senses external dsRNA that has been endocytosed or it may sense replicating RNA that enters the endosomal compartment. In any event, upon binding dsRNA, TLR3 signals through a protein called TRIF. In contrast, RIG-1, which has an RNA helicase domain that binds dsRNA, is present in the cytosol and therefore detects cytoplasmic dsRNA. Upon binding dsRNA, RIG-1 interacts with a protein called MAVS (it also has other names),which is attached to the mitochondrial membrane by means of a C-terminal transmembrane anchor. In either case, signaling proceeds through a number of intermediate steps with

TABLE 10.6   Characteristics of the Interferons

Characteristics	Type I		Type II
	INF-α	INF-β	IFN-γ
Alternative name	Leucocyte IFN	Fibroblast IFN	Immune IFN
Location	All cells	All cells	T-lymphocytes
Inducing agent	Viral infection or dsRNA	Viral infection or dsRNA	Antigen or mitogen
Number of species (number of genes)	14 (man), 22 (mouse)	1	1
Chromosomal location of gene	9 (man), 4 (mouse)	9 (man), 4 (mouse)	12 (man), 10 (mouse)
Number of introns	None	None	Three
Size of IFN protein	165–166 aa	166 aa	146 aa, dimerizes
Receptors	Receptor for both IFN-α and INF-β consists of 2 polypeptides: IFN-α R1 and INF-α R2, encoded on chromosome 21 (man) or 16 (mouse)		Receptor consists of two proteins: IFN-γR1 encoded on chromosome 6 (man) or 10 (mouse) and IFN-γR2 encoded on chromosome 21 (man) or 16 (mouse)
General functions	Anti-viral activity ↑ MHC class I[a]	Anti-viral activity ↑ MHC class I	Macrophage activation ↑ MHC class I ↑ MHC class II on macrophages NK cell activation, some antiviral activity ↓ MHC class II on B cells ↓ IgE, IgG production by B cells

[a] ↓ = downregulate; ↑ = upregulate.

Source: Data for this table came from Mims *et al.* (1993) Figure 12.9 and Fields *et al.* (1996) Table 3 on p. 378.

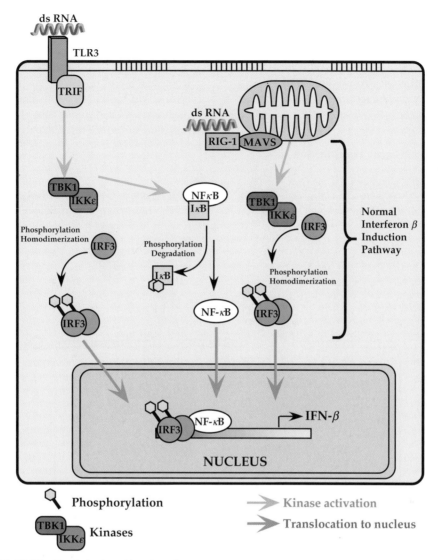

FIGURE 10.17   The induction of interferon β. External double-stranded RNA signals through Toll-like receptor 3 (TLR3) to TRIF, and internal dsRNA signals to MAVS through the *r*etinoic-acid *i*nducible *g*ene I (RIG-I) protein. MAVS is the *m*itochondrial *a*ntiviral *s*ignaling protein. MAVS and TRIF both can activate the kinases TBK1 and IKKε, leading to the phosphorylation and homodimerization of IRF3, and the phosphorylation and degradation of IκB, and enabling the translocation of IRF3 and NFκB to the nucleus. In the nucleus activated IRF3 cooperates with NFκB to induce IFN-β transcription. Adapted from Figure 1 in Freundt and Lenardo (2005).

the final result that transcription factors NFκB and IRF3 are activated (Fig. 10.17). These then translocate to the nucleus and initiate transcription of a type I IFN gene.

Transcription of type I IFNs is regulated at both the transcriptional and posttranscriptional levels. Regulation at the transcriptional level is complex and incompletely understood. At the posttranscriptional level, IFN mRNAs contain destabilization sequences in the 3' nontranslated region and have short half-lives. Thus, shutoff of IFN synthesis occurs fairly quickly once the gene is no longer transcribed.

Almost all cell types produce IFN-α or -β, and thus induction of these cytokines follows infection of almost any

cell by any virus. However, some viruses are much more efficient at the induction of IFN than others, and it has been postulated that viruses may differ in the extent of production of dsRNA, or perhaps in the production of products that activate TLRs other than TLR3.

In contrast to type I IFNs, production of IFN-γ is largely restricted to T cells and NK cells. Production is induced by agents that promote T-cell activation. These include exposure to antigens to which the organism has been presensitized or exposure to IL-12, a cytokine produced by monocytes and macrophages after infection by bacteria or protozoa. Other activators of T cells are also known that induce synthesis of IFN-γ.

Many other cytokines are also induced following infection by viruses. These include TNF-α, TNF-β, IL-1, IL-2, IL-4, IL-5, IL-6, IL-8, and GM-CSF (*g*ranulocyte-*m*acrophage *c*olony-*s*timulating *f*actor). The spectrum of cytokines induced is different for different viruses. Some of the mechanisms by which many viruses interfere with IFN or other cytokines are described later.

Interferon Receptors and Signal Transduction

IFNs effect their responses by binding to specific receptors on the cell surface. These receptors are composed of more than one polypeptide chain, at least one of which is an integral membrane protein that contains a cytoplasmic domain. The type I IFN receptor is present in low abundance (around 10^3 receptors per cell) on all major types of cells. It has a very high affinity for IFN (dissociation constant about 10^{-10} M).

Upon binding of type I or type II IFNs to their cognate receptors, tyrosine kinases associated with the cytoplasmic domains are activated. In the case of type I IFNs, the kinases activated are called TYK2 and JAK1. Upon activation, they phosphorylate transcription factors STAT1 and STAT2 which then heterodimerize, recruit IRF-9 (p48), and translocate to the nucleus, where they stimulate the transcription of a large number of genes (Fig. 10.18). Most genes that are induced by type I IFN have an upstream element referred to as ISRE, the

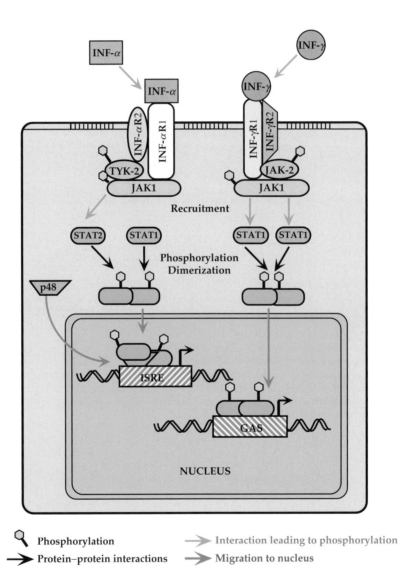

FIGURE 10.18　Overlapping signal transduction pathways used for gene induction by IFN-α and IFN-γ, which bind to different cellular receptors. Binding to their respective receptors leads to tyrosine phosphorylation of JAK1 and either TYK2 or JAK2. These in turn phosphorylate STAT1 and STAT 2 proteins. The phosphorylated STAT proteins dimerize, migrate to the nucleus, where they form complexes and bind to **ISRE** (**i**nterferon **s**timulated **r**esponse **e**lement) or **GAS** (IFN-**g**amma **a**ctivation **s**ite), which are present upstream of interferon inducible genes, resulting in transcription of these genes. Drawn from data in Fields *et al.* (1996) p. 379, Kalvakolanu (1999), and Nathanson *et al.* (1996) p. 123.

interferon stimulated response element. Induction of genes controlled by this element is very rapid and happens within minutes after treatment with IFN. Activation is transient, as one or more of the proteins induced act in a feedback loop to repress the continued transcription of these genes. In all, literally hundreds of genes are affected by IFN. The production of many genes is stimulated by IFN induction, but, in contrast, many genes are instead repressed upon IFN induction. A few of the genes induced are listed in Table 10.7.

In the case of type II IFN, the tyrosine kinases activated are JAK1 and JAK 2. STAT1 is phosphorylated, homodimerizes, and translocates to the nucleus. The genes that are induced by IFN-γ are under the control of several different regulatory elements, one of which is referred to as GAS, the IFN-*gamma activation site.* Induction by IFN-γ is slower and gene transcription continues for a longer period of time after induction. The complex of genes induced by type I and type II IFNs are overlapping because the tyrosine kinases associated with the type I and type II receptors and the transcription factors that are phosphorylated are overlapping (Fig. 10.18). In addition, many genes contain *cis*-acting regulatory elements that respond to both type I and type II IFN-induced transcriptional activators. A partial list of genes induced by IFN-γ is also given in Table 10.7.

Biological Effects of Interferons

Type I or type II IFN induces the expression of many different genes that have important biological effects. One effect is to stimulate the adaptive immune response. Both type I and type II IFNs induce increased production of class I MHC molecules, thus leading to enhanced surveillance by CTLs. Type II IFN also leads to increased production of class II MHC in macrophages, which are important players in the humoral response. The MHC response is augmented by the induction of genes in the MHC cluster that encode components of the proteasome. These cause the proteasome to become more active in producing peptides suitable for presentation by MHC molecules. Production of the TAP transporter system is upregulated, so that increased quantities of peptides are transferred across the ER membrane for binding by MHC molecules. Either type of IFN leads to the activation of monocytes and macrophages, the activation of natural killer cells, the activation of CTLs, and the modulation of the synthesis of Ig by B cells, all important for the immune response. IFN-γ also inhibits the growth of nonviral intracellular pathogens and induces the increased expression of Fc receptors in monocytes. Fc receptors bind to a conserved domain present in the heavy chain of all antibodies. This conserved domain is not involved in antibody binding but serves to interact with cells expressing Fc receptors and to bind the C1q component of complement. Thus, antibodies bound to an infected cell or to a pathogen can recruit cells such as monocytes or recruit complement to kill the pathogen or virus infected cell.

Both type I and type II IFNs are also key players in the innate immune defense against viruses. They are pyrogenic, inducing fever. High temperatures inhibit the replication of

TABLE 10.7   Genes Induced by Interferons

Protein	Induced by[a]			Inducible element	Functions/phenotype
	IFN-α	IFN-β	IFN-γ		
(2′-5′) (A$_n$) synthetase[b]	+++	+++	+	ISRE	(2′-5′) (A$_n$) synthesis/induction of antiviral state, esp. antipicornavirus
p68 Kinase (PKR)	+++	+++	+	ISRE	Protein kinase/induction of antiviral state
Indoleamine 2,3-dioxygenase	+	+	+++		Tryptophan degradation
γ56	+	+	+++		Trp-tRNA synthetase
GBP/g57	+	+	+++		Guanylate binding
MxA	+++	+++	+		Inhibits replication of influenza and VSV
IRF1/ISGF2	++	++	++		Transcription factor
IRF2	++	++			Transcription factor
MHC class I	+++	+++	+++		Upregulation of antigen presentation
MHC class II			++	Not ISRE nor GAS	Upregulation of antigen presentation
RING 12			+++		Proteosome subunit
RING 4	+++	+++			Putative TAP
β2 microglobulin	+++	+++	+++		MHC light chain

[a] The strength of the induction is indicated by the number of plus signs.
[b] Full induction also requires dsRNA.
Source: Adapted from Nathanson *et al.* (1996) p. 124.

many viruses or other pathogens. It has been proposed that the major symptoms of influenza infection, which include high fever as well as muscle aches and pains, are due to the induction of IFN rather than to virus replication per se. Both IFNs also induce what has been called the antiviral state, described next, in which cells are less susceptible to or resistant to infection by viruses.

Thus, the induction of IFNs is important for the control of viral infection at more than one level. Induction of the antiviral state results in lower yield of virus and the stimulation of immune responses leads to a more effective and faster clearance of the viral infection. Of interest is the fact that both types of IFNs inhibit cell growth, and treatment with IFN or with inducers of IFN has been effective in the treatment of at least some types of cancer.

The Antiviral State

Expression of the genes induced by interferon results in the establishment of the antiviral state in cells, in which viruses fail to replicate or replicate to much lower titers. The antiviral state is multifaceted and interference with the replication of viruses may occur at different stages of their replication cycle. The effect, therefore, depends on both the virus and the host cell. Furthermore, interference can occur at more than one stage of replication for some viruses, and such viruses tend to be more sensitive to the activities of interferon than others.

Interference with the virus replication cycle may occur very early, during penetration and uncoating of the virus. This occurs with SV40 and the retroviruses, for example. The proteins that are responsible for interference at these early stages of infection are unknown. For some viruses the transcription of the infecting viral genome is inhibited, examples being influenza, vesicular stomatitis virus, and herpes simplex virus. A host protein known as Mx, which is induced by interferon, is responsible, at least in part, for interfering with the transcription of influenza virus. In mice that are unable to produce the Mx protein, as is the case for most inbred strains of mice, infection with influenza produces a fatal outcome, demonstrating the importance of this gene. At a later stage in the infection cycle, interferon treatment results in reduced translation of many viral mRNAs, and the mechanisms by which this occurs are described next. Finally, some interferon-induced products interfere with virus assembly. This occurs with the retroviruses, vesicular stomatitis virus, and herpes simplex virus, but the proteins responsible are unknown.

Interference with Translation of Viral mRNAs

Two distinct pathways induced by IFN result in interference with the translation of viral mRNA. These are illustrated schematically in Fig. 10.19. Both of these pathways require that dsRNA be present for them to be active. Thus, not only is dsRNA a primary inducer of IFN synthesis, but these two pathways induced by IFN are also dependent on dsRNA for their activation. This dependence on dsRNA suggests that it is a common product in viral infection, whether the virus is DNA or RNA, but is not normally present in uninfected cells.

One inhibition pathway involves 2′-5′ oligo(A) synthetases, 2′-5′ OS. These synthetases, of which several are known, some of which have different subcellular locations, polymerize ATP into oligoadenylates that are joined in a 2′-5′ linkage. 2′-5′ Oligo(A) synthetase is induced by IFN but requires dsRNA as a cofactor for activity. 2′-5′ Oligo(A), in turn, is a cofactor for a latent ribonuclease, RNase L, which is present in all animal cells, but is inactive in the absence of 2′-5′ oligo(A). Once activated by its cofactor, which is bound tightly enough so that the two can be coimmunoprecipitated, RNase L can cleave single-strand mRNAs. 2′-5′ Oligo(A) is hydrolyzed by a phosphodiesterase present in cells, and activation of RNase L is transient.

Thus, RNase L results in a degradation of mRNAs only in those cells in which dsRNA is present. The picornaviruses are particularly sensitive to the action of RNase L, and it has been well established that activation of RNase L is a primary mechanism by which IFN interferes with the replication of these viruses. Other viruses seem to be less affected by RNase L, and the possible role of this enzyme in combating other virus infections remains to be determined.

The second pathway leading to the inhibition of translation of viral mRNAs involves a protein kinase known as PKR. PKR is present at low levels in most cells, but its concentration increases on IFN induction. It requires dsRNA for activity. On binding dsRNA, PKR is autophosphorylated on serine and threonine residues. Once activated, it phosphorylates the translation initiation factor, eIF-2. Phosphorylated eIF-2 cannot be recycled and protein synthesis is shut down. The importance of PKR for inhibiting viral replication is shown by the fact that several viruses encode inhibitors of its activity (see later). In particular, it is believed that reoviruses, adenoviruses, vaccinia virus, vesicular stomatitis virus, and influenza virus, among others, are sensitive to the effects of this enzyme.

PKR binds dsRNA and thus is another sensor for dsRNA within the cytoplasm of the cell. Whether it has any activity upon activation by dsRNA other than shutdown of protein synthesis is not clear, but PKR is known to have important roles in the regulation of cellular processes in addition to its antiviral activities. For example, it has a role in regulating several signal transduction cascades. In association with other factors, PKR can activate NFκB, which is proinflammatory. It also has a role in the activation of certain MAP kinases and a stress-related protein kinase. It is involved in the regulation of the cell cycle and if PKR is nonfunctional, unregulated cell proliferation and neoplastic transformation occur.

A. Development of the antiviral state

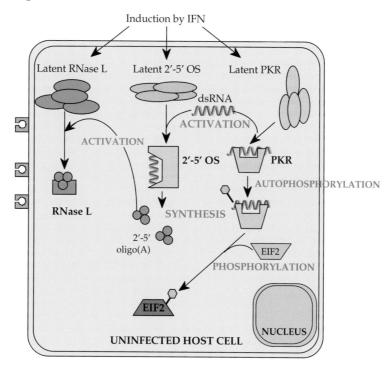

B. Virus attempts to replicate in a cell in the antiviral state

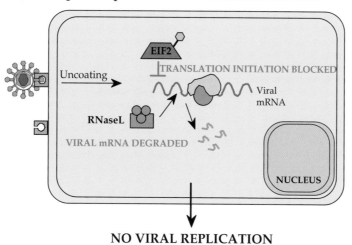

FIGURE 10.19 The antiviral state. (A) Development of the antiviral state begins with the action of interferon on an uninfected cell. The result of the signal transduction cascade shown in Figure 10.18 is the induction of expression of up to 100 genes, of which three are shown: RNase L, the 2′-5′ oligo(A) synthetase (2′-5′ OS), and the dsRNA-dependent protein kinase (PKR). These proteins are latent until they are activated by viral infection. PKR and 2′-5′ oligo(A) synthetase are activated by dsRNA that is produced during viral infection. Once activated, PKR autophosphorylates itself, and then phosphorylates EIF2. The activated synthetase makes trimeric oligonucleotides which in turn activate RNase L. (B) Phosphorylated EIF2 and activated RNase L are characteristic of the "antiviral state" in which a eukaryotic cell is refractory to infection by a wide variety of viruses. Phosphorylated EIF2 cannot serve to initiate translation of mRNA by ribosomes and activated RNase L degrades mRNAs, both viral and cellular, so protein synthesis stops. Without protein synthesis no virus replication can take place, but the inhibition of protein synthesis is transient and the cell may recover. Adapted from Nathanson *et al.* (1996) Figure 6.8 on p. 125.

Degradation of mRNA and shutoff of protein synthesis are drastic events that represent a defense of last resort. The dependence of both pathways on dsRNA means that only cells that contain dsRNA, presumably only virus-infected cells, will shut down as a result of these pathways. To simplify somewhat, one effect of IFN is to prepare the cell for instant shutdown if it becomes infected, while leaving uninfected cells essentially alone.

Therapeutic Uses of Interferon

Since the discovery of IFN some 50 years ago, there has been great hope and expectation that IFN therapy would be useful for the treatment of viral infections. In early studies, IFN was induced in patients by injection of dsRNA, but quantities of recombinant IFN-α suitable for injection are now available. In general, IFN-α therapy has been disappointing for treatment of viral infections in humans, although it is useful in a number of diseases (Table 10.5). Infection with hepatitis B or hepatitis C viruses often results in chronic infection. In at least some of these chronically infected patients, the virus load can be decreased, or the disease may go into remission, upon long-term treatment with recombinant IFN-α. However, side effects of IFN treatment often limit the dose and duration of treatment that can be used. IFN-α has also been useful in treatment of infection with papilloma viruses. More recently, combination therapy in which IFN-α has been combined with other drugs that interfere with virus replication, such as ribavirin, which depletes GTP pools in cells, have often proved more successful than treatment with IFN-α alone, and such methods for treatment of chronic virus infection continue to evolve.

During clinical trials with IFN-α, it was found that this IFN is useful for the treatment of at least some cancers, including hairy-cell leukemia and AIDS-related Kaposi's sarcoma. It is assumed that this control is based on the inhibition of cell growth caused by IFNs. Clinical trials using IFN-α, as well as other cytokines, for treatment of other viral diseases and other neoplasias continue and it is to be expected that further uses for these agents will be found (Table 10.5).

Gene Silencing

It has been known for many years that plants use RNA molecules 21 to 25 nucleotides in length, sometimes referred to as RNAi (i for interference), to silence genes. The discovery in 1999 that a similar mechanism exists in animals, in this case in the worm *Caenorhabditis elegans*, and that this mechanism is used to control a developmental pathway in the worm, created a great deal of excitement and resulted in a Nobel Prize for Andrew Fire and Craig Mello (see Table 1.1). Since 1999 a great deal of work has elucidated the details of the mechanisms involved and extended the findings to many other animals, including mammals. Attempts by many

investigators and by biotech companies to use these mechanisms to specifically turn off viral genes as a method to control viruses have followed. It is now clear that almost every cell can produce such interfering RNAs. These RNAs are used in the determination of the fate of the cell, in protecting organisms from transposons and genomic rearrangements, in physiological regulation, and in brain morphogenesis. The fact that several viruses have been found to encode products that interfere with gene silencing (described later in this chapter) indicates that the gene-silencing pathway is also important for the control of at least some viruses.

There are two sources of these small RNAs used for gene silencing. One source is long dsRNA, whether supplied externally or synthesized within the cell, which gives rise to RNAs called siRNAs (for *s*hort *i*nterfering RNAs). The long dsRNAs can be shortened by RNaseIII but are ultimately cleaved by the ribonuclease DICER into short dsRNAs that are typically 21 nucleotides long with 2 nucleotide overhangs at each 3′ end. These short molecules are then delivered to the RISC (*R*NA-*i*nduced *s*ilencing *c*omplexes) where they interact with one or more argonaute proteins (Ago1, Ago2). In the RISC complex the "sense" strand is selectively removed, and the resulting "guide strand" ssRNA binds to the complementary portion of mRNA from the gene to be silenced. Depending upon the gene, the organism, and the degree of sequence identity between the siRNA and the target sequence the silencing may be accomplished by cleavage of the mRNA or by repression of translation of the mRNA. The siRNA may also be used to repress the transcription of a gene through chromosome modification, but this process is not understood at present. Figure 10.20 illustrates the production and use of siRNA to silence genes.

The second source of interfering RNAs is ssRNA that is transcribed from the genome of many (all?) organisms, including humans. These RNAs form long hairpins of imperfect complementarity, which are processed by a nuclear enzyme called Drosha to form pre-micro RNA (pre-miRNA) of about 70 nts. These are exported to the cytoplasm by the protein Exportin5 where they are cleaved by DICER into dsRNAs about 21 nucleotides long, and delivered to RISC. Here the sense strand is removed and the resulting miRNA performs the same function as do siRNAs. miRNAs, however, often contain one or more mismatches with their targets, unlike siRNAs, and thus miRNAs are more likely to function as translational repressors rather than causing mRNA degradation. However, some miRNAs do cause mRNA degradation, and some siRNAs cause the inhibition of translation, so that the functions of these two RNAs are overlapping, as are the pathways that lead to their formation (Fig. 10.20). There are approximately 400 miRNA genes in humans, grouped by sequence into 350 families. Many of these are encoded in the 3′ NTRs of mammalian genes and a small fraction of them are tissue specific. Each miRNA can have many targets, since for most genes a 7-nucleotide

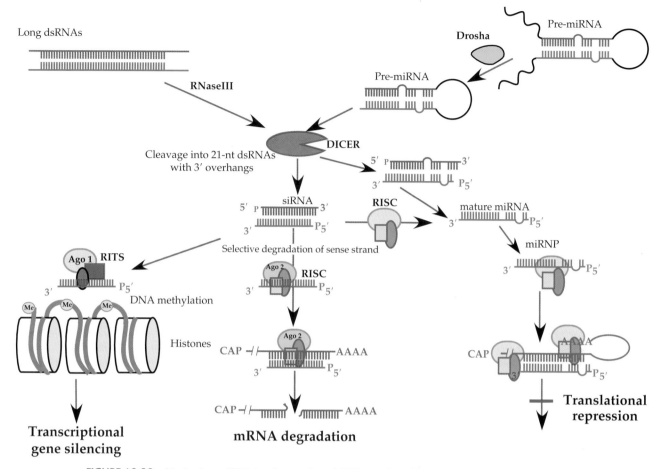

FIGURE 10.20   Mechanisms of RNA interference. Long dsRNAs are cleaved by RNaseIII and subsequently processed by DICER into siRNAs, which are short dsRNAs of 21–22 nts with 2-nt overhangs on the 3 termini. siRNAs then enter the *R*NA-*i*nduced *s*ilencing *c*omplex (RISC) where the sense strand is selectively degraded. miRNAs are encoded as incompletely self-complementary hairpins in viral and cellular genomes. They are cleaved by Drosha in the nucleus and the pre-miRNA exported to the cytoplasm by Exportin5. There they are also processed by DICER. siRNAs silence genes by binding in the context of RISC to mRNAs that are exactly complementary and causing mRNA degradation. miRNAs often contain some mismatches and generally exert their effects by inhibiting translation. siRNAs can also enter *R*NA-*i*nduced *t*ranscriptional *s*ilencing complexes (RITS) which recruit an enzyme to methylate DNA in chromatin, turning it into inactive heterochromatin. Both RITS and RISC contain argonaute proteins (Agos). The figure is adapted from those in Novina and Sharp (2004) and Schütz and Sarnow (2006).

match is sufficient for silencing. It is clear that these miRNAs are important for the development of the organism and for cell regulation, but the many functions of these RNAs have as yet to be unraveled.

There are ongoing attempts to develop gene silencing as a new antiviral strategy, and numerous trials have been made using a variety of viruses. Table 10.8 summarizes results in both cell culture and in experimental animals for a number of viruses. For many viruses, including HIV, polio, and foot-and-mouth disease virus, treatment of cells with siRNA either previous to or concomitantly with viral infection reduced viral replication significantly. For the experiments in cell culture, uptake of siRNA was usually induced by treatment with lipofectamine or other polyca-

tions, although expression of the RNA by expression vectors in the cell has been used in a few cases. These initial experiments highlight some of the problems with adapting siRNA to therapeutic use. First, many cells do not take up small RNAs easily or efficiently and expression vectors may be required, which pose another set of problems. Second, small RNAs have a short half-life in the average cell. Attempts to extend the half-life have included altering the siRNAs in a number of ways, such as 2′-O methylation of the bases or covalent attachment of cholesterol at the 3′ terminus. Third, it is unknown how specific siRNA or miRNAs would be for their intended targets and whether cellular mRNAs that might share some sequence identity with the gene to be silenced might be downregulated in a

TABLE 10.8   Viruses Suppressed by siRNA

Virus	Cell type or organism	Delivery protocol	Genes targeted	Reduction in viral yield
Viral replication in cultured cells				
HIV	Astroglioma cells	Lipofectamine	Polymerase, *Nef*	70–80%
HIV	Human primary T cells	Lentiviral vector	CCR5	???
	Cultured cell lines	Lipofectamine	Five viral genes and LTR	???
Poliovirus	HeLa Cells	Lipofectamine	Capsid, polymerase	90–95%
HRSV	Cultured cell lines	Lipofectamine	???	???
FMDV	BHK-21 cell lines	Lipofectamine	5′NTR, 3′NTR, VPg, VP4, Pol	90–99.9%
Flock house virus	*Drosophila* cell line	Endogenous cellular siRNA	???	Nearly 100%
Vaccinia	HeLa cells	???	E3L-specific siRNA	97%
Influenza	MDCK, CEF, chicken eggs	Plasmids	NP, PA	???
LCMV	Cultured cells	Recombinant adenovirus vector	L, Z1	Cures chronically infected cells
Viral replication in whole animals				
VSV	*C. elegans*	Gene products of *rde*-1, *rde*-4	???	???
Influenza	Mice (lung)	Inhale siRNA with polycations	???	99%
HRSV and parainfluenza	Mice (lung)	Inhale siRNA without transfection reagents	Essential viral genes	Protected from simultaneous viral challenge
HSV-2	Mice (vagina)	Oligofectamine	UL27 and UL29	Protected from challenge up to 3 hours later
Hepatitis B	Mice (intravenous)	siRNA in liposomes	???	???
Coxsackie B3	Mice (intravenous)	Hydrodynamic injection	???	Reduces virus by 6 logs
Japanese encephalitis	Mice (intracranial)	Lipid complexed lentiviral vector	cd-loop domain of envelope protein	Protects against both JE and WN encephalitis

Abbreviations: HIV, human immunodeficiency virus; LTR, long terminal repeat; HRSV, human respiratory syncytial virus; FMDV, foot and mouth disease virus; LCMV, lymphocytic choriomeningitis virus; VSV, vesicular stomatitis virus; HSV-2, herpes simplex virus 2; JE, Japanese encephalitis virus; WN, West Nile virus.
Source: Data from McManus and Sharp (2002), Lecellier *et al.* (2005), and Dykxhoorn *et al.* (2006).

"bystander effect." Obviously, such bystander downregulation could potentially be deleterious.

More detailed experiments with HIV in cell culture have used not only siRNA exogeneously supplied but also lentiviral vectors expressing siRNAs to target a variety of HIV genes, the 5′ LTR, and even the HIV coreceptor CCR5 (referred to in Table 10.8). What has become apparent from these studies is that HIV mutates rapidly under pressure from siRNA inhibition and resistant variants arise quickly. Even single nucleotide substitutions in the center of the core recognition sequence can make the virus resistant to silencing. Thus, resistant variants can arise that have an unaltered amino acid sequence because they possess only silent mutations. Similar results with picornaviruses have implied that any successful RNAi strategy must involve multiple siRNAs directed at a variety of targets so as to reduce the probability that resistant variants will quickly arise.

Preliminary experiments in mice have also been performed to explore ways of delivering siRNAs to appropriate tissues in sufficient quantities to effect viral silencing (Table 10.8). Some success has been achieved for viruses affecting the lungs, for lung tissue appears to be able to take up RNA even without transfection reagents such as lipofectamine. Similarly, viruses affecting the genital tract can be suppressed by siRNA inserted into the vagina with oligofectamine. Hepatitis B infection has been treated with siRNA introduced intravenously in liposomes. Other intravenous approaches have so far been less successful, although a Coxsackie B3 infection in mice could be suppressed by a million-fold after siRNA was introduced by hydrodynamic injection. Hydrodynamic injection, however, carries significant risk of cardiac failure and is not an option for human treatment. Thus at the current time, siRNA appears to be a molecule with much therapeutic potential but no proven efficacy.

VIRAL COUNTERDEFENSES

Mammals have evolved elaborate defenses to ward off infection by viruses. Viruses, in turn, have evolved counterdefenses that allow them to persist and continue to infect mammals. These counterdefenses vary from the simple to the elaborate. An overview of these mechanisms is given in Table 10.9.

The simplest counterdefense is to shut down the host cell rapidly after infection and to produce progeny virus very rapidly. Rapid shutoff of the host cell may prevent it from synthesizing IFN or other cytokines required for function of both the innate and adaptive immune systems. It will also shut off production of MHC class I molecules required for recognition of an infected cell by CTLs. Rapid growth allows the virus to go through several rounds of virus production before sufficient cytokine production has occurred and cells have responded to establish the antiviral state, and before the adaptive immune system can gear up. Many viruses take this approach, in particular RNA viruses, which have a limited coding capacity. As one example, soon after infection with poliovirus, the cap-binding protein required for cap-dependent initiation of protein synthesis is cleaved and host cap-dependent protein synthesis ceases. Translation of poliovirus

TABLE 10.9 Major Strategies Used by Viruses to Evade Host Defenses

Rapid shutdown of host macromolecular synthesis

Evasive strategies of viral antigen production

- Restricted gene expression; virus remains latent with minimal or no expression of viral proteins
- Infection of sites not readily accessible to the immune system
- Antigenic variation; antigenic epitopes mutate rapidly

Interference with MHC class I antigen presentation (see Table 10.10)

- Downregulation of transcription of MHC class I molecules
- Downregulation of transcription of TAP
- Interference with proteolysis to produce peptide epitopes
- Binding to TAP to inhibit its function
- Retention of MHC class I molecules within the cell
- Destabilization and degradation of MHC class I molecules

Interference with natural killer cell function

Interference with MHC class II antigen presentation

Interference with antiviral cytokine function (see also Table 10.13)

- Encoding of viral homologues of cellular regulators of cytokines
- Neutralization of cytokine activities
- Production of soluble cytokine receptors

Interference with interferons (see also Tables 10.12 and 10.15)

- Interference with IFN induction
- Interference with IFN signaling (effects on PKR, STAT1 and STAT2, etc.)

Inhibition of apoptosis (see also Table 10.11)

RNA is not affected since it utilizes a cap-independent initiation process. This elegant mechanism for specifically interfering with host protein synthesis does not appear to be the sole pathway used by the virus. Additional mechanisms also contribute to shutdown of host protein synthesis. Shutoff of host protein synthesis not only prevents the synthesis of IFN, for example, but it also frees up the translation machinery of the cell for the virus.

As a second example, the alphaviruses also rapidly shut off host protein synthesis. As for poliovirus, more than one mechanism appears to be used. One mechanism is an interference with the cell Na–K pump, which results in an increase in the Na^+ concentration and a decrease in the K^+ concentration in the cell. This inhibits translation of cellular mRNAs, but has no effect on translation of viral mRNAs.

These first two examples refer to plus-strand RNA viruses. As a third example, the minus-strand RNA virus vesicular stomatitis virus also grows rapidly and profoundly inhibits cellular protein synthesis. This shutoff is effected by the M protein, which has been shown to be a potent inhibitor of the transcription and transport of cellular mRNA.

Some DNA viruses also inhibit host protein synthesis shortly after infection. A component of the herpes simplex virion that enters the cell during infection inhibits host-cell protein synthesis, beginning very soon after infection. As infection progresses, additional proteins contribute to a more profound inhibition of host-cell macromolecular synthesis.

Although shutoff of host protein synthesis impairs the host defense systems, most viruses, even viruses with limited coding capacity, also use additional mechanisms to impede the host immune system. Viruses with smaller genomes may encode only one or two proteins that interfere with host immunity, whereas viruses with large genomes, such as herpes simplex virus, have the luxury of encoding a large number of proteins that interfere with host defenses.

Virus Evasion of the Adaptive Immune System

More than 50 viral proteins have been identified that modulate the defense mechanisms used by mammals to control viral infection and to eliminate the viruses once infection is initiated. Most have been described only within the last decade or so, and more viral mechanisms to interfere with host defenses will surely be discovered in the future. The most elaborate evasion mechanisms are utilized by viruses that contain larger genomes, and thus can afford to encode many proteins for this purpose.

Interference with the Expression of Peptides by Class I MHC Molecules

Many viruses downregulate the expression of class I MHC molecules on the surface of infected cells, thereby rendering the cells invisible to CTLs. A partial summary of the proteins used and the mechanisms involved is given in Table 10.10. Rapid shutdown of the cell can accomplish this, and this appears to occur following infection by picornaviruses, for example. However, the picornaviruses also utilize additional mechanisms to interfere with antigen presentation. Poliovirus protein 3A interacts with the membranes of the endoplasmic reticulum to prevent transport of proteins from the ER to the Golgi apparatus. This prevents the transport of MHC bearing polio-specific peptides to the cell surface. This mechanism, which can only be used by viruses that are not enveloped and that do not secrete proteins during the replication cycle, has the added advantage that previously transported MHC remains at the cell surface to ward off NK cells. Interestingly, foot-and-mouth disease virus also prevents transport of proteins to the cell surface but it is viral protein 2BC that is responsible.

More elaborate mechanisms that interrupt presentation of antigens by MHC are used by many viruses, including the lentiviruses, the adenoviruses, the herpesviruses, and the poxviruses. The lentivirus HIV-1 uses two different mechanisms to downregulate class I MHC expression. The Tat protein interferes with the transcription of mRNA for class I MHC molecules, thereby leading to reduced synthesis of class I molecules. The Nef protein downregulates the surface expression of class I MHC molecules by relocalizing them to the *trans*-Golgi network. It does this through interactions with another cellular protein called PACS-1, which is involved in sorting proteins to the trans-Golgi network.

Adenoviruses prevent the presentation of peptides by class I MHC molecules by one of two different methods. These are illustrated schematically in Fig. 10.21. Some adenoviruses, such as Ad2, produce an integral membrane protein of 19 kDa, a product of the E3 gene, which binds tightly to class I MHC molecules in the lumen of the ER. This prevents their export to the cell surface. Ad12, in contrast, produces a product from the E1 gene that interferes with the transcription of genes in the MHC locus. Transcription of the genes for class I molecules, the genes for TAP, and the genes for the MHC-encoded components of the proteasome is inhibited. Thus, synthesis of new class I molecules is inhibited, and the production and transport of peptides required for MHC presentation cannot be upregulated.

Several poxviruses are known to downregulate the expression of class I MHC at the surface of the infected cell. One mechanism used appears to be the production of proteins that bind MHC and target it for degradation.

The herpesviruses have learned to coexist with the immune system, enabling them to establish lifelong infections. They encode a wide variety of gene products that interfere with the presentation of peptides by class I MHC molecules. Several of these are illustrated schematically in Fig. 10.22. Human cytomegalovirus (HHV-5) encodes four different proteins that interfere with class I presentation by interacting with cellular proteins that form components of the presentation pathway. One protein, called US3, binds to class I MHC molecules and retains them in the ER. Two other proteins, US2 and US11, independently cause class I MHC molecules to be immediately recycled to the

TABLE 10.10   Some Viruses That Alter Antigen Presentation by Class I MHC Molecules

Interference level	Virus family	Virus	Virus protein	Mechanism
Downregulation of MHC class I expression at cell surface	*Adenoviridae*	Ad 2	E3-19K	Viral protein binds to class I molecules and keeps them in the ER
		Ad12	E1A	Inhibits transcription of class I mRNA
	Herpesviridae	HHV-5	US2, US11	Gene products lead to degradation of class I molecules
		HHV-5	UL18	Binds to β2 microglobulin
	Picornaviridae	PV	3A	Inhibits secretory pathway
	Retroviridae	HIV	Tat	Inhibits transcription of MHC class I mRNA
			Nef	Downregulates surface expression of class I proteins
Alteration of antigen processing	*Adenoviridae*	Ad12	E1A	Inhibits transcription of TAP1 and TAP2
	Herpesviridae	HHV-1	ICP47	Binds to TAP and prevents transport of antigenic peptide to MHC class I
		HHV-5	US6	Blocks transport by TAP
		HHV-4	EBNA-1	EBNA-1 protein with Gly-Ala repeat is not processed by proteasome
Alteration of spectrum of antigens presented	*Hepadnaviridae*	HepB	Ag epitopes	Epitopes mutate so that they are no longer recognized by CTLs

Virus abbreviations: Ad 2, adenovirus 2; Ad12, adenovirus 12; HHV-5, human cytomegalovirus; PV, poliovirus; HIV, human immunodeficiency virus; HHV-1, herpes simplex virus; HHV-4, Epstein-Barr virus; HepB, hepatitis B virus.

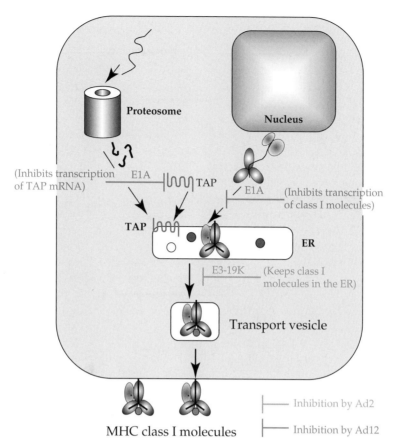

FIGURE 10.21   Mechanisms by which adenoviruses inhibit peptide presentation by class I MHC molecules on infected cells.

cytoplasm following synthesis, where they are degraded by the proteasome. The recycling of class I molecules to the cytoplasm is thought to occur by speeding up the kinetics of the normal cellular turnover pathway that recycles products in the ER back to the cytoplasm. A fourth HCMV protein, US6, binds to TAP and prevents transport of peptides into the lumen of the ER. Herpes simplex viruses (HSV) also encodes a protein (ICP47) that blocks transport of peptides by TAP. Interestingly, the HSV ICP47 binds to TAP from the cytoplasmic side of this protein, which spans the ER membrane, whereas HCMV US6 binds from the lumenal side. Thus, these proteins represent independent solutions to the problem of preventing transport of peptides across the ER membrane by TAP. Loss of TAP transport prevents the presentation of herpes peptides at the cell surface.

A fifth HCMV protein, UL83, is also involved in preventing the presentation of peptides by class I MHC molecules. This protein phosphorylates an early HCMV protein within an important T-cell epitope. Phosphorylation prevents the processing and presentation of this epitope.

Finally, a HCMV protein, UL18, binds β2 microglobulin. UL18 is homologous to MHC class I molecules (the virus

obviously swiped the gene from its host at some time in the past) and binds peptide. Its function is not well understood. It interacts with monocytes and presumably interferes with their function. It may have other functions as well. In any event, binding of β2 microglobulin reduces the pool available for formation of class I MHC molecules.

Mouse CMV also downregulates expression of class I molecules but uses different mechanisms to do so. An early gene, m06, encodes a protein that binds to MHC class I molecules in the ER. The bound MHC class I molecules exit the ER and pass through the Golgi apparatus. Rather than continuing on to the cell surface, however, they are redirected to lysosomes where they are degraded.

Interference with the Activity of Complement

Infected cells can be killed by a complement-mediated pathway in which antibodies first bind to viral antigens present on the surface. Virus neutralization by antibodies is also enhanced by complement. Herpesviruses express proteins that interfere with complement activation. HSV expresses a protein on the cell surface, called gC, that binds complement

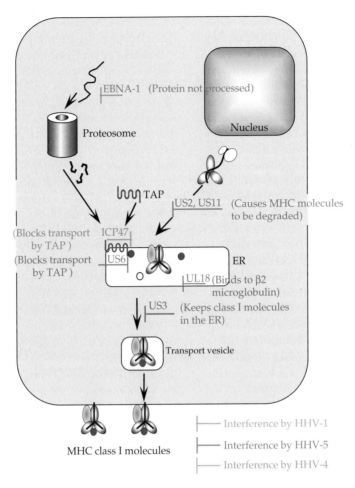

FIGURE 10.22   Mechanisms by which **herpesviruses** interfere with the immune defenses of the host cell, by altering peptide presentation by class I MHC molecules. Mechanisms used by human cytomegalovirus (HHV-5) are shown in red, by herpes simplex virus (HHV-1) in orange, and by Epstein–Barr virus (HHV-4) in ochre.

component C3b, thereby preventing initiation of the complement cascade. HSV also expresses a molecule, consisting of a complex of viral proteins gE and gI, that has an Fc receptor. It binds IgG through the Fc domain, which serves to block complement activation. HCMV also expresses an Fc receptor.

Poxviruses also interfere with the complement system in a variety of ways. The best understood mechanism is the secretion of a protein by vaccinia-infected cells called VCP (*v*accinia *c*omplement control *p*rotein). This protein binds both C3b and C4b, preventing the activation of the complement cascade by either the alternative or classical pathway. VCP appears to have other functions as well, and other poxvirus proteins appear to interfere with complement by less well understood mechanisms.

Latent Infections That Avoid CTL Surveillance

The herpesviruses establish latent infections that persist for the life of the host. The alphaherpesviruses express no protein in the latently infected cell and, furthermore, establish

a latent infection in neurons, which are immunologically privileged and express only low levels of class I MHC molecules. Epstein–Barr virus does express a protein in latently infected cells but this protein contains a glycine–alanine tract that interferes with processing by the proteasome, rendering this protein invisible to the class I pathway.

Infection of Cells of the Immune System to Thwart Immune Response

Another mechanism for evasion of the immune system is to infect immune effector cells. HIV lytically infects CD4$^+$ T$_H$ cells, which leads to a profound immunosuppression because these cells are required to mount an immune response. Similarly, measles virus infects many cells of the immune system, including T and B lymphocytes and monocytes. Infection by measles virus results in immunosuppression that lasts for some weeks after infection. As another example, the herpesvirus Epstein–Barr virus infects B cells. Other viruses are also known that infect T cells, B cells, macrophages, or other cells that are important for the immune response. In addition, if a virus infects the thymus early in the life of the animal, while the immune system is still developing, immune tolerance may result such that the virus will not be recognized as foreign. This enables the virus to establish an infection that lasts for the life of the animal.

Rapid Drift

Some viruses evade the adaptive immune system by rapid drift in the antigens that are recognized by it. Most viruses, but especially RNA viruses, undergo rapid evolution, and immune pressure can cause their sequence to drift. Viruses that are able to undergo reassortment, of which influenza virus is the classic example, are able to produce new forms very rapidly that have altered antigenic epitopes. Viruses such as the lentiviruses, of which HIV is the best known, establish a chronic infection in their host, and during this infection these viruses undergo continuing genetic drift, which may be responsible, at least in part, for their persistence.

Prevention of Killing by Natural Killer Cells

Downregulation of class I MHC expression renders a cell more sensitive to lysis by NK cells, the backup mechanism that eliminates cells that do not express class I molecules on their surface. The cytomegaloviruses (CMVs) resist NK cell activity, however, even though they severely downregulate the expression of MHC. In cells infected by CMVs, the molecules that are required for the stimulation of the killing activity of NK cells are downregulated by mechanisms that are poorly understood. The CMV protein UL18 is also important in some way. Mouse CMV deleted for this gene replicates poorly in mice, and NK cells appear to be responsible for the poor replication.

HIV also resists NK activity. Here the mechanism seems to be a selective interference with the expression of class I MHC molecules. Human cells are protected from NK killing primarily by HLA-C and HLA-E molecules. HIV-1 selectively downregulates expression of HLA-A and HLA-B molecules, which are the human class I molecules recognized by the majority of CTLs. It does not affect expression of HLA-C or HLA-E molecules. Thus, the infected cells are resistant to killing by NK cells. They remain sensitive to killing by CTLs, but their sensitivity to CTLs is greatly reduced.

Virus Counterdefenses against Apoptosis

Infection of an animal cell by a virus often results in apoptosis of the cell, or would result in apoptosis if the virus did not block its induction. NK cells or CTLs induce apoptosis as a way of killing infected cells. However, the replication activities of the virus itself are often apoptosis inducing. For example, DNA viruses deregulate the cell and cause it to enter S phase, which is required for optimal replication of the viral DNAs. This activity would normally induce apoptosis mediated by p53. Other aspects of viral replication also have the potential to induce apoptosis.

For viruses that replicate rapidly, such as many RNA viruses, apoptosis does not seem to inhibit the production of virus. It may actually lead to increased virus production, perhaps because of less competition for the resources of the infected cell. Apoptosis may even result in more rapid spread of viruses to neighboring cells because of cell fragmentation and uptake of the fragments by neighboring cells with less inflammation than would occur otherwise, since apoptosis itself does not induce inflammation. For viruses that replicate more slowly, however, such as most DNA viruses, premature apoptosis results in significant declines in virus yield. Thus, many viruses have evolved ways to inhibit or delay apoptosis in infected cells. An overview of mechanisms used by viruses in different families to interfere with apoptosis is given in Table 10.11. Viruses may interfere with the caspase activation pathway or with intracellular signaling that leads to apoptosis, produce anti-apoptotic proteins or proteins that regulate the activities of p53, or interfere with apoptosis in still other ways.

Production of Serpins

Many of the poxviruses produce proteins that inhibit the activity of caspases. These proteins are related to serpins (*ser*ine *p*rotease *in*hibitors), which are small proteins that serve as substrates for serine proteases, but which remain bound to the protease after cleavage and block their activity. Serpins are important in the regulation of inflammatory responses. The poxvirus serpins inhibit caspases, which are cysteine proteases. They appear to act similarly to host serpins in that they are cleaved by the caspase but remain bound to it and render it inactive. These viral products, of which crmA of cowpox virus is an example, appear to block apoptosis induced by any pathway requiring activation of caspases. An overview of poxvirus interference with apoptotic pathways is shown in Fig. 10.23.

Interference with Fas or TNF Signaling

The poxvirus rabbit myxoma virus encodes a homologue of the TNF receptor (TNFR), called T2. This protein inhibits the interaction of TNF-α with TNFR and thus prevents activation of the apoptotic pathway by TNF-α (Fig. 10.23). Some other poxviruses also produce TNFRs to neutralize TNF-α activity. Another approach is taken by the poxvirus molluscum contagiosum virus, which encodes proteins named MC159/160 that have death effector domains similar to those present in a cellular protein called FADD. FADD associates with caspase-8 and recruits it to activated Fas or TNF receptors in the cell surface, resulting in the activation of caspase-8. The viral proteins disrupt this interaction and prevent the activation of caspase-8 signaled by Fas ligand or TNF (Fig. 10.23).

TABLE 10.11   Viruses That Interfere with Apoptosis

Virus family	Virus	Viral protein	Mode of interference
Poxviridae	Cowpox	crmA	Serpin homologue, inhibits proteolytic activation of caspases
	Vaccinia	SPI-2	crmA homologue, inhibits activation of caspases
	Myxoma	M11L, T2	T2 is a homologue of TNFR, and inhibits interaction of TNA-α with TNFR; M11L has a novel function
	Molluscum contagiosum	M159, 160	Has death domains like FADD, inhibits FADD activation of caspase-8
Asfarviridae	African swine fever	LMW5-HL	Homologue of Bcl-2
Herpesviridae	HHV-1	γ34.5 gene	Prevents shutoff of protein synthesis in neuroblastoma cells
	SaHV-1	ORF 16 product	Homologue of Bcl-2
	HHV 8	KS bcl-2	Homologue of Bcl-2
		K13	vFLIPS, prevents activation of caspases by death receptors
	HHV-4 (latent)	LMP1	Upregulates transcription of Bcl-2 and A20 mRNAs; inhibits p53-mediated apoptosis
	HHV-4 (lytic)	BHFR1	Inhibits p53 activity; has some sequence similarity to Bcl-2
	HHV-5	IE-1, IE-2	Downregulates transcription of p53 mRNA
	Gammaherpesvirinae	Viral FLIPs	Inhibits signaling from death domains to caspases
Polymaviridae	SV40	Large T antigen	Binds to and inactivates p53
Papillomaviridae	HPVs	E6	Binds to p53 and targets it for ubiquitin-mediated proteolysis
Adenoviridae	Adenovirus	E1B-55K	Binds to and inactivates p53
		E3-14.7K	Interacts with caspase-8
		E3 10.4K/14.5K	Blocks caspase-8 activation by destruction of Fas
		E4 orf 6	Binds to and inactivates p53
		E1B 19K	Functional homologue of Bcl-2; interacts with Bax, Bi, and Bak
Baculoviridae	AcMNPV	p35	Forms a complex with caspases; inhibits caspase-mediated cell death
		IAP	Like FLIPs, inhibits activation of caspases
Hepadnaviridae	HepB	pX	Binds to p53
Flaviviridae	HepC	Core protein	Represses transcription of p53 mRNA

Virus name abbreviations: HHV-8, Kaposi's sarcoma herpesvirus; HHV-5, human cytomegalovirus; HHV-1, herpes simplex virus; HHV-4, Epstein–Barr virus; SaHV-1, Saimiriine herpesvirus 1; HPV, human papillomaviruses; AcMNPV, *Autographa californica* nucleopolyhedrovirus; HepB, hepatitis B virus; HepC, hepatitis C virus.
Source: Adapted from reviews by O'Brien (1998) and Tortorella *et al.* (2000).

Several herpesviruses also produce proteins that interfere with the activation of caspase-8 through its interactions with FADD. These viral FLIPs thus inhibit apoptosis in a manner similar to the molluscum contagiosum virus MC159/160.

Another mechanism to interfere with Fas-induced activation of the apoptotic pathway is simply to reduce the number of Fas molecules at the cell surface. The adenoviruses produce two proteins, known as RIDα and RIDβ, that cause Fas at the cell surface to be internalized and degraded. The RID proteins also inhibit apoptosis induced by TNF, but the mechanism is not known. The importance of the TNF-α pathway is shown by the fact that the adenoviruses produce four different proteins, found in different parts of the infected cell, that antagonize the effects of TNF-α.

Production of Homologues of Bcl-2

Bcl-2 is a cellular protein that inhibits apoptosis. Several herpesviruses and adenoviruses, as well as at least one poxvirus (fowlpox virus) and one asfarivirus (African swine fever virus), produce homologues of Bcl-2, probably obtained originally from the host, that act as anti-apoptotic agents. The adenovirus homologue, called E1B-19K, is multifunctional. It appears to inhibit apoptosis not only as a Bcl-2 homologue but also by interfering with FADD-mediated activation of the caspase pathway. An overview of herpesvirus interference with the induction of apoptosis is shown in Fig. 10.24 and of adenovirus interference in Fig. 10.25.

The herpesvirus EBV has evolved another solution to the production of Bcl-2. It does not encode a Bcl-2 homologue, but produces a protein that upregulates the production of cellular Bcl-2.

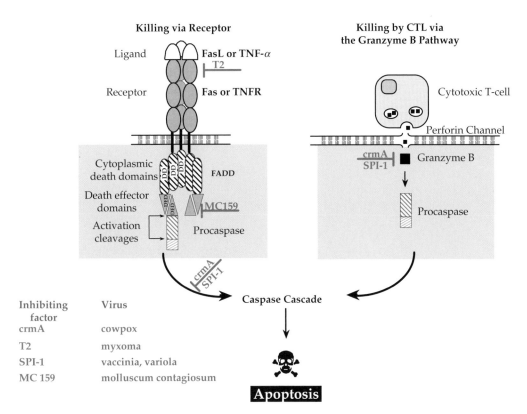

Killing via Receptor

Killing by CTL via the Granzyme B Pathway

Inhibiting factor	Virus
crmA	cowpox
T2	myxoma
SPI-1	vaccinia, variola
MC 159	molluscum contagiosum

FIGURE 10.23 **Poxviruses** inhibit apoptosis by interfering with the ligand–receptor pathway and by interfering directly with the caspase cleavage pathway. The latter mechanisms blocks induction of apoptosis by the granzyme B pathway (right) as well as by receptor signaling (left). Points of interference by products of cowpox, myxoma, vaccinia, and molluscum contagiosum viruses are shown in red. Drawn from data in Turner and Moyer (1998).

Control of p53 Concentrations

Many of the DNA viruses induce cellular DNA synthesis, which leads to an increase in p53 concentration. This antitumor protein plays a central role in the control of the cell cycle, and one of its functions is as a transcriptional activator. Its concentration in normal cells is controlled by rapid turnover and by its activation of the transcription of a gene called mdm-2, which also requires interaction with p300. The mdm-2 protein binds p53 and inhibits its ability to act as a transcription factor, thus regulating its own synthesis. This feedback loop is inactivated by DNA viruses in the process of stimulating the cell to enter S phase and p53 accumulates. High concentrations of p53 induce apoptosis, probably as a result of its transcriptional activities. Many DNA viruses resist p53-induced apoptosis by sequestering p53 or otherwise interfering with its function as a transcriptional activator, or by causing it to be rapidly degraded. These include at least some herpesviruses, adenoviruses, papillomaviruses, and hepatitis B virus. Examples include the SV40 T antigen, the adenovirus E1B-55K protein (see Fig. 10.25), the hepatitis B virus pX, two proteins encoded by Epstein–Barr virus (called EBNA-5 and BZLF1) which bind p53, two proteins (called IE1 and IE2) encoded by cytomegalovirus that interfere with p53-directed transcription (see also Fig. 10.24), and the papil-lomavirus product E6, which binds p53 and causes it to be rapidly degraded. Many of these products were described in Chapter 7.

Other Viral Proteins That Oppose Apoptosis

There are yet other viral proteins that cause the infected cell to resist apoptosis by mechanisms that are not well understood. In the poxviruses these include proteins that contain ankyrin repeats, a protein encoded by molluscum contagiosum virus that is 75% identical to human glutathione peroxidase (and may therefore prevent accumulation of oxidizing agents that can provoke apoptosis), a protein that acts at the mitochondrial checkpoint, as well as still other proteins. The multiple mechanisms used by poxviruses to inhibit apoptosis illustrate the importance of preventing apoptosis by the infected cell. Herpesviruses and adenoviruses also produce proteins in addition to those described before that inhibit apoptosis, again showing the importance of controlling the apoptotic pathway after infection by these viruses. It is interesting that the adenoviruses are so efficient at suppressing apoptosis that they encode another protein, the adenovirus death protein, which leads to cell death and release of virus. This protein, which is produced only late

Killing due to External Stimuli

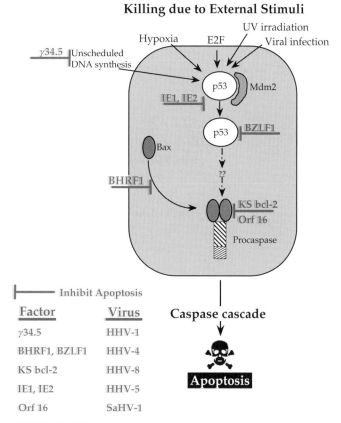

FIGURE 10.24 Various members of the **herpesvirus** family inhibit different steps in p53-mediated apoptosis. They interfere with Bax (an inducer of apoptosis), encode Bcl-2 homologues, and interact with p53 itself. The gammaherpesviruses also interfere with Fas-mediated signaling (not shown) by encoding products called viral FLIPs that interact with death domains and prevent signaling (similar to the action of poxvirus MC159 illustrated in Fig. 10.23). HHV-1, herpes simplex virus; HHV-4, Epstein–Barr virus; HHV-5, human cytomegalovirus; HHV-8, Kaposi's sarcoma herpesvirus; SaHV-1, Saimirine herpesvirus 1. Adapted from Hill and Masucci (1998).

after infection, may simply allow apoptosis to proceed, or the protein may initiate some other pathway that results in the death of the cell.

Evasion of the Antiviral State

Many viruses encode products that specifically interfere with the activation of the PKR pathway that leads to shutdown of protein synthesis. Several different mechanisms are used: synthesis of competitive RNA that binds PKR but does not activate it; synthesis of products that sequester dsRNA; production of proteins that bind PKR and prevent it from phosphorylating eIF2α; competitive inhibitors of the eIF2α substrate; activation of cellular proteins that inhibit or degrade PKR; production of products that result in dephosphorylation of eIF2α (Table 10.12). An overview of the viruses that use these various mechanisms is presented in

the following paragraphs. The fact that such a large number of viruses belonging to many different families have evolved such a diverse set of mechanisms to counter this pathway illustrates the importance of this pathway in controlling viral infections in vertebrates.

Several viruses, including adenoviruses, Epstein–Barr virus, and hepatitis C virus, encode small RNA products that bind to PKR in lieu of dsRNA, but which do not activate PKR. The adenoviral RNA is called VAI RNA and was the first of these products to be described. PKR has two dsRNA binding domains. The current model for the activation of PKR posits that dsRNA must be bound by both domains for activating autophosphorylation to occur, and VAI RNA seems to be incapable of binding to both sites simultaneously. After activation, dimerization of the enzyme is required for further phosphorylation *in trans* to occur, and whether RNA binding is required for this is not known. Regardless of mechanism, the net result is that the viral RNAs act as inhibitors of the cofactor dsRNA and prevent activation of PKR.

Many viruses, including vaccinia virus, reoviruses, influenza viruses, rotaviruses, and at least two herpesviruses, encode protein products that bind dsRNA, thus making it unavailable as a cofactor for activation of PKR. Furthermore, vaccinia virus appears to limit the production of dsRNA by using an arrangement of genes and stop transcription signals that reduces transcription from both strands, at least early after infection. Sequestering of dsRNA also inhibits the induction of IFN, since dsRNA is a primary inducer of interferon, as well as the activation of RNase L.

Many viruses interfere with the activity of PKR directly (Table 10.12). They encode products that bind PKR and prevent it from phosphorylating eIF2α. These products may inhibit the dimerization of PKR required for its activity, or may sequester the protein, or may bind to inactive or activated protein and prevent it from functioning, or may recruit cellular proteins that inhibit PKR function. Of interest is the fact that such proteins are not confined to viruses of vertebrates but one is also found in some insect baculoviruses.

Competitive inhibitors that are homologues of eIF2α are produced by some viruses, including vaccinia, HIV, and ranaviruses of frogs. These pseudosubstrates inhibit the phorphorylation of authentic eIF2α.

Poliovirus induces the degradation of PKR. Although the virus encodes more than one protease, it is thought that the virus induces a cellular protease to degrade PKR.

Finally, herpesvirus, papillomavirus, and SV40 viruses promote the dephosphorylation of eIF2α. They encode products that interact with cellular phosphatases to induce them to dephosphorylate eIF2α, thus maintaining it in an active state.

The RNase L pathway is also inhibited by viral products that bind dsRNA or by decoy RNAs. However, herpes simplex virus targets RNase L activity directly. It makes an analogue of 2′-5′ oligo(A) that binds to RNase L but which

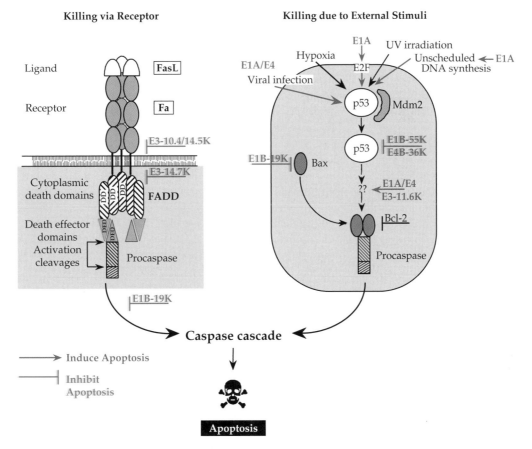

FIGURE 10.25 **Adenoviruses** have been shown to both induce apoptosis (blue arrows and text) and to encode factors which inhibit apoptosis (red symbols and text). These moieties interfere with "killing via receptor" and with the p53-mediated pathway of apoptosis. Drawn from data in Chinnadurai (1998).

does not activate the enzyme. By preventing the binding of authentic 2′-5′ oligo(A), the analogue prevents the activation of RNase L.

Viral Counterdefenses against Cytokines and Chemokines

The cytokines and chemokines are powerful regulators of both innate and adaptive immune defenses. Because of the importance of these agents in the regulation of the immune response, and because of their potential effectiveness against viral infection, many viruses have devised methods to disrupt their activities. These include the encoding of homologues or analogues of cytokines and chemokines or of their receptors. Some of these were acquired from the host at some time in the past and modified to meet the purposes of the virus, whereas others are viral products that have evolved to interact with the cytokine system. Because the networks of cytokine and chemokine interactions are complex, the pathways by which the viral products exert their effects are often poorly understood. It is in the interest of the virus to divert the immune responses in directions that not only favor

virus growth in the present host but also the persistence of the virus in nature. These two are often antagonistic, and compromises are required. Viruses also produce products that defeat specific aspects of the innate response induced by cytokines. Of these, products that evade the antiviral state induced by IFN were described earlier. A partial listing of strategies used by some viruses to modulate cytokine activities is given in Table 10.13, and a listing of defense molecules encoded by poxviruses is given in Table 10.14.

Interference with Signal Transduction Pathways

Adenoviruses, herpesviruses, poxviruses, hepatitis C virus, hepatitis B virus, and most or all minus-strand RNA viruses make products that interfere with signal transduction by the dsRNA sensors that lead to the production of IFN or with signal transduction by the IFN receptor on binding of its ligand. The mechanisms used by the minus-strand viruses are summarized in Table 10.15.

Hepatitis C virus establishes a chronic infection in the majority of infected individuals. One reason that the virus is so successful in establishing chronic infections is that it

TABLE 10.12   Viral Regulation of PKR

Mechanism of action	Virus family	Virus	Viral gene product
dsRNA binding proteins	*Poxviridae*	Vaccinia	E3L
	Reoviridae	Reovirus	σ3
		Group A rotavirus	NSP5?
		Group C rotavirus	NSP3
	Orthomyxoviridae	Influenza	NS1
	Herpesviridae	Herpes simplex (HHV-1)	Us11
		Epstein–Barr (HHV-4)	SM
RNA inhibitors	*Herpesviridae*	Epstein–Barr (HHV-4)	EBER RNA
	Adenoviridae	Adenovirus	VAI RNA, VAII RNA?
	Flaviviridae	Hepatitis C	IRES
PKR interaction	*Flaviviridae*	Hepatitis C	NS5A, E2
	Orthomyxoviridae	Influenza	p58
	Poxviridae	Vaccinia	E3L
	Herpesviridae	Herpes simplex (HHV-1)	Us11
		Human herpesvirus-8	vIRF-2
Competitive inhibitor	*Poxviridae*	Vaccinia	K3L
	Retroviridae	HIV	Tat
PKR degradation	*Picornaviridae*	Poliovirus	Cellular protease (activated by ??)
eIF2α phosphatase	*Herpesviridae*	Herpes simplex (HHV-1)	γ34.5
	Papillomaviridae	HPV type 18	E6
	Polyomaviridae	SV-40	Large-T antigen

Abbreviations: HIV, human immunodeficiency virus; HPV, human papillomavirus; SV-40, simian virus 40.
Source: Adapted from Table 1 in Langland *et al.* (2006).

prevents TLR3 or RIG-1 from signaling the presence of dsRNA. The viral NS3/4A protease cleaves both TRIF, an intermediate in the TLR3 signaling pathway, and MAVS, an intermediate in the RIG-1 signaling pathway (see Fig. 10.17). Each of these proteins is cleaved once at a specific site, and once cleaved they are no longer active in the signaling pathway. The result is that dsRNA cannot induce production of IFN following infection by the virus. Other TLR pathways are not known to be affected, and IFN induction can occur via such other TLR receptors, but the fact that HCV specifically targets the dsRNA signaling pathways illustrates the importance of this mechanism for induction of IFN following viral infection.

A number of paramyxoviruses are also known to interfere with this signaling pathway. The V proteins of several paramyxoviruses interfere at some undefined stage, and the W protein of Nipah virus blocks the activity of IRF-3 in the nucleus.

Many viruses interfere with the signal transduction pathways activated upon binding of IFN to its receptor on the cell surface, thereby preventing the induction of gene expression normally caused by IFN. As described before, activation of the transcription factor STAT1 by IFN-γ and its subsequent dimerization, and the activation of the transcription factors STAT1 and STAT2 by IFN-α/β and their subsequent heterodimerization, are critical for the transcription of genes activated by the activities of the IFNs (Fig. 10.18). Human cytomegalovirus produces a protein called IE1-72kDa that forms a complex with STAT1 and STAT2 in the nucleus of the cell and prevents their association with IRF-9, thereby preventing these transcription factors from functioning. Most of the minus-strand RNA viruses are also known to inhibit the STAT1–STAT2 pathways. The V proteins of several paramyxoviruses cause either STAT1 or STAT2 to be degraded by the host proteasome. SV5 and mumps viruses cause the degradation of STAT1, which inactivates both the type I and type II IFN responses. Human PIV2 V protein targets STAT2, and thus only the type I IFN response is ablated.

Other paramyxoviruses inhibit IFN action by binding to STAT1 or STAT2 to block their activity. Nipah virus V protein causes STAT1 and STAT2 to aggregate in the nucleus. The C proteins of Sendai virus, of which there are four, produced by in-frame translation starting from four different start codons, all bind to STAT1 to block its activity. In some cell lines, binding of the two larger C proteins leads to degradation of STAT1, but in all cells the binding of C proteins is sufficient to inhibit its activity.

Other minus-strand RNA viruses inhibit the activity of IFN by mechanisms that are not as yet established. A summary of the proteins involved is shown in Table 10.15.

Production of Cytokine-Binding Proteins

A number of viruses make proteins that bind to cytokines. Many of these cytokine-binding proteins are homologues of cellular cytokine receptors and have certainly been acquired from the host. Most of these viral proteins are secreted from the cell as soluble proteins that neutralize the activity of cytokines by binding to them in a nonproductive fashion. Others function at the surface of the infected cells.

Various poxviruses encode receptors for IFN-γ, IFN-α/β, IL-1, IL-6, IL-8, and TNF (Table 10.14). Most poxviruses

TABLE 10.13   Virus Manipulation of Cytokine Signaling

Virus family	Virus	Cellular target or homologue	Viral factor	Mode of action
Herpesviridae	HHV-5	TNF receptor	UL144	Unknown function, retained intracellularly
		Chemokine receptors	US28	Competitive CC-chemokine receptor, sequesters CC-chemokines
	HHV-8	Type 1 IFNs	vIRF K9	Blocks transcription activation in response to IFN
			vIRF-2	May modulate expression of early inflammatory genes
		Virus-encoded chemokines	vMIP-I, vMIP-II	T_H-2 chemoattractant, chemokine receptor antagonist
	HHV-1	Type 1 IFNs	$\gamma_1$34.5	Reverses IFN-induced translation block
		RNase L	2'-5' (A)	RNA analogue, inhibits RNase L
	HHV-4	Type 1 IFNs	EBNA-2	Downregulates IFN-stimulated transcription
		PKR	EBER-1	Blocks PKR activity
		Chemokine receptors	BARF-1	Secreted, sequesters CSF-1
		IL-10	BCRF-1	IL-10 homologue, antagonizes T_H-1 responses
Adenoviridae	Adenovirus	Type 1 IFNs	E1A	Blocks IFN-induced JAK–STAT pathway
		PKR	VA 1 RNA	Blocks PKR activity
		TNF-α	E3 proteins	Various mechanisms
Hepadnaviridae	HepB	Type 1 IFNs	Terminal protein	Blocks IFN signalling
Flaviviridae	HepC	PKR	E2	Inhibits PKR activation in response to type 1 IFN
Retroviridae	HIV	PKR	TAR RNA	Recruits cellular PKR inhibitor TRBP
Poxviridae	See Table 10.14			

Abbreviations: HHV-5, human cytomegalovirus; TNF, tumor necrosis factor; HHV-8, Kaposi's sarcoma herpesvirus; HHV-1, herpes simplex virus; HHV-4, Epstein–Barr virus; IFN, interferon; PKR, dsRNA-dependent protein kinase; CSF-1, colony stimulating factor; HIV, human immunodeficiency virus.
Source: Adapted from Table 3 of Tortorella *et al.* (2000).

secrete a receptor for IFN-γ that is distantly related to the human receptor. This receptor neutralizes the activity of IFN-γ and presumably functions to prevent IFN-γ-induced events. The potential efficacy of interference with IFN-γ is shown by the receptor secreted by rabbit myxoma virus. This virus causes an infection of European rabbits that has a 99% fatality rate (Chapter 7). Mutants that lack the IFN-γ receptor cause nonfatal illness in these rabbits.

Many poxviruses also produce a receptor for IFN-α/β and for TNF-α. TNF-α is a cytokine that has multiple roles in the control of virus-infected cells. It plays a role in apoptosis, but is also important in noncytolytic clearing of virus infection. The TNF-α receptor encoded by rabbit myxoma virus is multifunctional. It is secreted in part and binds TNF-α to neutralize it. However, it is partially retained within the cell where it interferes with signal transduction that induces apoptosis. The importance of TNF-α in control of viral infections is shown by the fact that some poxviruses produce two different TNF receptors to neutralize its activity, and by the fact that adenoviruses produce four different proteins, found in different parts of the infected cell, that antagonize the effects of TNF.

IL-1, IL-6, and IL-8 are also neutralized by gene products encoded by various poxviruses. Deletion of any of these poxvirus genes usually results in an attenuation of virus growth in experimental animals.

Poxviruses make a number of proteins that bind various chemokines. Neutralization of the activity of chemokines results in damping the inflammatory response to viral infection. Some of these chemokine-binding proteins are soluble proteins and some are expressed on the surface of the infected cell. Rabbit myxoma virus, for example, produces two proteins that bind chemokines. One protein binds with high affinity to a subset of chemokines called CC-chemokines. The second protein is the IFN-γ receptor described before. The rabbit myxoma virus IFN-γ receptor, but not that of other poxviruses, binds a number of chemokines through their heparin-binding domains. This binding is of low affinity. Thus, this protein is multifunctional, and binding of IFN-γ and chemokines are independent of one another.

Many beta- and gammaherpesviruses also produce cytokine or chemokine receptors. Human CMV, for example, produces four chemokine receptors, whereas HHV-8 produces one. Epstein–Barr virus encodes a receptor for the cytokine macrophage colony-stimulating factor (CSF-1).

In most cases, deletion of viral genes that interfere with cytokine or chemokine activity attenuates the virus in experimental animals. However, in some cases deletion of such

TABLE 10.14   Pox Defense Molecules

System	Target	Virus	Gene	Viral protein	Properties	Cellular homologue
Complement	C4B and C3B	Vaccinia Variola	C3L D15L	VCP SPICE	4 SCRs, secreted, binds, and inhibits C4B and C3B, binds heparan sulfate	C48-binding protein
	? ?	Vaccinia Variola	B5R B6R	42-kD protein	4 SCRs, EEV class I membrane glycoprotein, for virus egress	Complement control proteins
Interferon	Type 1 IFN	Vaccinia	B18R	IFNα/β BP	Binds to and inhibits IFN-α	IFN receptor
	PKR	Vaccinia Variola Swinepox	K3L C3L K3L	K3L protein	Binds PKR, inhibits phosphorylation of eIF-2α, leads to IFN resistance	
	dsRNA	Vaccinia Variola	E3L E3L	E3L protein	Binds dsRNA, nuclear localization, inhibits activation of PKR, leads to IFN resistance	
	IFN-γ	Myxoma Vaccinia Variola Swinepox	T7 B8R B8R C61	vIFN-γR	Secreted, binds and inhibits IFN-γ	IFN-γ receptor
Cytokine receptor homologues		Vaccinia Cowpox	B15R B14R	vIL-1β receptor	Secreted glycoprotein, binds, and inhibits IL-1β	IL-1β receptor
		Myxoma Vaccinia Variola Cowpox	T2 G2R C22L crm B	vTNF R (CrmB)	Secreted, binds and inhibits TNF-α, TNF-β	TNF receptor
		Variola Cowpox MCV	B6L C8L MC054	vIL-18BP	Secreted, binds soluble IL-18, inhibits induction of IFNγ	hIL-18 BP
Cytokine homologues		Orf	???	vIL-10	Secreted, binds to IL-10 receptors on inflammatory cells, blocks activation	IL-10
CC-chemokine inhibitors		Myxoma Vaccinia Variola Cowpox	p35	T1 and T7	Secreted, bind to CC-chemokines, inhibit binding of chemokines to heparan sulfate	None
Serine protease inhibitors		Myxoma Cowpox Vaccinia	M152R *crmA* B14R	Serp-2 Crm A SPI-1	Block serine proteases intracellularly, reduce inflammation, inhibit ICE, prevent apoptosis	SERPIN
		Myxoma		Serp-1	Secreted, binds to proteases like plasmin, thrombin	SERPIN

Abbreviations: MCV, molluscum contagiosum; SCR, 60 amino acid sequence called: "Short consensus repeat"; EEV, extracellular enveloped virions; PKR, dsRNA-dependent protein kinase; IFN, interferon; SERPIN, serine protease inhibitor superfamily; VCP, vaccinia complement control protein; BP, binding protein.
Source: Adapted from reviews by Tortorella *et al.* (2000), by Smith and Kotwal (2002), and by Seet *et al.* (2003).

genes leads to an increase in the virulence of the virus. The increased virulence appears to result from a more severe inflammatory response to virus infection.

Secretion of Virokines

Many beta- and gammaherpesviruses secrete cytokine or chemokine analogues, called virokines. The full range of activities of these products is unknown, but it appears that for many, perhaps all, virus replication is enhanced by the expression of increased amounts of the cytokine or chemokine, in contrast to the ablation of the activity of other cytokines or chemokines by virus-encoded receptors described in the previous section. As examples, HHV-8 produces three chemokines and one cytokine (IL-6), and Epstein–Barr virus encodes a homologue of IL-10. The chemokines may serve to attract target cells, since these viruses infect B cells. IL-6 and IL-10 are necessary for the growth of B cells

TABLE 10.15   Viral Interferon Antagonists of Negative-Strand RNA Viruses

Virus			Viral protein involved	Effect of antagonist
Family	Genus	Species		
Orthomyxoviridae	*Influenzavirus A*	FLUAV	**NS1**	Inhibit IFN induction; downregulate IRF-3, IRF-7, NF-κB; inhibit activation of PKR
	Thogotovirus	THOV	**ML**[*]	Blocks transcriptional activation of IFN
Paramyxoviridae	*Rubulavirus*	Mumps, SV5	**V**	Targets STAT1 for proteasome degradation
	Henipavirus	Nipah	**V**	Inhibits IFN signaling
	Respirovirus	Sendai	**C proteins**	Inhibit IFN signaling
	Pneumovirus	HRSV	**NS1 and NS2**	Mechanism unknown; both proteins needed
Rhabdoviridae	*Vesiculovirus*	VSV	**M**	Inhibits cellular mRNA synthesis and translation, hence inhibits IFN synthesis
Filoviridae	*Ebolavirus*	ZEBOV	**VP35**	Inhibits IFN induction
Bunyaviridae	*Bunyavirus*	BUNV	**NSs**	Inhibits IFN induction
	Tospovirus	TSWV	**NSs**	Inhibits RNA silencing in plants

Abbreviations: FLUAV, influenzavirus A; THOV, Thogotovirus; ML[*] is a 38 amino acid carboxyl terminus extended form of M from alternative splicing; HRSV, human respiratory syncytial virus; VSV, vesicular stomatitis virus; ZEBOV, Zaire ebolavirus; BUNV, buyamwera virus; TSWV, tomato spotted wilt virus.
Source: Data from Garcia-Sastre (2004).

and presumably serve to expand the target cell population. Furthermore, as described earlier, these cytokines skew the immune response toward a B-cell response, helped by T_H-2 cells, and away from a CTL response, helped by T_H-1 cells. Thus, while expanding the number of host cells available for infection by the virus, these cytokines also depress the number of CTLs that control the infected cell population.

Viral Use and Misuse of Gene-Silencing Pathways

As has been noted earlier, for almost all defense mechanisms of a host cell or organism against assault by viruses, the viruses have evolved counterdefenses, and the RNA interference pathway is no exception. Some viruses encode *suppressors* of *R*NA *s*ilencing or SRS proteins. Others encode miRNAs that appear to jam the pathway and prevent it from interfering with virus replication. Still other viruses go a step further and encode miRNAs that use the gene-silencing pathway to regulate their lifecycle or to silence cellular defense mechanisms. An overview of the little that is currently known in a rapidly developing field is given in Table 10.16.

Viruses in four different families have been shown to encode SRS proteins. Studies have been done in plants, insect cells, and mammalian cells, and it appears that SRS proteins are widely used in the virus world. These SRS proteins have been shown to inhibit degradation of viral RNA during virus infection.

Using improved algorithms for predicting miRNA genes in nucleotide sequences, viral genomes have been searched for likely miRNA sequences. It seems probable that viruses might use RNAi strategies to downregulate cellular genes, especially cellular genes involved in defending the host from viral invasion. miRNA sequences have been found in several different families of viruses and many of these predicted pre-miRNA sequences have been shown to be expressed in virus-infected cells and, at least in some cases, to be efficient substrates for processing by DICER. A sampling of such miRNAs is shown in Table 10.16. Although the cellular targets for most miRNAs that have been identified are not known, in three cases: the herpes simplex virus latency transcript, the adenovirus VAI and VAII RNAs, and miRNAs encoded by SV40, the function of the miRNAs has been established, as described next.

For herpes simplex virus, it has long been known that only a single viral transcript, from the *LAT* gene, was essential for maintenance of latent infection of neurons. The gene appeared to be transcribed into a 2-kb long-lived RNA that was retained in the nucleus, but no protein product was ever identified. It was shown that *LAT* promoted neuronal survival and inhibited apoptosis of infected neurons. Recent studies have shown that *LAT* mRNA contains a dsRNA stem-loop structure that the host cell recognizes as a pre-miRNA and processes to miR-LAT. miR-LAT functions as a gene silencer and promotes degradation of two cellular mRNAs, those for TGF-β and SMAD3, both of which are involved in induction of apoptosis. Thus, miR-LAT keeps

TABLE 10.16   Viral Defenses against Host RNA Interference

Virus family	Virus	Cell type or organism	Gene product	Effect
Viruses that encode SRS proteins				
Nodaviridae	Flock house	*Drosophila* cell lines, transgenic plants	Protein B2	
	Nodamura	Insect cells, mammalian cells	Protein B2	
Poxviridae	Vaccinia	*Drosophila* cell lines	E3L	Inhibit degradation of viral RNAs
Retroviridae	HIV	Human primary T cells	Tat	
Bunyaviridae	Tomato spotted wilt	Plants	NSs	
	LaCrosse	Human 293T cell line	NSs	
Viruses that encode miRNAs				
Adenoviridae	Adenovirus	Susceptible cells	VAI and VAII	Inhibit DICER
Polyomaviridae	SV-40	Cultured cells	Complements to tumor antigen mRNAs	Downregulate large T and small t
Retroviridae	HIV	Cultured cells	miRNAs in *env* and *nef*	???
Herpesviridae	Herpes simplex (HHV-1, 2)	Neurons	miR-LAT	Inhibit apoptosis
	Epstein–Barr (HHV-4)	Burkitt's lymphoma cell line	5 miRNAs	Maintain latency?
	Human cytomegalovirus (HHV-5)		Cultured cells	5 expressed miRNAs ???
	Kaposi's (HHV-8)	???	Several miRNAs	???

Abbreviations: SRS, *s*uppressors of *R*NA *s*ilencing. Data largely from Schütz and Sarnow (2006).

the latently infected neuron alive until such time as the virus reactivates and enters productive infection. Genes for miRNAs have been found in a number of other herpes viruses (see Table 10.16) and these miRNAs may also be involved in maintaining the latent or persistent infections established by all herpesviruses.

Adenoviruses produce small RNAs of ~160 nucleotides called VAI and VAII. Transcription is by the host RNA polymerase III. As described earlier, one function of VAI is to bind to PKR and prevent its activation by other dsRNA. VAI also interferes with the gene-silencing pathway, as does VAII. They are exported from the nucleus by Exportin5 and both VAs can be processed by DICER to VA-miRNAs, whose targets are unknown. It has been proposed that adenovirus VAI and VAII, which are produced in large amounts, suppress the host RNAi defense by saturating Exportin5, DICER, and RISC, but there could also be specific repression of genes by VA-miRNAs.

SV40 encodes two miRNAs whose activity appears paradoxical at first glance. They selectively downregulate production of SV40 large T and small t antigens. Downregulation of these antigens does not result in reduced viral yield, however, but downregulation did lead to reduced activation of cytotoxic T cells and helps the virus evade the host immune system.

Since the field of RNA interference is so new, it is likely that many of the RNAi-mediated interactions between hosts and viruses have yet to be discovered. The large genomes of DNA viruses such as the poxviruses and herpesviruses have lots of room to encode multiple 70–80 nucleotide hairpins

that could join the plethora of other counterdefense mechanisms these viruses possess. But even viruses with much smaller genomes could potentially encode such small RNAs, as shown by the examples of LaCrosse virus or HIV.

INTERACTIONS OF VIRUSES WITH THEIR HOSTS

The interaction of viruses with their hosts is intimate and the product of a long period of evolution during which viruses coevolved with their hosts. Humans cannot survive without a functioning immune system to protect them from viruses. However, this is the result of the long evolutionary history during which hosts and viruses adapted to one another, because viruses in turn cannot survive without their hosts. The example of rabbit myxoma virus demonstrates that the virulence of a virus diminishes if it kills or otherwise incapacitates too large a proportion of its hosts too rapidly (see Figs. 7.9 and 7.10). We can even speculate that the exceptional virulence of the influenza virus responsible for the 1918 pandemic might have been made possible because of active warfare ongoing at the time. The virus is so very virulent and incapacitates its victims so rapidly that we might expect such a virus to not persist and spread in the population. But very ill and dying soldiers continued to be moved about and the virus could continue to spread, perhaps could even spread more readily, even if it rapidly incapacitated its hosts. On the other hand, the many examples of ways in which viruses modify the immune and cytokine defenses of the host in order to replicate demonstrate that viruses are capable of evolving more virulent forms if it is to their advantage. The end result is an interplay in which viruses and their hosts exist in an uneasy equilibrium punctuated by the emergence of new viruses or the spread of new epidemics accompanied by changes in the immune system that protect against these viruses.

The virulence of a virus for its host depends in part on the epidemiology of the virus, how it gets from one host to another. The herpesviruses set up a lifelong infection in which they are effectively transferred once per generation. It is in the virus's interest not to incapacitate the host so that the host can pass it on 20 or 40 or 60 years later, and herpesviruses cause minor illness or no illness in most humans. On the other hand, arboviruses must cause a viremia (virus circulating in the blood) high enough to infect an insect taking a blood meal. Because many of these viruses are RNA viruses that encode relatively few functions to ablate the immune response, rapid and vigorous replication is required to establish the viremia before immunity is established, and this is often harmful to the host because many cells are killed in the process. To take another example, respiratory viruses that are transmitted as aerosols or in respiratory secretions must produce enough virus in the respiratory tract so that respiratory droplets expelled by coughing or sneezing will contain sufficient virus to infect a person nearby. These viruses are transmitted in epidemics that can spread rapidly and that require close contact between individuals, and one infected individual can infect dozens or even hundreds of others in a very short time. Thus these viruses need be transmissible only over a short period. Sexually transmitted viruses have different hurdles to overcome. Because the potential for sexual transmission is usually infrequent and one person interacts with a limited number of others, these viruses need to establish infections that last for long periods of time and that do not incapacitate the infected individual, at least not early in the infection.

The close interplay between viruses and their hosts means that the study of viruses continues to tell us much about the hosts. We now know much more about the adaptive immune system, the cytokine system, and apoptosis because of recent studies that started with viruses. Continuing studies on viruses have told us much about the function of regulatory genes and cancers. We are confident that the study of viruses will continue to teach us much about human biology.

FURTHER READING

General Summary of the Vertebrate Immune Response

Biron, C. A., and Sen, G. C. (2006). Innate responses to viral infection. Chapter 9 in: *Fields Virology, Fifth Edition* (D. M. Knipe and P. M. Howley, Eds. in chief), Philadelphia, Lippincott Williams & Wilkins, pp. 249–278.

Braciale, T. J., Hahn, Y. S., and Burton, D. R. (2006). Adaptive immune responses to viral infection. Chapter 10 in: *Fields Virology, Fifth Edition* (D. M. Knipe and P. M. Howley, Eds. in chief), Philadelphia, Lippincott Williams & Wilkins, pp. 279–326.

Cresswell, P., Ackerman, A. L., Giodini, A., Peaper, D. R., and Wearsch, P. A. (2005). Mechanisms of MHC class I-restricted antigen processing and cross-presentation. *Immunol. Rev.* **207**: 145–157.

Janeway, C. A., Travers, P., Walport, M., et al. (2004). *Immunobiology: The Immune System in Health and Disease, 6th Edition,* London, Elsevier Science.

Kuby, J. (1997). *Immunology.* New York, W. H. Freeman and Co.

Interferons and Other Cytokines

Buck, C. B., Day, P. M., Thompson, C. D., et al. (2006). Human α-defensins block papillomavirus infection. *Proc. Natl. Acad. Sci. U.S.A.* **103**: 1516–1521.

Landolfo, S., Gribaudo, G., Angeretti, A., et al. (1995). Mechanisms of viral inhibition by interferons. *Pharmacol. Ther.* **65**: 415–442.

Perry, A. K., Chen, G., Zheng, D., Tang, H., and Cheng, G. (2005). The host type I interferon response to viral and bacterial infections. *Cell Res.* **15**: 407–422.

Weber, F., Wagner, V., Rasmussen, S. B., Hartmann, R., and Paludan, S. R. (2006). Double-stranded RNA is produced by positive-strand RNA viruses and DNA viruses but not in detectable amounts by negative-strand RNA viruses. *J. Virol.* **80**: 5059–5064.

How Viruses Interfere with the Immune Response

Alejo, A., Ruiz-Argüello, M. B., Ho, Y., et al. (2006). A chemokine-binding domain in the tumor necrosis factor receptor from variola (smallpox) virus. *Proc. Natl. Acad. Sci. U.S.A.* **103**: 5995–6000.

Blair, G. E., and Hall, K. T. (1998). Human adenoviruses: evading detection by cytotoxic T lymphocytes. *Semin. Virol.* **8**: 387–398.

Conzelmann, K.-K. (2005). Transcriptional activation of alpha/beta interferon genes: interference by nonsegmented negative-strand RNA viruses. *J. Virol.* **79**: 5241–5248.

Davis-Poynter, N. J., and Farrell, H. E. (1998). Human and murine cytomegalovirus evasion of cytotoxic T lymphocyte and natural killer cell-mediated immune responses. *Semin .Virol.* **8**: 369–376.

Deitz, S. B., Dodd, D. A., Cooper, S., Parham, P., and Kirkegaard, K. (2000). MHC I-dependent antigen presentation is inhibited by poliovirus protein 3A. *Proc. Natl. Acad. Sci. U.S.A.* **97**: 13790–13795.

Esteban, D. J., and Buller, R. M. K. (2005). Ectromelia virus: the causative agent of mousepox. *J. Gen. Virol.* **86**: 2645–2659.

Gale, M. J., and Katze, M. G. (1998). Molecular mechanisms of interferon resistance mediated by viral-directed inhibition of PKR, the interferon-induced protein kinase. *Pharmacol. Ther.* **78**: 29–46.

García-Sastre, A. (2004). Identification and characterization of viral antagonists of Type I interferon in negative-strand RNA viruses. *Curr. Top. Microbiol. Immunol.* **283**: 249–280.

Hilleman, M. R. (2004). Strategies and mechanisms for host and pathogen survival in acute and persistent viral infections. *Proc. Natl. Acad. Sci. U.S.A.* **101**: 14560–14566.

Kalvakolanu, D. V. (1999). Virus interception of cytokine-regulated pathways. *Trends Microbiol.* **7**: 166–171.

Kash, J. C., Goodman, A. G., Korth, M. J., and Katze, M. G. (2006). Hijacking of the host-cell response and translational control during influenza virus infection. *Virus Res.* **119**: 111–120.

Katz, M. G., He, Y., and Gale, M., Jr. (2002). Viruses and interferon: a fight for supremacy. *Nature Rev. Immunol.* **2**: 675–687.

Klotman, M. E., and Chang, T. L. (2006). Defensins in innate antiviral immunity. *Nature Rev. Immunol.* **6**: 447–456.

Lachmann, P. J., and Davies, A. (1997). Complement and immunity to viruses. *Immunol. Rev.* **159**: 69–77.

Langland, J. O., Cameron, J. M., Heck, M. C., Jancovish, J. K., and Jacobs, B. L. (2006). Inhibition of PKR by RNA and DNA viruses. *Virus Res.* **119**: 100–110.

Leib, D. A., Machalek, M. A., Williams, B. R. G., Silverman, R. H., and Virgin, H. W. (2000). Specific phenotypic restoration of an attenuated virus by knockout of a host resistance gene. *Proc. Natl. Acad. Sci. U.S.A.* **97**: 6097–6101.

Li, X.-D., Sun, L., Seth, R. B., Pineda, G., and Chen, Z. J. (2005). Hepatitis C virus protease NS3/4A cleaves mitochondrial antiviral signaling protein off the mitochondria to evade innate immunity. *Proc. Natl. Acad. Sci. U.S.A.* **102**: 17717–17722.

Lichtenstein, D. L., Toth, K., Doronin, K., Tollefson, A. E., and Wold, W. S. M. (2004). Functions and mechanisms of action of the adenovirus E3 proteins. *Int.. Rev. Immunol.* **23**: 75–111.

Mahalingam, S., Meanger, J., Foster, P. S., and Lidbury, B. A. (2002). The viral manipulation of the host cellular and immune environments to enhance propagation and survival: a focus on RNA viruses. *J. Leukocyte Biol.* **72**: 429–439.

Paulus, C., Krauss, S., and Nevels, M. (2006). A human cytomegalovirus antagonist of type I IFN-dependent signal transducer and activator of transcription signaling. *Proc. Natl. Acad. Sci. U.S.A.* **103**: 3840–3845.

Poole, E., He, B., Lamb, R. A., Randall, R. E., and Goodbourn, S. (2002). The V proteins of simian virus 5 and other paramyxoviruses inhibit induction of interferon-β. *Virology* **303**: 33–46.

Rosengard, A. M., Liu, Y., Nie, Z., and Jimenez, R. (2002). Variola virus immune evasion design: Expression of a highly efficient inhibitor of human complement. *Proc. Natl. Acad. Sci. U.S.A.* **99**: 8808–8813.

Seet, B. T., Johnston, J. B., Brunetti, C. R., *et al.* (2003). Poxviruses amd immune evasion. *Annu. Rev. Immunol.* **21**: 377–423.

Shchelkunov, S. N., Totmenin, A. V., Safronov, P. F., *et al.* (2002). Analysis of the monkeypox virus genome. *Virology* **297**: 172–194.

Smith, G. L., Symons, J. A., and Alcami, A. (1998). Poxviruses: interfering with interferon. *Semin. Virol.* **8**: 409–418.

Smith, S. A., and Kotwal, G. J. (2002). Immune response to poxvirus infections in various animals. *Crit. Rev. Microbiol.* **28**: 149–185.

Tortorella, D., Gewurz, B. E., Furman, M. H., *et al.* (2000). Viral subversion of the immune system. *Annu. Rev. Immunol.* **18**: 861–926.

Apoptosis

Ashkenazi, A., and Dixit, V. M. (1998). Death receptors: signaling and modulation. *Science* **281**: 1305–1308.

Brydon, E. W. A., Morris, S. J., and Sweet, C. (2005). The role of apoptosis and cytokines in influenza virus morbidity. *FEMS Microbiol. Rev.* **29**: 837–850.

Raff, M. (1998). Cell suicide for beginners. *Nature* **396**: 119–122.

How Viruses Interfere with Apoptosis

Chinnadurai, G. (1998). Control of apoptosis by human adenovirus genes. *Semin. Virol.* **8**: 399–408.

Hill, A. B., and Masucci, M. G. (1998). Avoiding immunity and apoptosis: manipulation of the host environment by herpes simplex virus and Epstein-Barr virus. *Semin. Virol.* **8**: 361–368.

McFadden, G., and Barry, M. (1998). How poxviruses oppose apoptosis. *Semin. Virol.* **8**: 429–442.

Miller, L. K., and White, E. (Eds.) (1998). Apoptosis in virus infection. *Semin. Virol.* **8**: No. 6.

O'Brien, V. (1998). Viruses and apoptosis. *J. Gen. Virol.* **79**: 1833–1845.

Taylor, J. M., and Barry, M. (2006). Near death experiences: Poxvirus regulation of apoptotic death. *Virology* **344**: 1319–150.

Toll-like Receptors

Finberg, R. W., and Kurt-Jones, E. A. (2004). Viruses and toll-like receptors. *Microbes Infect.* **6**: 1356–1360.

Finberg, R. W., Knipe, D. M., and Kurt-Jones, E. A. (2005). Herpes simplex virus and Toll-like receptors. *Viral Immunol.* **18**: 457–465.

Häcker, H., Redecke, V., Blagoev, B., *et al.* (2006). Specificity in Toll-like receptor signalling through distinct effector functions of TRAF3 and TRAF6. *Nature* **439**: 204–210.

Iwasaki, A., and Medzhitov, R. (2004). Toll-like receptor control of the adaptive immune responses. *Nature Immunol.* **5**: 987–995.

Kawai, T., and Akira, S. (2006). TLR signaling. *Cell Death Differ.* **13**: 816–825.

Lee, J., Wu, C. C. N., Lee, K. J., *et al.* (2006). Activation of anti-hepatitis C virus responses via Toll-like receptor 7. *Proc. Natl. Acad. Sci. U.S.A.* **103**: 1828–1833.

López, C. B., Yount, J. S., and Moran, T. M. (2006). Toll-like receptor-independent triggering of dendritic cell maturation by viruses. *J. Virol.* **80**: 3128–3134.

Meylan, E., and Tschopp, J. (2006). Toll-like receptors and RNA helicases; two parallel ways to trigger antiviral responses. *Mol. Cell* **22**: 561–569.

O'Neill, L. A. J. (2005). Immunity's early-warning system. *Sci. Am.* **January**: 38–45.

Schröder, M., and Bowie, A. G. (2005). TLR3 in antiviral immunity: key player or bystander? *Trends Immunol.* **26**: 462–468.

Takeda, K., Kaisho, T., and Akira, S. (2003). Toll-like receptors. *Annu. Rev. Immunol.* **21**: 335–376.

Vaccines

Arvin, A. M., and Greenberg, H. B. (2006). New viral vaccines. *Virology* **344**: 240–249.

Bianchi, E., Liang, X., Ingallinella, P., *et al.* (2005). Universal influenza B vaccine based on the maturational cleavage site of the hemagglutinin precursor. *J. Virol.* **79**: 7380–7388.

de Filette, M., Jou, W. M., Birkett, A., *et al.* (2005). Universal influenza A vaccine: optimization of M2-based constructs. *Virology* **337**: 149–161.

Donnelly, J. J., Wahren, B., Liu, M. A., and Wahren, B. (2005). DNA vaccines: progress and challenges. *J. Immunol.* **175**: 633–639.

Gerhard, W., Mozdzanowska, K., and Zharikova, D. (2006). Prospects for universal influenza virus vaccine. *Emerg. Infect. Dis.* **12**: 569–574.

Graham, B. S., and Crowe, J. E., Jr. (2006). Immunization against viral diseases. Chapter 15 in: *Fields Virology, Fifth Edition* (D. M. Knipe and P. M. Howley, Eds. in chief), Philadelphia, Lippincott Williams & Wilkins, pp. 487–538.

Hirao, L., and Weiner, D. B. (2006). Advancing DNA vaccine technology. *Microbiol. Today* **33(1)**: 24–27.

Jaffe, S. (2005). New rotavirus vaccines on the horizon. *The Scientist* **February 15**: 37–39.

Lai, C.-J., and Monath, T. P. (2003). Chimeric flaviviruses: novel vaccines against dengue fever, tick-borne encephalitis, and Japanese encephalitis. *Adv. Virus Res.* **61**: 469–509.

Liu, M. A., Wahren, B., and Hedestam, G. B. K. (2006). DNA vaccines: recent developments and future possibilities. *Human Gene Ther.* **17**: 1051–1061.

Mason, P. W., Shustov, A. V., and Frolov, I. (2006). Production and characterization of vaccines based on flaviviruses defective in replication. *Virology* **351**: 432–443.

Robinson, H. L., and Weinhold, K. J. (2006). Phase I clinical trials of the National Institutes of Health vaccine research center HIV/AIDS candidate vaccines. *J. Infect. Dis.* **194**: 1625–1627.

Tesh, R. G., Travassos da Rosa, A. P. A., Guzman, H., Araujo, T. P., and Xiao, S.-Y. (2002). Immunization with heterologous flaviviruses protective against fatal West Nile encephalitis. *Emerg. Infect. Dis.* **8**: 245–251.

RNAi

Berkhout, B., and Haasnoot, J. (2006). The interplay between virus infection and the cellular RNA interference machinery. *FEBS Lett.* **580**: 2896–2902.

He, L., and Hannon, G. J. (2004). MicroRNAs; small RNAs with a big role in gene regulation. *Nature Rev. Genet.* **5**: 522–531.

Liu, J. D., Carmell, M. A., Rivas, F. V., *et al.* (2004). Argonaute 2 is the catalytic engine of mammalian RNAi. *Science* **305**: 1437–1441.

Nair, V., and Zavolan, M. (2006). Virus-encoded microRNAs: novel regulators of gene expression. *Trends Microbiol.* **14**: 169–175.

Pfeffer, S., Zavolan, M., Grässer, F. A., *et al.* (2004). Identification of virus-encoded microRNAs. *Science* **304**: 734–736.

Schütz, S., and Sarnow, P. (2006). Interaction of viruses with the mammalian RNA interference pathway. *Virology* **344**: 151–157.

11

Gene Therapy

INTRODUCTION

Molecular genetic studies during the last decades have led to an enormous increase in our understanding of the molecular biology of the replication of viruses. The complete nucleotide sequences of many virus genomes have been determined. Information on the origins required for the replication of these genomes, the promoters used to express the information within them, and the packaging signals required for packaging progeny genomes into virions have been established for many. The mechanisms by which viral mRNAs are preferentially translated have been explored. Together with methods for cloning and manipulating viral genomes, this information has made possible the use of viruses as vectors to express foreign genes. In principle, any virus can be used as a vector, and systems that use a very wide spectrum of virus vectors have been described. DNA viruses were first developed as vectors, since it is possible to manipulate the entire genome in the case of smaller viruses, or to use homologous recombination to insert a gene of interest in the case of larger viruses. Recent developments now make it possible to manipulate the entire genomes of even very large DNA viruses as artificial chromosomes, which potentially makes the use of these large genomes for expression of foreign proteins even more appealing. When complete cDNA clones of RNA viruses were obtained, it became straightforward to rescue plus-strand viruses from clones because the viral RNA itself is infectious, and many such viruses have been used to express proteins. The use of minus-strand RNA viruses as vectors was delayed because the virion RNA itself is not infectious, but recent developments has made it possible to rescue virus from cloned DNA by using coexpression of the appropriate viral proteins in a transfected cell. As a consequence, minus-strand RNA viruses have also joined the club of expression systems receiving intense study. Retroviruses have also been widely used because of their capacity to integrate into a host chromosome and potentially express foreign proteins indefinitely.

A sampling of expression systems and their uses is given here to illustrate the approaches that are being followed. Every virus system has advantages and disadvantages as a vector, depending on its intended use. One of the more exciting uses has been the development of viruses as vectors for gene therapy, that is, to correct genetic defects in humans. In the most general sense, gene therapy involves transfer of genetic information into a cell, tissue, organ, or organism with the goal of improving the clinical outcome, either by curing a disease, or alleviating an underlying condition in a patient. Although results have been disappointingly slow in coming, such systems offer great promise. This use represents an example of taking these infectious agents that have been the source of much human misery and developing them for the betterment of mankind. Such expression systems have a wide variety of other potential uses, however. Efforts to engineer viruses to kill cancer cells are also receiving attention. Viruses that express foreign proteins have potential uses in the engineering of new vaccines against other pathogens. Finally, viral expression systems have proved very useful in the expression of proteins in cell culture that can be used for various studies.

VIRUS VECTOR SYSTEMS

A representative sampling of viruses that are being developed as vectors is described next in order to illustrate some of the strengths and weaknesses of the different systems. The viruses used in most clinical trials to date have been the poxviruses, the adenoviruses, and the retroviruses, and these are described here. Several other virus systems that may be used in the future for treatment of humans, or that are useful for other purposes, are also described.

Vaccinia Virus

Vaccinia virus is a poxvirus with a large dsDNA genome of 200 kb (Chapter 7). Until recently, this genome was too big to handle in one piece in a convenient fashion, and homologous recombination has been used to insert foreign genes into it. The large size of the viral genome, however, does mean that very large pieces of foreign DNA can be inserted, while leaving the virus competent for independent replication and assembly. Another advantage of the virus is that it has been used to vaccinate hundreds of millions of humans against smallpox. Thus, there is much experience with the effects of the virus in humans. Although the vaccine virus did cause serious side effects in a small fraction of vaccinees, highly attenuated strains of vaccinia have been developed for use in gene therapy by deleting specific genes associated with

virulence. A new approach to the use of poxvirus vectors has been the development of nonhuman poxviruses, such as canarypox virus, as vectors. Canarypox virus infection of mammals is abortive and essentially asymptomatic, but foreign genes incorporated into the canarypox virus genome are expressed in amounts that are sufficient to obtain an immunologic response.

A variety of approaches have been used to obtain recombinant vaccinia viruses that express a gene of interest, but only the first such method to be used, and one that remains in wide use, is described here. This method is illustrated in Fig. 11.1. The thymidine kinase (TK) gene of vaccinia virus is nonessential for growth of the virus in tissue culture. Furthermore, deletion of the TK gene results in attenuation of the virus in humans, which is a desirable trait. Finally, the TK gene can be either positively or negatively selected

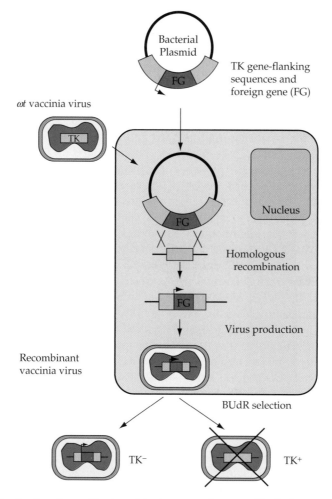

FIGURE 11.1   Construction of recombinant vaccinia virus expression vector. The foreign gene (red) is inserted into a bacterial plasmid adjacent to a vaccinia promoter (black arrow), flanked with sequences from the vaccinia thymidine kinase gene (turquoise). Plasmid DNA is transfected into cells infected with wild-type (TK+) vaccinia virus. Recombinant progeny from homologous recombination are all TK−, due to interruption of the TK gene, and can be selected by growth in bromodeoxyuridine (BUdR), since incorporation of BUdR into TK+ vaccinia is lethal. These TK− vaccinia will infect normally and express the foreign gene under control of the vaccinia promoter. Adapted from Strauss and Strauss (1997) Figure 2.25 on p. 115.

by using different media for propagation of the virus. The starting point is a plasmid clone that contains a copy of the TK gene that has a large internal deletion. In the region of the deletion a vaccinia virus promoter is inserted upstream of a polylinker. The gene of interest is inserted into the polylinker using standard cloning technology. Thus, we have the foreign gene downstream of a vaccinia promoter, and the entire insert is flanked by sequences from the vaccinia TK gene. The plasmid containing the cloned TK gene with its foreign gene insert is transfected into cells that have been infected by wild-type vaccinia virus. Homologous recombination between the TK gene in the virus and the TK-flanking sequences in the plasmid occurs with a sufficiently high frequency that a reasonable fraction of the progeny have the gene of interest incorporated. These viruses have an inactive TK gene (they are TK⁻), because the TK gene has been replaced by the deleted version containing the inserted foreign gene. The next step, then, is to select for viruses that are TK⁻ by growing virus in the presence of bromodeoxyuridine (BUdR). An active TK enzyme will phosphorylate BUdR to the monophosphate form, which can be further phosphorylated by cellular enzymes to the triphosphate and incorporated into the viral nucleic acid during replication. Incorporation of BUdR is lethal under the appropriate conditions, and thus viruses that survive this treatment are those in which the TK gene has been inactivated.

It is usually necessary to select among the surviving progeny for those that possess the gene of interest, because inactivation of the TK gene can occur spontaneously through deletion or mutation. Selection can be accomplished by a plaque lift hybridization assay in which virus in plaques is transferred to filter paper. Virus plaques on the filter paper are probed with radiolabeled hybridization probes specific for the inserted gene. Virus in plaques that hybridize to the probe is recovered and further passaged. In this way a pure virus stock that will express the gene of interest can be isolated.

Herpesviruses

The herpesviruses also have a large DNA genome that is capable of accommodating large inserts of foreign DNA. HSV-1, in particular, has been studied as an expression vector. Recombination has been used to insert foreign genes and to delete virus genes involved in lytic growth or toxicity. Because HSV-1 is neurotropic, it has been considered as a possible vector for the control or eradication of neural cancers. The viral DNA does not integrate, and the virus is capable of infecting nonreplicating neurons and being maintained in a latent state, properties that suggest it could be used for this purpose. It might also be used as an expression vector that could produce protein for long periods in neurons, and as such might be useful for the treatment of spinal nerve injury, for example, or for pain therapy.

Baculoviruses

The baculoviruses are insect viruses that have a large DNA genome capable of accommodating large DNA inserts. Foreign DNA is inserted by recombination and selection of appropriate viruses. They have been widely used to express high levels of protein in eukaryotic cells (insect cells in this case) that can be used, for example, in crystallization trials to determine protein structure, or to produce protein for immunization of animals, and other uses. Recent studies have suggested that baculoviruses might be useful for gene therapy in humans. The viruses will infect a number of human cells resulting in expression of proteins of interest, but the viruses are nonpathogenic in humans, suggesting a level of safety in their possible application. Whether problems associated with these viruses, such as low levels of expression and the rejection of them by the immune system, can be overcome remains to be determined.

Adenoviruses

Adenovirus infections of humans are common and normally cause only mild symptoms. Deletion of virulence genes from adenovirus vectors further attenuates these viruses. In addition, adenovirus vaccines have been used by the military for some years and, therefore, some experience has been gained in the experimental infection of humans by adenoviruses, although gene therapy trials use a different mode of delivery of adenovirus vectors. Because of their apparent safety, adenoviruses have been developed for use as vectors in gene therapy trials or for vaccine purposes. Two approaches have been used. In one, infectious adenoviruses have been produced that express a gene of interest. In the second approach, suicide vectors are produced that can infect a cell and express the gene of interest, but which are defective and cannot produce progeny virus. Suicide vectors cannot spread to neighboring cells, and the infection is therefore limited in scope and in duration.

The genome of adenoviruses is dsDNA of 36 kbp (Chapter 7). Thus, the genome is smaller than that of poxviruses or other large DNA viruses such as the herpesviruses and the baculoviruses and can accommodate correspondingly smaller inserts. However, inserts large enough for most applications can be accommodated. The genome is small enough that the virus can be reconstituted from DNA clones. Such an approach is inconvenient, however, and homologous recombination is often used to insert the gene of interest into the virus genome.

The foreign gene is inserted into the region occupied by either the adenovirus E1 or E3 genes, one or both of which are deleted in the vector construct. Virus lacking E1 cannot replicate, and such viruses form suicide vectors. For gene therapy, suicide vectors are normally used so as to prevent the spread of the infection. To prepare the stock of virus lacking E1, the virus must be grown in a cell line that expresses E1. An overview of this process is shown in Fig. 11.2. The complementing cell line, which produces E1 constitutively, supplies the E1 needed for replication of the defective adenovirus. The cells are transfected with the defective adenovirus DNA and a full yield of progeny virions results. The progeny virus is defective and cannot replicate in normal cells, but it can be amplified by infection of the complementing cell line. On introduction of the virus into a human, the virus will infect cells and express the foreign gene, but the infection is abortive and no progeny virus

is formed. The stock of defective virus must be tested to ensure that no replication-competent virus is present, since such virus can arise by recombination between the vector and the E1 gene in the complementing cell line.

Adenoviruses with only E3 deleted are often used to express proteins for vaccine purposes. These E3-deleted viruses possess intact E1 and will replicate in cultured cells and in humans, but are attenuated. Because the virus replicates, expression of the immunizing antigen persists for a long time and a good immune response usually results.

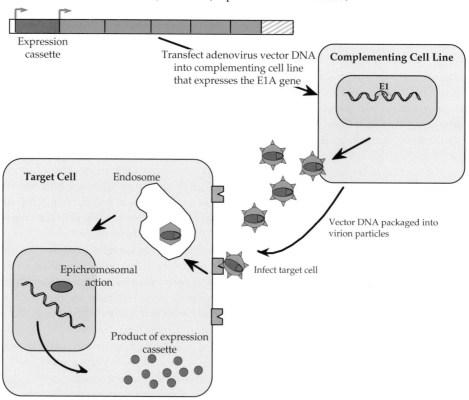

A. Wild-type Adenovirus Genome

B. Adenovirus vector DNA (E1, E3 deleted, expression cassette inserted)

FIGURE 11.2   Generation of a nonreplicating adenovirus expression vector. From the wild-type adenovirus genome, the E1 and E3 genes are removed. The E1 genes are replaced with an expression cassette. This adenovirus DNA is transfected into a complementing cell line that produces E1 protein. The transfection produces particles that are able to infect cells, but which are E1⁻ and nonreplicating. The DNA genome is delivered to the nucleus where it functions as an epichromosome and directs expression of the inserted foreign gene. Adapted from Crystal (1995).

The procedure for insertion of the gene of interest by homologous recombination resembles that used for the pox-viruses. The gene is inserted into a plasmid containing flanking sequences from the E1 or E3 region, and transfected into cells infected with adenovirus. Recombinant viruses containing the gene of interest are selected and stocks prepared. It is also possible to transfect cells with the E1 or E3 expression cassette together with DNA clones encoding the rest of the adenovirus genome, in which case homologous recombination results in the production of virus. In the case of insertions into E1, cells that express E1 must be used to produce the recombinant virus.

Adeno-Associated Viruses

Adeno-associated viruses (AAVs) have a single-stranded DNA genome of 4.7 kb (Chapter 7). They normally require coinfection of a cell by a helper virus, usually an adenovirus or a herpesvirus. They are being developed as expression vectors because they are not pathogenic in humans and because they normally integrate into the host-cell genome in a specific region, thus minimizing the problems of insertional mutagenesis. The genome size is small enough to be readily manipulated as a DNA clone, but the small size also limits the amount of DNA that can be inserted and therefore the applicability of the virus for gene transfer experiments. A related problem is that for expression studies, the genome is normally deleted for the AAV genes with only the ends that function as promoter sites retained. However, site-specific integration requires the activity of the Rep protein. Nonetheless, the system is sufficiently attractive that efforts to develop AAV as a gene therapy vector continue and the virus has been used in a number of clinical trials, as described later.

Retroviruses

Retrovirus-based expression systems offer great promise because the retroviral genome integrates into the host-cell chromosome during infection and, in the case of the simple retroviruses at least, remains there as a Mendelian gene that is passed on to progeny cells on cell division. Thus, there is the potential for permanent expression of the inserted gene of interest. The essential components of a retrovirus vector are the long terminal repeats (LTRs), the packaging sequences known as ψ, the primer-binding site, and the sequences required for jumping by the reverse transcriptase during reverse transcription to form the dsDNA copy of the genome (Chapter 6).

The process of creating and packaging a retrovirus-based expression cassette is illustrated in Fig. 11.3. A packaging cell line is created that expresses the retroviral *gag*, *pol*, and *env* genes, but whose mRNAs do not contain the packaging signal and so cannot be packaged. The vector DNA/RNA is created by modifying a DNA clone of a retrovirus to contain the gene of interest in place of the *gag–pol–env*

genes. In the process, all of the essential *cis*-acting signals required for packaging, reverse transcription, and integration are retained. The foreign gene can be under the control of the LTRs, or it can be under the control of another promoter positioned in the insert upstream of it. The resulting DNA clone is transfected into the packaging cell line, and a producer cell line isolated that expresses the vector DNA as well as the helper DNA. Vector RNA transcribed from the vector DNA is packaged into retroviral particles, using the proteins expressed from the helper DNA. These particles are infectious and can be used to infect other cells or to transfer genes into a human. On infection of cells by the packaged vector, the vector RNA is reverse transcribed into DNA that integrates into the host-cell chromosome, where it can be expressed under the control of the promoters that it contains. The limitation on the size of the insert is about 10 kb, the upper limit of RNA size that can be packaged.

Although murine leukemia viruses are not known to cause disease in man, it has been found that these viruses will cause tumors in immunosuppressed subhuman primates. Thus, it is thought to be essential that there be no replication-competent virus in stocks used to treat humans. Replication-competent virus can arise during packaging of the vector by recombination between the vector and the retroviral sequences used to produce Gag–Pol–Env. At the current time, preparations of packaged vectors are screened to ensure that replication-competent viruses are not present. Efforts are being made to reduce the incidence of recombination during packaging in order to simplify the procedure. One approach is to develop vectors that have very little sequence in common with the helper sequences, in order to reduce the incidence of homologous recombination. A second approach is to separate the Gag–Pol sequences from the Env sequences in the helper cell. In this case, recombination between three separate DNA fragments in the producer cell (that encoding Gag–Pol, that encoding Env, and sequences in the vector) are required in order to give rise to replication-competent retrovirus.

In gene therapy trials that use retroviruses, it has been found that the expression of the foreign gene in humans is often downregulated after a period of months. Attempts are being made to identify promoters that will not be downregulated. Different promoters might be required for different uses, and promoters that target transcription to particular cell types would be useful.

A major problem with retroviral vectors is that simple retroviruses will only infect dividing cells. Although they enter cells and are reverse transcribed into DNA, the DNA copy of the genome can enter the nucleus only during cell division. In many gene therapy treatments, it is desirable to infect stem cells in order to maintain expression of the therapeutic gene indefinitely. Because stem cells divide relatively infrequently, it is difficult to infect a high proportion of them by vectors used to date. Attempts are being made to identify methods to stimulate stem cells to divide during *ex vivo* treatment,

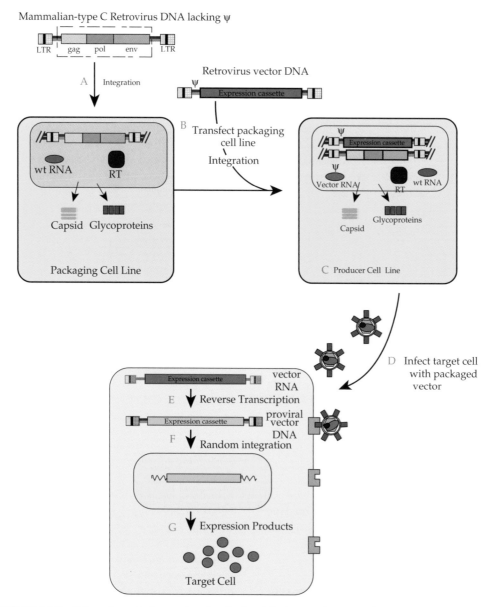

FIGURE 11.3   Scheme for producing a packaged replication-defective retrovirus expression vector. (A) A "packaging" cell line is generated by introduction of DNA encoding *gag*, *pol*, and *env* genes into the chromatin of a fibroblast cell line. This DNA lacks the packaging signal Ψ and RNA transcribed from it is not packaged. (B) The packaging cell line is transfected with a second retroviral DNA, in which the foreign gene (expression cassette) replaces *gag*, *pol*, and *env*, but which has an intact Ψ packaging signal and intact LTRs, to form a "producer" cell line. This producer cell line (C) releases vector particles containing the expression cassette genome packaged with the proteins from the helper genome. (D) These particles enter target cells via specific cell surface receptors, (E) are reverse transcribed, (F) randomly integrate, and (G) produce expression products. Adapted from Dunbar (1996).

so that a larger fraction of them can be infected. A second approach is to develop lentivirus vectors. Lentiviruses, which include HIV, can infect nonreplicating cells and could potentially infect nondividing stem cells during *ex vivo* treatment. Lentivirus vectors could also be useful for therapy involving other nondividing cells, such as neurons.

It would be of considerable utility to be able to target retroviruses to specific cells. One possible approach to this is to replace all or part of the external domains of the retroviral surface glycoprotein with a monoclonal antibody that is directed against an antigen expressed only on the target cells. In principle, this approach is feasible, but whether it can be developed into something practical is as yet an open question. If specific cells could be infected, it would allow protocols in which the therapeutic gene would be expressed only in cells where it would be most useful. It would also allow the specific killing of cells such as tumor cells or HIV-infected cells. For example, the retrovirus could express a gene that rendered the cell sensitive to toxic drugs such as BUdR. A retrovirus vector that expressed such a gene could also be useful for conventional gene therapy, because it would allow the infected cells to be killed if the infection process threatened to get out of hand.

Alphaviruses

The genomes of plus-strand RNA viruses are self-replicating molecules that replicate in the cytoplasm, and they can express very high levels of protein. These properties make them potentially valuable as expression vectors.

The alphaviruses possess a genome of single-strand RNA of about 12 kb (Chapter 3). Their genomes can be easily manipulated as cDNA clones, and infectious RNA can be transcribed from these clones by RNA polymerases, either *in vivo* or *in vitro*. RNA transcribed *in vitro* can be transfected into cells and give rise to a full yield of virus, whereas RNA transcribed *in vivo* will begin to replicate and produce virus. The structural proteins are made from a subgenomic mRNA, making it easy to insert a foreign gene under the control of the subgenomic promoter. Two approaches that have been used are illustrated in Fig. 11.4. In one approach, a second subgenomic promoter is inserted into the genome downstream of the structural proteins, or between the structural proteins and the nonstructural proteins (Fig. 11.4C). Two subgenomic mRNAs are transcribed,

one for the structural proteins and the second for the gene of interest. The size of the insert must be relatively small, on the order of 2000 nucleotides or less, because longer RNAs are not packaged efficiently. However, this system has the advantage that the resulting double subgenomic virus is an infectious virus that can be propagated and maintained without helpers.

A second approach is to delete the viral structural proteins and replace them with the gene of interest. In this case, there is room for an insert of about 5 kb that will still allow the resulting replicon to be packaged. The replicon is capable of independent replication, and transcription of a subgenomic messenger results in expression of the gene of interest. The replicon constitutes a suicide vector. It cannot be packaged unless the cells are coinfected with a helper to supply the structural proteins, or unless a packaging cell line that expresses the viral structural proteins is used.

Alphavirus replicons can be extremely efficient in expressing a foreign gene. In some cases as much as 25% of the protein of a cell can be converted to the foreign protein expressed by the replicon over a period of about 72 hours. Wild-type replicons are cytolytic in vertebrate cells, inducing apoptosis, and the infection dies out. However, replicons have been produced with mutations in the replicase proteins that are not cytolytic and will produce the protein of interest

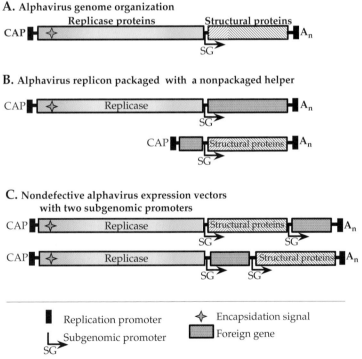

FIGURE 11.4 Alphavirus expression vectors. (A) The genome organization of a typical alphavirus with the location of the promoters for replication and production of subgenomic RNA as well as the RNA-packaging signal indicated. (B) A simple alphavirus replicon. The structural proteins of the virus have been replaced with the foreign gene to be expressed. If packaging of the replicon is required, the structural proteins of the virus are supplied on a DI RNA lacking a packaging signal. (C) Packaged expression vectors with two subgenomic promoters. These constructs are unstable if the foreign gene is much larger than 2 kb. Adapted from Strauss and Strauss (1994) Figure 23.

indefinitely. Thus, a wide sprectrum of choices is available, and the system chosen can be adapted to the needs of a particular experiment or treatment.

Viral expression systems would be more useful if they could be directed to specific cell types. An approach that uses monoclonal antibodies to direct Sindbis virus to specific cells has been described. Protein A, produced by *Staphylococcus aureas*, binds with high affinity to IgG. It is an important component of the virulence of the bacterium because it interferes with the host immune system. The IgG-binding domain of protein A has been inserted into one of the viral glycoproteins. Virions containing this domain are unable to infect cells using the normal receptor. However, the virus will bind IgG monoclonal antibodies. If an antibody directed against a cell surface component is bound, the virus will infect cells expressing this protein at the cell surface. Thus, this system has the potential to direct the virus to a specific cell type. One of the advantages of this approach is that the virus, once made, can be used with many different antibodies and thus directed against a variety of cell types. This approach is potentially applicable to any enveloped virus, and perhaps to nonenveloped viruses as well.

A modification of the alphavirus system is to use a DNA construct containing the replicon downstream of a promoter for a cellular RNA polymerase, rather than using packaged RNA replicons. On transfection of a cell with the DNA, the replicon RNA is launched when it is transcribed from the DNA by cellular enzymes. Once produced, the RNA replicates independently

and produces the subgenomic mRNA that is translated into the gene of interest. As described in Chapter 10, naked DNA can be used to transfect muscle cells and perhaps other cells.

Polioviruses

Plus-strand viruses that do not produce subgenomic mRNAs, such as the picornaviruses and flaviviruses, present different problems for development as vectors. The translated product from the gene of interest must either be incorporated into the polyprotein produced by the virus and provisions made for its excision, or tricks must be used to express the gene of interest independently. Two approaches with poliovirus will be described as examples of how such viruses might be used as vectors.

Poliovirus replicons have been constructed by deleting the region encoding the structural proteins and replacing this sequence with that for a foreign gene. The foreign gene must be in phase with the remainder of the poliovirus polyprotein, and the cleavage site recognized by the viral 2A protease is used to excise the foreign protein from the polyprotein. Because the poliovirus replicon lacks a full complement of the structural genes (it is a suicide vector), packaging to produce particles requires infection of a cell that expresses the poliviral structural proteins by some mechanism. A construct that uses this approach to express the cytokine tumor necrosis factor alpha (TNF-α) is illustrated in Fig. 11.5. A poliovirus "infectious

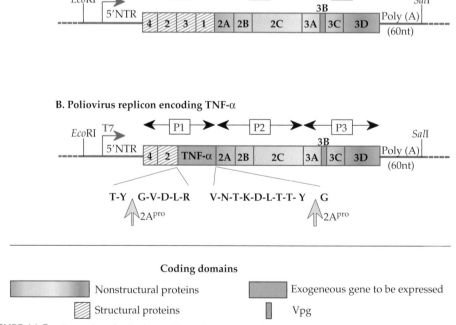

FIGURE 11.5   Generation of poliovirus replicons for expression of foreign genes in motor neurons. Based on an earlier construct to express interleukin-2 via a poliovirus replicon, the gene for wild-type murine tumor necrosis factor alpha (TNF-α) was positioned between the VP0 and 2A proteins of poliovirus, replacing VP3 and VP1. It was flanked on either side by sites for cleavage by the poliovirus 2A protease. These constructs were injected into transgenic mice expressing the poliovirus receptor, and expression of murine TNF-α was monitored. Adapted from Bledsoe *et al.* (2000).

clone" in which a DNA copy of the viral genome is positioned downstream of a promoter for T7 RNA polymerase is modified by replacing the genes for VP3 and VP1 with the gene for TNF-α. Recognition sites for the poliovirus 2A protease are positioned on both sides of the TNF-α gene. The TNF-α protein is produced as part of the poliovirus polyprotein, and cleaved from the polyprotein by the 2A protease. Packaged replicons were used to infect transgenic mice that expressed the polio receptor (Chapter 1). One of the interests of this system is that poliovirus exhibits an extraordinary tropism for motor neurons in the central nervous system (CNS) (Chapter 3). The packaged replicons, on introduction into the CNS, infected only motor neurons, and therefore the foreign gene was expressed only in motor neurons. Such replicons may be useful to treat CNS diseases in which motor neurons are affected.

A second approach to the use of poliovirus replicons is to use a second internal ribosome entry site (IRES) (Chapter 1) to initiate the synthesis of the nonstructural proteins. If the foreign gene replaces the structural genes, it will be translated from the 5' end of the genome. If the poliovirus nonstructural genes are placed downstream of a second IRES, internal initiation at this IRES results in production of a polyprotein for the nonstructural proteins. This approach is similar to the approach shown in Fig. 3.3, where the structural proteins are replaced by a gene of interest.

Coronaviruses as Expression Vectors and Vaccine Candidates

The coronaviruses have long been considered as potential candidates for experimental expression vectors or as candidate vaccines. The viruses have the largest nonsegmented RNA viral genomes, up to 31 kb (see also Chapter 3), which is both an advantage in that they could potentially accommodate a large amount of heterologous nucleic acid, and a disadvantage due to the difficulties of manipulating large RNA molecules. The essential genes are arranged 5'-replicase-S-E-M-N-3', and are interspersed with a number of nonessential genes which are group specific. Recent studies have revealed a number of characteristics which make them even more attractive as vector candidates. The first was that the deletion of the nonessential genes is sufficiently attenuating that no further mutations in the essential genes are required to produce an avirulent virus. Second, as the precise domains on the S protein which interact with the species-dependent cellular receptors were determined, it was found that both species and tissue specificity could be altered by relatively minor changes in the sequence of S. Third, it is possible to rearrange the linear order of the genes, and while this altered the relative amounts of the products, it reduces the possibility that the vector (vaccine) could undergo recombination with field strains. Fourth, it is possible to insert heterologous genes anywhere in the genome by simply incorporating

a cassette comprised of the gene of interest preceded by the specific intergenic sequence of the parental virus.

The generation of a recombinant MHV (mouse hepatitis virus) encoding *Renilla* luciferase is shown in Figure 11.6 to illustrate the strategies employed. These include maintenance of a replication defective genome as a bacterial plasmid under the control of a T7 promoter, transcription of RNA *in vitro*, electroporation of RNA into feline cells previously infected with a murine coronavirus engineered to infect feline cells, *in vivo* recombination, and selection for recombinants on murine cells. Due to the high frequency of recombination in coronaviruses, many recombinants are unstable and not suitable for use as vaccines. However, the location of the foreign gene within the genome, the particular coronavirus used, and the identity of the particular heterologous gene all have significant effects on stability, and coronaviruses may yet prove to be useful for targeted gene delivery.

Rhabdoviruses and Other Negative Strand Viruses

In minus-strand RNA viruses, the genomic RNA is not itself infectious. Ribonucleoprotein containing the N, P, and L genes is required for replication of the viral RNA, and thus for infectivity, and only recently have methods been devised to recover virus from cDNA clones. A schematic diagram of how virus can be recovered from DNA clones of the rhabdovirus vesicular stomatitis virus (VSV) (Chapter 4) is shown in Fig. 11.7. A cell is transfected with a set of cDNA clones that together express N, P, and L as well as the genomic or antigenomic RNA. The antigenomic RNA usually works better, probably because it does not hybridize to the mRNAs being produced from the plasmids. Encapsidation of the antigenomic RNA by N, P, and L to form nucleocapsids allows it to replicate and produce genomic RNA that is also encapsidated. Synthesis of mRNAs from the genomic RNA, together with continued replication, results in a complete virus replication cycle and production of infectious progeny virus that have as their genome the RNA supplied as a cDNA clone. The yield of infectious virus is small, but sufficient to isolate individual plaques and thus obtain viruses from the cDNA clones.

The ability to rescue virus from a cDNA clone makes it possible to manipulate the viral genome. Since the rhabdovirus genome is transcribed into multiple mRNAs, one for each gene, and the transcription signals recognized by the enzyme are well understood, it is relatively simple to add or delete genes. A modified VSV that was produced by using DNA clones is illustrated in Fig. 11.8. In this VSV, the surface glycoprotein present on the VSV particle, called G, has been deleted and replaced with CD4, the cell surface protein that is used as a receptor by HIV. In addition a new gene has been inserted, the gene encoding the HIV coreceptor CXCR4, so that the virus now contains six genes. The virions produced

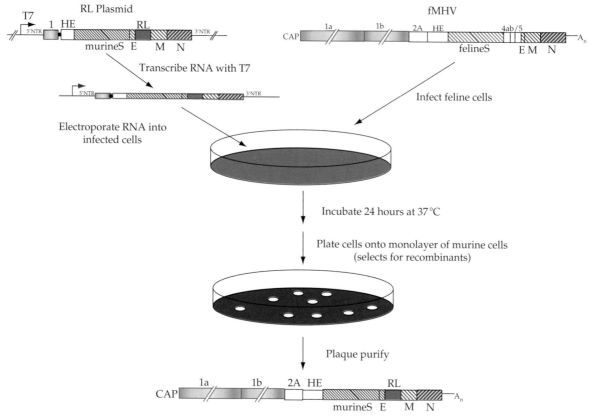

FIGURE 11.6   Construction of a coronavirus expression plasmid. An attenuated and nonreplicating plasmid from MHV is constructed in which the 5' end of the ORF 1 sequence is fused to the last 28 codons of HE. In addition, all of the accessory protein genes are deleted, and the gene for *Renilla* luciferase (RL) inserted. RNA is transcribed *in vitro* from this plasmid using T7 polymerase and the RNA electroporated onto feline cells which have been infected with MHV virus in which the spike S protein has been mutated to recognize receptors on feline cells (fMHV). After 24 hours these cells are plated out onto a monolayer of murine cells to select for coronaviruses that have undergone recombination between the plasmid and the helper virus. Recombinants are plaque-purified and stocks grown.

are unable to infect the cells normally infectable by VSV because they lack the G protein. However, because they contain the HIV receptor and coreceptor on their surface, they do infect cells that express the HIV glycoprotein on their surface, such as cells infected by HIV. Since VSV is a lytic virus, the HIV-infected cells are killed.

USE OF VIRUSES AS EXPRESSION VECTORS

Viruses have been widely used as vectors to express a variety of genes in cultured cells. This use is of long standing and has led to important results. Of perhaps more interest are efforts to develop viruses as vectors for medical purposes. The manipulation of virus genomes to develop new vaccines is very promising. Although no licensed human vaccines have been introduced using this technology, clinical trials are ongoing and it is to be expected that

several such vaccines will be licensed in the near future. There is also expectation that viruses will be useful as vectors for gene therapy, and numerous clinical trials are taking place. The results to date have been disappointing, but the promise remains.

Expression of Proteins in Cultured Cells

The use of viruses to express foreign genes in cultured cells is well established and only a few examples are cited to illustrate the range and purpose of such use.

Hepatitis C virus (HCV) does not grow in cultured cells to titers sufficient to allow studies on the expression of viral proteins. The only experimental model for the virus is the chimpanzee, which severely restricts the number and nature of experiments that can be done. Thus, most of what we know about the expression of the HCV genome has been obtained through expression of parts of

A. Rhabdovirus genome organization

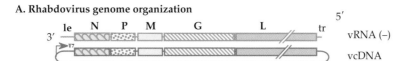

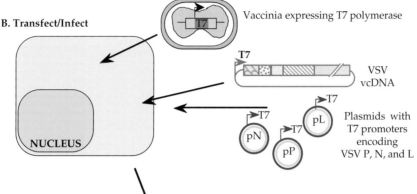

C. Expression of VSV replicase proteins (L, P) and N protein under the control of T7 promoter

D. Transcription of VSV vRNA from cDNA by T7 polymerase; replication of vcRNA, vRNA, and transcription of VSV mRNAs by VSV replicase/transcriptase

E. Translation of viral proteins and assembly of infectious virus.

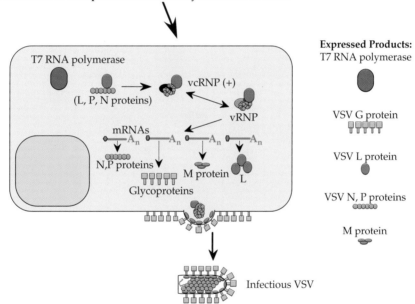

FIGURE 11.7   Rescuing infectious VSV virions from cDNA clones. (A) The rhabdovirus genome organization, and a schematic of a cDNA clone containing the genome sequence (cDNA copy of vRNA). (B) A susceptible cell is infected with vaccinia virus expressing the T7 RNA polymerase, and transfected with four separate plasmids: the genome plasmid from which plus strand anti-genome RNA is transcribed, and 3 individual plasmids expressing VSV N protein, P protein, and L protein, all under the control of T7 promoters. (C), (D), (E) Steps in the synthesis of infectious virus are described within the figure. Infectious virions bud from the cell and can infect a new susceptible cell. Adapted from Conzelmann and Meyers (1996).

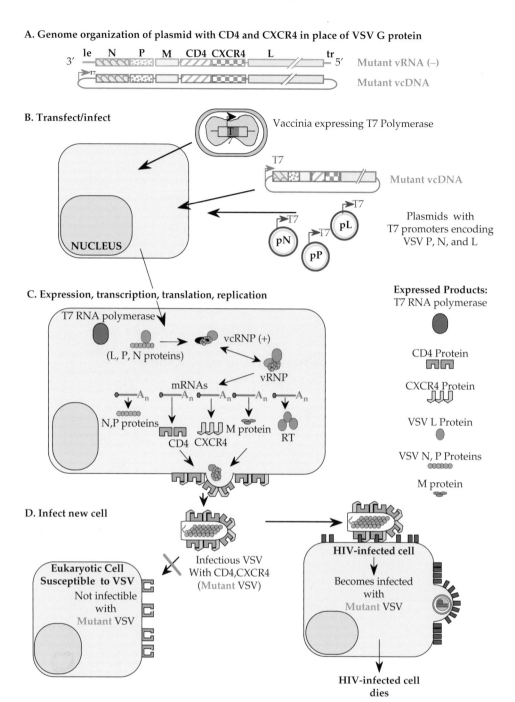

A. Genome organization of plasmid with CD4 and CXCR4 in place of VSV G protein

B. Transfect/infect

Vaccinia expressing T7 Polymerase

Mutant vcDNA

Plasmids with
T7 promoters encoding
VSV P, N, and L

NUCLEUS

C. Expression, transcription, translation, replication

T7 RNA polymerase

(L, P, N proteins)

vcRNP (+)

vRNP

mRNAs

N,P proteins

CD4 CXCR4 M protein

RT

Expressed Products:
T7 RNA polymerase

CD4 Protein

CXCR4 Protein

VSV L Protein

VSV N, P Proteins

M protein

D. Infect new cell

Infectious VSV
With CD4,CXCR4
(Mutant VSV)

**Eukaryotic Cell
Susceptible to VSV**

Not infectible
with
Mutant VSV

HIV-infected cell

Becomes infected
with
Mutant VSV

**HIV-infected cell
dies**

FIGURE 11.8   Producing a mutant VSV targeted to kill HIV-infected cells. (A) Genome of a rhabdovirus in which the glycoprotein G gene has been replaced with sequences encoding CD4 and CXCR4, the HIV primary receptor and coreceptor, and a schematic of a cDNA clone containing the genome sequence (cDNA copy of vRNA). (B) A susceptible cell is infected with vaccinia virus expressing the T7 RNA polymerase, and four separate plasmids: the mutant genome plasmid from which full-length vc (plus strand) RNA is transcribed, and three individual plasmids expressing VSV N protein, P protein, and L protein, all under the control of T7 promoters. (C) Plus-strand mutant vcRNA is transcribed and encapsidated with N, P, and L. The RNP then replicates and both viral proteins and CD4 and CXCR4 are expressed from individual mRNAs transcribed from genome sense RNPs. Virions bud from the cell and (D) cannot infect a new susceptible cell as before in Fig. 11.7, but can infect an HIV-infected cell expressing HIV *env* proteins on its surface. Infection with VSV is cytolytic and the HIV-infected cell dies. Adapted from Conzelmann and Meyers (1996).

the genome by virus vectors, often by recombinant vaccinia virus. These studies have resulted in an understanding of the two viral proteases within the HCV genome, the processing pathway through which the polyprotein translated from the genome is processed, the function of the viral IRES, and the function of the viral replicase, among other results. The use of virus vectors means that such studies on HCV can be conveniently conducted in mammalian cells under conditions that are related to the natural growth cycle of the virus.

Norwalk virus is another virus for which there is no cell culture system. The virus can be grown only in human volunteers, again limiting the range of studies that can be done. Virus particles isolated from the stools of infected volunteers are often degraded and difficult to purify to homogeneity. Thus, structural studies of infectious virus have been limited. Expression of cDNA copies of the structural proteins of the virus in baculovirus vectors has allowed the production of large amounts of viral structural proteins that spontaneously assemble into virus-like particles. These virus-like particles have been studied by cryoelectron microscopy, and detailed information on the structure of the virus has been obtained in this way.

Baculoviruses are also widely used to prepare large amounts of protein for crystallographic studies. Such studies require 20 mg or more of protein, and the baculovirus system can be used to prepare such quantities. An advantage of the system is that the protein is made in a eukaryotic cell, which can be important for obtaining the protein folded into its correct three-dimensional conformation. Also of importance is the use of secretion sequences in the constructs that lead to the secretion of the protein from the infected cell, making it easier to purify the protein.

Even for viruses for which cell culture systems exist, the use of virus vectors that express to higher levels can be advantageous. There are cell culture systems in which rubella virus will grow and plaque, and there is a full-length cDNA clone of rubella virus from which infectious RNA can be recovered. However, the cell culture systems produce only low amounts of virus proteins, especially of the nonstructural proteins, and it has been difficult to study the expression and processing of the nonstructural polyprotein. Expression of the nonstructural region of rubella virus in vaccinia virus vectors or in Sindbis virus vectors has allowed the production of much larger quantities of the polyprotein precursor. This has been used to determine the processing pathways, the identification of the virus nonstructural protease, and the identification of the cleavage sites that are cleaved by this protease.

As a final example, vaccinia virus vectors and Sindbis virus vectors have been used to map T-cell epitopes for a number of viruses (Chapter 10). For this, defined regions of a viral protein are expressed in order to determine whether a particular T-cell epitope lies within that region.

Viruses as Vectors to Elicit an Immune Response

Much effort is being put into the development of viruses as agents to immunize against other infectious agents, including other viruses. Such an approach has a number of advantages. There is a large body of experience in the use of attenuated or avirulent viruses as vaccines. Many of these, such as vaccinia virus or the yellow fever 17D virus, both of which have been used to immunize many millions of people, can be potentially developed as vectors to express other antigens, such as those of HCV or HIV. Use of a live virus as a vector to express antigens of other pathogens has many of the advantages of live virus vaccines. This includes the fact that only low initial doses are required, and therefore the expense of vaccine production may be less; that subsequent virus replication leads to the expression of large amounts of the antigen over an extended period of time, and the antigen folds in a more or less native conformation; and that a full range of immunity, including production of CTLs as well as of humoral immunity, usually develops.

No human vaccines have been licensed that use such recombinant viruses, but there are ongoing clinical trials of several potential vaccines. Several trials of candidate vaccines against HIV have been conducted that use vaccinia virus or retrovirus vectors to express the HIV surface glycoprotein. These trials have been moderately successful in the sense that immune responses to HIV glycoprotein were obtained, but these immune responses were not particularly vigorous and it is not known if the immune response is protective. HIV is able to persist in infected patients despite a vigorous immune response, and sterilizing immunity might be required. Further, the HIV surface glycoprotein is highly glycosylated and neutralizing antibodies are difficult to obtain. Studies in monkeys with related vaccines against simian immunodeficiency virus have given mixed results. In most such trials, immune responses were generated, but these were not fully protective. One recent trial did generate a protective response, however, giving hope that continued efforts in this direction will ultimately work out. Recent studies with anti-HIV drugs given very soon after infection found that limiting the replication of the virus early appears to allow the generation of a protective immune response in at least some patients. Although such studies remain preliminary, they do suggest that a nonsterilizing immune response that restricts virus replication early might prove to be protective.

Other clinical trials have also tested poxviruses as vectors. Vaccinia virus has been used in an attempt to immunize against Epstein–Barr virus, and canarypox virus has been used as a vector for potential immunization against rabies virus.

Although no licensed human vaccines use poxvirus vectors, veterinary vaccines that are based on poxvirus vectors are in use. One such vaccine consists of vaccinia virus that expresses the rabies surface glycoprotein. This vaccine has been used to immunize wildlife. The recombinant vaccinia viruses are spread in baits that are eaten by wild animals that serve as reservoirs of the virus, such as skunks, raccoons, foxes, and coyotes. This approach has been useful in limiting the spread of rabies in wildlife populations. Other poxvirus-based vaccines include vaccinia virus vectors to protect cattle against vesicular stomatitis virus and rinderpest virus, and to immunize chickens against influenza virus; pigeonpox virus vectors to immunize chickens against Newcastle disease virus; fowlpox virus vectors to immunize chickens against influenza, Newcastle disease, and infectious bursal disease viruses; a capripox virus vector to immunize pigs against pseudorabies virus; and a canarypox virus vector used to immunize dogs against canine distemper virus. Thus, it should be possible to develop human vaccines based on poxvirus vectors.

In a quite different approach, clinical trials of a novel vaccine against Japanese encephalitis (JE) virus have been conducted. JE is a scourge in parts of Asia, causing a large number of deaths and neurological sequelae in people that survive the encephalitis (Chapter 3). Vaccines in widespread use are inactivated virus vaccines, and the difficulties in preparing the large amounts of material required and delivering it to large segments of the population are significant. An attenuated virus vaccine, SA14-14-2, has been prepared in China by passing the virus in cultured cells and in rodent tissues. This vaccine is safe but overattenuated, so that the effectiveness is only 80% after a single dose. In contrast, the yellow fever virus (YF) 17D vaccine has an effectiveness of virtually 100% after a single dose, and immunity is long lasting, probably lifelong. A candidate JE vaccine has been developed that consists of the 17D strain of YF virus in which the prM and E genes have been replaced with those of JE, as illustrated in Fig. 11.9. Four chimeric viruses were tested. The JE structural proteins were taken from either the virulent Nakayama strain or from the attenuated SA-14-14-2 strain. In both cases, chimeras containing all three structural proteins from JE were tested

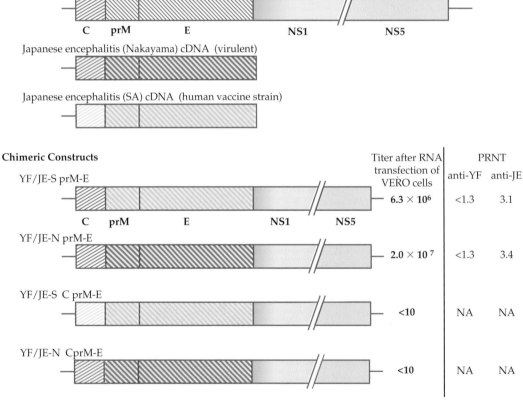

FIGURE 11.9 Construction of yellow fever/Japanese encephalitis chimeric viruses. Starting with the full-length cDNA clone for 17D yellow fever virus, a number of chimeric viruses were constructed in which the M and E proteins were replaced with those of different strains of Japanese encephalitis virus. However, when C, M, and E of JE were put into the yellow fever clone, no viable virus was obtained. Both prME chimeras grew well in tissue culture, and were neutralized by anti-JE antiserum. YF/JE-S prM-E was attenuated, and did not kill adult mice by intracerebral inoculation, but YF/JE-N prM-E was neurovirulent. PRNT is the log reciprocal of the dilution yielding 50% plaque reduction neutralization, based on 100 PFU on LLV-MK cells, using either YF or JE hyperimmune ascitic fluid. Adapted from Chambers et al. (1999).

as well as chimeras that contained only prM and E from JE. Chimeras containing C, prM, and E from JE were not viable, whereas chimeras containing only prM and E from JE were viable and grew well in culture (Fig. 11.9).

The viable chimeras were first tested in mice. The chimera containing the Nakayama strain proteins caused lethal encephalitis in mice, as does the YF 17D virus (even though it is safe for use in humans). However, the chimera containing prM and E from the attenuated JE strain was fully attenuated in mice and did not cause illness. The fully attenuated chimera was chosen for testing in monkeys, and was found to be safe and to protect monkeys against challenge with JE virus.

Clinical trials of this candidate vaccine have taken place in humans. The vaccine appears to be safe and more effective than the JE vaccines now in use. Furthermore, this approach is applicable to other flaviviruses, such as the dengue viruses, for which no licensed vaccines exist, or West Nile virus, which spread recently to the Americas where it caused a number of fatal cases of human encephalitis. Recombinant YF 17D expressing the prM and E proteins of all four serotypes of dengue viruses and recombinant viruses expressing prM and E of West Nile virus are also in clinical trials with encouraging results.

Yet another possible approach to developing new generations of vaccines using the power of biotechnology is to attenuate a virus by making changes in the laboratory that are expected to cripple the virus. Such an approach can be used with virtually any virus. A candidate vaccine strain of dengue virus has been constructed by making deletions in the 3′ nontranslated region of the genome that attenuate the virus, and such viruses are being tested in early trials.

GENE THERAPY

A number of genetic diseases result from the failure to produce a specific protein, usually due to a single defective gene. One of the more exciting possible uses for virus vectors is for the expression of a missing protein as a cure for the genetic defect associated with its absence. Some of these "monogenic diseases" that might be curable through the use of gene therapy are listed in Table 11.1. For successful treatment, expression of the missing protein must be long-term and preferably lifelong, the levels of protein produced must be sufficient to alleviate the symptoms of the disease, the protein must be expressed in or translocated to those cells that require the normal protein for function, and infection with the virus vector must be free of disease symptoms. Because of the requirement for long-term expression, viruses whose DNA integrates into the host chromosome, such as the simple retroviruses as well as the lentiviruses, and adeno-associated viruses, offer the most promising system for many diseases. To date, several hundred patients have been treated with vectors based on Moloney murine leukemia virus in clinical trials. Clinical trials have also been conducted that use adenovirus. adeno-associated virus, poxvirus, and herpesvirus vectors (Fig. 11.10).

Clinical trials in humans, which require extensive prior testing in animals, are divided into three phases. Phase I involves relatively few, usually healthy individuals. The objective of Phase I trials is to test the safety of a vaccine or treatment as well as the dosage that is tolerated, and the individuals are closely monitored during the trial. Phase II trials involve more individuals and test the efficacy of the treatment, and patients are again closely monitored. If a treatment

TABLE 11.1   Genetic Defects That Are Candidates for Gene Therapy

Disease	Defect	Incidence	Target cells
Severe combined immunodeficiency (SCID)	Adenosine deaminase (ADA) in 25% of SCID patients	Rare	Bone-marrow cells or T lymphocytes
Hemophilia ⟨A B⟩	Factor VII deficiency Factor IX deficiency	1:10,000 males 1:30,000 males	Liver, muscle, fibroblasts or bone marrow cells
Familial hypercholesterolemia	Deficiency of low-density lipoprotein (LDL) receptor	1:1 million	Liver
Cystic fibrosis	Faulty transport of salt in lung epithelium	1:3000 Caucasians	Airways in the lungs
Hemoglobinopathies thalassemias	(Structural) defects in the α or β globin gene	1:600 in certain ethnic groups	Bone marrow cells that are precursors to red blood cells
Gaucher's disease	Defect in the enzyme glucocerebrosidase	1:450 in Ashkenazi Jews	Bone marrow cells, macrophages
α1 antitrypsin deficiency inherited emphysema	Lack of α1 antitrypsin	1:3500	Lung or liver cells
Duchenne muscular dystrophy	Lack of dystrophin	1:3500 males	Muscle cells
Xeroderma pigmentosa	Impaired DNA repair, leading to severe sensitivity to sunlight	Rare	Fibroblasts

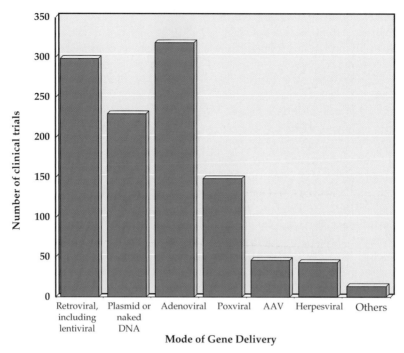

FIGURE 11.10   Vectors used in gene therapy trials as of January 2007. The "others" category includes flavivirus (5), measles virus (3), Newcastle disease virus (1), poliovirus (1), Semliki Forest virus (1), Sendai virus (1) and SV40 (1). Data are from Gene Therapy Clinical Trials Worldwide Web site of the *Journal of Gene Medicine* at: http://www.wiley. co.uk/genetherapy/clinical/.

passes both of these tests, Phase III trials can begin in which thousands of individuals are treated to test the efficacy of treatment. Virtually all of the clinical trials in gene therapy conducted to date are Phase I or Phase II; fewer than 2% of trials have progressed to Phase III (Fig. 11.11), and no gene therapy treatments have been licensed to date.

In addition to the possible treatment of genetic defects, virus vectors may also be useful for the treatment of a number of acquired diseases. These include cancer, HIV infection, Parkinson's disease, injuries to the spinal cord, and vascular diseases such as restenosis and arteriosclerosis. A partial listing of acquired diseases that have been suggested as candidates for gene therapy is given in Table 11.2, and the number of trials for a number of different conditions is shown in Fig. 11.12. Despite the large efforts to use gene therapy in clinical settings, the progress has been disappointingly slow, and many of the trials have been aborted due to unforeseen adverse consequences.

Nevertheless, as infectious clones of viruses continue to be developed, a large body of research is being devoted to construction of vectors, especially now to second and third generation vectors, as the problems associated with the initial systems are becoming clear. A comparison of various virus systems that are being considered for gene therapy is shown in Tables 11.3 and 11.4. Naked DNA has also been used in a recent trial for coronary artery disease, and the properties of this system are included in Table 11.4. Most of the modern vectors have had more and more of the dispensable viral genes deleted. Deletion of these genes reduces pathogenicity, and prevents the production of immunogenic viral

antigens. Often only the gene of interest and the viral transcriptional regulatory elements are left, and to prepare the vectors for use in trials, all other functions must be supplied by a helper virus or a packaging cell line. Another advantage of such stripped down vectors is the fact that it is improbable that the vector can recombine with wild-type viruses, either exogenous or endogenous, to cause disease.

A partial listing of clinical trials that attempt to treat several different genetic defects by using virus vectors to deliver specific genes is given in Table 11.5. There have been few successes to date and the table is more of a compendium of the variety of genes and diseases, as well as the variety of delivery schemes, that are being examined. Also included in the table is an impending trial for the vaccination of humans against HPIV-3 using a bovine virus.

Retrovirus Vectors to Genetically Mark Cells

Retroviruses have been used in a number of clinical trials to genetically tag cells. Although this use does not fall within the narrow definition of gene therapy, it does provide background experience in the use of retrovirus vectors in humans. One such use has been in bone marrow transplantation for leukemia. Severe forms of leukemia can sometimes be treated by ablation of the hematopoietic system with chemotherapy and/or X-rays in order to kill all tumor cells, followed by reconstitution of the system by transplantation of bone marrow from a compatible donor. Although often successful, the leukemia sometimes recurs and it is desirable

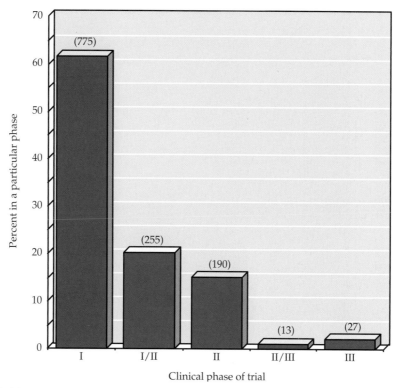

FIGURE 11.11 Percent of gene therapy clinical trials that are in each phase as of January 2007. The numbers above the bars are the actual number of trials. Data from Gene Therapy Clinical Trials Worldwide from the Web site of the *Journal of Gene Medicine* at: http://www.wiley.co.uk/genetherapy/clinical/.

to know whether it recurs because of incomplete destruction of the patient's leukemic cells or whether the donor cells are the source of the leukemia. Experiments in which the donor cells have been tagged using retroviruses that express a marker gene have been used to answer this question, which is important for the design of transplantation protocols.

Gene Therapy for ADA Deficiency

Patients who lack the enzyme adenosine deaminase (ADA) will die early in life unless treated. Lack of ADA results in the failure to clear adenosine from the body and, consequently, the accumulation of adenosine in cells throughout the body.

Adenosine is toxic at high concentrations, producing a variety of symptoms. The most serious symptom results from the extreme sensitivity of T cells to elevated adenosine concentrations. Loss of T cells results in SCID, *s*evere *c*ombined *i*mmu-*no*deficiency. Both CTL responses (which are T-cell based) and humoral responses (which require T-helper cells) are impaired. People with SCID syndrome are unable to mount an immunologic response to infectious agents, and SCID is invariably fatal early in life unless treated in some way. ADA deficiency accounts for about 25% of SCID syndromes in humans.

SCID can be treated by bone marrow transplantation if a suitable donor can be found. In the case of SCID due to ADA deficiency, weekly or twice weekly injections of ADA

TABLE 11.2 Some Acquired Diseases That Are Candidates for Gene Therapy

Disease	Defect	Incidence	Target cells
Cancer	Many causes, including genetic and environmental	1 million/year in United States	Variety of cancer cell types, in liver, brain, pancreas, breast, kidney
Neurological diseases	Parkinson's, Alzheimer's spinal-cord injury	1 million Parkinson's and 4 million Alzheimer's patients in the United States	Neurons, glial cells, Schwann cells
Cardiovascular	Restenosis, arteriosclerosis	13 million in United States	Arteries, vascular endothelial walls
Infectious diseases	AIDS, hepatitis B	Increasing numbers	T cells, liver, macrophages
Rheumatoid arthritis	Autoimmune inflammation of joints	Increasing numbers with aging population	Intra-auricular delivery and expression of IL-1 and TNF-α inhibitors

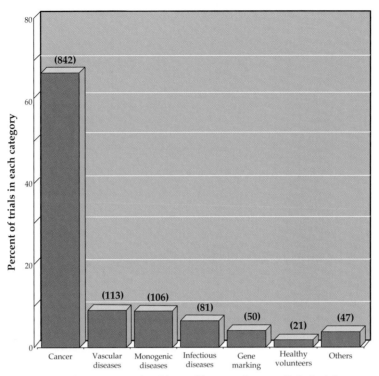

Conditions currently addressed by gene therapy clinical trials

FIGURE 11.12   Percent of gene therapy trials which are directed toward various conditions as of January 2007. The numbers above the bars are the actual number of trials. Data from Gene Therapy Clinical Trials Worldwide from the Web site of the *Journal of Gene Medicine* at: http://www.wiley.co.uk/genetherapy/clinical/.

TABLE 11.3   Comparison of Properties of Various Vector Systems Based on RNA Viruses and Retroviruses

Features	Simple retroviral	Lentiviral	Alphaviral	Coronaviral	Negaviral[a]
Maximum insert size	7–7.5 kb	7–7.5 kb	5 kb	2.7 kb	~4 kb[b]
Concentration in viral particles/ml	$>10^8$	$>10^8$	$>10^9$	$>10^8$	$>10^9$ for VSV, PIVs 10–1000 fold lower
Route of gene delivery	*Ex vivo*	*Ex/in vivo*	*In vivo*	*In vivo*	*In vivo*
Integration	Yes	Yes	No	No	No
Duration of expression *in vivo*	Shorter than theorized	Long	Short	Variable	Not known
Stability	Good	Not tested	Good	Dependent on background	Good
Ease of preparation scale up	Pilot scale up to 20–50 liters	Not known	Not known	Not known	Not known
Preexisting host immunity	Unlikely	Unlikely, except in AIDS patients	No	Unlikely	Unlikely
Safety problems	Insertional mutagenesis?	Insertional mutagenesis?	Few	Recombination with wild strains	—
Other advantages	—	Replicate in nondividing cells	—	—	Recombination virtually unknown; naturally atttenuated viruses exist

[a] Includes consideration of VSV based and PIV based vectors.
[b] Longer inserts are tolerated but vectors are too attenuated *in vivo* to be useful.
Source: Verma and Somia (1997), Jolly (1994), and Bukreyev *et al.* (2006).

TABLE 11.4 Comparison of Properties of Various Vector Systems Based on DNA Viruses

Features	Vectors based on DNA viruses					Naked/lipid DNA
	Adenoviral	AAV	Herpesviral	Vaccinia	Baculovirus	
Maximum insert size	7.5 kb	<4 kb	~30 kb	25–75 kb	>38 kb	Unlimited size
Concentration in viral particles/ml	>10^{10}	>10^{12}	>10^{10}	10^7–10^9	>10^8	No limitation
Route of gene delivery	*Ex/in vivo*	*Ex/in vivo*	*Ex vivo*	*Ex/in vivo*	*Ex/in vivo*	*Ex/in vivo*
Integration	No	Yes/no	No	No	No	Very poor
Duration of expression *in vivo*	Short	Long	Short/long in CNS?	Short	Short	Short
Stability	Good	Good	Unknown	Good	Good	Very good
Ease of preparation scale up	Easy to scale up	Difficult to purify, difficult to scale up	Not yet tried	Vaccine production facilities exist	Easy to scale up	Easy to scale up
Immunologic problems	Extensive	Not known	Not known	Extensive	Not known	None
Preexisting host immunity	Yes	Yes	Yes	Diminishing as unvaccinated population grows	Unlikely	No
Safety problems	Inflammatory response, toxicity	Inflammatory response, toxicity	Neurovirulence? Insertional mutagenesis	Disseminated vaccinia in immuno-compromised hosts	None?	None?

Source: Verma and Weitzman (2005), Jolly (1994), Boulaiz *et al.* (2005), Hu (2006).

mixed with polyethylene glycol (PEG) have been used to successfully treat about 60 patients in whom bone marrow transplantation cannot be used because of the lack of compatible donors. Of these, about 10 patients have also been treated with retroviral vectors that express ADA. In these experiments, T cells were taken from the patient (or in the case of three newborns, umbilical cord cells were used), infected *ex vivo* with the retrovirus vector using a number of different cell culture and infection protocols, and the cells reinfused into the patient. Many of the patients continue to produce ADA from the vector several years after treatment. However, all of the patients continue to receive ADA–PEG injections, which is known to be an effective treatment. Although some patients who have received retroviral therapy have been partially weaned from the supplementary ADA–PEG, it appears that some of these, and perhaps all, do not produce enough ADA to be cured. Thus, although no cures were effected in these early trials, the results were encouraging and suggested that future protocols might be more successful. Two areas of retroviral therapy that needed improvement were to increase the efficiency with which stem cells are infected, and the need to prevent the retroviral promoter from being downregulated.

A more recent trial involving two SCID-ADA patients in Italy was more successful (Table 11.5). Two infants for whom no compatible donor existed and for whom no PEG-ADA treatment was available were treated with improved retroviral therapy. Both patients developed functional immune systems and no adverse events have been reported. The improved results appear to arise from improved protocols as well as to selection in the patient for lymphoid progenitor cells that expressed adequate amounts of ADA.

Treatment for SCID Caused by IL2RG Deficiency

SCID disease can also be caused by a failure to produce the receptor for the cytokine interleukin-2 (Chapter 10). Thirteen SCID patients with this deficiency were treated in two different clinical trials with retroviruses that expressed the defective gene (Table 11.5). The results illustrate the highs and the lows of gene therapy trials. All 13 patients developed functional immune systems, and the trials at first appeared to be a complete success. However, three of the patients later developed T-cell leukemia and one has died of the leukemia. The leukemia was at first suspected to arise from insertional mutagenesis, a chronic worry with vectors that insert into the host DNA, and many gene therapy trials that used retroviral vectors were suspended. Recent studies indicate that the disease is not due to insertional mutagenesis, however, but rather due to the oncogenic potential of the IL2RG gene itself, as studies have shown that overexpression of this gene in mice results in leukemia.

TABLE 11.5   Recent Human Gene Therapy Trials

Disease	Therapeutic gene	Total patients	Vector/ promoter	Method of delivery	Outcome
Monogenic diseases					
OTC deficiency	OTC cDNA	18	E1E4 deleted adenovirus/CMV	*In vivo* injection to hepatic artery	No clinical benefit; **1 death**
Factor IX deficiency (hemophilia B)	Modified Factor IX gene	8	rAAV2/ CMV	*In vivo* intramuscular	No sustained clinical benefit
		7	rAAV2/APOE-SERPINA1	*In vivo* injection to hepatic artery	Transient Factor IX expression, no sustained benefit
SCID-X1 (French trial)	IL2RG cDNA	10	Retrovirus/MLV-LTR	*Ex vivo* transduction of CD34⁺ cells	9 patients developed functional immune system; **3 developed T-cell leukemia, 1 death**
SCID-X1 (British trial)	IL2RG cDNA	4	GALV-pseudotyped retrovirus/MFG-LTR	*Ex vivo* transduction of CD34⁺ cells	4 patients developed functional immune system; no adverse results reported in first 2 years
SCID-ADA (Italian trial)	ADA cDNA	2	Retrovirus/MLV-LTR	*Ex vivo* transduction of CD34⁺ cells	2 patients developed functional immune system; no adverse results reported in first 4 years
CGD	GP91PHOX	2	Retrovirus/SFFV-LTR	*Ex vivo* transduction of CD34⁺ cells	2 patients developed functional neutrophils and **clonal myelo-proliferation (cancer)**
Duchenne muscular dystrophy	Dystrophin	9	Plasmid/CMV	Intramuscular injection	Low dystrophin expression in 6 of 9, no adverse effects (Phase I)
	Microdystrophin	?	AAV/CMV	Intramuscular biceps injection	Phase I/II ongoing
	2OMeAOs	9	Oligonucleotide	Injection into extensor digitorum brevis	Successful non-human primate study; human Phase I/II not yet opened for patient recruitment
Vaccine for infectious disease					
HPIV-3 vaccine	F and HN surface glycoproteins	—	BPIV-3	Infection	Successful non-human primate study; human Phase I trial not yet opened for patient recruitment

APOE-SERPINA1, α1-antitrypsin promoter linked to APOE enhancer; BPIV-3, bovine parainfluenzavirus 3; CGD, chronic granulomatous disease; CMV, cytomegalovirus promoter; GALV, gibbon ape leukemia virus; HPIV, human parainfluenzavirus; LTR, long terminal repeat; MFG, derivative of MLV; MLV, Moloney murine leukemia virus; OTC, ornithine transcarbamylase; rAAV, recombinant adeno-associated virus serotype 2; SCID-ADA, severe combined immunodeficiency secondary to adenosine deaminase deficiency; SCID-IX, severe combined immunodeficiency secondary to mutations in the IL2RG gene; SFFV, Friend mink cell spleen focus-forming virus.
Source: Data for this figure came from Porteus *et al.* (2006), Foster *et al.* (2006), Bukreyev *et al.* (2006).

Cystic Fibrosis

Cystic fibrosis results from loss of the cystic fibrosis transmembrane conductance regulator (CFTR), which regulates epithelial transport of ions and water. Although lack of this protein results in damage to the epithelium in many parts of the body, the most serious manifestation is lung disease accompanied by chronic bacterial infection of the airways. Clinical trials using adenoviral vectors, which infect respiratory epithelium, to express CFTR in the lungs have been conducted. The first such studies were encouraging, but a more recent trial that was carefully controlled found no relief of symptoms. Inflammation produced by the high doses of adenovirus used in trials is also a problem. It is difficult to get efficient delivery to the lung, especially through the thick mucus that is a characteristic of cystic fibrosis. In addition, the expression needs to continue for the life of the patient, which means either a very stable (integrated?) gene being expressed, or a system of repeated administration of vector that can be tolerated without immunologic consequences. Lentiviruses have been proposed as an attractive vector, but there have been concerns about probable pathogenesis due to the lentivirus itself. However, for this disease, the most promising mode of gene delivery so far developed has been DNA compacted into nanoparticles with polycations, in particular PEG-polylysine, nicknamed "polyplexes." A clinical trial in humans used CFTR polyplexes, and the Phase I trial showed no adverse effects. For this disease a nonviral approach may well be the best solution.

Duchenne Muscular Dystrophy

Duchenne muscular dystrophy (DMD) is a severe muscle wasting disorder due to the lack of functional dystrophin protein. It occurs in 1/3500 male births. There are a number of difficulties in attempting to cure DMD with gene therapy, including the size of the protein, which is encoded in a cDNA of 11kb, and the need to deliver the vector to a large proportion of the body mass, that is to all of the striated muscles and cardiac muscle. Several approaches have been tried, and the first human clinical Phase I trial has just been completed (see Table 11.5). In this study, a plasmid containing the entire gene under the control of a CMV promoter was injected intramuscularly. Although only low levels of dystrophin expression were observed in 6 out of 9 patients, there was no evidence for adverse reactions and the trial was considered a success.

A second approach has been to attempt to introduce the gene in an AAV vector, especially in one of the many human isolates to which most of the population show no preexisting immunity, or into a nonhuman AAV. However, here the size of the gene is a problem, but it has been shown that the dystrophin protein contains a large number of repeated elements (Fig. 11.13) and that "mini-dystrophin" and "micro-dystrophin" are functionally active in the mouse model for the disease, the *mdx* mouse. Low-pressure intravenous injection of AAV6 expressing micro-dystrophin into a mouse could transfect 90% of muscle, but high titers of the AAV6 vector were required. A Phase I/II clinical trial has been initiated for delivery of micro-dystrophin in AAV under the control of a CMV promoter into human patients, but there are no results as yet.

A third approach attacks the specific nature of the genetic defect. It has been shown that 75% of DMD is caused by a frameshift mutation in one of the exons, such that no dystrophin is produced. Since severely truncated dystrophin (like micro-dystrophin) can be functional (Fig. 11.13), therapy to get rid of the exon causing the problem is being developed. Modified antisense oligonucleotides (AOs) can be used to alter the splicing pattern of the gene such that the exon in which the frameshift occurs is skipped, thereby restoring the reading frame. Following success of an AO with a 2′O-methyl-phosphorothiolate backbone (2OMe AOs) to restore function in the *mdx* mouse, two clinical Phase I trials have been initiated, one in the Netherlands using 2OMe AOs and one in Great Britain using morpholino AOs. Since both are targeting the exclusion of exon 51, it will be possible to directly compare the two chemically modified AOs.

Rheumatoid Arthritis

Rheumatoid arthritis is a chronic, progressive inflammatory disease of the joints. An estimated 5 million people in the United States suffer from it. There is no cure. Drug therapies are used that ameliorate the symptoms, but most of these drugs have side effects and cannot be taken indefinitely. If the disease progresses far enough, joint replacement may be required. The disease is associated with the release of inflammatory cytokines in the affected joints. Clinical trials have started that use retroviruses to deliver the gene for an anti-arthritic cytokine gene to the joints. The gene encodes the interleukin (IL)-1 receptor antagonist, which inhibits the biological actions of both IL-1α and IL-1β. It is hoped that such treatment might damp out the disease or at least keep it from progressing.

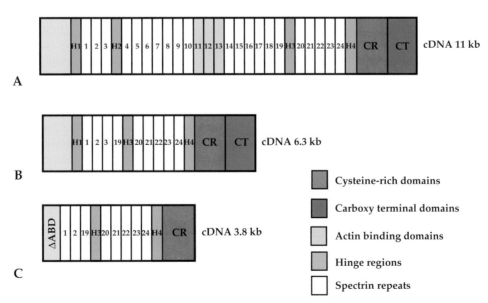

FIGURE 11.13   Forms of the dystrophin protein. (A) Full-length dystrophin containing all functional domains. (B) Mini-dystrophin from a Duchenne muscular dystrophy patient who was only mildly impaired. (C) Micro-dystrophin engineered for delivery in AAV vectors. Adapted from Figure 1 in Foster *et al.* (2006).

A Gene Therapy Failure

Patients who have deficiencies in enzymes that participate in the urea cycle have increased concentrations of ammonia in the blood. High concentrations of ammonia result in various symptoms, which can include behavioral disturbances or coma. Severe deficiencies in these enzymes result in early death, but moderate deficiencies can result in delayed appearance of symptoms and may be partially controlled by diet. One such enzyme is ornithine transcarbamylase (OTC), which is found on the X chromosome. Deficiencies in OTC are therefore more common in males than in females.

Gene therapy trials that use virus vectors recently received a major setback when a relatively fit 18-year-old male with an inherited deficiency for OTC died 4 days after an adenovirus vector was injected into his liver. A high dose of adenovirus (4×10^{10}) that expressed OTC was injected in an effort to achieve adequate levels of enzyme production. The virus unexpectedly spread widely and a systemic inflammatory response developed, inducing a fever of 40.3°C. He went into a coma, his lungs filled with fluid, and he died of asphyxiation. This unfortunate result makes clear the possible drawbacks to experimental treatments and the difficulties in designing protocols that allow an adequate margin of safety while trying to achieve a clinically relevant result.

Treatment of Restenosis

A gene therapy trial in patients with heart disease gave very encouraging results. Although this study did not involve virus vectors, a brief description will be given since it serves as an incentive for continuation of gene therapy trials. Coronary artery disease is common in older people. Angioplasty or bypass surgery is used to open clogged arteries, but in many patients the arteries close up again (a process called restenosis). Thirteen patients with chronic chest pain who had failed angioplasty or bypass surgery or both were injected in the heart muscle with DNA encoding vascular endothelial growth factor. This factor promotes the growth of blood vessels, a process called angiogenesis. Two months after treatment, all patients exhibited an improvement in vascularization of damaged areas of the heart, as shown by imaging and mapping studies. All patients reported a decrease in disease symptoms, and all had an improved performance in treadmill tests. Although the number of patients is small, the uniformly positive results are encouraging.

Viruses as Anticancer Agents

A large number of clinical trials have examined the possibility of using viruses as anticancer agents. Table 11.6 lists a number of trials that were active in the year 2000, to give a flavor of what is being tried and the number of malignacies that are being considered as candidates for treatment using gene therapy approaches. These trials are all Phase I or I/II, but more than 1000 patients were participating in the trials listed in the table. As shown in Fig. 11.12, the development of gene therapy approaches for the control of cancers continues to attract much effort, such that the majority of clinical trails to date have been directed against cancers. Although progress has been painfully slow and disappointing, the prospect remains that effective treatment may yet be obtained for at least some cancers using such approaches.

In most of the trials in Table 11.6, viruses are used to express proteins that control the growth of tumors or that are toxic to tumor cells. A number of different cytokines are being tried as antitumor agents, such as IFN-γ, IL-2, TNF, and GM-CSF. Another approach is to try to repair the defective regulatory gene in the tumor cell, which is often p53. Many other gene products are also being tested. The viruses used to express these products include the retroviruses, the adenoviruses, or the poxviruses. More recent trials have used additional viruses as well, in particular the herpesviruses and the adeno-associated viruses. Herpes simplex type 1 would seem to be particularly appropriate for control of brain tumors, because the virus is neurotropic but sets up a latent infection in neurons. One idea would be to engineer herpes to express a protein that is only toxic in dividing cells but which would be nontoxic in mature, nondividing neurons. The table also lists a number of trials that use lipofection to introduce the gene of interest into target cells.

Further afield, thought is being given to the possibility of using viruses to express proteins that are overexpressed in tumor cells in an attempt to stimulate the immune system to respond by killing tumor cells. This is in essence an attempt to vaccinate a person against a tumor. For this approach to succeed, an antigen overproduced by a tumor cell, such as a melanoma cell, must be identified, inserted into a suitable vector, and the person with the tumor infected with the virus vector in an attempt to stimulate the immune system. In principle, this approach may be feasible, but only time will tell whether it is in fact practical.

Another approach is to try to direct the virus, more or less specifically, to infect the tumor cells, so that upon infection the cells are killed. Cell death might result either because the virus itself is cytolytic or because the virus expresses a protein that renders the cell sensitive to a toxic agent such as BUdR. A number of the trials listed in Table 11.6 use the TK gene for this, since cells that express TK are sensitive to BUdR. One possible approach is to engineer the virus so that its surface glycoprotein expresses a monoclonal antibody directed against an antigen expressed only on the tumor cell, while at the same time causing the virus to be unable to infect cells that do not express the antigen. Experiments have established the possibility of this approach, at least in principle, with viruses such as the alphaviruses. Another approach was illustrated by the experiments with VSV to design a virus that could infect only HIV-infected cells.

TABLE 11.6   Clinical Trials of Gene Transfers for Cancer Therapy in the United States in 2000

Location	Gene[a]	Vector	Number of trials	Number of patients	Phase[b]
Brain cancers					
Neuroblastoma	IFNγ	Retrovirus	1	4	I
	IL-2	Retrovirus	1	12	I
	IL-2	Adenovirus	1	6	I
Central nervous system	TK	Adenovirus	2	22	I
Pediatric tumor	TK	Retroviral producing cells	1	2	I
Adult brain tumor	TK	Retroviral producing cells	1	15	I
Ovarian cancer	HSV-TK	Adenovirus	1	10	I
	TK	Retroviral producing cells	3	42	I
	BRCA-1	Retrovirus	1	40	I/II
	p53	Adenovirus	1	16	I
Small cell lung cancer	IL-2+NeoR	Lipofection	1	8	I
	Anti-sense to k-*ras*	Retrovirus	1	9	I
	p53	Adenovirus	2	59	I/II
Prostate cancer	GM-CSF	Retrovirus	1	8	I/II
	PSA	Poxvirus	1	3	I
	HSV-TK	Adenovirus	1	18	I
Breast cancer	BRCA-1	Retrovirus	1	21	I
	E1A	Lipofection	1	16	I
	MDR-1+NeoR	Retrovirus	4	39	I
	CD80	Lipofection	1	15	I
	CEA	Poxvirus	4	53	I
	CEA	RNA transfer	1	30	I
Melanoma	GM-CSF	Gene gun	1	17	I
	GM-CSF	Retrovirus	2	29	I
	HLA-B7/β2 m	Lipofection	8	165	I/II
	IL-2+NeoR	Retrovirus	5	115	I
	IFNγ	Retrovirus	3	91	I
	TNF+NeoR	Retrovirus	1	12	I/II
	MART-1	Adenovirus	1	33	I
	MART-1	Poxvirus	2	16	I
	gp100	Poxvirus	1	19	I
	gp100	Adenovirus	1	7	I
	CD80	Lipofection	1	17	I
Miscellaneous carcinomas	p53	Adenovirus	1	26	I
	HLA-B7/β2m	Lipofection	4	76	II
	IL-2	Lipofection	1	11	I
	CEA	Poxvirus	1	8	I
Lymphomas and solid tumors	IL-2	Retrovirus	2	29	I
	TK	Retrovirus	1	11	I
	IL-12+NeoR	Retrovirus	1	31	I
Bladder cancer	p53	Adenovirus	1	5	I
Colo/rectal, renal, and liver cancers	GM-CSF	Retrovirus	1	18	I
	HLA-B7/β2m	Lipofection	4	53	I/II
	CD	Adenovirus	1	6	I
	IL-4	Retrovirus	1	18	I
	TNF+NeoR	Retrovirus	1	12	I

(Continued)

[a]Abbreviations: IFNγ, interferon gamma; IL-2, interleukin 2; TK, thymidine kinase (sometimes used with bromodeoxyuridine); HSV-TK, herpes simplex thymidine kinase, often coupled with gancyclovir treatment; BRCA-1, breast cancer 1, early onset; PSA, puromycin-sensitive aminopeptidase; CEA, carcinoembryonic antigen; GM-CSF, granulocyte-macrophage colony stimulating factor; MDR-1, multi-drug resistance protein 1 (used to insert chemotherapy-resistance genes into the hematopoetic lineage); CD80, protein involved in T-cell activation; CD, cytosine deaminase; TNF, tumor necrosis factor.

[b]Definitions of phases in a clinical trial: Phase I usually fewer than 100 healthy volunteers, primarily to gauge adverse reactions, and to determine optimal dose and best route of administration. Phase II are generally pilot efficacy studies usually involving 200–500 volunteers randomly assigned to control and study groups. Phase II will test for immunogenicity in the case of vaccines, and duration of expression and amelioration of symptoms for gene therapy. Note that none of these trials have proceeded beyond Phase II, and most are in Phase I.

Source: Wiley Journal of Gene Medicine/Clinical Trial Database at: http://www.wiley.co.uk/genetherapy/clinical/.

The possible use of herpesviruses to control brain tumors, especially gliomal tumors, has been cited earlier. Simple retroviruses are also being examined for this purpose. These viruses can only replicate in dividing cells. Thus, they should be able to infect only tumor cells in the brain, since most neuronal cells are terminally differentiated and do not divide. If the retroviruses express a protein that renders the cells sensitive to a toxin, it might be possible to kill replicating cells and therefore only the tumor cells.

FURTHER READING

Infectious Clones

Bredenbeek, P. J., Kooi, E. A., Lindenbach, B., et al. (2003). A stable full-length yellow fever virus cDNA clone and the role of conserved RNA elements in flavivirus replication. J. Gen. Virol. 84: 1261–1268.

Bridgen, A., and Elliott, R. M. (1996). Rescue of a segmented negative-strand RNA virus entirely from cloned complementary DNAs. Proc. Natl. Acad. Sci. U.S.A. 93: 15400–15404.

Conzelmann, K.-K., and Meyers, G. (1996). Genetic engineering of animal RNA viruses. Trends Microbiol. 4: 386–393.

Domi, A., and Moss, B. (2002). Cloning the vaccinia virus genome as a bacterial artificial chromosome in Escherichia coli and recovery of infectious virus in mammalian cells. Proc. Natl. Acad. Sci. U.S.A. 99: 12415–12420.

Komoto, S., Sasaki, J., and Taniguchi, K. (2006). Reverse genetics system for introduction of site-specific mutations into the double-stranded RNA genome of infectious rotavirus. Proc. Natl. Acad. Sci. U.S.A. 103: 4646–4651.

Neumann, G., and Kawaoka, Y. (2004). Reverse genetics systems for the generation of segmented negative-sense RNA viruses entirely from cloned DNA. Curr. Top. Microbiol. Immunol. 283: 43–60.

Rice, C. M., Levis, R., Strauss, J. H., and Huang, H. V. (1987). Production of infectious RNA transcripts from Sindbis virus cDNA clones: Mapping of lethal mutations, rescue of a temperature sensitive marker, and in vitro mutagenesis to generate defined mutants. J. Virol. 61: 3809–3819.

Schnell, M. J., Mebatsion, T., and Conzelmann, K.-K. (1994). Infectious rabies viruses from cloned cDNA. EMBO J. 13: 4195–4203.

Shi, P.-Y., Tilgner, M., Lo, M. K., Kent, K. A., and Bernard, K. A. (2002). Infectious cDNA clone of the epidemic West Nile virus from New York City. J. Virol. 76: 5847–5856.

Thumfart, J. O., and Meyers, G. (2002). Feline calicivirus: recovery of wild-type and recombinant viruses after transfection of cRNA or cDNA constructs. J. Virol. 76: 6398–6407.

Nonviral Vectors for Gene Therapy

Boulaiz, H., Marchal, J. A., Prados, J., Melguizo, C., and Aránega, A. (2005). Non-viral and viral vectors for gene therapy. Cell. Mol. Biol. 51: 3–22.

Cohen, E. P., de Zoeten, E. F., and Schatzman, M. (1999). DNA vaccines as cancer treatment. American Scientist 87: 328–335.

Viruses as Vectors for Heterologous Antigen Expression

Basak, S., McPherson, S., Kang, S., et al. (1998). Construction and characterization of encapsidated poliovirus replicons that express biologically active murine interleukin-2. J. Interferon Cytokine Res. 18: 305–313.

Bledsoe, A. W., Jackson, C. A., McPherson, S., et al. (2000). Cytokine production in motor neurons by poliovirus replicon vector gene delivery. Nature Biotechnol. 18: 964–969.

Bukreyev, A., Skiadopoulos, M. H., Murphy, B. R., and Collins, P. L. (2006). Nonsegmented negative-strand viruses as vaccine vectors. J. Virol. 80: 10293–10306.

de Haan, C. A. M., Haijema, B. J., Boss, D., Heuts, F. W. H., and Rottier, P. J. M. (2005). Coronaviruses as vectors: stability of foreign gene expression. J. Virol. 79: 12742–12751.

Gillet, L., Daix, V., Donofrio, G., et al. (2005). Development of bovine herpesvirus 4 as an expression vector using bacterial artificial chromosome cloning. J. Gen. Virol. 86: 907–917.

Mackett, M., Smith, G. L., and Moss, B. (1982). Vaccinia virus: a selectable eukaryotic cloning and expression vector. Proc. Natl. Acad. Sci. U.S.A. 79: 7415–7419.

Monath, T. P., Levenbook, I., Soike, K., et al. (2000). Chimeric yellow fever virus 17D-Japanese encephalitis virus vaccine: Dose-response effectiveness and extended safety testing in rhesus monkeys. J. Virol. 74: 1742–1751.

Pennathur, S., Haller, A. A., MacPhail, M., et al. (2003). Evaluation of attenuation, immunogenicity and efficacy of a bovine parainfluenza virus type 3 (PIV-3) vaccine and a recombinant chimeric bovine/human PIV-3 vaccine vector in rhesus monkeys. J. Gen. Virol. 84: 3253–3261.

Perri, S., Greer, C. E., Thudium, K., et al. (2003). An alphavirus replicon particle chimera derived from Venezuelan equine encephalitis and Sindbis viruses is a potent gene-based vaccine delivery vector. J. Virol. 77: 10394–10403.

Pinto, V. B., Prasad, S., Yewdell, J., et al. (2000). Restricting expression prolongs expression of foreign genes introduced into animals by retroviruses. J. Virol. 74: 10202–10206.

Schneider, U., Bullough, F., Vongpunsawad, S., et al. (2000). Recombinant measles viruses efficiently entering cells through targeted receptors. J. Virol. 74: 9928–9936.

Shen, Y., and Post, L. (2006). Virus vectors and their applications. Chapter 16 in: *Fields Virology, Fifth Edition* (D. M. Knipe and P. M. Howley, Eds. in chief), Lippincott Williams & Wilkins, pp. 539–564.

Van Craenenbroeck, K., Vanhoenacker, P., and Haegeman, G. (2000). Review: Episomal vectors for gene expression in mammalian cells. *Eur. J. Biochem.* **267**: 5665–5678.

Gene Therapy

Athanasopoulos, T., Fabb, S., and Dickson, G. (2000). Gene therapy vectors based on adeno-associated virus: characteristics and applications to acquired and inherited diseases. *Int. J. Mol. Med.* **6**: 363–375.

Crystal, R. (1995). Transfer of genes to humans: early lessons and obstacles to success. *Science* **270**: 404–410.

Fischer, A., Deist, F. L., Hacein-Bey-Abina, S., *et al.* (2005). Severe combined immunodeficiency: A model disease for molecular immunology and therapy. *Immunol. Rev.* **203**: 98–109.

Foster, K., Foster, H., and Dickson, J. G. (2006). Gene therapy progress and prospects: Duchenne muscular dystrophy. *Gene Therapy* **13**: 1677–1685.

Hu, Y.-C. (2006). Baculovirus vectors for gene therapy. *Adv. Virus Res.* **68**: 287–320.

Jackson, C. A., Messinger, J., Peduzzi, J. D., Ansardi, D. C., and Morrow, C. D. (2005). Enhanced functional recovery from spinal cord injury following intrathecal or intramuscular administration of poliovirus replicons encoding IL-10. *Virology* **336**: 173–183.

Klimstra, W. B., Williams, J. C., Ryman, K. D., and Heidner, H. W. (2005). Targeting Sindbis virus-based vectors to Fc receptor-positive cell types. *Virology* **338**: 9–21.

Kohn, D. B., Hershfield, M. S., Carbonaro, D., *et al.* (1998). T lymphocytes with a normal ADA gene accumulate after transplantation of transduced autologous umbilical cord blood CD34+ cells in ADA-deficient SCID neonates. *Nature Med.* **4**: 775–780.

Le Bec, C., and Douar, A. M. (2006). Gene therapy progress and prospects-Vectorology: design and production of expression cassettes in AAV vectors. *Gene Ther.* **13**: 805–813.

Marchetto, M. C. N., Correa, R. G., Menck, C. F. M., and Muotri, A. R. (2006). Functional lentiviral vectors for xeroderma pigmentosum gene therapy. *J. Biotechnol.* **126**: 424–430.

Mohan, R. R., Sharma, A., Netto, M. V., Sinha, S., and Wilson, S. E. (2005). Gene therapy in the cornea. *Prog. Retin. Eye Res.* **24**: 537–559.

Porteus, M. H., Connelly, J. P., and Pruett, S. M. (2006). A look to future directions in gene therapy research for monogenic diseases. *PLOS Genet.* **2**: 1285–1292.

Russell, W. C. (2000). Update on adenovirus and its vectors. *J. Gen. Virol.* **81**: 2573–2604.

Shen, Y., and Nemunaitis, J. (2006). Herpes simplex virus 1 (HSV-1) for cancer treatment. *Cancer Gene Ther.* **13**: 975–992.

Shen, Y., and Post, L. (2006). Virus vectors and their applications. Chapter 16 in: *Fields Virology, Fifth Edition* (D. M. Knipe and P. M. Howley, Eds. in chief), Philadelphia, Lippincott Williams & Wilkins, pp. 539–564.

Thorne, S. H., Bartlett, D. L., and Kirn, D. H. (2005). The use of oncolytic vaccinia viruses in the treatment of cancer: A new role for an old ally? *Curr. Gene Ther.* **5**: 429–443.

Verma, I. M., and Weitzman, M. D. (2005). Gene therapy: twenty-first century medicine. *Annu. Rev. Biochem.* **74**: 711–738.

Walther, W., and Stein, U. (2000). Viral vectors for gene transfer: a review of their use in the treatment of human diseases. *Drugs* **60**: 249–271.

Woods, N. B., Bottero, V., Schmidt, M., von Kalle, C., and Verma, I. (2006). Gene therapy: Therapeutic gene causing lymphoma. *Nature* **440**: 1123.

Ziady, A. G., and Davis, P. B. (2006). Current prospects for gene therapy of cystic fibrosis. *Curr. Opin. Pharmacol.* 6: 515–521.

References Used for Figures and Tables

CHAPTER 1

Tables

TABLE 1.2   Fauquet, C. M., Mayo, M. A., Maniloff, J., Desselberger, U., and Ball, L. A. (Eds.), (2005). *Virus Taxonomy, the Eighth Report of the International Committee on Taxonomy of Viruses*, Academic Press, New York.

Figures

FIGURE 1.1   *Morbidity and Mortality Weekly Report* (*MMWR*) (1999). Vol. 48, #29, p. 621.

FIGURE 1.2   World Health Organization Web site: http://www.who.int/infectious-disease-report/pages/graph 5.html and World Health Report 2003 at: http://www.who.int/whr/2003/en/.

FIGURE 1.3   Lattanzi, M., Rappuoli, R., and Tonini T. (2006). The future of vaccines. *Microbiol. Today* **33**: 8–11.

FIGURE 1.4   Adapted from Fields, B. N., Knipe, D. M., and Howley, P. M. (Eds.), (1996). *Fields Virology, 3rd Edition*, Lippincott-Raven, New York, p. 1788 and Coffin, J. M., Hughes, S. H., and Varmus, H. E. (Eds.) (1997). *Retroviruses,* Cold Spring Harbor Laboratory Press, Cold Spring Harbor, NY, pp. 76–82.

FIGURE 1.6   Redrawn from Fields, B. N., Knipe, D. M., and Howley, P. M. (Eds.), (1996). *Fields Virology, 3rd Edition*, Lippincott-Raven, New York, p. 1361.

FIGURE 1.7   Adapted from Figure 6 in Koshiba, T., and Chan, D. C. (2003). The prefusogenic intermediate of HIV-gp41 contains exposed c-peptide regions. *J. Biol. Chem.* **278**: 7573–7579.

FIGURE 1.8   Adapted from Mims, C. A., Playfair, J. H. L., Roitt, I. M., Wakelin, D., and Williams, R. (1993). *Medical Microbiology,* Mosby Europe Limited, London, p. 2.3.

FIGURE 1.9   Adapted from Flint, S. J., Enquist, L. W., Krug, R. M., Racaneillo, V. R., and Skalka, A. M. (2000). *Principles of Virology, Molecular Biology, Pathogenesis, and Control,* ASM Press, Washington DC, Figures. 9.8, 9.16, 9.10, and 9.9, respectively.

FIGURE 1.10   Adapted from Mims, C. A., Playfair, J. H. L., Roitt, I. M., Wakelin, D., and Williams, R. (1993). *Medical Microbiology,* Mosby Europe Limited, London, p. 2.3 and Strauss, E. G., and Strauss, J. H. (1997). Eukaryotic RNA viruses, a variant genetic system. In: *Exploring Genetic Mechanisms,* (M. Singer and P. Berg, Eds.), University Science Books, Sausalito, CA, Figure 2.2, p. 76.

FIGURE 1.12   Adapted from Strauss, E. G., and Strauss, J. H. (1997). Eukaryotic RNA viruses, a variant genetic system. In: *Exploring Genetic Mechanisms*, (M. Singer and P. Berg, Eds.), University Science Books, Sausalito, CA, Figure 2.3, p. 77.

FIGURE 1.13   Adapted from Fields, B. N., Knipe, D. M., and Howley, P. M. (Eds.), (1996). *Fields Virology, 3rd Edition*, Lippincott-Raven, New York, p. 1786 and Coffin, J. M., Hughes, S. H., and Varmus, H. E. (Eds.) (1997). *Retroviruses,* Cold Spring Harbor Laboratory Press, Cold Spring Harbor, NY, p. 8.

FIGURE 1.15   Scheme derived from Fields, B. N., Knipe, D. M., and Howley, P. M. (Eds.), (1996). *Fields Virology, 3rd Edition*, Lippincott-Raven, New York, p. 2709.

FIGURE 1.16   Adapted from Wimmer, E., Hellen, C. U. T., and Cao, X. (1993). Genetics of poliovirus. *Annu. Rev. Genet.* **27**: 353–436. (Figure on p. 374).

FIGURE 1.17   Adapted from Goff, S. P. (1997). Retroviruses: infectious genetic elements. In: *Exploring Genetic Mechanisms,* (M. Singer and P. Berg, Eds.), University Science Books, Sausalito, CA, Figure 3.15, p. 156 and Fields, B. N., Knipe, D. M., and Howley, P. M. (Eds.), (1996). *Fields Virology, 3rd Edition*, Lippincott-Raven, New York, p. 577.

FIGURE 1.18   Adapted from Fields, B. N., Knipe, D. M., and Howley, P. M. (Eds.), (1996). *Fields Virology, 3rd Edition*, Lippincott-Raven, New York, p. 566.

FIGURE 1.19   From Matthews, D. A., Smith, W. W., Ferre, R. A., *et al.* (1994). Structure of the human rhinovirus 3C protease reveals a trypsin-like polypeptide fold, RNA-binding site, and means for cleaving precursor polyprotein. *Cell* **77**: 761–771 Figure 1 ribbon diagram; Ding, J., McGrath, W. J., Sweet, R. M., and Mangel, W. F. (1996) Crystal structure of the human adenovirus proteinase with its 11 amino acid cofactor. *EMBO J.* **15**: 1778–1783; Rutenber, E., Fauman, E. B., Keenan, R. J., Fong, S., Furth, P. S., Ortiz de Montellano, P. R., Meng, E., Kuntz, I. D., DeCamp, D. L., and Salto, R., (1993). Structure of a non-peptide inhibitor complexed with HIV-1 protease. Developing a cycle of structure-based drug design. *J. Biol. Chem.* **268**: 15343–15346.

CHAPTER 2

Figures

FIGURE 2.1   Granoff, A., and Webster, R. G. (Eds.) (1999). *The Encyclopedia of Virology, Second Edition,* Academic Press, San Diego, a copy of Figure 5 on p. 401 (Vol. 1).

FIGURE 2.2   Murphy, F. A., Fauquet, C. M., Bishop, D. H. L., Ghabrial, S. A., Jarvis, A. W., Martelli, G. P., Mayo, M. A., and Summers, M. D.

(Eds.) (1995). *Virus Taxonomy,* Sixth Report of the International Committee on Taxonomy of Viruses, Springer-Verlag, Vienna, Austria, p. 434.

FIGURE 2.3   From Plate 31 in color in Vol. 3 of Granoff, A., and Webster, R. G. (Eds.) (1999). *The Encyclopedia of Virology, Second Edition,* Academic Press, San Diego, originally from Johnson, J. (1996). Review: Functional implications of protein-protein interactions in icosahedral viruses. *Proc. Natl. Acad. Sci U.S.A.* **93**: 27–33.

FIGURE 2.4   Adapted from Plate 32 in Vol. 3 of Granoff, A., and Webster, R. G. (Eds.) (1999). *The Encyclopedia of Virology, Second Edition,* Academic Press, San Diego.

FIGURE 2.5   Most of the images are taken from Baker, T. S., Olson, N. H., and Fuller, S. D. (1999). Adding the third dimension to virus life cycles: three-dimensional reconstruction of icosahedral viruses from cryo-electron micrographs. *Microbiol. Mol. Biol. Rev.* **63**: 862–922, except the images of Ross River virus and of dengue virus, which were kindly provided by Drs. R. J. Kuhn and T. S. Baker.

FIGURE 2.6   From Figure 4 in Johnson, J. E. (1996). Review: Functional implications of protein-protein interactions in icosahedral viruses. *Proc. Natl. Acad. Sci. U.S.A.* **93**: 27–33.

FIGURE 2.7   Adapted from Granoff, A., and Webster, R. G. (Eds.) (1999). *The Encyclopedia of Virology, Second Edition,* Academic Press, San Diego, p. 287.

FIGURE 2.9   Adapted from Kolatkar, P. R., Bella, J., Olson, N. H., Bator, C. M., Baker, T. S., and Rossmann, M. G. (1999). Structural studies of two rhinovirus sertotypes complexed with fragments of their cellular receptor. *EMBO J.* **18**: 6249–6259.

FIGURE 2.10   From Figure 1 in Stehle, T., Gamblin, S. J., Yan, Y., and Harrison, S. C. (1996). The structure of simian virus 40 refined at 3.1Å resolution. *Structure* **4**: 165–182; and Fields, B. N., Knipe, D. M., and Howley, P. M. (Eds.), (1996). *Fields Virology, 3rd Edition,* Lippincott-Raven, New York, Colorplate 4.

FIGURE 2.11   From Plate 17, Granoff, A., and Webster, R. G. (Eds.) (1999). *The Encyclopedia of Virology, Second Edition,* Academic Press, San Diego.

FIGURE 2.12   Adapted from Fields, B. N., Knipe, D. M., and Howley, P. M. (Eds.), (1996). *Fields Virology, 3rd Edition,* Lippincott-Raven, New York, p. 80; Stewart, P. L., Burnett, R. M., Cyrklaff, M., and Fuller, S. D. (1991). Image reconstruction reveals the complex molecular organization of adenovirus. *Cell* **4**: 145–154; Athappilly, F. K., Murali, R., Rux, J. J., Cai, Z., and Burnett, R. M. (1994). The refined crystal-structure of hexon, the major coat protein of adenovirus type 2, at 2.9 angstrom resolution. *J. Mol. Biol.* **242**: 430–455.

FIGURE 2.13   Adapted from Murphy, F. A., Fauquet, C. M., Bishop, D. H. L., Ghabrial, S. A., Jarvis, A. W., Martelli, G. P., Mayo, M. A., and Summers, M. D. (Eds.) (1995). *Virus Taxonomy,* pp. 51, 60, 55.

FIGURE 2.14   Panels A and B are adapted from Strauss, J. H., Strauss, E. G., and Kuhn, R. J. (1995). Budding of alphaviruses. *Trends Microbiol.,* **3**: 346–350, Figures 4b and 3.

FIGURE 2.15   Adapted from Figure 5 in Zhang, Y., Corver, J., Chipman, P. R., Zhang, W., Pletnev, S., Sedlak, D., Baker, T. S., Strauss, J. H., Kuhn, R. J., and Rossmann, M. G. (2003). Structures of immature flavivirus particles. *EMBO J.* **22**: 2604–2613.

FIGURE 2.16   Figure 6 from Mukhopadhyay, S., Zhang, W., Gabler, S., Chipman,R., Strauss, E. G., Strauss, J. H., Baker, T. S., Kuhn, R. J., and Rossmann, M. G. (2006). Mapping the structure and function of the E1 and E2 glycoproteins in alphaviruses. *Structure* **14**: 63–73, reprinted with permission.

FIGURE 2.17   Adapted from Figure 2 in Lescar, J., Roussel, A., Wien, M. W., Navaza, J., Fuller, S. D., Wengler, G., Wengler, G., and Ray, F. A. (2001). The fusion glycoprotein shell of Semliki Forest virus: an icosahedral assembly primed for fusogenic activation at endosomal pH. *Cell* **105**: 137–148, reprinted with permission.

FIGURE 2.18   Panels (A) and (C) were provided by Richard J. Kuhn; panel B was reprinted from Figure 3c in Kuhn, R. J., Zhang, W., Rossmann,

M. G., Pletnev, S. V., Corver, J., Lenches, E., Jones, C. T., Chipman, P. R., Strauss, E. G., Baker, T. S., and Strauss, J. H. (2002). Structure of dengue virus: implications for flavivirus organization, maturation, and fusion. *Cell* **108**: 717–725; panel (D) was adapted from Figure 3b in Zhang, Y., Corver, J., Chipman, P. R., Zhang, W., Pletnev, S., Sedlak, D., Baker, T. S., Strauss, J. H., Kuhn, R. J., and Rossmann, M. G. (2003). Structures of immature flavivirus particles. *EMBO J.* **22**: 2604–2613; panel E is reprinted from Figure 1a in Zhang, W., Chipman, P. R., Corver, J., Johnson, P. R., Zhang, Y., Mukhopadhyay, S., Baker, T. S., Strauss, J. H., Rossmann, M. G., and Kuhn, R. J. (2003). Visualization of membrane protein domains by cryo-electron microscopy of dengue virus. *Nature Struct. Biol.* **10**: 907–912; panel F is adapted from Figure 4a in Strauss, J. H., Strauss, E. G., and Kuhn, R. J. (1995). Budding of alphaviruses. *Trends Microbiol.* **3**: 346–350.

FIGURE 2.19   Panel B is reprinted from Figure 4a in Zhang, W., Chipman, P. R., Corver, J., Johnson, P. R., Zhang, Y., Mukhopadhyay, S., Baker, T. S., Strauss, J. H., Rossmann, M. G., and Kuhn, R. J. (2003). Visualization of membrane protein domains by cryo-electron microscopy of dengue virus. *Nature Struct. Biol.* **10**: 907–912.

FIGURE 2.20   Panel A was adapted from Murphy, F. A., Fauquet, C. M., Bishop, D. H. L., Ghabrial, S. A., Jarvis, A. W., Martelli, G. P., Mayo, M. A., and Summers, M. D. (Eds.) (1995). *Virus Taxonomy* p. 114.

FIGURE 2.21   Adapted from Coffin, J. M., Hughes, S. H., and Varmus, H. E. (Eds.) (1997). *Retroviruses,* Cold Spring Harbor Laboratory Press, Cold Spring Harbor, NY, (A) is from p. 2, (B) is from p. 30.

FIGURE 2.23   Panel (C) is from Simpson, R. W., and Hauser, R. E. (1966). Structural components of vesicular stomatitis virus. *Virology* **29**: 654, reproduced in Fenner, F., McAuslan, B. R., Mims, C. A., Sambrook, J., and White, D. O. (Eds.) (1974). *The Biology of Animal Viruses,* Academic Press, New York, p. 127, plate 3-20b; Panel (D) is from Birdwell, C. R., and Strauss, J. H. (1974). Maturation of vesicular stomatitis virus: electron microscopy of surface replicas of infected cells. *Virology* **69**: 587–590; (F) and (G) are from Murphy, F. A., Fauquet, C. M., Bishop, D. H. L., Ghabrial, S. A., Jarvis, A. W., Martelli, G. P., Mayo, M. A., and Summers, M. D. (Eds.) (1995). *Virus Taxonomy,* Sixth Report of the International Committee on Taxonomy of Viruses, Springer-Verlag, Vienna, Austria, p. 289.

FIGURE 2.24   Diagrams from Fenner, F., and Nakano, J. H. (1988) *Poxviridae:* the poxviruses. In: (E. H. Lennette, P. Halonen, and F. A. Murphy, Eds.), *The Laboratory Diagnosis of Infectious Diseases: Principles and Practice,* Vol. II. Springer-Verlag, New York, pp. 177–210.

FIGURE 2.25   Original sources: (A) Roizman, B. (1969) *Curr. Top. Microbiol. Immunol.* **49**: 1–79. as shown on p. 93 of Atlas; (B) Murphy, F. A., Webb, P. A., Johnson, K. M., Whitfield, S. G., *et al.* (1969). *J. Virol.* **4**: 535, as shown on p. 329 of the Atlas. (C) Strauss, J. H., Strauss, E. G., and Kuhn, R. J. (1995). *Trends Microbiol.* **3**: 346–350. (D) Higashi, N., Arimura, H., Matsumoto, A., and Nagatomo, Y. (1976). Septième Congrés International de Microscopie Electronique, Grenoble Vol. III, p. 295, as seen in the Atlas, p. 194; (E) Compans, R. W., and Choppin, P.W. (1973). Orthomyxoviruses and paramyxoviruses, Chap. 11 in the Atlas, p. 229; (F) Compans, R. W., Holmes, J. V., Dales, S., and Choppin, P. W. (1966). *Virology* **42**: 880, as shown in the Atlas p. 227; [Atlas= *Ultrastructure of Animal Viruses and Bacteriophages:* an atlas, (1973). (A. J. Dalton and F. Haguenau, Eds.), Academic Press, New York, pages as cited above.]

CHAPTER 3

Tables

All taxonomy tables throughout are according to the new data in Fauquet, C. M., Mayo, M. A., Maniloff, J., Desselberger, U., and Ball, L. A. (Eds.), (2005). *Virus Taxonomy, the Eighth Report of the International Committee on Taxonomy of Viruses,* Academic Press, New York.

TABLE 3.6 *Summary of Notifiable Diseases-United States, 2004, MMWR* Vol. 53, #53.

TABLE 3.9 Adapted from Granoff, A., and Webster, R. G. (Eds.) (1999). *Encyclopedia of Virology, 2nd Edition*, Academic Press, San Diego, p. 442.

Figures

FIGURE 3.1 Adapted from Yamashita, T., Ito, M., Kabashima, Y., Tsuzuki, H., Fujiura, A., and Sakae, K. (2003). Isolation and characterization of a new species of kobuvirus associated with cattle *J. Gen. Virol.* **84**; 3069–3077 and Fauquet, C. M., *et al*. (Eds.) (2005). *Virus Taxonomy*, the Eighth Report of the International Committee on Taxonomy of Viruses, Academic Press, pp. 757–778.

FIGURE 3.2 Adapted from Murphy, F. A., Fauquet, C. M., Bishop, D. H. L., Ghabrial, S. A., Jarvis, A. W., Martelli, G. P., Mayo, M. A., and Summers, M. D. (Eds.) (1995). *Virus Taxonomy*, Sixth Report of the International Committee on Taxonomy of Viruses, Springer-Verlag, Vienna, Austria, p. 300; and Yamashita, T., Sakae, K., Tsuzuki, Y., Ishikawa, N., Takeda, N., Miyamura, T., and Yamazaki, S. (1998). Complete nucleotide sequence and genetic organization of Aichi virus, a distinct member of the *Picornaviridae* associated with acute gastroenteritis in humans. *J. Virol.* **72**: 8408–8412.

FIGURE 3.3 Adapted from Molla, A., Paul, A. V., Schmid, M., Jang, S. K., and Wimmer, E. (1993). Studies on dicistronic polioviruses implicate viral proteinase 2A^pro in RNA replication *Virology* **196**; 739–747; and Lu, H.-H., Li, X., Cuconati, A., and Wimmer, E. (1995). Analysis of picornavirus 2A^pro proteins: separation of proteinase from translation and replication functions. *J. Virol.* **69**: 7445–7452.

FIGURE 3.4 Data came from *MMWR* (1997). Vol. 46, #4 p. 79; Nathanson, N. Ahmed, R., Gonzalez-Scarano, F., Griffin, D. E., Holmes, K. V., Murphy, F. A., and Robinson, H. L. (Eds.) (1996). *Viral Pathogenesis*, Lippincott-Raven, Philadelphia, p. 556; updated with data from *MMWR Summary of Notifiable Diseases, United States-2003* Vol. 52, #42.

FIGURE 3.5 From Halstead, Lauro S. (1998). Post-Polio Syndrome. *Scientific American* **278**: 42–47, with the permission of the author and the publisher.

FIGURE 3.7 (A) From the WHO Web site: www.who.int/gpv-surv/graphics/NY graphics/global polio 98.htm; (B) from *MMWR* Vol. **49**, #16, p. 352.

FIGURE 3.8 Wild poliovirus importations—West and Central Africa, January 2003–March 2004. *MMWR* Vol. 53, #20; 433–435; Resurgence of wild poliovirus type 1 transmission: consequences of importation—21 countries, 2002–2005. *MMWR* Vol. **55**, #6, 145–150.

FIGURE 3.9 *MMWR* Vol. **52**, #54, *Summary of Notifiable Diseases-United States, 2003.*

FIGURE 3.10 Innis, B. L., Snitbhan, R., Kunasol, P., Laorakpongse, T., Poopatanakool, W., Kozik, C. A., Suntayakorn, S., Suknuntapong, T., Safary, A., Tang, D. B., and Boslego, J. W., (1994). Protection against hepatitis A by an inactivated vaccine. *JAMA* **271**: 1328–1334.

FIGURE 3.11 Adapted from Fauquet, C. M., Mayo, M. A., Maniloff, J., Desselberger, U., and Ball, L. A. (Eds.) (2005). *Virus Taxonomy, the Eighth Report of the International Committee on Taxonomy of Viruses*, Academic Press, New York, p. 847.

FIGURE 3.12 Adapted from Fauquet, C. M., Mayo, M. A., Maniloff, J., Desselberger, U., and Ball, L. A. (Eds.) (2005). *Virus Taxonomy, the Eighth Report of the International Committee on Taxonomy of Viruses*, Academic Press, New York, p. 854.

FIGURE 3.13 Adapted from Fields, B. N., Knipe, D. M., and Howley, P. M. (Eds.) (1996). *Fields Virology, 3rd Edition*, Lippincott-Raven, New York, p. 2838.

FIGURE 3.14 Adapted from Tsarev, S. A., Binn, L. N., Gomatos, P. J., Arthur, R. R., Monier, M. K., van Cuyck-Gandre, H., Longer, C. F.,

and Innis, B. L. (1999). Phylogenetic analysis of hepatitis E virus isolates from Egypt. *J. Med. Virol.* **57**: 68–74.

FIGURE 3.15 Adapted from Fauquet, C. M., Mayo, M. A., Maniloff, J., Desselberger, U., and Ball, L. A. (Eds.) (2005). *Virus Taxonomy, the Eighth Report of the International Committee on Taxonomy of Viruses*, Academic Press, New York, p. 860.

FIGURE 3.16 Adapted from Figure 34 in Strauss, J. H., and Strauss, E. G. (1994). The alphaviruses: gene expression, replication and evolution. *Microbiol. Rev.* **58**: 491–562.

FIGURE 3.17 Adapted from Figures 2B and 4B from Luers, A. J., Adams, S. D., Smalley, J. V., and Campanella, J. J. (2005). A phylogenomic study of the genus Alphavirus employing whole genome comparison. *Comp. Funct. Genom.* **6**: 217–227.

FIGURE 3.18 Adapted from Figure 5 in Strauss, J. H., and Strauss, E. G. (1994). The alphaviruses: gene expression, replication and evolution. *Microbiol. Rev.* **58**: 491–562.

FIGURE 3.19 Adapted from Figures 7 and 9 in Strauss, J. H., and Strauss, E. G. (1994). The alphaviruses: gene expression, replication and evolution. *Microbiol. Rev.* **58**: 491–562.

FIGURE 3.20 Adapted from Figure 18 in Strauss, J. H., and Strauss, E. G. (1994). The alphaviruses: gene expression, replication and evolution. *Microbiol. Rev.* **58**: 491–562.

FIGURE 3.21 Adapted from Figure 3 in Frey, T. K., Gard, D. L., and Strauss, J. H. (1979). Biophysical studies on circle formation by Sindbis virus 49S RNA. *J. Mol. Biol.* **132**: 1–18.

FIGURE 3.22 Adapted from Monath, T. P. (1988). *The Arboviruses*, Vol. I, CRC Press, Boca Raton, FL, p. 91.

FIGURE 3.23 From Monath, T. P. (1988). *The Arboviruses* **1**: 129.

FIGURE 3.24 Data from Fields, B. N., Knipe, D. M., and Howley, P. M. (Eds.) (1996). *Fields Virology, 3rd Edition*, Lippincott-Raven, New York, p. 875 using data from Tsai, and later data from *MMWR Summary of Notifiable Diseases, 1998, 1999, 2000, 2001, 2002*, and *2003*.

FIGURE 3.25 Fauquet, C. M., Mayo, M. A., Maniloff, J., Desselberger, U., and Ball, L. A. (Eds.) (2005). *Virus Taxonomy, the Eighth Report of the International Committee on Taxonomy of Viruses*, Academic Press, New York, Figure 4, p. 1007.

FIGURE 3.26 From *MMWR*, 1998, Vol. **46**, #54, *Summary of Notifiable Diseases in the United States for 1997* and later years.

FIGURE 3.27 Adapted from Figures 2, 4, and 5 of the *Flaviviridae* in: Fauquet, C. M., Mayo, M. A., Maniloff, J., Desselberger, U., and Ball, L. A. (Eds.) (2005). *Virus Taxonomy, the Eighth Report of the International Committee on Taxonomy of Viruses*, Academic Press, New York, pp. 981–998. The data on the F protein of Hepatitis C virus which is produced from the core protein sequence by ribosomal frameshifting came from Xu, Z., Choi, J., Yen, T. S. B., Lu, W., Strohecker, A., Govindarajan, S., Chien, D., Selby, M. J., and Ou, J.-H. (2004). Synthesis of a novel hepatitis C protein by ribosomal frameshift. *EMBO J.* **20**: 3840–3848.

FIGURE 3.28 Adapted from Billoir, F., de Chesse, R., Tolou, H., de Micco, P., Gould, E.A., and de Lamballerie, X. (2000). Phylogeny of the genus Flavivirus using complete coding sequences of arthropod-borne viruses and viruses with no known vector. *J. Gen Virol.* **81**: 781–790.

FIGURE 3.29 Redrawn from Figures 3, 4, and 6 in Strauss, J. H., and Strauss, E. G. (1996). Current advances in yellow fever research. In: *Microbe Hunters Then and Now*, (H. Koprowski and M. B. A. Oldstone, Eds.), Medi-Ed Press, Bloomington, IL pp. 113–137.

FIGURE 3.30 Adapted from Figure 5 in Alvarez, D. E., Lodiero, M. F., Ludueña, S J., Pietrasanta, L. I., and Gamarnik, A. V. (2005). Long range RNA-RNA interactions circularize the dengue virus genome *J. Virol.* **79**: 6631–6643.

FIGURE 3.31 Adapted from Figure 7 in Hahn, C. S., Hahn, Y. S., Rice, C. M., Lee, E., Dalgarno, L., Strauss, E. G., and Strauss, J. H. (1987). Conserved elements in the 3′ untranslated region of flavivirus RNAs and potential cyclization sequences. *J. Mol. Biol.* **198**: 33–41.

FIGURE 3.32 From data at: http://www.who.int/immunization monitoring/data/data_subject/en/index.html.

FIGURE 3.33 Adapted from *MMWR (1993)*, Vol. **42**, RR-1, Figure 1, p. 2.

FIGURE 3.34 Data came from Fields, B. N., Knipe, D. M., and Howley, P. M. (Eds.) (1996). *Fields Virology*, *3rd Edition*, Lippincott-Raven, New York, and from *MMWR, Summary of Notifiable Diseases for 1996, 1997, 1998, 1999, 2000, 2001, 2002, 2003.*

FIGURE 3.35 Adapted from Porterfield, J. S. (Ed.) (1995). *Exotic Viral Infections*. Chapman & Hall Medical, London, p. 207.

FIGURE 3.36 http://www.reliefweb.int/rw/RWB.NSF/db900LargeMaps/SKAR-64GDV4?OpenDocument.

FIGURE 3.37 Data from Meyers, G., and Thiel, H.-J. (1996). Molecular characterization of pestiviruses. *Adv. Virus Res.* **47**: 53–118.

FIGURE 3.38 Adapted from de Vries, A. A. F., Horzinek, M. C., Rottier, P. M., and de Groot, R. J. (1997). The genome organization of the *Nidovirales*: Similarities and differences between arteri-, toro-, and coronaviruses. *Semin. Virol.* **8**: 33–47, with permission.

FIGURE 3.39 Adapted from from de Vries, A. A. F., Horzinek, M. C., Rottier, P. M., and de Groot, R. J. (1997). The genome organization of the *Nidovirales*: Similarities and differences between arteri-, toro-, and coronaviruses. *Semin. Virol.* **8**: 33–47, with permission; and den Boon, J. A., Snijder, E. J., Chirnside, E. D., de Vries, A. A. F., Horzinek, M. C., and Spaan, W. J. M. (1991). Equine arteritis virus is not a togavirus but belongs to the coronavirus-like superfamily. *J. Virol.* **65**: 2910–2920.

FIGURE 3.40 Redrawn from Cowley, J. A., and Walker, P. J. (2002). The complete genome sequence of gill-associated virus of *Penaeus monodon* prawns indicates a gene organization unique among nidoviruses. *Arch. Virol.* **147**: 1977–1987.

FIGURE 3.41 Adapted from Figure 35 in Strauss, J. H., and Strauss, E. G. (1994). The alphaviruses: gene expression, replication and evolution. *Microbiol. Rev.* **58**: 491–562.

FIGURE 3.42 Adapted from Strauss, E. G., and Strauss, J. H. (1997). Eukaryotic RNA viruses: a variant genetic system. In: *Exploring Genetic Mechanisms*, (M. Singer and P. Berg, Eds.), University Science Books, Sausalito, CA, Figure 2.12, p. 94.

FIGURE 3.43 Original data in Koonin, E. V., and Dolja, V. V. (1993). Evolution and taxonomy of positive-strand RNA viruses—Implication of comparative analysis of amino acid sequences. *Crit. Rev. Biochem. Mol. Biol.* **28**: 375–430. Figure was redrawn as Figure 36 in Strauss, J. H., and Strauss, E. G. (1994). The alphaviruses: gene expression, replication and evolution. *Microbiol. Rev.* **58**: 491–562.

FIGURE 3.44 Data from: Weaver, S. C., Kang, W. L., Shirako, Y., Rümenapf, T., Strauss, E. G., and Strauss, J. H. (1997). Recombinational history and molecular evolution of Western equine encephaloymyelitis complex alphaviruses. *J. Virol.* **71**: 613–623; Takeda, N., Tanimura, M., and Miyamura, K. (1994). Molecular evolution of the major capsid protein VP1 of enterovirus 70. *J. Virol.* **68**: 854–862; Saitou, N., and Nei, M. (1986). Polymorphism and evolution of influenza A viruses. *Mol. Biol. Evol.* **3**: 57–74. Also used in Strauss, J.H and Strauss, E.G (1988). Evolution of RNA Viruses *Annu. Rev. Microbiol.* **42**: 657–683.

CHAPTER 4

Tables

TABLE 4.4 Adapted from Granoff, A., and Webster, R. G. (Eds.) (1999). Tables of Respiratory Viruses. In: *Encyclopedia of Virology*, *2nd Edition*, Academic Press, San Diego, pp. 1493 and 1494.

TABLE 4.8 Adapted in part from Fields, B. N., Knipe, D. M., and Howley, P. M. (Eds.) (1996). *Fields Virology*, *3rd Edition*, Lippincott-Raven, New York, Table 2, p. 1355 and recent information from Fauquet, C. M., Mayo, M. A., Maniloff, J., Desselberger, U., and Ball, L. A. (Eds.) (2005). *Virus*

Taxonomy, the Eighth Report of the International Committee on Taxonomy of Viruses, Academic Press, New York, p. 683.

TABLE 4.12 Adapted in part from Fields, B. N., Knipe, D. M., and Howley, P. M. (Eds.), (1996). *Fields Virology*, *3rd Edition*, Lippincott-Raven, New York, Table 1, p. 1522; Porterfield, J. S. (Ed.) (1995). *Exotic Viral Infections*. Chapman & Hall Medical, London, Table 11.1, p. 228; and recent information from Fauquet, C. M., Mayo, M. A., Maniloff, J., Desselberger, U., and Ball, L. A. (Eds.) (2005). *Virus Taxonomy, the Eighth Report of the International Committee on Taxonomy of Viruses*, Academic Press, New York, pp. 725–734.

TABLE 4.13 This table includes data from Nathanson, N., Ahmed, R., Gonzalez-Scarano, F., Griffin, D. E., Holmes, K. V., Murphy, F. A., and Robinson, H.L. (Eds.) (1996). *Viral Pathogenesis*, Lippincott-Raven, Philadelphia, Table 32.1, p. 780.

Figures

FIGURE 4.1 Redrawn from Strauss, E. G., Strauss, J. H., and Levine, A. J. (1996). Virus evolution. In: *Fields Virology*, 3rd Edition (B. N. Fields, D. M. Knipe, P. M. Howley, *et al.* (Eds.), Lippincott-Raven, Philadelphia, Figure 5 on p.168.

FIGURE 4.2 Data for this figure came from Rose, J. K., and Schubert, M. (1987). Rhabdovirus genomes and their products. In: *The Rhabdoviruses* (R. R. Wagner, Ed.), Plenum Press, New York, pp. 129–166, Figure 3.

FIGURE 4.3 From Smith, J. S., Orciari, L. A., and Yager, P. A. (1995). Molecular epidemiology of rabies in the United States. *Semin. Virol.* **6**: 387–400 and additional data from *MMWR Summaries of Notifiable Diseases, 1996 and 2003*.

FIGURE 4.4 From *MMWR* (1997). Vol. **45**, p. 1119, updated with data from CDC found at: www.rabavert.com/caserc.html.

FIGURE 4.5 Adapted from Chua, K. B., Bellini, W. J., Rota, P. A., Harcourt, B. H., Tamin, A., Lam, S. K., Ksiazek, T. G., Rollin, P. E., Zaki, S. R., Shieh, W. J., Goldsmith, C. S., Gubler, D. J., Roehrig, J. T., Eaton, B., Gould, A. R., Olson, J., Field, H., Daniels, P., Ling, A. E., Peters, C. J., Anderson, L. J., and Mahy, B. W. J. (2000). Nipah virus: a recently emergent deadly paramyxovirus. *Science* **288**: 1432–1435.

FIGURE 4.7 From E. G. Strauss and Strauss, J. H. (1991). RNA viruses: genome structure and evolution. *Curr. Opin. Genet. Dev.* **1**: 485–493, Figure 1, with additional data from Chua, K. B., Bellini, W. J., Rota, P. A., Harcourt, B. H., Tamin, A., Lam, S. K., Ksiazek, T. G., Rollin, P. E., Zaki, S. R., Shieh, W. J., Goldsmith, C. S., Gubler, D. J., Roehrig, J. T., Eaton, B., Gould, A. R., Olson, J., Field, H., Daniels, P., Ling, A. E., Peters, C. J., Anderson, L. J., and Mahy, B. W. J. (2000). Nipah virus: a recently emergent deadly paramyxovirus. *Science* **288**: 1432–1435.

FIGURE 4.8 From *MMWR, Summary of Notifiable Diseases-1996*, Vol. **45**, p. 45 and the comparable summaries for 1998, 2001, and 2003 and *MMWR (2006)*, Vol. **55**, p. 1152.

FIGURE 4.9 Data from Black, F. L. (1966). Measles endemicity in insular populations: critical community size and its evolutionary implication. *J. Theoret. Biol.* **11**: 207–211.

FIGURE 4.10 From *MMWR, 1998, Summary of Notifiable Diseases-1998*, Vol. **47**, No. 53, p. 48, and the comparable summary for 2003 and *MMWR (2006)*, Vol. **55**, p. 1348.

FIGURE 4.11 Adapted from Fauquet, C. M., Mayo, M. A., Maniloff, J., Desselberger, U., and Ball, L. A. (Eds.) (2005). *Virus Taxonomy, the Eighth Report of the International Committee on Taxonomy of Viruses*, Academic Press, New York, Figure 3, p. 652.

FIGURE 4.12 Adapted from Fauquet, C. M., Mayo, M. A., Maniloff, J., Desselberger, U., and Ball, L. A. (Eds.) (2005). *Virus Taxonomy, the Eighth Report of the International Committee on Taxonomy of Viruses*, Academic Press, New York, Figure 2, p. 617.

FIGURE 4.13 Adapted from Strauss, E. G., and Strauss, J. H. (1997). Eukaryotic RNA viruses: a variant genetic system. In: *Exploring Genetic*

Mechanisms (M. Singer and P. Berg, Eds.), University Science Books, Sausalito, CA, Figure 2.19, p. 105.

FIGURE 4.15 Panels (A) and (B) were adapted from Neumann, G., Brownlee, G. G., Fodor, E., and Kawaoka, Y. (2004). Orthomyxovirus replication, transcription, and polyadenylation. *Curr. Top. Microbiol. Immunol.* **283**: 121–143 and Panel C was adapted from Figure 1 in Catchpole, A. P., Mingay, J., Fodor, E., and Brownlee, G. G. (2003). Alternative base pairs attenuate influenza A virus when introduced into the duplex region of the conserved viral RNA promoter of either the NS or the PA gene. *J. Gen. Virol.* **84**: 507–515.

FIGURE 4.16 Reid, A. H., Fanning, T. G., Janczewski, T. A., Lourens, R. M., and Taubenberger, J. K. (2004). Novel origin of the 1918 pandemic influenza virus nucleoprotein gene. *J. Virol.* **78**: 12462–12470, Figure 2.

FIGURE 4.17 Redrawn from Fields, B. N., Knipe, D. M., and Howley, P. M. (Eds.) (1996). *Fields Virology, 3rd Edition*, Lippincott-Raven, New York, p. 1421; with updated information from Thompson, W. W., Shay, D. K., Weintraub, E., Brammer, L., Cos, N., Anderson, L. J., and Fukuda, K. (2003). Mortality associated with influenza and respiratory syncytial virus in the United States. *JAMA* **289**: 179–186.

FIGURE 4.18 Data from Crosby, A. W. (1989). *America's Forgotten Pandemic: the Influenza of 1918*. Cambridge, England, Cambridge University Press, pp. 22–24, with updated information from Thompson, W. W., Shay, D. K., Weintraub, E., Brammer, L., Cos, N., Anderson, L. J., and Fukuda, K. (2003). Mortality associated with influenza and respiratory syncytial virus in the United States. *JAMA* **289**: 179–186.

FIGURE 4.19 Adapted from *ASM News*, July 1999.

FIGURE 4.20 Adapted from Fanning, T. G., Slemons, R. D., Reid, A. H., Janczewski, T. A., Dean, J., and Taubenberger, J. K. (2002). 1917 Influenza virus sequences suggest that the 1918 pandemic virus did not acquire its hemagglutinin directly from birds. *J. Virol.* **67**: 7860–7862, Figure 1.

FIGURE 4.22 Adapted from Figure 8 in Kukkonen, S. K. J., Vaheri, A., and Plyusnin, A. (2005). L Protein, the RNA-dependent RNA polymerase of hantaviruses. *Arch. Virol.* **150**: 533–556.

FIGURE 4.24 Adapted from Peters, C. J. (1998a). Hantavirus pulmonary syndrome in the Americas. In: *Emerging Infections 2* (W. M. Scheld, W. A. Craig, and J. M. Hughes, Eds.), ASM Press, Washington, DC, pp. 17–64, Figure 2.

FIGURE 4.25 Adapted from Porterfield, J. S. (Ed.) (1995). *Exotic Viral Infections,* Chapman & Hall Medical, London, p. 276 and data from Lee, H. W. (1996). Hantaviruses: an emerging disease. *Philos. J. Microbiol. Infect. Dis.* **25**(1): S19–S24.

FIGURE 4.26 Adapted from Peters, C. J. (1998a). Hantavirus pulmonary syndrome in the Americas. In: *Emerging Infections 2* (W. M. Scheld, W. A. Craig, and J. M. Hughes, Eds.), ASM Press, Washington, DC, Figure 1 and Table 3, and data from Yates, T., Mills, J. N., Parmenter, C. A., Ksiazek, T. G., Parmenter, R. R., Vande Castle, J.R., Calisher, C. H., Nichol, S. T., Abbott, K. D., Young, J. C., Morrison, M. L., Beaty, B. J., Dunnum, J. L., Baker, R. J., Salazar-Bravo, J., and Peters, C. J. (2002). The ecology and evolutionary history of an emergent disease: Hantavirus pulmonary syndrome. *Bioscience* **52**: 989–998, and data from the Pan American Health Organization Web site at: http://www.paho.org/.

FIGURE 4.27 Drawn from data in Clegg, J. C. S., Wilson, S. M., and Oram, J. D. (1990). Nucleotide sequence of the S RNA of Lassa virus (Nigerian Strain) and comparative analysis of arenavirus gene products. *Virus Res.* **18**: 151–164; Lukashevich, I. S., Djavani, M., Shapiro, K., Sanchez, A., Ravkov, E., Nichol, S. T., and Salvato, M. S. (1997). The Lassa fever virus L gene: nucleotide sequence, comparison, and precipitation of a predicted 250 kDa protein with monospecific antiserum. *J. Gen. Virol.* **78**: 547–551.

FIGURE 4.28 Adapted from Peters, C. J. (1998b). Hemorrhagic fevers: how they wax and wane, In *Emerging Infections 1* (W. M. Scheld, D.

Armstrong, and J. M. Hughes, Eds.), ASM Press, Washington, DC, pp. 15–25, Figure 1.

CHAPTER 5

Tables

TABLE 5.4 Information in this table can be found in Joklik, W. K., and Roner, M. R. (1996). Molecular recognition in the assembly of the segmented reovirus genome. *Prog. Nuc. Acid Res.* **53**: 249–281.

Figures

FIGURE 5.1 Adapted from Fauquet, C. M., Mayo, M. A., Maniloff, J., Desselberger, U., and Ball, L. A. (Eds.) (2005). *Virus Taxonomy, the Eighth Report of the International Committee on Taxonomy of Viruses*, Academic Press, New York. Figure 3, p. 453.

FIGURE 5.2 Adapted from Duncan, R. (1999). Extensive sequence divergence and phylogenetic relationships between the fusogenic and nonfusogenic orthoreoviruses: a species proposal. *Virology* **260**:316–328; and additional data from Fauquet, C. M., Mayo, M. A., Maniloff, J., Desselberger, U., and Ball, L. A. (Eds.) (2005). *Virus Taxonomy, the Eighth Report of the International Committee on Taxonomy of Viruses*, Academic Press, New York, p. 465.

FIGURE 5.3 Fields, B. N., Knipe, D. M., and Howley, P. M. (Eds.) (1996). *Fields Virology, 3rd Edition*, Lippincott-Raven, New York, p. 1559.

FIGURE 5.4 Redrawn from a figure in Niebert, M. L., and Fields, B. N. (1995). Early steps in reovirus infection of cells. In: *Cell Receptors for Animal Viruses*, Cold Spring Harbor Laboratory Press Cold Spring Harbor, NY, pp. 341–364; reprinted in Fields, B. N., Knipe, D. M., and Howley, P. M. (Eds.) (1996). *Fields Virology, 3rd Edition*, Lippincott-Raven, New York, p. 1562.

FIGURE 5.5 Adapted from Fields, B. N., Knipe, D. M., and Howley, P. M. (Eds.) (1996). *Fields Virology, 3rd Edition*, Lippincott-Raven, New York, Figure 6, p. 1706.

FIGURE 5.6 Adapted from Kapikian, A. Z. (1993). Viral gastroenteritis. *JAMA* **269**: 627–630.

FIGURE 5.7 Glass, R. I. (2006). New hope for defeating rotavirus. *Scientific American* **294**, #4: 46–55.

FIGURE 5.8 Drawn from data from Barnes, G. L., Uren, E., Stevens, K. B., and Bishop, R. F. (1998). Etiology of acute gastroenteritis in hospitalized children in Melbroune, Australia, from April 1980 to March 1993. *J. Clin. Microbiol.* **35**: 133–138, and Brandt, C. D., Kim, H.W., Rodriguez, W. J., Arrobio, J. O., Jeffries, B. C., Stallings, E. P., Lewis, C., Miles, A. J., Chanock, R. M., Kapikian, A. Z., and Parrott, R. H. (1983) Pediatric viral gastroenteritis during eight years of study. *J. Clin. Microbiol.* **18**: 71–78.

FIGURE 5.9 Adapted from Török, T. J., Kilgore, P. E., Clarke, M. J., Holman, R. C., Bresee, J. S., and Glass, R. I. (1997). Visualizing geographic and temporal trends in rotavirus activity in the United States, 1991 to 1996. *Pediatr. Infect. Dis. J.* **16**: 941–946.

FIGURE 5.10 Adapted from Fields, B. N., Knipe, D. M., and Howley, P. M. (Eds.) (1996). *Fields Virology, 3rd Edition*, Lippincott-Raven, New York, p. 1736.

FIGURE 5.11 Original source: Tsai, T. F. (1991). Arboviral infections in the United States. *Infect. Dis. Clin. North Am.* **5**: 73–102; reprinted in Fields, B. N., Knipe, D. M., and Howley, P. M. (Eds.) (1996). *Fields Virology, 3rd Edition*, Lippincott-Raven, New York, p. 1753.

FIGURE 5.12 Redrawn from Bowen, G. S. (1988). Colorado Tick Fever. In: *The Arboviruses* (T. P. Monath, Ed.), Vol. II, pp. 159–176, Figure 4.

CHAPTER 6

Tables

TABLE 6.2   Adapted from Coffin, J. M., Hughes, S. H., and Varmus, H. E. (Eds.) (1997). *Retroviruses*, Cold Spring Harbor Laboratory Press, Cold Spring Harbor, NY, Table 2, p. 38.

TABLE 6.3   Adapted from Coffin, J. M., Hughes, S. H., and Varmus, H. E. (Eds.) (1997). *Retroviruses*, Cold Spring Harbor Laboratory Press, Cold Spring Harbor, NY, Table 1 on p. 36 and information from Fauquet, C. M., Mayo, M. A., Maniloff, J., Desselberger, U., and Ball, L. A. (Eds.) (2005). *Virus Taxonomy, the Eighth Report of the International Committee on Taxonomy of Viruses*, Academic Press, New York, pp. 421–440.

TABLE 6.4   Adapted from Levy, J. A. (1994) *HIV and the Pathogenesis of AIDS*, ASM Press, Washington, DC, Table 1.4, p. 8.

TABLE 6.5   Adapted from Fields, B. N., Knipe, D. M., and Howley, P. M. (Eds.) (1996). *Fields Virology, 3rd Edition*, Lippincott-Raven, New York, p. 309.

TABLE 6.6   Adapted from Coffin, J. M., Hughes, S. H., and Varmus, H. E. (Eds.) (1997). *Retroviruses*, Cold Spring Harbor Laboratory Press, Cold Spring Harbor, NY, Table 1, p. 597.

TABLE 6.7   Data from the web site: www.unaids.org/epidemic_update/report/.

TABLE 6.8   From Eickbush, T. (1994) The evolution of retroelements. In: *The Evolutionary Biology of Viruses* (S. S. Morse, Ed.), Raven Press, New York, pp. 121–157, Table 1, p. 122.

TABLE 6.9   Fauquet, C. M., Mayo, M. A., Maniloff, J., Desselberger, U., and Ball, L. A. (Eds.) (2005). *Virus Taxonomy, the Eighth Report of the International Committee on Taxonomy of Viruses*, Academic Press, New York, pp. 397–420.

TABLE 6.10   Fields, B. N., Knipe, D. M., and Howley, P. M. (Eds.) (1996). *Fields Virology, 3rd Edition*, Lippincott-Raven, New York, Table 1, p. 2708, using taxonomy from Fauquet, C. M., Mayo, M. A., Maniloff, J., Desselberger, U., and Ball, L. A. (Eds.) (2005). *Virus Taxonomy, the Eighth Report of the International Committee on Taxonomy of Viruses*, Academic Press, New York, pp. 373–384.

Figures

FIGURE 6.1   Adapted from Coffin, J. M., Hughes, S. H., and Varmus, H. E. (Eds.) (1997). *Retroviruses*, Cold Spring Harbor Laboratory Press, Cold Spring Harbor, NY, Figure 6 on page 43; and Fields, B. N., Knipe, D. M., and Howley, P. M. (Eds.) (1996). *Fields Virology, 3rd Edition*, Lippincott-Raven, New York, p. 1769.

FIGURE 6.2   Adapted from Goff, S. P. (1997). Retroviruses: infectious genetic elements. In: *Exploring Genetic Mechanisms* (M. Singer and P. Berg, Eds.), University Science Books, Sausalito, CA, Figure 3.5, p. 143.

FIGURE 6.3   Adapted from a combination of Fields, B. N., Knipe, D. M., and Howley, P. M. (Eds.) (1996). *Fields Virology, 3rd Edition*, Lippincott-Raven, New York, p. 1792; Goff, S. P. (1997). Retroviruses: infectious genetic elements. In: *Exploring Genetic Mechanisms* (M. Singer and P. Berg, Eds.), University Science Books, Sausalito, CA, Figure 3.6 on p. 145; and Coffin, J. M., Hughes, S. H., and Varmus, H. E. (Eds.) (1997). *Retroviruses*, Cold Spring Harbor Laboratory Press, Cold Spring Harbor, NY, Figure 2, p. 123.

FIGURE 6.4   Original sources: Redrawn from a combination of Hindmarsh, P., and Leis, J. (1999). Retroviral DNA Integration, *Microbiol. Mol. Biol. Rev.*, **63**: 836–843; Goff, S. P. (1997). Retroviruses: infectious genetic elements. In: *Exploring Genetic Mechanisms* (M. Singer and P. Berg, Eds.), University Science Books, Sausalito, CA, Figure 3.10, p. 145; and Coffin, J. M., Hughes, S. H., and Varmus, H. E. (Eds.) (1997). *Retroviruses*, Cold Spring Harbor Laboratory Press, Cold Spring Harbor, NY, Figure 8, p. 185.

FIGURE 6.5   Composite of Figures 4, 5, 7, 8, 9, and 11 in Chapter 6 of Coffin, J. M., Hughes, S. H., and Varmus, H. E. (Eds.) (1997). *Retroviruses*, Cold Spring Harbor Laboratory Press, Cold Spring Harbor, NY.

FIGURE 6.6   This figure is a composite of Goff, S. P. (1997). Retroviruses: infectious genetic elements. In: *Exploring Genetic Mechanisms* (M. Singer and P. Berg, Eds.), University Science Books, Sausalito, CA, Figure 3.16, p. 157; and Coffin, J. M., Hughes, S. H., and Varmus, H. E. (Eds.) (1997). *Retroviruses*, Cold Spring Harbor Laboratory Press, Cold Spring Harbor, NY, Figures on pp. 45, 269, 795, and 799.

FIGURE 6.7   Coffin, J. M., Hughes, S. H., and Varmus, H. E. (Eds.) (1997). *Retroviruses*, Cold Spring Harbor Laboratory Press, Cold Spring Harbor, NY, pp. 44 and 798.

FIGURE 6.8   Coffin, J. M., Hughes, S. H., and Varmus, H. E. (Eds.) (1997). *Retroviruses*, Cold Spring Harbor Laboratory Press, Cold Spring Harbor, NY, Figure 10, p. 56.

FIGURE 6.9   Redrawn from Coffin, J. M., Hughes, S. H., and Varmus, H. E. (Eds.) (1997). *Retroviruses*, Cold Spring Harbor Laboratory Press, Cold Spring Harbor, NY, Figure 5, p. 37; and Fields, B. N., Knipe, D. M., and Howley, P. M. (Eds.) (1996). *Fields Virology, 3rd Edition*, Lippincott-Raven, New York, p. 1776.

FIGURE 6.10   Redrawn from Coffin, J. M., Hughes, S. H., and Varmus, H. E. (Eds.) (1997). *Retroviruses*, Cold Spring Harbor Laboratory Press, Cold Spring Harbor, NY, Figure 12, p. 226.

FIGURE 6.11   Redrawn from Coffin, J. M., Hughes, S. H., and Varmus, H. E. (Eds.) (1997). *Retroviruses*, Cold Spring Harbor Laboratory Press, Cold Spring Harbor, NY, p. 803.

FIGURE 6.12   Adapted from Coffin, J. M., Hughes, S. H., and Varmus, H. E. (Eds.) (1997). *Retroviruses*, Cold Spring Harbor Laboratory Press, Cold Spring Harbor, NY, p. 244.

FIGURE 6.13   Redrawn from Coffin, J. M., Hughes, S. H., and Varmus, H. E. (Eds.) (1997). *Retroviruses*, Cold Spring Harbor Laboratory Press, Cold Spring Harbor, NY, pp. 5 and 504; and from Fields, B. N., Knipe, D. M., and Howley, P. M. (Eds.) (1996). *Fields Virology, 3rd Edition*, Lippincott-Raven, New York, p. 1776.

FIGURE 6.14   Redrawn from Coffin, J. M., Hughes, S. H., and Varmus, H. E. (Eds.) (1997). *Retroviruses*, Cold Spring Harbor Laboratory Press, Cold Spring Harbor, NY, Figure 21, p. 544.

FIGURE 6.15   Adapted from Coffin, J. M., Hughes, S. H., and Varmus, H. E. (Eds.) (1997). *Retroviruses*, Cold Spring Harbor Laboratory Press, Cold Spring Harbor, NY, Figure 5, p. 600.

FIGURE 6.16   Redrawn from Mellors, J. W., Rinaldo, C. R., Jr., Gupta, P., White, R. M., Todd, J. A., and Kingsley, L. A. (1996). Prognosis in HIV-1 infection predicted by the quantity of virus in plasma. *Science* **272**: 1167–1170.

FIGURE 6.17   Redrawn from Schwartlander, B., Garnett, G., Walker, N., and Anderson, R. (2000). AIDS special report—AIDS in a new millennium. *Science* **289**: 64–67; and data from *2006 Report on the Global AIDS Epidemic*, from the UNAIDS Programme.

FIGURE 6.18   Adapted from UNAIDS publications, Slide 12 of the *AIDS in Africa* series on the Web site www.unaids.org.

FIGURE 6.19   Data from *MMWR Summary of Notifiable Diseases, United States-2003*.

FIGURE 6.20   *2006 Report on the Global AIDS Epidemic*, from the UNAIDS Programme.

FIGURE 6.21   Data from http://www.cdc.gov/hiv/graphics/images/.

FIGURE 6.22   *Thailand's Response to HIV/AIDS: Progress and Challenges* (2004). United Nations Development Programme report. Figure 2, p. 2.

FIGURE 6.23   Data were from CDC, National Center for Injury Prevention and Control and National Vital Statistics System.

FIGURE 6.24   Redrawn from Coffin, J. M., Hughes, S. H., and Varmus, H. E. (Eds.) (1997). *Retroviruses*, Cold Spring Harbor Laboratory Press, Cold Spring Harbor, NY, Figure 17, p. 411 and information from Fauquet, C. M., Mayo, M. A., Maniloff, J., Desselberger, U., and Ball, L. A. (Eds.) (2005). *Virus Taxonomy, the Eighth Report of the International Committee on Taxonomy of Viruses*, Academic Press, New York, p. 418.

FIGURE 6.25   Adapted from Eickbush, T. H. (1994). The origin and evolutionary relationships of retroelements. In: *The Evolutionary Biology of Viruses* (S. S. Morse, Ed.), Raven Press, New York, pp. 121–157.

FIGURE 6.26 Data from Yen, T. S. B. (1998). Posttranscriptional regulation of gene expression in hepadnaviruses. *Semin. Virol.* **8**: 319–326; Hu, J., and Seeger, C. (1997). RNA signals that control DNA replication in Hepadnaviruses. *Semin. Virol.* **8**: 205–211; and Fields, B. N., Knipe, D. M., and Howley, P. M. (Eds.) (1996). *Fields Virology*, *3rd Edition*, Lippincott-Raven, New York, p. 2706.

FIGURE 6.27 Modified from Locarnini, S. A., Civitico, G. M., and Newbold, J. E. (1996). Hepatitis B: new approaches for antiviral chemotherapy. *Antiviral Chem. Chemother.* **7**: 53–64.

FIGURE 6.28 Redrawn from Buckwold, V. E., and Ou, J.-H. (1999). Hepatitis B virus C-gene expression and function: the lessons learned from viral mutants. *Curr. Topics Virol.* **1**: 71–81, Figure 2.

FIGURE 6.29 From WHO Program on Diseases and Vaccines Web page: http://www.cdc.gov/ncidod/diseases/hepatitis/slides.

FIGURE 6.30 Data from Globocan 2002 at: http://www-dep.iarc.fr/.

FIGURE 6.31 From http://www.who.int/immunization_monitoring/en/globalsummary/timeseries/tscoveragehepb3.htm.

CHAPTER 7

Tables

TABLE 7.5 Adapted from Fields, B. N., Knipe, D. M., and Howley, P. M. (Eds.) (1996). *Fields Virology*, *3rd Edition*, Lippincott-Raven, New York, Table 3, p. 2645; with additional information from Goebel, S. J., Johnson, G. P., Perkus, M. E., Davis, S. W., Winslow, J. P., and Paoletti, E. (1990). The complete sequence of vaccinia virus. *Virology* **179**: 247–266.

TABLE 7.6 Adapted from Fields, B. N., Knipe, D. M., and Howley, P. M. (Eds.) (1996). *Fields Virology*, *3rd Edition*, Lippincott-Raven, New York, Table 2, p. 2643; with additional information from Goebel, S. J., Johnson, G. P., Perkus, M. E., Davis, S. W., Winslow, J. P., and Paoletti, E. (1990). The complete sequence of vaccinia virus. *Virology* **179**: 247–266.

TABLE 7.9 Adapted from Fields, B. N., Knipe, D. M., and Howley, P. M. (Eds.) (1996). *Fields Virology*, *3rd Edition*, Lippincott-Raven, New York, Table 2, p. 2433.

TABLE 7.11 Adapted from Fields, B. N., Knipe, D. M., and Howley, P. M. (Eds.) (1996). *Fields Virology*, *3rd Edition*, Lippincott-Raven, New York, p. 2122.

TABLE 7.13 Adapted from Fields, B. N., Knipe, D. M., and Howley, P. M. (Eds.) (1996). *Fields Virology*, *3rd Edition*, Lippincott-Raven, New York, p. 2030 and additional data from Walker, D. L., and Frisque, R. J. (1986). The biology and molecular biology of JC virus. In: *The Papovaviridae* (N. P. Salzman, Ed.), Plenum Press, New York, Vol. 1, pp. 327–377.

TABLE 7.15 Adapted from Fields, B. N., Knipe, D. M., and Howley, P. M. (Eds.) (1996). *Fields Virology*, *3rd Edition*, Lippincott-Raven, New York, Table 1, pp. 2048–2049, Tables 3 and 4, p. 2085 and additional data from Alani, R. M., and Münger, K. (1998). Human papillomaviruses and associated malignancies. *J. Clin. Oncol.* **16**: 330–337.

Figures

FIGURE 7.1 Adapted from Fields, B. N., Knipe, D. M., and Howley, P. M. (Eds.) (1996). *Fields Virology*, *3rd Edition*, Lippincott-Raven, New York, p. 2638.

FIGURE 7.2 Adapted from Fenner, F., Wittek, R., and Dumbell, K. R. (Eds.) (1988). *The Orthopoxviruses*, Academic Press, New York.

FIGURE 7.3 Redrawn from data from Traktman, P. (1990). The enzymology of poxvirus DNA replication. *Curr. Topics Microbiol. Immunol.* **163**: 93–123; and Moyer, R. M., and Graves, R. L. (1981). Mechanism of cytoplasmic orthopox DNA replication. *Cell* **27**: 391–401.

FIGURE 7.4 (A) Adapted from Afonso, C. L., Tulman, E. R., Delhon, G., Lu, Z., Viljoen, G. J., Wallace, D. B., Kutish, G. F., and Rock, D. L. (2006).

Genome of crocodilepox virus. *J. Virol.* **80**: 4978–4991. (B) Adapted from Chen, N. H., Danila, M. I., Feng, Z. H., Buller, R. M. L., Wang, C. L., Hang, X. S., Lefkowitz, E. J., Upton, C. (2003). The genomic sequence of ectromelia virus, the causative agent of mousepox. *Virology* **317**: 165–186, Figure 3.

FIGURE 7.5 From Fenner, F., Henderson, D. A., Arita, I., Jezek, A., and Ladnyi, I. D. (1988). *Smallpox and its Eradication*, WHO Geneva, pp. 15 and 57.

FIGURE 7.6 From Fenner, F. (1983). The Florey Lecture, 1983. Biological Control, as exemplified by smallpox eradication and myxomatosis. *Proc. R. Soc. London B* **218**: 259–285.

FIGURE 7.7 From *Morbidity and Mortality Weekly Report* (1997), Vol. **46**, p. 306.

FIGURE 7.8 From *Morbidity and Mortality Weekly Report* (2003), Vol. **54**, p. 642.

FIGURE 7.9 Drawn from data in Fenner, F. (1983). Biological control as exemplified by smallpox eradication and myxomatosis. *Proc. R. Soc. London B* **218**: 259–285.

FIGURE 7.10 Drawn from data from Fenner, F. (1983). The Florey Lecture, 1983. Biological Control, as exemplified by smallpox eradication and myxomatosis. *Proc. R. Soc. London B* **218**: 259–285.

FIGURE 7.11 Adapted from Fauquet, C. M., Mayo, M. A., Maniloff, J., Desselberger, U., and Ball, L. A. (Eds.) (2005). *Virus Taxonomy, the Eighth Report of the International Committee on Taxonomy of Viruses*, Academic Press, New York, Figure 5, p 211.

FIGURE 7.12 Adapted from McGeoch, D. J., Dolan, A., and Ralph, A. C. (2000). Toward a comprehensive phylogeny for mammalian and avian herpesviruses *J. Virol.* **74**; 10401–10406.

FIGURE 7.13 Redrawn from Fauquet, C. M., Mayo, M. A., Maniloff, J., Desselberger, U., and Ball, L. A. (Eds.) (2005). *Virus Taxonomy, the Eighth Report of the International Committee on Taxonomy of Viruses*, Academic Press, New York, p. 195 and Gompels, U. A., Nicholas, J., Lawrence, G., Jones, M., Thomson, B. J., Martin, M. E. D., Efstathiou, S., Craxton, M., and Macaulay, H. A. (1995). The DNA sequence of human herpesvirus-8: structure, coding content, and genome evolution. *Virology* **209**: 29–51.

FIGURE 7.14 Redrawn from Fields, B. N., Knipe, D. M., and Howley, P. M. (Eds.) (1996). *Fields Virology*, *3rd Edition*, Lippincott-Raven, New York, p. 2245.

FIGURE 7.15 Redrawn from data in Roizman, B., and Sears, A. E. (1990). The replication of herpes simplex viruses. In: *The Human Herpesviruses* (B. Roizman, C. Lopez, and R. J. Whitley, Eds.), Raven Press, New York, pp. 11–68.

FIGURE 7.16 Drawn from data from Nahmias, A. J, Lee, F. K., and Bechman-Nahmias, S. (1990). Sero-epidemiological and sociological patterns of herpes simplex virus infection in the world. *Scand. J. Infect. Dis. Suppl.* **69**: 19–30 (1990) and Johnson, R. E., Nahmias, A. J., Magder, L. S., Lee, F. K., Brooks, C., and Snowden, C. (1989). A seroepidemiological survey of the prevalence of herpes simplex virus type 2 infection in the United States. *N. Engl. J. Med.* **321**: 7–12.

FIGURE 7.19 Data from Fields, B. N., Knipe, D. M., and Howley, P. M. (Eds.) (1996). *Fields Virology*, *3rd Edition*, Lippincott-Raven, New York, pp. 2399–2402.

FIGURE 7.20 Redrawn from a figure in Wold, W. S. M., and Golding, L. R. (1991). Region E3 of adenovirus: a cassette of genes involved in host immunosurveillance and virus-cell interactions. *Virology* **184**: 1–8.

FIGURE 7.21 Adapted from Fields, B. N., Knipe, D. M., and Howley, P. M. (Eds.) (1996). *Fields Virology*, *3rd Edition*, Lippincott-Raven, New York, p. 2119 and Ginzberg, H. S. (1984). An overview. In: *The Adenoviruses* (H. S. Ginzberg, Ed.), Plenum Press, New York, Figure 1, p 2.

FIGURE 7.22 Adapted from Berg, P., and Singer, M. (1997). Mammalian DNA viruses: Papoviruses as models of cellular genetic function and oncogenesis. In: *Exploring Genetic Mechanisms* (M. Singer and P. Berg, Eds.), University Science Books, Sausalito CA, Figure 1.35, p. 61.

FIGURE 7.23 Data from Strauss, S. E. (1984). Adenovirus infections in humans. In: *The Adenoviruses* (H. Z. Ginsberg, Ed.), Plenum Press, New York, pp. 451–496.

FIGURE 7.24 From Fox, J. P., Brandt, C. D., Wasserman, F. E., Hall, C. E., Spigland, I., Kogon, A., and Elveback, L. R. (1969). The Virus Watch Program: a continuing surveillance of viral infections in metropolitan New York families. VI. Observations of adenovirus infections: virus excretion patterns, antibody response, efficiency of surveillance, patterns of infection and relation to illness. *Am. J. Epidemiol.* **89**: 25–50.

FIGURE 7.25 From Dingle, J., and Langmuir, A. D. (1968). Epidemiology of acute respiratory diseases in military recruits. *Am. Rev. Resp. Dis.* **97**: 1–65.

FIGURE 7.26 Redrawn completely from Brady, J. N., and Salzman, N. P. (1986). The papovaviruses: general properties of polyoma and SV40. In: *The Papovaviridae*, (N. P. Salzman, Ed.), Plenum Press, New York, Vol. **1**, pp. 1–26, Figures 2 and 3.

FIGURE 7.27 Adapted from Berg, P., and Singer, M. (1997). Mammalian DNA viruses: Papoviruses as models of cellular genetic function and oncogenesis. In: *Exploring Genetic Mechanisms* (M. Singer and P. Berg, Eds.), University Science Books, Sausalito CA, Figures 1.23 and 1.24; and Fields, B. N., Knipe, D. M., and Howley, P. M. (Eds.) (1996). *Fields Virology, 3rd Edition*, Lippincott-Raven, New York, Figure 6, p. 2011.

FIGURE 7.28 Drawn from data in DiMaio, D., Lai, C.-C., and Klein, O. (1998). Virocrine transformation: the intersection between viral transforming proteins and cellular signal transduction pathways. *Annu. Rev. Microbiol.* **52**: 397–421.

FIGURE 7.29 From Fields, B. N., Knipe, D. M., and Howley, P. M. (Eds.) (1996). *Fields Virology, 3rd Edition*, Lippincott-Raven, New York, Figure 6, p. 2039.

FIGURE 7.30 Adapted from Nathanson, N., Ahmed, R., Gonzalez-Scarano, F., Griffin, D. E., Holmes, K. V., Murphy, F. A., Robinson, H. L. (Eds.) (1996). *Viral Pathogenesis*, Lippincott-Raven, Philadelphia, p. 27, and Fields, B. N., Knipe, D. M., and Howley, P. M. (Eds.) (1996). *Fields Virology, 3rd Edition*, Lippincott-Raven, New York, pp. 2051 and 2052.

FIGURE 7.31 Data from McBride, A. A., Byrne, J. C., and Howley, P. M. (1989). E2 polypeptides encoded by bovine papilloma virus type-1 form dimers through the common carboxyl-terminal domain—transactivation is mediated by the conserved amino-terminal domain. *Proc. Natl. Acad. Sci. U.S.A.* **86**: 510–514.

FIGURE 7.32 Adapted from Nathanson, N., Ahmed, R., Gonzalez-Scarano, F., Griffin, D. E., Holmes, K. V., Murphy, F. A., Robinson, H. L. (Eds.) (1996). *Viral Pathogenesis*, Lippincott-Raven, Philadelphia, p. 273; and taxonomic data from Fauquet, C. M., Mayo, M. A., Maniloff, J., Desselberger, U., and Ball, L. A. (Eds.) (2005). *Virus Taxonomy, the Eighth Report of the International Committee on Taxonomy of Viruses*, Academic Press, New York, pp. 239–253.

FIGURE 7.33 Adapted from Heegaard, E. D., and Brown, K. E. (2002). Human parvovirus B19. *Clin. Microbiol. Rev.* **15**: 485–505, Figure 2. (with permission); Mouw, M. B., and Pintel, D. J. (2000). Adeno-associated virus RNAs appear in a temporal order and their splicing is stimulated during coinfection with adenovirus. *J. Virol.* **74**: 9878–9888, Figure 1 (with permission).

FIGURE 7.34 Adapted from Fields, B. N., Knipe, D. M., and Howley, P. M. (Eds.) (1996). *Fields Virology, 3rd Edition*, Lippincott-Raven, New York, p. 2175.

FIGURE 7.35 Adapted from Brister, J. R., and Muzyczka, N. (2000). Mechanism of rep-mediated adeno-associated virus origin nicking. *J. Virol.* **74**: 7762–7771.

FIGURE 7.36 From Parrish, C. R. (1997). How canine parvovirus suddenly shifted host range. *ASM News* **63**: 307–311.

CHAPTER 8

Figures

FIGURE 8.1 Adapted from Figure 1B in Eaton, B. T., Broder, C. C., Middleton, D., and Wang, L.-F. (2006). Hendra and Nipah viruses: different and dangerous. *Nature Rev. Microbiol.* 4: 23–35.

FIGURE 8.2 From data from the World Health Organization at: http://www.who.int/csr/don/archive/disease/severe_acute_respiratory_syndrome/en/.

FIGURE 8.3 Data from Porterfield, J. S. (Ed.) (1995). *Exotic Viral Infections*, Chapman & Hall Medical, London, p. 320; and later data from Georges-Courbot, M.-C., Sanchez, A., Lu, C.-Y., Baize, S., Leroy, E., Lansout-Soukate, J., Tévi-Bénissan, C., Georges, A. J., Trappeir, S. G., Zaki, S. R., Swanepoel, R., Leman, P. A., Rollin, P. E., Peters, C. J., Nichol, S. T., and Ksiazek, T. G. (1997). Isolation and phylogenetic characterization of Ebola viruses causing different outbreaks in Gabon. *Emerg. Infect. Dis.* **3**: 59–62; Peters, C. J., and Khan, A. S. (1999). Filovirus diseases. *Curr. Topics Microbiol.* **235**: 85–95, and news bulletins from the World Health Organization (2000–2005) at: http://www.who.int/disease-outbreak-news/.

FIGURE 8.7 Adapted from Figure 1 in Wang, E., Ni, H.., Xu, R., Barrett, A. D. T., Watowich, S. J., Gubler, D. J., and Weaver, S. C. (2000). Evolutionary relationships of endemic/epidemic and sylvatic dengue viruses. *J. Virol.* **74**; 3227–3234.

FIGURE 8.9 Adapted from Siquiera, J. B., Martelli, C. M. T., Coelho, G. E., Simplicio, A. C. da R., and Hatch, D. L. (2005). Dengue and dengue hemorragic fever, Brazil, 1981–2002. *Emerg. Infect. Dis.* **11**: 48–53.

FIGURE 8.10 Goncalvez, A. P., Escalante, A. A., Pujol, F. H., Ludert, J. E., Tovar, D., Salas, R. A., and Liprandi, F. (2002). Diversity and evolution of the envelope gene of dengue virus type 1. *Virology* **303**: 110–119.

FIGURE 8.11 Panel (A) was redrawn from Whetter, L. E., Ojukwu, I. C., Novembre, F. J., and Dewhurst, S. (1999). Pathogenesis of simian immunodeficiency virus infection. *J. Gen. Virol.* **80**: 1557–1568. Panel (B) is adapted from Figure 2 in Sharp, P. M., Shaw, G. M., and Hahn, B. H. (2005). Simian immunodeficiency virus infection of chimpanzees. *J. Virol.* **79**: 3891–3902.

FIGURE 8.12 Adapted from Figure 1 in Sharp, P. M., Shaw, G. M., and Hahn, B. H. (2005). Simian immunodeficiency virus infection of chimpanzees. *J. Virol.* **79**: 3891–3902.

FIGURE 8.13 Adapted from Figure 4 in Sharp, P. M., Shaw, G. M., and Hahn, B. H. (2005). Simian immunodeficiency virus infection of chimpanzees. *J. Virol.* **79**: 3891–3902.

CHAPTER 9

Tables

TABLE 9.3 Adapted from Granoff, A., and Webster, R. G. (Eds.) (1999). *Encyclopedia of Virology, 2nd Edition*, Academic Press, San Diego, p. 1389.

Figures

FIGURE 9.1 Adapted from Figure 1 in Strauss, J. H., and Strauss, E. G. (1997). Recombination in alphaviruses. *Semin. Virol.* **8**: 85–94.

FIGURE 9.2 Adapted from Whelan, S. P. J., and Wertz, G. W. (1997). Defective interfering particles of vesicular stomatitis virus: Functions of the genomic termini. *Semin. Virol.* **8**: 131–139; and from Figure 1 in Brian, D. A., and Spaan, W. J. M. (1997). Recombination and coronavirus defective interfering RNAs. *Semin. Virol.* **8**: 101–111.

FIGURE 9.3 Adapted from Granoff, A., and Webster, R. G. (Eds.) (1999). *Encyclopedia of Virology, 2nd Edition*, Academic Press, San Diego, Figure 2, p. 373.

FIGURE 9.4 Adapted from Flores, R., Di Serio, F., and Hernández, C. (1997). Viroids: the noncoding genomes. *Semin. Virol.* **8**: 65–73.

FIGURE 9.5 Adapted from Flores, R., Di Serio, F., and Hernández, C. (1997). Viroids: the noncoding genomes. *Semin. Virol.* **8**: 65–73; and Pelchat, M., Lévesque, D., Ouellet, J., Laurendeau, S., Lévesque, S., Lehoux, J., Thompson, D. A., Eastwell, K. C., Skrzeczkowski, L. J., and Perreault, J. P. (2000). Sequencing of peach latent mosaic viroid variants from nine North American peach cultivars shows that this RNA folds into a complex secondary structure. *Virology* **271**: 37–45.

FIGURE 9.6 Adapted from Fields, B. N., Knipe, D. M., and Howley, P. M. (Eds.) (1996). *Fields Virology*, *3rd Edition*, Lippincott-Raven, New York, p. 2826.

FIGURE 9.7 Adapted from Fields, B. N., Knipe, D. M., and Howley, P. M. (Eds.) (1996). *Fields Virology*, *3rd Edition*, Lippincott-Raven, New York, p. 2825.

FIGURE 9.8 Adapted from Modahl, L. E., and Lai, M. M. C. (2000). Hepatitis delta virus: The molecular basis of laboratory diagnosis. *Crit. Rev. Clin. Lab. Sci.* **37**: 45–92; and from Lai, M. M. C. (2005). RNA replication without RNA-dependent RNA polymerase: surprises from hepatitis delta virus. *J. Virol.* **79**: 7951–7958.

FIGURE 9.9 Adapted from Lai, M. M. C. (2005). RNA replication without RNA-dependent RNA polymerase: surprises from hepatitis delta virus. *J. Virol.* **79**: 7951–7958.

FIGURE 9.10 Redrawn from data in Prusiner, S. (1998). Prions. *Proc. Natl. Acad. Sci. U.S.A.* **95**: 13363–13383; and Riek, R., Hornemann, S., Wider, G., Billeter, M., Glockshuber, R., and Wüthrich, K., (1996). NMR structure of the mouse prion protein domain PrP(121–231). *Nature* **382**: 180–183.

FIGURE 9.11 Adapted from Figure 1g in Anderson, R. M., Donnelly, C. A., Ferguson, N. M., Woolhouse, M. E. J., Watt, C. J., Udy, H. J., MaWhinney, S., Dunstan, S. P., Southwood, T. R. E., Wilesmith, J. W., Ryan, J. B. M., Hoinville, L. J., Hillerton, J. E., Austin, A. R., and Wells, G. A. H. (1996). Transmission dynamics and epidemiology of BSE in British cattle. *Nature* **382**: 779–788.

FIGURE 9.12 Adapted from Figure 1a in Anderson, R. M., Donnelly, C. A., Ferguson, N. M., Woolhouse, M. E. J., Watt, C. J., Udy, H. J., MaWhinney, S., Dunstan, S. P., Southwood, T. R. E., Wilesmith, J. W., Ryan, J. B. M., Hoinville, L. J., Hillerton, J. E., Austin, A. R., and Wells, G. A. H. (1996). Transmission dynamics and epidemiology of BSE in British cattle. *Nature* **382**: 779–788.

FIGURE 9.13 Data for this figure is present in Nathanson, N., Wilesmith, J., Wells, G. A., and Griot, C. (1999). Bovine spongiform encephalopathy and related diseases. In: *Prion Biology and Diseases* (S. Prusiner, Ed.), Cold Spring Harbor Laboratory Press, Cold Spring Harbor, NY, Figure 7, p. 449; and Monthly CJD Statistics, from the Department of Health of the United Kingdom.

FIGURE 9.14 From Riek, R., Hornemann, S., Wider, G., Billeter, M., Glockshuber, R., and Wüthrich, K., (1996). NMR structure of the mouse prion protein domain PrP(121–231). *Nature* **382**: 180–183.

FIGURE 9.15 Redrawn from data in Weismann, C. (1996). Molecular biology of transmissible spongiform encephalopathies. *FEBS Lett.* **389**: 3–11; Riek, R., Hornemann, S., Wider, G., Billeter, M., Glockshuber, R., and Wüthrich, K., (1996). NMR structure of the mouse prion protein domain PrP(121–231). *Nature* **382**: 180–183.; and Prusiner, S. (1998). Prions. *Proc. Natl. Acad. Sci. U.S.A.* **95**: 13363–13383.

FIGURE 9.16 Adapted from Hegde, R. S., Mastrianni, J. A., Scott, M. R., DeFea, K. A., Tremblay, P., Torchia, M., DeArmond, S. J., Prusiner, S. B., and Lingappa, V. R. (1998). A transmembrane form of the prion protein in neurodegenerative disease. *Science* **279**: 827–834.

FIGURE 9.17 Adapted from Figure 10 of Wickner, R. B., Taylor, K. L., Edskes, H. K., Maddelein, M.-L., Moriyama, H., and Roberts, B. T. (1999). Prions of *Saccharomyces* and *Podospora* ssp.: protein-based inheritance. *Microbiol. and Mol. Biol. Rev.* **63**: 844–861.

CHAPTER 10

Tables

TABLE 10.1 Data from Janeway, C. A., Travers, P., Walport, M., and Capra, J. D. (1999). *Immunobiology: the Immune System in Health and Disease*, Elsevier Science, Burlington, MA, pp. 62, 93, 158.

TABLE 10.2 Data for this table came from Granoff, A., and Webster, R. G. (Eds.) (1999). *Encyclopedia of Virology*, *2nd Edition*, Academic Press, San Diego, p. 1862, from Fields, B. N., Knipe, D. M., and Howley, P. M. (Eds.) (1996). *Fields Virology*, *3rd Edition*, Lippincott-Raven, New York, p. 371, and from Arvin, A. M., and Greenberg, H. B. (2006). New viral vaccines. *Virology* **344**: 240–249.

TABLE 10.3 Data for this table came from Granoff, A., and Webster, R. G. (Eds.) (1999). *Encyclopedia of Virology*, *2nd Edition*, Academic Press, San Diego, p. 1862; and Fields, B. N., Knipe, D. M., and Howley, P. M. (Eds.) (1996). *Fields Virology*, *3rd Edition*, Lippincott-Raven, New York, p. 371.

TABLE 10.4 Iwasaki, A., and Medzhitov, R. (2004). Toll-like receptor control of the adaptive immune responses. *Nature Immun.* **5**: 987–995; O'Neill, L. A. J. (2005). Immunity's early warning system. *Scientific American* January, pp. 38–45; Takeda, K., Kaisho, T., and Akira, S. (2003). Toll-like Receptors. *Annu. Rev. Immunol.* **21**: 335–376, Kawai, T., and Akira, S. (2006). TLR Signaling. *Cell Death Differ.* **13**: 816–825.

TABLE 10.5 Data for this table came from Mims, C. A., Playfair, J. H. L., Roitt, I. M., Wakelin, D., and Williams, R. (1993). *Med. Microbiol.*, Mosby Europe Limited, London, p. 37.3 and Figure 7.13; and Kuby, J. (1997). *Immunology*, *3rd Edition*, W. H. Freeman and Company, New York, pp. 318 and 319.

TABLE 10.6 Data for this table came from Mims, C. A., Playfair, J. H. L., Roitt, I. M., Wakelin, D., and Williams, R. (1993). *Med. Microbiol.*, Mosby Europe Limited, London, Figure 12.9; and Fields, B. N., Knipe, D. M., and Howley, P. M. (Eds.) (1996). *Fields Virology*, *3rd Edition*, Lippincott-Raven, New York, Table 3, p. 378.

TABLE 10.7 Adapted from Nathanson, N. Ahmed, R., Gonzalez-Scarano, F., Griffin, D. E., Holmes, K. V., Murphy, F. A., and Robinson, H. L. (Eds.) (1996). *Viral Pathogenesis,* Lippincott-Raven, Philadelphia, p. 124.

TABLE 10.8 Data from McManus, M. T., and Sharp, P. A., (2002). Gene silencing in mammals by small interfering RNAs. *Nature Rev. Genet.* **3**: 737–747; Lecellier, C.-H., Dunoyer, P., Arar, K., Lehmann-Che, J., Eyquem, S., Himber, C., Saïb, A., and Voinnet, O. (2005). A cellular microRNA mediates antiviral defense in human cells. *Science* **308**: 557–560; and Dykxhoorn, D. M., Palliser, D., and Lieberman, J. (2006). The silent treatment: siRNAs as small molecule drugs. *Gene Ther.* **13**: 541–552.

TABLE 10.11 Data for this table came from reviews by O'Brien, V. (1998). Viruses and apoptosis. *J. Gen. Virol.* **79**; 1833–1845; and Tortorella, D., Gewurz, B. E., Furman, M. H., Schust, D. J., and Ploegh, H. L. (2000). Viral subversion of the immune system. *Annu. Rev. Immunol.* **18**: 861–926.

TABLE 10.12 Adapted from Table 1 in Langland, J. O., Cameron, J. M., Heck, M. C., Jancovich, J. K., and Jacobs, B. L. (2006). Inhibition of PKR by RNA and DNA viruses. *Virus Res.* **119**: 100–110.

TABLE 10.13 Adapted from Table 3 of Tortorella, D., Gewurz, B. E., Furman, M. H., Schust, D. J., and Ploegh, H. L. (2000). Viral subversion of the immune system. *Annu. Rev. Immunol.* **18**: 861–926.

TABLE 10.14 Data for this table came from reviews by Tortorella, D., Gewurz, B. E., Furman, M. H., Schust, D. J., and Ploegh, H. L. (2000). Viral subversion of the immune system. *Annu. Rev. Immunol.* **18**: 861–926; Smith, S. A., and Kotwal, G. J. (2002). Immune response to poxvirus infectins in various animals. *Crit. Rev. Microbiol.* **28**: 149–185; and Seet, B. T., Johnston, J. B., Brunetti, C. R., Barrett, J. W., Everett, H., Cameron, C., Sypula, J., Nazarian, S. H., Lucas, A., and McFadden, G. (2003). Poxviruses and immune evasion. *Annu. Rev. Immunol.* **21**: 377–423.

TABLE 10.15 From data presented in García-Sastre, A. (2004). Identification and characterization of viral antagonists of Type I interferon in negative-strand viruses. *Curr. Topics Microbiol. Immunol.* **283**: 249–280.

TABLE 10.16 Schütz, S., and Sarnow, P. (2006). Interaction of viruses with the mammalian RNA interference pathway. *Virology* **344**: 151–157.

Figures

FIGURE 10.1   Adapted from Mims, C. A., Playfair, J. H. L., Roitt, I. M., Wakelin, D., and Williams, R. (1993). *Med. Microbiol.*, Mosby Europe Limited, London p. 5.8.

FIGURE 10.2   Panel (B) from Bjorkman, P. J., Saper, M. A., Samraoui, B., Bennett, W. S., Strominger, W. L., and Wiley, D. C. (1987). Structure of the human class I histocompatibility antigen, HLA-A2. *Nature* **329**: 506–512, as reprinted in Kuby, J. (1997). *Immunology, 3rd Edition*, W. H. Freeman and Company, New York, p. 229.

FIGURE 10.3   Adapted from Chen, J., and Alt, F. W. (1997). Generation of antigen receptor diversity. In: *Exploring Genetic Mechanisms* (M. Singer and P. Berg, Eds.), University Science Books, Sausalito CA, Figures 7.6 and 7.7, pp. 344 and 345.

FIGURE 10.4   Adapted from Kuby, J. (1997). *Immunology, 3rd Edition*, W. H. Freeman and Company, New York, Figure 11.6, p. 269.

FIGURE 10.5   Adapted from Chen, J., and Alt, F. W. (1997). Generation of antigen receptor diversity. In: *Exploring Genetic Mechanisms* (M. Singer and P. Berg, Eds.), University Science Books, Sausalito CA, pp. 344 and 345; and data from Kuby, J. (1997). *Immunology, 3rd Edition*, W. H. Freeman and Company, New York, pp. 275–277.

FIGURE 10.6   Adapted from Fields, B. N., Knipe, D. M., and Howley, P. M. (Eds.) (1996). *Fields Virology, 3rd Edition*, Lippincott-Raven, New York, p. 351.

FIGURE 10.7   Adapted from Kuby, J. (1997). *Immunology, 3rd Edition*, W. H. Freeman and Company, New York, p. 113.

FIGURE 10.8   Adapted from Gally, J. (1993). Structure of immunoglobulins. In: *The Antigen* (M. Sela, Ed.), Academic Press, New York, p. 209.

FIGURE 10.9   Adapted from Chen, J., and Alt, F. W. (1997). Generation of antigen receptor diversity. In: *Exploring Genetic Mechanisms* (M. Singer and P. Berg, Eds.), University Science Books, Sausalito CA, Figures 7.2, p. 340 and Figure 7.8 on p. 345.

FIGURE 10.10   Adapted from Kuby, J. (1997). *Immunology, 3rd Edition*, W. H. Freeman and Company, New York, Figures 7.4 and 7.5 on pp. 172 and 173.

FIGURE 10.11   From Kuby, J. (1997). *Immunology, 3rd Edition*, W. H. Freeman and Company, New York, Figure 16.19, p. 398.

FIGURE 10.12   Adapted from Kuby, J. (1997). *Immunology, 3rd Edition*, W. H. Freeman and Company, New York, Figure 7.17, p. 185; and Figure 4.18 in Janeway, C. A., Travers, P., Walport, M., and Capra, J. D. (2004). *Immunobiology: The Immune System in Health and Disease. 6th Edition*, Elsevier Science, London.

FIGURE 10.13   From maps on the Web site of WHO international at: http://www.who.int/vaccines-surveillance/graphics/NY_graphics/meainc.htm and at: http://www.who.int/vaccines-surveillance/graphics/NY_graphics/meacovmap.htm.

FIGURE 10.14   Redrawn from Garenne, M., Leroy, O., Beau, J. P., and Sene, I. (1991). Child-mortality after high-titer measles vaccines: Prospective study in Senegal. *Lancet* **338**: 903–907.

FIGURE 10.15   Original figure using data from Ashkenazi, A., and Dixit, V. M. (1998). Death receptors: Signaling and modulation. *Science* **281**: 1305–1308; and Raff, M. (1998). Cell suicide for beginners. *Nature* **396**: 119–122.

FIGURE 10.16   Adapted from Figure 1 on p. 340 in Griffin, D. E. (1999). Cytokines and chemokines. In: *Encyclopedia of Virology, 2nd Edition* (A. Granoff and R. G. Webster, Eds.), Academic Press, San Diego, pp. 339–344.

FIGURE 10.17   Adapted from Figure 1 in Freundt, E. C., and Lenardo, M. J. (2005). Interfering with interferons: hepatitis C virus counters innate immunity. *Proc. Natl. Acad. Sci. U.S.A.* **102**: 17539–17540.

FIGURE 10.18   Drawn from data in Fields, B. N., Knipe, D. M., and Howley, P. M. (Eds.) (1996). *Fields Virology, 3rd Edition*, Lippincott-Raven, New York, p. 379; Kalvakolanu, D. V. (1999). Virus interception of cytokine regulated pathways. *Trends in Microbiol.* **7**: 166–171; and Nathanson, N. Ahmed, R., Gonzalez-Scarano, F., Griffin, D. E., Holmes, K.

V., Murphy, F. A., and Robinson, H. L. (Eds.) (1996). *Viral Pathogenesis*, Lippincott-Raven, Philadelphia, p. 123.

FIGURE 10.19   Redrawn from Nathanson, N. Ahmed, R., Gonzalez-Scarano, F., Griffin, D. E., Holmes, K. V., Murphy, F. A., and Robinson, H. L. (Eds.) (1996). *Viral Pathogenesis*, Lippincott-Raven, Philadelphia, Figure 6.8 on p. 125.

FIGURE 10.20   Schütz, S., and Sarnow, P. (2006). Interaction of viruses with the mammalian RNA interference pathway. *Virology* **344**: 151–157; and Novina, C. D., and Sharp, P. A., (2004). The RNAi revolution. *Nature* **430**: 161–164.

FIGURE 10.23   Drawn from data in Turner, P. C., and Moyer, R. W. (1998). Control of apoptosis by poxviruses. *Semin. Virol.* **8**: 453–469.

FIGURE 10.24   Drawn from data in Hill, A. B., and Masucci, M. G. (1998). Avoiding immunity and apoptosis: manipulation of the host environment by herpes simplex virus and Epstein-Barr virus. *Semin. Virol.* **8**: 361–368.

FIGURE 10.25   Drawn from data in Chinnadurai, G. (1998). Control of apoptosis by human adenovirus genes. *Semin. Virol.* **8**: 399–408.

CHAPTER 11

Tables

TABLES 11.3   Verma, I. M., and Somia, N. (1997). Gene therapy—promises, problems and prospects. *Nature* **389**: 239–242; Jolly, D. (1994). Viral vector systems for gene-therapy. *Cancer Gene Ther.* **1**: 51–64; Bukreyev, A., Skiadopoulos, M. H., Murphy, B. R., and Collins, P. L. (2006). Nonsegmented negative-strand viruses as vaccine vectors. *J. Virol.* **80**: 10293–10306.

TABLE 11.4   Verma, I. M., and Weitzman, M. D. (2005). Gene therapy: twenty-first century medicine. *Annu. Rev. Biochem.* **74**: 711–738; Jolly, D. (1994). Viral vector systems for gene-therapy. *Cancer Gene Ther.* **1**: 51–64; Boulaiz, H., Marchal, J. A., Prados, J., Melguizo, C., and Aránega, A. (2005). Non-viral and viral vectors for gene therapy. *Cell. Mol. Biol.* **51**: 3–22; Hu, Y.-C. (2006). Baculovirus vectors for gene therapy. *Adv. Vir. Res.* **68**: 287–320.

TABLE 11.5   Porteus, M. H., Connelly, J. P., and Pruett, S. M. (2006). A look to future directions in gene therapy research for monogenic diseases. *PLOS Genet.* **2**: 1285–1292; Foster, K., Foster, H., and Dickson, J. G. (2006). Gene therapy progress and prospects: Duchenne muscular dystrophy. *Gene Ther.* **13**: 1677–1685; Bukreyev, A., Skiadopoulos, M. H., Murphy, B. R., and Collins, P. L. (2006). Nonsegmented negative-strand viruses as vaccine vectors. *J. Virol.* **80**: 10293–10306.

TABLE 11.6   Wiley Journal of Gene Medicine/Clinical Trial Database at: http://www.wiley.co.uk/genetherapy/clinical/.

Figures

FIGURE 11.1   Adapted from Strauss, E. G., and Strauss, J. H. (1997). Eukaryotic RNA viruses: a variant genetic system. In: *Exploring Genetic Mechanisms*, (M. Singer and P. Berg, Eds.), University Science Books, Sausalito, CA, Figure 2.25 on p. 115.

FIGURE 11.2   Adapted from Crystal, R. G. (1995). Transfer of genes to humans—early lessons and obstacles to success. *Science* **270**: 404–410.

FIGURE 11.3   Adapted from Dunbar, C. E. (1996). Gene transfer to hemapoietic stem cells: implications for gene therapy of human disease. *Annu. Rev. Med.* **47**: 11–20.

FIGURE 11.4   Adapted from Strauss, J. H., and Strauss, E. G. (1994). The Alphaviruses: Gene expression, replication, and evolution. *Microbiol. Rev.* **58**: 491–562; Figure 23.

FIGURE 11.5   Adapted from Bledsoe, A. W., Jackson, C. A., McPherson, S., and Morrow, C. D. (2000). Cytokine production in motor neurons by poliovirus replicon vector gene delivery. *Nature Biotechnol.* **18**: 964–969.

FIGURE 11.7 AND 11.8   Adapted from Conzelmann, K.-K., and Meyers, G. (1996). Genetic engineering of animal RNA viruses. *Trends Microbiol.* **4**: 386–393.

FIGURE 11.9   Adapted from Chambers, T. J., Nestorowicz, A., Mason, P. W., and Rice, C. M. (1999). Yellow-fever/Japanese encephalitis chimeric viruses: construction and biological properties. *J. Virol.* **73**: 3095–3101.

FIGURES 11.10, 11.11, AND 11.12   data from Wiley Journal of Gene Medicine/Clinical Trial Database at: http://www.wiley.co.uk/genetherapy/clinical/.

FIGURE 11.13   Adapted from Figure 1 in Foster, K., Foster, H., and Dickson, J. G. (2006). Gene therapy progress and prospects: Duchenne muscular dystrophy. *Gene Ther.* **13**: 1677–1685.

Index